MOTOR APPLICATION AND MAINTENANCE HANDBOOK

OTHER McGRAW-HILL HANDBOOKS OF INTEREST

Baumeister · MARKS' STANDARD HANDBOOK FOR MECHANICAL ENGINEERS
Bovay · HANDBOOK OF MECHANICAL AND ELECTRICAL SYSTEMS FOR BUILDINGS
Brady and Clauser · MATERIALS HANDBOOK
Brater and King · HANDBOOK OF HYDRAULICS
Chopey and Hicks · HANDBOOK OF CHEMICAL ENGINEERING CALCULATIONS
Croft, Watt, and Summers · AMERICAN ELECTRICIANS' HANDBOOK
Dudley · GEAR HANDBOOK
Fink and Beaty · STANDARD HANDBOOK FOR ELECTRICAL ENGINEERS
Harris · SHOCK AND VIBRATION HANDBOOK
Hicks · STANDARD HANDBOOK OF ENGINEERING CALCULATIONS
Hicks · STANDARD HANDBOOK OF CONSULTING ENGINEERING PRACTICE
Juran · QUALITY CONTROL HANDBOOK
Karassik, Krutzsch, Fraser, and Messina · PUMP HANDBOOK
Kurtz · HANDBOOK OF ENGINEERING ECONOMICS
Maynard · INDUSTRIAL ENGINEERING HANDBOOK
Optical Society of America · HANDBOOK OF OPTICS
Pachner · HANDBOOK OF NUMERICAL ANALYSIS APPLICATIONS
Parmley · STANDARD HANDBOOK OF FASTENING AND JOINING
Peckner and Bernstein · HANDBOOK OF STAINLESS STEELS
Perry and Green · PERRY'S CHEMICAL ENGINEERS' HANDBOOK
Raznjevic · HANDBOOK OF THERMODYNAMIC TABLES AND CHARTS
Rohsenow and Hartnett · HANDBOOK OF HEAT TRANSFER
Schwartz · METALS JOINING MANUAL
Seidman and Mahrous · HANDBOOK OF ELECTRIC POWER CALCULATIONS
Shand and McLellan · GLASS ENGINEERING HANDBOOK
Smeaton · SWITCHGEAR AND CONTROL HANDBOOK
Transamerica Delaval, Inc. · TRANSAMERICA DELAVAL ENGINEERING HANDBOOK
Tuma · ENGINEERING MATHEMATICS HANDBOOK
Tuma · TECHNOLOGY MATHEMATICS HANDBOOK
Tuma · HANDBOOK OF PHYSICAL CALCULATIONS

MOTOR APPLICATION AND MAINTENANCE HANDBOOK

ROBERT W. SMEATON, Editor
EDITOR, *Switchgear and Control Handbook*
REGISTERED PROFESSIONAL ENGINEER, Wisconsin
MEMBER, WSPE, IEEE, AND ESM

Second Edition

McGRAW-HILL BOOK COMPANY
New York St. Louis San Francisco Auckland Bogotá
Hamburg Johannesburg London Madrid Milan
Mexico Montreal New Delhi Panama
Paris São Paulo Singapore
Sydney Tokyo Toronto

Library of Congress Cataloging-in-Publication Data

Motor application and maintenance handbook.

Includes index.
1. Electric motors—Handbooks, manuals, etc.
I. Smeaton, Robert W.
TK2514.M58 1986 621.46′2 86-7330
ISBN 0-07-058448-6

1234567890 DOC/DOC 8932109876

ISBN 0-07-058448-6

The editors for this book were Harold B. Crawford and Ingeborg M. Stochmal, the designer was Mark E. Safran, and the production supervisor was Sally Fliess. It was set in Times Roman by University Graphics, Inc.

Printed and bound by R. R. Donnelley & Sons Company.

CONTENTS

Contributors vii

Preface ix

1. Motor Load and Mechanical Considerations 1-1
2. Electric Power Supply and System Considerations 2-1
3. Basic Motor Types 3-1
4. Large Polyphase Induction Motors (Over 680 Frame Size) 4-1
5. Synchronous Motors 5-1
6. Ac Three-Phase Motors (680 Frame and Smaller) 6-1
7. Fractional-Horsepower ac Motors 7-1
8. Direct-Current Motors 8-1
9. Small Specialty Motors 9-1
10. Special-Application Motors 10-1
11. Insulation 11-1
12. Motor Bearings 12-1
13. Bearing and Shaft Seals 13-1
14. Lubrication 14-1
15. Flywheels 15-1
16. Foundations, Grouting, and Alignment 16-1
17. Preventive Maintenance 17-1
18. Refurbishing Old Motors 18-1
19. Emergency Care of Damaged Motors 19-1
20. Motor Noise 20-1

Index follows Section 20

CONTRIBUTORS

Annis, R. R., P.E., *Vice President, Electrical Engineering, John Oster Manufacturing Company, Milwaukee, Wis.* (SECTION 9)

Blakey, R. C., *Manager, Apparatus Repair, Westinghouse Electric Corporation, Phoenix, Ariz. (retired)* (SECTION 17)

Cobb, C. F., *Chief Engineer, Electrical Engineering Section, Allis-Chalmers Manufacturing Company, Norwood, Ohio (retired)* (SECTION 10, PART 3)

Daley, L. T., P.E., *Commander USNR, Philadelphia Naval Ship Yard, Naval Ships System Command, Philadelphia, Pa.* (SECTION 19)

Dumper, W. C., P.E., *Station Design Division Engineer, Inside Plant Bureau, Consolidated Edison Company, New York, N.Y. (retired)* (SECTION 2)

Gibbs, C. R., *Director of Service, Allis-Chalmers Manufacturing Company, Milwaukee, Wis. (retired)* (SECTION 16)

Gregory, L. P., *Senior Application Engineer, Allis-Chalmers Manufacturing Company, Norwood, Ohio* (SECTION 10, PART 6)

Horrell, R. F., P.E., *Chief Engineer, Induction Machines, Electric Machinery Manufacturing Company, Minneapolis, Minn. (retired)* (SECTIONS 3 AND 10, PART 2)

Kaiser, A. N., *Manager of Product Engineering, Prestolite Division of Eltra Corporation, Toledo, Ohio (retired)* (SECTION 10, PART 4)

Koons, H. O., P.E., *Chief Engineer, Waukesha Bearing Corporation, Waukesha, Wis. (retired)* (SECTION 12)

Kuehlthau, J. L., P.E., *Consulting Engineer, New Berlin, Wis. (retired)* (SECTION 11)

Lefferts, W. G., *Senior Engineer, Process Equipment and System Division, Allis-Chalmers Manufacturing Company, Milwaukee, Wis. (retired)* (SECTION 1)

Petermann, J. E., P.E., *Senior Engineer, Motor-Generator Department, Allis-Chalmers Manufacturing Company, Milwaukee, Wis.* (SECTION 12)

Sargent, J. W., *Senior Engineer, Motor-Generator Department, Allis-Chalmers Manufacturing Company, Milwaukee, Wis.* (SECTION 11)

Sellers, J. F., *Chief Engineer, DC Machines, Allis-Chalmers Manufacturing Company, Milwaukee, Wis. (retired)* (SECTIONS 8 AND 15)

Szabo, E. J., P.E., *Manager of Engineering, Regular Products Division, Leece-Neville Company, Cleveland, Ohio (retired)* (SECTION 10, PART 4)

Tsivitse, P. J., Ph.D, P.E., *Chief Engineer, Rotating Machinery Group, Reliance Electric Company, Cleveland, Ohio (retired)* (SECTION 10, PART 5)

Veinott, C. G., D.Eng., P.E., *Machinery Consultant and Chief Engineering Analyst, Reliance Electric Company, Cleveland, Ohio* (SECTION 7)

Wiedener, E. L., *Professor Emeritus, Milwaukee School of Engineering, Milwaukee, Wis.* (SECTION 18)

The persons listed above were the contributors for the first edition. Their material was updated by others because they were unavailable for this edition.

The following contributors updated their original material.

Bartheld, R. G., *Manager, Technology Transfer, Siemens Energy and Automation, Inc., Atlanta, Ga.* (SECTION 20)

Booser, E. R., P.E., *Turbine Technology Laboratory, General Electric Company, Schenectady, N.Y.* (SECTION 14)

Burke, R. R., P.E., *Manager, Mechanical Engineering, Siemens Energy and Automation, Inc., Bradenton, Fla.* (SECTION 13)

Dobbins, P. J., *Electrical Design and Development Engineer, Large AC Motor and Generator Department, General Electric Company, Schenectady, N.Y. (retired)* (SECTIONS 6 AND 19).

Moore, R. C., *Senior Engineer, Motor-Generator or Department, Allis-Chalmers Manufacturing Company, Milwaukee, Wis. (retired); Distinguished Lecturer, Milwaukee School of Engineering* (SECTIONS 4, 5, AND 15)

Nickley, A. P., *Supervisor, Motor and Motor-Generator Branch, Electrical Division, Naval Sea Systems Command, Washington, D.C.* (SECTION 10, PART 7)

Poland, H. O., *Senior Design Engineer, AC Motor and Generator Engineering, Westinghouse Electric Corporation, East Pittsburgh, Pa.* (SECTION 10, PART 8)

The following persons are new contributors for this edition.

Adolphson, E. J., *Senior Insulation Engineer, Motor-Generator Department, Siemens Energy and Automation, Inc., Bradenton, Fla.* (SECTION 11)

Harms, H. B., P.E., *Senior Research Engineer, Motor Business Group, General Electric Company, Fort Wayne, Ind.* (SECTION 10, PART 9)

Moser, P. R., P.E., *Chief Product Engineer, Balder Electric Company, Fort Smith, Ark.* (SECTION 3)

Shively, R. A., *Senior Application Engineer, U.S. Electrical Motors Division, Emerson Electric Company, Prescott, Ariz.* (SECTION 2)

Soukup, G. C., P.E., *Manager, Large Motor Engineering, U.S. Electrical Motors Division, Emerson Electric Company, Prescott, Ariz.* (SECTION 2)

Stiffler, W. G., *Manager of Engineering, Rotating Machinery Group, Reliance Electric Company, Cleveland, Ohio* (SECTIONS 7 AND 10, PART 5)

Wagner, P. D., *Manager, Advanced Technology, Industrial Motor Division, Siemens Energy and Automation, Inc., Little Rock, Ark.* (SECTION 10, PARTS, 1, 6, AND 10)

PREFACE

Modern motors are much smaller and lighter than motors of only a few years ago. They are precisely designed to exact ratings and are the result of new higher-quality materials, improved manufacturing techniques, and computer-derived optimum designs.

To obtain maximum efficiency and reliability from these motors, considerable care must be given to their application, control, and protection.

This book was written as a ready reference for engineers and technicians to use when motor-application problems are being considered. Machinery operators and plant-maintenance personnel too will find valuable information on motor installation, preventive maintenance, and repair.

Through this Handbook, eminently qualified engineers give the reader the benefit of their many years of practical experience. All thirty-three contributors have become nationally known through their published books, technical articles, engineering papers, or through committee activities in their engineering societies.

Sections 1, 2, and 3 offer basic information governing the choice of motor for a given application. Section 1 covers the first consideration in motor application—the mechanical load. It includes a comprehensive table of machine-load characteristics needed for the selection of a proper drive and control. Section 2 gives the electric-power-supply and system considerations, and Section 3 explains the standard motor terminology and gives general background information on motor selection.

Other sections of the book cover in detail information on the application of specific sizes and types of motors, their components, their installation, maintenance, or repair. Section 20 deals with motor-noise problems, an important consideration in today's installations.

Standards and Codes. Industrial motor standards are continually improved and updated by the addition of new material or by changes in existing material. It is therefore recommended that those working with the application and maintenance of electric motors avail themselves of the latest applicable standards and codes and that changes be noted as they are made.

Numbers as referred to in the text of this book will change as new sections and paragraphs are added to these standards. However, readers who become familiar with current standards and their numbering systems can readily locate new references. The more important of these standards may be briefly described as follows:

a. NEMA standards are voluntary standards of the National Electrical Manufacturers Association and represent general practice in the industry. They define a product, process, or procedure with reference to nomenclature, composition, construction, dimensions, tolerances, operating characteristics, performance, quality, rating, and testing. Specifically, they cover such matters as frame sizes, torque classifications, and basis of rating.

b. IEEE standards are concerned with fundamentals such as basic standards for temperature rise, rating methods, classification of insulating materials, and test codes.

c. ANSI standards are national standards established by the American National Standards Institute, which represents manufacturers, distributors, consumers, and others concerned. ANSI standards may be sponsored by any responsible body and may become national standards only if a consensus of those having substantial interest is reached. Standards may cover a wide variety of subjects, such as dimensions, specifications of materials, methods of test, performance, and definition of terms. ANSI standards frequently are those previously sponsored and adopted by NEMA, IEEE, etc. The chief motor and generator standard of ANSI is C50, "Rotating Electric Machinery," which is substantially in agreement with current NEMA standards.

d. National Electrical Code (NEC) is an ANSI standard sponsored by the National Fire Protection Association for the purpose of safeguarding persons and buildings from electrical hazards arising from the use of electricity for light, heat, power, and other purposes. It covers wiring methods and materials, protection of branch circuits, motors and control, grounding, and recommendations regarding suitable equipment for each classification.

e. Underwriters Laboratories, Inc., is an independent testing organization which examines and tests devices, systems, and materials with particular reference to life, fire, and casualty hazards. It develops standards for motor and control for hazardous locations through cooperation with manufacturers. It has several different services by which a manufacturer can indicate compliance with Underwriters Laboratories standards. Such services are utilized on motors only in the case of explosionproof and dust-ignition-proof motors where label service is used to indicate to code-enforcing authorities that motors have been inspected to determine their adherence to Underwriters Laboratories standards for motors for hazardous locations.

f. Federal Specification CC-M-641 for integral-horsepower ac motors has been issued by the federal government to cover standard motors for general government uses. Standard motors meet these specifications, but other federal specifications issued by various branches of the government for specific use may require special designs.

g. World standards, that is, standards similar to our NEMA standards, have been established in other countries. These standards specify dimensions, classes of insulation, and in some cases horsepower ratings.

Since U.S. industrial standards dealing with motor dimensions and horsepower ratings have not been changed to the metric system, and will not be changed in the foreseeable future, these dimensions and ratings remain in the U.S. customary system in this edition. Conversions can be made using the following factors:

1 ft = 0.3048 m
1 in = 25.4 mm
1 hp = 0.7457 kW

These standards are important to all motor users because motor manufacturers follow them when designing, manufacturing, and testing motors. Only when motors are operated within their ratings will they perform satisfactorily.

This Handbook will help the reader to understand and apply these standards when using or specifying motors.

ACKNOWLEDGMENTS

The editor wishes to express his most sincere appreciation to the many engineers who gave freely of their advice and encouragement during the planning, organizing, and revising of this book, and to David L. Hendrickson, who prepared the illustrations and graphs for reproduction.

Robert W. Smeaton

SECTION 1

MOTOR LOAD AND MECHANICAL CONSIDERATIONS

W. G. Lefferts*

FOREWORD 1-2
CONSTANT-SPEED LOADS 1-2
1. Motor Load 1-2
ADJUSTABLE-SPEED LOADS 1-4
2. Constant Torque 1-4
3. Constant Horsepower 1-4
4. Variable Torque 1-4
5. Common Machine Characteristics 1-4
6. Duty Cycle 1-5
7. Multiple Drives 1-16
UNUSUAL TORQUE REQUIREMENTS 1-17
8. Rotary Kilns 1-17
9. Grinding Mills 1-18
10. Vibrating Screens, Feeders, Conveyors, and Oscillating-Grate Coolers 1-19
11. Repetitive Machines 1-19
12. Overhauling Loads 1-20
13. Balling Drums 1-20
14. Wash Mills 1-20
15. Positioning Drives 1-20
16. Cranes, Hoists, and Freight or Personnel Elevators 1-20
17. Flywheel Machines 1-20
REFERENCES 1-21

*Senior Engineer, Process Equipment and System Division, Allis-Chalmers Manufacturing Company, Milwaukee, Wis. (retired); Member, IEEE.

FOREWORD

The first step in selecting a motor is to determine the load to be driven and its torque characteristics. Along with these characteristics, information is required on the motor operating conditions, which include ambient environment, mounting, how the load is driven, and the load-speed characteristics.

This section provides means of evaluating pertinent motor-load and operating considerations prior to applying a motor.

CONSTANT-SPEED LOADS

1. Motor Load. The term "motor load" usually means the horsepower required by the driven machine. "Shaft horsepower" is a mechanically oriented synonym. Since 1 hp = 33,000 ft·lb/min, the motor load in hp = ft·lb/33,000. For rotating machinery ft·lb is the force required to turn the shaft multiplied by the number of feet through which the force moves; the ft·lb term becomes torque in lb·ft = radius × 2π × lb × r/min. Therefore,

$$\text{Motor load in hp} = \frac{\text{radius} \times 2\pi \times \text{lb} \times \text{r/min}}{33{,}000} = \frac{\text{radius} \times \text{lb} \times \text{r/min}}{5250}$$

The relationship is established by the following formula:

$$\text{hp} = \frac{\text{lb}\cdot\text{ft} \times \text{r/min}}{5250}$$

where lb·ft is the torque, or the effort required to turn the load. Motor load is therefore best described as the torque required by the load. In addition, the torque requirement is usually dependent upon the speed at which the torque is required. To describe a motor load accurately, consideration must be given to the torque requirements under the following conditions.

a. Breakaway. This torque is normally defined by motor manufacturers as "locked-rotor" torque. It is the torque required to start a shaft turning. Methods of bearing lubrication and types of lubricants have a pronounced effect on this torque requirement. Some loads are harder to start turning than others. Consider, for example, the case of grinding mills of the types called ball mills, rod mills, or pebble mills. For generations these grinding mills were built with grease-lubricated bearings. After standing idle overnight they were difficult to start on cold mornings. The motor had to supply at least 150 percent of "full-load" running torque to start the mill. Nowadays mill bearings are pressure-lubricated to float the trunnions on a film of oil before the motor is energized. As a result the torque required to start a modern mill may be 70 to 90 percent of full-load torque. Another case where breakaway torque must be carefully considered is in crushers whose crushing chambers are full of material. To allow a crusher shaft to start turning when its bearings are dry and its chambers are full of material, more than 200 percent locked-rotor torque may be required. High-torque-jog reversing capabilities have been provided to permit starting some crushers under such conditions.

b. Accelerating or pull-up. This torque, usually expressed in percent of running torque, is required to accelerate the load from standstill to full speed. It is the torque required not only to overcome friction, windage, and product loading, but also to overcome the inertia of the machine. The torque required by a machine may not be constant after the machine has started to turn. This type of load is particularly characteristic of fans and centrifugal pumps and also of certain machine tools.

Another type of load different from pumps and fans but which also demands high accelerating torque is that imposed by vibrating screens whose vibration results from rotating eccentric weights. The ratio of accelerating to full-speed running torque may reach 3:1 in some screen designs.

In considering the torque required by a machine during acceleration, its maximum torque required is the most significant. The minimum accelerating torque capability of the driving motor must exceed the maximum accelerating torque required by the machine. Special consideration must be given to the selection of the motor to assure that it will have the necessary thermal capacity to bring the machine to full speed. During the period of acceleration one-half of the energy input to the motor is absorbed by the motor rotor circuit while the other half is stored in the driven machine. In larger cage motors of conventional design the temperature rise of the rotor may limit its ability to accelerate a high-inertia load to its full speed.

High-inertia drives, including a wide range of machinery, usually with heavy or relatively high-speed rotating elements, are treated in more detail in Art. 5.

c. Pull-in. This torque is of importance in the use of a synchronous motor to drive a machine. As far as the driven machine is concerned, it corresponds to the accelerating-torque requirement.

d. Peak. The peak torque is the maximum momentary torque that a machine may require from its driving motor. The peak torque required by a load is directly related to the breakdown, or pull-out, torque for its driving motor. High peak-torque requirements for brief periods of time are available from the breakdown torque of an induction motor assisted by the inertia of the rotating system. If the peak-torque requirement is of any appreciable duration, however, it is necessary that the breakdown torque exceed the peak-load requirement. The term "pull-out" torque usually refers to the maximum running torque of a synchronous motor. Inertial energy cannot assist the pull-out torque in carrying a peak-torque requirement of the load unless the motor is of the nonsalient-pole or the reluctance type. It is therefore necessary in practically every instance to consider specifying a synchronous motor whose breakdown or pull-out torque exceeds the peak running torque of the load.

For some basic-materials-processing machinery, such as rock or ore crushers, the peak-torque requirements may be 300 percent or more of the full-load running-torque capabilities of their drive motors. In the case of gyratory crushers the breakdown torques of the driving motors should exceed those peaks to prevent stalls since the inertias of their rotating systems are relatively small. Jaw crushers, such as punch presses, store most of their energy in flywheels to supply their peak-torque requirements. Their driving motors are quite small, horsepowerwise, but they should be of the high-slip induction type to permit their flywheels to deliver their energy upon demand.

It is apparent from these considerations that once a load has been brought up to speed, the motor must be sized both on a thermal basis and on a torque basis to match the load.

ADJUSTABLE-SPEED LOADS

Additional mechanical requirements of motor loads must be considered when machines require adjustable-speed drives. Load characteristics require further definitions to describe them adequately.

2. Constant Torque. The term "constant torque" refers to loads whose horsepower requirements vary linearly with changing speeds during normal operation. Most machinery for processing basic materials falls in this category. Conveyors of nearly all types, some crushers, some coolers, some rock-grinding machines, rotary kilns, and a host of other machines require practically the same torque whether running fast or slowly when operating normally. Constant-displacement pumps delivering liquid against a constant head at varying speeds are another prime example of a constant-torque load.

3. Constant Horsepower. Such a load absorbs the same amount of horsepower regardless of its speed during normal operation. Machine tools normally used for production purposes are prime examples of constant-horsepower loads. Reel motors also fall into this class under certain operating conditions, and some types of earth-moving equipment may be constant-horsepower loads. For this type of load, torque requirements decrease as the speed increases. Traction loads frequently approach the characteristics of constant horsepower.

4. Variable Torque. A load whose torque requirements vary with speed in any fashion other than those mentioned in Arts. 2 and 3 is in this class. Centrifugal pumps and blowers, whose torque requirements increase approximately as the square of the speed, are examples of this type of load. Some mixers are also variable-torque loads.

The type of solid-materials feeder having a rotating vibration generator that operates near its critical speed presents a torque ratio at low speed compared with that at high speed that is approximately the fifth root of the speeds.

Most loads, when abnormally operated, present variable-torque characteristics to their drives.

5. Common Machine Characteristics. Table 1 presents the load characteristics of a large variety of common types of machines found in industrial plants. In making this tabulation, it was assumed that in those cases where the load inertia in lb·ft^2 is more than six times the normal Wk^2 of an electric-motor drive, special drive characteristics are required. This ratio is especially true where the drive is to be an internal-combustion engine or a turbine. NEMA Standard MG1-20.42 lists normal load Wk^2 capabilities for certain induction motors. These characteristics have been used as a guide in arriving at the ratios listed. Normal load Wk^2 for small motors (1 through 200 hp) have been estimated in arriving at the tabulated inertia ratios. The Wk^2 ratio listed in the table tends to be the maximum encountered in the tabulated types of loads and may give conservative results.

The inertia of a load is generally expressed as Wk^2 in lb·ft^2, where W is the weight of the rotating parts in pounds and k is the radius of rotation of the center of gravity of the rotating parts (usually referred to as the radius of gyration). In general k is expressed in feet. The difference between the torque (including friction) required by the load to perform its function, such as transporting material, pumping water, mixing, etc., and the torque supplied by the drive is applied to accelerate the load. The greater the accelerating torque is, the less the time required for acceleration. Consequently less thermal dissipation is required from the drive. The relationship is expressed by

$$t = \frac{Wk^2(n_2 - n_1)}{308T_a}$$

where t = accelerating time, s
n_2 = final speed, r/min
n_1 = initial speed, r/min
T_a = accelerating torque available from drive, lb·ft

Where the accelerating torque from the drive is not constant through the accelerating speed range (such as that exerted by an electric motor), the average torque should be determined and used as T_a. To refer Wk^2 from a load turning at a lower speed to a drive rotating at a higher speed, the load Wk^2 should be multiplied by the low speed squared divided by the drive speed squared. Belted or geared drives would require this computation. Likewise, in the case of a slow-speed prime mover, the Wk^2 of the load would be multiplied by the load speed squared divided by the prime-mover speed squared. In general a standard motor and control should accelerate its load from rest to full speed in 30 s. Also in the table certain load characteristics which should be recognized in specifying the drive are indicated by a note to "see text." These are categories as follows.

6. Duty Cycle. Duty cycle includes frequent starts, plugging stops, reversals, or stalls. These characteristics are usually involved in batch-type processes and may include tumbling barrels, certain cranes, shovels and draglines, dampers, gate- or plow-positioning drives, drawbridges, freight and personnel elevators, press-type extractors, some feeders, presses of certain types, hoists, indexers, boring machines, cinder-block machines, keyseating, kneading, car-pulling, shakers (foundry or car), swaging and washing machines, and certain freight or passenger vehicles. The list is not all-inclusive. The drives for these loads must be capable of absorbing the heat generated during the duty cycles. Adequate thermal capacity would be required in slip couplings, clutches, or motors to accelerate or plug-stop these drives or to withstand stalls. It is the product of the slip speed and the torque absorbed by the load per unit of time that generates heat in these drive components. All the events that occur during the duty cycle generate heat, which the drive components must dissipate.

Since motors are rated by horsepower, designs have been standardized to include locked-rotor, accelerating, and breakdown torques (Art. 1). It is seen that each of the three torques listed must be sufficiently large, or a motor cannot handle its load. To assure that a motor is not overpowered, the duty cycle of the load must be considered. A load whose torque requirement varies in accordance with a repetitive cycle would have its motor horsepower determined on the basis of the internal heat generated in the motor after it has reached full speed. Core loss, windage, and friction losses are considered constant no matter what load the motor is carrying. The copper losses, however, are a function of the torque output of the motor, and so determine its horsepower rating. They are proportional to the square of the line current to the motor. The rms horsepower concept is therefore generally used to calculate the motor rating where the speed remains essentially constant. As an example, a gyratory crusher is being fed by dump trucks which arrive alternately from two separate quarries. The time required for each

TABLE 1. Load Characteristics of Various Machines

Load description	Load torques, % full-load drive torques			Inertia ratio (see notes)	Ambient	Environment	Mounting	How driven?	Adjustable speed range, max	Remarks
	Breakaway	Accelerating	Peak running							
Actuators:										
Screw-down (rolling mills)	200	150	125	1	A	D	R	L	2:1	Could be intermittently operated
Positioning	150	110	100	1	3C	5	RV	L	6:1 S	See text
Agitators:										
Liquid	100	100	100	2	AW	FJ	RV	B	3:1	Can be direct-connected
Slurry	150	100	100	2	AW	CDJ	RV	B	....	Settling of solids when idle may cause difficult restarting
Barrels, tumbling (foundry)	50	150	100	3	A	DJ	R	B		
Bars, boring, rotary kiln	75	125	100	4	A	DJ	R	B	4:1	Reversing required
Beaters:										
Standard	110	120	100	4	AW	DJ	R	B		
Breakers	110	120	120	4	AW	5	R	B		
Blowers, centrifugal:										
Valve closed	30	50	40	60	3	5	R	BLT	3:1	Some applications would require constant speed
Valve open	40	110	100	60	3	5	R	BLT	3:1	Same as above
Blowers, positive-displacement, rotary, bypassed	40	40	100	2	AW	DJ	R	L		
Breakers, flake, starting loaded	150	110	100	2	A	DEJ	R	B		
Calenders, textile or paper	75	110	100	3	AW	J	R	L		
Card machines, textile	100	110	100	2	A	E	R	L		
Centrifuges (extractors)	40	60	125	50	AW	DJ	R	L	....	Starting unloaded*
Chippers, wood, starting empty	50	40	200	100 max	AW	D	R	T		
Compactors, solids	100	110	125	1	A	DEJ	R	L	2:1	Constant speed may be used
Compressors, axial-vane, loaded	40	100	100	8	AW	DJ	R	L		
Compressors, reciprocating, start unloaded	40	50	100	10	A	DJ	R	BL		
Converters, copper, loaded	150	150	125	8	AH	D	R	L	4:1 S	See text
Conveyors, belt (loaded)	110	130	100	4	A	CDJ	R	B	....	Inertia depends on load (see text)

Conveyors, drag (or apron) . . .	100	150	100	2	AH	D	R	B	. . . .	Starting loaded. Inertia depends on load (see text)
Conveyors, screw (loaded)	150	100	100	1	AH	CDEJ	R	B		
Conveyors, shaker-type (vibrating)	50	150	75	6	AH	DJ	I	B	. . . .	See text
Coolers, hot solids, rotary (loaded)	175	140	100	2	AH	DJ	R	L		
Coolers, grate, reciprocating (loaded)	50	125	75	1	A	DJ	R	B	3:1	
Coolers, grate, oscillating (loaded)	50	100	40	1	A	DJ	R	B	1.5:1	
Coolers, grate, traveling (stoker-type)	100	110	100	1	AH	D	R	L	6:1	Starting loaded—torque-limiting drive is desirable
Cranes, traveling:										
Bridge motion	100	300	100	4	AH	CDJ	R	L	10:1	Drives must be suited to duty cycle and service. Hoisting inertia depends on load
Trolley motion	100	200	100	4	AH	CDJ	R	L	10:1	
Hoist motion	50	200	100	. . . .	AH	CDJ	R	L	10:1	
Crushers, gyratory:										
Starting unloaded	50	60	300	2	AC	DJ	R	B	. . . .	Large crushers are usually direct-connected
Choke-fed	100	200	300	2	AC	DJ	R	B	. . . .	Reverse jogging may be necessary to start
With feeder	100	150	150	2	AC	DJ	R	B		
Crushers, jaw:										
Starting unloaded	50	100	200	10	AC	DJ	R	B	. . . .	Usually started unloaded*
Choke-fed	. . .	. . .	200	10	AC	DJ	R	B		
Crushers, pulverizing (hammer-mill)	50	100	150	25	A	CDJ	R	L	. . . .	Usually started unloaded*
Crushers, roll:										
Starting unloaded	50	50	. . .	10	A	DJ	R	B	. . . .	*
Choke-fed, loaded	200	200	150	10	A	DJ	R	B		
Cutter bars, balling drum	50	150	150	25	A	DJ	R	L	2:1	See text
Cutter heads, dredge	50	125	150	2	W	. . . . .	R	L	S	
Dampers, fan, centrifugal, cold	200	200	100	1	AC	CDEJ	RV	L	S	See text*
Dampers, fan, centrifugal, hot	400	300	100	1	AH	CDJ	RV	L	S	See text*
Drawbridges	100	125	100	10	AW	. . . . .	R	L	. . . .	Drive coordination required*
Draw presses (flywheel)	50	50	200	10	A	D	R	B	. . . .	High inertia*
Drill presses	25	50	150	2	A	D	R	B		

TABLE 1. Load Characteristics of Various Machines (*Continued*)

Load description	Load torques, % full-load drive torques			Inertia ratio (see notes)	Ambient	Environment	Mounting	How driven?	Adjustable speed range, max	Remarks
	Breakaway	Accelerating	Peak running							
Drums, balling (ore)	50	125	100	4	A	D	R	B	2:1	See text
Dryers, rotary (rock or ore)	50	150	100	4	AH	D	R	B		
Dryers, grain	50	100	90	2	AH	DJ	R	L		
Edgers (starting unloaded)	40	30	200	10	A	DE	R	B	....	High inertia*
Elevators, bucket (starting loaded)	150	175	150	2	AHC	CDJ	R	L	....	Antirollback required
Elevators, freight (loaded)	100	125	100	4	ACW	DFJ	R	L	S	
Elevators, man lift	50	125	100	1	AH	5	R	L		
Elevators, personnel (loaded)	110	150	100	4	AHC	5	R	L	10:1	Speed range required during acceleration and deceleration
Escalators, stairways (starting unloaded)	50	75	100	2	A	.....	R	L		
Extractors (press type)	50	150	150	1	3	5	RV	L		
Extruders (rubber or plastic)	100	150	100	1	3	5	R	L		
Fans, centrifugal, ambient:										
Valve closed	25	60	50	25	ACW	5	R	BT	2:1	*
Valve open	25	110	100	25	ACW	5	R	BT	2:1	*
Fans, centrifugal, hot gases:										
Valve closed	25	60	100	60	AH	DJ	R	R	3:1	*
Valve open	25	200	175	60	AH	DJ	R	B	3:1	Peak running overload torque occurs when handling colder gases
Fans, propeller, axial-flow	40	110	100	25	AH	DJ	R	BT	....	High inertia*
Feeders, belt (loaded)	100	120	100	2	3	5	R	B	10:1 S	
Feeders, distributing, oscillating drive	100	150	100	4	AW	CDJ	R	B	6:1 S	Starting loaded
Feeders, screw, compacting rolls	100	100	100	1	AW	CDEJ	RV	L	S	Starting loaded, torque limited
Feeders, screw, filter-cake	100	100	100	1	3	DJ	R	B	3:1 S	Starting loaded

Feeders, screw, dry	150	100	100	1	AH	CDEJ	R	B	3:1 S	Starting loaded
Feeders, slurry, ferris-wheel	110	100	75	2	AW	DJ	R	B	3:1 S	Starting loaded
Feeders, table	125	110	100	2	A	DJ	R	L	6:1 S	Starting loaded
Feeders, vane-type	150	80	75	1	AW	CDEJ	R	L	6:1 S	Starting loaded
Feeders, vibrating, magnetic	100	100	100	....	AH	CDEJ	R	L	3:1 S	Starting loaded. No rotating member
Feeders, vibrating, motor-driven	50	150	100	4	AH	CDEJ	RV	L	3:1 S	Starting loaded
Forge presses	25	50	150	10	AH	D	R	B	....	High inertia*
Frames, spinning, textile	50	125	100	2	A	E	R	B		
Furnaces, holding, copper	150	125	100	4	AH	D	R	L	4:1 S	Overhauling load
Gates, diverting, solids	200	125	100	1	3	CDJ	R	L	S	
Gates, locks, hydraulic	25	200	200	2	W	.....	R	L	S	*
Generators, electric, flywheel-type	50	100	400	100	A	.....	R	L	....	High inertia*
Generators, electric, general use	25	30	150	3	AH	D	R	L		
Generators, electroplating	25	30	100	3	AW	F	R	L		
Generators, welding	30	50	200	3	AHC	D	RV	L	....	Peak torque required when arc is struck
Grates, indurating (preheater)	100	110	200	3	AHC	DJ	R	L	4:1 S	
Grates, stoker (furnace)	75	110	100	1	AH	D	R	L	2:1 S	
Grinders, metal	25	50	100	2	A	D	RV	LB	....	Starting unloaded
Grinders, pulp or meat	40	50	150	2	AW	F	R	L	....	Starting unloaded
Grinders, pulp-magazine type	50	50	150	5	AW	CJ	R	L	....	Starting unloaded
Grinders, pulp, pocket-type	40	30	150	5	AW	CJ	R	L	....	Starting unloaded
Hammers, power, flywheel	50	50	150	10	A	D	R	B	....	High inertia*
Hoists, skip	100	150	100	10	A	D	R	L	6:1	*
Hydropulpers	125	125	150	1	W	.....	R	L		
Indexers	150	200	150	2	A	D	R	L	....	*
Ironers, laundry (mangles)	50	50	125	1	3	.....	R	L	S	*
Jointers, woodworking	50	125	125	1	A	E	R	L		
Jordans, plug out	50	50	150	9	AW	.....	R	L	....	*
Kilns, rotary (loaded)	200	125	125	4	AH	DJ	R	B or L	6:1 S	See text
Log washers, rock or ore (loaded)	75	125	150	1	AW	J	R	B		
Looms, textile, without clutch	125	125	150	2	A	E	R	L	....	*
Machines, boring (loaded)	150	150	100	2	A	D	R	L	6:1	Constant-hp drives may be required
Machines, bottling	50	50	100	2	AW	.....	R	B		

TABLE 1. Load Characteristics of Various Machines (*Continued*)

Load description	Load torques, % full-load drive torques			Inertia ratio (see notes)	Ambient	Environment	Mounting	How driven?	Adjustable speed range, max	Remarks
	Breakaway	Accelerating	Peak running							
Machines, briquetting	100	125	150	2	A	CDEJ	R	L		
Machines, buffing, automatic	50	75	100	2	A	DJ	R	L		
Machines, cinder-block, vibrating	50	150	70	4	AW	DJ	I	B	....	Frequent starting and plugging required
Machines, keyseating	25	50	100	2	A	D	R	B		
Machines, kneading	50	150	175	1	AW	J	R	L		
Machines, polishing	50	75	100	2	A	DJ	R	L		
Mills, attrition (starting unloaded)	100	60	120	10	AW	CDEJ	R	L	....	High inertia*
Mills, autogenous, grinding (prelubricated)	90	140	100	10	A	DJ	R	L	....	See text, high inertia
Mills, ball, grinding (prelubed bearings)	90	130	100	6	AW	CDJ	R	L	....	See text, high inertia
Mills, ball, grinding (dry bearings)	140	130	100	6	AW	CDJ	R	L or B	....	See text, high inertia
Mills, B & W, coal (loaded)	150	110	100	3	A	CDE	R	L		
Mills, bowl, Raymond, coal (loaded)	130	120	100	5	A	CDE	R	L	....	Grinding mill and exhauster coupled together
Mills, Bradley-Hercules (loaded)	150	110	100	3	A	CDEJ	R	L		
Mills, flour, grinding	50	75	100	6	A	E	R	B	....	High inertia*
Mills, pan	125	125	150	4	A	D	R	B		
Mills, rod or tube, grinding (prelubed bearings)	90	120	100	6	AW	CDJ	R	L	....	See text
Mills, rod or tube, grinding (dry bearings)	140	130	100	6	A	CDJ	R	B		
Mills, rolling metal:										
Billet, skelp and sheet, bar	50	30	200	1	A	D	R	L	1.5:1	Accurate speed control required
Brass and copper finishing	120	100	200	1	A	D	R	L	1.5:1	See text

Brass and copper roughing	40	30	200	1	A	D	R	L	1.5:1	See text
Merchant mill trains	50	30	200	1	A	D	R	L	1.5:1	See text
Plate	40	30	250	1	A	D	R	L	1.5:1	See text
Reels, wire or strip	100	100	100	2	A	D	R	L	20:1	See text. Torque controlled
Rod	90	50	200	1	A	D	R	L	1.5:1	See text
Sheet and tin (cold rolling)	150	110	200	1	A	D	R	L	1.5:1	See text
Strip, hot	40	30	200	1	A	D	R	L	1.5:1	See text
Structural and rail finishing	40	30	200	1	A	D	R	L	1.5:1	See text
Structural and rail roughing	40	30	250	1	A	D	R	L	1.5:1	See text
Tube	50	30	200	1	A	D	R	L	1.5:1	See text
Tube piercing and expanding	50	30	250	1	A	D	R	L	1.5:1	See text
Tube reeling	50	30	200	1	A	D	R	L	10:1	See text
Mills, rubber	100	100	200	1	A	CE	R	L	....	*
Mills, saw, band	50	75	200	100	A	D	R	B	....	High inertia*
Mills, stamp	50	75	150	10	A	D	R	B	....	High inertia*
Mills, wash	25	30	100	1	A	.....	R	L	....	*
Mixers, banbury	125	100	250	1	A	CE	R	L	....	*
Mixers, concrete	40	50	100	2	AW	D	R	L		
Mixers, dough	100	125	100	1	AW	E	RV	L		
Mixers, liquid	100	100	100	2	AW	EFJ	RV	B or L		
Mixers, sand, centrifugal	50	100	100	4	A	D	R	B		
Mixers, sand, screw	100	100	100	1	A	D	R	B		
Mixers, slurry	100	125	100	1	A	DJ	RV	B		
Mixers, solids (mullers)	100	125	175	1	A	CDJ	RV	BT		
Pans, pelletizing, ore	50	100	100	6	A	DJ	RV	B	2:1	
Planers, metalworking	50	150	150	4	A	D	R	L	4:1	Plugging and reversing service required
Planers, woodworking	50	125	150	10	A	D	R	B	....	High inertia*
Plasticators	125	100	250	1	A	5	R	L		
Plows, conveyor, belt (ore)	150	150	200	1	A	DJ	RV	L	....	See text
Positioners, indexing (machine tool)	50	200	100	4	A	D	RV	L	....	See text
Presses, brick	100	175	150	4	A	DJ	R	B		
Presses, drill, production, automatic	50	60	125	1	A	D	RV	BL	....	Frequent starts with plugging
Presses, pellet (flywheel)	50	75	150	10	A	D	R	B	....	High inertia*
Presses, printing, production-type	100	150	150	4	A	.....	R	L	20:1	*
Presses, punch (flywheel)	50	75	100	10	A	D	R	B	....	See text, high inertia*

TABLE 1. Load Characteristics of Various Machines (*Continued*)

Load description	Load torques, % full-load drive torques			Inertia ratio (see notes)	Ambient	Environment	Mounting	How driven?	Adjustable speed range, max	Remarks
	Breakaway	Accelerating	Peak running							
Presses, punch (no flywheel)	10	40	150	1	A	D	R	B	....	See text, high inertia*
Pug mill (solids mixing)	150	125	100	1	A	CDJ	RV	T	....	Solids may "set up" on emergency shutdown
Puller, car	150	110	100	25	A	D	R	L	....	Inertia depends on number of cars
Pumps, adjustable-blade, vertical	50	40	125	1	AW	.....	RV	T	....	Unloaded start
Pumps, centrifugal, discharge open	40	100	100	1	AW	FJ	RV	T	....	Loaded start
Pumps, oil-field, flywheel	50	200	200	10	AC	D	R	B	....	*
Pumps, oil, lubricating	40	150	150	1	AC	D	R	L	....	Cold oil can cause drive overloads
Pumps, oil, fuel	40	150	150	1	AC	D	R	L	....	Peak torque caused by more viscous oils
Pumps, propeller	40	100	100	1	AW	F	RV	LT	....	Handling nonviscous fluids
Pumps, reciprocating, positive-displacement	40	30	150	1	AW	.....	R	B	....	Starting dry, handling nonviscous fluids
Pumps, reciprocating, positive-displacement	40	30	20	1	AW	.....	R	B	....	Bypassed, handling nonviscous fluids
Pumps, reciprocating, positive-displacement	100	100	150	4	AW	.....	R	B	....	3 cylinder, not bypassed, handling a nonviscous fluid
Pumps, screw-type, started dry	40	30	100	1	AW	F	R	L	....	Handling nonviscous fluids
Pumps, screw-type, primed, discharge open	40	100	100	1	AW	F	R	L	....	Handling nonviscous fluids
Pumps, slurry-handling, discharge open	100	100	100	1	AW	D	R	B		
Pumps, turbine, centrifugal, deep-well	50	100	100	2	AW	.....	RV	L		
Pumps, vacuum (paper-mill service)	60	100	150	4	AW	.....	R	L		

Pumps, vacuum (other applications)	40	60	100	4	A	....	R	L		
Pumps, vacuum, reciprocating	40	60	150	10	A	....	R	B	....	Starting unloaded*
Pumps, vane-type, positive-displacement	100	150	150	1	A	DJ	R	L	....	Viscous fluids may overload drive
Rolls, bending	150	150	100	2	A	D	R	L		
Rolls, compacting (loaded)	100	110	125	1	A	DJ	R	L		
Rolls, crushing (sugarcane)	50	110	125	2	AW	J	R	L		
Rolls, flaking	30	50	100	2	A	E	R	B		
Sanders, woodworking, disk or belt	30	50	100	1	A	D	R	L or B		
Saws, band, metalworking	30	50	100	4	A	D	R	B		
Saws, circular, metal, cutoff	25	50	150	6	A	D	R	L		
Saws, circular, wood, production	50	30	150	10	A	E	R	B	....	High inertia*
Saws, edger (see Edgers)										
Saws, gang	60	30	150	10	A	D	R	B	....	High inertia*
Saws, trimmer	40	30	150	10	A	D	R	B	....	High inertia*
Screens, centrifugal, paper-mill	50	100	100	50	AW	FJ	R	L	....	High inertia*
Screens, centrifugal (centrifuges)	40	60	125	50	AW	DJ	R	L	....	High inertia*
Screens, rotary, stone (trommel)	70	100	100	1	A	DJ	R	B		
Screens, vibrating	50	150	70	6	3	5	I	B	....	See text
Separators, air (fan-type)	40	100	100	15	A	CJ	R	L	....	High inertia
Shakers, foundry or car	50	150	70	6	AHC	5	K	B	....	See text
Shears, flywheel-type	50	50	120	10	A	D	R	B	....	See text*
Shovels, dragline, hoisting motion	50	150	100	4	A	CDJ	R	L	6:1	See text on Cranes*
Shovels, dragline, platform motion	50	100	100	4	A	CDJ	R	L	4:1	See text on Cranes*
Shovels, large, digging motion	50	200	200	3	ACW	CDJ	R	L	10:1	See text on Cranes*
Shovels, large, platform motion	50	100	100	4	ACW	CDJ	R	L	4:1	See text on Cranes*
Shredders (see Crushers, pulverizing)										
Sifters, shaker-type	50	100	70	3	A	EJ	IV	B		
Stokers, traveling-grate-type	50	110	100	1	A	CD	R	L	2:1	Torque-limiting drive is desirable
Swagers	100	110	150	1	A	C	R	L		

TABLE 1. Load Characteristics of Various Machines (*Continued*)

Load description	Load torques, % full-load drive torques			Inertia ratio (see notes)	Ambient	Environment	Mounting	How driven?	Adjustable speed range, max	Remarks
	Breakaway	Accelerating	Peak running							
Tension-maintaining drives	100	100	100	1	AW	DE	R	L	10:1	See text
Textile machinery	50	100	90	2	A	EJ	R	L	....	See text
Tools, machine	100	150	100	2	A	D	R	L	4:1	See text
Tools, machine, broaching, automatic	50	150	150	1	A	D	R	L	....	Frequent starts and plugged stops
Tools, machine, lathe, metal, production	50	200	200	2	A	D	R	L	....	Frequent starts, plugs and changes in depth of cut
Tools, machine, mill, boring, production, metal	100	125	100	4	A	D	R	L	20:1	Inertia depends on work on table
Tools, machine, milling, production	100	100	100	1	A	D	R	L	4:1	
Tools, machine, planer, production, metal (see Planers)										
Tools, machine, shaper, metal, automatic	50	75	150	2	A	D	R	B		
Vehicles, freight	200	200	200	50	AW	DJ	R	L	3:1	Special engineering required
Vehicles, passenger	100	400	200	25	AW	DJ	R	L	10:1	Special engineering required
Walkways, mechanized	50	50	100	2	A	.....	R	L		
Washers, laundry	25	75	100	4	AW	.....	R	B		
Winches	125	150	100	4	A	D	R	L	S	Drive must be coordinated with service
Wood hogs	60	100	200	30	AW	.....	R	L	.....	High inertia*

* Special drive may be required.

Notes for Table 1

Inertia Ratio. Load inertia compared within normal inertial capability of its drive motor (see elaboration below).

Ambient

A = High altitude, which for motors may be in excess of 3300 ft (1000 m) above sea level.
H = High temperatures, which for motors may be in excess of 40°C (104°F).
C = Extreme cold, −40°C (−40°F) or lower. Such temperatures can affect lubricants to prevent starts.
W = High humidity, continuous exposure to atmospheres of 100 percent relative humidity and/or frequent "hose-downs."
3 = Where A, H, and W may exist either simultaneously or individually at different times.

Environment

C = Atmosphere heavy with carbon dust such as occasionally exists in rubber plants or coal- and coke-handling facilities.
D = "Dirty," or atmospheres containing abrasive dusts.
E = Explosive atmospheres (dust or gas).
F = Atmospheres containing relatively high concentrations of acid fumes.
J = Atmospheres containing quantities of chemical dusts which may be corrosive or gummy after exposure to high humidity.
5 = All of the above.

Mounting

R = Bolted or securely fastened to a metallic base so that whatever vibration is generated by the driven machine is transmitted directly to the motor.
I = Mounting provides some degree of vibrational isolation between driven machine and motor.
K = Hinged mounting, usually tensioning belt, at least partially, with weight of motor.
V = Vertical mounting can be required.

How Driven?

B = Usually belted to motor.
L = Usually direct-connected to motor.
T = Direct-connected with axial thrust placed on motor bearings.

Adjustable Speed Range. Ratios indicate maximum ranges employed on equipment indicated.
S = Starting and stopping replace adjustable speed range.

Note. Any one application could not meet all conditions listed in the notes above.

truck to dump is 1 min. Table 2 shows the horsepower required, the times during which it is required, and the method of calculation.

According to the calculation on the table there are two NEMA motor ratings that can be used, viz., 150 and 200 hp. If other conditions are suitable, a motor with a rating of 150 hp with a 1.15 service factor should be satisfactory because 150 × 1.15 = 1.72 hp. The breakdown torque in horsepower of a 150-hp NEMA B or NEMA C motor is 150 × 1.9 = 261 hp. This peak torque is also more than adequate. If the altitude, ambient temperature, or motor enclosure is such that the standard motor would have a unit service factor, the 200-hp NEMA standard motor would be required.

TABLE 2. Typical Horsepower and Time Duty Cycle for Crusher Motor

	Hp	Time, s	Hp × min
Quarry 1	200	45	30,000
Idle	10	15	25
Quarry 2	175	45	22,968
Idle	10	15	25
Total			53,018
Total ÷ 2 min			26,509
Square root of result = rms hp			163 hp

7. Multiple Drives. Certain loads require more than one drive. Usually multiple drives for the same machine must be synchronized for torque and speed output. Certain belt conveyors which are long or inclined or which transport very dense material may require several torque-balanced drives to prevent rupture of the belting. Large traveling cranes may have multimotor bridge drives which are torque- and speed-synchronized to prevent skew. Some large gyratory and roll crushers require multiple drives because unwieldy mechanical rotating elements would otherwise be required. Multiple drives may be used because of economics or maintenance requirements, or because there is no other way to do the job. Examples of multiple drives used because of economics are rotary kilns, grinding mills, and vehicles. Some loads which require multiple drives because of their very nature are belt conveyors, large traveling cranes, roll crushers, metal-rolling mills, and certain types of paper and textile machinery.

Multiple drives in paper mills which handle the same product are typically the paper machine with its drying rolls and reels whose drives must be closely torque- and speed-controlled to prevent rupture of the paper being made. The spinning of yarn in a textile mill requires like consideration. Lack of close torque and speed control of any drive handling the yarn will result in its rupture. Metal-rolling mills, however, may have the different rolls through which the product passes torque- and speed-controlled for still another reason. The forming and dimensioning of the product may also be accomplished by properly torquing the different drives.

Some large vibrating screens and feeders use multiple drives because single drives large enough to do the job would be so heavy and bulky that other parts of the machine would have to be made inordinately large and heavy to accommodate them.

UNUSUAL TORQUE REQUIREMENTS

8. Rotary Kilns. Undoubtedly the largest and heaviest piece of rotating equipment is the rotary kiln. Some of the largest require over 1000 hp and rotate at less than 2 r/min. Sleeve-type support bearings are practically dry of lubricant after the kiln has been at rest. Thus an unusually high breakaway torque is required. Were the load in the kiln not at rest at the bottom of the kiln at breakaway, the drive might not be able to start the kiln. Because of ring formation or unsymmetrical coating thickness, continuous running torques may periodically exceed full load of the drive. Thermal shielding of the prime mover (if a motor) is also desirable because of the high ambient temperatures usually encountered. Figure 1 shows how the shafting, gearing, and coupling torque rises under the initial torque loading from the motor to about 175 percent load torque before the kiln actually turns. Torque oscillations then occur until the kiln accelerates nearly to operating speed. The load does not break away until the crest of the torque curve near the 880,000-lb·in ordinate is reached. Rotary kilns, like some other equipment in the rock- and ore-processing industries, require their drives to be sized empirically on the basis of experience. The normal running torque cannot always be foretold. Many kilns are overpowered and some have been underpowered.

Some materials being processed have a steeper angle of repose than others and tend to absorb heat more readily than those materials that lie like water in the bottom of the kiln. Kilns operated with a steep angle of repose have a tendency to require more torque than their counterparts. Some kilns may also tend to require more torque at lower speeds than they do when operating at their maximum speed. This knowledge generally engenders a philosophy of oversizing the drive to ensure that enough torque is available when required. Since a hot kiln cannot be stopped indefinitely, it is usual to have a source of standby power to rotate it intermittently until it has cooled. This so-called emergency drive should be capable of devloping about 200 percent of maximum running torque to ensure that it will turn the kiln when required to do so.

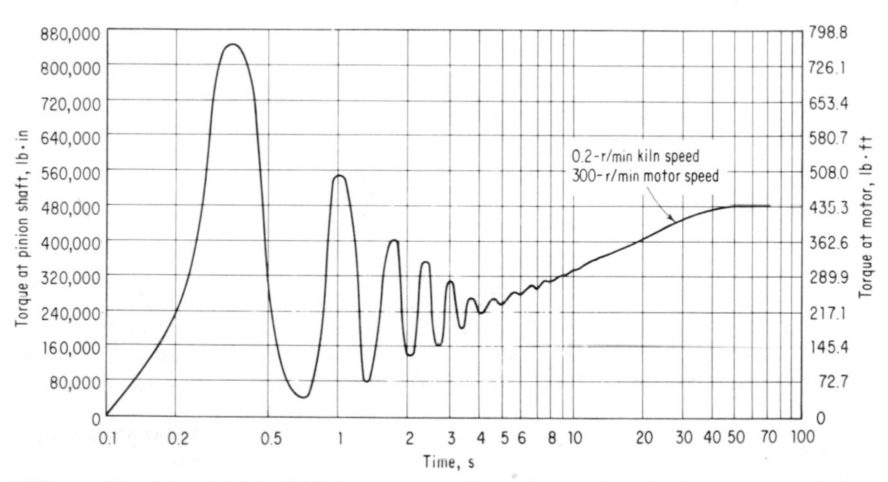

FIG. 1. Torque at starting of a 13- by 17-ft rotary kiln, started with load in kiln after a 2-h shutdown.

9. Grinding Mills. In some industries, such as mining, cement, and ore beneficiation, the largest block of energy is used in grinding. To reduce the grinding cost per ton grinding mills have been increasing in size to 6000 hp. Undoubtedly they will become larger. Since large mills are motor loads (usually synchronous), the maximum size becomes a question of power-distribution economics and mechanical design (Ref. 1). These mills all have high-pressure-lubricated bearings which reduce breakaway-torque requirements. Essentially the requirements for grinding mills consuming 1000 hp or more are the same no matter whether they are autogenous, ball, rod, or tube type. Some mills requiring less than 1000 hp may have oil- or grease-lubricated bearings, which are essentially dry when starting. Such mills may require as much as 140 percent breakaway torque. The accelerating and running torques are the same, except for the autogenous mill, whose Wk^2 is higher than that of either the ball, rod, or tube mills.

The electric-current inrush to the motors has developed problems which have engendered some devious schemes in attempts at solution. The most economical and efficient motor is the salient-pole full-voltage-starting synchronous machine. When a ball mill with this type of motor is started, torque pulsations occur, as shown in Fig. 2. Since the mill itself rotates at about 20 r/min, the gearing and other mechanical parts are subjected to severe stresses synchronously with these torque pulsations. Likewise electric surges occur at the same time.

To eliminate some of these undesirable side effects of starting the load directly connected to this type of motor a friction clutch may be installed between the mill gearing and the motor. The clutch is engaged to drive the mill after the motor is at full speed. Clutch engagement must be controlled to prevent pulling the motor out of synchronism with its electric-power supply. Figure 3 shows the variation

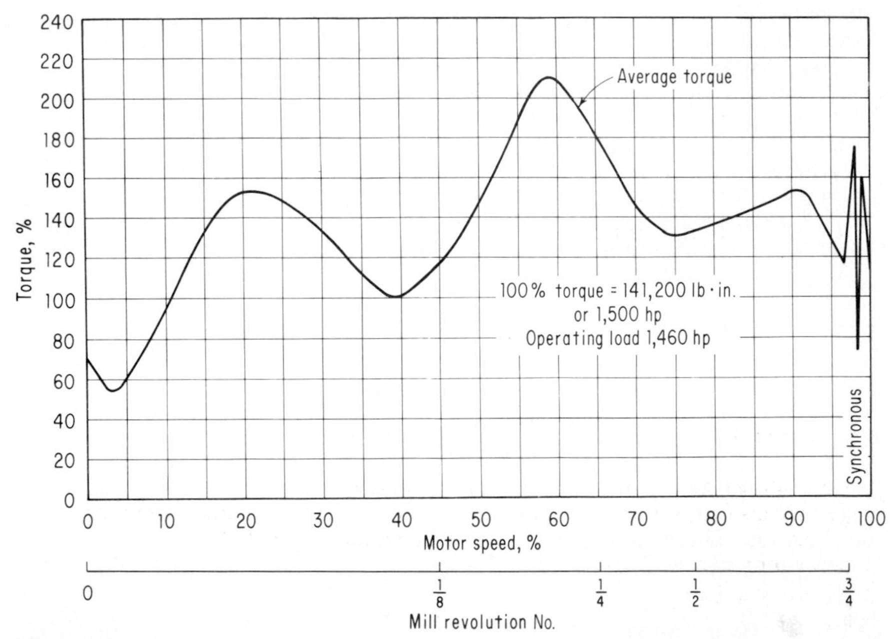

FIG. 2. Starting-torque pulsations when starting a 1500-hp salient-pole full-voltage-starting synchronous motor driving a grinding mill.

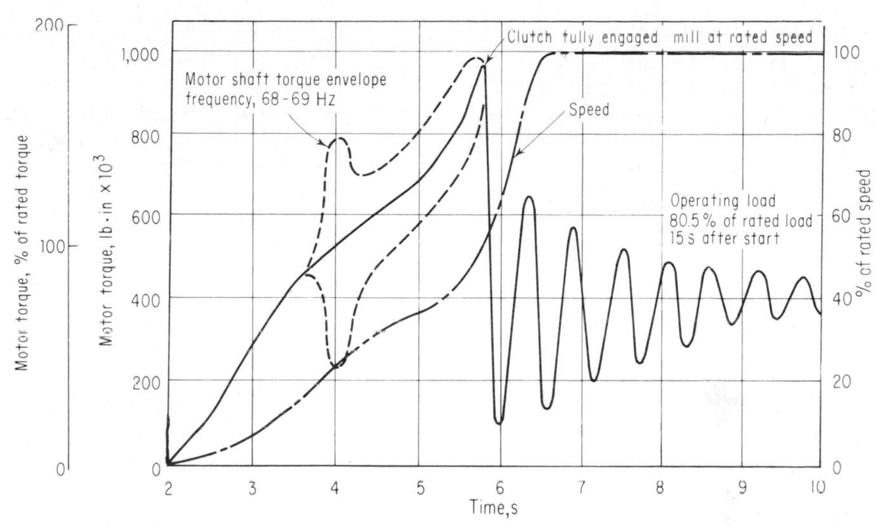

FIG. 3. Starting-torque pulsations are greatly reduced when a clutch is used that is engaged after the motor is up to speed.

of torque required by the mill with respect to time and the instantaneous speeds also plotted with respect to time. The displacement of one curve with respect to the other is not intentional. The test data just plotted that way. Peak pull-out torque required by this mill and clutch is about 180 percent of full-load motor torque. Acceleration of the mill from standstill to full speed occurs in less than 4 s. If the clutch were inflated less rapidly, the peak accelerating torque would be reduced. The accelerating time would be lengthened, of course.

10. Vibrating Screens, Feeders, Conveyors, and Oscillating-Grate Coolers. Some of these machines operate at or near their critical speeds, and others (such as the screens) operate considerably above their critical speeds. Vibration is produced by the rotation of eccentric weights. Enough torque must be available from the drive to cause the weights to rotate. Torque requirements then decrease as the speed approaches the critical speed of the machine. The torque required to raise the speed beyond the critical speed increases rapidly and then becomes relatively constant.

The torque required from the drive is further increased by the amount of power required to transport whatever material is being handled by the machine. Those machines in this category which operate near or at their critical speeds, such as feeders, conveyors, and coolers, require operating torques usually less than half of those required for acceleration from standstill.

11. Repetitive Machines. These include banbury, dough mixers, plasticators, large shovels (digging), tension-maintaining drives, some textile machinery, and machine tools. Machines operating on a batch basis usually have definitely predictable duty cycles. During parts of these duty cycles, torque requirements may increase considerably. The drives, of course, must be able to supply the maximum torque but do not have to bear a continuous rating at this maximum torque. Because of this characteristic, an economic advantage can be taken in selecting suitable drives for these machines.

12. Overhauling Loads. Copper converters, cranes, cutter bars, elevators, escalators, hoists, conveyor-belt plows, vehicles, and winches are examples of overhauling loads. The drives for these machines must be able to brake their loads during parts of their duty cycles. Braking may be accomplished with friction, magnetic, or fluid brakes as separate units, if the prime mover does not have the means of braking. Some of these machines are also required to hold their loads stationary. Cutter bars act as pendulums. The overhauling load on the downswing must not cause an appreciable increase in speed. A full copper converter beginning to pour its charge will require more torque to hold the load than that required when the converter is almost empty.

13. Balling Drums. To obtain the optimum retention time for making the best pellets, it may be necessary to adjust the slope of a balling drum. Changes in slope require different torques. The drum manufacturer may recommend a drive which approaches constant horsepower (i.e., greater torque availability at lower speeds) to compensate for the torque requirements. Balling-drum operators require minimum deviation of speed from their settings for both the drum and its cutter bar. To effect the desired speed regulation, control of their drives should be held within 1 percent from no load to full load.

14. Wash Mills. An exceptionally small accelerating torque is required for this drive. The clay is agitated with beams, suspended from chains. Too much acceleration will cause the beams to strike the walls and cause damage. Torque-limit acceleration is most suitable for this application.

15. Positioning Drives. Dampers, diverting gates, indexers, cinder-block vibrating machines, plows, and positioners all require frequent starting and stopping under maximum-torque conditions and therefore present special problems. The duty cycles of the cinder-block vibrators and certain diverting gates and indexers can be calculated. Those of dampers, plows, and other positioners, such as speed-adjusting scoop tubes for adjusting speed outputs of liquid couplings whose speeds are under automatic control, are unknown. The worst conditions of operation should therefore govern the design of these drives. Indexing duty cycles can usually be calculated when used on production machines performing repetitive operations. When indexers are used on equipment designed for other tasks, it is usually best to design for the worst condition, which is usually continuous starting and stopping for 8 h per day.

16. Cranes, Hoists, and Freight or Personnel Elevators. The duty cycles of these machines can be calculated and the drives designed accordingly. If any of this type of machine has its function changed, its drive may also have to be changed. A comprehensive treatment and calculations of the duty cycles of cement-plant overhead storage cranes are contained in Ref. 2.

17. Flywheel Machines. The purpose of the flywheel is, of course, to deliver part of its inertia to supply peak-torque demands of the machine (see Sec. 15). It is therefore necessary for its driving prime mover to slow down upon increases in torque requirements but still be able to restore the flywheel to its original (maximum) speed between load peaks. On machines of this type, and especially on jaw crushers and punch presses, even the peak torques are limited by springs or pneumatic cylinders. Since the machines in this category do have flywheels, particular attention must be given in each case to the torque requirements and the time required to accelerate these flywheels during starting (Ref. 3). Most therefore fall into the category of high-inertia loads.

REFERENCES

1. A. R. Olds, Jr., K. E. Olsen, and I. M. Watson, "More on Large Grinding Mill Drives—Background and Test," presented at the IEEE Cement Industry Conference (Apr. 1964).
2. Leo E. Swanson, "Requirements for Cement Plant Overhead Cranes from a User's Viewpoint," presented at the AIEE Cement Industry Subcommittee Technical Conference (Milwaukee, Wis., 1960).
3. Charles C. Libby, *Motor Selection and Application* (McGraw-Hill, New York, 1960).
4. Leonard Anderson, *Electric Machines and Transformers* (Reston Publishing, Reston, Va., 1980).

SECTION 2

ELECTRIC POWER SUPPLY AND SYSTEM CONSIDERATIONS

William C. Dumper*

George C. Soukup†

Russell A. Shively‡

FOREWORD 2-2

GENERAL 2-3

1. Factors to Consider for Proper Motor Application . . 2-3

SYSTEM-VOLTAGE CONSIDERATIONS 2-3

2. System-Voltage Variations 2-4
3. Effect of Voltage Variation on Equipment 2-6

SYSTEM-FREQUENCY CONSIDERATIONS 2-10

4. System-Frequency Variations 2-10
5. Use of Higher Frequencies 2-10
6. Use of Direct Current 2-11

MOTOR STARTING 2-11

7. Manual and Magnetic Starters 2-11
8. Classes of Motor Starters 2-12
9. Combination-Type Starters 2-12
10. Use of Circuit Breakers for Motor Starting 2-12
11. Standard Starter Ratings 2-12

MOTOR-STARTING METHODS 2-18

12. Squirrel-Cage Induction Motors 2-18
13. Wound-Rotor Induction Motors 2-22
14. Synchronous Motors 2-23
15. Dc Motors 2-23

*Station Design Division Engineer, Inside Plant Bureau, Consolidated Edison Company, New York, N. Y. (retired); Registered Professional Engineer (N.Y.); Member, IEEE.

†Manager, Large Motor Engineering, U.S. Electrical Motors Division, Emerson Electric Company, Prescott, Ariz.; Registered Professional Engineer (Wisc.).

‡Senior Application Engineer, U.S. Electrical Motors Division, Emerson Electric Company, Prescott, Ariz.

NOTE: The section was written by William C. Dumper for the first edition and reviewed and updated by George C. Soukup and Russell A. Shively.

VOLTAGE-DROP DETERMINATION 2-24
16. Steady-State Voltage-Drop Calculation 2-24
17. Voltage Drop Due to Motor Starting 2-25

CONTRIBUTIONS OF MOTORS TO SYSTEM FAULTS 2-36
18. Synchronous-Motor Behavior during Faults 2-36
19. Induction-Motor Behavior during Faults 2-37
20. Approximate Motor-Reactance Values 2-37
21. Dc Motor Behavior during Faults 2-38

POWER-FACTOR IMPROVEMENT 2-41
22. Disadvantages of Low Power Factor 2-41
23. Induction-Motor Power Factor 2-41
24. Synchronous-Motor Power Factor 2-42
25. Use of Synchronous Motors for Power-Factor Correction 2-43
26. Use of Induction Motors and Capacitors 2-44
27. Capacitor Limitations for Unit Switching 2-44
28. Suggested Maximum Capacitor Values 2-46
29. Released System Capacity 2-46

MOTOR PROTECTION 2-46
30. Protective Features of Motor-Starting Equipment . . 2-46
31. Stator-Winding Fault Protection 2-53
32. Stator-Winding Overheating Protection 2-55
33. Rotor Protection 2-55
34. Protection against Improper Operating Conditions . 2-57
35. Essential-Service Motors 2-58
36. Surge Protection 2-59
37. Lightning Protection 2-60

REFERENCES 2-63
BIBLIOGRAPHY 2-63

FOREWORD

The proper application of electric motors demands thorough consideration of many factors in regard to both the driven-load requirements and the characteristics of the electric-supply system. Failure to consider fully the effect of motor operation upon the system can result in improper operation of other connected utilization equipment. On the other hand, failure to consider system characteristics properly may limit the performance designed and built into the motor by the manufacturer for the particular application. This section is devoted to the proper application of electric motors from the electric-system viewpoint.

GENERAL

1. Factors to Consider for Proper Motor Application. In general, the following factors should receive consideration in electric-motor application. It is not intended that this list be all-inclusive. Peculiarities of certain applications may require consideration of additional items. Conversely, many of the factors listed below may require little or no consideration in some applications.

1. The available system voltages and frequencies. Are they compatible with reasonable motor design and cost in the horsepower and speed ratings required?
2. The types of other equipment presently connected to the system or envisioned for future connection.
3. The voltage spread that can be tolerated for each type of equipment without impairing its performance or useful life.
4. The effect of motor starting and operation on voltage levels in other parts of the system. What will be the frequency of motor starting and the accelerating period during which the starting current affects the system?
5. The expected variations in motor terminal voltage and frequency due to the operation of other system equipment or the motor itself. Will these variations result in inadequate torque capabilities, improper speed, excessive heating, reduced life, etc., of the motor?
6. Limitations placed on starting current by the power-supply system, utility service, or local generation. Is reduced-voltage starting equipment required to keep the system disturbance within acceptable limits?
7. The possibility that pulsating driven-load requirements may cause cyclic variations in the motor input, the disturbing effects of which will be reflected throughout the system.
8. The effect of motor power factor and efficiency on system losses and operation. Is the choice of synchronous motors or capacitors suitable for system power-factor correction or increased efficiency for large low-speed drives? (Sec. 5.)
9. The magnitude and duration of fault current contributed by the motor during system-fault conditions.
10. Changes in protective devices or interrupting equipment to provide proper protection for motor or system.

SYSTEM-VOLTAGE CONSIDERATIONS

The basic pattern of voltage identification followed in the United States is shown in Table 1. Rated generator voltage, transformer secondary voltage, and motor and control voltages are tabulated for each nominal system voltage. In general, the rated generator and transformer secondary voltages are the same value as the nominal system voltage. Motor and control rated voltages are lower than the nominal system voltages in almost all cases to compensate for system-voltage drops. Similar practice is followed in regard to voltage ratings of other utilization equipment.

TABLE 1. Preferred Voltage Ratings for Equipment

Preferred nominal system voltage	Generator rated voltage	Transformer secondary rated voltage	Motor and control rated voltage
Single-phase systems and single-phase components applied on three-phase systems			
120	120	120	115
120/240	120/240	120/240	230
208Y/120 (3-phase)	125/216Y (3-phase)	216Y/125 (3-phase)	115*
460Y/265 (3-phase)	277/480Y (3-phase)	480Y/277 (3-phase)	
Three-phase low-voltage systems			
208Y/120	125/216Y	216Y/125	208
230	139/240Y	240Y/139 or 240	230
460Y/265	277/480Y	480Y/277	460
460	277/480Y	480Y/277	460
575	346/600Y	600Y/346 or 600	575
Three-phase medium-voltage systems			
2,400	1,388/2,400Y	2,400Y/1,388 or 2,400	2,300
4,160	2,400/4,160Y	4,160Y/2,400	4,000
4,800	2,770/4,800Y	4,800Y/2,770 or 4,800	4,600
6,900	3,980/6,900Y	6,900Y/3,980	6,600
7,200	3,980/6,900Y	7,200Y/4,160	6,600
12,000	7,210/12,500Y	12,000Y/6,920	11,000
7,200/12,470Y	7,210/12,500Y	12,470Y/7,200	
13,200	7,970/13,800Y	13,200Y/7,610	13,200
13,800	7,970/13,800Y	13,800Y/7,970	13,200
14,400	8,320/14,400Y	13,800Y/7,970	13,200

*Line-to-neutral rating.

2. System-Voltage Variations. Electric-power systems consist of circuit elements which contain impedance. Current flow through these impedances produces voltage drops. As a result, voltage levels throughout the entire system and at utilization-equipment terminals vary with system operating conditions. These voltage variations are classified in four categories: voltage level, voltage regulation, voltage flicker, and voltage balance.

a. Voltage level. This is the difference between the maximum and minimum voltages experienced at any particular system location. Voltage level includes voltage drops in branch circuits, secondary feeders, and supply transformers due to system load and also the voltage variation in the primary supply system. The latter variation may be completely independent of the degree of loading on the particular plant or secondary system under consideration.

b. Voltage regulation. This variation is expressed as the percentage decrease in voltage from no load to full load at the particular system location under consideration.

c. Voltage flicker. This is the abrupt or rapid cyclic variation in voltage generally detectable by changes in illumination intensity. Assessment of the annoyance caused by voltage flicker is a difficult task. Many variables enter into the determination of what level of flicker is perceptible or objectionable to any individual or group of individuals.

Figure 1 shows the relationship between the threshold-of-perception and the threshold-of-objection curves (see Ref. 1) plotted in voltage (120-V basis) change as a function of flicker frequency. A large variation in the magnitude of voltage change with flicker frequency is apparent. A voltage change in the order of 1 V or less is considered as objectionable at a frequency of 7 Hz. Also shown are two curves of electric-utility flicker limitations. The main difference between these curves is the number of people involved. The lower curve applies to systems that supply lighting which affects large groups of people, such as theaters, office buildings, hotels, and large stores. The upper curve is typical of the higher limit applicable to residences, small apartments, stores, and industrial establishments in general. It is again stressed that these are general limits. The specific limitations for the particular application should be determined.

d. Voltage balance. This is the difference between phase voltages measured at the equipment terminals or anywhere in the system. The voltage unbalance (or negative-sequence voltage) in percent may be defined as

$$\text{Percent voltage unbalance} = 100 \times \frac{\text{maximum voltage deviation from average voltage}}{\text{average voltage}}$$

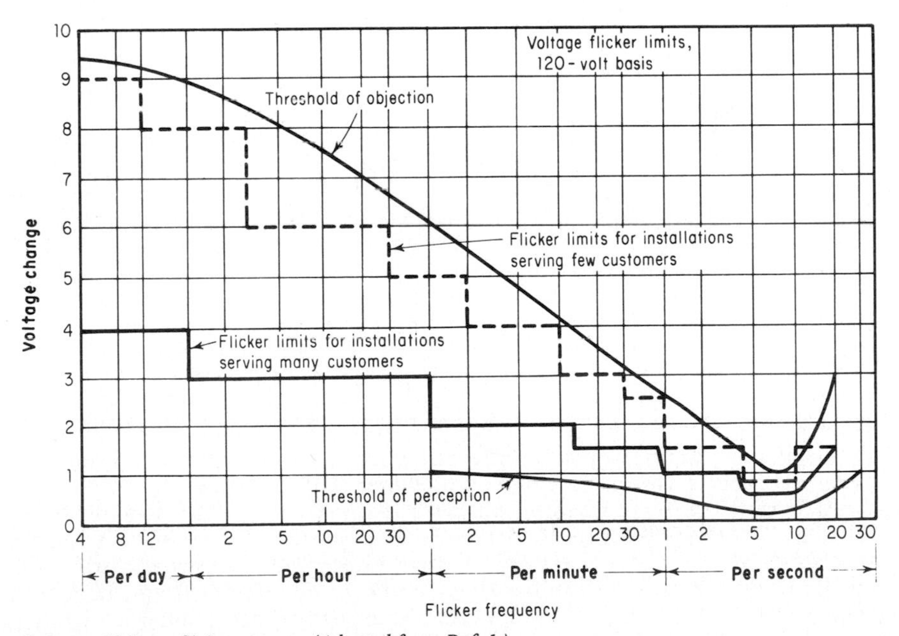

FIG. 1. Voltage-flicker curves. *(Adapted from Ref. 1.)*

Example. With voltages of 220, 215, and 210 V, the average is 215, the maximum deviation from the average is 5, and the percent unbalance is $100 \times 5/215$ 2.3 percent. The effect of unbalanced voltages is equivalent to the introduction of negative-sequence voltage having a rotation opposite to that occurring with balanced voltage. This can produce current in the motor considerably in excess of that present in balanced-voltage conditions. These currents will also be unbalanced, but of a much greater magnitude than the voltage unbalance. Typically current unbalance will be six to ten times the equivalent voltage unbalance. In addition to increased losses, locked-rotor and breakdown torque will be decreased and full-load speed slightly reduced. Normally motors operating on systems where the voltage is unbalanced should be derated (Fig. 2).

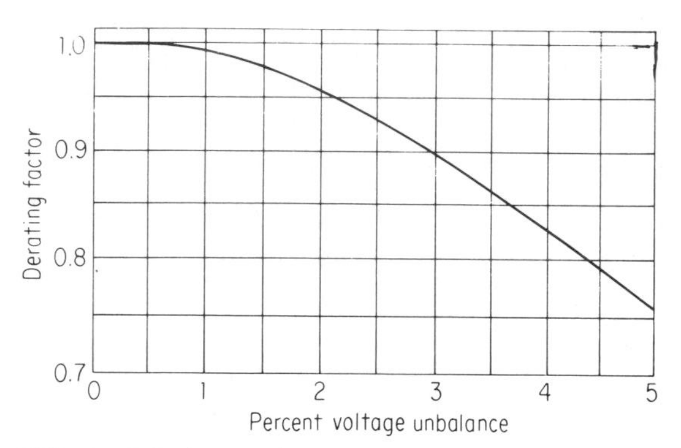

FIG. 2. Polyphase squirrel-cage induction-motor derating factor versus unbalanced voltage. *(Ref. 2.)*

3. Effect of Voltage Variation on Equipment. Motors and other utilization equipment are designed for operation at rated voltage. A loss in performance or life is experienced if other than rated voltage is applied to the equipment terminals. The ultimate effect of voltage variation is a function of the design of the equipment and the magnitude of the variation. Recognizing the fact that constant rated terminal voltage may not be maintained, various standardized bodies have allowed certain voltage tolerances within which the equipment will operate satisfactorily but not necessarily within guaranteed performance values.

a. Effect on motors. A maximum voltage variation of ± 10 percent of rated is allowed for satisfactory operation of electric motors. Table 2 shows the general effect of voltage variation on induction-motor characteristics. Synchronous-motor performance is generally affected in the same manner, with the exception of the pull-out torque, which is a direct function of the voltage. Table 3 shows the general effects of voltage variation on dc motor characteristics.

The major effects of motor operation at reduced voltages are increased losses, increased temperature rises, and reduction in starting and maximum running torques. Operation at voltages higher than rated produces greater starting and running torques, higher starting current, and decreased power factor. In general, there is less adverse effect on motor performance at terminal voltages slightly in excess of rated values than for voltages less than rated.

TABLE 2. General Effect of Voltage and Frequency Variation on Induction-Motor Characteristics

Variation	Starting and max running torque	Synchronous speed	% slip	Full-load speed	Efficiency			Power factor			Full-load current	Starting current	Temp rise, full load	Max overload capacity	Magnetic noise, no load in particular
					Full load	¾ load	½ load	Full load	¾ load	½ load					
Voltage variation:															
120% voltage	Increase 44%	No change	Decrease 30%	Increase 1.5%	Increase 1 point	Decrease ½ point	Decrease 2 points	Decrease 10 points	Decrease 16 points	Decrease 25 points	Decrease 11%	Increase 25%	*	*	Noticeable increase
110% voltage	Increase 21%	No change	Decrease 17%	Increase 1%	Increase ½–1 point	Practically no change	Decrease ½ point	Decrease 3 points	Decrease 5 points	Decrease 9 points	Decrease 7%	Increase 12%	Decrease 5°C	Increase 21%	Increase slightly
Function of voltage	$(\text{Voltage})^2$	Constant	$(\text{Voltage})^{-2}$	(Synchronous speed slip)								Voltage		$(\text{Voltage})^2$	
90% voltage	Decrease 19%	No change	Increase 23%	Decrease 1½%	Decrease 2 points	Decrease 1 point	No change	No change	Increase 1 point	Increase 4–5 points	Increase 11%	Decrease 10–12%	Increase 15°C	Decrease 25%	Decrease slightly
Frequency variation:															
105% frequency	Decrease 10%	Increase 5%	Practically no change	Increase 5%	No change	No change	No change	Slight increase	Slight increase	Slight increase	No change	Decrease 5–6%	Decrease slightly	Increase slightly	Decrease slightly
Function of frequency	$(\text{Frequency})^{-2}$	Frequency		(Synchronous speed slip)								$(\text{Frequency})^{-1}$			
95% frequency	Increase 11%	Decrease 5%	Practically no change	Decrease 5%	No change	No change	No change	Slight decrease	Slight decrease	Slight decrease	Increase slightly	Increase 5–6%	Increase slightly	Decrease slightly	Increase slightly

*Temperature rise may increase significantly and overload capacity may not be possible. Consult motor manufacturer for conditions outside of NEMA limits.

Note. This table shows general effects. For a 200-hp motor, effects can vary significantly for other horsepower, speed, and voltage combinations. *(Ref. 2.)*

TABLE 3. General Effect of Voltage Variation on dc Motor Characteristics

Voltage variation	Starting and max run torque	Full-load speed	Efficiency			Full-load current	Temp rise, full load	Max overload capacity	Magnetic noise
			Full load	¾ load	½ load				
Shunt-wound									
120% voltage	Increased 30%	110%	Slight increase	No change	Slight decrease	Decrease 17%	Main field increase. Commutator field and armature decrease	Increased 30%	Slight increase
110% voltage	Increased 15%	105%	Slight increase	No change	Slight decrease	Decrease 8.5%	Main field increase. Commutator field and armature decrease	Increased 15%	Slight increase
90% voltage	Decreased 16%	95%	Slight decrease	No change	Slight increase	Increase 11.5%	Main field decrease. Commutator field and armature increase	Decreased 16%	Slight decrease
Compound-wound									
120% voltage	Increased 30%	112%	Slight increase	No change	Slight decrease	Decrease 17%	Main field increase. Commutator field and armature decrease	Increased 30%	Increased slightly
110% voltage	Increased 15%	106%	Slight increase	No change	Slight decrease	Decrease 8.5%	Main field increase. Commutator field and armature decrease	Increased 15%	Increased slightly
90% voltage	Decreased 16%	94%	Slight decrease	No change	Slight increase	Increase 11.5%	Main field decrease. Commutator field and armature increase	Decreased 16%	Decreased slightly

Starting current is controlled by starting resistor.
Note. This table shows general effects, which will vary somewhat for specific ratings.

b. Recommended voltage spread for motors. Table 4 shows the recommended voltage spread for motors and other devices rated 600 V and below. Table 5 gives the recommended voltage spread for higher-voltage motors. It should be noted that the voltage spreads in Table 5 are biased toward voltages higher than rated. This is done to take advantage of the less adverse effect on motor performance as noted above.

TABLE 4. Voltage Nomenclature for Low-Voltage Systems

	Voltage	% of system voltage	Voltage spread, %	
Single-phase				
Favorable zone, max voltage	125	104.2	+4.2	12.5
System nominal voltage and transformer voltage rating	120	100.0		
Motor voltage rating	115	95.0		
Favorable zone, min voltage	110	91.7	−8.3	
Favorable zone, max voltage	250	104.2	+4.2	12.5
System nominal voltage and transformer voltage rating	240	100.0		
Motor voltage rating	230	95.9		
Favorable zone: min voltage	220	91.7	−8.3	
Three-phase				
Favorable zone, max voltage and transformer voltage rating	216	103.8	+3.8	9.1
System nominal voltage	208	100.0		
Motor voltage rating	208	100.0		
Favorable zone: min voltage	197	94.7	−5.3	
Favorable zone, max voltage and transformer voltage rating	240	104.3	+4.3	13.0
System nominal voltage	230	100.0		
Motor voltage rating	230	100.0		
Favorable zone, min voltage	210	91.3	−8.7	
Favorable zone, max voltage and transformer voltage rating	480	104.3	+4.3	13.0
System nominal voltage	460	100.0		
Motor voltage rating	460	100.0		
Favorable zone, min voltage	420	91.3	−8.7	
Favorable zone, max voltage and transformer voltage rating	600	104.3	+4.3	13.0
System nominal voltage	575	100.0		
Motor voltage rating	575	100.0		
Favorable zone, min voltage	525	91.3	−8.7	

Adapted from Ref. 1.

TABLE 5. Voltage Nomenclature for Medium-Voltage Systems (Ref. 1)

	Voltage	% of system voltage	Voltage spread, %	
Favorable zone, max voltage	2,450	102.1	+2.1	10.4
System nominal voltage and transformer voltage rating	2,400	100.0		
Motor voltage rating	2,300	95.9		
Favorable zone, min voltage	2,200	91.7	−8.3	
Favorable zone, max voltage	4,240	102.1	+2.1	10.4
System nominal voltage and transformer voltage rating	4,160	100.0		
Motor voltage rating	4,000	96.1		
Favorable zone, min voltage	3,810	91.7	−8.3	
Favorable zone, max voltage	4,900	102.1	+2.1	10.4
System nominal voltage and transformer voltage rating	4,800	100.0		
Motor voltage rating	4,600	95.9		
Favorable zone, min voltage	4,400	91.7	−8.3	
Favorable zone, max voltage	7,050	102.1	+2.1	10.4
System nominal voltage and transformer voltage rating	6,900	100.0		
Motor voltage rating	6,600	95.7		
Favorable zone, min voltage	6,320	91.7	−8.3	
Favorable zone, max voltage	14,100	102.1	+2.1	10.4
System nominal voltage and transformer voltage rating	13,800	100.0		
Motor voltage rating	13,200	95.7		
Favorable zone, min voltage	12,630	91.7	−8.3	

SYSTEM-FREQUENCY CONSIDERATIONS

The standard frequency for ac electric motors in the United States is 60 Hz. In the past, 25-Hz systems found wide usage in traction and steel-mill applications. The modern trend is away from 25-Hz systems to 60-Hz systems. Few of the 25-Hz systems still exist.

4. System-Frequency Variations. The effects of system-frequency variation on induction motors are shown in Table 2. In general, the frequency stability of utility power systems is excellent and frequency variations are not of any great concern in regard to motor operation. Power-supply systems fed from isolated or local generation, however, may be subject to larger frequency variations. In these cases the expected frequency variations must be considered in the motor application. NEMA standards permit a frequency variation of ±5 percent from rated. Within this range motors will normally operate satisfactorily but not necessarily within guaranteed performance values.

5. Use of Higher Frequencies. The use of higher frequencies in industrial plants has been increasing in recent years. The machine-tool industry has created a need for higher motor speeds for grinding, polishing, spinning, and other metalworking operations. Higher speeds can be obtained by using the more common higher frequencies of 120, 180, 360, 420, and 900 Hz. Since the motor size and weight are reduced with speed increase, the use of higher frequencies is attractive for small power hand tools. Power systems on aircraft have been standardized at 400 Hz to take advantage of smaller and lighter motors and generators.

Higher frequencies are readily obtained from converters. Advances in the field of solid-state devices are enabling frequency conversion without the use of rotating equipment. These variable-frequency devices allow adjustment of the motor and driven equipment speeds with a corresponding improvement in system efficiency over conventional methods, for example, adjusting the output of a pump by varying the speed versus the use of a discharge valve.

Systems employing higher frequencies require special design considerations. Since reactance is a direct function of frequency, voltage variations will be considerably higher than those experienced on similar 60-Hz systems. Skin effects will also produce higher power losses in conductors at these elevated frequencies.

6. Use of Direct Current. Dc power systems experienced a decline with the development of ac equipment and its advantages of transformation from one voltage level to another. Speed control inherent in dc motors has resulted in an increase in their usage in recent years to meet the requirements of modern manufacturing processes where accurate control of motor speed is necessary.

Present-day usage of direct current retains the advantages of ac transmission to the point of application where transformation and rectification equipment is used to provide direct current. Recent advances in solid-state equipment, such as silicon diodes, have permitted the construction of compact rectifiers with capabilities in the thousands of kilowatts. The generation of direct current by prime movers or the use of motor-generator sets is rapidly becoming obsolete.

MOTOR STARTING

As defined in NEMA Standard ICS 1-1978 (Ref. 3), "an electric motor controller is a device or a group of devices which serve to govern, in some predetermined manner, the electric power to the motor." Also, "a starter is an electric controller for accelerating a motor from rest to normal speed." The starter is the important connecting link between the motor and the electric-supply system.

7. Manual and Magnetic Starters. Starters range in complexity from simple manual across-the-line devices to more sophisticated controllers permitting proper starting and stopping with minimum disturbance to the system or adverse effect on the driven equipment. In the smaller motor sizes, manual starters are available. This type of starter requires that the operator furnish the necessary operating force. When starting and stopping are relatively infrequent, manual starters can provide a satisfactory installation. However, the modern trend is toward the use of magnetic motor starters in which the operating force is furnished by electromagnetic action. This permits remote operation of the motor by means of a push button from a convenient location. When a motor is part of an integrated control process requiring starting and stopping in response to process variables such as flow, pressure, and temperature, remote control is a necessity. Motor starters associated with a particular process often are grouped together in compact motor-control centers for convenience in power-supply arrangement (Ref. 9).

8. Classes of Motor Starters. NEMA has divided motor starters into Classes A through E. The more commonly used Classes A, B, and E are described (Ref. 3) as follows:

a. Class A. "Alternating current air-break and oil-immersed manual or magnetic controllers for service on 600 volts or less and capable of interrupting operating overloads up to and including 10 times their normal motor rating but not short circuits or faults beyond operating overloads."

b. Class B. "Direct current air-break manual or magnetic controllers for service on 600 volts or less and capable of interrupting operating overloads but not short circuits or faults beyond operating overloads."

c. Class E. "Alternating current air-break and oil-immersed magnetic controllers for service on voltages from 2200 to 4600 volts and capable of interrupting short circuits or faults beyond operating overloads." This class is subdivided into Classes E1 and E2, the former utilizing contacts for starting and interrupting while the latter employs fuses for interrupting.

d. Classes C and D. These two classes are, respectively, ac and dc controllers capable of interrupting fault currents beyond operating overloads. These devices are not too widely used, and when class designations for 600 V and less are not stated, it is assumed that Class A or B is meant.

9. Combination-Type Starters. Short-circuit protection extending from the maximum fault current available from the power-supply system down to fault-current magnitudes of ten times the full-load rating of the starter must be provided by a device on the line side of the starter. This is commonly accomplished by the use of combination-type starters for 600 V and less, where the ability to interrupt fault currents is applied to the combination as a whole. Combination starters are manufactured with fault protection in the form of a fused disconnect switch, an air circuit breaker, or a coordinated combination of fuses and a circuit breaker. Asymmetrical interrupting ratings up to 100,000 A can be obtained with combination starters.

10. Use of Circuit Breakers for Motor Starting. Motor starters are not manufactured for induction motors larger than 3500 hp or synchronous motors larger than 4000 hp at unity power factor or for nominal system voltages above 7200 V. Motor applications in excess of these limits require circuit breakers for starting. In contrast to motor starters, circuit breakers are not designed for repetitive operation. A circuit breaker, however, does provide short-circuit protection. In many cases where repetitive operation is not required, considerable savings can be achieved through the use of low-voltage circuit breakers instead of combination starters for horsepower ratings as low as 200 hp at 208 V or 400 hp at 460 V. In the higher nominal system voltages of 2400 and 4800 V Class E starters are considerably less expensive than circuit breakers and should be used whenever possible, as determined by the horsepower rating. From the standpoint of maintenance and reduction in fire hazard air circuit breakers have an advantage over oil circuit breakers, especially for indoor applications.

11. Standard Starter Ratings. Tables 6 through 12 list the standard starter horsepower and current ratings for squirrel-cage and wound-rotor induction motors, synchronous motors, and dc motors. The continuous-current rating shown is the maximum rms current, in amperes, which the controller may be expected to carry continuously without exceeding the allowable temperature rises.

TABLE 6. Manually Operated Full-Voltage Controllers (Ref. 3)

Size of controller	hp at 3-phase 110 volts	3-phase 208/230 volts	3-phase 460/575 volts	Single-phase 115 volts	Single-phase 230 volts
M-0	1½	2	3	1	1½
M-1	3	5	7½	1½	3
M-1P	...	...	...	3	5

TABLE 7. Full-Voltage Magnetic Controllers for Single-Speed Squirrel-Cage Induction Motors (Ref. 3)

Ratings for single-phase full-voltage magnetic controllers for nonplugging and nonjogging duty

Size of controller	Continuous current rating, A	Single-phase hp at 115 volts	Single-phase hp at 230 volts	Service-limit current rating, A
00	9	⅓	1	11
0	18	1	2	21
1	27	2	3	32
1P	36	3	5	42
2	45	3	7½	52

Ratings for polyphase single-speed full-voltage magnetic controllers for nonplugging and nonjogging duty

Size of controller	Continuous-current rating, A	3-phase hp at 110 volts	3-phase hp at 208/230 volts	3-phase hp at 460/575 volts	Service-limit current rating, A
00	9	¾	1½	2	11
0	18	2	3	5	21
1	27	3	7½	10	32
2	45	...	15	25	52
3	90	...	30	50	104
4	135	...	50	100	156
5	270	...	100	200	311
6	540	...	200	400	621
7	810	...	300	600	932
8	1,215	...	450	900	1,400
9	2,250	...	800	1,600	2,590

TABLE 8. Reduced-Voltage, Part-Winding, and Wye-Delta Magnetic Controllers for Single-Speed Squirrel-Cage Induction Motors (Ref. 3)

Ratings for single-phase reduced-voltage general-purpose reversing or nonreversing magnetic controllers

Size of controller	Continuous-current rating, A	Single-phase hp at 115 volts	Single-phase hp at 230 volts	Service-limit current rating, A
1	27	2	3	32
2	45	3	7½	52

Ratings for polyphase reduced-voltage general-purpose reversing or nonreversing magnetic controllers

Size of controller	Continuous-current rating, A	3-phase hp at 110 volts	3-phase hp at 208/230 volts	3-phase hp at 460/575 volts	Service-limit current rating, A
1	27	3	7½	10	32
2	45	...	15	25	52
3	90	...	30	50	104
4	135	...	50	100	156
5	270	...	100	200	311
6	540	...	200	400	621
7	810	...	300	600	932
8	1,215	...	450	900	1,400
9	2,250	...	800	1,600	2,590

Ratings for three-phase, full-voltage, nonreversing, and nonjogging magnetic controllers for motors suitable for part-winding starting[a]

Size of controller[b]	Controller continuous-current rating per winding, A	Contactor size, line[c] (1M and 2M)	Controller locked-rotor rating per winding, A, at 208/230 volts	460 volts	575 volts[d]	Controller service-limit current rating per winding, A
1PW	27	1	140	88	70	32
2PW	45	2	255	210	168	52
3PW	90	3	500	418	334	104
4PW	135	4	835	835	668	156
5PW	270	5	1,670	1,670	1,334	311

Ratings for three-phase, reduced-voltage, nonreversing, and nonjogging magnetic controllers for motors suitable for part-winding starting[a]

Size of controller[b]	Controller continuous-current rating per winding, A	Contactor size: Line[c] (1M and 2M)	Contactor size: Accelerating	Controller locked-rotor rating per winding, A, at 208/230 volts	460 volts	575 volts[d]	Controller service-limit current rating per winding, A
1PW	27	1	1	140	88	70	32
2PW	45	2	2	255	210	168	52
3PW	90	3	3	500	418	334	104
4PW	135	4	4	835	835	668	156
5PW	270	5	4[e]	1,670	1,670	1,334	311

TABLE 8. Reduced-Voltage, Part-Winding, and Wye-Delta Magnetic Controllers for Single-Speed Squirrel-Cage Induction Motors (Ref. 3) (*Continued*)

Ratings for three-phase, wye-delta, nonreversing, and nonjogging magnetic controllers for either open- or closed-circuit transition

Size of controller	Controller continuous-current rating, A	Contactor size			Controller locked-rotor rating, A, at			Controller service-limit current rating, A
		1M and 2M[f]	S or 1S	28[g]	208/230 volts	460 volts	575 volts[d]	
1YD	47	1	1	. . .	242	152	121	55
2YD	78	2	2	. . .	441	363	291	90
3YD	156	3	3	. . .	866	725	578	179
4YD	233	4	4	. . .	1,445	1,445	1,155	270
5YD	467	5	4	. . .	2,510	2,510	2,000	538
6YD	935	6	5	. . .	5,020	5,020	4,000	1,075
7YD	1,400	7	6	. . .	8,660	8,660	6,930	1,610
8YD	2,100	8	7	. . .	13,000	13,000	10,400	2,420
9YD	3,900	9	8	. . .	23,200	23,200	18,500	4,480

[a]These ratings are established on a per winding basis because, on part-winding starting, the locked-rotor current is considerably in excess of 50 percent of the total locked-rotor-current rating of the motor. The percentage varies over a wide range, depending on motor design.

[b]Data shown for 1PW to 5PW only. Controllers are available up to size 9PW.

[c]Each contactor 1M and 2M is intended to carry one half of the motor running current, which is assumed to be equally divided between its two windings (see Fig. 3).

[d]This column has been approved as Suggested Standard for Future Design.

[e]The use of a size 4 accelerating contactor on a size 5PW controller is suitable when the contactor is shunted by its associated line contactor.

[f]Each contactor 1M and 2M carries only 0.577 times the motor line current when the motor is switched to the delta (running) connections (see Fig. 3).

[g]Contactor 28 should be capable of interrupting the current in the transition resistor circuit.

TABLE 9. Class A Controllers for Polyphase Wound-Rotor Induction Motors (Ref. 3)

Ratings for primary contactors of magnetic controllers for reversing and nonreversing duty

Controller size	Continuous-current rating, A	3-phase hp at			Service-limit current rating, A
		110 volts	208/230 volts	460/575 volts	
1	27	3	7½	10	32
2	45	. . .	15	25	52
3	90	. . .	30	50	104
4	135	. . .	50	100	156
5	270	. . .	100	200	311
6	540	. . .	200	400	621
7	810	. . .	300	600	932
8	1,215	. . .	450	900	1,400
9	2,250	. . .	800	1,600	2,590

TABLE 10. Controllers for Synchronous Motors (Ref. 3)

Horsepower and current ratings of low-voltage controllers for synchronous motors (nonplugging or nonjogging duty)

Size of controller	hp ratings: 230 volts power factor 1.0	230 volts power factor 0.8	460–575 volts power factor 1.0	460–575 volts power factor 0.8	Continuous-current rating, A
2	20	15	30	25	45
3	40	30	60	50	90
4	60	50	125	100	135
5	125	100	250	200	270
6	250	200	500	400	540
7	350	300	700	600	810
8	500	450	1,000	900	1,215
9	1,000	800	2,000	1,600	2,250

TABLE 11. High-Voltage Class E Controllers (Ref. 3)

Horsepower ratings

Size of controller	2,200–2,400 volts: Synchronous motors 100% power factor	2,200–2,400 volts: Synchronous motors 80% power factor	2,200–2,400 volts: Induction motors	4,000–4,600 volts: Synchronous motors 100% power factor	4,000–4,600 volts: Synchronous motors 80% power factor	4,000–4,600 volts: Induction motors	Continuous-current rating, A*
H2	900	700	700	1,500	1,250	1,250	180
H3	1,750	1,500	1,500	3,000	2,500	2,500	360

Interrupting ratings†

Size of controller	Volts	Range of system voltages at which interrupting rating applies: Max	Min	Interrupting ratings, 3-phase available symmetrical short-circuit kVA, 50 or 60 Hz: Class E1	Class E2	Continuous-current rating, A
H2	2,300	2,500	2,200	25,000 and 50,000	150,000	180
H2	4,600	5,000	4,000	25,000 and 50,000	250,000	180
H3	2,300	2,500	2,200	50,000	150,000	360
H3	4,600	5,000	4,000	50,000	250,000	360

The contactors of Class E2 controllers shall have the ability to interrupt operating overloads up to and including ten times the normal full-load currents of the motors for which the controllers are rated.

*Rating for oil circuit breaker; values are higher for air or vacuum circuit breakers.

†When connected to a circuit having the available symmetrical short-circuit kVA shown, Class E controllers shall interrupt the currents resulting from a short circuit at the load terminals.

TABLE 12. DC General-Purpose and Machine-Tool Class B Controllers (Ref. 3)

Horsepower ratings of across-the-line manually operated controllers

Size	8-h open rating, A	hp at 115 volts	hp at 230 volts
0	15	1	1½
1	25	1½	2

Horsepower ratings of full-voltage magnetic controllers

Size	8-h open rating, A	hp at 115 volts	hp at 230 volts
0	15	1	1
1	25	1½	2

Horsepower ratings of reduced-voltage magnetic controllers

Size of controller	8-h open rating, A	115 volts		230 volts		575 volts	
		hp rating	No. of accelerating contactors	hp rating	No. of accelerating contactors	hp rating	No. of accelerating contactors
1	25	3	1	5	1		
2	50	5	2	10	2	20	2
3	100	10	2	25	2	50	2
4	150	20	2	40	2	75	3
5	300	40	3	75	3	150	4
6	600	75	3	150	4	300	5
7	900	110	4	225	5	450	6
8	1,350	175	4	350	5	700	6
9	2,500	300	5	600	6	1,200	7

The service-limit current rating shown is the maximum rms current, in amperes, which the controller may be expected to carry for protracted periods in normal service. At these service-limit ratings temperature rises may exceed those obtained by testing the controller at its continuous-current rating. The ultimate trip current of motor protective devices used shall not exceed the service-limit rating. The 8-h rating of the contactor in Class B dc controllers is based upon the current-carrying capacity for 8 h, starting with clean contacts and free ventilation and with full voltage on the operating coil without exceeding any of the established limits.

With the exception of Class E controllers, all ratings are for the controller when mounted in any type of enclosure. The standard Class E controller is furnished as a complete, totally enclosed, self-supporting electrically coordinated unit. The controllers shown in Tables 6 through 12 should not be used with motors whose full-load current or horsepower ratings exceed the listed values of continuous current or horsepower.

It is recommended that NEMA Standard ICS 1-1978 be consulted for applications involving multispeed, plugging, jogging, etc., operation. This publication also contains much valuable information in regard to definitions, engineering information, duty cycles, ratings. etc.

MOTOR-STARTING METHODS

The electric-motor starter is the important connecting link between the motor and the electric-supply system. Properly applied, the motor starter can reduce the effects of motor starting on the system and on the driven load. The various methods available for starting squirrel-cage and wound-rotor induction motors, synchronous motors, and dc motors are described in the following paragraphs (Ref. 9).

12. Squirrel-Cage Induction Motors. Starting can be accomplished by either full-voltage, reduced-voltage, or reduced-inrush methods. Reduced-voltage starters include autotransformer, primary-resistor, and primary-reactor types. Reduced-inrush starters include part-winding and wye-delta types. A comparison of the various methods showing performance and relative costs is shown in Table 13. Figure 3 shows schematic diagrams for reduced-voltage and reduced-inrush starting of squirrel-cage induction motors.

a. Full-voltage starting. This is the simplest and lowest-cost method of starting. It produces maximum starting torque and minimum acceleration time. It also produces the maximum disturbance to the electric distribution system. The motor and driven load must be designed to withstand the high torques developed by this method of starting. Starting-equipment maintenance is a minimum because of the simplicity of this method.

b. Autotransformer starting. The principal advantage of this starting method is the high value of torque produced per unit of starting current. The motor current is reduced in proportion to the voltage applied to the motor terminals. The line current, however, is reduced in proportion to the square of the motor-terminal voltage because of the autotransformer action. The cost of an autotransformer starter is generally the highest of all starting methods. It has the advantage of torque and inrush-current adjustment in the field by simple tap selection. Autotransformers are generally provided with two or three taps.

Autotransformer starters can be obtained with either open or closed transition. Open transition has the disadvantage of possible transient switching-current peaks at the instant of transfer to full system supply voltage. Closed-transition starting eliminates transient current peaks since the motor is never disconnected from the system. Autotransformer starting can be accomplished with either two- or three-winding transformers. The two-winding arrangement results in a current-voltage unbalance between phases. The transformer rating in an autotransformer starter is based on a duty cycle in accordance with the application. The transformer windings are removed from the circuit once starting has been accomplished.

c. Primary resistor starting. This method employs series resistors in each phase of the motor primary circuit. The value of the resistance is reduced in one or more steps to meet inrush requirements until full voltage is applied to the

TABLE 13. Comparison of Motor-Starting Methods for Squirrel-Cage Induction Motors

Type of starter	Motor terminal voltage	Motor-starting torque	Motor-starting current	Line starting current	Torque efficiency	Approx relative cost 50 hp 460 volt†
Full voltage.	1.0	1.0	1.0	1.0	1.0	1.0
Reduced voltage:						
Autotransformer:						
80% voltage tap.	0.8	0.64	0.8	0.64*	1.0	5.5
65% voltage tap.	0.65	0.42	0.65	0.42*	1.0	(closed
50% voltage tap.	0.5	0.25	0.5	0.25*	1.0	transition)
Primary resistor:						
80% terminal voltage.	0.8	0.64	0.8	0.8	0.8	
65% terminal voltage.	0.65	0.42	0.65	0.65	0.65	5.0
50% terminal voltage.	0.5	0.25	0.5	0.5	0.5	
Primary reactor.	Values the same as for primary resistor starter as listed above					Generally applied at higher voltages
Reduced inrush:						
Part-winding 50%:						
Low-speed motors. . . .	1.0	0.5	0.5	0.5	0.5	
High-speed motors. . .	1.0	0.5	0.7	0.7	0.7	3.2
Part-winding 75%:						
Low-speed motors. . . .	1.0	0.75	0.75	0.75	1.0	
Wye-delta.	1.0	0.33	0.33	0.33	1.0	4.0–6.0

*Does not include transformer magnetizing current, which is usually less than 25 percent of motor full-load current.

†At higher horsepower ratings, such as 500 to 1000 hp at 460 V, the spread in relative costs tends to narrow to approximately a 1.0 to 2.5 range with roughly proportionate decreases for the various methods. Specific costs should be obtained from the manufacturer. In some horsepower ratings the actual costs may be contrary to the relative values shown. Use of 460-V motors at horsepowers above 500 is not recommended.

Low-speed motors are considered to be 514 r/min and below.

All values shown are per unit. One per unit is equal to motor rated terminal voltage, rated starting current, and rated starting torque. Torque efficiency is motor-starting torque divided by motor-starting current on a per unit basis.

If the line voltage applied to the motor-starter terminals is not equal to the motor rated terminal voltage, the tabulated values can be adjusted as follows:

Multiply the motor-starting torque values by (line per unit voltage)^2.
Multiply the motor-starting current values by (line per unit voltage).
Multiply the line starting current values by (line per unit voltage).
Multiply the torque-efficiency values by (line per unit voltage).

motor terminals. Starting-torque magnitudes are high but the torque efficiency is low, being equal to the per unit value of voltage appearing at the motor terminals. Closed transition is inherent with this method of starting. Motor terminal voltage increases automatically during acceleration as the line current decreases.

Although transient switching-current peaks are not a problem, large values of starting current can be experienced depending upon the timing of the resistance-reduction steps. The inrush current of a normal induction motor remains at a fairly high level until relatively high speeds are attained. If the step reductions are

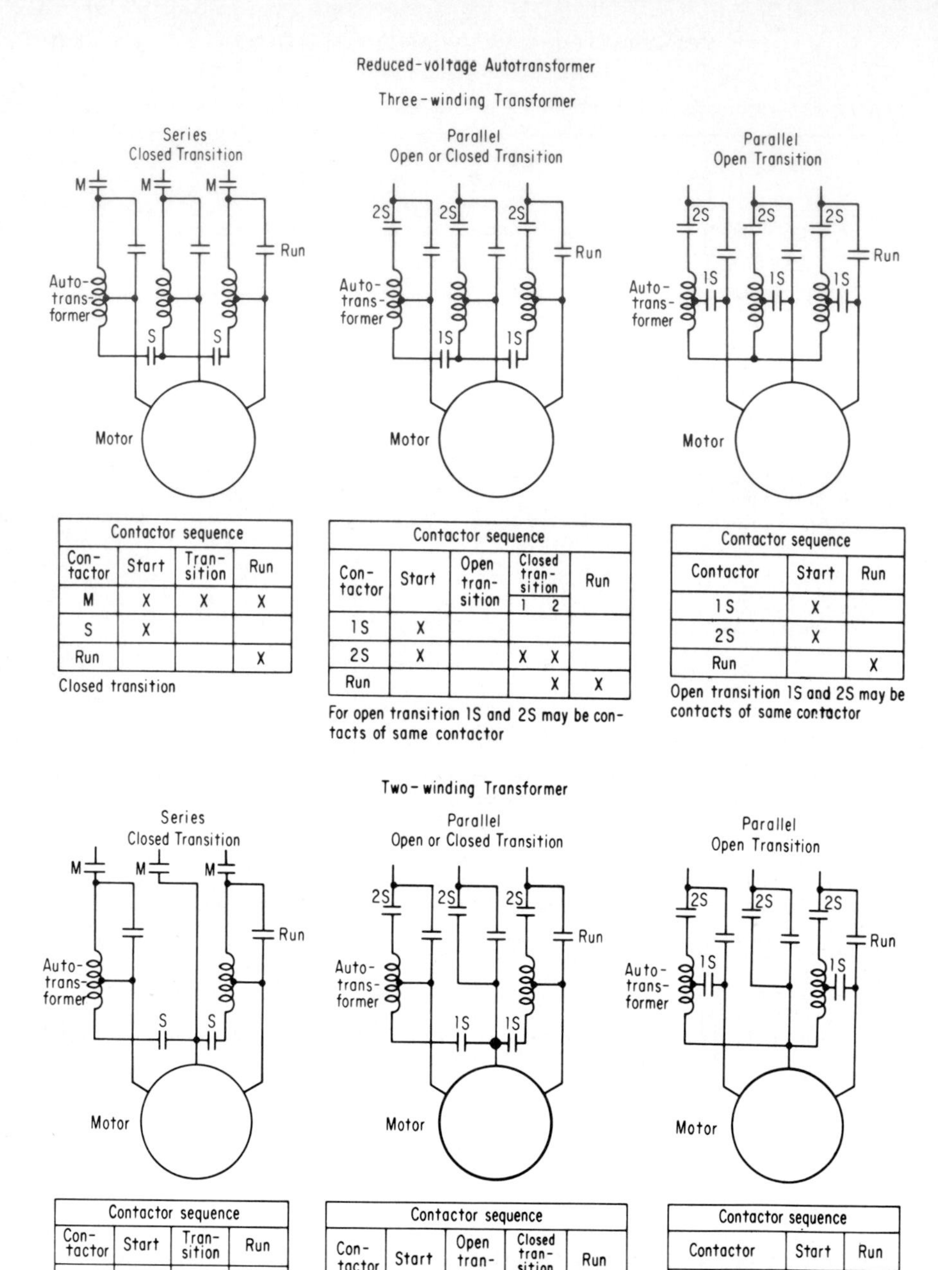

Three-winding transformer, series, closed transition:

Contactor sequence			
Contactor	Start	Transition	Run
M	X	X	X
S	X		
Run			X

Closed transition

Three-winding transformer, parallel, open or closed transition:

Contactor sequence					
Contactor	Start	Open transition	Closed transition 1	Closed transition 2	Run
1S	X				
2S	X		X	X	
Run				X	X

For open transition 1S and 2S may be contacts of same contactor

Three-winding transformer, parallel, open transition:

Contactor sequence		
Contactor	Start	Run
1S	X	
2S	X	
Run		X

Open transition 1S and 2S may be contacts of same contactor

Two-winding transformer, series, closed transition:

Contactor sequence			
Contactor	Start	Transition	Run
M	X	X	X
S	X		
Run			X

Two-winding transformer, parallel, open or closed transition:

Contactor sequence					
Contactor	Start	Open transition	Closed transition 1	Closed transition 2	Run
1S	X				
2S	X		X	X	
Run				X	X

For open transition 1S and 2S may be contacts of same contactor

Two-winding transformer, parallel, open transition:

Contactor sequence		
Contactor	Start	Run
1S	X	
2S	X	
Run		X

Open transition 1S and 2S may be contacts of same contactor

FIG. 3. Typical methods of starting squirrel-cage induction motors. *(Ref. 3.)*

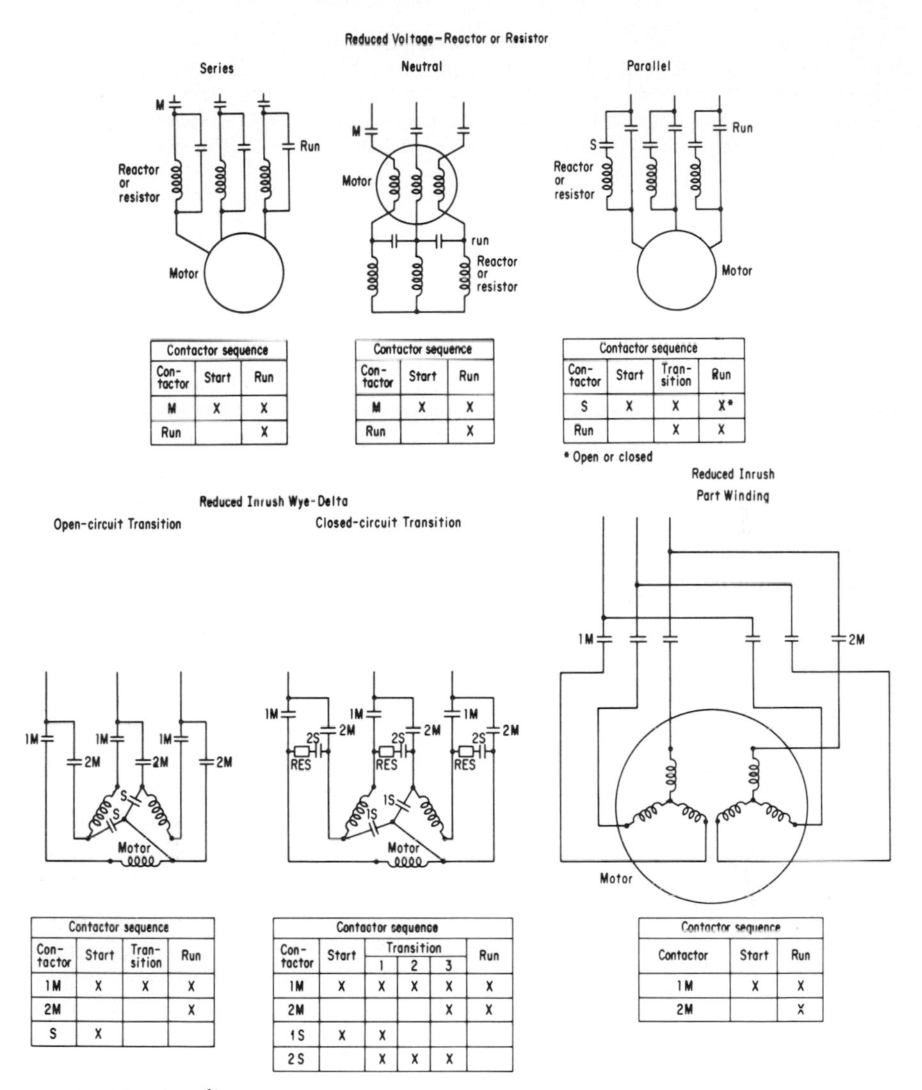

Contactor sequence		
Con-tactor	Start	Run
M	X	X
Run		X

Contactor sequence		
Con-tactor	Start	Run
M	X	X
Run		X

Contactor sequence			
Con-tactor	Start	Tran-sition	Run
S	X	X	X*
Run		X	X

* Open or closed

Contactor sequence			
Con-tactor	Start	Tran-sition	Run
1M	X	X	X
2M			X
S	X		

Contactor sequence					
Con-tactor	Start	Transition 1	Transition 2	Transition 3	Run
1M	X	X	X	X	X
2M				X	X
1S	X	X			
2S		X	X	X	

Contactor sequence		
Contactor	Start	Run
1M	X	X
2M		X

FIG. 3. (*Continued.*)

quickly accomplished, there may not be a great reduction in the maximum inrush current. In many cases the limiting factor is not the total current magnitude but the incremental current increase allowed by the utility. In this situation, the starter may have performed its function well if all the resistance steps are short-circuited before breakaway is achieved, provided that excessive motor heating does not occur.

d. Primary reactor starting. This method is quite similar to resistor starting. It is generally more applicable to larger motors and at voltages above 600 V. There is little power consumption in the reactor as contrasted to the resistor method. The reactor method does not improve the starting power factor of the supply system since the starting current is largely reactive in the first place. However, the improvement in motor power factor with speed increase automatically raises the motor terminal voltage during acceleration. Starting characteristics can be adjusted by tap selection. Increment starting requires separate reactors for each step since a portion of a single reactor cannot be short-circuited in the same manner as a resistor.

e. Part-winding starting. This method is attractive in its simplicity and is generally the least expensive of the reduced starting-current methods. Transition is the closed type. Torque efficiency is low for high-speed (above 514-r/min) motors; it approaches unity for low-speed motors. Motor torque-speed curves tend to dip at certain speeds, especially one-half speed, when operated on part winding. The possibility that the motor may not accelerate to rated speed because of torque dips requires careful consideration. This method often is not suitable for starting high-inertia loads or equipment requiring relatively large torques during acceleration. While motors with multiple parallel windings, such as dual-voltage motors, can be successfully applied, it is recommended that the motor be designed specifically for part-winding starting. Factors such as winding stresses, torque dips, and heating should be considered in the motor design. It should be kept in mind that the primary purpose for utilizing a part-winding start motor is to minimize starting current inrush and its resulting system disruption. Because of this, some motors, by design, may not rotate when first energized. As soon as the system voltage has recovered, generally in just a few seconds, the motor should be fully energized. The motor will then accelerate to full speed. Motors should be energized on part winding only long enough for the system voltage to recover. Excessive operation under part-winding conditions may result in damage to the motor due to the higher current present in the portion of the winding that is energized (Ref. 2).

f. Wye-delta starting. The starting torque is only one-third of the value at rated voltage. The torque efficiency is unity. This method is desirable only where low starting torque is acceptable. The motor must run with a delta connection. Both closed- and open-transition types are available. Starting duty cycles are limited by motor heating for open transition. Closed transition utilizes resistors to keep the motor energized during transfer to the running connection. In the latter case, as with other similar types of reduced-current starters, the motor control itself may impose restrictions on the starting time or frequency. The cost of open-transition wye-delta starters tends to be less than that of autotransformer or primary-resistor starters. However, closed-transition starters may be more expensive than autotransformer starters in some ratings.

13. Wound-Rotor Induction Motors. In contrast to the squirrel-cage induction motor, limitation of the starting current is accomplished by insertion of a resistance in the secondary or rotor circuit of a wound-rotor induction motor. The starting torque can be varied from a fraction of rated full-load torque to breakdown torque by proper selection of the external resistance value. A wound-rotor induction motor is capable of producing rated full-load torque at standstill with rated full-load current. When low starting current, high starting torque, and smooth acceleration are required, a wound motor merits consideration. Because of the complexity of rotor construction and the motor-control equipment, a wound-rotor induction-motor installation is considerably more expensive than an equivalent rated squirrel-cage induction motor.

Automatic starting of wound-rotor induction motors can be accomplished by three methods: (1) current-limit acceleration, (2) secondary-frequency acceleration, or (3) definite-time acceleration. Wound-rotor induction motors are very seldom started with the rotor short-circuited because of the high-starting-current, low-starting-circuiting torque, and poor accelerating-torque characteristics under this condition. All three methods involve short-circuiting the secondary resistance in steps as the motor accelerates. As their names indicate, each method utilizes a different variable to initiate step changes. Current-limit acceleration requires that the current fall to a predetermined value before the next step. Secondary-frequency acceleration utilizes frequency-sensitive relays in the secondary circuit to determine the speed at which each step is accomplished. These two methods provide torque limitation and smooth acceleration. Definite-time acceleration short-circuits the starting resistor on the basis of a predetermined time schedule regardless of the motor speed or starting-current magnitude. This method requires coordination with the actual starting cycle to prevent large torques and starting-current peaks.

14. Synchronous Motors. Conventional synchronous motors are not capable of starting without auxiliary means such as a squirrel-cage winding or a separate starting motor. Starting motors are generally of the wound-rotor induction-motor type and are a practical solution where very large synchronous motors, such as frequency changers, must be started from a limited-capacity system. Another application is the pumped-storage unit, which may require a starting motor as large as 18,000 hp or more.

Starting a synchronous motor equipped with an amortisseur or squirrel-cage winding involves the same considerations as starting an induction motor, with the additional requirement of field application at the proper time. The synchronous-motor rotor winding is closed through a starting resistor to protect against insulation breakdown due to induced voltages. Lightly loaded synchronous motors, such as those on motor-generator sets, can be started and synchronized without elaborate field application relays. Installations having large inertias or high pull-in torques require the use of a field application relay. Generally, the slip-frequency current induced in the rotor winding is used to select the proper time for dc field application. The proper choice of a starting-resistance value can increase the pull-in torque by its beneficial effect on the speed-torque curve near synchronous speed.

15. Dc Motors. Except for the very smallest sizes, 2 hp maximum at 230 V, dc motors require reduced-inrush starting. This is accomplished by inserting resistance in the armature circuit to limit the current. As the motor accelerates, the resistance is either manually or automatically reduced in steps. Acceleration can be controlled by (1) current-limit, (2) counter-emf, or (3) definite-time methods. Current limit requires that the inrush current drop to a predetermined level before the next step. In a constant-flux machine, such as a shunt motor, the counter emf generated in the armature is a measure of the speed and can be used to initiate resistance-reduction steps. Definite-time control accelerates the motor on a time schedule regardless of current magnitude or armature speed. This method provides the greatest assurance of achieving breakaway from standstill but requires coordination with the starting cycle to avoid large values of inrush current and torque.

Dc motors can also be started from a variable-voltage source. In this manner the voltage can be increased as the motor accelerates to meet the required inrush current and torque limitations. This method eliminates the need for a starting resistance in the armature circuit.

VOLTAGE-DROP DETERMINATION

Voltage-drop determination for motor applications can be divided into the steady-state and motor-starting categories. Formulas and typical equipment characteristics are included here for steady-state calculations. Graphic data are included for determination of voltage drops due to motor starting. Data for the determination of the minimum voltage level and the restored voltage magnitude are included for those applications where generators are involved. The calculation of the transient duration associated with motor starting is beyond the scope of this section. A large part of this material was adapted from Ref. 4. This reference and other publications contain graphic data in much greater detail which reduce the effort involved in the calculation of system voltage drops.

Of necessity, the generator, motor, transformer, cable, etc., data included here are typical values. Whenever possible, actual values should be used for the particular case under investigation. These typical data are suitable for estimating purposes or for use when actual characteristics are unavailable.

16. Steady-State Voltage-Drop Calculation. The voltage drop in a power system may be calculated by selecting the formula most suitable as to accuracy desired and the voltage which is known, such as the load-end or source-end voltage of the circuit.

a. Formulas for voltage-drop calculation. In the following formulas the voltages and voltage drops are on a line-to-neutral basis. To obtain the line-to-line voltage drop in a three-phase system, multiply the line-to-neutral voltage drop by $\sqrt{3}$. For single-phase systems the line-to-line voltage drop is obtained by multiplying the line-to-neutral voltage drop by 2.

Under some conditions, an answer with a negative sign can be obtained from the following formulas. In such cases the answer should be interpreted as showing that the load voltage is higher than the source voltage. These cases will be rare, however, since the great majority of systems will have load voltages which are lower than the source voltage.

The nomenclature for the formulas is as follows:

E = line-to-neutral voltage drop
E_s = line-to-neutral voltage at source end
E_r = line-to-neutral voltage at load end
θ = angle whose cosine is the load power factor
I = line current
R = resistance of circuit, Ω
X = reactance of circuit, Ω (by convention, inductive reactance is positive and capacitive reactance is negative)
$\cos\theta$ = load power factor in decimals
$\sin\theta$ = load reactive factor in decimals (by convention, $\sin\theta$ is positive for lagging-power-factor loads and negative for leading-power-factor loads)

The following formulas are exact. If E_r is known,

$$E = [(E_r \cos\theta + IR)^2 + (E_r \sin\theta + IX)^2]^{1/2} - E_r$$

If E_s is known,

$$E = E_s + IR \cos \theta + IX \sin \theta - [E_s^2 - (IX \cos \theta - IR \sin \theta)^2]^{1/2}$$

For most practical purposes, the following approximate formula is of sufficient accuracy:

$$E = I(R \cos \theta + X \sin \theta)$$

From the phasor diagram (Fig. 4), it can be seen that the approximate formula is close enough for most applications. In practical cases the angle between E_s and E_r will be small, and it approaches zero as the power factor of the load approaches that of the supply system.

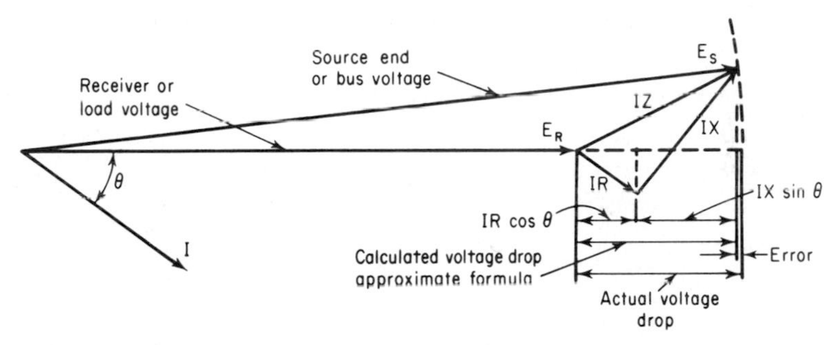

FIG. 4. Typical phasor diagram showing voltage relationships in power-supply system. *(Ref. 4.)*

b. Typical equipment impedance data. All reactance data are based on a frequency of 60 Hz. Typical impedance values for 600- and 5000-V cables in magnetic ducts are shown in Tables 14 and 15 respectively. Correction factors for nonmagnetic ducts are shown in Tables 16 and 17. Tables 18 through 20 contain approximate impedance data for single-phase distribution transformers, load-center-type transformers, and power transformers 500 kVA and above. Tables 21 and 22 contain resistance and reactance data for busway circuits.

17. Voltage Drop Due to Motor Starting. The voltage behavior at the motor terminals and throughout the supply system during motor starting is more difficult to determine than for steady-state operation. When a motor is started from a synchronous generator source, the voltage variation is greatly dependent upon the initial generator loading, the relative size of the motor, and the type of voltage regulator. Figure 5 shows typical results of starting a motor whose inrush kVA is equal to the generator rating for several starting conditions. It can be seen that the restored voltage during the starting period is dependent upon the initial load and the response of the exciter and voltage-regulating system. Calculation of the restored voltage value may be important for synchronous-motor applications to determine if sufficient pull-in torque can be developed. The value of restored voltage should also be used where starting times are long, such as in accelerating high-inertia fans. The minimum voltage magnitude should be determined to investi-

TABLE 14. Approximate Resistance, Reactance, and Impedance of 600-V Cables in Magnetic Ducts (Ref. 4)

Cable size	Three single-conductor cables per duct, ohms per 100 ft			Three-conductor cable including interlocked armor cable, ohms per 100 ft		
	*R**	*X*	*Z*	*R**	*X*	*Z*
No. 14 AWG....	0.3135	0.00765	0.3135	0.3135	0.00468	0.31352
No. 12 AWG....	0.1972	0.00710	0.1972	0.1972	0.00456	0.19720
No. 10 AWG....	0.1240	0.00687	0.1240	0.1240	0.00448	0.12410
No. 8 AWG.....	0.0779	0.00638	0.0782	0.0779	0.00427	0.07460
No. 6 AWG.....	0.0498	0.00598	0.0500	0.0493	0.00391	0.04899
No. 4 AWG.....	0.0318	0.00551	0.0322	0.0312	0.00362	0.03140
No. 2 AWG.....	0.0203	0.00513	0.0209	0.0197	0.00344	0.02000
No. 1 AWG.....	0.0163	0.00500	0.0171	0.0157	0.00342	0.01606
No. 1/0 AWG...	0.0131	0.00495	0.0140	0.0125	0.00340	0.01296
No. 2/0 AWG...	0.0106	0.00490	0.0117	0.0100	0.00336	0.01054
No. 3/0 AWG...	0.00860	0.00486	0.00986	0.00800	0.00333	0.00866
No. 4/0 AWG...	0.00700	0.00482	0.00850	0.00640	0.00327	0.00721
250 MCM.......	0.00608	0.00480	0.00778	0.00547	0.00322	0.00632
300 MCM.......	0.00520	0.00474	0.00704	0.00460	0.00316	0.00557
350 MCM.......	0.00461	0.00469	0.00658	0.00400	0.00310	0.00510
400 MCM.......	0.00419	0.00462	0.00625	0.00354	0.00304	0.00469
500 MCM.......	0.00359	0.00450	0.00575	0.00292	0.00295	0.00412
750 MCM.......	0.00280	0.00438	0.00520	0.00208	0.00284	0.00346

* Based on 75°C.

gate possible dropout of undervoltage devices, contactors, and associated equipment. The minimum voltage level also determines the degree of light flicker.

a. Minimum generator voltage. The curves of Fig. 6 show the minimum generator voltage as a function of motor-starting kVA in percent of rated generator kVA. The motor-starting kVA is that which would be drawn if the generator voltage were maintained at rated value. The actual kVA will be less than that at rated voltage because of the drop in the generator. This effect is taken into account in the curves of Fig. 6. Three families of curves are shown for three speed ranges. Limiting conditions are represented by the curves for generators with no voltage regulators and those equipped with electronic or high-speed regulators. Intermediate curves are a function of the excitor response as related to a performance factor. The performance factor K is the exciter response in volts per second divided by the exciter voltage for rated generator voltage at rated load multiplied by the generator open-circuit field time constant in seconds. Approximate values of K are given in Fig. 7 and are based upon the use of a self-excited exciter controlled by a direct-acting rheostatic voltage regulator. Multipliers to allow for variation of exciter response with generator initial load are included. The initial load is considered to be of the type which draws constant current during the voltage disturbance, such as a mixture of partially loaded induction motors and lighting.

TABLE 15. Approximate Resistance, Reactance, and Impedance of 5000-V Cables in Magnetic Ducts (Ref. 4)

Cable size	Three single-conductor cables per duct, ohms per 100 ft			Three-conductor cable including interlocked armor cable, ohms per 100 ft		
	*R**	*X*	*Z*	*R**	*X*	*Z*
No. 14 AWG....	0.3135	0.00969	0.3135	0.3135	0.006664	0.3291
No. 10 AWG....	0.1240	0.00850	0.1240	0.1240	0.005745	0.1241
No. 8 AWG.....	0.0779	0.00788	0.0781	0.0779	0.005308	0.07808
No. 6 AWG.....	0.0498	0.00748	0.0503	0.0493	0.004941	0.04944
No. 4 AWG.....	0.0318	0.00681	0.0325	0.0312	0.004619	0.03154
No. 2 AWG.....	0.0203	0.00623	0.0212	0.0197	0.004366	0.02017
No. 1 AWG.....	0.0163	0.00588	0.0173	0.0157	0.003964	0.01619
No. 1/0 AWG...	0.0131	0.00567	0.0143	0.0125	0.003792	0.01304
No. 2/0 AWG...	0.0106	0.00545	0.0119	0.0100	0.003677	0.01061
No. 3/0 AWG...	0.00860	0.00535	0.0101	0.00800	0.003631	0.008785
No. 4/0 AWG...	0.00700	0.00529	0.00877	0.00640	0.003585	0.007335
250 MCM.......	0.00609	0.00525	0.00802	0.00547	0.003562	0.006527
300 MCM.......	0.00520	0.00519	0.00735	0.00460	0.003518	0.005791
350 MCM.......	0.00461	0.00514	0.00690	0.00400	0.003477	0.005299
400 MCM.......	0.00419	0.00506	0.00657	0.00354	0.003436	0.004923
500 MCM.......	0.00359	0.00495	0.00611	0.00292	0.003344	0.004439
750 MCM.......	0.00280	0.00474	0.00551	0.00208	0.003088	0.003723

* Based on 75°C.

TABLE 16. Correction Factors for Nonmagnetic Ducts (Three-Conductor Cables) (Ref. 4)

Factor for correcting reactances, all sizes of cable	Factors for correcting resistances	
	No. 14 to No. 00 AWG	No. 0000 AWG to 750 MCM
0.87	1.0	0.98

Determine correct *Z* from corrected values of *X* and *R*. No correction is required for interlocked armor.

TABLE 17. Correction Factors for Nonmagnetic Ducts (Single-Conductor Cables) (Ref. 4)

Factor for correcting reactances, all sizes of cable	Factors for correcting resistances				
	No. 14 to No. 8 AWG	No. 6 to No. 0 AWG	No. 00 to 250 MCM	300 to 500 MCM	750 MCM
0.8	1.0	0.96	0.93	0.83	0.72

TABLE 18. Approximate Resistance, Reactance, and Impedance of Single-Phase Distribution Transformers (Ref. 4)

kVA	High voltage: 2,400/4,160Y volts and 2,400/4,800/8,320Y volts Low voltage: 120/240, 240/480, 600 volts—60 Hz			High voltage: 7,200/12,470Y volts Low voltage: 120/240, 240/480, 600 volts—60 Hz		
	% R	% X	% Z	% R	% X	% Z
3 5	1.7	1.5	2.3	2.2	1.7	2.8
10 15 25	1.5	1.7	2.3	1.6	1.6	2.3
37½ 50	1.3	2.2	2.6	1.3	2.0	2.4
75 100	1.2	2.3	2.6	1.2	3.5	3.7
167	1.1	3.8	4.0	1.0	3.6	3.7
250 333 500	1.0	4.7	4.8	1.0	5.1	5.2

TABLE 19. Approximate Reactance of Load-Center-Type Transformers (Ref. 4)

kVA range	(Three-phase) 15-kV max primary voltage 600-volt max secondary voltage, % reactance on own kVA base*
112½–150	3.0
225–500	5.0
750–2,000	5.5

*Percent resistance on own kVA base is approximately 1.5 percent for 150 kVA and below and varies from approximately 1 down to 0.8 percent on ratings above 150 kVA.

TABLE 20. Approximate Impedance of Power Transformers (above 500 kVA) (Ref. 4)

Insulation class, kV		Impedance at kVA base equal to 55°C rating of largest capacity winding for	
High voltage	Low voltage	Self-cooled or water-cooled rating, %	Forced-oil-cooled rating, %
15 or lower	15 or lower	5½	6¾
25	15 or lower	5½	8¼
34.5	15 or lower	6	9
46.0	15 or lower	6½	9¾
69.0	15 or lower	7	10½
92.0	15 or lower	7½	11¼
115.0	15 or lower	8	12
138.0	15 or lower	8½	12¾

For high-voltage insulation classes intermediate of those given, use the impedance of the next higher listed insulation class.

For transformers with a load-ratio control add 0.5 percent to the values listed above except in those cases in which a lower impedance has been specified.

The percent resistance on the base given above ranges from 1.0 down to 0.06.

TABLE 21. Resistance of Typical Copper Busway Circuits (Ref. 4)

Current capacity of busway, A	Resistance, ohms per 1,000 ft
250	0.114
400	0.033
600	0.023
800	0.016
1,000	0.012
1,350	0.0096
1,600	0.0073
2,000	0.0055

TABLE 22. Reactance of Typical Three-Phase Low-Voltage Copper Busway Circuits (Ref. 4)

Percent reactance of 1000 circuit feet on a 1000-kVA base

	System voltage, V		
Busway rating, A	240	480	600
Plug-in type:			
Up to 600	98.8	24.7	15.8
601–1000	62.4	15.6	10.0
Low-impedance type:			
Up to 600	45.2	11.4	7.3
601–1000	17.2	4.3	2.7
1350–1600	10.8	2.7	1.7
2000	7.6	1.9	1.2

b. Restored voltage. The curves of Fig. 8 may be used for estimating the generator voltage attained after the regulator has acted to apply maximum excitation current to the generator during the starting of a squirrel-cage induction motor or a synchronous motor. Initial generator load is assumed to be of the constant-current type with 0.8 power factor.

c. Initial generator voltage and load considerations. As shown in Table 1, the voltage ratings of motors are not the same as the generator voltage ratings in many cases for a given nominal system voltage. If, for example, a 460-V motor is started directly from a 480-V generator operating initially at its rated voltage, the motor inrush kVA must be adjusted directly as the square of the motor terminal voltage prior to using Figs. 6 and 8. The restored voltage values cannot exceed the initial voltage setting of the regulator. Refer-

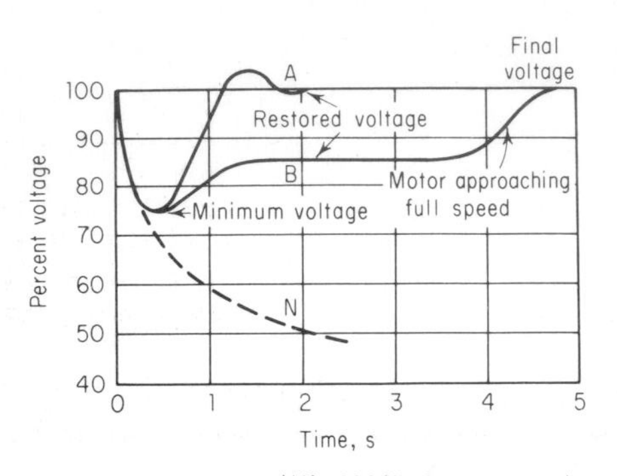

A - No initial load on generator
B - 50 % initial load on generator
N - No regulator

FIG. 5. Typical generator-voltage behavior. *(Ref. 4.)*

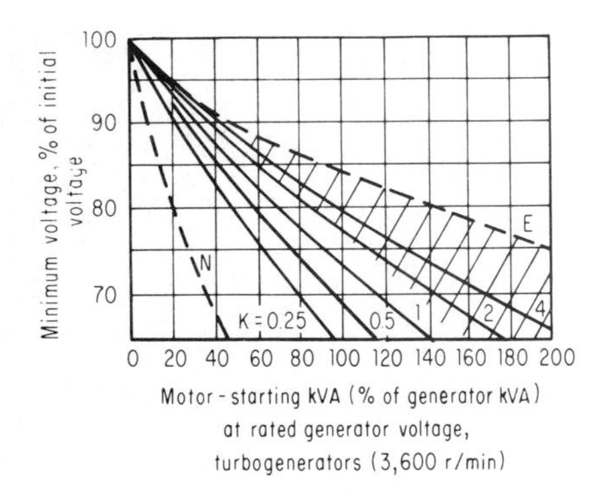

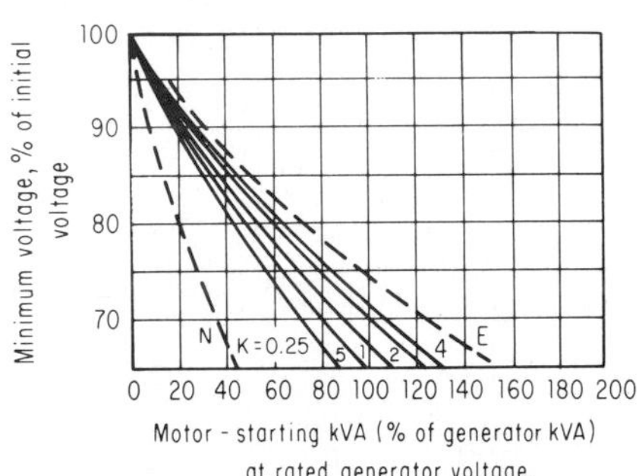

Initial load assumed constant-current type
K- Performance factor (Fig. 7)
N- With no voltage regulator
E- With electronic exciter
In shaded areas, with high initial loads, minimum voltage is the lower of the values from Figs. 6 and 8.

Generator reactances are taken to be as follows:

Generator speed, r/min	Motor-starting kVA, %	Effective transient reactance x'_d, %	Synchronous reactance x_d, %
3,600	50	25	120
3,600	100	18	120
3,600	150	16	120
1,800-600	0-150	25	120
Below 600	0-150	35	120

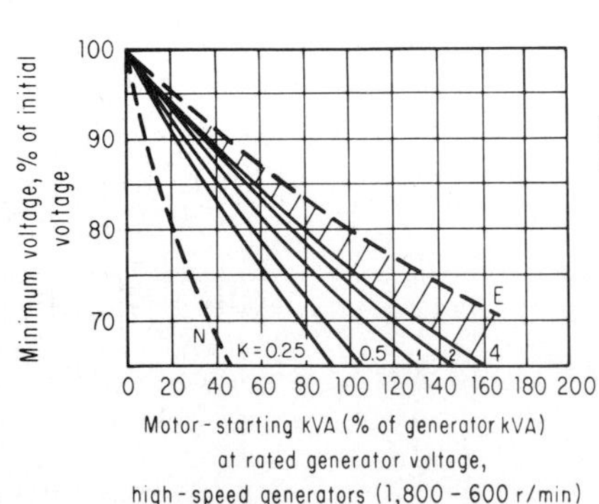

FIG. 6. Generator minimum voltage. *(Ref. 4.)*

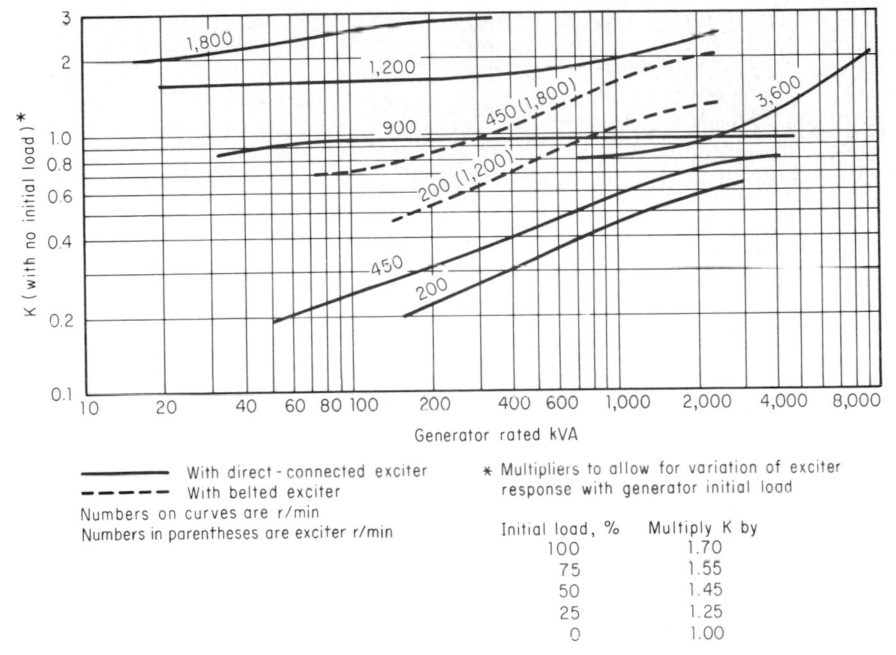

FIG. 7. Typical values of performance factor K for ac generators. *(Ref. 4.)*

ring to Fig. 8, if the initial setting of the regulator is 90 percent, all restored voltage values become horizontal lines at 90 percent of rated voltage rather than at 100 percent voltage.

If the generator initial load is not of the constant-current type but consists entirely of heavily loaded induction motors, the voltage disturbance will be more severe than shown in Figs. 6 and 8. Heavily loaded induction and synchronous

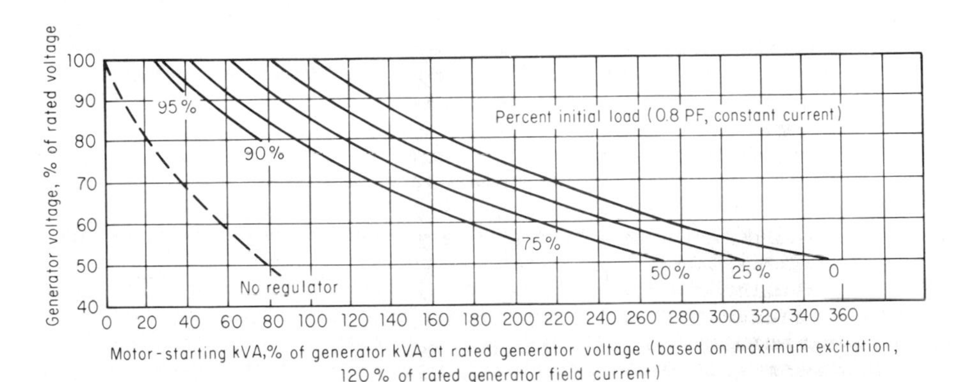

FIG. 8. Restored generator voltage. *(Ref. 4.)*

motors tend to be a constant-kVA-type load since the current increases as the terminal voltage decreases. On the other hand, if the initial system load consists mainly of a resistive load, the current will decrease with voltage reduction and the disturbance will not be as severe as shown in Figs. 6 and 8.

Figure 9 provides a method of adjusting the motor-starting kVA to an equivalent value to compensate for the initial heavy loading of induction motors. This permits the initial loading to be considered as being the constant-current type.

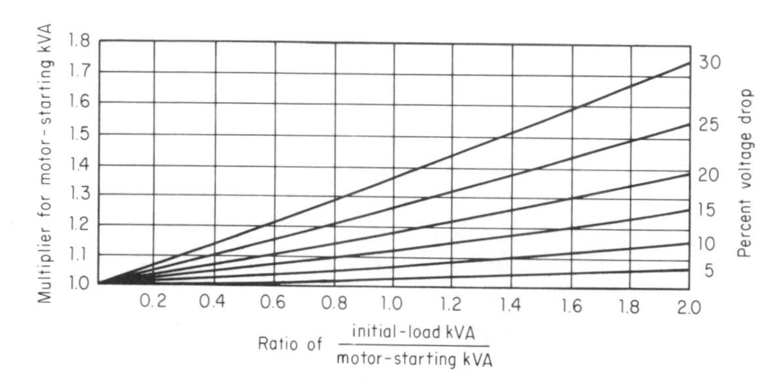

FIG. 9. Approximate effect of initial load consisting of fully loaded induction motors. *(Ref. 4.)*

The following example demonstrates use of Fig. 9. Assume that the constant-current voltage drop of a generator is determined to be 25 percent from Fig. 6 or 8 if a 100 percent motor-starting load is applied in addition to a 50 percent initial load. If the initial load consists of heavily loaded induction motors, enter Fig. 9 with the 25 percent voltage drop and a 0.5 ratio of initial-load kVA to motor-starting kVA. A multiplier of 1.13 for motor-starting kVA is obtained from the curve. The generator voltage drop is then redetermined using this larger motor-starting kVA.

d. Motor-starting power factor. The approximate starting power factors of typical squirrel-cage induction motors are shown in Fig. 10. Wound-rotor induction motors have a starting power factor of approximately 0.8.

e. Distribution-system voltage drops. The voltage drops in lines, cables, and transformers are frequently more important than the generator voltage drop during motor starting. When the total rating of the connected generation is large in comparison with the motor being started, the generator voltage drop will be small and quickly corrected by regulators. As the point of motor application becomes more remote from the source of generation, the effect of generator voltage drop becomes less and less of a factor. The voltage drop in a supply transformer can become the limiting consideration during motor starting in many instances.

f. Determination of transformer voltage drops. The voltage drop through typical transformers during the starting of squirrel-cage induction or synchronous motors can be estimated from Fig. 11. Motor-starting kVA is the value that would be obtained if rated transformer secondary voltage were maintained. The secondary voltage on starting of the motor, in percent of the initial secondary voltage, is

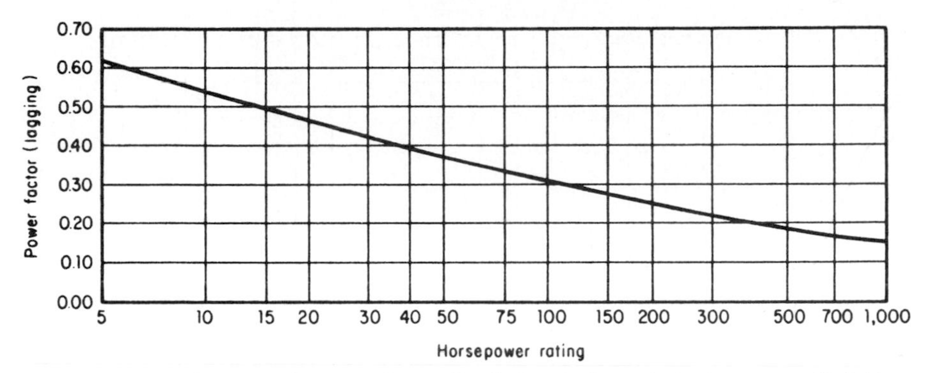

FIG. 10. Approximate starting power factor of typical squirrel-cage induction motors. *(Ref. 4.)*

plotted as a function of the motor-starting kVA in percent of transformer kVA rating. Figure 11 neglects the effect of primary voltage drops. Methods for taking these effects into account are described later.

The curves of Fig. 11 were prepared on the basis that the initial load, if any, draws constant current during the motor-starting voltage disturbance. If the initial load consists of a constant-kVA type, the motor-starting kVA should be adjusted to an equivalent value, as previously described through use of Fig. 9.

The curves of Fig. 11 apply for motor-starting power factors in the usual range of 10 to 40 percent. For wound-rotor motors with a starting power factor of about 80 percent, the drop will be approximately 70 percent of the values shown. These curves also do not take into account any effect of automatic tap-changing apparatus on the transformer, which may serve to restore the voltage level during the motor-starting interval.

g. Voltage drops in cables and overhead lines. The curves of Figs. 12 and 13 may be used to estimate the voltage drops in lines and cables when starting squirrel-cage induction and synchronous motors. In Fig. 12 the percent of initial load end voltage is shown as a function of the loading factor M for typical lines and cables. Figure 13 applies to the specific case where M is equal to 1. Since the curves of Fig. 12 are practically straight lines, the value of voltage drop may be obtained from Fig. 13 by multiplying by the value of M. The circled points of Figs. 12 and 13 are corresponding circuit configurations.

The power factor of the motor-starting load is considered to be 30 percent. For conductor sizes above No. 0 AWG, variations over the range of 20 to 40 percent

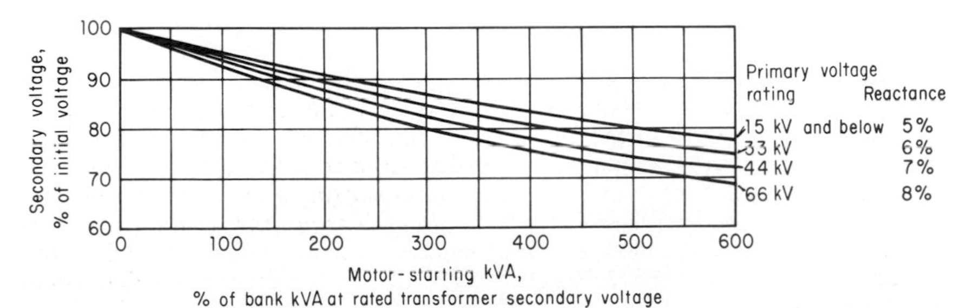

FIG. 11. Transformer secondary voltage. *(Ref. 4.)*

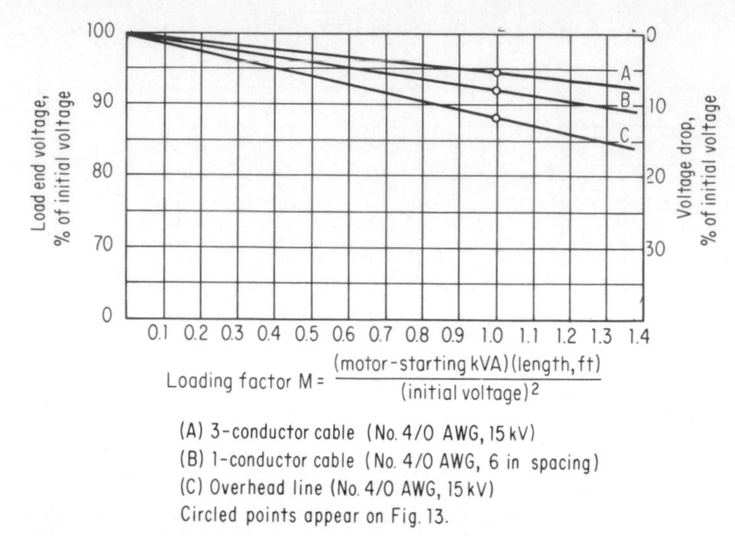

FIG. 12. Variation of voltage drop with loading factor M for typical lines and cables. *(Ref. 4.)*

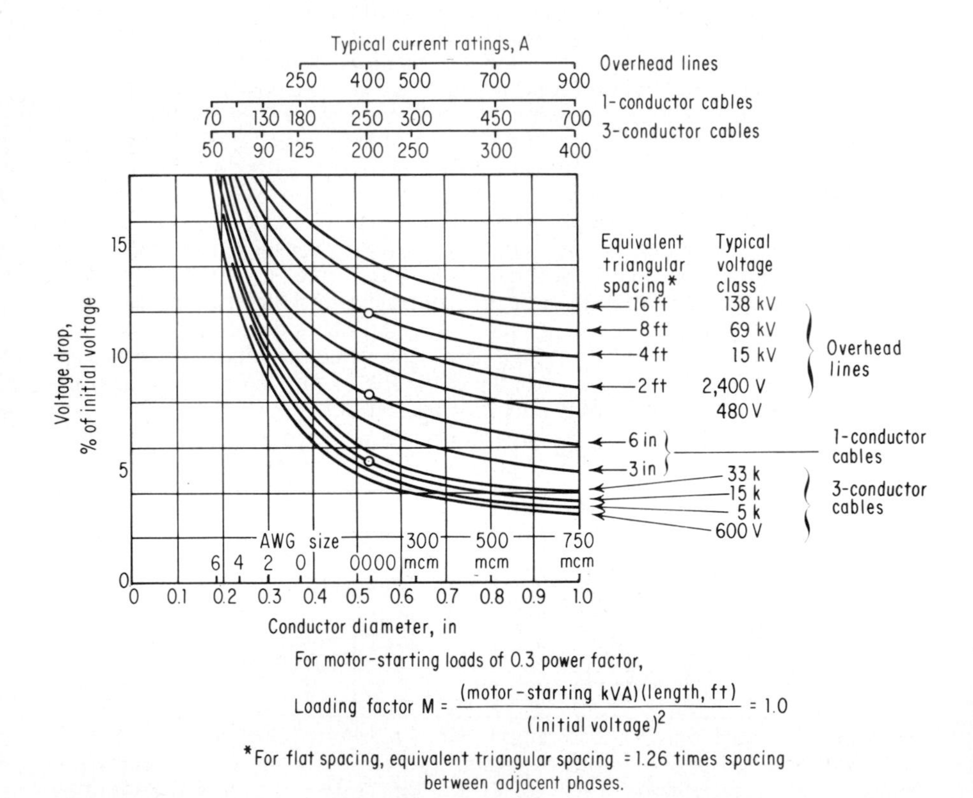

FIG. 13. Voltage drop in lines and cables with loading factor M being unity. *(Ref. 4.)*

will not have an important effect on the voltage drop. The curves are shown for a frequency of 60 Hz. For other frequencies, the voltage drop will vary directly as the frequency for conductor sizes above No. 0 AWG. For conductor sizes smaller than this, the variation will be less than directly proportional to frequency.

h. Voltage drop of reactors. The voltage drop in current-limiting reactors may be estimated from Fig. 11 during motor starting. Current-limiting reactors are usually rated as having a certain percent reactance on a specified system-kVA and system-voltage base. The motor-starting kVA of Fig. 11 should be that drawn at the specified system voltage expressed in percent of the specified system kVA. If the percent reactance of a reactor does not lie between 5 and 8 percent, the motor-starting kVA should be adjusted to an equivalent value and the secondary voltage read from the percent reactance curve used as the basis of adjustment.

i. Voltage drop of power systems. Motors are frequently supplied from power systems consisting of complex networks for which calculation of the voltage drop would be difficult. The voltage drop may be estimated, however, if the short-circuit kVA is known at the point of power delivery.

When motor-starting kVA is drawn from a system, the voltage drop in percent of the initial voltage is approximately equal to 100 times the motor-starting kVA divided by the sum of this kVA and the short-circuit kVA. The motor-starting kVA used should be that drawn by the motor if the initial system voltage were maintained. For example, if a 1000-hp motor has a starting kVA of 5000 at the initial system voltage and the system short-circuit kVA is 50,000, the voltage drop will be approximately

$$\frac{5000 \times 100}{5000 + 50{,}000} = 9\% \text{ of initial voltage}$$

In many systems the short-circuit kVA varies over a wide range, depending upon the number of parallel lines in service, system interconnections, etc. While the highest short-circuit kVA is of interest for circuit interruption, the minimum short-circuit kVA should be used for voltage-drop calculations since it gives the highest value. Table 23 shows typical power-system short-circuit kVA ranges. A corresponding variation in the voltage drop produced by a certain motor-starting kVA will occur, depending upon the chosen value within the short-circuit kVA range. In those cases where this voltage is of importance, the actual minimum short-circuit kVA should be used in the calculation.

TABLE 23. Power-System Short-Circuit kVA

System voltage	Usual range of short-circuit kVA
2,400	15,000–150,000
4,160	25,000–250,000
6,900	50,000–500,000
13,800	100,000–1,000,000
23,000	150,000–1,500,000
34,500	150,000–1,500,000
69,000	150,000–1,500,000
115,000	250,000–2,500,000

The method of calculating voltage drop given above should be used for systems consisting mainly of transformers, transmission lines, reactors, and cables remote from the source of generation. It is not applicable where the short-circuit kVA is appreciably affected by generator reactances.

j. Method of combining system voltage drops. Often a motor is supplied through a series combination of transformers, overhead lines, cables, and generators. A rough estimate of the total voltage drop may be obtained by adding the motor-starting voltage drops as determined in the foregoing paragraphs. This simple addition method results in a pessimistic answer since the addition of series impedance tends to decrease the current supplied to the motor (Ref. 4).

Where a more accurate voltage-drop determination is required, the following procedure is suggested:

1. Determine the voltage drop in the circuit element nearest the motor, neglecting the other elements. For example, for a motor supplied from a generator, transformer and cable in series, determine the drop in the cable first.
2. Multiply the motor-starting kVA by the ratio of the load-end voltage to the initial voltage of the cable just determined.
3. Using this new value of motor-starting kVA, determine the voltage drop in the next circuit element. In the example selected, this drop is the transformer drop.
4. Multiply the motor-starting kVA by the product of the ratio of the load-end voltage to the initial voltage of the cable and the ratio of the secondary voltage to the initial secondary voltage of the transformer.
5. Using this new value of motor-starting kVA, determine the voltage drop in the next circuit element. In the example selected, this drop is the generator voltage drop.
6. Continue the process until all series circuit elements have been considered.
7. Multiply the initial voltage at the motor by the product of the ratios of final voltage to initial voltage of all the circuit elements. This result is the final voltage at the load.

CONTRIBUTIONS OF MOTORS TO SYSTEM FAULTS

The basic sources of system-fault current are synchronous and induction rotating machines. The contributions of synchronous and large induction motors must be considered in system-fault studies along with the fault contributions of generators.

The fault contribution of a rotating machine can be considered as being furnished from an infinite bus and limited by the machine reactance. The magnitude of fault current decreases with time following the fault initiation. The reactance of the machine thus appears to be a variable quantity. Approximate reactance data are included to permit calculation of the short-circuit currents produced by synchronous and induction motors.

18. Synchronous-Motor Behavior during Faults. Synchronous motors behave in much the same manner during fault conditions as do synchronous generators. The rotational energy stored in the motor rotor and the driven load act as the prime mover. Excitation is maintained by the separate source of excitation current. The variable reactance of a synchronous motor following fault initiation is defined in the same terms as that of a synchronous generator.

This variable reactance can be divided into three separate reactances of increasing magnitude, namely, the subtransient reactance x''_d, the transient reactance x'_d, and the synchronous reactance x_d. The particular value to use in fault calculations is a function of the elapsed time from initiation of the fault. The value of subtransient reactance determines the value of fault current during the first cycle. This current value decays exponentially during the first few cycles to the value determined by the transient reactance. The value of transient short-circuit current, in turn, decays exponentially in about 30 to 120 cycles to the steady-state value determined by the synchronous reactance. The above values of fault current are further diminished following the fault by the reduction in generated voltages as the motor speed decreases.

19. Induction-Motor Behavior during Faults. The main flux in an induction motor is maintained by the system to which it is connected. During fault conditions the system voltage is reduced to a very low value. The decay of flux linkages within the motor generates a voltage in the stator which, in turn, produces short-circuit current to the fault. The stored energy in the rotating masses acts as the prime mover.

The duration of the short-circuit current is short, however, because of the rapid decay of flux. Usually the contribution has disappeared in a few cycles, as can be seen in Fig. 14. This oscillogram shows the short-circuit current produced by a lightly loaded 150-hp 460-V three-phase 60-Hz 720-r/min wound-rotor induction motor faulted at its terminals. The rotor was short-circuited during this test to simulate a squirrel-cage induction motor. The fault contribution of a wound rotor operating with an external resistance in the rotor circuit may be of little or no importance because of the reduced time constant. It is necessary to check the individual application before neglecting a wound-rotor induction-motor contribution.

Since the duration of short-circuit current is on the order of a few cycles, an induction motor is considered to have only a subtransient reactance value x''_d. The initial value of symmetrical fault current is approximately equal to the full-voltage motor-starting current.

20. Approximate Motor-Reactance Values. The following 60-Hz reactance values are expressed in per unit values based upon the motor kVA rating. Base kVA ratings are determined as follows:

$$\text{Exact base kVA of motor} = \sqrt{3}EI/1000$$

where E and I are nameplate voltage and current ratings.

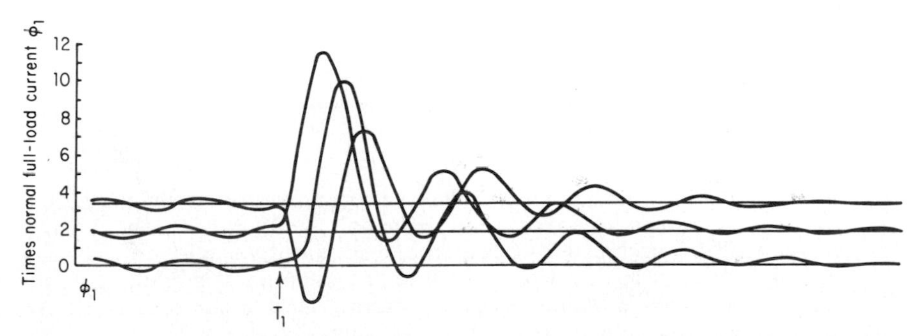

FIG. 14. Traces of oscillograms of short-circuit currents produced by an induction motor running at light load. *(Ref. 4.)*

The approximate kVA base of the motor when full-load currents are not known is

Induction-motor kVA base = hp rating
0.8-power-factor synchronous-motor kVA base = hp rating
1.0-power-factor synchronous-motor kVA base = 0.8 × hp rating

Exact motor-reactance values should be used when available from the motor manufacturer for fault-current-contribution studies. Subtransient reactances should be used in calculating momentary-withstand ratings and interrupting duties for devices which open in a few cycles, such as fuses. Transient reactances should be used for determining interrupting duties for circuit breakers that open after the first few cycles, such as 5- and 8-cycle breakers. In all fault-current-contribution studies the effect of dc offset must also be considered in addition to the symmetrical component.

a. Large induction motors. The subtransient reactance can be calculated as

$$\text{Per unit } x''_d = \frac{\text{rated full-load current}}{\text{locked-rotor current at rated voltage}}$$

This reactance value will generally lie between 0.14 and 0.25, with a mean of 0.17 per unit.

b. Large synchronous motors. Table 24 shows the approximate values of both x''_d and x'_d as a function of motor speed for motors in the several hundred to several thousand horsepower range.

TABLE 24. Approximate Reactance Values of Large Synchronous Motors (Ref. 4)

No. of poles	Percent x_d''		Percent x_d'	
	Range	Mean	Range	Mean
6	10–20	15	15–30	23
8–14	15–25	20	20–40	30
16 or more	25–45	30	25–60	40

c. Groups of small motors. Short-circuit contributions from a group of small motors based upon the installed kVA can be estimated from Table 25. The proportion of synchronous and induction motors should be known or estimated. The relationship between installed motor kVA and energized kVA should be known or estimated to make a study meaningful.

21. Dc Motor Behavior during Faults. The transient armature current of a faulted dc machine rises rapidly to a maximum value and then decays slowly to a sustained value. If the machine is self-excited, the sustained value is zero. The calculation of dc machine fault-current contributions is complicated by the variability and nonlinearity of machine parameters during short-circuit conditions. An approximate method utilizing nameplate data and having a range of accuracy from −20 to +30 percent is presented below. In general, the results determined

TABLE 25. Reactances Based on kVA of Installed Motors (Ref. 4)

Item	Motor ratings and connections	Subtransient reactance x_d'' %	Transient reactance x_d' %
1	600 volts or less—induction	28	
2	600 volts or less—synchronous (items 1 and 2 include motor leads)	21	29
3	600 volts or less—induction	34	
4	600 volts or less—synchronous (items 3 and 4 include motor leads and step-down transformers)	27	35
5	Motors above 600 volts—induction	20	
6	Motors above 600 volts—synchronous	15	25
7	Motors above 600 volts—induction	26	
8	Motors above 600 volts—synchronous (items 7 and 8 include step-down transformers)	21	31

by this method will be on the high side. It is assumed that a flashover has not occurred. This method is covered in more detail in Ref. 5.

This method employs the use of two equivalent circuits to calculate the initial rate of rise of the armature current and its peak value. These two circuits cannot be combined to calculate an overall transient response since each is applicable only for its specific purpose. The rate of decay from the peak-current value to the sustained value is determined by the time constant of the exciting windings, which is on the order of 0.1 to 1.0 s.

a. Calculation of initial rate of rise of armature current. The initial rate of rise of armature current for a terminal fault can be approximated from the following formula:

$$\frac{di_a}{dt} = \frac{PN_1E_0}{19.1C_x} \qquad \text{per unit A/s}$$

where P = number of poles
N_1 = base speed, r/min
E_0 = per unit armature voltage before short circuit
C_x = a characteristic constant for a particular class of machine

Values of C_x are listed in Table 26. The equivalent circuit employed in the initial rate-of-rise calculation consists of a voltage source representing the machine-generated voltage at the initial speed and an inductance equivalent to the armature-circuit inductance.

TABLE 26. Characteristic Inductance Constant for dc Motors (Ref. 5)

Motor type	Approx value of C_x	$\frac{1}{19.1\,C_x}$
Without pole-face windings........	0.4	0.13
With pole-face windings...........	0.1	0.52

If a short-circuited dc machine behaved as a constant-parameter circuit there would be little difficulty in describing the complete transient. Unfortunately, the inductance and equivalent armature-circuit resistance change because of saturation and eddy currents during the transient. Tests indicate, however, that the current will rise at a constant rate as determined by the above equation until approximately two-thirds of the maximum value is reached.

b. Calculation of peak armature current. The equivalent circuit describing the machine for peak-armature-current determination consists of a voltage source and a resistance. The voltage source again represents the machine-generated voltage prior to short circuit. It is assumed that the initial speed remains constant until the peak current is reached. The resistance is an equivalent internal value which represents all the factors tending to limit the flow of current from the machine. This equivalent or effective value of internal armature-circuit resistance is designated as r'_d.

Values of r'_d expressed in the per unit system for general-purpose machines, shunt- or compound-wound, constant- or adjustable-speed, with or without pole-face windings, continuous-duty, and having a voltage rating in the range from 115 to 750 V are shown in Fig. 15. The initial condition is no load and r'_d is expressed as a function of a horsepower-r/min product. The value of r'_d for low-voltage machines usually will be nearer the top curve of Fig. 15. The r'_d value for high-voltage machines will generally lie closer to the lower curve. The effect of load current for a machine operating as a motor prior to the fault can be introduced

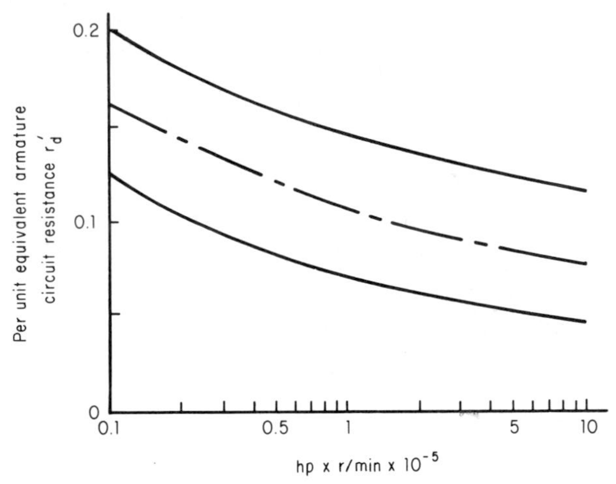

FIG. 15. Per unit equivalent armature-circuit resistance generalized curve for dc machines. *(Ref. 5.)*

by subtracting this value from the calculated short-circuit current obtained from E_0 and r'_d.

c. Groups of dc motors. The value of equivalent armature resistance obtained from Fig. 15 and the per unit generated voltage can be used to represent an equivalent circuit for each dc motor on the system. It is important that resistance external to the machines be included in this equivalent circuit. If the external resistance reflected into each machine exceeds approximately twice the value of r'_d

obtained from Fig. 15, the motor must be considered overloaded rather than short-circuited. When calculating the magnitude of short-circuit current it must be remembered that the circuit representation is for the peak value. If the short-circuit contributions of all motors in the system reach a peak at approximately the same time, the overall accuracy should be commensurate with the accuracy of the r'_d values. If there are large variations in the time to peak current, improved accuracy can be obtained by the more exact methods in Ref. 5. Also included are calculation methods for system contributions from a machine that has flashed over at the armature. The contribution to the system is less in this case than if the machine had failed at its terminals.

POWER-FACTOR IMPROVEMENT

Industrial-plant power factors are generally lagging because of the exciting currents required by induction motors, transformers, fluorescent lighting, induction-heating furnaces, etc. Power-factor improvement can be obtained through the use of synchronous motors or capacitors at the proper locations.

22. Disadvantages of Low Power Factor. Low power factor has an adverse effect upon system operation. This fact applies to both industrial and utility power systems. For this reason, the rate structures of many utilities contain power-factor clauses which penalize consumers with low-power-factor loads. The savings from improvement in power factor can be calculated from the daily load chart of the plant and the particular rate structure involved. It is not uncommon for capacitors to pay for themselves in a period of a few years.

Low power factor should be avoided for three reasons. First, since circuits and circuit elements tend to be more reactive than resistive, reactive components of current produce larger voltage drops than an equal resistive component of current. System-voltage regulation suffers and additional voltage-regulating equipment may be required for satisfactory operation.

The second disadvantage of low power factor is the inefficient utilization of system equipment due to the increased current flow per unit of real power transmitted. This larger current magnitude produces additional heating in system equipment and, in effect, derates these components. Power-factor correction will release this system capacity and permit increased loading without installation of additional distribution equipment.

A third disadvantage is the cost of the increased losses throughout the system. These losses vary as the square of the current and also inversely as the power factor squared. The reduction in system losses can result in an annual gross return of as much as 15 percent of the investment in power-factor-improvement equipment.

23. Induction-Motor Power Factor. The approximate full-load power factors of squirrel-cage induction motors are shown in Fig. 16. The power factors of larger and higher-speed motors are much higher than for small and low-speed ratings. A typical power-factor characteristic for a medium-size and medium-speed induction motor is shown in Fig. 17. The rapid deterioration of motor power factor with reduced loading is evident. The power factor at partial loading is caused by the essentially constant exciting current, which is independent of load. In many applications induction motors do not operate at full load since margin is included

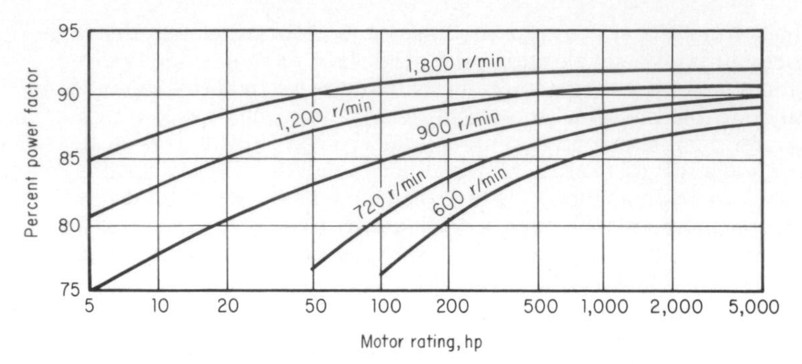

FIG. 16. Approximate full-load power factors of three-phase 60-Hz general-purpose squirrel-cage induction motors. *Note:* The values given indicate a general relationship only. Power factors of individual motor designs may vary widely.

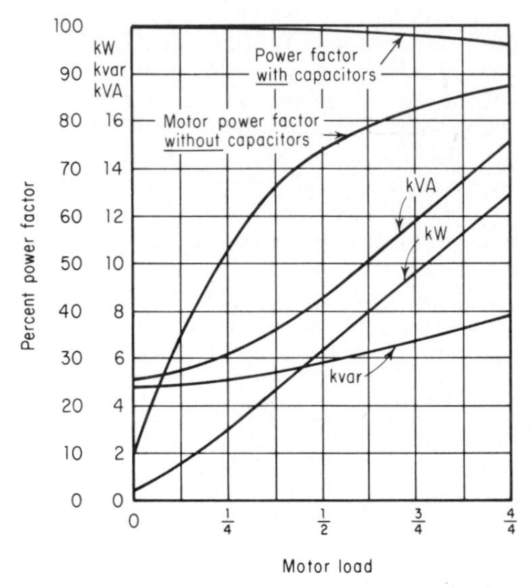

FIG. 17. Motor characteristics for typical medium-size and medium-speed induction motor. *(Ref. 4.)*

for overloads and possible future load increases due to expansion. Where possible, "overmotoring" should be avoided to minimize associated power-factor losses.

24. Synchronous-Motor Power Factor. A synchronous machine, unlike the asynchronous or induction machine, is capable of furnishing its own excitation requirement. An overexcited synchronous machine, whether it be a motor or a generator, furnishes reactive power to the system to which it is connected. Conversely, an underexcited synchronous machine draws reactive power or exciting current from the system. The dividing line between overexcited and underexcited operation is the value of field current required to produce unity-power-factor operation at any particular load.

The standard power-factor ratings of synchronous motors are unity and 0.8 leading power factor. Because of the increased armature and field currents associated with 0.8-power-factor operation, the 0.8-power-factor motor is larger and more expensive than an equally rated unity-power-factor motor. The ability of unity- and 0.8-power-factor motors to furnish reactive power is shown in Fig. 18. The 0.8-power-factor motor is capable of furnishing considerable reactive power throughout its entire load range. A unity-power-factor motor furnishes reactive power only at reduced loading. The curves shown in Fig. 18 are for rated full-load excitation.

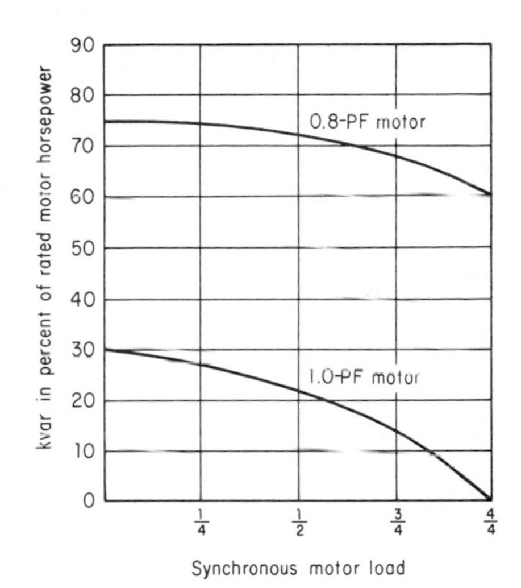

FIG. 18. Curves showing approximate kvar values supplied by synchronous motors with rated excitation.

25. Use of Synchronous Motors for Power-Factor Correction. Synchronous motors find the greatest application in low-speed drives, such as compressors. Low-speed applications are generally considered to be those operating at 600 r/min and below. It is in these applications that synchronous motors can provide very economical power-factor correction.

In general, the cost of a synchronous motor may be less than that of an induction motor when the ratio of horsepower to speed is greater than 1. The use of synchronous motors for power-factor correction in horsepower-speed ratings favoring less expensive induction motors requires careful economic study. In many cases induction motors and capacitors can provide lower first cost and reduced maintenance expense when compared with synchronous motors.

Another factor to consider is the relative efficiencies of synchronous and induction motors. Unity-power-factor synchronous motors are more efficient than induction motors of equal rating. Synchronous motors rated 0.8 power factor are less efficient than unity-power-factor motors, but they are comparable with induction-motor efficiencies up to the higher-speed intermediate-horsepower ratings. As rated speed is reduced, the synchronous motor becomes increasingly attractive from an efficiency standpoint when compared with an induction motor.

The improvement in power factor resulting from a synchronous-motor addition is demonstrated in the following example. Assume a plant load of 500 kW at 0.8 power factor lagging. The phasor relationships are shown in Fig. 19. Assume that a new drive requiring 200 hp is to be added to the plant. From Fig. 18 it can be determined that at full load a 0.8-power-factor synchronous motor will provide 120 kVA of reactive power. The resultant power factor would increase to 0.935. Improvement of the power factor to unity would require an additional 255 kvar. In general, it is not economic to improve the power factor beyond 0.95. The installation of a less expensive unity-power-factor motor would improve the

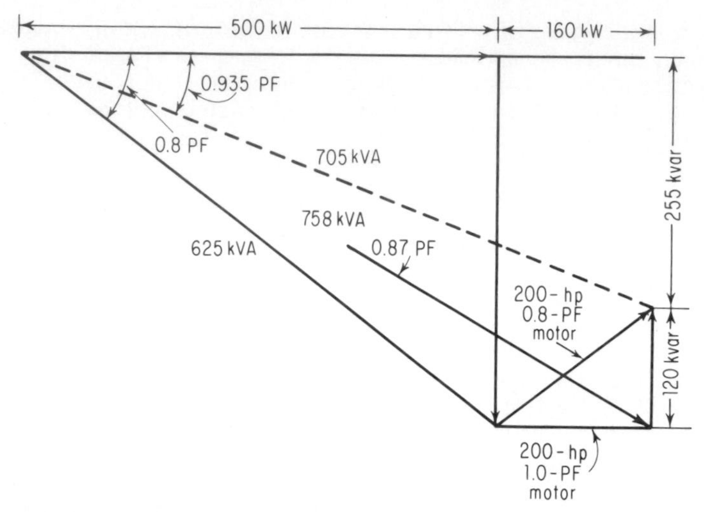

FIG. 19. Phasor diagram showing system power-factor correction by means of synchronous-motor addition.

power factor to 0.87 at full load. Operation of these two motors while essentially unloaded would result in system power factors of 0.91 and 0.85, respectively.

26. Use of Induction Motors and Capacitors. Static capacitors provide a simple means of correcting the induction-motor power factor. Figure 20 shows three possible capacitor connections for use with induction motors. The most common practice is to switch the capacitor and the motor as a unit, as shown in Fig. 20*a* or *b*. When capacitors are applied on the load side of the overload relay, the reduction in current through the overload relay must be taken into account. This method is more applicable to new motor installations since the proper relay can be purchased initially. Capacitor location on the source side of the overload relay is more applicable to older installations where only capacitors are being added. No changes in the existing overload relays are required. When it is desired to leave the capacitor permanently connected to the system, the connection in Fig. 20*c* can be used. This method eliminates the need for a separate capacitor switching device.

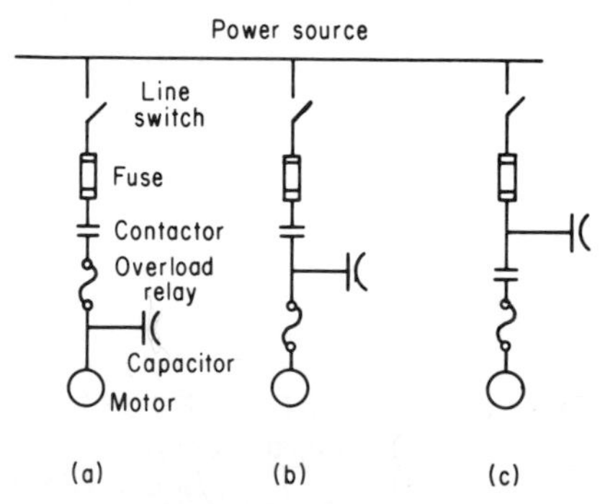

FIG. 20. Electrical location of capacitors when used with induction motors for power-factor improvement. *(Ref. 1.)*

27. Capacitor Limitations for Unit Switching. When switching a capacitor–induction-motor combination, there are two considerations which limit the capacitor kvar that can be applied. These two factors are the allowable overvoltage due to self-excitation and the permissible transient torque. These limitations do not apply to the arrangement shown in Fig. 20*c* since the capacitor is not switched with the motor and the plant is usually large compared to the switched load.

a. Self-excitation overvoltage. The rotational energy stored in the motor rotor and the drive continues to rotate the motor following disconnection from the source of supply. Normally the voltage generated during this period collapses rapidly since the required source of excitation is not present. However, if a capacitor is switched with the motor, a source of excitation at the motor terminals is provided and high voltages can be induced in the stator. The magnitude of the induced voltage is dependent upon the value of capacitance. It is possible to generate voltages on the order of 135 to 175 percent of rated when using capacitance sufficient to optimize the full-load power factor. The use of the motor no-load saturation curve for determination of the overvoltage value is shown in Fig. 21. The assumption is made that the motor speed remains constant. High-inertia drives are capable of maintaining self-excitation for a period of several minutes. Generally, however, a speed reduction of 15 percent eliminates self-excitation because of the decrease in generated voltage and the increase in capacitive reactance with decreasing frequency.

b. Transient torques. As described above, an induction motor can operate as an induction generator immediately after disconnection from the source. Reenergization of the motor during this self-excited period can produce severe transient electrical torques in much the same manner as a synchronous machine connected

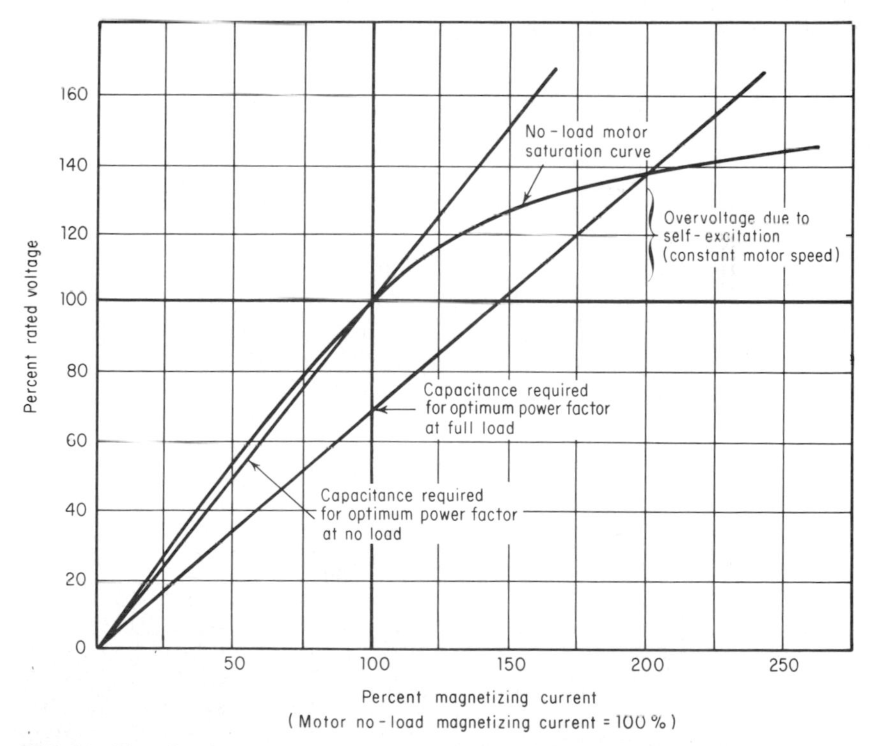

FIG. 21. Use of no-load saturation curve to determine overvoltage due to switching motor-capacitor combinations.

to the system out of phase. Transient torques on the order of 20 times rated full-load torque can be produced when switching with large values of capacitance. The capacitor has an additional effect in that it increases the time constant of voltage decay. In general, switching with motor residual voltages which have decayed to 25 percent of rated is satisfactory.

28. Suggested Maximum Capacitor Values. The suggested capacitor values for open motors manufactured after 1956, NEMA Design B or equivalent, are given in Tables 27 and 28. Similar tabulations for totally enclosed motors and NEMA Design C motors are contained in Ref. 1. The capacitor values shown are conservative; they will prevent overvoltages due to self-excitation and will limit transient torques to low values.

The listed values, when multiplied by 1.1, are suitable for wound-rotor induction-motor applications. For motor ratings not shown, a conservative rule is that the capacitor current should not exceed the motor magnetizing current, which is essentially the no-load current. This will usually result in a full-load power factor of not less than 95 percent. A typical power-factor curve as a function of motor load is shown in Fig. 17. The suggested value of capacitance for magnetizing-current compensation has been utilized in this case.

If a 200-hp induction motor were installed in the example shown in Fig. 19 instead of a 200-hp unity-power-factor synchronous motor, the values of capacitance required for correction would be as follows. From Table 27 and assuming a 600-r/min motor, it can be seen that approximately 75 kvar of capacitance is required to compensate for the no-load magnetizing current. If the resultant full-load power factor of the capacitor-motor combination is 0.97, an additional 40 kvar is required for unity-power-factor operation. Power-factor correction equivalent to a 0.8-power-factor synchronous motor requires an additional 120 kvar. In order not to produce overvoltages or transient torques, any additional capacitance values above 75 kvar must be applied with separate switching or be left connected to the system as shown in Fig. 20*c*.

29. Released System Capacity. Figure 22 can be used to determine the percent system capacity released, thus permitting additional system loading through the installation of capacitors. This figure can also be used in conjunction with synchronous-motor applications if the kW component of the motor load is included in the plant load and power factor prior to use of the curves. The kvar of the motor can then be treated as a capacitor.

MOTOR PROTECTION

The basic principles of motor protection described here apply to both small and large motors. It is recommended that the National Electrical Code be consulted for protection requirements for very small motors. The protection requirements for fire-pump motors are covered in Publication 20 of the National Fire Protection Association. In addition to the protection of stator and rotor windings, protection of motors against improper operating conditions, bearing protection, and surge protection are also discussed.

30. Protective Features of Motor-Starting Equipment. Motor starters and circuit breakers can be furnished with a variety of protective devices as described below.

TABLE 27. Suggested Switched Capacitor Ratings for Low-Voltage Induction Motors (Ref. 1) — IEEE Red Book

230-, 460-, and 575-V motors, enclosure open—including dripproof and splashproof, NEMA Design B and larger motors of similar design manufactured in 1956 or later

Induction-motor hp rating	Motor speed, r/min, and number of poles: 3,600 / 2		1,800 / 4		1,200 / 6		900 / 8		720 / 10		600 / 12	
	kvar	% AR	kvar	% AR	kvar	% AR	kvar	% AR	kvar	% AR	kvar	% AR
2	1	16	1	20	1	22	1	24				
3	1	10	1	16	1	21	2	24				
5	1	9	2	16	2	21	2	21				
7½	1	8	2	13	2	15	4	21	5	29		
10	2	8	2	13	4	15	5	21	5	25	5	30
15	4	8	4	13	5	15	5	13	5	25	5	25
20	4	7	5	9	5	12	5	13	10	25	10	25
25	4	7	5	9	5	11	5	11	10	25	10	25
30	5	7	5	7	5	11	10	11	10	15	10	19
40	5	5	5	7	10	11	10	11	10	15	10	19
50	5	5	10	7	10	9	15	11	20	15	25	19
60	5	5	10	7	10	9	15	11	20	15	30	19
75	10	5	10	7	15	9	15	9	30	15	40	19
100	15	5	20	7	25	9	30	9	40	15	45	17
125	15	5	20	7	30	9	30	9	45	15	50	17
150	15	5	25	6	30	9	40	9	50	13	60	17
200	40	5	40	6	45	8	50	9	70	13	75	17
250	45	5	50	6	50	8	70	9	75	12	90	17
300	50	5	50	6	70	8	70	9	75	11	105	17
350	50	5	50	5	70	8	80	9	80	11	105	17
400	60	5	60	5	70	8	100	9	100	11	110	17
450	60	5	70	5	70	6	100	9	100	11	110	17
500	70	5	90	5	90	6	110	9	120	11	120	17

These data are representative for three-phase 60-Hz general-purpose induction motors.

For standard 60-Hz wound-rotor induction motors operating at 60 Hz, the following data should be used:

$$\text{kvar} = 1.1 \text{ of the kvar values listed}$$
$$\%\text{AR} = 1.05 \text{ of the }\%\text{AR values listed}$$

The listed kvar values give the rating of the capacitors. This value is approximately equal to the motor no-load magnetizing kvar values.

%AR is the percent reduction in full-load line current due to capacitors. A capacitor located on the motor side of the overload relay reduces current through the relay. Therefore a smaller relay may be necessary. The motor-overload relay should be selected on the basis of the motor full-load nameplate current reduced by the present reduction in line current (%AR) due to capacitors. If a capacitor of lower kvar rating is used, the actual %AR will be approximately proportional to

$$\text{Listed }\%\text{AR} \times \frac{\text{actual capacitor rating}}{\text{kvar value in tables}}$$

In general, the capacitance added should not exceed the motor no-load magnetizing kVA for switched applications to prevent overexcitation from occurring.

Warning. Do not exceed the motor manufacturer's specified power-factor-correction capacitor rating. To do so may result in overexcitation, which can cause higher than designed torques, currents, and transient voltages. These can lead to increased safety hazards to both equipment and personnel. Refer to NEMA Standard MG 2 1977 or National Electric Code Article 460 for additional information and safety precautions (Ref. 10).

TABLE 28. Suggested Switched Capacitor Ratings for Medium-Voltage Induction Motors (Ref. 1)

2300- and 4000-V motors, enclosure open—including dripproof and splashproof, normal starting torque and current, NEMA Design B and larger motors of similar design manufactured in 1956 or later

Induction-motor hp rating	Motor speed, r/min, and number of poles											
	3,600 / 2		1,800 / 4		1,200 / 6		900 / 8		720 / 10		600 / 12	
	kvar	% AR	kvar	% AR	kvar	% AR	kvar	% AR	kvar	% AR	kvar	% AR
100	20	7	25	10	25	11	25	11	30	12	45	17
125	30	7	30	9	30	10	30	10	30	11	45	15
150	30	7	30	8	30	8	30	9	30	11	60	15
200	30	7	30	6	45	8	60	9	60	10	75	14
250	45	7	45	5	60	8	60	9	75	10	90	14
300	45	7	45	5	75	8	75	9	75	9	90	12
350	45	6	45	5	75	8	75	9	75	9	90	11
400	60	5	60	5	60	6	90	9	90	9	90	10
450	75	5	60	5	75	6	90	8	90	8	90	8
500	75	5	75	5	90	6	120	8	120	8	120	8
600	75	5	90	5	90	5	120	7	120	8	135	8
700	90	5	90	5	90	5	135	7	150	8	150	8
800	90	5	120	5	120	5	150	7	150	8	150	8

See footnote to Table 27.

The selection of the proper device depends upon the application and the relaying philosophy of the protection engineer (Ref. 9).

a. Fault protection—motor starters. Motor-starter contactors are generally limited in interrupting duty to not more than ten times rated current. This limitation permits the contactor to interrupt overloads and locked-rotor currents. It does not, however, provide interrupting ability for faults on the motor side of the contactor in excess of ten times rated current. Protection above this value is generally provided by a fused disconnect switch or a circuit breaker on the line side of the contactor. Circuit breakers used for this application are usually of the low-voltage molded-case type. When these protective devices are furnished as an integral assembly with the contactor, the unit is known as a combination starter.

The choice of fuses or a circuit breaker for motor-starter fault protection is largely a matter of individual preference. Each of these devices has certain advantages over the other. Fuses are low in initial cost, simple and compact, are available in higher interrupting ratings than circuit breakers, and can be obtained with current-limiting features. On the other hand, fuses require replacement after operation, the failure of one fuse can result in single-phase operation of three-phase motors, replacement fuses of proper rating must be kept in stock, and fuse coordination with other circuit-protective devices may be difficult or impossible.

b. Fault protection—circuit breakers. When circuit breakers are used as motor-starting devices, fault-current-interrupting ability is inherent in the breaker up to

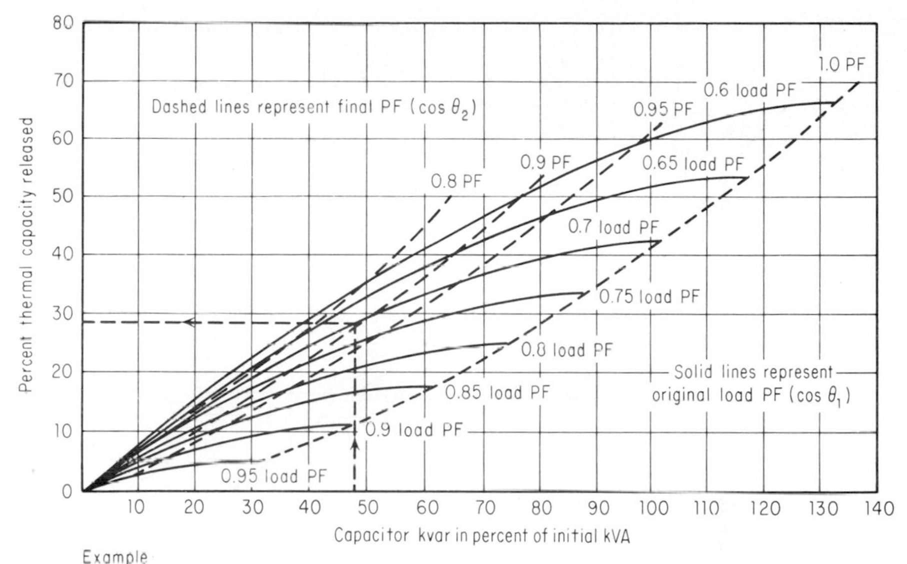

Example

If a plant has a load of 1,000 kVA at 70 percent power factor and 480 kvar of capacitors are added, the system electric capacity released is approximately 28.5 percent; that is, the system can carry 28.5 percent more load (at 70 percent power-factor) without exceeding the kVA before the power factor was improved. The final power factor (cos θ_3) of the original load plus the additional load is approximately 90 percent.

FIG. 22. Percent capacity released and approximate combined-load power factor with capacitors. *(Ref. 1.)*

its interrupting rating. This rating should not be less than the maximum fault-current capability of the system at the point of application.

Circuit breakers can be obtained with direct-acting trips that operate the breaker by mechanical action. This method has the advantage of not requiring a separate source of tripping power. The majority of direct-acting trips are used on low-voltage (600 V and below) air circuit breakers. They also can be obtained on circuit breakers of relatively low interrupting capacity, rated above 600 V. In this case, the trip mechanisms are usually operated from current transformers rather than being directly connected in series with the primary circuit.

Direct-acting trips can be obtained with instantaneous, long-time-delay, and short-time-delay characteristics in various combinations. These devices are generally of the adjustable type, but fixed-setting devices are sometimes furnished. Fault protection is obtained through use of the instantaneous-trip device or, in some cases, the short-time-delay device. The typical time-current characteristic curves of these various devices are shown in Fig. 23. A typical coordination of selected devices from Fig. 23 is shown in Fig. 24. The maximum and minimum curves represent the extremes of acceptable performance for these devices and include the effects of manufacturing tolerances, material variations, etc. The performance of a particular tripping device will be within a much narrower tolerance band.

Circuit breakers above 600-V rating are usually operated by means of relays and a reliable source of tripping power. This method permits a greater selection

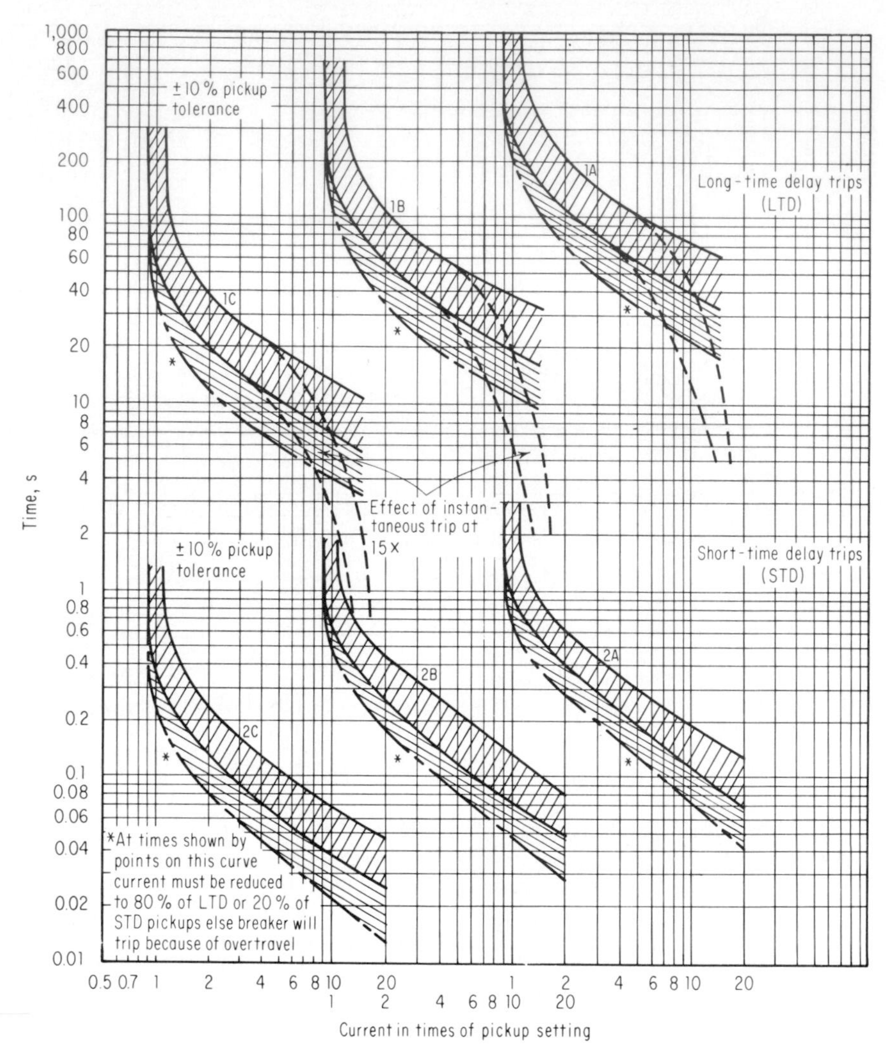

FIG. 23. Typical time-current characteristics of direct-acting trips on 600-V air circuit breakers. *(Ref. 4.)*

of protective features and more selective coordination with other protective devices on the system. Relays are also more accurate than direct-acting trip devices.

c. Overheating protection—motor starters. Motor starters are equipped with overload relays of either the magnetic or the thermal type. To permit motor starting, magnetic relays utilize dashpots for time delay. The thermal type has inherent time delay. Three overload devices are provided in motor starters for three-phase motors. Single-phase motors require an overload device in only one lead.

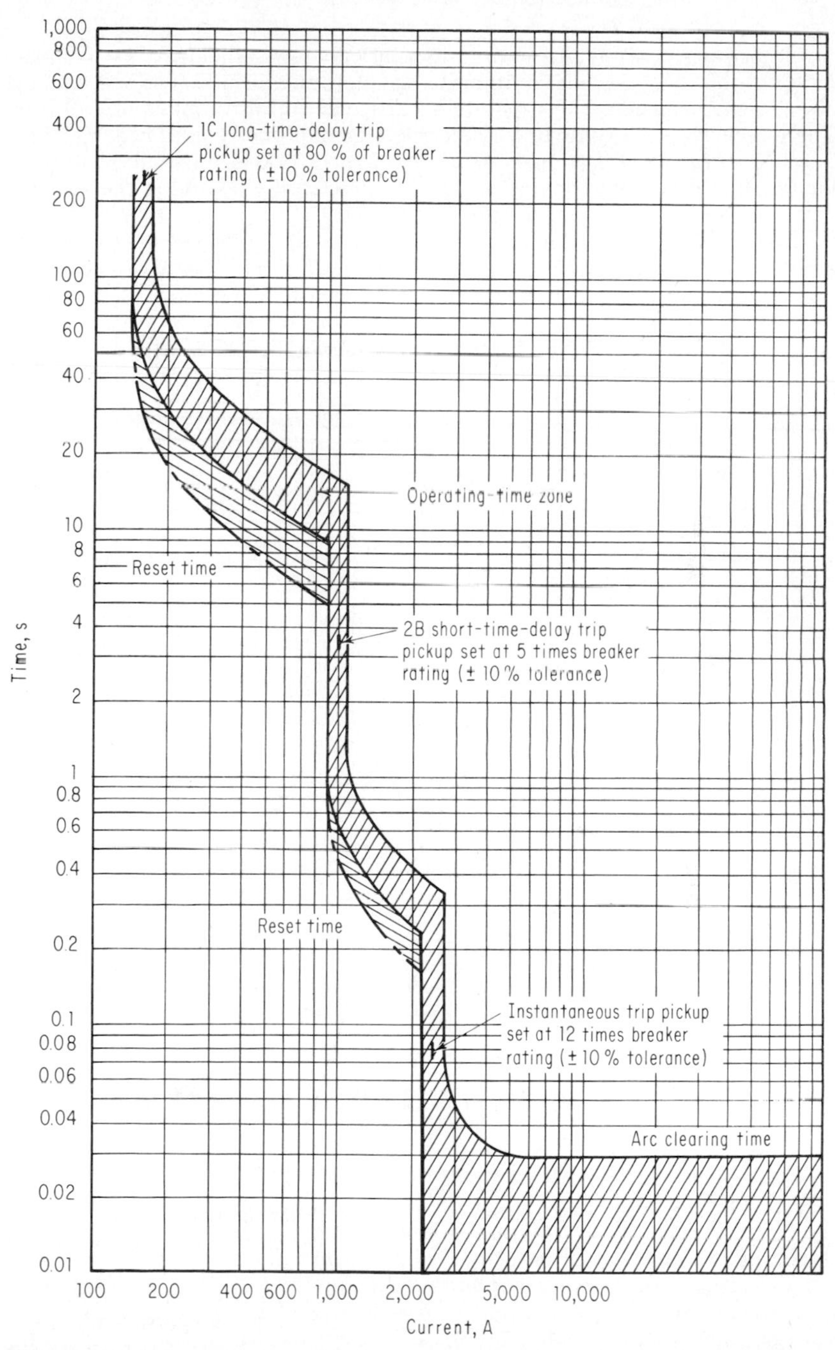

FIG. 24. Typical coordination of instantaneous, short-time-delay, and long-time-delay trips on 600-V air circuit breakers. *(Ref. 4.)*

Motor starters are manufactured in relatively few standard sizes. To adapt a motor starter to a particular motor size, an appropriate-sized overload relay must be selected. Magnetic relays require a particular coil for a given size of motor. Thermal relays are coordinated by the selection of properly sized heater elements.

d. Overheating protection—circuit breakers. Overload protection can be provided by direct-acting trips as previously described. Generally, the long-time-delay trip device is used for overload protection.

Protective relays in conjunction with a source of tripping power are generally used with higher-voltage circuit breakers. Figure 25 shows typical protective-relay and motor-heating characteristics. It can be seen that the replica-type thermal-overload-relay characteristic more nearly matches the motor-heating curve than does the inverse-time relay. Thermal relays also reflect the immediate thermal history of the motor and the ambient temperature at the relay location.

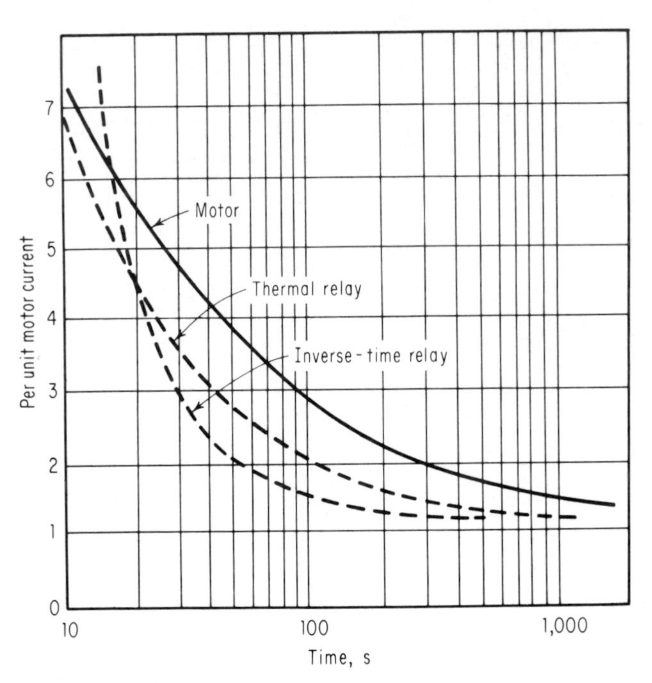

FIG. 25. Typical time-current characteristics for a motor and protective devices.

e. Undervoltage protection. The contactor in a motor starter is maintained in the closed position by an interlock which shunts the start button, thus allowing its release. A power failure or a large voltage drop will result in practically instantaneous opening of the contactor and motor shutdown. To restart the motor it is necessary to press the start button again. This type of operation, known as undervoltage protection, eliminates the hazard of unexpected motor operation upon return of normal voltage.

Automatic motor-control systems may incorporate a form of protection known as undervoltage release. An automatic device such as a float switch or pressure switch maintains the contactor in the closed position as dictated by system requirements. Upon voltage failure the contactor will open, and it will close again when normal voltage returns. The hazard of unexpected starting is not so great in this case since it is not expected that an operator will be in attendance at the motor location. Also, unexpected motor starting is the rule rather than the exception since it is in response to an unseen system variable.

Unnecessary shutdowns can be avoided by using a time-delay undervoltage device in which the control circuit of the starter is maintained closed for a period of several seconds. This arrangement allows the motor to ride through voltage dips of short duration. Circuit breakers are obtainable with adjustable direct-acting time-delay trip devices in the low-voltage ratings. Undervoltage inverse-time relays generally are used with circuit breakers rated above 600 V.

31. Stator-Winding Fault Protection. Protection of the stator windings from fault conditions is achieved through use of phase-overcurrent, ground-overcurrent, and differential relaying.

a. Phase overcurrent. Instantaneous phase-overcurrent relays and instantaneous direct-acting trips are generally set to operate slightly above locked-rotor current. Inverse-time relays and fuses should be selected to operate at no more than four times rated motor current with adequate time delay to permit the flow of motor inrush current during starting.

b. Ground current. Motors can be protected from ground faults through the use of instantaneous and inverse-time ground-overcurrent relays. Ground-fault relays connected in the wye of the three-current transformer secondaries will sense only the unbalanced ground-current flow. Ground-fault currents will be of different magnitude than phase-fault currents. Ground-fault relays can be set much more sensitively than phase relays. They also are not subject to motor inrush current.

Inverse-time ground relays should be adjusted to pick up at no more than about 20 percent of rated current or about 10 percent of the maximum ground-fault current, whichever is smaller. Instantaneous ground-relay pickup should be set from about 2.5 to 10 times rated current. In many three-phase applications adequate fault-current protection is obtained with instantaneous overcurrent relays in two phases and one inverse-time ground-overcurrent relay (Ref. 6). Figure 26 illustrates a method of obtaining sensitive ground-fault protection with instantaneous overcurrent relays.

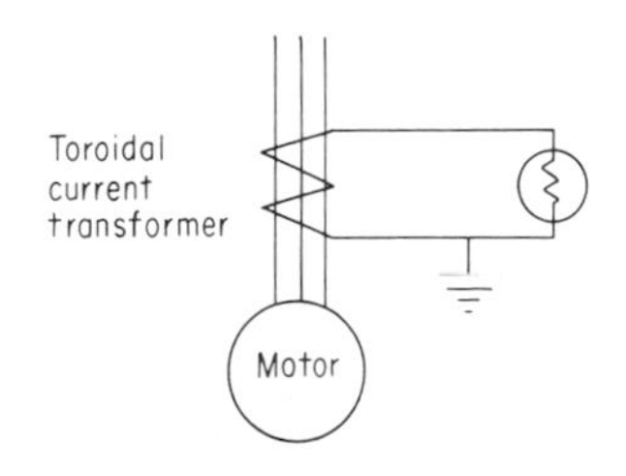

FIG. 26. Typical ground-fault protection with instantaneous overcurrent relay.

c. Differential protection. Differential protection is commonly used on motors larger than 500 hp at less than 600 V or over 1500 hp for voltages greater than 600 V. It provides for matching the current that enters the motor winding with that leaving. It cannot detect turn-to-turn faults but responds to ground-fault or phase-to-phase-type failures. Its chief advantage is its sensitivity and fast response, which minimize internal damage that might otherwise occur. Coordination with other overcurrent relays in the system is not required.

Figure 27 depicts both a six- and a three-current-transformer method of differ-

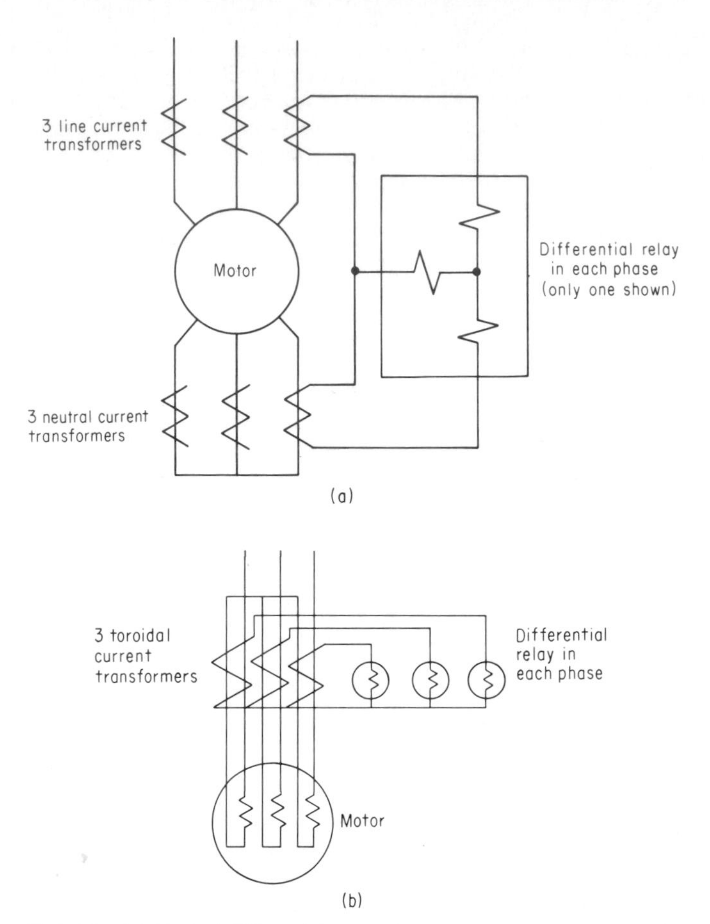

FIG. 27. (*a*) Typical differential-protection relay circuit. (*b*) Differential relay protection using three current transformers.

ential protection. The six-current-transformer method shown allows mounting three of the current transformers in the switchgear to include the motor cables in the protection zone for a wye-connected motor. Delta-connected motors require all six current transformers to be located in the motor terminal box. Today most differential protection uses the three-current-transformer or zero-sequence self-balancing method. Whether wye- or delta-connected, both the incoming and the outgoing motor leads for each phase pass through a doughnut-type current transformer of typically a 50:5 or a 100:5 ratio. No current-transformer output is normally present until a fault occurs. This method has the advantage of lower cost and less required coordination (Ref. 7).

Split-phase protection is a form of differential protection that can be applied when there are parallel circuits within each phase. In this case the current transformers in the two parallel circuits or groups of parallel circuits are balanced against each other. In this manner turn-to-turn faults can be detected. Split-phase protection is generally not so fast as differential relaying since time delay must be introduced to prevent operation during transient unbalance. The relay used is an inverse-time overcurrent relay rather than an instantaneous percentage differential relay.

32. Stator-Winding Overheating Protection. All motors should have protection against overheating. Complete protection for three-phase motors requires an overload element in each phase. As a result of an open circuit in a wye-delta transformer supply one phase of the motor may carry twice the current of the other two phases, as shown in Fig. 28. Continued operation under this condition will result in failure of the motor. In spite of this, in older controls motors rated

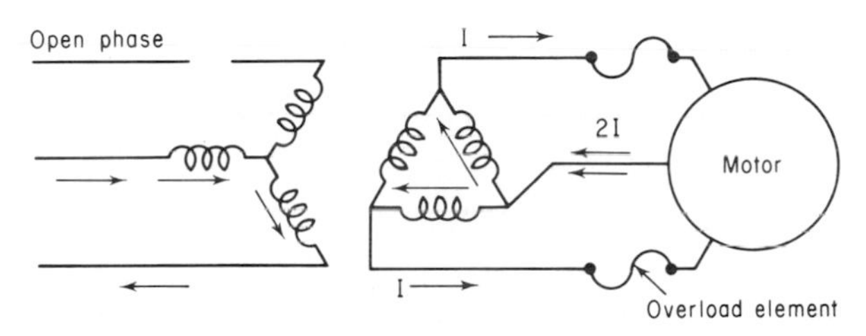

FIG. 28. Effect of open phase in primary of wye-delta transformer upon motor operation with only two-phase protection.

approximately 1500 hp and less usually had overload elements in two phases only. Control standards now require overload protection in all three phases (Ref. 9).

Protective devices for motors with no service factor should be set to trip at approximately 115 percent of rated motor current. Motors with a 115 percent service factor should have protective devices set to trip at about 125 percent of rated motor current. For motors with special short-time overload ratings or other service factors, the manufacturer should be consulted for proper settings.

Stator-temperature detectors are usually provided in motors rated 1500 hp and larger. These devices consist of a copper resistance coil embedded in the stator slot between the top and bottom coil sides. The resistance of the coil is usually 10 Ω at 25°C, although 100 and 200 Ω are also quite common. The resistance of the coil is a function of the temperature in the slot. A Wheatstone-bridge circuit as shown in Fig. 29 permits monitoring this temperature at a remote location. This device can also be used to operate a single relay to protect against overload. A current-balance relay should also be provided when a single-resistance temperature detector is used.

Smaller motors may utilize bimetallic switches or nonlinear resistance devices attached to or buried in the end windings for winding-temperature indication or protection. It must be remembered that the end windings may not be the hottest part of the stator winding. Also, any device that is applied to the surface of the insulation and depends upon heat flow through the insulation may not give a true indication of the highest winding temperature. Depending upon the voltage class of the motor, the actual hot-spot temperature may be 10 to 20°C higher than indicated by the temperature-sensing device.

33. Rotor Protection. Squirrel-cage induction motors, wound-rotor induction motors, and the amortisseur windings of synchronous motors require protection

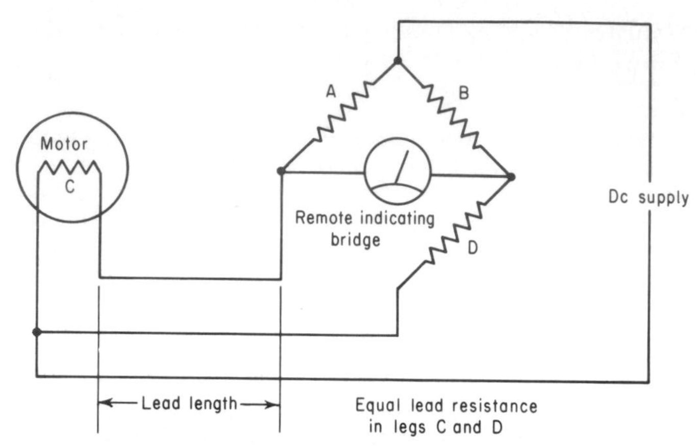

FIG. 29. Stator-resistance temperature detector monitoring circuit showing method of canceling effect of lead resistance.

against overheating. Field ground protection may also be required on large and important synchronous motors.

a. Rotor-overheating protection. In general, squirrel-cage overheating protection can be obtained from the thermal or inverse-time overcurrent relays used for stator protection as previously described. From the typical curves shown in Fig. 25 it can be seen that overload relays on high-inertia drives, requiring long acceleration times, may trip during starting. Applications such as these require special time-current characteristics to match the accelerating conditions. In those cases where stator protection is in the form of a stator-resistance temperature detector, proper rotor protection requires the addition of a single thermal or inverse-time overcurrent relay, or rate of use thermostat.

Wound-rotor protection may not be adequate with the above devices. Each case should be considered separately because of the range in rotor operating conditions. The manufacturer can be of assistance in determining proper protection.

Synchronous-motor amortisseur-overheating protection should be provided during starting. An exception would be a motor driving a generator. Such a motor operates unloaded during starting and should accelerate to rated speed without difficulty assuming proper terminal voltage. Amortisseur-overheating protection is best provided by a time-delay thermal-overload relay in the field discharge circuit (Ref. 6).

b. Field ground protection. Synchronous-motor field windings operate ungrounded. A single ground fault does not cause any operating difficulty. The first ground fault does, however, increase the stress to ground at other points in the winding. Also, the probability of a second ground fault is increased. A second ground fault will short-circuit part of the winding, and the unbalanced flux distribution may result in severe vibration.

Ground-fault protection can be provided by an overvoltage relay and an ac or dc voltage source connected between the field winding and ground. To ground the rotor effectively and not rely on a path through the bearing oil film, a shaft brush may be required. Care must be taken not to short out any bearing insulation provided to prevent shaft-current flow. In general, field ground-fault protection is provided only for large and important motors and may be only of an alarm nature.

34. Protection against Improper Operating Conditions. A number of improper operating conditions can result in damage to the motor. The degree of protection provided against these conditions is a function of the importance of the motor, its value and type of service.

a. Undervoltage protection. With the exception of essential-service motors, all ac motors should have undervoltage protection. This protection may take the form of single-phase protection for smaller motors and polyphase undervoltage protection for large motors above 1500 hp. Protection equipment should have inverse-time-delay characteristics to prevent shutdown during system transients and possible voltage reduction during starting.

b. Current balance. Imbalance in motor line circuits on polyphase systems can result from an open lead in the motor connections, an open circuit on the primary side of a wye-delta supply transformer, or a system-voltage unbalance.

An open lead in one phase of the motor connections will result in single-phase operation. Once started, a polyphase motor will continue to operate on a single-phase supply with excessive current in two leads. Overload protection in two leads will protect against this condition, provided the motor is not lightly loaded. For this added protection, a phase-balance relay is required.

As discussed under stator-winding-overheating protection and shown in Fig. 28, an open primary lead in a wye-delta supply transformer can cause one phase to carry twice the current in the other two phases. Stator-overheating protection with sensing elements in two leads or a relay operated from a single-resistance temperature detector will not provide adequate protection against this type of operation.

Unbalanced polyphase system voltages can result in quite large unbalanced motor currents. For example, voltages of 102, 100, and 98 percent applied to the primary of a delta-connected transformer can result in motor currents of 115, 100, and 85 percent when supplied from a wye-connected secondary. Inadequate overheating protection is obtained from two overload elements or a single temperature sensor. Three overload elements are now required (Ref. 3).

A current-balance relay capable of operating on a 25 percent or less unbalance or a negative-sequence relay will protect against these modes of operation.

c. Phase sequence. This is a safety requirement of the National Electrical Code for motors in equipment transporting people or in other applications where the motor must not be started in reverse, such as an exhaust fan in a mine. It prevents starting and operation of motors should the phase sequence become reversed.

d. Loss of excitation. A synchronous motor which experiences loss of excitation will continue to run as an induction motor if it is provided with an amortisseur winding. If the load is light, a synchronous motor may remain in synchronism because of salient torques. When excitation is lost or severely reduced, the motor draws its magnetizing requirements from the system. The ill effect of this type of operation on the system is compounded in those cases where a synchronous motor was installed for power-factor correction and low system voltages may result.

Loss-of-excitation protection can be provided with a time-delay reset undercurrent relay adjusted to respond to a low value of field current. Inherent loss-of-excitation protection is obtained with provision of amortisseur-overheating protection, loss-of-synchronism protection, and stator-overheating protection (Ref. 6).

e. Loss of synchronism. All synchronous motors which start under load should be provided with loss-of-synchronism protection. This protection can be used to remove excitation and unload the motor and reapply the motors when proper conditions exist. A second approach is to trip the motor supply. Synchronous motors used in frequency-changer sets should also be provided with loss-of-synchronism protection. For a motor driving a dc generator, loss-of-synchronism protection can be simulated by the combination of undervoltage, loss-of-excitation, and dc-generator overcurrent protection (Ref. 6).

f. Overspeed. This type of protection is applicable to dc motors which may reach excessive speed because of loss of load or excitation. It should also be provided on any motor that may be driven above its safe limit by the equipment to which it is coupled. One example is an induction motor used as an induction generator, where the driving equipment can cause potentially high overspeed. The overspeed protection should be arranged to trip the motor supply, and in the case of dc motor-generator sets to trip the dc generator.

g. Bearing protection. Antifriction bearings generally are more adaptable to vibration protection rather than temperature protection. The failure of an antifriction bearing usually is preceded well in advance by an increase in vibration level and audible noise level. Sleeve bearings can be protected with temperature relays arranged to alarm or trip at excessive temperatures. A probe inserted in a well in the bearing shell close to the babbitt surface senses the bearing temperature. The probe can be a resistance temperature detector, a thermocouple, or a vapor-filled bulb. Remote indication from a resistance temperature detector, which can be obtained as shown in Fig. 29, permits monitoring any number of bearings at a central location.

Sleeve bearings generally are of the self-lubricated type with oil rings. Two-pole motors in the larger ratings require flood or forced oil lubrication. A pressure switch with alarm contacts can be used for indication of pressure failure. Forced-oil-lubricated bearings are also equipped with oil rings for temporary lubrication until pressure can be restored or the motor safely shut down.

35. Essential-Service Motors. Unnecessary tripping of motors must be avoided if such tripping would result in the shutdown of a vital generating unit or the interruption of a process with large financial loss or damage. In some cases the motor may be expendable to a certain degree to maintain operation or until an orderly or convenient shutdown can be arranged. Each case must be considered on its own merits and the consequences of continued operation or shutdown evaluated.

Stator-fault protection on essential-service motors may be reduced to instantaneous phase-overcurrent relays, differential relays, and inverse-time and instantaneous ground-overcurrent relays. These relays provide tripping of the motor for fault conditions only. An ungrounded or high-resistance grounded system can provide continued operation with a ground fault if conditions are critical enough to do so.

Stator-overheating protection can be provided by long-time inverse-time overcurrent relays connected to an alarm. Tripping of the motor is then left to the judgment of the operator. Locked-rotor protection, if required, can be obtained with instantaneous relays set to pick up at about 250 percent of rated current. The contacts of these relays should be connected in series with the inverse-time overcurrent relays to provide automatic tripping. This arrangement produces an alarm only if the current is less than 250 percent and long-time inverse-time overcurrent protection for currents in excess of 250 percent. If it is desired to protect the motor against damage but take advantage of the maximum motor thermal capability between 115 and 250 percent of rated motor current, thermal tripping relays may be used. These relays should be set to operate as far as possible beyond the alarm point but not so far as to require motor repair before reuse (Ref. 6).

36. Surge Protection. The basic impulse insulation level (BIL) of rotating machines is limited by the dry type of insulation used, space restrictions, and mechanical, thermal, and cost considerations. In comparison with other types of electrical equipment the BIL of rotating machinery is very low. For example, the BIL rating of a 5-kV oil-immersed distribution transformer or a 4.16-kV oil circuit breaker is 60 kV. The BIL rating of a 4-kV motor is considerd to be on the order of 13 kV. This level is equal to 1.25 times the peak value of the 1-min high-potential proof test, $1.25 \times \sqrt{2}\,(2 \times \text{rated voltage} + 1000)$ V.

An ac-motor stator winding can be considered to be a short transmission line. It exhibits the properties of surge impedance, electrical length, and terminal reflection and refraction phenomena normally associated with a transmission line. A surge voltage impressed upon the motor terminals stresses both the ground insulation and the turn-to-turn insulation. In general, the highest stress is across the first few turns of the motor winding since the surge is attenuated as it passes through the winding. It is possible, however, to have stresses due to reflected waves at ungrounded or high-impedance grounded neutral connections or midway between terminals of delta-connected windings under certain conditions.

Typical surge-impedance values for a transmission line lie between 400 and 500 Ω. Cables have a typical surge-impedance value in the order of 30 to 40 Ω. The surge impedance of ac-motor stator windings varies between approximately 150 and 1500 Ω. The upper end of this range is usually associated with higher-voltage motors having multiturn coils and series-connected windings. Typical velocity-of-propagation values are 1000 ft/μs for transmission lines, 600 ft/μs for cables, and 45 ft/μs for the slot portion of motor stator windings. The velocity in the end-turn region will be much higher than that in the slot portion.

Surge voltages can be generated by lightning strokes, switching operations, lightning arrester action, fault conditions, forced-current-zero devices such as current-limiting fuses, vacuum circuit breakers, and simple energization of cable-connected motors. In general, switching surges are less than two or three times rated line-to-neutral peak voltage.

When a surge is impressed on a motor terminal, a portion of the wave is reflected along the incoming conductor back toward the source and a portion is refracted into the motor winding. This effect is most severe when the motor surge impedance is much larger than that of the incoming conductor, as is the case for a cable-connected motor. The refracted voltage can approach twice the incoming voltage level for motors with high surge impedances. Even with a relatively low motor surge-impedance value of 300 Ω and a cable of 30 Ω, the refracted voltage is 1.8 times the incoming surge-voltage value.

The magnitude of the turn-to-turn stress caused by this refracted wave is dependent upon the steepness of the wavefront. Figure 30 shows a typical surge waveform. In order to gain some idea of the voltage impressed across a single turn, the following situation will be considered. Assume that a surge voltage of twice the peak line-to-neutral voltage (6500 V) is refracted into a 4-kV motor with a core length of 25 in. Assume also that the rise time of the wavefront is 1 μs and that the velocity of propagation is 500 in/μs. The electrical length of the iron portion of one turn is therefore 0.1 μs. This results in a turn-to-turn stress of approximately 650 V when taken from the wavefront.

The wire insulation used in motors is generally either enamel or fused double Da-Glas, which requires that the surge turn voltage be limited to 600 to 1000 V, depending upon the thickness of the insulation used. Repeated exposure to surge voltages above these values may reduce the life of the insulation and result in motor failure. In practice it has been found that wavefronts with a rise time of 10 μs or longer will not cause excessive stresses between turns of motor-stator wind-

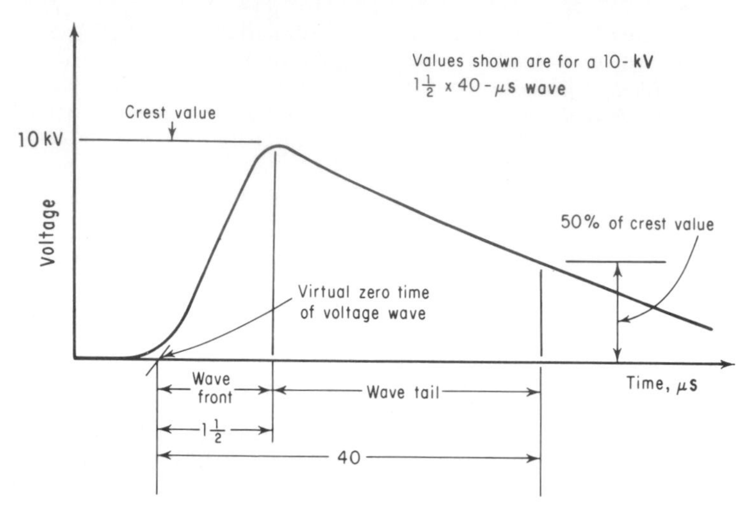

FIG. 30. Typical surge wave.

ings. The use of surge capacitors, as in Table 29, will reduce the rate of rise of an incoming wave. The capacitors must be connected across each phase to ground at the incoming cable terminals. The leads from this point to the motor should be kept as short as possible, 3 ft or less. Recommended surge capacitor values are 1.0 μF for motors on 600-V systems and below and 0.5 μF for motors on 2400- to 6900-V systems.

37. Lightning Protection. In addition to surge capacitors to reduce the steepness of the wavefront, machines connected to circuits exposed to lightning require further protection in the form of lightning arresters. Station-type arresters designed specifically for rotating machines are available in ratings of 3 to 27 kV. For protection of motors and other equipment on low-voltage circuits (110 to 125 V) a 175-V arrester is available. The ratings of protective equipment for three-phase machines are shown in Table 29.

Motors connected directly to exposed overhead lines can be protected as shown in Fig. 31. This arrangement consists of a distribution-type arrester 1500 to 2000 ft (450 to 600 m) out on the line and a station-type arrester across the surge capacitor for protection of the machine ground insulation. The surge capacitor and the arrester should be located at the machine terminals for maximum effectiveness. If they cannot be located at the machine terminals, they should be connected ahead of the machine on the incoming lines rather than on separate leads from the machine terminals. An overhead ground wire extending for a distance of 2000 ft (600 m) from the plant will provide additional protection against direct strokes on lines in the immediate vicinity of the plant.

Figure 32 shows the arrangement of lightning-protective equipment for a motor connected to an exposed overhead line through a transformer. The station-type arrester provides protection for the transformer and eliminates the need for a distribution-type arrester out on the line. The arrester across the motor terminals protects the windings from the electrostatic and electromagnetic transient voltages transmitted through the transformer. The value of transient voltage is highest for wye-wye transformers with both neutrals grounded and autotrans-

TABLE 29. Protective Equipment for Three-Phase ac Rotating Machines (Ref. 4)

	For installation at machine terminals or on machine bus†						For installation 1,500 to 2,000 ft (450 to 600 m) out on directly connected exposed overhead lines		
	Protective capacitors			Station-type arresters			Distribution-type arresters		
				Voltage rating			Voltage rating		
Machine voltage rating (phase-to-phase)	Voltage rating	Micro-farads per pole	Single-pole units required	Ungrounded or resistance grounded system	Effectively grounded system	Single-pole units required	Ungrounded or resistance grounded system	Effectively grounded system	Single-pole units required
0–650	0–650	1.0	3*	650	650	3*	650	650	3
2,400	2,400	0.5	3*	3,000	3,000	3	3,000	3,000	3
4,160	4,160	0.5	3*	4,500	3,000‡	3	6,000	3,000‡	3
6,900	6,900	0.5	3	7,500	6,000	3	9,000	6,000	3
13,800	13,800	0.25	3	15,000	12,000	3	18,000	10,000	3

*A single three-pole unit is commonly used.

†For single-phase machines the same recommendations apply except that only two single-pole units are required if neither line is grounded and only one (on the ungrounded line) if one line is grounded.

‡The use of 3,000-volt arresters on a 4,160-volt system requires an X_0/X_1 ratio less than that necessary to make the system "effectively grounded."

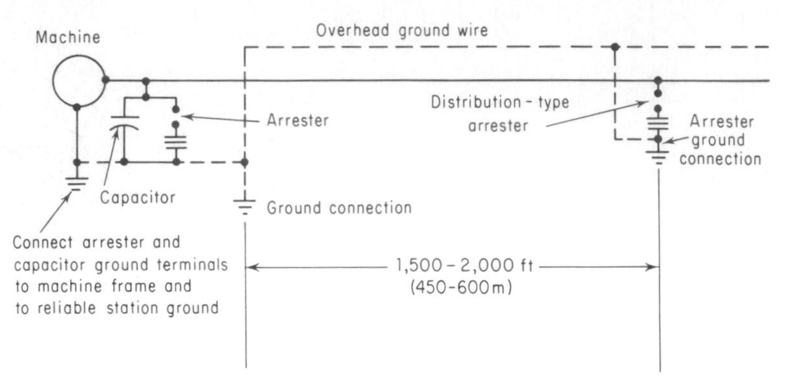

FIG. 31. Lightning-protective equipment for motor connected to exposed overhead line. *(Ref. 4.)*

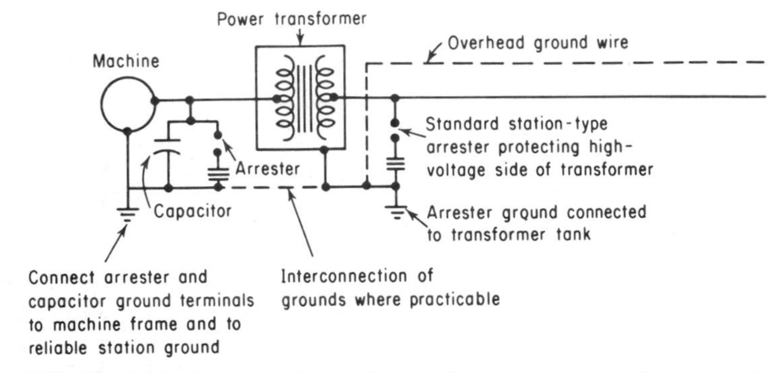

FIG. 32. Lightning-protective equipment for motor connected to exposed overhead line through a transformer.

formers. Wye-delta-connected transformers are more effective in reducing the transient-voltage magnitude. In smaller installations the less expensive distribution-type arrester may be used on the transformer primary in place of the station-type arrester.

The protective scheme for a motor connected to an exposed overhead line through a reactor or regulator is shown in Fig. 33. This same arrangement can be used where a cable is the connecting link instead of the reactor. In this case the cable sheath should be grounded to the arrester grounds on each end of the cable. Since the surge impedance of a cable is generally so much lower than that of an overhead line, the cable itself will reduce the incoming voltage. The arrester at the line end of the cable may be eliminated if the strength of the cable and the termination equipment is sufficient for the expected voltage. Where the length of cable is considerable it may be possible to omit the arrester at the machine because of the attenuation of the surge voltage.

Low-voltage motors, 600 V and less, have higher dielectric strengths than higher-voltage motors on a relative basis. Low-voltage motors connected to

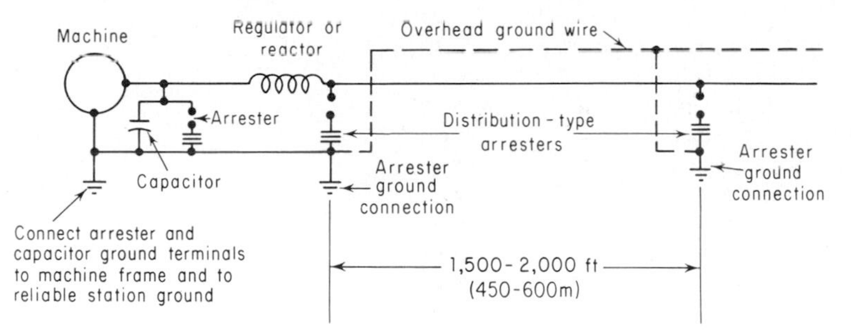

FIG. 33. Lightning-protective equipment for motor connected to exposed overhead line through a reactor or regulator.

exposed lines through transformers usually require no lightning-protective equipment if proper protection has been provided for the transformer. Lightning arresters are required for low-voltage motors connected directly to exposed overhead lines.

REFERENCES

1. "Recommended Practice for Electric Power Distribution for Industrial Plants" (IEEE Red Book), IEEE Standard 141-1976.
2. "Motors and Generators," ANSI-NEMA Standard MG 1-1978.
3. "Industrial Control and Systems," ANSI-NEMA Standard ICS 1-1978.
4. D. Beeman, *Industrial Power Systems Handbook* (McGraw-Hill, New York, 1955).
5. "Proposed Guide for Determination of Short-Circuit Characteristics of Direct-Current Machinery," AIEE Publication 66 (July 1957).
6. C. R. Mason, *The Art and Science of Protective Relaying* (Wiley, New York, 1956).
7. "Induction Motor Protection," IEEE Standard 288 (Feb. 1969).
8. *Ac Rotating Machine Protection* (General Electric Co., Schenectady, N.Y., May 1984).
9. Robert W. Smeaton, Ed., *Switchgear and Control Handbook* (McGraw-Hill, New York, 1986).
10. "Safety Standard for Construction and Guide for Selection, Installation and Use of Electric Motors and Generators," ANSI-NEMA Standard MG 2-1977.

BIBLIOGRAPHY

W. W. Lewis, *The Protection of Transmission Systems against Lightning* (Wiley, New York, 1950).

D. R. Shoults, C. J. Rife, and T. C. Johnson, *Electric Motors in Industry* (Wiley, New York, 1942).

M. Liwschitz-Garik and C. C. Whipple, *Electric Machinery,* vols. 1 and 2 (Van Nostrand, Princeton, N.J., 1946).

SECTION 3
BASIC MOTOR TYPES

Raymond F. Horrell*
Paul R. Moser†

FOREWORD . . . 3-2
SYMBOLS . . . 3-3
GENERAL INTRODUCTION . . . 3-4
CONSTRUCTION FEATURES . . . 3-5
1. Active Iron . . . 3-5
2. Windings . . . 3-5
3. Inactive Iron . . . 3-5
4. Bearings . . . 3-5
PERFORMANCE OF MOTORS . . . 3-6
5. Torque . . . 3-6
6. Speed . . . 3-8
7. Power . . . 3-10
8. Losses and Efficiency . . . 3-10
9. Excitation and Power Factor . . . 3-12
10. Direction of Rotation . . . 3-14
11. Acceleration and Deceleration . . . 3-15
12. Electric-Motor Controller . . . 3-16
13. Braking . . . 3-16
14. External Braking . . . 3-16
15. Internal Braking . . . 3-17
PROTECTION OF MOTOR PARTS . . . 3-23
16. Type and Degree of Protection . . . 3-23
17. Factors Used to Determine Protection . . . 3-23
18. Enclosures . . . 3-25
19. Winding-Insulation Protection . . . 3-27
20. Causes of Overheating . . . 3-27
21. Protection from Overheating . . . 3-28

*Chief Engineer, Induction Machines, Electric Machinery Manufacturing Company, Minneapolis, Minn. (retired); Registered Professional Engineer (Minn.); Senior Member, IEEE.

†Chief Product Engineer, Baldor Electric Company, Fort Smith, Ark.; Registered Professional Engineer (Ohio); Member, IEEE.

NOTE: The section was written by Raymond E. Horrell for the first edition and reviewed and updated for this edition by Paul R. Moser.

22. Current-Responsive Protectors 3-28
23. Temperature-Responsive Protectors 3-29
24. Thermal Protectors 3-30
25. Protection from Humidity 3-30
26. Protection of Mechanical Parts 3-30
27. Other Protective Devices 3-31

DC MOTORS 3-32
28. Basic Equations 3-33
29. Construction 3-33
30. Shunt-Wound Motors 3-33
31. Series-Wound Motors 3-34
32. Compound-Wound Motors 3-35
33. Effect of Voltage Variation 3-36
34. Efficiency of dc Motors 3-36

DC-MOTOR CONTROL 3-37
35. Starters 3-37
36. Controllers 3-37
37. Control Systems 3-37
38. Protection 3-37

AC MOTORS 3-37

INDUCTION MOTORS 3-38
39. Basic Equations 3-38
40. Construction 3-38
41. Single-Phase Induction Motors 3-38
42. Polyphase Induction Motors 3-44

INDUCTION-MOTOR CONTROL 3-53
43. Starters 3-54
44. Controllers 3-55
45. Protection 3-55

SYNCHRONOUS MOTORS 3-55
46. Basic Equations 3-55
47. Construction 3-55
48. Nonexcited Motors 3-57
49. Dc-Excited Motors 3-57

SYNCHRONOUS-MOTOR CONTROL 3-66
50. Starters 3-66
51. Controllers 3-66

REFERENCES 3-67

FOREWORD

This section consists of three parts. The first part is a general introduction covering basic similarities in construction, performance, and protection of motor types. These are discussed in greater detail in the other two parts.

The second part covers three fundamental types of dc motors and their differences in operating characteristics and performance. General control functions and protective features are listed.

The third part covers two types of ac motors: induction and synchronous. Operating characteristics, generally available control functions, and protective features are listed.

By referring to the tables and charts, one may readily find one or more types of motors which will accomplish the desired performance for a particular application. Reference to other applicable sections of the handbook will supply further detailed information for making the final choice.

SYMBOLS

C	= compressor factor
d	= displacement angle or angle of lag between stator and rotor fields of a dc-excited synchronous motor, electrical degrees
E	= potential or voltage of power supply, volts, kilovolts, etc.
EFF	= efficiency of motor, per unit
EFF_1, EFF_2	= efficiency of motor, percent
f	= frequency of power supply, hertz
I	= current, amperes
I_a	= armature current of dc motor, amperes
I_2	= secondary current of ac motor, amperes
$K, K_1, K_2, K_3, K_4, K_5, K_p, k_1, k_2$	= constants
kW	= useful power, kilowatts
kVA	= apparent power, kilovolt-amperes
kvar	= reactive kilovolt-amperes
N	= speed, revolutions per minute
ΔN	= change in speed, revolutions per minute
N_s	= synchronous speed, revolutions per minute
P	= number of poles
PF	= power factor, per unit
P_r	= synchronizing coefficient, kilowatts per electrical radian
R	= resistance, ohms per phase
R_a	= armature resistance of dc motor, ohms
R_1	= primary resistance, ohms per phase
R_2	= secondary resistance, ohms
R_{21}	= secondary resistance referred to primary, ohms per phase
S, S_1, S_2	= slip, percent

$S_{r/min}$ = slip, revolutions per minute
t = acceleration or deceleration time, seconds
T = torque, lb·ft, oz·ft, etc.
T_F = frictional torque, lb·ft, oz·ft, etc.
T_L = load torque imposed on motor, lb·ft, oz·ft, etc.
T_M = motor torque, lb·ft, oz·ft, etc.
T_0 = induction-motor open-circuit time constant, voltage, seconds
T'_0 = induction-motor open-circuit time constant with capacitance, voltage, seconds
T_{ac} = induction-motor current time constant, ac component, seconds
T_{dc} = induction-motor current time constant, dc component, seconds
Wk^2 = inertia, lb·ft^2
W_a = energy absorbed in rotor windings during acceleration, kilowatt-seconds
W_s = energy absorbed in stator windings during acceleration, kilowatt-seconds
X_c = capacitive reactance, ohms per phase
X_L = locked-rotor reactance, ohms per phase
X_m = magnetizing reactance, ohms per phase
X_1 = stator reactance, ohms per phase
X_2 = rotor reactance referred to stator, ohms per phase
ω = constant equal to $2\pi f$
ϵ = constant equal to 2.7182818+
π = constant pi, equal to 3.14159+
Φ = magnetic flux

GENERAL INTRODUCTION

Three forms of energy conversion are accomplished by rotating machinery: mechanical to electrical, one type of electrical to another, and electrical to mechanical. Large steam turbogenerators, waterwheel generators, and smaller driven generator sets are the most familiar means of converting mechanical energy to electrical energy. Rotary converters, motor-generator sets, and frequency changers convert ac energy to dc energy, or vice versa, or change the frequency of ac power. Voltage regulators are used to control the voltage of electrical energy. Transformers convert voltage from one value to another. Electric motors are the machines most frequently used to convert electrical energy into mechanical energy.

CONSTRUCTION FEATURES

All types of electrical apparatus mentioned above have three common elements of construction. They are: active iron to carry magnetic flux, windings to carry electric current, and inactive iron or other suitable material to support and provide some degree of protection to the other two elements. Significant differences in the number, arrangement, and construction of the windings and the active iron elements produce diverse performance characteristics.

1. Active Iron. In electric motors, the active iron generally consists of two cylindrical laminated-iron cores, one within the other, and separated by an air gap. Magnetic flux passing from one core to the other across the air gap forms the magnetic circuit. Usually the inner core will rotate and the outer core remains stationary. In special cases, the reverse may be true or both cores may rotate. (In transformers, both cores are stationary and the air gap is eliminated, thereby leaving one core physically.) Most laminations are thinly coated with either an organic or an inorganic insulating material to provide interlaminar resistance which reduces losses and heating caused by stray currents.

2. Windings. The conductors, usually copper, of electric motors are located near the surface of the active-iron elements (cores) in the vicinity of the air gap. They may be located in slots or encircle a salient-pole structure and are usually electrically insulated from the iron cores. The conductors are made into coils with external end connections and insulated from the iron cores. The coils within each core are connected to form the several individual circuits or windings which carry independent currents. Distribution of the currents flowing in the windings around the surfaces of the active iron determines the operation of the motor. In Sec. 11 the different systems used to insulate the windings from the iron elements are described.

3. Inactive Iron. The inactive iron generally surrounds the active iron and the windings in the form of a frame and some type of end enclosure. The frame serves to anchor the stationary elements to a foundation. In the case of direct current, the frame (sometimes called yoke) may also carry magnetic flux. The end enclosure may contain the bearings which properly position the rotating element with respect to the stationary element. Sometimes bearings are separated from the end enclosure and mounted in pedestals which are mounted on either sole plates or bed plates. The frame and the end enclosure also provide various degrees of protection to the active iron and windings. In addition they serve to direct the ventilating air efficiently for cooling the parts.

4. Bearings. There are sleeve-type and antifriction-type bearings. Sleeve bearings for horizontal motors are not normally designed to carry external thrust loads. Antifriction bearings may be either ball- or roller-bearing types and are not normally designed to carry external thrust loads. For vertical motors, either sleeve or antifriction bearings may be designed to carry external thrust loads. Kingsbury-type thrust bearings are used to carry high external thrust loads. Sleeve bearings are oil-lubricated, whereas antifriction bearings may be either oil- or grease-lubricated. Cooling for the bearing may be provided by circulating the oil through suitable coolers. Bearing types and their applications are described in Sec. 12.

PERFORMANCE OF MOTORS

5. Torque. The torque is the force that tends to produce rotation, the turning moment of tangential effort. Torque in electric motors is the turning force developed by the motor. It may also be referred to as the resistance to the turning force exerted by the driven load.

Torque is expressed in units of force and distance to represent the turning moment. The units normally used are pound-feet, ounce-inches, or gram-centimeters, depending upon the magnitude of the torque and the unit system of expression used. A preferred method of expression is percent of rated full-load torque.

The torque of electric motors can be described in several definitive terms (Refs. 2, 3), most of which are illustrated in Fig. 1.

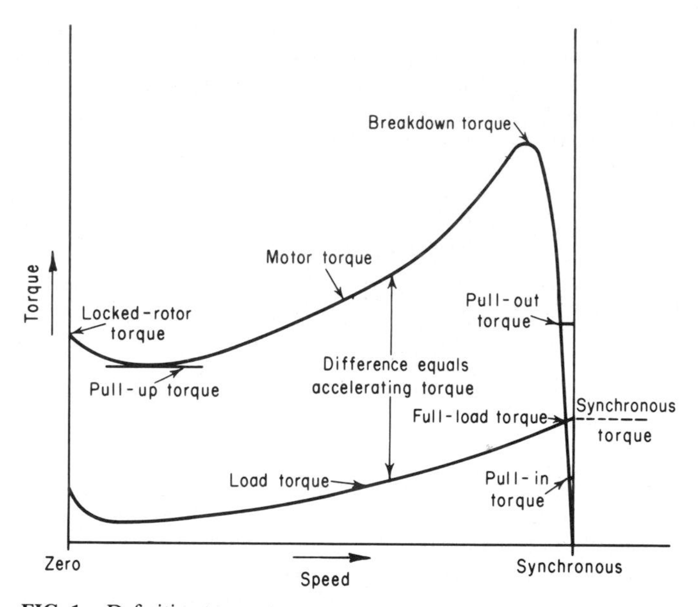

FIG. 1. Definitive torque terms.

a. Locked-rotor torque is the minimum torque developed by the motor for all angular positions of the rotor at the instant of rated-power application to the motor primary winding circuits. This torque is sometimes referred to as breakaway or starting torque.

b. Accelerating torque is the torque developed with rated-power input during the period from standstill (or zero speed) to full rated speed. The term "accelerating torque" is frequently referred to and used as the net accelerating torque, which is that positive torque available beyond the torque required by the load.

c. Breakdown torque, sometimes called maximum torque, is the maximum torque developed at rated-power input without an abrupt change in speed.

d. Pull-up torque is the minimum torque developed with rated-power input during the period of acceleration from standstill to the speed at which breakdown torque occurs or, if the motor does not have a definite breakdown torque, up to full rated speed.

e. Pull-in torque is the torque developed during the transition from slip speed to synchronous speed. For the synchronous motor it is the maximum constant torque with which the motor will pull its connected load inertia into synchronism, with rated power input, when its field excitation is applied.

f. Reluctance torque is a pulsating torque which has a net average value of zero and a frequency of twice line frequency. Reluctance torque affects the pull-in and pull-out torques of motors and is characteristic of motors having salient-pole construction.

g. Synchronous torque is the steady-state torque developed during synchronous operation with rated-power input applied. It is the torque available to drive the load.

h. Full-load torque is the torque necessary to produce rated output at rated speed with rated-power input. The other definitive torques are usually expressed in percent of this rated full-load torque.

i. Pull-out torque is the maximum torque a motor can develop without stalling. This torque is frequently referred to as breakdown or maximum torque. For synchronous motors, it is the maximum sustained torque developed for 1 min at synchronous speed with rated-power input and normal excitation.

j. Braking torque is that torque which tends to decelerate the motor to some lower speed or to zero speed.

k. Definitive motor torques. Table 1 is arranged to show which of the definitive torques are normally associated with the basic types of motors. In this table, a yes indicates the term is well defined by usage and standards, and appears in publications concerning the particular type of motor. A no indicates that the term either is not applicable to that type of motor or is not defined in standards. Special cases, to be pointed out, may permit the use of the term where a no appears. Different terms or names are sometimes used for the definitive torques. The more common names are shown in the table.

TABLE 1. Definitive Motor Torques

Definitive torques	DC	AC		
		Induction		Synchronous
		Single-phase	Polyphase	
Locked-rotor torque (starting, breakaway)	No	Yes	Yes	Yes
Accelerating torque (net accelerating)	Yes	Yes	Yes	Yes
Pull-up torque	No	Yes	Yes	Yes
Pull-in torque	No	No	No	Yes
Synchronous torque	No	No	No	Yes
Reluctance torque	No	No	No	Yes
Full-load torque	Yes	Yes	Yes	Yes
Pull-out torque	No	Yes	Yes	Yes
Breakdown torque	No	max Yes	max Yes	No

Yes indicates well-defined usage of the definitive torque term.

No indicates the term either is not applicable or is not normally used in conjunction with that particular type of motor.

6. Speed. Speed is a measure of the rate of motion. For electric motors, speed designates the number of revolutions of the shaft with respect to time. The unit normally used is revolutions per minute (r/min). The speed at which torque is delivered by electric motors forms the basis of motor application, and for this reason several definitive speed terms (Refs. 2, 3) common to all types of motors have been established by definition and usage.

a. Full-load speed is the rated speed at which rated full-load torque is delivered with rated-power input.

b. Constant speed indicates that the normal operating speed is constant, or practically constant, for a specified range of torque. For example, a synchronous motor operates at constant speed, whereas low-slip induction motors and shunt-wound dc motors operate at practically constant speed over their load range.

c. Synchronous speed indicates that the speed is "in step" or in synchronism with the frequency of the power supply. For ac motors, synchronous speed may be found by the formula

$$N_s = \frac{120f}{P} \tag{1}$$

where N_s = synchronous speed, r/min
f = frequency of power system, Hz
P = number of poles for which motor is operating

d. Adjustable speed indicates that the speed may be varied gradually over a considerable range, but remains practically unaffected by load for each adjustment. For example, a dc shunt-wound motor with field-resistance control may be designed for considerable range of speed adjustment.

e. Varying speed indicates that the speed varies with the load, usually decreasing as the load increases. For example, series-wound, repulsion, and induction motors with high slip have varying-speed characteristics.

f. Adjustable varying speed indicates that the speed may be varied gradually and that the speed will then vary over a considerable range with change in load. For example, a compound-wound dc motor with field control or a wound-rotor induction motor with rheostatic speed control offers adjustable varying-speed characteristics.

g. Multispeed indicates that more than one definite constant-speed rating, each being practically independent of the load, can be obtained by appropriate arrangement and connection of the windings of the motor, for example, a dc motor with two armature windings or an induction motor with windings which can be reconnected for different pole groupings. The speed of multispeed permanent-split capacitor and shaded-pole motors will vary with the load.

In some cases, two-frame construction may be used. Two motors having different speeds are suitably coupled together on a common base. Synchronous multispeed motors are usually of two-frame construction.

h. Slip speed (sometimes called slip r/min) is the difference between synchronous speed and the actual rotor speed.

i. Overspeed is that range of speed which exceeds rated full-load speed. The term is sometimes used to indicate the maximum speed at which a motor may operate without physical damage.

j. Reduced speed indicates that the motor is operating at some specified speed less than rated full-load speed.

k. Reversed speed, reversible, and similar terms are used to indicate that the direction of rotation may be reversed by appropriate arrangement or connection of motor windings.

l. Critical speed refers to the speed at which the shaft and rotor element responds unfavorably to certain exciting forces. There are two types of critical speed, and there may be more than one speed which is critical for each type. Safe operation is possible above or below each critical speed. One type of critical speed, called torsional critical speed, occurs in the case of pulsating torques, such as exist in compressors and in circuit-breaker reclosure. Operation at one of the torsional critical speeds may result in severe vibration and even in shaft breakage. The second type of critical speed, called rotational critical speed, is associated with the length-to-diameter proportions of the shaft and rotor. Operation at one of the rotational critical speeds results in severe vibration.

m. Definitive motor speed terms. Table 2 is arranged to show which of the definitive speed terms are normally associated with the basic types of motors. In this table, a yes indicates that the term is well defined by standards and general usage, and appears in published literature for the particular type of motor

TABLE 2. Definitive Motor Speed Terms

Definitive speed terms	DC	AC		
		Induction		Synchronous
		Single-phase	Polyphase	
Full-load speed	Yes	Yes	Yes	Yes
Constant speed	Sometimes	Essentially	Essentially	Yes
Synchronous speed	No	Reference	Reference	Yes
Adjustable speed	Yes	No	Wound-rotor	No
Variable speed	Yes	No*,‡	No*,†,‡	No*,‡
Multispeed	No	Some types	Yes§	Yes§
Slip speed	No	Yes	Yes	No
Overspeed	Yes	Yes	Yes	Yes
Critical speed:				
Torsional	No	No	Yes	Yes
Rotational	Yes	Yes	Yes	Yes
Critical speed probable:				
Torsional	Not common	No	No	Compressor drives
Rotational	Not common	Not common	Large high-speed drives	No
Reduced speed	Yes	No	Wound-rotor	No
Reversed speed	Yes	Some types	Yes	Yes

Yes indicates well-defined usage of the definitive speed term.

No indicates the term either is not applicable or is not normally associated with that particular type of motor.

* The motor itself does not produce variable output speed. When some form of speed changer is combined with the motor in a unit frame, variable output speed may be obtained from the combination.

† A water rheostat used in the secondary circuit of the wound-rotor motor will give variable speed.

‡ Variable speed may result if a variable power-supply frequency is applied to the motor.

§ Multispeed motors are usually one-frame construction if induction type, two-frame construction if synchronous type.

involved. A no indicates that the term either is not applicable to that type of motor or is not generally associated with that particular type of motor. Other explanatory notes and wording are used in the table to point out the more common meaning of the definitive terms. These items are more fully explained in other sections of the handbook.

7. Power. Mechanical work is done whenever motion results from the action of force. The amount of work done is the product of the force times the distance through which the body moves along the line of action of the force, that is, pound-feet. The amount of work done is also a measure of the energy transferred or transformed. The time required to do the work determines the rate of working, or the rate at which work is done. The time rate of working or of transforming energy is known as power.

a. Electric power is the rate of doing work as represented by the amount of current I flowing under and in phase with a potential of E volts, or $I \times E$. For electric motors, the electric power may be represented generally by the formula

$$\text{Power} = k_p EI \qquad \text{watts} \tag{2}$$

where k_p is a constant representative of the type of machine involved. Electric power is usually expressed in watts, kilowatts, megawatts, or microwatts, depending upon the amount of power involved.

b. Mechanical power may be measured by the number of grams weight that can be raised 1 cm/s, or by the number of pounds raised 1 ft/s. James Watt introduced the unit of power, known as horsepower, commonly used in rating motors with respect to mechanical power. One horsepower is equal to 33,000 ft·lb/min (550 ft·lb/s). In electric motors, the torque (turning moment) is acting through the shaft, which rotates at some rate of speed (r/min). The power, or rate of working, may be found in the formula

$$\text{Power} = \frac{TN}{5252} \qquad \text{hp} \tag{3}$$

where torque T is in lb·ft and speed N in r/min. The relationship given by the formula applies to all types of motors.

Electric and mechanical power are related by the formula

$$\text{hp} = \frac{\text{watts}}{746} = \frac{\text{kW}}{0.746} = \frac{TN}{5252} \tag{4}$$

Table 3 indicates the general formulas for torque, speed, electric power, and mechanical power for the three general basic types of motors. The constants K_1, K_2, K_3, and K encompass the design parameters that apply to the particular type of machine. The magnetic flux represented by Φ is used in connection with dc motors. More exact equations and theory may be found in textbooks (Refs. 2, 3, 9–12, 15, 16).

c. Synchronizing power has been defined by ANSI standards as the power at synchronous speed corresponding to the torque developed at the air gap between the armature and the field with rated-power input. This torque tends to restore the rotor to no-load position with respect to the line voltage.

8. Losses and Efficiency. Certain losses will occur in the motor elements during operation. Each loss represents the power diminution in a circuit element corre-

TABLE 3. Generalized Equations for Basic Types of Electric Motors

	DC	AC	
		Induction	Synchronous
Torque T	$K_1 \Phi I_a$	$K_2 I_2 R_2 / S$	$K_3 I^2 R_2$
Speed N	$E - I_a R_a / (K_4 \Phi)$	$(120\, f/P) - S_{rpm}$	$120\, f/P$
Power:			
Watts	EI EFF	KEI EFF PF	KI EFF PF
hp	EI EFF/746	KEI EFF PF/746	KEI EFF PF/746
hp	$TN/5{,}252$	$TN/5{,}252$	$TN/5{,}252$

Note. See the list of symbols at the beginning of this section for definition of the symbols used.

sponding to the conversion of electric power into heat by some form of resistance. The presence of these losses causes a reduction in mechanical power delivered by the electric motor relative to the electric-power input. The efficiency of power conversion, known as motor efficiency, is represented by the formula

$$\text{Efficiency} = \frac{\text{output}}{\text{input}} = \frac{\text{input} - \text{losses}}{\text{input}} \tag{5}$$

Efficiency may be expressed in per unit terms as given by the formula but is often expressed as percent, which is obtained by multiplying per unit values by 100.

a. Electric-motor losses are associated with the three common elements of construction. There are losses in the circuits that carry current, in the circuits that carry flux, and in those parts that help support and ventilate or cool the other two parts. Table 4 indicates the losses which are associated with electric motors. Those that are normally considered to vary with load are noted. The definitive terms as applied to the three basic motor types are indicated with a yes in the three right-hand columns. Whereas the various losses are common to all types of motors depending upon their common features of construction, the definitive terms used may vary with the basic types of machines, especially for those circuits which carry current. Generally, the losses associated with the current circuits are called I^2R losses, and for motors using brushes they include brush, brush contact, and brush frictional losses. The losses associated with the active-iron elements are generally called core losses. Other losses are associated with the motor bearings, ventilating fans or blowers, and excitation or control circuits. Bearing frictional losses are usually combined with the rotor and integral-fan windage losses and called friction and windage loss. The ventilating loss of separately mounted blowers, rheostat loss, and exciter loss may or may not be charged to the motor, depending upon the conditions given in one of several test codes (Refs. 12, 19, 20, 23) for the particular type of motor. The stray, stray-load, or load loss is any additional loss that is not included in the other losses. It includes eddy-current loss in primary winding, some active-iron loss, loss in some of the inactive-iron parts, and some secondary-winding loss, all of which are proportional to the square of the stator current.

b. Electric-motor efficiencies are usually given at one-half, three-fourths, and rated load. Values for other loads and for specified conditions may usually be

TABLE 4. Electric-Motor Losses

Loss	Vary with load	Present in motor type		
		DC	Induction	Synchronous
I^2R loss (windings):				
Armature I^2R	Yes	Yes		
Series field I^2R	Yes	Yes		
Shunt field I^2R	No	Yes		
Primary or stator I^2R	Yes		Yes	Yes
Field (dc field) I^2R	Yes			Yes
Secondary or rotor I^2R	Yes		Yes	
Brush loss	Yes	Yes	Yes[a]	Yes[b,c]
Brush contact loss	Yes	Yes	Yes[a]	Yes[b,c]
Brush friction loss	Yes	Yes	Yes[a]	Yes[b]
Core loss (active iron)	DC only	Yes	Yes	Yes
Other losses:				
Bearing friction	No[d]	Yes	Yes	Yes
Rotor windage (includes any fan loss)	No[e]	Yes	Yes	Yes
Rheostat loss	Yes	Sometimes	No[f]	Sometimes
Exciter loss	Yes	Sometimes	No	Sometimes
Ventilating loss	No	Sometimes	Sometimes	Sometimes
Stray (stray-load or load) loss	Yes	Yes	Yes	Yes

See text for further discussion on each basic motor type.
[a]For wound-rotor type only.
[b]Unless excited by some form of brushless.
[c]Varies with field excitation.
[d]Additional losses caused by external thrust losses are not usually charged to the motor.
[e]Varies with speed on self-ventilated motors.
[f]Secondary controlled resistance loss not normally charged to motors.

obtained from the motor manufacturer. The efficiency rises rapidly from zero at no load to a maximum value and then diminishes gradually over the useful-load range. Efficiencies are higher for larger horsepower ratings and for higher speed ratings. Maximum efficiency will usually occur at lesser load values for the higher horsepowers and the higher speeds. Efficiencies will also vary with variations of other factors, such as impressed voltage, frequency, and speed-torque characteristics desired. These are discussed more fully in the text for each basic type of motor. Figure 2 indicates the representative shape of the efficiency curve versus load.

9. Excitation and Power Factor. A magnetic field must be established in electric motors from some source of electric power. This source of power excitation may be in the form of either a permanent magnet or an electromagnet. The permanent-magnet form requires no excitation current. The electromagnet form requires either direct current or alternating current which flows through one or more of the separate individual windings. The arrangement of the windings and the interaction of the magnetic fields set up by the currents in the different windings determine the type of motor and its operating characteristics.

All basic types of motors have primary windings which carry the load or primary current. This current is called armature current in the case of dc motors, and primary, armature, or line current in ac motors. The value of this load current varies with the load requirement relative to the amount of energy transformation.

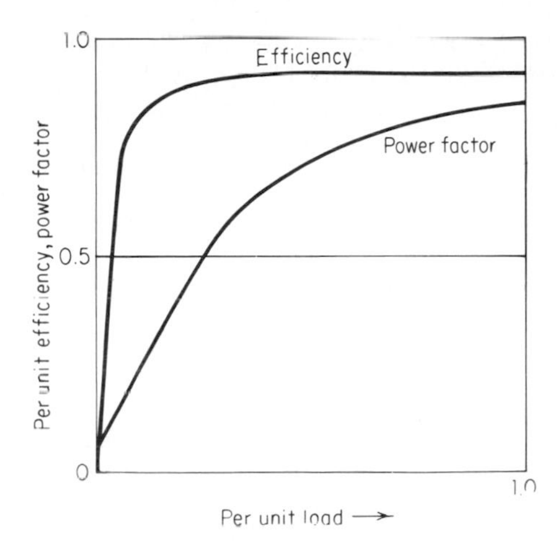

FIG. 2. Typical efficiency and power-factor curves illustrating variance with load. Shape and knee of the curves differ with motor size and speed. Efficiency curve is representative of most motor types. Power-factor curve is representative of ac induction-motor types.

The secondary windings are generally known as the excitation windings. These windings are independent of the main windings, differ in physical form, and carry the secondary or excitation current. The current may be supplied by the primary current or by a separate dc or ac power source. This secondary excitation current is independent of load or the amount of energy transformation.

If the motor *is* separately excited, the secondary or exciting current is called field current, which is usually direct current supplied by batteries or by some form of dc generator or converter. Dc motors and ac synchronous motors are the most familiar types of motors using this form of excitation.

In the case of induction-type motors, the excitation current is obtained from the primary power source and the excitation or magnetizing current flows in the primary winding. The secondary currents and their associated fields are induced by transformer action between the primary and secondary windings. The primary magnetizing current in this case is usually called no-load current.

The primary input current which flows in the primary windings of an ac motor has two components. One component is in phase with the voltage and is called load component. The power associated with this current is measured in watts, kilowatts, etc., and varies with the load or the amount of energy transferred.

The second component lags the voltage by 90° and is called reactive component. It is measured in terms of reactive watts, reactive kilowatts, etc. The net interchange of energy due to the reactive current is zero.

a. The power factor of an ac motor is numerically equal to the cosine of the angle of lag of the primary input current with respect to its voltage. With certain windings and excitation conditions, the current may lead the voltage, as in the case of the synchronous motor with overexcitation or underloaded with rated field excitation. The power factor may also be defined as the ratio of the active power (watts) to the apparent power (volt-amperes).

The power factor may vary with load, as in the case of the induction-type motor, or it may be controlled by adjusting the field current (or excitation), as in the case of synchronous motors having separate dc excitation. Figure 2 shows a representative power-factor curve of induction-type motors.

The armature current of a dc motor is "in phase" with the voltage, and hence its power factor is unity.

b. Basic motor power formulas. Table 5 shows the relationship of the various definitive terms associated with electrical volts, current, power, efficiency, and power factor. In the table, voltage is defined as the line-to-line voltage across the motor primary terminals, as distinguished from line-to-neutral voltage. Current is defined as the current flowing in any one main primary lead as distinguished from any neutral lead. Kilowatts (kW) and kilovolt-amperes (kVA) as used in this table are also most commonly used in application work but may be converted into other units with appropriate conversion factors.

10. Direction of Rotation. A standard direction of rotation (Ref. 3) is specified for some types of motors. Facing the end of the motor opposite the drive or shaft extension, the standard direction of rotation is counterclockwise for all ac single-phase motors, all synchronous motors, and all universal motors.

When motors have double shaft extensions, or when two or more machines are mechanically coupled together, the standard may not apply to all units.

A standard direction of rotation is not given for polyphase induction motors because many applications require both directions of rotation. Also, the phase sequence of the power source is not always known.

Where a standard direction of rotation does not apply, the direction should be shown on the dimension outline drawing. If the phase sequence is known, the motor manufacturer can check the direction of rotation for that phase sequence

TABLE 5. Electric-Motor Power Formulas

$$\text{Apparent power kVA} = k_1 \times \text{VA}/1000$$
$$\text{Active power, kW} = k_1 \times \text{VA} \times k_2/1000 = \text{kVA} \times k_2$$
$$\text{Power factor, PF} = \text{kW/kVA}$$
$$\text{Reactive kilovolt-amperes,* kvar} = k_1 \times \text{VA}\,\sqrt{1 - \text{PF}^2}/1000 = \text{kVA}\,\sqrt{1 - \text{PF}^2} = \text{kVA}\,\sqrt{1 - k_2^2}$$
$$\text{Horsepower, hp} = k_1 \times \text{VA} \times \text{EFF} \times k_2/746$$
$$\text{Current, } I = \text{hp} \times 746/(k_1 \times \text{V} \times \text{EFF} \times k_2) = \text{kW} \times 1000/(k_1 \times \text{V} \times k_2) = \text{kVA} \times 1000/(k_1 \times \text{V})$$

where k_1 = 1 for single-phase and dc motors
= 2 for two-phase four-wire ac motors; for two-phase three-wire ac motors, the current in the common wire is 1.41 ($\sqrt{2}$) times the current in either of the other two wires
= 1.73 ($\sqrt{3}$) for three-phase ac motors

PF = motor power factor expressed in per unit (decimal) form = 1 for dc motors
EFF = motor efficiency expressed in per unit (decimal) form
V = line voltage
A = line current
k_2 = power factor (PF) for ac motors
= 1 for dc motors

*Not applicable to dc motors.

and identify the motor primary leads accordingly. A small nameplate affixed to the motor will indicate the correct connection to the power source to obtain the specified direction of rotation.

Some motors, usually those with peripheral rotor speeds in excess of 10,000 ft/min, may use high-efficiency unidirectional fans, which move the ventilating air through the motor with minimum windage loss. The motor dimension outline and the plate affixed to the motor will indicate the correct direction of rotation. If it becomes desirable to reverse the direction of rotation of these motors, the fans should be interchanged end for end, and the rotor rebalanced. Fans which have slanted, curved, or propeller-type blades are unidirectional, whereas fans with radial blades may usually be operated in either direction.

Multispeed motors and motors which require any rearrangement of the motor terminal-lead connections for control purposes will rotate in the same direction for each of the connection arrangements unless specifically provided otherwise.

11. Acceleration and Deceleration. A motor must be able to accelerate or decelerate (brake) its load to the desired operating speed when required without damage or adverse effect to itself, the load, or the power source. Occasionally it may be desirable to reverse the rotation of the motor, and at times rapidly and frequently. The motor will perform these functions when properly designed and when operated with the proper control.

The motor design must provide adequate net torque to accelerate or decelerate the load under the conditions specified. The design must also provide sufficient thermal capacity in the windings to absorb and/or dissipate the losses associated with these functions. The factors involved in successful acceleration or deceleration are:

Motor-accelerating torque at operating input conditions

Total motor and load inertia Wk^2

Motor thermal capacity

Frequency and number of accelerations and brakings with respect to time

Ability to dissipate the losses for the entire duty cycle

a. The number of accelerations and brakings may be limited if these requirements exceed the maximum capability provided for in the motor design relative to net torque, thermal capacity, or cooling ability. The motor manufacturer should be consulted concerning the motor's capabilities.

b. The time of acceleration or deceleration is important and often vital to the overall application intended for the motor. Disturbances on the system supplying power must be kept to a minimum. Load requirements involving load changes and continuity must be performed. Braking to a standstill rapidly or even suddenly is often required as a personnel-safety feature. The time of acceleration or deceleration may be calculated by the general formula

$$t = \frac{Wk^2 \, \Delta N}{308T} \tag{6}$$

where t = time of acceleration or deceleration, seconds

Wk^2 = total motor and load inertia being accelerated or braked, all referred to the motor shaft rated speed, $\text{lb} \cdot \text{ft}^2$

ΔN = change of speed, r/min

T = net acceleration or deceleration torque, $\text{lb} \cdot \text{ft}$

For acceleration, the net motor torque is

$$T = T_M - T_L - T_F \tag{7}$$

and for deceleration,

$$T = T_M + T_L + T_F \tag{8}$$

where T_M = motor torque, lb·ft
T_L = load torque, lb·ft
T_F = frictional torque, lb·ft

All torque values are average torque acting over the change-of-speed (ΔN) range.

12. Electric-Motor Controller. The basic motor functions of start, acceleration, retardation or deceleration, reversing, stop, etc., are performed through the use of electric-motor controllers (Refs. 8, 24). The electric-motor controller is a device or a group of devices which govern in some predetermined manner the electric power delivered to the motor. (See Sec. 1, Arts. 7 through 15.)

13. Braking. Braking of an electric motor is a control function which dissipates the kinetic energy of the motor and any connected load during the period of retardation (deceleration) of the motor speed from a high value to either a lower value or zero. The dissipation of kinetic energy may be either external to the motor or internal to the motor windings (Ref. 7).

14. External Braking. External braking is provided by some form of brake which is coupled to the motor (Ref. 7). It is applicable to all motor types and may require some form of end-bracket modification, special shaft extension, and suitable control. Friction braking employs a shoe- or disk-type brake coupled to the motor. The brake may be actuated mechanically, electrically, hydraulically, or pneumatically with proper control elements. Eddy-current braking employs a magnetic coupling unit which is direct-connected to the motor. The eddy currents which provide the braking load are produced by excitation (usually direct current) and motor rotation. Hydraulic braking employs a form of hydraulic pump which is coupled to the motor. Braking torque is produced by motor rotation. Magnetic-particle braking employs a magnetic-particle-type coupling which is direct-connected to the motor. The coupling unit may be either dry or fluid. Braking torque is developed with excitation, usually direct current. Advantages of external braking include:

Holding action, when required in addition to deceleration, may be obtained with the proper control system.

Heating is not produced in the motor.

The disadvantages of external braking include:

Tension (countertorque) cannot be provided.

Space is required in addition to that required for the motor.

A special shaft extension may be required.

Inertia is added to the motor-rotating system by the brake-rotating element.

15. Internal Braking. Internal braking torque for retardation and deceleration is produced electrically by the currents flowing in the windings in much the same way as motor torque is produced (Ref. 7). The methods of producing internal or electrical braking torque can be classified into two fundamental groups: countertorque and generating. The countertorque group develops torque which tends to rotate the motor rotor in the opposite direction to that which existed before the application of the braking operation. This type of braking torque exists whether the motor is rotating or not. Energy is dissipated within the motor elements. In the generating group, the torque is developed because the motor is rotating. Dynamic braking torque is developed if the motor is rotating below normal speed. Energy is dissipated within the motor or in a connected load. However, the torque developed by this method becomes zero at zero speed. Regenerative braking torque is developed if the motor is rotating above the normal operating speed (above synchronous speed in ac motors). The motor is always connected to the power system and the generated power is then returned to that system. If many similar motors are tied to a power system, considerable power savings can be attained. Energy is dissipated within the motor. The advantages of internal braking include:

Deceleration and tension may be provided by using the proper control system.

Excellent overspeed control is available for overhauling types of load.

Positive retardation exists.

Retardation torque is controllable for speed or torque regulation.

Inertia is not added to the rotating system.

Extra space is not always required.

The disadvantages of external braking include:

Holding power is not provided.

Heating is produced in the motor.

Power is always required for braking.

Current magnitudes are on the same order as acceleration currents.

Dissipation of heat decreases as speed decreases.

Under certain conditions, the current magnitudes, the heating produced in the motor, and the reduction of heat dissipation may be limited by the motor winding elements (thermal capacity of conductors, commutators, slip rings, etc.). In such cases, larger motors using forced ventilation may be required.

Many of the advantages of internal and external braking may be obtained by combining more than one method of braking. Table 6 indicates the methods of braking most generally available for the basic types of motors. Figure 3 compares the typical braking curves obtained by several methods of dynamic and regenerative braking.

a. Plugging indicates that countertorque braking is obtained by reversing or reconnecting the power input leads of certain types of motors which may be running at any speed. Plugging provides simple and fast braking action, which is severe unless limited by resistors in series with the motor windings. Motor windings may require additional bracing. The power system must supply peak plugging current. The motor will reverse the direction of rotation unless removed from the line at zero speed by an operator, a zero-speed switch, or a timer. Losses are high

TABLE 6. Electric-Motor Braking Methods

Method of braking	DC	AC		
		Induction		Synchronous
		Single-phase	Polyphase	
Internal:				
Plugging, countertorque	Shunt series permanent magnet	Repulsion universal	Yes	Yes
Regenerative	Yes	Yes	Yes	No
Dynamic:				
AC excitation	No	No	Yes	Yes
DC excitation	No	Most types	Yes	No
DC with resistors	Yes	Universal only	Universal only	Yes
Capacitor excitation	No	No	Yes	No
Combination*	No	Most types	Yes	No
External:				
Friction braking	Yes	Yes	Yes	Yes
Eddy-current	Yes	Yes	Yes	Yes
Hydraulic braking	Yes	Yes	Yes	Yes
Magnetic-particle braking	Yes	Yes	Yes	Yes

* A combination of capacitor, resistor, and rectifier.

during plugging, reaching three times the acceleration loss for a squirrel-cage motor (four times if the motor is allowed to accelerate to full speed in the reverse direction). Figure 4 indicates typical braking speed-torque curves for a plugging operation. The value of braking torque can be varied by any means that varies the current flowing in the windings.

b. Regenerative braking indicates that dynamic braking torque is obtained if a motor is driven at higher than normal or synchronous speed. This method of braking is applicable to all types of motors except ac synchronous motors. The motor is not disconnected from the power supply and is loaded by feeding generated power back into the power-supply system. The braking torque produced is high, and the losses within the motor are approximately equal to the acceleration losses for an ac motor if braked from twice synchronous speed. Regenerative braking is of no value for stopping a motor. It is used for limiting the speed of overhauling loads and for positive deceleration from high to low speed on adjustable-speed or multispeed motors. It is more effective on dc motors at reduced speeds if the field strength is increased. However, controls may become more expensive, and special field windings may be required at the lower speeds. Regenerative braking may cause damage to the motor if high-inertia loads are connected to the motor. Figure 5 indicates typical braking curves for a three-phase squirrel-cage induction motor, a shunt-wound dc motor, and a compound-wound dc motor. The value of braking torque can be varied by any means that varies the current flowing in the windings. The speed at which the maximum braking torque of the squirrel-cage motor occurs can be varied by changing the motor power-supply frequency.

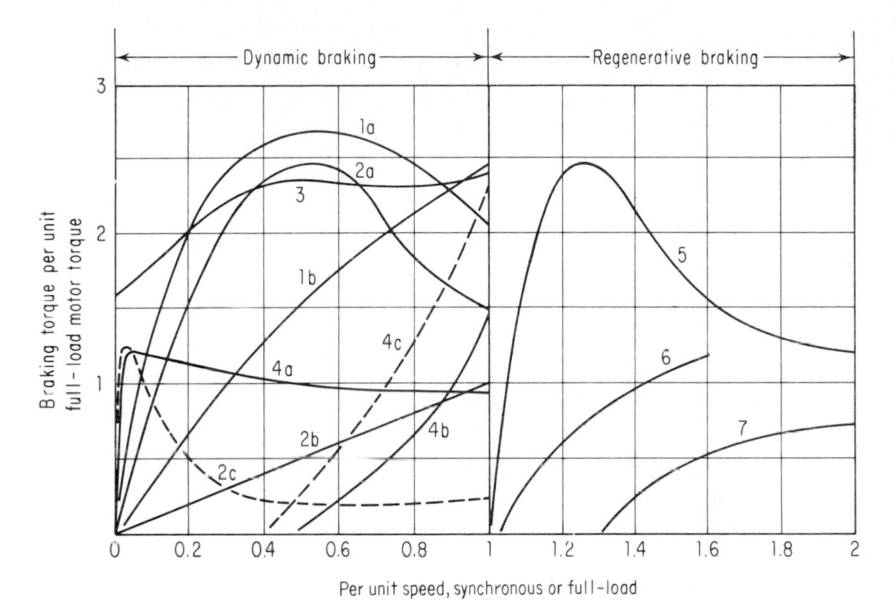

FIG. 3. Comparison of typical dynamic and regenerative braking curves. Dynamic braking with resistors having low resistance (1*a*) and high resistance (1*b*), for ac excitation, squirrel-cage motor with special winding (2*a*) and wound-rotor motor (2*b*), and for dc excitation, three-phase induction motor (2*c*); for plugging with 100 percent applied voltage (3); for capacitor-rectifier-resistor combination (4*a*), capacitor only (4*b*), and capacitor with loading resistors (4*c*). Regenerative braking for three-phase induction motor (5); for shunt-wound dc motor (6); for compound-wound dc motor (7).

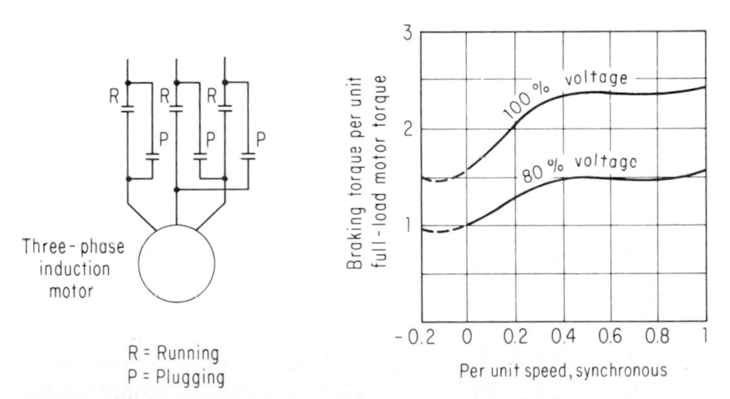

FIG. 4. Plugging (braking). Braking-torque curves shown are typical for three-phase induction motors. Torque values will vary with motor rating, design parameters, and applied voltage and may be more constant over plugging range than illustrated. Zero-speed switch is used to prevent reversal of motor rotation. Contactors *R* and *P* must not be closed at the same time.

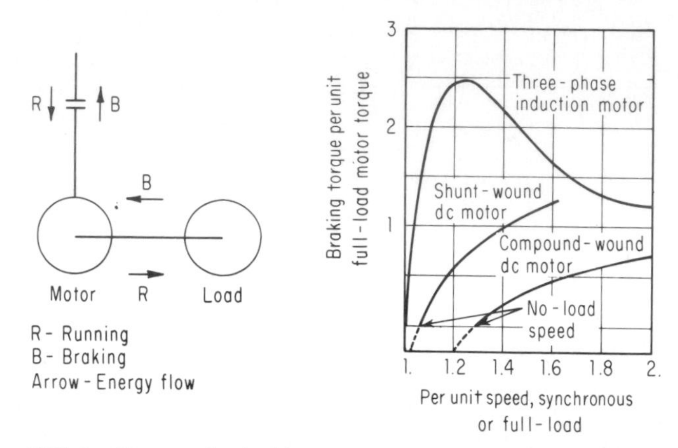

FIG. 5. Regenerative braking. Motors driven above their synchronous speed (ac motors) or rated full-load speed (dc motors) by their load for any reason will develop braking torque by returning energy to the power source.

c. Dynamic braking by ac excitation indicates that braking torque is obtained by applying single-phase ac power to one phase of a motor while another phase is short-circuited. This connection produces a pulsating magnetic field instead of a revolving field, and no torque is produced to drive the load. Instead, a current is generated in the rotor windings and associated external resistance as the rotor revolves in the pulsating field. The current circulating in the winding and its external resistance produces braking torque. Dynamic braking by ac excitation is applicable to polyphase induction motors and is simple, comparatively inexpensive, and suitable for overhauling loads. Dissipation of losses within the motor winding may require a larger motor thermally. Braking torque may be varied by changing the value of resistance in the rotor circuits. The torque decreases to zero as the speed decreases to zero.

A two-phase multipolar braking winding which is not inductively coupled to the main stator winding may be superimposed on the regular stator winding. Power is supplied to one phase from the line while the second phase is short-circuited to produce braking torque. This torque is quite similar to that produced by dc excitation. These braking windings are usually rated for intermittent duty. The torque may be varied by changing the excitation drawn from the line. Figure 6 indicates the braking torque typical of dynamic braking by ac excitation with or without the special braking windings.

d. Dynamic braking by dc excitation indicates that braking torque is obtained by applying dc power to one phase of a motor after the motor has been disconnected from its power source. Braking is obtained because the motor is loaded by the induced current which flows through the rotor windings. This method of braking is applicable to polyphase induction motors and to most single-phase ac motor types. The braking torque is low at the higher speeds and increases to a high maximum value as the speed is reduced, but drops rapidly to zero at zero speed. The losses are approximately the same as the acceleration losses. The value of braking torque may be varied by changing the dc value and by using high-resistance rotors. For rapid deceleration, a high dc excitation is required. Figure 7 indicates typical braking curves obtained with different excitation values.

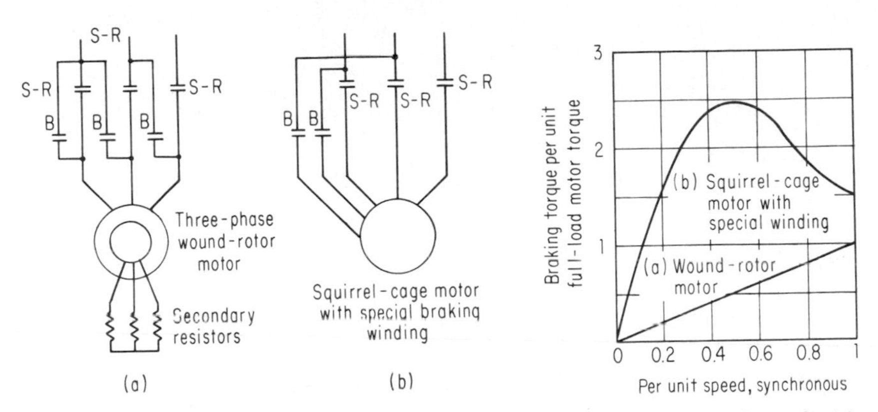

FIG. 6. Dynamic braking by ac excitation. Typical braking-torque curves are shown for (*a*) three-phase wound-rotor motor and (*b*) squirrel-cage induction motor with special braking winding. *S* is starting, *R* is running, and *B* is braking contact.

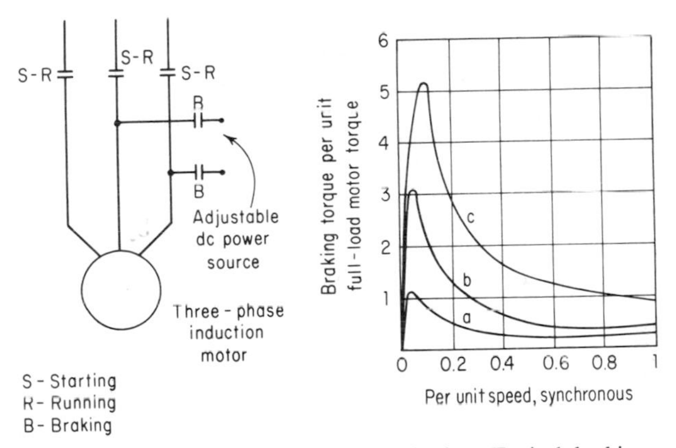

FIG. 7. Dynamic braking by dc excitation. Typical braking-torque curves for three-phase induction motor using dc braking current equal to (*a*) ac full load, (*b*) two times full load, and (*c*) four times full load.

e. Dynamic braking with resistors indicates that dynamic braking torque is obtained by operating the motor as a generator which supplies power to a resistor load after the motor has been disconnected from the power supply. Energy is dissipated in the resistor load and in the motor windings. This method of braking is applicable to most dc motors, to ac and dc universal motors, and to ac synchronous motors. Some form of field excitation, usually direct current, must be applied to the field winding of all motor types except permanent-magnet types. Braking is not available if excitation is removed for any reason, and there is no holding force available at zero speed. Deceleration is smooth and less severe than with plugging. The value of braking torque may be varied by changing the value of either the load resistors or the braking excitation current. Figure 8 indicates the typical braking torques obtained with high- and low-resistance load resistors.

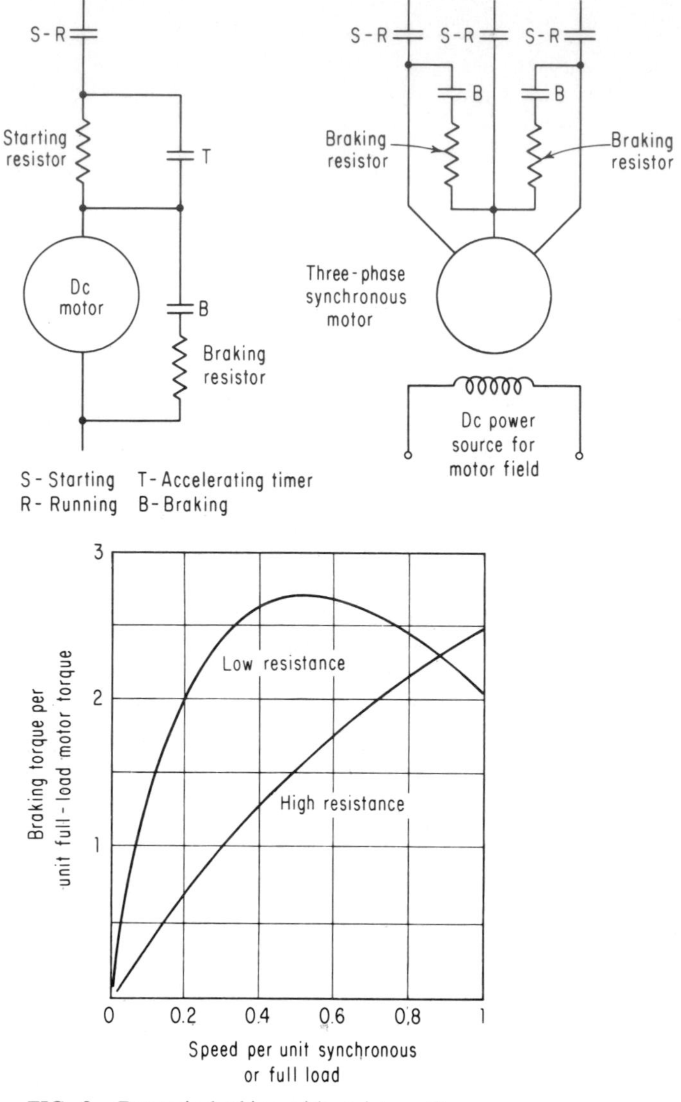

FIG. 8. Dynamic braking with resistors. Curves illustrate typical variations in braking torque which can be obtained by adjusting resistor values. Torque values may be varied by adjusting field excitation.

f. Dynamic braking by capacitors indicates that dynamic braking torque is obtained by generator action when the motor is disconnected from the line and loading excitation current is supplied by capacitors connected across the line. The braking torque can be increased by using resistors as a load. Energy is dissipated in the motor windings and in the load resistors. This method is applicable to three-phase induction motors. When the capacitors are connected across the line during normal-run operation, power-factor correction is provided. The braking

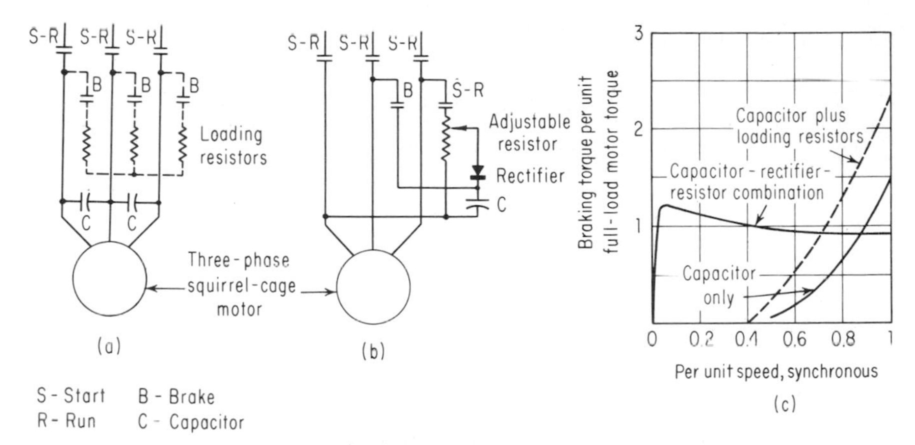

FIG. 9. Dynamic braking by capacitors. Typical braking-torque curves for dynamic braking by capacitors, for capacitors plus loading resistors, and for capacitor-rectifier-resistor combination.

torque is not dependent upon the power-system voltage, and it requires practically no maintenance but drops to zero at approximately one-third speed. Capacitors connected across two or three phases (Fig. 9*a*) give the most consistent results. The value of capacitance used must not exceed that needed to correct the motor no-load power factor to unity because damage to the winding insulation from high transient voltage would occur. Figure 9*c* shows the braking torque typical of dynamic braking by capacitors with and without load resistors.

g. Dynamic braking by capacitor-rectifier-resistor combination. Dynamic braking by a capacitor-rectifier-resistor combination indicates that dynamic braking is obtained by generator action with a form of dc excitation. A capacitor which is charged through a rectifier circuit when the motor is running discharges direct current through the motor windings to produce braking torque when the motor is disconnected from the line (Fig 9*b*). This method is applicable to three-phase induction motors. It differs from dynamic braking by capacitor in that dc power is discharged into the motor windings. It differs from dynamic braking by dc excitation in that a constant dc voltage is applied. The capacitor discharge provides maximum voltage at maximum motor speed. The dc voltage and motor speed decrease together to give more nearly constant braking torque. The dc excitation is available until the capacitor has discharged. Capacitor values chosen to match the motor and load characteristics can result in a discharge time equal to the deceleration time. The energy stored in the capacitor is all that is required for the braking power supply. Large capacitor ratings are required. Figure 9*c* shows the braking torque typical of dynamic braking by the capacitor-rectifier-resistor method.

PROTECTION OF MOTOR PARTS

16. Type and Degree of Protection. All electric-motor elements (active iron, inactive iron, and windings) require some type of protection to ensure that the motor will remain in operation safely and economically (Refs. 8, 24). The degree of protection depends upon the particular combination of service conditions and the importance of the application. The protection may be in the form of an enclo-

TABLE 7. Protection of Electric Motors

General cause of deterioration or damage	*General type of protection**
Corrosive gases, vapors, chemical fumes, and solid material which may cause corrosion	Winding insulation and proper enclosure
Atmospheric conditions: high humidity, rain, snow, steam, salt air, unusual ambients, wind-blown foreign materials such as dirt, dust, sand, twigs, gritty and conducting dust	Winding insulation and proper enclosure
Explosive atmospheres, gases, vapors, dust	Proper enclosure
Flammable gases, combustible dust, and other similar materials	Proper enclosure
Electrical considerations:	
Level of potential to ground	Class of insulation
Excessive temperatures	Voltage class of insulation. Winding-protection devices
Mechanical considerations:	
Corrosion	Suitable paints
Loss of bearing lubrication	Bearing-protection devices
Bearing and shaft currents	Bearing insulation
Excessive overspeeds	Overspeed switches
Vibration, unbalance, misalignment	Proper installation and maintenance

* The degree of protection depends upon the specific motor part subjected to damage and upon the specific cause. Where more than one type of protection is shown, the choice of whether one or both types are required depends upon the severity of the cause of damage.

sure, a warning by sound or light, or disconnection of the motor from the power source before damage can occur. Damage occurs because of deterioration and electrical breakdown of the insulation system used on the windings, because of a mechanical-component failure, or from a combination of the two. Both the electrical and the mechanical deterioration and breakdown may be either the cause or the result of a motor failure. The user must determine how extensive the protective features should be for a particular application. Table 7 indicates the general causes of motor damage and the general forms of protection available.

17. Factors Used to Determine Protection. These are given by groups in the following paragraphs.

a. Motor-design characteristics and parameters. The type of motor, its rated horsepower, voltage, speed, cost, service factor, and the thermal capacity of the windings are the most important motor items to be considered. The motor manufacturer must be consulted in regard to the thermal capacity.

b. Motor-starting method and acceleration conditions. The voltage at the motor terminals, the load inertia, the load-torque requirements during acceleration, and the duty cycle or frequency of starts contribute to chances of stalling, long acceleration times, and excessive heating of the windings. The motor manufacturer should be consulted as to the best design for any unusual conditions.

c. Power-supply system. The type of utility feeder, whether manual or automatic reclosure or transfer is used, the speed of reclosure or transfer, voltage unbalance for any reason, the type of circuits (underground or open-wire), the possibility of surges (switching, lightning, transient), the grounding system used (resistance, solid, ungrounded neutral), and whether fed from feeder or individual transformer must be considered. It is desirable to consult the utility and the motor manufacturer when making these considerations.

d. Service conditions. Service conditions merit special attention, and where critical and unusual risk is involved, extreme care should be used in selecting motors for successful operation and satisfactory service (Ref. 3). The motor manufacturer should be consulted about how to provide the best protection needed to minimize the risk involved.

1. Service conditions more favorable than usual for successful motor operation include:

   Operation at rated voltage and frequency

   Application of the motor to a machine whose loads and duty cycle are accurately known and cannot be exceeded

2. Usual service conditions (Ref. 3) in which little risk is encountered in successful motor operation include:

   Operation in an ambient temperature not exceeding that for which the motor was designed, for example, 40°C

   Operation within allowable frequency variation for ac motors, normally ±5 percent of rated frequency

   Operation within allowable voltage variation, for example, ±10 percent of rated voltage for all motors except universal motors, in which case ±6 percent is permissible

   Operation within allowable combined voltage and frequency variation, for example, up to 10 percent above or below the rated voltage and the rated frequency, provided that the frequency variation does not exceed 5 percent

   Operation at an altitude for which the motor was designed, for example, 3300 ft or less

   Operation in locations or supplementary enclosures which do not interfere with motor ventilation

   Operation with solid mounting and with belt, chain, or gear drive in accordance with good practice

3. Service conditions less favorable than usual for which the risk is great and for which extreme care should be exercised in selecting motors and motor protection for successful operation include (Ref. 3):

   Operation of motor in locations having chemical vapors and fumes, inflammable gases, and oil vapor

   Operation of motor in damp locations, or where steam, salt atmosphere, snow, and high humidity exist

   Operation of motor in locations where combustible dust, dust of explosives, and explosive gases may exist

   Operation of motor in locations containing lint, sand, twigs, or other debris, gritty or conducting dust, lampblack, and cement dust

   Operation in extreme ambient temperatures, below 10°C or above 50°C, or at high altitudes

   Operation in poorly ventilated rooms, pits, or small confining enclosures

   Operation at speeds in excess of overspeed limits, at voltages and frequencies beyond the normal limits, or with unbalanced voltages

   Operation under conditions of abnormal shock or vibration caused by external sources

e. Commercial codes and specifications. Various organizations sponsor codes which relate to motor protection. The principal one is the National Fire Protection Association (NFPA), which sponsors the National Electric Code (NEC).

Code NFPA 70 is also an American National Standard, ANSI/NFPA70. Devices conforming to the intent of the NEC are "approved" by Underwriters Laboratories (U/L). The Institute of Electrical and Electronics Engineers (IEEE) and the National Electrical Manufacturers Association (NEMA) set maximum temperature limits for the various classes of winding-insulation systems.

18. Enclosures. The type of motor enclosure used provides physical protection from external sources of motor damage (Ref. 3). There are several standard and well-defined types of enclosures, all of which vary in degree of enclosure and method of ventilating the motor.

a. An open enclosure is the simplest type of enclosure. It has ventilating openings which permit passage of external air over and around the motor windings, around and, in some cases, through ducts in the active-iron core stack. Openings are semiguarded or fully guarded when screens are added to part or all of the openings to limit access to the live or rotating parts of the motor.

b. A dripproof enclosure is an open enclosure whose openings are constructed in such a manner as to prevent drops of liquid or solid particles falling on the machine at any angle of 15° or less from the vertical from entering the motor. Guard screens may be added.

c. A splashproof enclosure is similar to the dripproof enclosure, except that an angle of 100° or less from the vertical applies.

d. An externally ventilated enclosure is an open enclosure which is ventilated by a separately motor-driven blower mounted on the motor enclosure. Guard screens may be added.

e. A pipe-ventilated enclosure is an open enclosure whose inlet ventilating openings are constructed to permit the connection of inlet-air pipes or ducts to them. It is called a forced-ventilated motor when the air is circulated through the motor enclosure by an external motor-driven blower.

f. A weather-protected type 1 enclosure is an open enclosure whose ventilating passages are constructed and arranged to minimize the entrance of rain, snow, and airborne particles to the live and rotating parts. Guard screens may be added.

g. A weather-protected type 2 enclosure is an open enclosure whose ventilating passages at both intake and discharge are constructed and arranged to permit high-velocity air and airborne particles to be discharged without entering the internal ventilating passages of the motor. Guard screens may be added.

h. A totally enclosed enclosure prevents free exchange of air between the inside and the outside of the enclosure. The enclosure is not airtight. Different methods of cooling can be used with this enclosure.

i. A totally enclosed nonventilated enclosure has no provisions for external cooling of the enclosed parts. The motor is cooled by heat radiation from the exterior surfaces to the surrounding atmosphere.

j. A totally enclosed fan-cooled enclosure provides for exterior cooling by means of a fan or fans integral with the machine, but external to the enclosing parts. When screens are used to cover the openings giving access to the fan or fans, the term "guarded" applies.

k. An explosionproof enclosure is a totally enclosed enclosure which is constructed to withstand an explosion of a specified gas or vapor which may occur within it. Should such an explosion occur, the enclosure will prevent the ignition or explosion of the gas or vapor which may surround the motor enclosure.

l. A dust-ignition-proof enclosure is totally enclosed and constructed to exclude ignitible amounts of dust or amounts that might affect the performance or rating of the motor. Any heat, arcs, or sparks generated or liberated inside the enclosure cannot cause ignition of specified dust accumulations or atmospheric suspensions thereof either on or in the vicinity of the motor enclosure. Excessive overloads, stalling, or excessive quantities of accumulated dust on the motor must be avoided to prevent overheating of the motor.

m. A waterproof enclosure is totally enclosed and constructed to exclude water applied externally from a hose. A stream of water from a 1-in nozzle delivering 65 gal/min of water from a 10-ft distance in any direction for a period of not less than 5 min is the common test. Leakage may occur around the shaft provided it is not allowed to enter the oil reservoir. A check valve or tapped hole in the lowest part of the frame is provided to drain the enclosure.

n. A totally enclosed pipe-ventilated enclosure is a totally enclosed enclosure whose inlet and outlet ventilating openings are constructed to permit connection of air pipes or ducts to them. It is called a forced-ventilated enclosure when the air is circulated through the enclosure by an external motor-driven blower.

o. A totally enclosed water-cooled enclosure is a totally enclosed enclosure whose cooling is by water circulated through conductors which come in direct contact with the motor parts. In special cases, the water itself may come in direct contact with the motor parts.

p. A totally enclosed water-air-cooled enclosure is a totally enclosed enclosure constructed to cool the motor by means of an air-to-water heat exchanger. The internal motor air is circulated by either integral rotor fans or separately driven blowers through and around the motor parts and through the exchanger. Circulating water through the heat-removal section of the heat exchanger removes the heat from the motor-ventilating air.

q. A totally enclosed air-to-air-cooled enclosure is a totally enclosed enclosure constructed to cool the motor by means of an air-to-air heat exchanger. The motor's internal air is circulated by either integral rotor fans or separately driven blowers through and around the motor parts and through the exchanger. External air circulated by separately driven blowers through the heat-removal exchanger removes the heat from the motor-ventilating air.

19. Winding-Insulation Protection. The main function of winding-insulation protection is to prevent uneconomical and excessive rates of electrical-insulation-system deterioration caused by excessive temperature (Refs. 8, 24). The deterioration rate varies with time and temperature for a given insulation system, as indicated in Sec. 11. Assuming that the proper design, materials, processes, workmanship, and other types of protection have been supplied, and if the insulation is not physically damaged by external forces, the deterioration rate of insulation systems will depend upon the insulation time-temperature characteristics. Severe overheating may result in immediate motor burnout. Also, overheating results in fire hazard and risk of electric shock to operating personnel.

20. Causes of Overheating (Ref. 8). The several causes of overheating are as follows.

a. Electric-power supply. For constant power output, operation at lower than rated voltage results in lower iron loss, reduced speed (nonsynchronous motors) and ventilation, higher currents and loss in the windings, and higher stray load loss. High voltage produces opposite effects. Lower or higher than rated frequencies on ac motors produces much the same effect as lower or higher voltages, respectively. Lower than rated voltage on synchronous motors increases the possibility of the motor pulling out of step. Individual motor designs vary in their ability to operate at other than rated voltage and frequency without excessive overheating. In general motors having higher temperature ratings at full load, those having short time ratings, and those of high-torque design are more sensitive to power-supply variations.

b. Sustained overloading. Operation at higher than rated load and service factor results in higher currents and increased losses in the windings. If the overload is large enough, the motor may stall, fail to start, or fail to accelerate to rated speed. Synchronous motors may pull out of step or out of synchronism. These conditions are quite severe, and whenever the thermal capacity of the winding is limited, destructive heating can occur in a matter of a few seconds. Heating is critical in the cage and damper windings under such conditions.

c. Sustained and excessive duty cycle. Duty cycles involving rapid repetitive starting, plugging, jogging, and reversals may result in accumulative heating which exceeds the thermal capacity of the windings. If the ventilation, either integral with the motor or separately provided, is inadequate, damage can occur. Time for speed recovery must be provided for high-slip ac motors driving high-inertia loads. High currents for short time intervals can be dangerous if repeated rapidly.

d. Loss of ventilation. Anything which reduces the rate of heat removal from the motor can contribute to dangerous overheating. Loss or reduction of ventilation reduces the rate of heat removal. The most common cause of loss of ventilation is clogging or blocking of the air ventilation paths to, into, and through the motor. This applies to the external ventilating system, when used, as well as to the internal ventilating system. Loss of cooling in the heat-removal section of a heat exchanger can occur if the temperature of the cooling medium increases or if the volume of the cooling medium decreases. Overheating from loss or reduction of ventilation can cause overheating without excessively high currents. The heating can be cumulative because the copper losses will increase with the higher winding resistance caused by the higher temperatures.

e. High ambient temperatures. Overheating can result if the motor is operated in a higher ambient temperature than intended. The rated temperature rise of the motor under rated operating conditions is a constant independent of the ambient temperature. The rise and the ambient temperatures added together should not exceed the rated total temperature of the insulation system provided in the motor. Confining enclosures or rooms from which heat cannot be effectively removed can result in higher than rated ambient temperatures. Loss of cooling in the heat-removal section of heat exchangers will result in higher temperature of the cooling air or gas entering the motor internally. Overheating can result without excessive motor currents.

f. Failure of electrical elements. Overheating can result from a failure of an electrical element within either the motor or its control equipment. Failure of the insulation system between turns in the coil, to ground, or between phases can result in excessive heating, although in most cases this type of failure results in shutdown. Single phasing, for any reason, can cause severe overheating in the remaining phases. Malfunction of switching of the motor circuits can produce excessive currents in some portions of the windings with resultant overheating. Intermittent short circuits between turns, poor contact joints, broken bars in cage windings, or unsatisfactory brush operation on either commutators or slip rings can result in overheating of the motor elements.

g. Failure of mechanical components. Loss of lubrication, faulty or broken belts and couplings, and misalignment of the motor and its driven load can result in overload heating of the motor. These items sometimes cause mechanical rubbing of rotating and stationary members of the motor with resultant local overheating. Mechanical failure of either the motor or its driven load could result in a stalled motor with resultant rapid overheating.

21. Protection from Overheating (Ref. 8). Motor protection from overheating is provided by sensing the motor line current, the internal motor temperature, or both. More than one type of protection can be provided. The choice will depend upon the probable cause of overheating, the motor size, the distance between the motor and its control, ambient-temperature variations, the type of load, and the degree of protection desired.

22. Current-Responsive Protectors (Refs. 2, 8, 24). Current-responsive protectors respond to motor-overload current and are usually located external to the motor, often at some distance away. They may function either as line-break devices or as control-circuit devices used to activate alarm or contactor circuits.

a. Time-lag fuses, which respond to heat by current flowing through them, are used to prevent sustained motor-overload current. They are simple and have sufficient time delay to prevent tripping on either starting or momentary overload conditions. They are not generally used to protect larger motors because their melting characteristics do not match the motor-heating characteristics.

b. Magnetic relays respond to the magnetic field of the motor line current. They are not affected by ambient temperatures and for this reason may be located at a different place from the motor.

c. Thermal relays respond to the heating produced by motor line current flowing through a resistance-type heating coil. The heat is either conducted or radiated to a heat-sensitive element which causes the relay to operate at a predetermined temperature. The melting-alloy relay, or solder-pot relay, uses an eutectic alloy which suddenly changes from a solid to a liquid at the selected temperature and allows the spring-loaded contacts to open. After the solder has cooled enough to become solid, the relay may be reset manually. The bimetallic relay uses two thermally responsive dissimilar metal sheets bonded together. The difference in the thermal coefficients of expansion for the two sheets causes the bimetallic sheet to bend, therby activating the contact mechanism of the relay. The sheet may be in the form of a disk which snaps from convex to concave or a strip which bends. The strip type may be manually reset, and the disk type can be either manually or automatically reset.

23. Temperature-Responsive Protectors (Refs. 2, 8, 24). Temperature-responsive protectors respond to the internal motor heat and operate either at a given value of temperature or at a rate-of-rise value. They are used as control-circuit devices to activate alarm or contactor circuits, although some may be used as line-break devices.

a. Thermostats are applied directly to the motor windings and are calibrated to operate within a narrow range of temperature. Their mass is made as small as possible in order to follow the motor temperature as closely as possible. However, they cannot follow a rapid rate of rise such as that caused by locked-rotor conditions. Bimetallic thermostats, operating on the same principle as the bimetallic relay element, are used to activate normally closed switches. The switches may operate as line-break devices for fractional-horsepower motor use and as control-circuit devices for the larger motors.

b. Resistance-type protectors are sensitive to the winding temperature and are used to activate a control-circuit device. A resistance-coil type commonly used is the resistance-temperature detector having either 10 or 120 Ω resistance at 25°C. These detectors are embedded between the coils in the slots and are used on motors above 500 hp. Thermistors are thermally sensitive semiconductor resistors whose resistance value lies between that of a conductor and that of an insulator. They exhibit extreme sensitivity to minute temperature changes. Their small mass enables them to follow temperature changes more closely than the bimetallic types. Negative-temperature-coefficient (NTC) thermistors have a high resistance, which decreases linearly as the temperature increases. Positive-temperature-coefficient (PTC) thermistors have essentially constant resistances until a predetermined critical temperature causes a sharp increase in resistance. The NTC-type thermistor is most commonly used. Both types require amplification of the resistance variation through an electronic circuit to actuate the control device. Several thermistors may be used in series or parallel.

c. Rate-of-rise protectors are anticipatory types used as sensors for control-circuit devices. A metal rod is enclosed in a tube in a manner which permits them to operate as a differential-expansion thermal element actuating a self-contained snap-action switch. For slow temperature changes, the rod and tube heat evenly and the switch opens at the predetermined temperature. For fast temperature changes, the tube heats faster than the rod and causes the switch to open at a lower temperature. This anticipatory action makes it possible to protect the motor for running and stalled conditions at rates of temperature rise of up to 16°C/s.

d. A thermocouple is composed of two dissimilar conductors joined so as to produce a thermoelectric effect. It may be used as a thermometer to measure temperature because variations in temperature will cause a variation of the joint-contact electromotive force. Thermocouples may be used with proper circuits and instruments to indicate the temperature of any motor element to which they may be attached. Temperatures may be read on some form of instrument or an alarm can be sounded when a predetermined temperature is reached. The motor may also be disconnected from the line if desired.

24. Thermal Protectors (Refs. 2, 8, 24). Devices responsive to both temperature and current are called thermal protectors. They are basically bimetallic thermostats combined with a heating element which is in series with the motor winding. Thermal protectors are used as line-break devices on 15 hp or less and as control-circuit devices on larger motors. When used, they must match the heating characteristics of the motor in order to provide both running and stalled protection.

25. Protection from Humidity. The winding may require protection if the motor is to be left idle for any length of time. The idle time will depend upon the motor environment. Either space heaters (energized when the motor is idle) mounted in the motor enclosure or current circulated from a separate source through the idle windings will prevent condensation of moisture on the electrical parts by keeping them 5°C above the ambient temperature. Power requirements are approximately equal to the stator core $D \times L$, where D is the outside diameter and L is the length.

26. Protection of Mechanical Parts. Prevention of bearing failures is one of the most important items of motor protection. Loss of lubrication for any reason can cause overheating of the bearings, leading eventually to bearing failure and to damage of other motor elements.

a. Bearing-temperature detectors similar to the resistance-coil winding protector can be used to detect overheating of the bearings. The detector, having 10 or 120 Ω resistance at 25°C, can be used with the same control-circuit devices as the stator protector. The bearing-temperature detector is placed against the bearing babbitt in sleeve bearings and against the outer race of a ball bearing. Dial-type and tube-type thermometers can be used in place of the detectors and control-circuit devices. The tube-type thermometer may also be provided with electrical contacts that can be used for control-circuit purposes.

b. Loss of oil pressure or flow in forced-lubricated bearings may be detected by devices placed in the oil lines. The devices are used to sound alarms when predetermined pressures, either high or low, are reached.

c. Shaft currents, which are caused by any dissymmetries in the motor magnetic circuits, can cause bearing failure by destroying the bearing and journal surfaces. They may also cause a breakdown of the oil. Current flows from the bearing journal through the oil film (sleeve bearings) to the bearing and its housing, the motor frame and foundation, to the other bearing housing, bearing, and oil film, to the shaft to complete the circuit. Insulation material may be placed at any convenient place which will interrupt the shaft-current flow. For instance, insulation may be placed between the motor feet or between the pedestal and the foundation. Oil pipelines, when used, should be insulated at one joint. Care should be taken not to short-circuit the insulation. Insulating the bearing from its housing is the preferred way of preventing shaft and bearing currents.

d. Liquid-level relays can be used in motor enclosures to detect an accumulation of liquids that might damage the insulation. The source of the liquid might be leaks in the cooling system, lubricating systems, or floods.

e. Water-flow meters with alarm contacts can be used to indicate inadequate flow of water to the air cooler or lubricating-oil cooler whenever these coolers are supplied.

f. End-thrust damage to bearings of horizontal motors may be prevented by using limited end-float couplings. Motor bearings are seldom designed to take thrust from the driven machine unless specifically requested.

g. Overspeed for any reason may be prevented by using a speed-limit switch which can be arranged to disconnect the motor from the line.

27. Other Protective Devices (Refs. 2, 24). There are several other protective or auxiliary devices which may be used to protect motors from damage from external and internal sources. Most of these offer winding protection from electric-power sources of damage.

a. Overload relays may be used to disconnect the motor from the line at a predetermined value of overcurrent. In some cases, an undercurrent relay may be used in a similar manner.

b. Phase-sequence or phase-reversal relays may be used to disconnect or prevent energization of a polyphase motor from the line if the phase sequence of the motor power supply is reversed for any reason.

c. Undervoltage relays (low voltage) may be used to disconnect and prevent reenergization of the motor from the line either if voltage is reduced to a predetermined value below rated voltage or if voltage failure occurs. Undervoltage protection is the most common type of protection used because low voltage is the most common hazard to satisfactory motor operation. If the relay is designed to permit reenergization of the motor upon reestablishment of voltage to the motor, the term "undervoltage release" (low-voltage release) applies. Overvoltage relays are also available for protection against higher than rated voltages. Some special relays will provide both overvoltage and undervoltage protection.

d. Frequency relays may be used to protect ac motors from overfrequency, underfrequency, or both, should this protection be necessary.

e. Open-phase relays. Single phasing, opening of one phase, and undervoltage in one phase of polyphase motors may be prevented by using open-phase relays. The open-phase relay functions whenever one or more phases of the power supply are opened for any reason. Starting open-phase protection prevents connecting the motor to the line until all phases are energized. Running open-phase protection removes the motor from the line if the current in one phase drops to zero. Phase-unbalance relays will disconnect the motor from the line if the phase unbalance and current reach preset values.

f. Differential relays (ratio-balance relays, biased relays, percentage-differential relays) operate whenever a predetermined vector difference of two or more similar electrical values is exceeded. They are used to prevent damage caused by short circuits in the stator windings of ac motors. Whereas phase-to-phase and phase-to-ground faults can be detected, short-circuited turns will not be detected. Differential protection is desirable protection for wye-connected ac motors connected to ungrounded systems.

g. Current-limiting power fuses having high interrupting capacity are used to protect against short circuits. Both low-voltage (600 V and less) and high-voltage (above 600 V) fuses are available. Close coordination between motor locked-rotor current, full-load current, and overload characteristics is required.

h. Surge arresters and capacitors. Surge voltages may occur on improperly grounded or ungrounded systems, because of fast reclosure or transfer on the power system, open-transition motor starting, a synchronous motor pulling out of synchronism with the system, and overhead lines exposed to lightning. Surge protection, especially important on larger motors at the higher voltages, can be supplied by surge arresters or surge capacitors connected between each motor terminal and ground.

i. Field-failure relays provide protection from loss of field current in the motor armature circuit by sounding an alarm, disconnecting the motor from the line, or both.

j. Field-protective relays provide protection against overheating of the field-excitation winding by reducing or removing the field excitation.

k. Field-discharge resistors are used to provide protection to the motor field windings by limiting the induced high voltages in the field windings whenever the field circuit is opened for any reason. They also provide for dissipation of the field energy under these conditions.

l. Pull-out protection prevents synchronous motors from running out of synchromism by either disconnecting the motor from the line or allowing the motor to resynchronize if possible.

m. Amortisseur-winding protection for synchronous motors may be provided by a thermal overcurrent relay, which senses the induced current which flows in the field-discharge resistor and is proportional to the amortisseur-bar current. Protection may also be obtained by a frequency-tuned induction voltage relay which is connected across the discharge resistor.

n. Incomplete-sequence protection. Protection from incomplete sequence of normal starting functions of motors and their associated controllers may be provided by a timing relay. The timing relay is energized when power is applied to the motor. After a preset interval of time, the relay will disconnect the motor from the line if acceleration to rated speed has not been completed.

DC MOTORS

Dc motors (Refs. 9–11) are used to convert dc energy into mechanical energy. They are well suited for use as either constant-speed or adjustable-speed motors. The three types (shunt, series, and compound-wound) which are available provide the range of speed-torque characteristics for applications which require:

Smooth acceleration, retardation, or deceleration

A carefully controlled rate of acceleration or retardation (deceleration)

Controlled speed changes, over small or wide ranges

Accurate and positive speed matching

Control or limit of torque or tension

28. Basic Equations. Table 3 lists the basic equations for dc motor operation. These equations apply equally to each of the three types of dc motors, which differ only in the arrangement of the field windings.

29. Construction. The main elements of the dc motor are the frame and the armature. The frame, also called yoke or field, serves as an enclosure and a support for the field windings, and forms a part of the magnetic-flux path. The armature is the rotating element of the motor and includes the shaft, the armature core to complete the magnetic-flux path, the armature windings, and the commutator. End brackets added to the frame complete the motor enclosure and contain the bearings which position the armature within the frame to obtain the proper air gap. A brushholder assembly is mounted either to the frame or to the end enclosures to position carbon composition brushes properly on the commutator.

a. The frame is constructed of steel and may be either cast or fabricated from plates. The laminated field poles are bolted to the inside of the frame. Insulated field windings are mounted on the poles. The windings are appropriately connected to produce alternate north and south flux poles. The flux flows through the pole body, yoke, air gap, and armature. Some dc motors use commutating-pole (interpole) assemblies which are similar to, and mounted between, the main field poles.

b. The armature is composed of a laminated-core assembly mounted on the shaft. The armature winding is suitably placed in slots in the core and the coil ends are terminated to appropriate commutator bars. Current flow in the armature winding produces a magnetic field which reacts with the field flux to produce torque.

c. Commutator and carbon composition brushes form a mechanical rectifier which ensures that the current flow through the armature winding is in the proper direction at any instant. Successful operation of the dc motor depends, to a large extent, upon the choice and location of the brushes and upon proper care of the commutator (Ref. 13).

30. Shunt-Wound Motors. Figure 10 gives the schematic diagram for the dc shunt-wound motor.

a. The main-field, or shunt-field, winding is connected in parallel with the armature winding across the dc power supply of constant voltage. A large number of turns of small wire size is used to carry the field current. The shunt-field winding is highly inductive and requires some means of limiting the discharge voltage should the field circuit be interrupted at any time.

b. The main-field, or shunt-field, current is usually less than 5 percent of rated full-load armature current. It is independent of the armature current.

c. Commutating field windings are supplied on many of the shunt-wound motors. When used, these windings consist of a few turns of large wire size placed on interpoles which are located between the main-field poles. Commutating windings are connected in

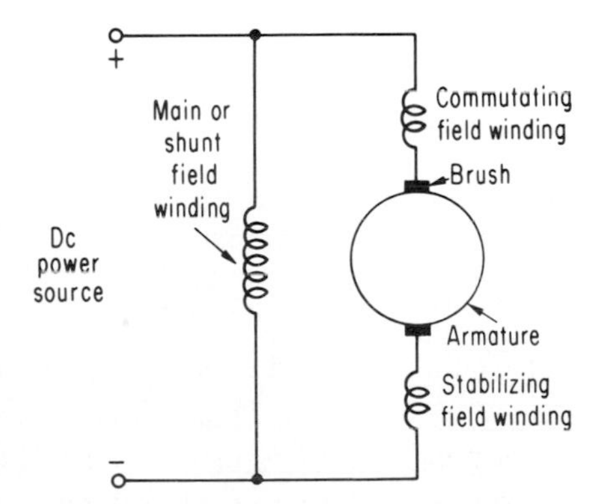

FIG. 10. Schematic diagram of shunt-wound dc motors.

series with the armature and carry armature current. The commutating field improves the commutating characteristics of the motor but has little effect upon the motor-performance characteristics otherwise.

d. Stabilizing field windings are supplied on some shunt-wound motors. When used, these windings consist of a few turns of large wire size placed on the main shunt-field poles with the shunt winding. Stabilizing windings are connected in series with the armature and carry armature current. The stabilizing field helps stabilize the running-performance characteristics of the motor.

e. Typical operating characteristics of the shunt-wound motor are shown in Fig. 11. These characteristics are well suited for flexible general use because field current and flux are independent of the armature or load current.

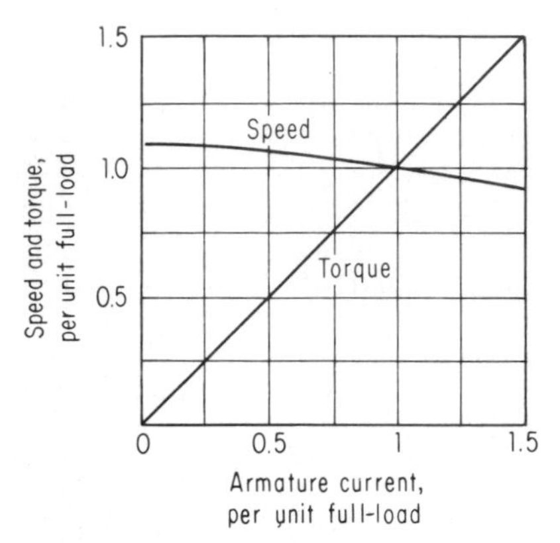

FIG. 11. Typical speed and torque curves for shunt-wound dc motors.

f. The speed of the shunt-wound motor can be changed by varying the shunt-field current or armature voltage. Normally there is very little drop in speed with increasing load, and rarely will the drop exceed 5 percent. Speed control by changing the armature resistance is unsatisfactory because the speed regulation is objectionable. Whenever the load changes slowly, the flux changes as a result of armature reaction and the speed will remain constant. However, if the load changes more rapidly than the self-induction of the field windings will allow the flux to change, then the speed will change rapidly. A speed range of approximately 4:1 with reasonable running stability is possible for loads up to full-load torque. Field failure will cause an unloaded motor to "run away" to excessively high speeds if the armature remains energized.

g. The starting, or locked-rotor, torque is inherently low, and for this reason the shunt-wound motor is not suitable for accelerating high-inertia loads.

31. Series-Wound Motors. Figure 12 gives the schematic diagram for the dc series-wound motor.

a. The main, or series, field winding is connected in series with the armature winding across the dc supply of constant voltage. A few turns of large wire size are required to produce the motor flux. The motor is simple and rugged.

b. Commutating field windings, as described under shunt-wound motors, are sometimes used on series-wound motors.

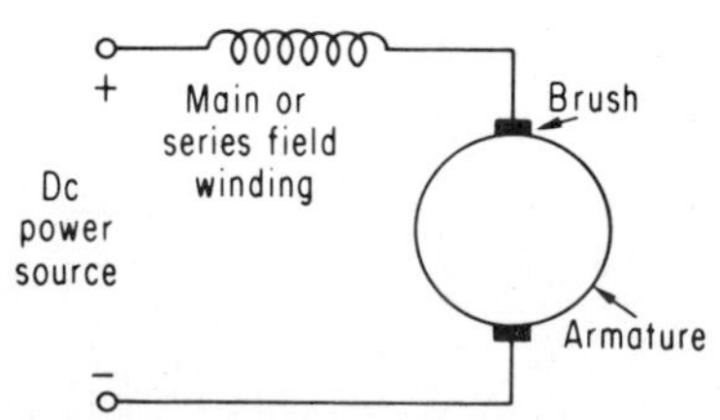

c. Typical operating characteristics of the series-wound motor are shown in Fig. 13. The series motor will develop high starting torque and is good for crane service and high-inertia loads. The torque will vary as the square of the armature current, neglecting saturation of the field poles, which

FIG. 12. Schematic diagram of series-wound dc motors.

reduces this relationship. The torque and speed are very sensitive to the load current (which is also the field current) because of the corresponding change in flux.

d. The speed of the series-wound motor may be adjusted by shunting out the series winding, short-circuiting some field turns, or inserting resistance in series with the field and/or armature. However, speed adjustment is not easily accomplished. This motor has one disadvantage in that it tends to "run away" at light loads. The overspeed can reach a destructive value if the load is suddenly removed. For this reason, positive direct couplings or gearing to the load is preferred to belting.

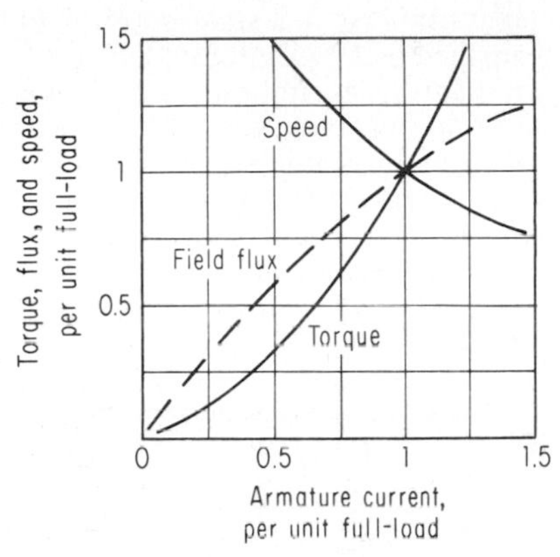

FIG. 13. Typical torque, field flux, and speed curves for series-wound dc motors.

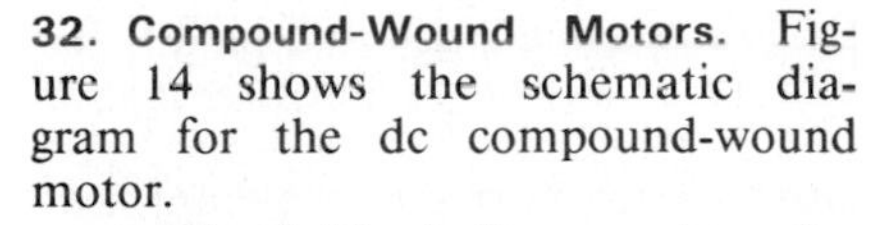

32. Compound-Wound Motors. Figure 14 shows the schematic diagram for the dc compound-wound motor.

a. The field winding consists of a shunt winding and a series winding. Each winding has turns and wire sizes similar to the shunt-wound and series-wound motor field windings. The proportion of the total flux supplied by the series winding determines the amount of "compounding." The amount of compounding can be varied to suit the speed characteristics desired. A strong series field will give speed characteristics approaching those of a series motor, whereas a weak series field will give characteristics approaching those of a shunt motor. Motors with series fields producing 40 to 75 percent of the total flux are frequently used, with a value of 50 percent most commonly provided. Compound motors having series fields producing 10 to 25 percent of the total flux (lightly compounded) are also used for some general industrial applications. Generally, then, the speed characteristics lie between those of shunt-wound and series-wound motors (Fig. 15). Compound-wound motors can be used when speed variation with load variation is permitted.

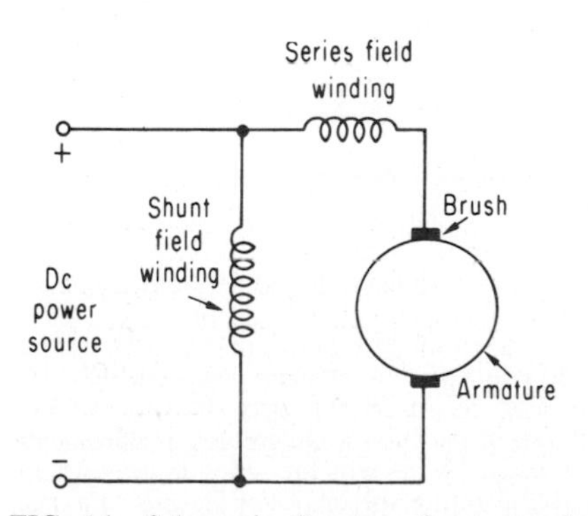

FIG. 14. Schematic diagram of compound-wound dc motors.

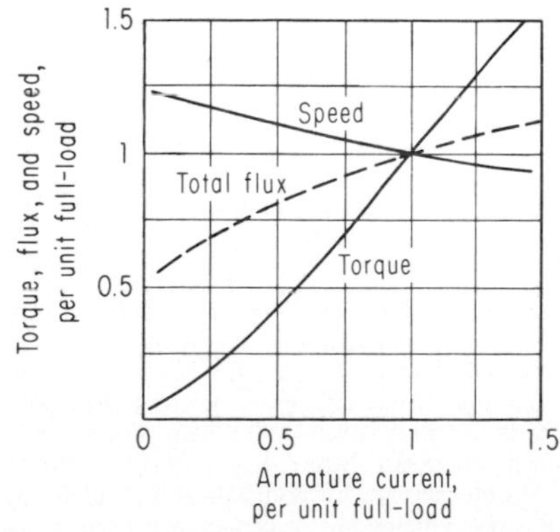

FIG. 15. Typical torque, total flux, and speed curves for compound-wound dc motors.

b. Commutating field windings as described under shunt-wound motors are sometimes used on compound-wound motors.

c. Indirect compounding may be supplied by a series field winding which is wound similarly to a small shunt-field winding. The current for this field winding is supplied by an unsaturated, constant-speed, separately driven exciter, which in turn draws its excitation from the main motor line or armature current. A rheostat and a reversing switch in the series exciter armature circuit provide variation in main-motor series-field strength and in polarity of the compounding. Constant-speed regulation over a given speed range can be obtained in this manner.

d. Differential compounding is obtained by a series field which opposes the shunt field. Proper proportion of the two field strengths will give constant speed from no load to full load. Also, an increase of speed with load is possible if desired. Differential compounding is used mostly on small motors.

e. The starting torque of the compound-wound motor is high, although not as high as that of a series-wound motor. The torque will increase rapidly with load because the series field will increase the flux. The speed will decrease rapidly for the same reason. However, the motor will not run away at light loads because of the shunt-field flux.

f. The speed of the compound-wound motor can be adjusted with a shunt-field rheostat over a 1½:1 speed range.

33. Effect of Voltage Variation. The general effect of voltage variation on shunt-wound and compound-wound dc motor characteristics is shown in Sec. 2, Table 3.

34. Efficiency of dc Motors. Figure 16 gives typical efficiency curves for shunt-wound and compound-wound 1150-r/min motors. The curves are necessarily general and will vary for specific ratings.

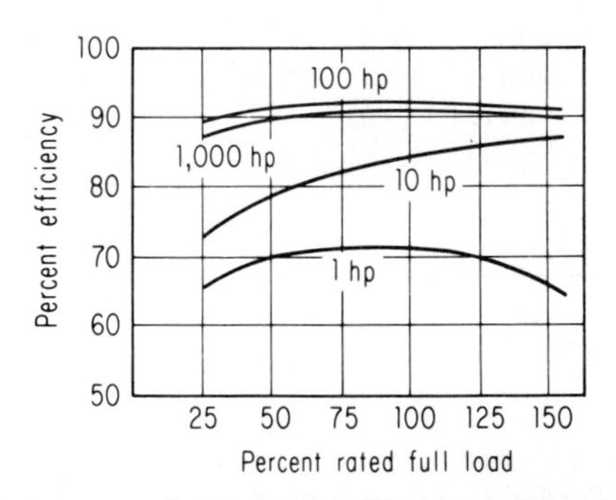

FIG. 16. Typical efficiency curves for standard 1150-r/min industrial-type dc motors. Generally, for motors 10 hp and less, efficiency will increase or decrease slightly as the base speed is increased or decreased, and peak efficiency will occur at 1750 r/min. For motors in the 10- to 100-hp range, maximum efficiency will occur at about 1150 r/min so that efficiency will decrease for speeds different from 1150 r/min. For the 1000-hp class, maximum efficiency will occur at 850 r/min, where it will be 1 to 2 percent higher than shown in the curves. Differences in efficiencies of shunt-, series-, or compound-wound motors are negligible, except for series motors where light-load efficiencies will decrease because of higher operating speeds. *(Westinghouse Electric Corporation, Buffalo, N.Y.)*

DC-MOTOR CONTROL

The various types of dc motors will provide the torque characteristics necessary to drive a particular load (Ref. 24). The control equipment will determine how the motor will operate to give the desired performance of the driven machine or load. NEMA Standard ICS 1 (Ref. 4) contains general information and specific standards on control equipment.

35. Starters. Starters are used to provide the basic functions of start, accelerate, and stop.

36. Controllers. Controllers are used to provide additional functions of speed reduction, braking, reversing, and variation in speed.

a. Speed reduction and variation, or reduced-speed operation, is generally obtained by reducing the voltage across the armature. Resistors placed in parallel (shunt) or in series with the armature are used for this purpose. This method is used mostly on shunt or light compound-wound motors to provide a slowdown before a stop (positioning) and for running at slow speed. One disadvantage of this method is the speed variation with load. When used on series-wound motors, this method will limit the overspeed tendencies, but speed will vary greatly with load and results are harder to predict.

b. Braking may be obtained by one of the methods listed in Table 6.

37. Control Systems. Systems involving extensive control equipment external to the motor are used where several motors are involved in one large operation or process. The system is used to provide coordinate and sequence speed matching, regulation of torque or tension, and interlocking. Motor and control manufacturers will help develop and apply control systems meeting the user's requirements (Ref. 24).

38. Protection. The control equipment will include devices for proper protection to the motor and personnel (Ref. 24). Table 7 indicates that most common types of motor protection are available for dc motors. Protection provided by enclosures and protection of the windings, insulation, and mechanical parts are readily available and will adequately protect the motor from most common sources of damage (Ref. 24).

AC MOTORS

Ac motors consist of a magnetic circuit which is separated by an air gap into two elements, one of which rotates with respect to the other. The stationary member, called stator, usually contains the primary windings. Current flowing in the primary winding will establish an alternating magnetic field. The rotating member, called rotor, usually contains the secondary windings. Current flowing in the secondary windings will establish a second magnetic field. The manner in which the secondary is constructed, together with the manner in which the currents are caused to flow, will establish the type of motor and its performance characteristics.

INDUCTION MOTORS

Induction motors (Refs. 15, 16) are used to convert ac energy into mechanical energy. They are suited for use as constant-speed, varying and reduced-speed (in some cases), reduced and reversed-speed (in some cases), and multispeed motors. The many types available cover a wide range of torque characteristics and may be designed to operate from many different power supplies having different combinations and values of phases, frequencies, and voltages.

39. Basic Equations. Table 3 gives the basic equations for induction-machine operation. The generalized equations apply to the many different types of induction motors.

40. Construction. The main elements of the induction motor are the stator and the rotor. The stator generally consists of a frame which serves as an enclosure and as a support for the stator magnetic laminated core. The core forms a part of the magnetic-flux path and supports the primary windings. The rotor is the rotating element of the motor and includes the shaft, the rotor magnetic core to complete the magnetic-flux path, and the secondary windings. End brackets added to the stator frame complete the motor enclosure and contain the bearings which position the rotor within the stator. The bearings of some large machines may be mounted in pedestals which are mounted on a common bedplate with the stator to obtain the proper air gap. Specific types of induction motors may include the use of slip rings or commutators with carbon composition brushes, short-circuiting switches, and starting windings.

a. The stator frame may be cast iron or steel, fabricated from steel plate, cast aluminum or extruded aluminum. The laminated core consists of laminations or segments made from electrical sheet steel having high permeability with low hysteresis and eddy-current losses. Insulated windings are placed in slots near the air gap in the stator core. The windings are designed and connected in accordance with the number of phases, the power-supply frequency and voltage, and the desired speed. The primary or stator currents produce alternate north and south flux poles. The flux flows through the stator core, across the air gap, through the rotor core, and back across the air gap to complete the circuit.

b. The rotor is composed of a laminated core assembly mounted on the shaft. The appropriate secondary windings are suitably placed in slots in the core near the air gap and are short-circuited or closed by means of external impedance. Current flow in the stator winding induces an opposing current flow in the secondary winding. The interaction of the induced motor-current field with the stator-current field produces the motor torque. The induced secondary current distinguishes the induction motor from other types of motors.

41. Single-Phase Induction Motors. The alternating current in the primary winding of a single-phase motor produces a pulsating, or stationary, field as opposed to a rotating field. The induced alternating single-phase current in the rotor windings produces a stationary field in the rotor. Both fields alternate in polarity but produce no starting torque by themselves. However, once the rotor is revolving at approximately one-half speed or more, the interaction between the primary field and the induced secondary field will produce sufficient torque for acceleration and load-carrying purposes.

Single-phase motors employ some device or means of providing, in effect, a rotating two-phase primary field. The interaction between this field and the associated induced secondary field will produce torque to start and accelerate the motor. The type of starting device or means employed determines the type of single-phase motor. The size and the application of the motor determine the type of motor needed. Single-phase motors are made in horsepower ratings from approximately 25 down to small fractional values.

a. Series-type single-phase motors are constructed in the same manner as a series-wound dc motor. The schematic diagram in Fig. 12 will apply. However, a well-laminated field-pole structure is required to reduce losses. The use of a series motor is limited to fractional sizes of approximately 1/150 to 3/4 hp and to very high speed (1500 to 15,000 r/min), or to intermittent-duty applications because of the low power factor and poor commutation at low speeds. Serious sparking generally occurs at the start, even when minimized by the use of high-resistance leads between the seondary (armature) windings and the commutator segments. The motor has varying speed characteristics. Speed increases with decreasing load and decreases with increasing load. Speed can be adjusted by series rheostat or special windings and controlled by an electric governor. The nominal starting current produces high starting torque. Series motors are usually designed for unidirectional rotation unless special windings are supplied.

b. A compensated series single-phase motor is a series-type single-phase motor to which a compensating winding (Fig. 17) has been added. The compensating winding is in series with the armature and produces a field which counteracts the armature field. A commutating winding placed in series with the main field will reduce the armature reaction in the commutating zone as in the dc motor. The power factor of a compensated series ac motor is much higher than that of a straight series ac motor. The motor has varying speed characteristics and is built in ratings of approximately 1/40 to 2½ hp at 2500 to 15,000 r/min. It cannot be reversed unless special windings are used.

c. A universal motor is a series-type single-phase motor which is designed to operate on either alternating or direct current. Because of poorer commutation on alternating current, these motors are built only in sizes of ½ hp and less. They may be either compensated or uncompensated. Uncompensated, they are useful in smaller sizes and higher speeds (3000 to 10,000 r/min) only. Compensated, they can be designed to have approximately the same speed-torque characteristics over the load range for all frequencies from zero to 60 Hz. They can easily be adapted to compact power tools and appliances. Figure 18 indicates the typical performance characteristics of a universal motor. The motor has varying speed characteristics and cannot be reversed unless special windings are provided.

d. A capacitor-start induction-run single-phase motor utilizes a capacitor and an auxiliary starting winding con-

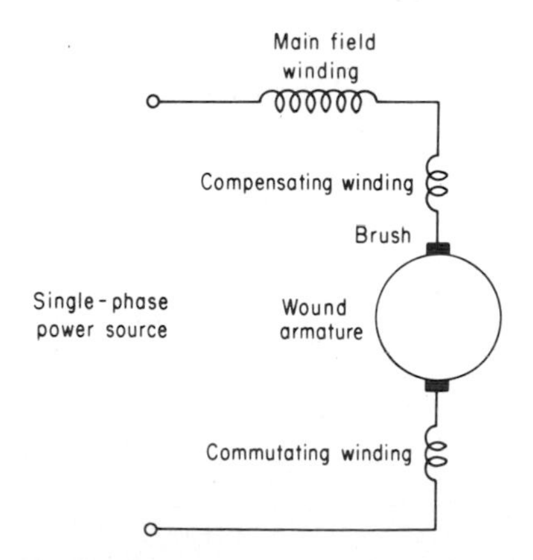

FIG. 17. Schematic diagram of compensated series single-phase ac motors.

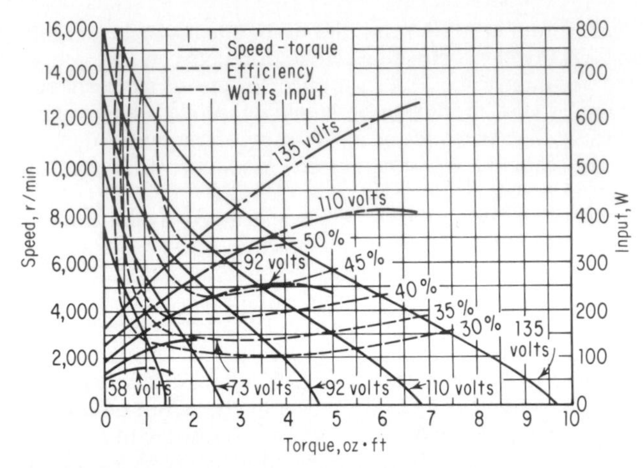

FIG. 18. Performance characteristics of universal motors. *(Ref. 5.)*

nected in parallel with the main-field winding, as shown in Fig. 19. The starting winding, in conjunction with the main field, produces a strong two-phase rotating field when connected to the line. A high torque per ampere is produced for start and acceleration. A centrifugal switch is used to disconnect the starting winding and capacitor from the line as the motor accelerates to rated speed. Figure 20 shows typical performance curves for this type of motor. Capacitor-start motors are manufactured in sizes from approximately ⅛ to 7½ hp in speeds up to 3450 r/min. Their high starting torque is useful on constant-speed loads, such as reciprocating compressors and vacuum pumps. The motor may be reversed at or near standstill. Reversal at full speed requires a special relay or switch. This motor can be supplied with two windings for two speeds.

e. A two-value capacitor (capacitor-start, capacitor-run) motor uses a capacitor and an auxiliary winding in parallel with the main-field winding. A starting capacitor and switch are placed in parallel with the auxiliary winding circuit, as shown in Fig. 21. This arrangement gives a high torque during start and acceleration. A centrifugal switch removes the starting capacitor from the line as the motor accelerates to rated speed. The auxiliary winding and running capacitor will give higher efficiency and power factor over the load range. Two-value capacitor motors are manufactured from approximately ½ to 25 hp. They are used on material-moving equipment such as conveyors, stokers, hoists, and compressors. The motor may be reversed at or near standstill.

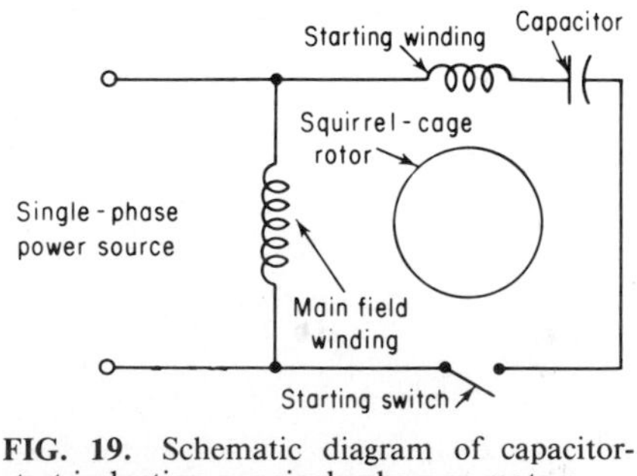

FIG. 19. Schematic diagram of capacitor-start induction-run single-phase ac motors.

f. A permanent-split-capacitor single-phase motor uses a permanent auxiliary winding with capacitor as shown in Fig. 22. This arrangement gives low starting current and low (approximately 100 percent full-load) starting

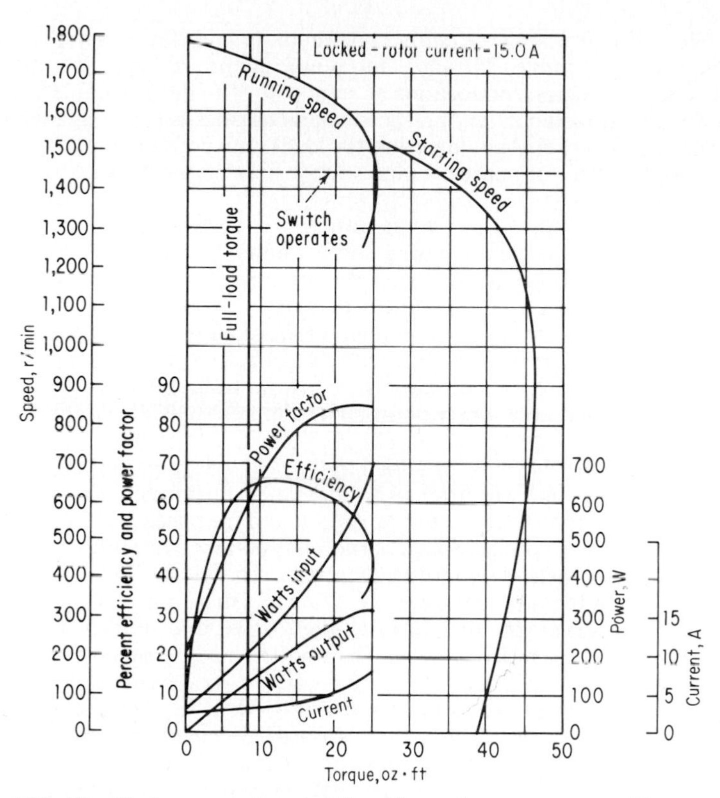

FIG. 20. Performance characteristics of capacitor-start motors. *(From Archer & Knowlton.* Standard Handbook for Electrical Engineers, *8th ed., McGraw-Hill, New York, 1949, Fig. 7-106, Sec. 7-255.)*

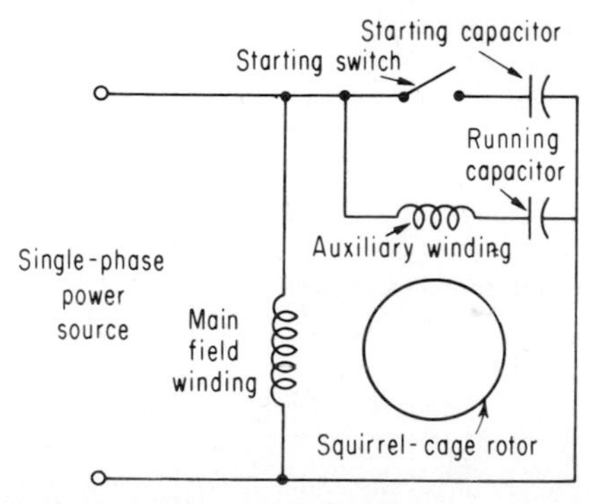

FIG. 21. Schematic diagram of two-value (capacitor-start, capacitor-run) single-phase motors.

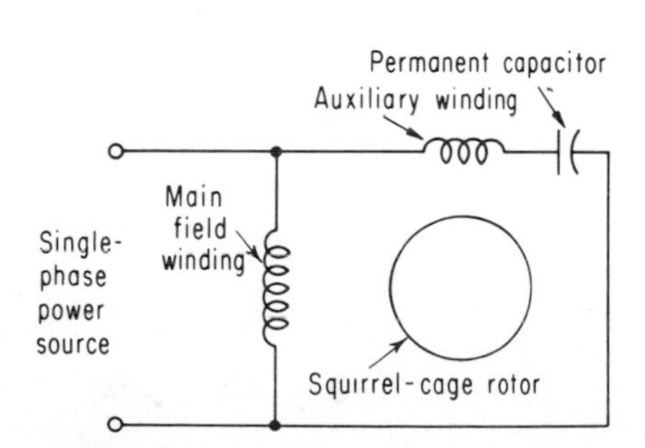

FIG. 22. Schematic diagram of permanent-split-capacitor single-phase motors.

torque. However, it operates quietly and smoothly and has essentially constant speed over the load range. Efficiency and power factor are high. The motor may be reversed by changing connections at standstill or under favorable conditions while running. Adjustable varying-speed characteristics are also available. Friction loads are favorable, but high-inertia loads are unfavorable. This type of motor is built in sizes of approximately 1/20 to 1 hp. A typical speed-torque curve is shown in Fig. 23.

g. A repulsion single-phase motor is similar to a series-wound dc motor in construction. It has a main-field winding and a wound armature with commutator and brushes. The brushes short-circuit the armature, as shown in Fig. 24. The brush setting will cause the rotor magnetic axis to be inclined from that of the main field. Changing the brush setting will produce adjustable-speed characteristics, will produce high starting torque with proper setting, or can cause the motor to run at speeds above synchronous speed. Commutation is very good at all speeds except those above synchronous. The simple repulsion motor is used for fan drives.

An extra set of brushes may be added in the field axis to permit armature excitation from another source or from a variable-ratio transformer. The performance obtained is similar to that of a shunt dc motor, and the speed may be adjusted by varying the excitation voltage. Speed variation is obtained by shifting the two sets of brushes together instead of varying the excitation voltage. This condition produces series-wound motor performance characteristics. This form of motor is used for variable-speed spinning, printing-press drive, and similar applications.

h. A repulsion-start induction motor is similar to the repulsion motor except that a commutator short-circuiting device is added. The motor starts and accelerates as a repulsion motor. As rated speed is approached the commutator bars are short-circuited, and the motor operates as a single-phase induction motor. The brushes are usually lifted off the commutator after acceleration. This motor combines high starting torque (300 to 600 percent) with constant-speed operation over the load range and is suitable for pump and compressor drives. Figure 25 shows typical performance characteristics of high-torque repulsion-start motors.

i. A repulsion induction motor is a repulsion motor to which a deeply embedded low-resistance squirrel-cage winding is added in the armature. The motor has

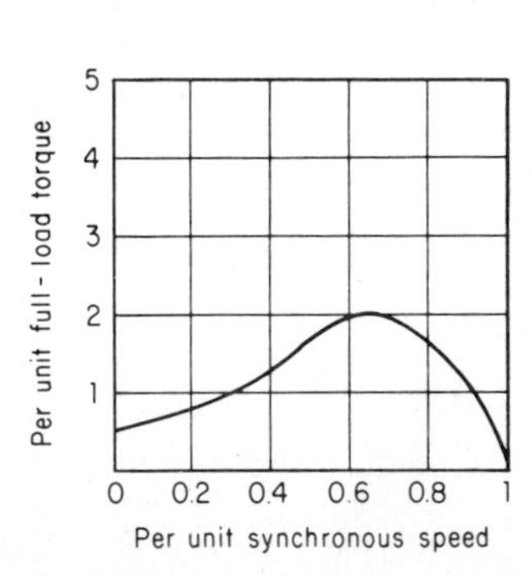

FIG. 23. Typical speed-torque curve for permanent-split-capacitor single-phase motors.

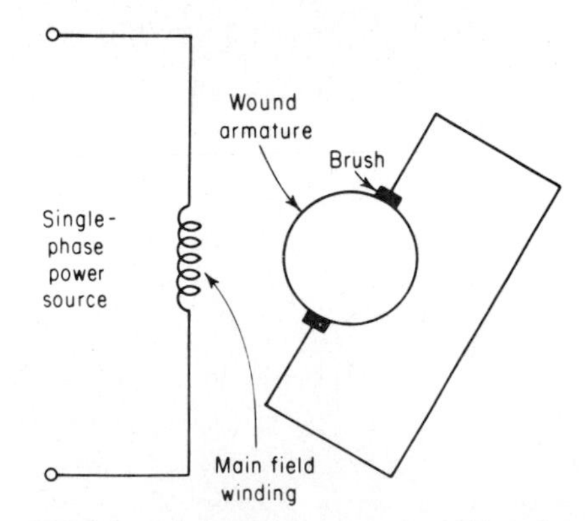

FIG. 24. Schematic diagram of repulsion single-phase motors.

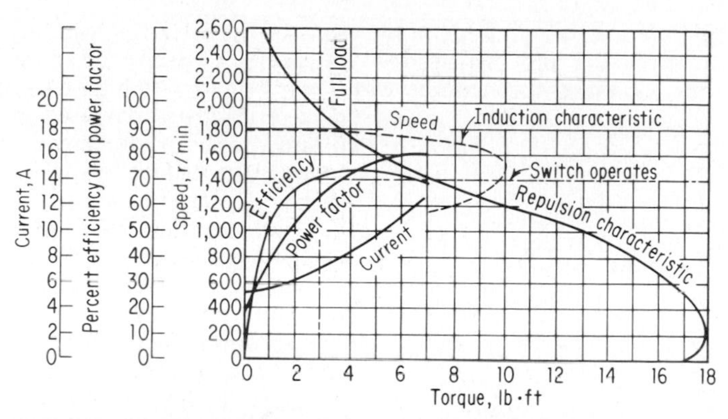

FIG. 25. Performance characteristics of high-torque repulsion-start induction motors. *(From Archer & Knowlton,* Standard Handbook for Electrical Engineers, *8th ed., McGraw-Hill, New York, 1949, Fig. 7-101, Sec. 7-255.)*

a high starting torque (repulsion start on commutated repulsion winding) combined with a high breakdown torque (squirrel-cage winding induction motor action). Variable- or constant-speed characteristics are obtained. The squirrel-cage winding improves the power factor up to as much as 10 percent above that of other repulsion motors. Figure 26 shows performance characteristics of normal-torque repulsion induction motors.

j. A shaded-pole single-phase motor uses a permanently short-circuited winding of high resistance mounted on the main pole, as shown in Fig. 27. This coil embraces approximately one-third of the pole pitch. The current induced in the coil, by the main-field flux it embraces, reduces this flux and causes it to lag in time phase. This "shading" effect is sufficient to produce enough torque to accelerate loads having low torque requirements. This type of motor has low efficiency and high slip and is used in ratings of approximately 30 W or less. The speed can be varied by using a series reactor to reduce the applied voltage. The motor cannot be reversed unless "shading" coils are supplied on both sides of the main-field poles. Some means of closing one coil and opening the other must be provided. Figure 28 shows performance characteristics of a shaded-pole motor.

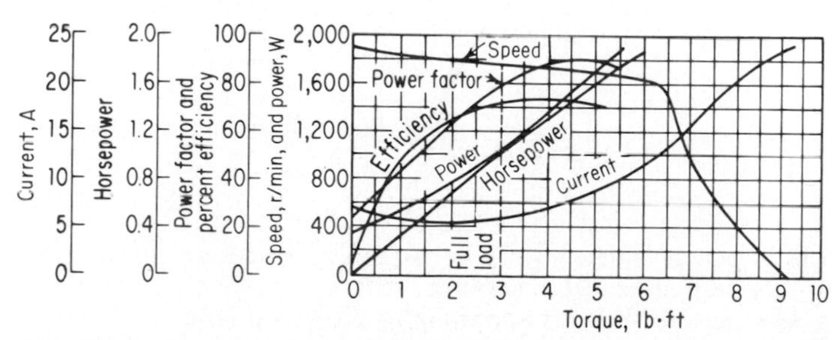

FIG. 26. Performance characteristics of normal-torque squirrel-cage repulsion motors. *(Ref. 5.)*

k. A split-phase single-phase motor uses an auxiliary starting winding which is connected in parallel and displaced 90° magnetically from the main field, as shown in Fig. 29. Either the auxiliary winding is wound with high-resistance wire or a series resistor is used to give the desired ratio of resistance and reactance. A centrifugal switch or electromagnetic relay is used to remove the auxiliary winding from the line as the motor approaches rated speed. The motor can be reversed at standstill by reversing the leads of either winding. Special designs with relays are required for reversal while running. The motor is suitable for operating constant-torque loads requiring low starting and accelerating torque. It is built in ratings of 1/20 to 1/2 hp, and two-speed ratings are also available. It is not suitable for use on frequent starts or high-inertia loads because of high losses in the high-resistance starting winding circuit. Figure 30 shows performance characteristics of resistance split-phase motors.

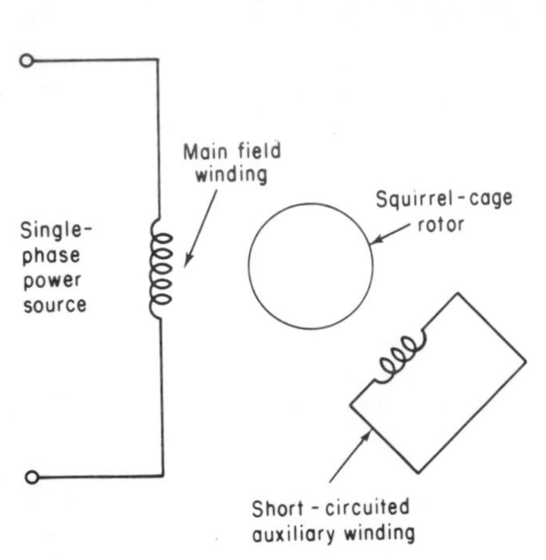

FIG. 27. Schematic diagram of shaded-pole single-phase motors.

42. Polyphase Induction Motors. The alternating current in the primary winding of a polyphase motor produces a revolving field. The short-circuited secondary windings carry the induced secondary current. The frequency of the secondary current is a function of the rotor slip speed. At zero speed, line-frequency currents flow in the secondary windings. At no-load speed, the frequency is usually less than 1 percent of the primary line frequency. The no-load speed of the polyphase induction motor is always less than synchronous speed in order to supply the no-load losses. The electromagnetic forces existing between the primary (stator) and the secondary (rotor) develop the torque to drive the load. This torque varies with the motor slip, as shown for the induction motor in Table 3. A typical speed-torque curve is shown in Fig. 1.

a. The efficiency of polyphase induction motors is shown in Figs. 31 and 32. Values for specific designs will vary somewhat from the values shown. Deductions are made for higher-voltage motors, as shown in Table 8, because of higher primary copper losses. Deductions are made for motors having higher than normal slip in accordance with the formula

$$\text{EFF}_2 = \text{EFF}_1 \frac{100 - S_2}{100 - S_1} \qquad (9)$$

where EFF_1 = percent efficiency of a normal-torque low-starting-current motor at desired load-torque point
S_1 = percent slip corresponding to EFF_1 load point
EFF_2 = percent efficiency of motor having higher than S_1 slip
S_2 = percent slip value for which efficiency is desired

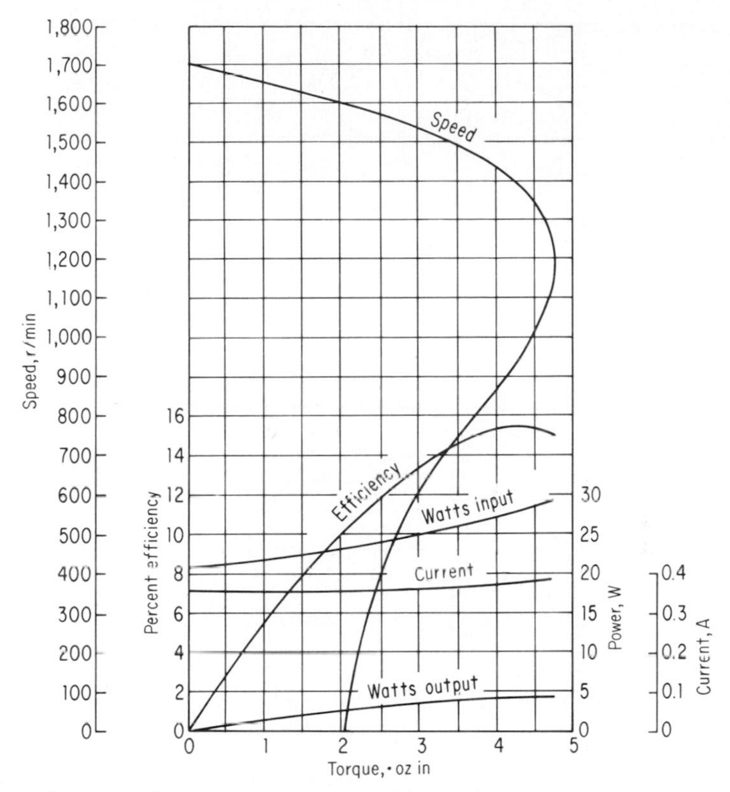

FIG. 28. Performance characteristics of shaded-pole motors. *(From Archer E. Knowlton,* Standard Handbook for Electrical Engineers, *8th ed., McGraw-Hill, New York, 1949, Fig. 7-99, Sec. 7-252.)*

This formula is very useful in estimating the efficiency of wound-rotor motors operating at reduced speeds. Table 9 indicates the general effect of voltage and frequency variation on motor efficiency.

b. The power factor of induction motors is shown in Figs. 33 and 34. Values for specific designs will vary somewhat from the values shown. The power factor will be lower for motors having larger than normal air gaps and for motors having intermittent- or short-time ratings. Over the load operating range, the power factor of a wound-rotor motor at a given torque or primary load current does not change if the speed is reduced. Table 9 indicates the general effect of voltage and frequency variation on motor efficiency.

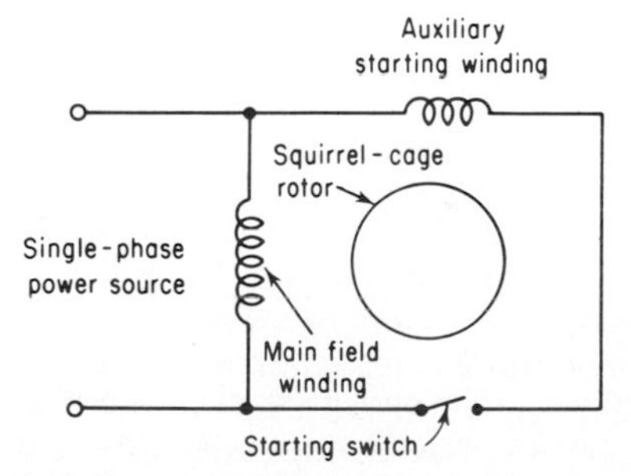

FIG. 29. Schematic diagram of split-phase single-phase motors.

c. Power-factor correction is desira-

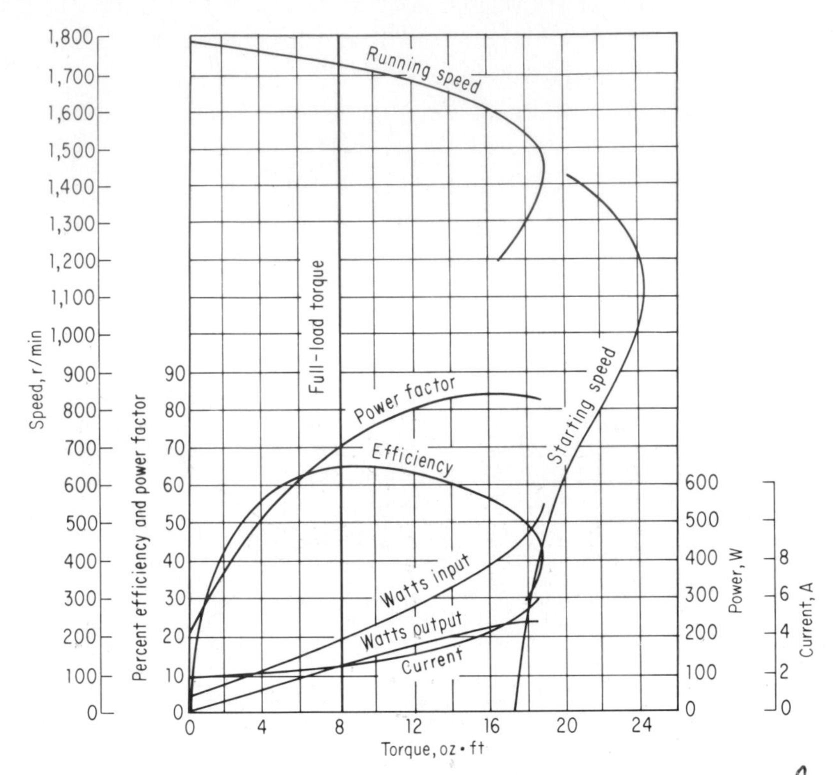

FIG. 30. Performance characteristics of resistance split-phase motors. *(From Archer & Knowlton,* Standard Handbook for Electrical Engineers, *8th ed., McGraw-Hill, New York, 1949, Fig. 7-100, Sec. 7-253.)*

ble in some instances to reduce the kvar requirements of the induction motor. This can be accomplished by adding capacitors across the motor terminals. *However, the amount of correction that can be safely made is limited to that amount needed to correct the motor no-load power factor to unity.* Amounts in excess of this value may cause insulation breakdown because of high transient voltage whenever the motor is disconnected from the power line. The no-load power factor generally ranges from approximately 4 to 5 percent for large high-speed motors to about 10 to 12 percent for smaller low-speed motors. The motor manufacturer will recommend the maximum safe value of capacitance that may be applied to a given motor.

d. The general effect of voltage and frequency variation on induction-motor performance is shown in Table 9. It must not be concluded that it would be safe to operate all motor designs at 110 percent overvoltage. The magnetic circuit of some short-time or intermittent rated and of Class B, F, or H insulated motors may saturate enough to cause overheating.

e. The voltage ratings for which induction motors can be built are limited by horsepower and speed ratings, as shown in Table 10. The mechanical strength of the primary coil as well as the electrical strength is set to a large degree by the

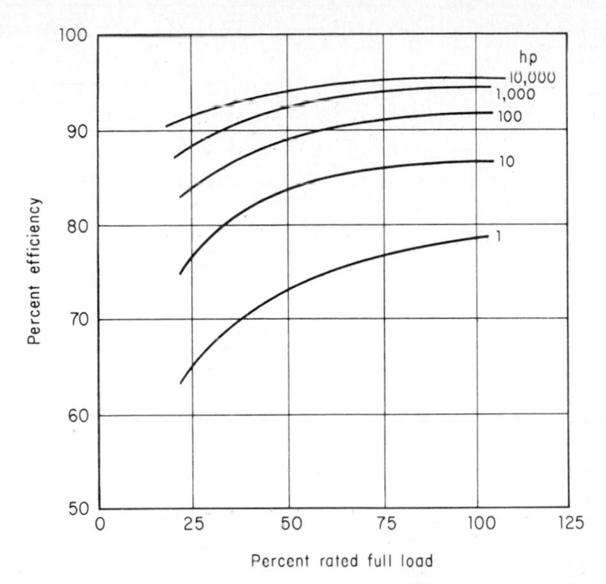

FIG. 31. Typical efficiency curves for 1800-r/min normal-torque low-starting-current polyphase induction motors. Values will vary for a specific design. Deductions for high voltages are made as shown in Table 9. Full-load efficiencies for other speeds may be taken from Fig. 32 and partial-load values extrapolated from curves above.

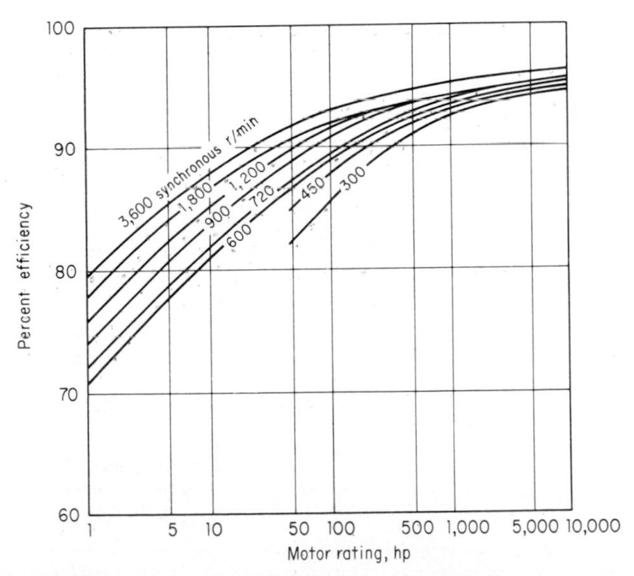

FIG. 32. Typical full-load efficiencies for normal-torque low-starting-current polyphase induction motors. Partial full-load values may be estimated as noted in Fig. 31.

TABLE 8. Approximate Deductions* for Load Efficiency (4/4, 3/4, 1/2) of High-Voltage (2301 to 13,200 V) Motors

hp range	Voltage range								
	2,301–5,000			5,001–7,000			7,001–13,200		
	4/4	3/4	1/2	4/4	3/4	1/2	4/4	3/4	1/2
250–800	0.3	0.5	1.4	0.5	0.7	1.5			
801–1,250	0.2	0.4	1.3	0.3	0.5	1.4			
1,251–5,000	0.1	0.3	1.2	0.2	0.4	1.3	0.5	0.7	1.5
5,001–10,000	0.0	0.2	1.0	0.1	0.2	1.1	0.2	0.4	1.3
10,001 and up	0.0	0.1	0.8	0.0	0.1	0.9	0.1	0.2	1.1

* Deductions apply to induction motors. Deductions are generally less for dc-excited synchronous motors.

TABLE 9. General Effect of Voltage and Frequency Variation on Induction-motor Characteristics

Characteristic	AC (induction) motors			
	Voltage		Frequency	
	110 %	90 %	105 %	95 %
Torque,* starting and max running	Increase 21 %	Decrease 19 %	Decrease 10 %	Increase 11 %
Speed:†				
Synchronous	No change	No change	Increase 5 %	Decrease 5 %
Full load	Increase 1 %	Decrease 1.5 %	Increase 5 %	Decrease 5 %
% slip	Decrease 17 %	Increase 23 %	Little change	Little change
Efficiency:				
Full load	Increase 0.5 to 1 point	Decrease 2 points	Slight increase	Slight decrease
¾ load	Little change	Little change	Slight increase	Slight decrease
½ load	Decrease 1 to 2 points	Increase 1 to 2 points	Slight increase	Slight decrease
Power factor:				
Full load	Decrease 3 points	Increase 1 point	Slight increase	Slight decrease
¾ load	Decrease 4 points	Increase 2 to 3 points	Slight increase	Slight decrease
½ load	Decrease 5 to 6 points	Increase 4 to 5 points	Slight increase	Slight increase
Current:				
Starting	Increase 10 to 12 %	Decrease 10 to 12 %	Decrease 5 to 6 %	Increase 5 to 6 %
Full load	Decrease 7 %	Increase 11 %	Slight decrease	Slight increase
Temp rise	Decrease 3 to 4°C	Increase 6 to 7°C	Slight decrease	Slight increase
Max overload capacity	Increase 21 %	Decrease 19 %	Slight decrease	Slight increase
Magnetic noise	Slight increase	Slight decrease	Slight decrease	Slight increase

*The starting and maximum running torque of ac induction motors will vary as the square of the voltage.

†The speed of ac induction motors will vary directly with the frequency.

Note. This table shows general effects, which will vary somewhat for specific ratings. (From Ref. 5.)

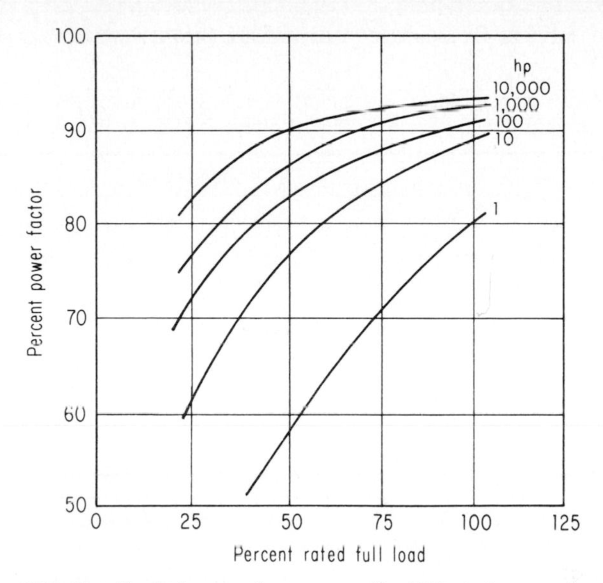

FIG. 33. Typical power-factor curves for 1800-r/min normal-torque low-starting-current polyphase induction motors. Values will vary for a specific design. Full-load power factors for other speeds may be taken from Fig. 34 and partial-load values extrapolated from curves above.

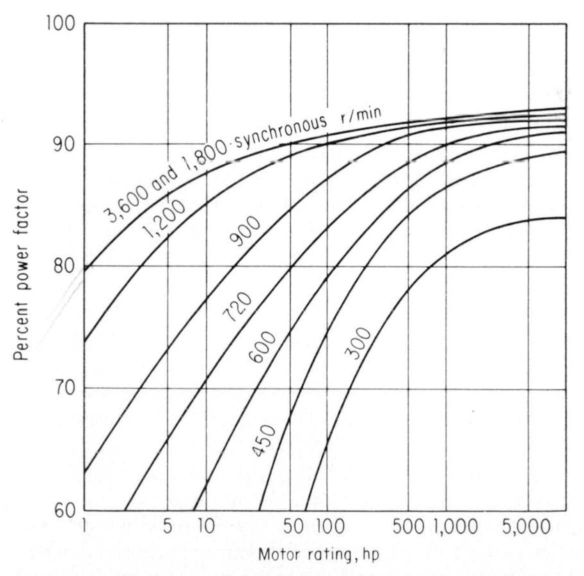

FIG. 34. Typical full-load power factors for normal-torque low-starting-current polyphase induction motors. Partial full-load values may be estimated as noted in Fig. 33.

TABLE 10. Recommended Minimum and Maximum Horsepower Ratings for Which Reliable Coil Designs Can Be Made for High-Voltage Ratings at Various Speed Ratings

	Voltage range		
	2,301–5,000	5,001–7,000	7,001–13,200
Condition	hp Synchronous r/min	hp Synchronous r/min	hp Synchronous r/min
Min hp possible	200 at 3,600 150 at 1,800 to 600	300 at 3,600 250 at 1,800 to 600	3,000 at 3,600* 3,500 at 1,800 to 600
Recommended† min hp	350 at 3,600 300 at 1,800 to 600	500 at 3,600 800 at 1,800 to 600	5,000 at 3,600 5,000 at 1,800 to 1,200 4,000 at 900 to 600
Recommended max hp	5,000 at 3,600 to 600		

*Motors for 11 to 13.2 kV, 3600 r/min not recommended.
†For those applications where extreme reliability is required.

number of turns and by the wire size. Table 10 reflects the most practical, reliable, and economical coil designs based on mechanical and electrical design considerations.

f. A wound-rotor induction motor uses insulated secondary windings which are arranged and connected for the same number of poles and phases as the primary winding. The secondary leads are short-circuited externally through collector rings and brush assemblies to complete the secondary circuit. Secondary resistance inserted into the secondary circuits determines the starting, accelerating, and running characteristics. Very flexible speed-torque characteristics are obtained by this arrangement, as shown in Fig. 35. The wound-rotor motor can be used to accelerate high-inertia, high-torque, or delicate loads that require a sequence of slow-speed steps to limit the current demand during acceleration, and to provide either adjustable-speed or reduced-speed operation. Reduced-speed operation of self-ventilated wound-rotor motors is limited to 70 percent rated speed on constant-torque applications or 50 percent speed with 40 percent rated horsepower. Rotor losses are external to the motor in proportion to the external rotor-circuit resistance. Successful operation of the wound-rotor motor depends upon proper slip-ring design and judicious choice of the brushes (Ref. 21).

g. The squirrel-cage induction motor uses a rotor having solid conductors instead of wire windings in the rotor-core slots. These conductors are short-circuited on each end of the rotor by means of solid low-resistance end rings. The resistance and the reactance of the rotor largely determine the speed-torque characteristics of the motor. Increasing the resistance increases the slip over the usable load range and increases the locked-rotor torque. The resistance is increased by decreasing the rotor slot or conductor size, by using conductor material having low conductivity, or by both. Increasing the reactance will decrease the torque obtained at a given speed, and the locked-rotor current drawn will be decreased. Each of the speed-torque curves shown in Fig. 35 may be matched with a squirrel-cage rotor design which is, of course, fixed and cannot be varied. All the rotor losses are internal to the motor and the rotor cage; therefore, the thermal capacity is limited. For this reason, high-torque high-slip designs are generally given inter-

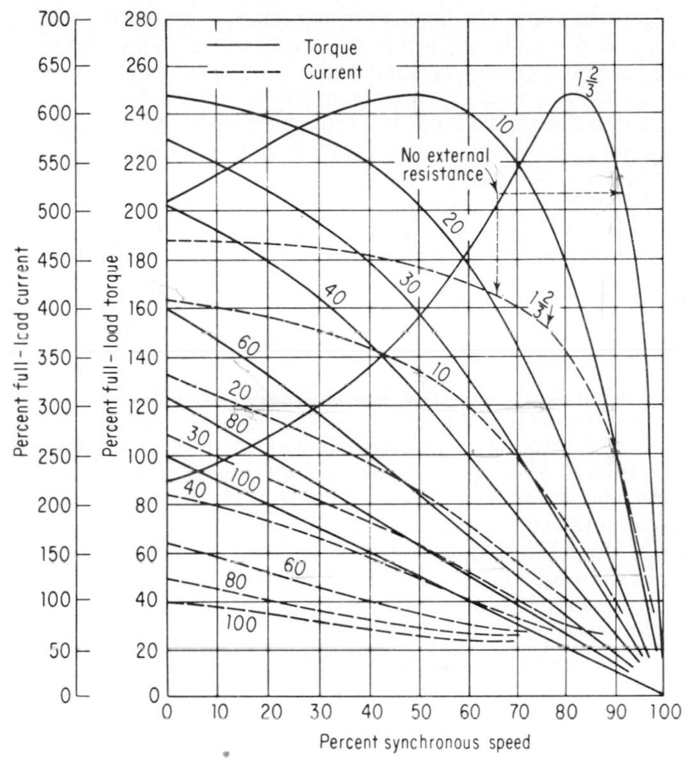

FIG. 35. Speed-torque and speed-current curves for typical wound-rotor induction motors. *(Ref. 5.)*

mittent-time ratings. In some cases, the number of successive starts of motors connected to high-inertia loads is limited to as few as two. The squirrel-cage-winding design provides a means of varying the torque per ampere to be obtained at the start and during acceleration. However, the motor impedance determines the starting current as well as the maximum torque. If high breakdown torque is desired, a high starting current must be accepted. Die-cast rotors are made by injecting molten aluminum or aluminum alloy into the rotor slots and end ring molds. The metal hardens upon cooling to form the short-circuited cage winding. Centrifugal-cast copper and copper alloys are used similarly to form a cage winding. The die-cast and centrifugal-cast rotors permit use of many sizes and configurations of slots to form the bar conductors. Brazed rotor cages consist of copper or copper-alloy bars driven through the rotor slots. The ends of the bars are short-circuited by brazing end rings of copper or copper alloy to them. The brazed-type rotor conductor may be round, square, rectangular, trapezoidal, or of any conveniently drawn shape. A single-cage rotor, meaning the rotor slot effectively contains only one conductor, is the most rugged and most used type of rotor-cage design. It may be designed to have high thermal capacity. Double-cage rotor designs, meaning the rotor slots effectively contain two conductors per slot, are used in special cases to obtain high starting or breakaway torque with low running slip and high load efficiency. One set of conductors having high resistance and low reactance is placed in the top of the slots, and a second set of conductors having low resistance and high reactance is placed in the bottom of the slots. The conductors of the two windings may or may not be joined, and either common or separate end rings may be used. The top cage winding has very limited thermal capacity and may "burn out" in less than a minute if the motor fails to start and accelerate rapidly, or if it stalls.

h. Multispeed induction motors. Squirrel-cage multispeed induction motors are available for those applications requiring more than one speed. However, there are practical and design limitations as to both the number of speeds and the speed ratios. Single-frame construction is used almost exclusively because the induced-current characteristic of the induction-motor rotor permits use of the rotor at more than one speed. Two-frame construction can rarely be justified economically. Constant-horsepower, constant-torque (rated horsepower varies with the speed ratio), or variable-torque (rated horsepower varies with the square of the speed ratio) characteristics are available. Multispeed motors are made with either one or two windings (Ref. 18). One-winding motors can be connected to give speed ratios of only 2:1. Two-winding motors have two separate windings which can be wound for any number of poles so that other speed ratios in addition to 2:1 can be obtained. Three speeds can be obtained if one of the windings is arranged for reconnection for two speeds, and four speeds can be obtained if both windings are reconnectable for two speeds. However, the size of the motor, the voltage rating, the type of coil used (random or form-wound), the type of winding used (four coil sides per slot or two coil sides interlaced), the number of speeds, and the speed ratios will limit the choice as to what is practical. The higher the speed ratio together with the number of speeds provided, the more difficult it becomes to find suitable stator and rotor slot combinations and stator winding designs.

Wound-rotor multispeed motors are rarely used because the rotor windings must be arranged for reconnection in the same manner as the stator windings and additional slip rings are required. Their use is limited economically. The speed ratio is limited to 2:1 whenever the secondary resistance is varied to obtain adjustable-speed operation at both speeds. If adjustable speed is needed on one speed only, two, three, or four speeds are available. All speeds except the one to be adjusted operate as a squirrel-cage motor for this arrangement.

i. Brush-shifting motors (Ref. 5) are used occasionally. Speed control of series brush-shifting motors is accomplished by shifting brushes instead of by varying the impressed voltage. The no-load speed is limited to approximately 150 percent of synchronous speed. Speed regulation is better than that of other forms of series motors. The shunt brush-shifting motor (Schräge motor) gives adjustable constant speed with good torque and efficiency. Shifting of brushes changes the speed. Sizes up to several hundred horsepower are available. The motor provides constant-torque characteristics. The Fynn-Weichsel motor utilizes fixed brushes instead of movable brushes as on the Schräge motor. This motor provides characteristics similar to the ordinary wound-rotor motor.

j. Time constants of induction motors (Ref. 22). The transients produced by induction motors under open- and short-circuit conditions require attention, especially when larger horsepower ratings are used with high-speed switchgear.

1. *The open-circuit time constant* T_0 is defined as the time that is required for the voltage to drop to $1/\epsilon$ (0.368) of the air-gap voltage upon deenergization of the motor with constant shaft speed and no electrical load. It may be calculated by

$$T_0 = X_m + \frac{X_2}{\omega R_{21}} \quad \text{seconds} \tag{10}$$

where X_m = magnetizing reactance, ohms per phase
X_2 = rotor reactance, ohms per phase referred to the primary
R_{21} = rotor resistance, ohms per phase referred to the primary
$\omega = 2\pi f$

If the motor is reenergized before the time T_0 has elapsed, very large transient currents and torques can result. It is recommended that the residual voltage be allowed to drop to 0.25 of rated voltage before the motor is reenergized. The addition of capacitors will extend the open-circuit time constant, which will be designated T_0', as follows:

$$T_0' = \frac{X_2 + X_m(X_1 - X_c)/(X_m + X_1 - X_c)}{\omega R_{21}} \quad \text{seconds} \tag{11}$$

where X_1 = stator reactance, ohms per phase
X_c = capacitive reactance, ohms per phase

2. *The short-circuit time constant* is defined as the time that is required for the current to drop to $1/\epsilon$ (0.368) of the line current which exists prior to a fault. The short-circuit transient current is composed of an ac component and a dc component. The ac time constant T_{ac} is found by

$$T_{ac} = \frac{X_L}{\omega R_{21}} \quad \text{seconds} \tag{12}$$

and the dc time constant T_{dc} is found by

$$T_{dc} = \frac{X_L}{\omega R_1} \quad \text{seconds} \tag{13}$$

where X_L = locked-rotor reactance, ohms per phase
R_1 = stator resistance, ohms per phase

The motor manufacturer should be consulted for the time-constant values for a specific design.

INDUCTION-MOTOR CONTROL

The various types of induction motors will provide the torque characteristics necessary to drive a particular load with definite performance characteristics. Control equipment will provide means for accelerating the motor, changing speed, reversing, and stop. NEMA Standard ICS 1 (Ref. 4) contains general information and specific standards on control equipment.

43. Starters. Starters are used to provide the basic functions of start, accelerate, and stop. Full-voltage full-winding start is the most popular method used on motors of almost any size. However, the driven equipment must be able to withstand the starting-torque shock, and the locked-rotor power requirements (kVA) must not exceed power-supply limits or voltage-dip limits. For such cases, other full-voltage methods or one of the reduced-voltage methods may be chosen, provided acceleration can be obtained without overheating the motor windings. Table 11 indicates the reduction in locked-rotor power and motor torque obtained by the various methods of starting.

The energy absorbed in the rotor windings during acceleration W_a is given by the expression

$$W_a = 0.231\,Wk^2 \left(\frac{N}{1000}\right)^2 \frac{T_M}{T} \quad \text{kWs} \tag{14}$$

where Wk^2 = total motor and load inertia being accelerated, lb·ft²
N = speed attained after acceleration, r/min
T_M = average motor torque over speed range, lb·ft
T = average net accelerating torque, lb·ft

The energy absorbed in the stator windings during acceleration W_s is given by the expression

$$W_s = W_a \frac{R_1}{R_{21}} \tag{15}$$

where R_1/R_{21} is the ratio of stator and rotor winding resistances. The ratio of T_M/T increases the losses in the windings very fast if the motor voltage is reduced to the point where very little torque is left to accelerate the load.

TABLE 11. Starting kVA and Motor Torques for Different Starting Methods Used on ac Motors

Method of starting	% rated volts at motor terminals	% of full-voltage kVA	% of full-voltage torque
Full-voltage:			
Full-winding	100	100	100
Part winding	100	60–80*	40–55*
Wye-delta start	57	33	33
Reduced-voltage:			
Autotransformer	80, 65, or 50	64, 42, or 25	64, 42, or 25
Series reactor or resistor	80 or 65	80 or 65	64 or 42
Rheostatic starter or controller†	100	100 or 150	100

*Part-winding start values vary depending upon winding-connection arrangements and number of poles. The motor manufacturer should be consulted as to acceleration capabilities on part winding, especially when used on induction motors. Adverse harmonic effects caused by omission of coils may prevent motor from starting or accelerating to full speed, thereby causing rapid and severe overheating in the energized portion of the stator windings. Transfer to full winding within 3 to 5 s is recommended for such cases.

†Used on wound-rotor induction motors only. The starting resistor limits the current to values shown to obtain 100 percent rated full-load torque at start (see Fig. 35).

44. Controllers. Controllers are used to provide additional functions of speed changing, reduction, braking, and reversing.

a. Speed changing of multispeed motors is accomplished by making the necessary changes in motor-lead connections (Refs. 15, 17, 18). These changes are made in the controller by appropriate contactor arrangements. Connections for these motors are given in Sec. 6.

b. Speed reduction of the wound-rotor induction motor is provided by the secondary (circuit) controller, which inserts the appropriate external resistance into the rotor winding circuits. See Fig. 35 for typical speed-torque curves of this arrangement.

c. Braking of induction motors may be obtained by one of the methods listed in Table 6.

d. Reversing of induction motors is accomplished by a controller that will make the necessary change of motor leads to the line. If the motor is plug-reversed, the losses in the windings will be four times the acceleration losses of one start.

e. Plug stop is accomplished by a controller that will make the necessary change of motor leads to the line and will disconnect the motor from the line when zero speed is reached. The losses in the windings for a plug stop will be three times the acceleration losses of one start.

45. Protection. The control equipment will include devices for proper protection of the motor and personnel. Table 12 indicates that almost any type of protection is available for induction motors. The protection provided by present insulation systems has reduced the need for extensive protection usually afforded by enclosures.

SYNCHRONOUS MOTORS

Synchronous motors (Refs. 5, 16) are available in sizes from subfractional to many thousand horsepower and are used to convert electrical energy into mechanical energy. They are suited for constant-speed drives, reversed-speed and inching service in some cases, and multispeed drives. They can provide reactive or leading kilovolt-amperes to the power source, thereby providing power-factor correction. The many types available cover a wide range of applications and are designed to operate from many different power supplies with many different combinations and values of phases, frequencies, and voltages.

46. Basic Equations. Table 3 gives the basic equations for synchronous-machine operation. The generalized equations apply to the many different types of synchronous machines.

47. Construction. The main elements of the synchronous motor are the stator (sometimes called primary or armature) and the rotor (sometimes called field or secondary). The stator generally consists of a frame which serves as an enclosure and as a support for the magnetic laminated core. The core forms a part of the magnetic-flux path and supports the primary windings. The rotor is the rotating element of the motor and includes the shaft and rotor magnetic-core structure to complete the magnetic-flux path. Rotors of large motors are usually built with

TABLE 12. Protection of Motor Parts

General type of protection	Types of motors		
	DC	AC induction	AC synchronous
Enclosures, protection by	Yes	Yes	Yes
Windings and insulation (protection from overheating):			
Current-responsive	Yes	Yes	Yes
Temperature-responsive	Yes	Yes	Yes
Thermal protectors	Yes	Yes	Yes
Humidity, protection from	Yes	Yes	Yes
Mechanical parts:			
Bearings	Yes	Yes	Yes
Lubrication	Yes	Yes	Yes
Shaft current	Yes	Yes	Yes
Overspeed	Yes	Yes	Yes
Auxiliary protective devices:			
Overload	Yes	Yes	Yes
Phase-sequence, phase-reversal	N.A.	Yes	Yes
Undervoltage, overvoltage	Yes	Yes	Yes
Frequency	N.A.	Yes	Yes
Open-phase	N.A.	Yes	Yes
Differential	N.A.	Yes	Yes
Current-limiting power fuses	.....	Yes	Yes
Surge arresters and capacitors	.....	Yes	Yes
Field failure	Yes	N.A.	Yes
Field protective	Yes	N.A.	Yes
Field discharge	Yes	N.A.	Yes
Pull-out	N.A.	N.A.	Yes
Amortisseur winding	N.A.	N.A.	Yes
Incomplete sequence	Yes	Yes	Yes

N.A. indicates the protection is not applicable to the particular type of motor.

field windings and amortisseur windings, but some very small motors are built without windings of any kind. End brackets added to the stator frame complete the motor enclosure and contain the bearings which position the rotor within the stator. The bearings of some large machines may be mounted in pedestals which are mounted on a common bedplate with the stator.

a. The stator frame may be cast iron or steel, or fabricated from steel plates. The laminated core stack consists of laminations or segments made from electrical sheet steel having high permeability with low hysteresis and eddy-current losses. Insulated windings are placed in slots near the air gap in the stator-core stack. The windings are designed and connected in accordance with the number of phases, the power-supply frequency and voltage, and the desired synchronous speed. The primary or stator currents produce alternate north and south flux poles, thereby setting up a revolving magnetic field.

b. The rotor is composed of a laminated core assembly mounted on a shaft. The appropriate windings, when used, may be placed in slots in the core near the air gap or be wirewound around salient-pole assemblies. Alternate north and south magnetic poles are formed by magnetic reluctance or by current flow in rotor windings. The same number of poles exists in the rotor as in the stator. The poles of the rotor lock "in step" with opposite-polarity poles of the revolving stator field, so that the rotor revolves at synchronous speed under all operative loads.

48. Nonexcited Motors. Nonexcited synchronous motors do not use rotor windings for separate excitation purposes. They are arranged to be self-starting by means common to single-phase induction motors, such as shaded-pole, split-phase, capacitor-start, and repulsion-start. They are built in sizes ranging from subfractional to approximately 100 hp. Efficiency, power factor, and torque are lower than for separately excited types.

a. Reluctance motors are the most commonly used of the nonexcited types. The smaller subfractional sizes use notched rotors to form the poles. They operate on the magnetic-reluctance principle. Larger motors use modified squirrel-cage rotors. Portions of the rotor magnetic steel are removed near the air gap to give a salient-pole effect. Both the bars and the steel are removed from a die-cast rotor, whereas the bars are sometimes left in the brazed-type rotor. Another modification of the die-cast rotor uses nonmagnetic barriers to channel the magnetic flux between poles below the slots. Cast aluminum is also used in the spaces between the salient-pole steel. This type, called synchronous induction, gives better performance than the other two reluctance types. Reluctance-type motors are built in sizes from subfractional to about 30 hp.

b. Hysteresis motors use solid metal rotors which may or may not be slotted or notched. They require no rotor windings. Poles are formed randomly on the rotor surface by hysteresis effect caused by eddy-current flow as the rotor locks in step. Hysteresis motors are suitable for applications requiring a precise constant-speed motor. They are built in subfractional-horsepower sizes.

c. Permanent-magnet motors use rotors in which permanent magnets are placed to increase the polar effect. These motors are used where operation at synchronous speed is absolutely necessary. They are built in sizes from 5 to 100 hp. Permanent-magnet motors are now used in smaller sizes in conjunction with solid-state control to obtain variable speed (Sec. 10, Part 9).

49. Dc-Excited Motors. Dc-excited motors are usually made with wirewound field windings mounted on salient-pole assemblies. As the motor is loaded, the rotor magnetic field will lag that of the stator field. The value of synchronous torque developed at synchronous speed depends upon the angle of lag, which is called the rotor-displacement angle, as shown in Fig. 36. This torque is the sum of the magnetic-reluctance torque and the torque which is developed by excitation or current in the field windings. The displacement angle is near zero at no load, approximately 30 to 35° at full load on unity-power-factor motors, and approximately 17 to 32° at full load on 0.8-leading-power-factor motors. Maximum torque occurs at approximately 70°. The motor will pull out of step and its operation will be unstable between 70 and 180° displacement angle.

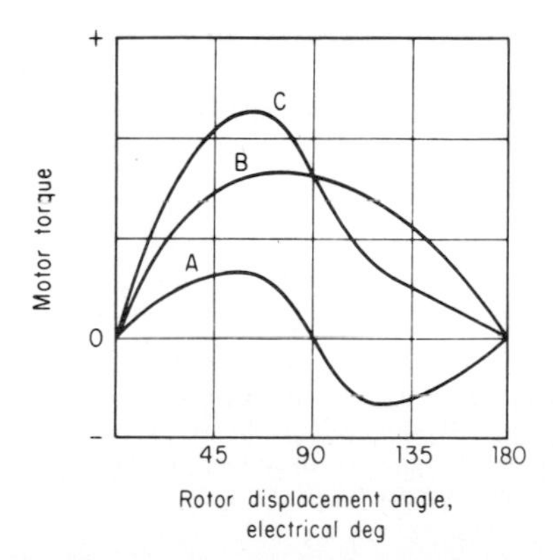

FIG. 36. Torque of dc-excited synchronous motors. *A*—reluctance torque; *B*—torque of excited salient pole; *C*—resultant or synchronous torque.

a. The synchronizing power is defined by NEMA (Refs. 2, 3) as the shaft power which corresponds to the torque developed between armature and field at the air gap when the motor is running at synchronous speed. The synchronizing coefficient P_r is expressed in kilowatts per electrical radian and is given by the expression

$$P_r = \frac{\text{kW}}{d} \tag{16}$$

where kW is the power which appears at the motor shaft and d is the rotor-displacement angle in radians at rated load, voltage, frequency, and power factor. The synchronizing coefficient indicates the "stiffness" of the magnetic coupling between stator and rotor. Its value is approximately 1.35 × hp for unity-power-factor motors and 1.8 × hp for 0.8-leading-power-factor motors.

b. Synchronization occurs whenever the rotor pulls into step with the stator magnetic field. The rotor-field polarity must be correct relative to the stator field, and the slip must be small enough to permit synchronization. A polarized field-frequency relay is used to apply excitation when polarities are favorable. The maximum slip which will allow synchronization depends upon the load and its inertia as shown by the expression

$$S = \frac{(\text{hp}\ K_5/Wk^2)^{1/2}}{N_s} \tag{17}$$

where S = slip at pull-in, percent of synchronous speed
N_s = synchronous speed of the motor
Wk^2 = sum of motor and load inertias
K_5 = a constant which takes efficiency, power factor, etc., into account

Figure 37 shows the effect of the inertia ratio upon the maximum slip from which the motor can pull into synchronism. The inertia ratio is the ratio of total load inertia to normal-load inertia and is used as the abscissa. Normal-load-inertia (Wk^2) values are established by NEMA (Ref. 3) using the formula

$$Wk^2 = \frac{3.75 \times 10^5 \times \text{hp}^{1.15}}{N_s^2} \tag{18}$$

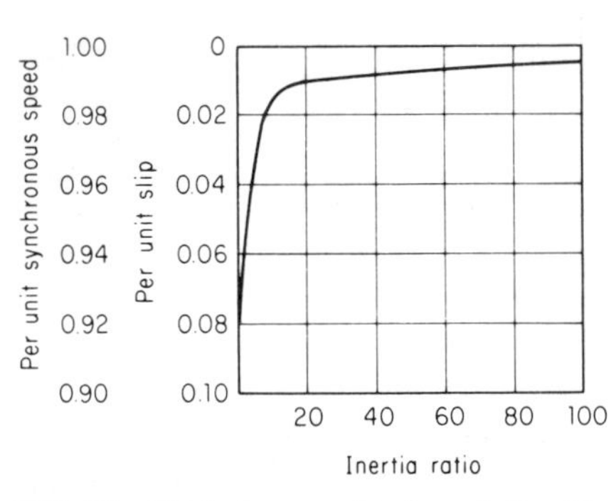

FIG. 37. Typical curve showing relationship between inertia ratio (load inertia/normal-load inertia) and maximum slip for pulling into step of synchronous motors.

Figure 38 indicates the relationship between normal-load-inertia values and rated horsepower at different speed ratings. The values of inertia ratio may vary from 1 for a centrifugal-pump drive to 100 for some chipper drives.

c. The pull-out torque of the synchronous motor is defined by NEMA as the maximum sustained torque the motor will develop at synchronous speed for 1 min when operating at rated voltage, frequency, and field excitation. The normal value of pull-out torque is about 150 percent of full-load torque for unity-power-factor motors and 200 to 225 percent for leading-power-factor motors. A sudden increase in load causes an

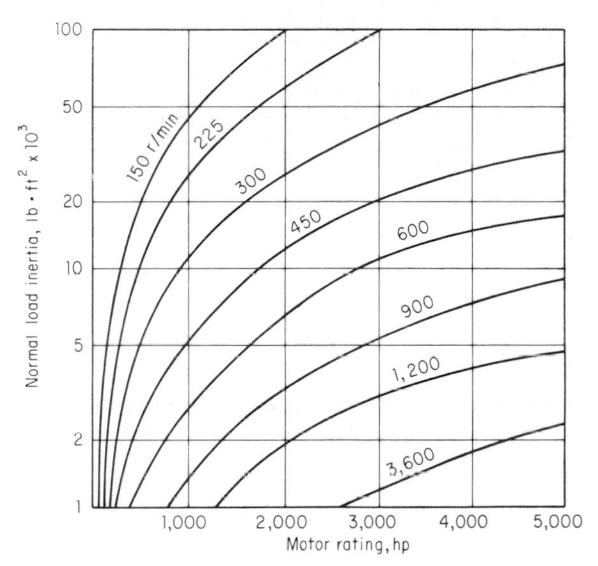

FIG. 38. Relationship of rated horsepower, speed, and normal-load inertia for dc-excited synchronous motors.

increase in the rotor-displacement angle and in the line current. This results in a transient increase in excitation and pull-out torque because of transformer action between the motor winding circuits. The increased values last only a few cycles but can be used to initiate an increase of excitation to help provide a higher maximum torque before the motor pulls out of synchronism.

d. The starting and accelerating torque of the dc-excited synchronous motor must be obtained by separate starting means because the torque developed in the pole faces by hysteresis and eddy currents and by magnetic reluctance is of little significant value for large-industrial-motor use. Normally the field winding is short-circuited through the field-discharge resistor, and an amortisseur winding is provided in slots in the pole faces. The torque typically developed by the field winding circuit is shown in Fig. 39. Note the dip in torque caused by the negative rotating-torque component. Increasing the value of the field-discharge resistance will reduce the "dip torque" effect and at the same time significantly affect the pull-in and locked-rotor torque, as shown in Fig. 40. The amortisseur winding develops most of the torque which is available for starting and acceleration. This winding is very similar to the brazed cage winding of induction motors. The number of bars varies in size, number, and location, and there may be common or separate end rings to form single or double cages as desired. Torque is developed by induction as in the induction motor. Figure 41 indicates the typical speed-torque-current curves developed by synchronous motors. In general, the torque per ampere cannot be varied as much as that of induction motors, but on the other hand, the starting current and torque are largely independent of the machine reactance and maximum torque.

e. Pulsations of armature current and power can become severe if the inertia of the rotating parts is too low. At synchronous speed, the rotor-displacement angle changes as load is applied or removed. Any pulsating load requiring variable torque during each revolution can cause undesirable pulsations. The rotor will

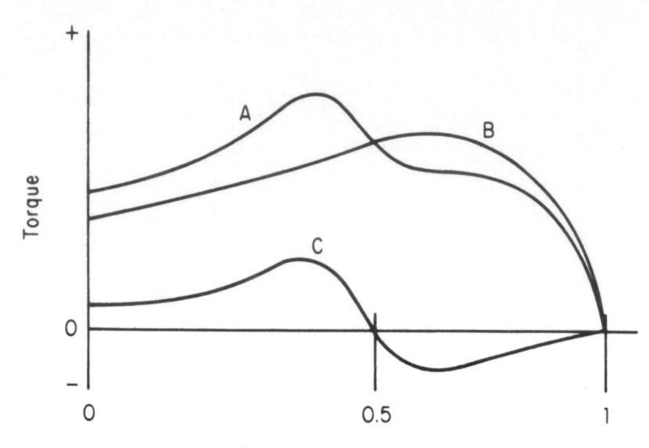

FIG. 39. Illustration of resultant torque *A*, showing positive rotating component *B* and negative rotating component *C*.

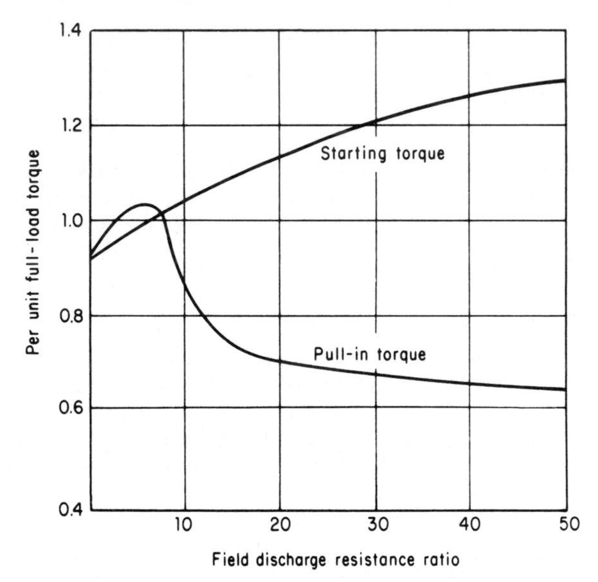

FIG. 40. Illustration of effect of field-discharge resistance ratio (resistance of field-discharge resistor to resistance of field winding) on starting and pull-in torques of dc-excited synchronous motors.

oscillate about the average displacement angle. Normally inertia is added to the rotor to limit current pulsations to a value not exceeding 66 percent of full-load current.

f. Natural frequency oscillations will occur if there is resonance between the frequency of load-torque impulses and the natural frequency of the motor on the power system. This condition will cause the percentage pulsation of current and power to be higher than the variations in torque impulses causing the pulsation.

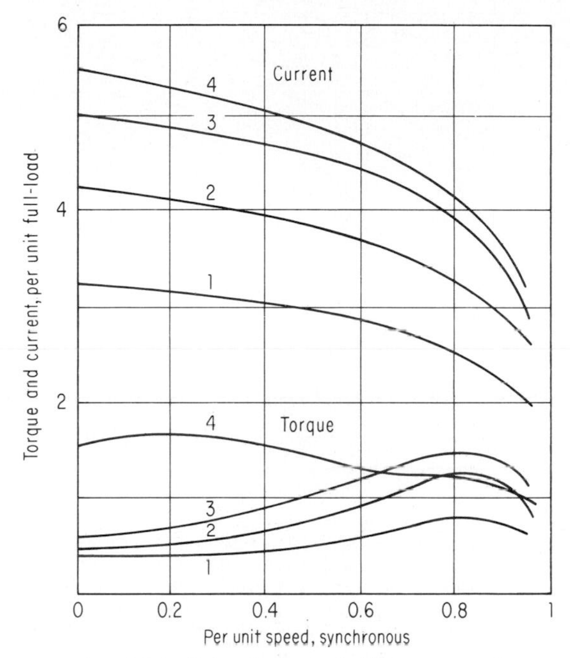

FIG. 41. Illustration of effect of different amortisseur winding constructions on torque and torque-current ratios of dc-excited synchronous motors. Effective resistance and reactance of the amortisseur winding are main factors that determine shape of curves and ratio of torque per ampere.

The undamped natural frequency (Refs. 3, 5) of a synchronous motor operating on a large (infinite) power source is given by the expression

$$\text{Natural frequency} = \frac{(P_r f / Wk^2)^{1/2} \times 35{,}200}{N_s} \tag{19}$$

The natural frequency should differ by 20 percent from the load-torque pulsations, varying with the number of torque pulsations per revolution.

g. The amount of inertia, or flywheel, effect (Ref. 5) needed to limit power pulsations can be calculated from the expression

$$Wk^2 = \frac{1.34 P_r f C}{(N_s/100)^4} \tag{20}$$

The term C (formerly X) is called compressor factor (Refs. 3, 5), and its value may be found from the expression

$$C = \frac{0.746\, Wk^2 N_s^4}{P_r f \times 10^8} = 9.25 \left(\frac{N_s}{f}\right)^2 \tag{21}$$

h. The efficiencies for synchronous motors operating at unity power factor are shown in Figs. 42 and 43. The efficiencies for motors operating at 0.8 power factor

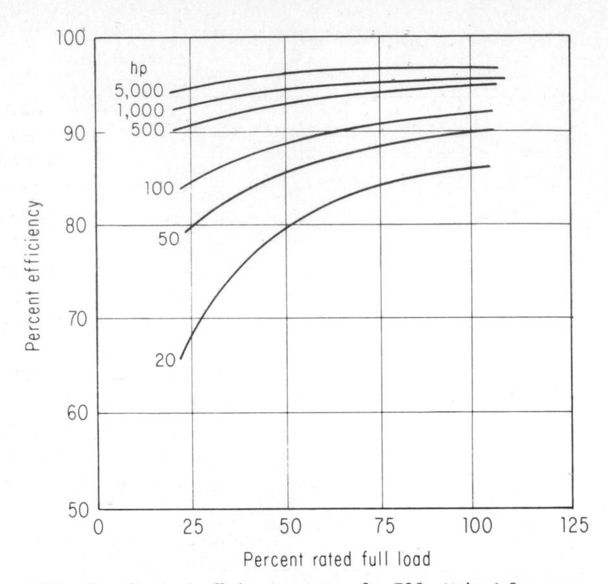

FIG. 42. Typical efficiency curves for 720-r/min 1.0-power-factor general-purpose dc-excited synchronous motors. Values may vary for a specific design. Deductions for high voltages are less than shown in Table 9. Full-load efficiencies for other speeds may be taken from Fig. 43 and partial-load values extrapolated from curves above.

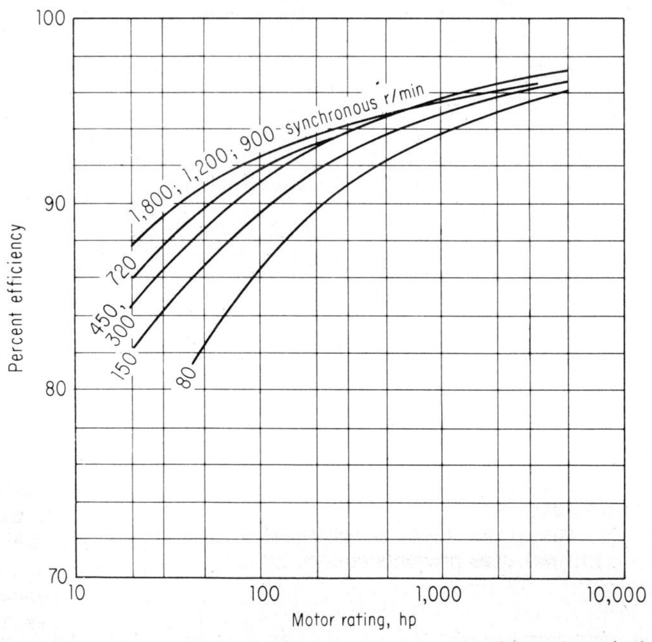

FIG. 43. Typical full-load efficiencies for 1.0-power-factor dc-excited synchronous motors. Partial full-load values may be estimated as noted in Fig. 42.

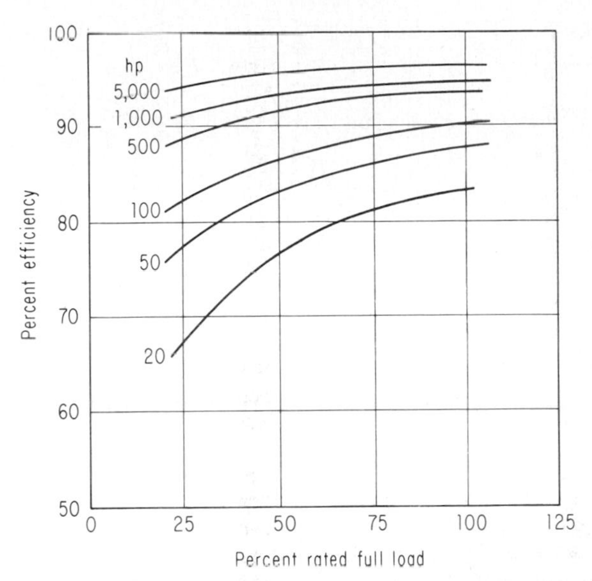

FIG. 44. Typical efficiency curves for 720-r/min 0.8-leading-power-factor general-purpose dc-excited synchronous motors. Values may vary for a specific design. Deductions for high voltages are less than shown in Table 9. Full-load efficiencies for other speeds may be taken from Fig. 45 and partial-load values extrapolated from curves above.

leading are shown in Figs. 44 and 45. Values for specific designs will vary somewhat from the values shown. Deductions are made for high-voltage motors, as shown in Table 8, because of high primary copper losses. The effect of voltage and frequency variation is similar to that shown in Table 9 for induction motors, except that the variations in slip and power factor do not apply.

i. The power factor of separately excited synchronous motors can be controlled within the motor design and load limits by varying the field excitation. Excitation may be furnished from any dc source such as a bus; a direct-connected, belted, or separately driven exciter; a motor dc generator set; or some form of silicon-rectifier unit such as a brushless exciter.

When operating at unity power factor, the excitation is adjusted to that value which is sufficient to just meet the motor requirements. Reactive current does not flow between the motor and the power system. The operating efficiency is high because the stator winding loss is minimum. However, motors having unusually high starting, pull-in, or pull-out torques have lower efficiencies when operating at unity power factor.

Increasing the field-excitation value (overexcitation) will cause the motor to operate with a leading power factor or to carry an overload at the rated power factor. The current drawn from the line leads the voltage, and the motor furnishes reactive current to the system. The motor efficiency is less and the pull-out torque is higher than for unity-powr-factor motors. Synchronous motors designed for 0.8 leading power factor are standard. They will deliver from 0.4 to 0.6 leading magnetizing kilovolt-amperes per horsepower depending upon the load carried. For this reason, 0.8-leading-power-factor synchronous motors are used extensively to

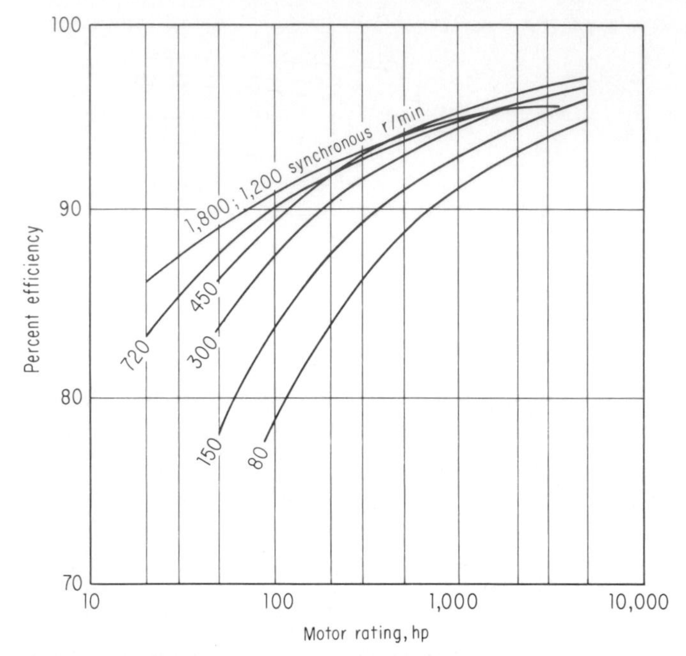

FIG. 45. Typical full-load efficiencies for 0.8-leading-power-factor dc-excited synchronous motors. Partial full-load values may be estimated as noted in Fig. 44.

offset the lagging magnetizing kilovolt-amperes drawn from the power system by induction motors, transformers, and other inductive apparatus. They have lower costs and higher efficiencies than induction motors, especially in the lower-speed range.

Decreasing the excitation (underexcitation) causes the motor to operate at a more lagging power factor. The current drawn lags the voltage, and the motor draws a more reactive current from the system. Synchronous motors are rarely designed for operation at lagging power factor.

The minimum current drawn from the line for each load condition occurs at unity power factor, as shown in Fig. 46. This occurs at some given value of excitation. For any load and at some given value of excitation, the power factor is the ratio of the minimum line current (at that load) to the line current at the given excitation value. For example, in Fig. 46 the power factor for one-half load at 95 percent of full-load 0.8-leading-power-factor excitation is approximately $50/75$, or 67 percent leading.

The reactive kilovolt-amperes (kvar) drawn by a synchronous motor may be calculated from a known load and the operating power factor by the expression

$$\text{kvar} = \frac{0.746 \times \text{hp}\,(1 - \text{PF}^2)^{1/2}}{\text{EFF} \times \text{PF}} \tag{22}$$

where EFF and PF are in per unit terms. The leading kvars available at part loads may be estimated from Fig. 47. The approximate leading kvar in percent of rated horsepower for various loads and various full-load power-factor values may be

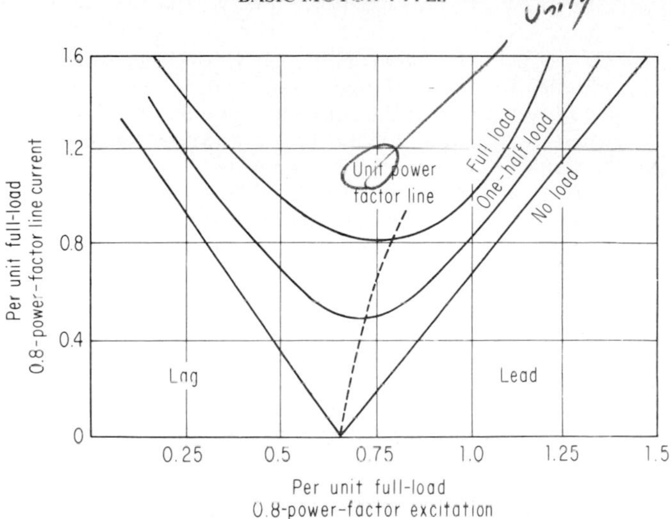

FIG. 46. Relationship of motor line current and excitation for full load, one-half load, and no load. Overexcitation causes motor to operate with leading power factor (to right of unity-power-factor line). Underexcitation causes motor to operate with lagging power factor (to left of unity-power-factor line).

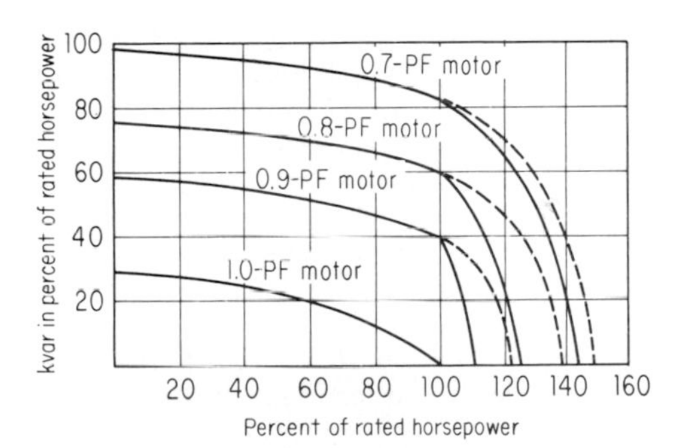

FIG. 47. Leading kvars of dc-excited synchronous motor operating at various loads and power-factor conditions. Solid curves are for reduced excitation at overload to maintain rated full-load motor current. Broken curves are for rated excitation to maintain rated pull-out torque.

determined from the curves. For instance, an 0.8-power-factor motor operating at 60 percent of full load will deliver approximately 70 percent of its horsepower rating (0.70 × hp) in leading kvars to the system. The curves assume the motor is operating at full-rated excitation from zero to 100 percent load, whereas above 100 percent load the excitation is reduced to maintain full-rated stator current and prevent overheating. However, reduced excitation will reduce the pull-out torque. Therefore full excitation may be required for peak overloads. The broken lines indicate the kvars for 100 percent excitation.

j. Multispeed synchronous motors are usually supplied as two-speed two-frame units. The two frames are mounted on a common bedplate. A common rotor for two speeds, such as is possible on the squirrel-cage induction motor, is not easily obtained because it is somewhat difficult to reconnect the salient poles for speed changes and the resulting efficiencies are lowered.

k. Time constants of synchronous motors The open- and short-circuit time constants of dc-excited synchronous motors depend upon the motor design and the specific conditions at the installation. The armature-winding time constant is usually one cycle or less in duration. The field-winding time constant depends upon whether the field-discharge resistor is involved. If so, the added resistance will materially reduce the value of the time constant. It is recommended that the motor manufacturer be consulted for the time constants for a specific motor design and application.

SYNCHRONOUS-MOTOR CONTROL

The various types of synchronous motors will provide torque characteristics necessary to drive a particular load. The nonexcited types require very little or no control equipment except some form of switch to connect or disconnect the motor and its power system.

50. Starters. Starters are used on dc-excited-type synchronous motors to provide the basic functions of starting, accelerating, stopping, and excitation adjustment. NEMA standards contain general information and specific standards on control equipment. This equipment will provide application of the dc excitation at the proper time for synchronization, remove the excitation should pull-out of synchronism occur, and provide for proper sequence of operations. See Sec. 2, Art. 14. Full-voltage full-winding start is the most popular method for most sizes, but full-voltage part-winding start is frequently used. Table 11 indicates the reduction in locked-rotor power and motor torque obtained by the various methods of starting. The heating caused by energy absorption in the windings of the synchronous motor is the same as that given for induction motors in Eqs. (14) and (15).

51. Controllers. Controllers are used to provide the additional functions of normal operation, speed changing, inching, reversing, and braking.

a. Speed changing of multispeed motors is accomplished by disconnecting one motor from the line and connecting the other to the line.

b. Inching, or spotting, of a synchronous motor and its driven unit can be accomplished by a controller that uses commutated direct current to provide slow-motion high-torque rotation of the rotor. The operator can control the rate and direction of inching at will.

c. Reversing of synchronous motors is accomplished by a controller that will make the necessary changes of stator (armature) and rotor (field) leads to their respective power sources.

d. Braking of synchronous motors may be accomplished by one of the methods shown in Table 6. A plug stop will produce losses equal to three times those produced during one start.

e. Protection. The control equipment will include devices for proper protection of the motor and personnel. Table 12 indicates that almost any type of protection is available. The protection provided by present insulation systems has reduced the need for extensive protection heretofore afforded by enclosures.

REFERENCES

General

1. Bernard Atkins, *The General Theory of Electrical Machines* (Chapman & Hall, London, 1957, reprinted 1959).
2. "Standard Dictionary of Electrical and Electronics Terms," ANSI/IEEE Standard 100-1984.
3. "Motors and Generators," ANSI/NEMA Standard MG 1-1978.
4. "Industrial Control and Systems," ANSI/NEMA Standard ICS 1-1978.
5. Donald G. Fink and H. Wayne Beaty, Eds., *Standard Handbook for Electrical Engineers,* 11th ed. (McGraw-Hill, New York, 1978).
6. "National Electrical Code," ANSI/NFPA No. 70-1984.
7. John C. Ponstingl, "Electrical Motor Braking," *Machine Design* (Jan. 12, 1956).
8. Jon Campbell, "Motor Protection," *Machine Design* (Aug. 13, 1964).

Dc Motors

9. Scott Hancock, *Design of Direct-Current Machines* (International Textbook Co., Scranton, Pa., 1934).
10. R. G. Kloeffler, J. L. Brenneman, and R. M. Kerchner, *Direct-Current Machinery* (Macmillan, New York, 1934).
11. M. Liwschitz-Garik, *Electric Machinery,* vol. 1: "Fundamentals and DC Machines" (Van Nostrand, Princeton, N.J., reprinted 1947).
12. "Guide on Test Procedures for DC Machines," IEEE Standard 113-1985.
13. P. Hunter-Brown, *Carbon Brushes and Electrical Machines,* 3d ed. (Morgan Crucible Co., Battersea Works, London, 1952).
14. W. C. Kalb, Ed., *Carbon Graphite, and Metal-Graphite Brushes* (National Carbon Co., St. Marys, Pa., 1946).

Ac Motors

15. P. L. Alger, *The Nature of Polyphase Induction Machines* (Wiley, New York, 1951).
16. M. Liwschitz-Garik, *Electric Machinery,* vol. 2: "AC Machines" (Van Nostrand, Princeton, N.J., reprinted 1947).
17. M. Liwschitz-Garik, *Winding Alternating-Current Machinery* (Van Nostrand, Princeton, N.J., 1950).
18. A. M. Dudley, *Connecting Induction Motors,* 3d ed. (McGraw-Hill, New York, 1936).
19. "Test Procedure for Polyphase Induction Motors and Generators," ANSI/IEEE Standard 112–1983.
20. "Test Procedure for Single-Phase Induction Motors," ANSI/IEEE Standard 114-1982.

21. R. F. Horrell, *Slip-Ring and Brush Application—Wound-Rotor Induction Machines* (Helwig Carbon Products Co., Milwaukee, Wis., 1962).

22. *Electrical Transmission and Distribution Reference Book,* 4th ed. (Westinghouse Electric Corp., East Pittsburgh, Pa., 1950).

23. "Test Procedures for Synchronous Machines," ANSI/IEEE Standard 115-1983.

Control of Electric Motors

24. Robert W. Smeaton, Ed., *Switchgear and Control Handbook,* 2d ed. (McGraw-Hill, New York, 1986).

25. Donald V. Richardson, *Handbook of Rotating Electric Machinery* (Reston Publishing, Reston, Va., 1980).

26. Balbir Singh, *Electric Machinery Design* (Advent Books, New York, 1981).

SECTION 4

LARGE POLYPHASE INDUCTION MOTORS (Over 680 Frame Size)

Robert C. Moore*

FOREWORD 4-2

GENERAL CHARACTERISTICS 4-2

1. Induction Motors 4-2
2. Horsepower Ratings 4-3
3. Frequency of Power Supply 4-3
4. Number of Phases 4-3
5. Standard Voltages 4-3
6. Insulation 4-5
7. Temperature Rise 4-5
8. Synchronous Speed 4-5
9. Classification by Speed 4-7
10. Classification by Enclosure 4-8
11. Motor Characteristics 4-9
12. Motor Tests 4-11

APPLICATION 4-12

13. Motor-Application Data 4-12
14. Starting Methods 4-15
15. Starting of Large Polyphase Wound-Rotor Motors 4-20
16. Speed Variation 4-21
17. Thermal and Electrical Protection of Motors 4-22
18. Combined Characteristics 4-24
19. Emergency Switching 4-26
20. Capacitors 4-28
21. Fault-Current Contributions 4-30
22. Frequency-Change Effects 4-33
23. Special Polyphase Induction Motors 4-35

REFERENCES 4-35

*Senior Engineer, Motor-Generator Department, Allis-Chalmers Manufacturing Company, Milwaukee, Wis. (retired); Registered Professional Engineer (Wis.); Senior Member, IEEE; Distinguished Lecturer, Milwaukee School of Engineering.

FOREWORD

The large polyphase induction motors treated in this section are of the types shown in Fig. 1. Types rarely used in the United States, the commutator type, for example, are not listed or treated. Motor ratings above 200 hp, approximately, in either horizontal or vertical types, are considered in this section. Standardizing bodies have not assigned frame sizes or dimensions to these motor ratings. The motors do, however, follow standards relating to horsepower, voltage, frequency, Wk^2 capability, etc.

Extensive use of standards is made in preparing this section. Misapplication of motors can result in interpreting specifications not related to standards. Thus standards provide for mutual understanding of motor mechanical and electrical technical terminology, motor characteristics, test procedures, and general motor specifications. They also promote motor-application economies in providing practical industrial ranges of horsepower, voltage, speed, etc.

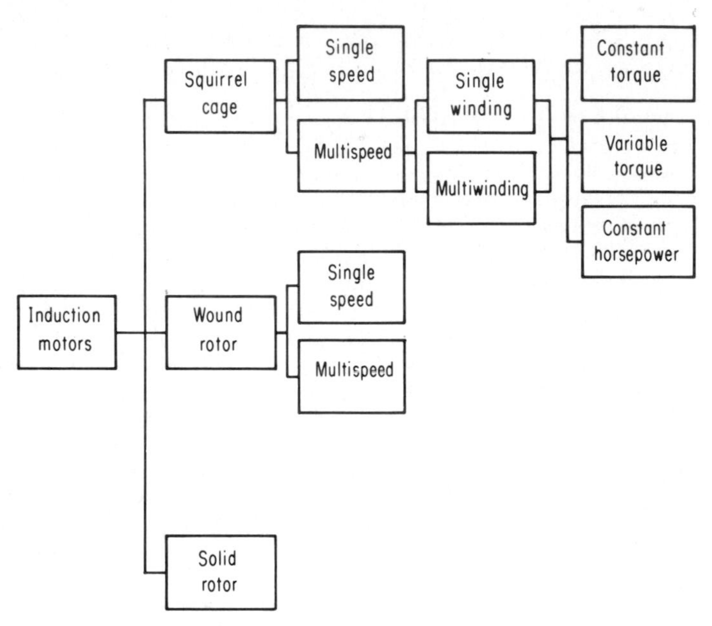

FIG. 1. Types of induction motors.

GENERAL CHARACTERISTICS

1. Induction Motors. Induction motors have two major mechanical members: one stationary and one rotating. The members are separated by a small air-gap clearance to avoid mechanical rubbing contact. On the stationary member a (primary) polyphase insulated winding connected to the main electric-power lines sets up a synchronously rotating magnetic field in the air gap. Such a rotating field induces currents in the rotor or secondary winding, provided the rotor runs more slowly than the synchronously rotating air-gap magnetic field. Thus torque is developed to turn the rotor and drive its connected load.

a. Squirrel-cage induction motor. The secondary, or rotor, winding is embedded in slots near the outer periphery of the rotor. The winding consists of (usually) uninsulated slot-embedded copper, copper alloy, or other suitable bar or rod materials. Extensions of the winding bars or rods beyond each end of the rotor laminated-core iron are connected together by short circuiting rings to provide current paths.

Torque, speed, thermal, etc., application requirements influence materials used in cage construction. When the cage-type rotor is built, the starting torque, current, and full-load speed are fixed.

b. Wound-rotor induction motor. The secondary, or rotor, winding has insulated coils disposed in slots near the outer periphery of the rotor. Coil connections produce a polyphase winding, normally three-phase. The three terminal, or lead, ends of the winding are connected to insulated rotor-mounted collector rings to provide, through brushes sliding on the rings, rotor-current conduction paths to external rotor circuits. External circuits may contain resistances or impedances. Adjusting the values of the external elements provides variation possibilities in motor torque and current at the start, or variation in motor speed during running operation under load.

c. Solid-rotor induction motor. The rotor of solid magnetic steel or steel shell provides the complete secondary winding. Induced-current penetration into the steel body is related to the strength of the inducing field and the inducing frequency. These phenomena, together with the low conductivity of steel, provide a motor having speed-torque characteristics similar to high-resistance rotor squirrel-cage induction motors.

Typical applications of polyphase induction motors are shown in Table 1.

2. Horsepower Ratings. The horsepower ratings standardized by NEMA (Ref. 1) are in increments as shown in Table 2. Listed ratings provide intervals to suit practical applications. Motor suppliers can provide nonstandard ratings. A price penalty may apply, however, when deviating from the standard listings.

3. Frequency of Power Supply. The power-supply frequency is standardized at 60 Hz and at 50 Hz (for export). In many foreign countries, 50-Hz generation is standard. In isolated locations, mostly foreign, frequencies other than those mentioned exist. Suppliers can provide induction motors for such applications; however, a price penalty may apply for odd- or non-standard-frequency design.

4. Number of Phases. Polyphase motors are wound three-phase almost without exception throughout the world.

5. Standard Voltages. Table 3 lists the standard voltages for large induction-motor stator windings. Motors of small hosepower ratings may use 115, 200, 230, 460, and 575 V. Larger motors may use higher voltages, such as 2300 or 4000 V, or even values such as 6600 or 13,200 V. These are general statements for normal applications, since as suggested in NEMA standards (Ref. 1), it is not practical to build motors of all horsepower ratings for all the standard voltages.

Low-horsepower motors for high-voltage or high-horsepower motors for low-voltage applications incur price penalties, if the motor can be designed at all. It may be more economical to purchase, for example, a 600-hp 13,200-V

TABLE 1. Applications of Polyphase Induction Motors

Application	Motor	Application	Motor
Air, gas compressors:		Condensate	1-F
Axial	1-F	Fire	1-F
Centrifugal	1-F	Pipeline	1-E,F
Reciprocating	1-E,G	Cooling tower	1-G
Rotary, vane	1-E	Sewage	2
Rotary (vacuum)	1-F	Axial flow	1-E
Coal:		Mixed flow	1-F;2
Mill, attrition	1-F,G	Hot well	1-F
Mill, bowl	1-F,G	Descaling	1-F,G
Conditioner	1-F,G	Reciprocating	1-F
Pulverizer (see mill)		Dredge	1-F,G
Crusher	1-F,G	Sand	1-F
Fans:		High lift	1-F;2
Forced draft	1-F,G	Rubber:	
Induced draft	1-E,F,G	Banbury	1-F,G;2;3
Gas recirculating	1-E,F,G	Extruder	1-F,G;3
Preheat	1-F,G	Mills	1-F,G
Sintering	1-G	Rolls	1-F,G;3
Mine	1-G	Strainer	1-F,G;3
Waste gas	1-F,G	Sugar:	
Exhauster (kiln)	1-F,G	Cane knife	1-F;2
Scrubber	1-F,G	Centrifugal	1-F;2
Cooling	1-F,G	Mill	2
Blower	1-F,G	Shredder	1-E
Purge	1-F,G	Wood:	
Primary air	3-G	Beater	1-D
Metals:		Chipper	1-F,G;2
Roughing rolls	2	Hydropulper	2
Piercing mill	2	Jordan	1-E,F
Drawbench	3	Refiner	1-E,F
Rolling mill	2	Shredder	1-E,F
Scalper	1-F	General:	
Pumps:		Ball, rod mill	1;2
Centrifugal	1-F	Crusher, stone	1-F,G
Charging	1-F	High-inertia drive starting	1;2;4
Boiler feed	1-F	Kiln	1-F,G
Circulating water	1-D,E	Mine hoist	2

Number designation: 1 = squirrel-cage, 2 = wound-rotor, 3 = multispeed, 4 = solid-rotor.
Letters indicate typical code letters for squirrel-cage motors. See Table 9.

TABLE 2. Standard Horsepower Ratings for Large Polyphase Induction Motors

hp range	hp increments
100–200	25
250–500	50
500–1,000	100
1,000–2,500	250
2,500–6,000	500
6,000–20,000	1,000
20,000–40,000	2,500
40,000–80,000	5,000
80,000–100,000	10,000

TABLE 3. Standard Voltages for Large Polyphase Induction Motors (Ref. 1)

Voltage*	hp
460 or 575	100–600
2,300	200–4,000
4,000 or 4,600	400–7,000
6,600	1,000–12,000
13,200	3,500–25,000

*60 Hz only.

1800-r/min induction motor with price penalty for high voltage than to provide a separate lower-voltage supply with required transformer, etc., for a nonpenalty low-voltage 600-hp motor. However, such applications should be investigated for price differentials.

6. Insulation. Insulation systems used in induction-motor windings are classified according to their thermal endurance for temperature-rating purposes (see Sec. 11). Four classes may be used in large polyphase induction motors: Classes A, B, F, and H. These classes correspond, in general, to IEEE insulating material classes 105, 130, 155, and 180, respectively. Hot-spot temperatures for Classes A, B, F, and H are 105, 130, 155, and 180°C, respectively. The temperature rise of induction motors with commonly used classes of insulation is given in Art. 7.

7. Temperature Rise. The temperature rise of continuously rated motors varies with the insulation classification and the type of mechanical enclosure. The temperature rise for a particular class of insulation is related to the thermal rating (Art. 6).

Advances in the development of insulating materials are relegating the once popular 40°C rise, Class A insulated induction motor to a minor role in application. Higher-temperature Class B insulated motors are being more widely applied, with resulting lighter and smaller frames. Advantages are smaller motor-space requirements and easier handling of lighter machines. Table 4 lists insulated-winding temperature rises for continuously rated polyphase induction motors of the different insulation types.

Motor operation at ambient temperatures and altitudes different from those listed in Table 4 may be encountered in some applications. Thus for higher ambients than 40°C and altitudes exceeding 3300 ft (1000 m) temperature-rise reductions should be made in accordance with standards recommendations (Ref. 1).

Large polyphase induction motors are occasionally encountered in special-service applications, such as short-time ratings and intermittent, periodic, or varying duty. A short-time-rated motor, for example, may be used for starting synchronous condensers, large vertical generators, etc. Temperature rises of the insulated windings for such special-service motors are generally higher than the values listed in Table 4. By agreement between manufacturer and user a temperature-rise value for a special-service motor may be assigned consistent with the class of insulation used, service conditions, and the expected life of the motor.

8. Synchronous Speed. For standardization purposes, synchronous-speed values of polyphase induction motors are recognized at 60 Hz (Ref. 1), from 2 to 32 poles, inclusive, and shown in Table 5. For any particular application, however, many other values of synchronous speed may be designed [see Eqs. (1) and (2)].

Synchronous speed, inflexible when related to the frequency-supply source, may be calculated from the following formula:

$$\text{Synchronous speed (r/min)} = \frac{7200}{P} \qquad \text{(60 Hz only)} \tag{1}$$

Equation (1) is arranged so that P is the number of poles for which the stator winding is wound. For example, setting $P = 4$ for a four-pole 60-Hz motor, the motor synchronous speed is, from Eq. (1), 1800 r/min. The equation is a special case, for 60 Hz, of a more general formula from which the synchronous speed

TABLE 4. Temperature Rise of Induction Motors (Ref. 1)

The observable temperature rise under rated load conditions of each of the various parts of the induction motor, above the temperature of the cooling air, shall not exceed the values given in the following table. The temperature of the cooling air* is the temperature of the external air as it enters the ventilating openings of the machine, and the temperature rises given in the table are based on a maximum temperature of 40°C for this external air. Temperatures shall be determined in accordance with the latest revision of the IEEE test procedure 112.

Machine part	Method of temperature determination	Temperature rise, °C Class of insulation			
		A	B	F	H
Insulated windings:					
All hp ratings	Resistance	60	80	105	125
1500 hp or less	Embedded detector†	70	90	115	140
Over 1500 hp:					
7000 V or less	Embedded detector†	65	85	110	135
Over 7000 V	Embedded detector†	60	80	105	125

Cores, cage, and mechanical parts such as collector rings, brushholders, and brushes may attain such temperatures as will not injure the machine in any respect.

60 Hz or 50 Hz (export).

*For totally enclosed water-air-cooled machines, the temperature of the cooling air is the temperature of the air leaving the coolers. On machines designed for cooling water temperatures up to 30°C, the temperature of the air leaving the coolers shall not exceed 40°C. For higher cooling-water temperatures, the temperature of the air leaving the coolers may exceed 40°C provided the temperature rises of the machine parts are limited to values less than those given in the table by the number of degrees that the temperature of the air leaving the coolers exceeds 40°C.

†Resistor or thermocouple in slot.

Note 1. Totally enclosed water-air-cooled machines are normally designed for the maximum cooling water temperature encountered at the location where each machine is to be installed.

Note 2. For motors which operate under prevailing barometric pressure and which are designed not to exceed the specified temperature rise at altitudes from 3300 ft (1000 m) to 13,000 ft (4000 m), the temperature rises, as checked by tests at low altitudes, shall be less than those listed in the foregoing table by 1 percent of the specified temperature rise for each 330 ft (100 m) of altitude in excess of 3300 ft (1000 m).

Note 3. Temperature rises in the foregoing table are based upon a reference ambient temperature of 40°C. However, it is recognized that induction motors may be required to operate in an ambient temperature higher than 40°C. For successful operation of the motors in ambient temperatures higher than 40°C, it is recommended that the temperature rises of the motors given in the foregoing table be reduced, as indicated below, for the ranges of ambient temperature given:

Ambient temperature, °C	*Values by which temperature rises in the foregoing table should be reduced*
Above 40 up to and including 50	10°C
Above 50 up to and including 60	20°C

TABLE 5. Synchronous-Speed Ratings, r/min, at 60 Hz (Ref. 1)

3,600	720	400	277
1,800	600	360	257
1,200	514	327	240
900	450	300	225

At 50 Hz speeds are five-sixths of the 60-Hz speeds.

Note. It is not practical to build motors of all horsepower ratings at all speeds.

may be calculated for any frequency f. Thus for any frequency f of the power-supply source,

$$\text{Synchronous speed (r/min)} = \frac{120f}{P} \tag{2}$$

Polyphase induction motors running unloaded operate very close to their designed synchronous speed. Under full load, cage-type motors operate at speeds from approximately ½ to 10 percent below synchronous speed. Actual full-load speed depends on motor design and drive-application requirements. At full load wound-rotor motors, with slip rings short-circuited, operate at speeds from approximately ½ to 3 percent below synchronous speed.

In induction motor design and application percent speed departure above or below synchronous speed is called slip (Ref. 2). Thus an induction motor operating at full load 2 percent below a synchronous speed of 1200 r/min has a full-load slip of 2 percent and an actual operating speed of 1200(1.00 − 0.02) = 1176 r/min. A formula commonly used to express slip S is

$$\text{Percent slip } S = \frac{\text{synchronous r/min} - \text{actual r/min}}{\text{synchronous r/min}} \times 100 \tag{3}$$

9. Classification by Speed. Large polyphase induction motors can be classified by speed as follows (Ref. 1):

1. Constant-speed motor
2. Adjustable varying-speed motor
3. Multispeed motor

a. Constant-speed motor. The speed of normal operation of a constant-speed motor is constant or practically constant.

The squirrel-cage motor is an example of such a motor. Squirrel-cage motors are widely applied to such drives as pumps of all types, fans, blowers, compressors, rubber mills, crushers, and pulverizers (Table 1). Speed variation of driven loads may be obtained if they are driven by cage-type motors through hydraulic or magnetic slip-type couplings.

b. Adjustable varying-speed motor. The speed of such a motor can be adjusted gradually. Once adjusted for a given load, it will vary to a considerable degree with a change in load.

The wound-rotor induction motor is an example of an adjustable varying-speed motor. In such a motor speed adjustment is usually made by placing resistance across the collector or slip rings, thus changing the total resistance of the rotor, or secondary, winding.

Wound-rotor induction motors may be used with certain types of pumps, mine hoists, mills (rolling, rubber crushers, coal etc.), and in other applications requiring speed reduction.

c. Multispeed motor. A multispeed motor can be operated at any of two or more definite speeds, each being practically independent of the load.

Either squirrel-cage or wound-rotor induction motors may be designed for multispeed operation. Squirrel-cage motors may have pole changing effected by simply reconnecting a single stator winding to produce two operating speeds. No rotor changes are required. The reconnection, according to the consequent-pole scheme, provides speed ratios of 2:1. Thus speeds of 1200/600 r/min, 900/450 r/min, or other combinations of the 2:1 ratio may be obtained. Stator-winding leads are brought out so that they may be conveniently reconnected for either speed. Wound-rotor stator windings can also be reconnected to obtain 2:1 speed ratios. In such designs, the rotor winding must also be arranged for connection to produce the change in the number of poles. Such motors may have six collector rings.

Speed ratios other than 2:1 for squirrel-cage motors are obtained by providing two separate stator windings. Each winding is designed for the number of poles to produce the desired speed [Eq. (1) or (2) applies]. Rotor design is suitable without change for the speeds desired. Squirrel-cage motors having two separate windings, each arranged for consequent-pole connection, can provide four different synchronous speeds.

Multispeed squirrel-cage induction motors may be used for such drives as wire-drawing machines, banbury mixers (rubber industry), and fans where different operating speeds must be used.

Multispeed consequent-pole wound-rotor induction motors have been designed and applied to drives. Construction costs of the motor, especially of the rotor, make such motors uneconomical compared with other types of speed-changing methods. Speed ratios other than 2:1, using a single stator winding, have been developed in recent years. The stator winding of such a motor has special design features such as to produce pole amplitude modulation (PAM). They are commonly referred to as PAM induction motors.

10. Classification by Enclosure. Large polyphase induction motors are classified by enclosure according to their mechanical structural embodiments (Ref. 1). General classifications may be made as follows:

1. Open motor
2. Totally enclosed motor

a. Open motors have ventilating openings through which external cooling air comes in contact with and passes over and around the machine windings. Open motors and their many variations are described in Sec. 3. Methods to conform to standards vary with manufacturers. The following list is for reference:

1. Dripproof motor
2. Splashproof motor
3. Semiguarded motor
4. Guarded motor
5. Dripproof fully guarded motor
6. Open externally ventilated motor
7. Open pipe-ventilated motor
8. Weather-protected motor, types I and II

b. Totally enclosed motors are so enclosed as to prevent the free exchange of air within and without the case, but the motors are not sufficiently enclosed to be termed airtight. Totally enclosed machines have many variations as described in Sec. 3. Methods to provide conformity with standards and to provide heat removal from motor-heated parts vary with manufacturers' designs. The following list is for reference:

1. Totally enclosed nonventilated motor
2. Totally enclosed fan-cooled motor
3. Explosionproof motor
4. Dust-ignition-proof motor
5. Waterproof motor
6. Totally enclosed pipe-ventilated motor
7. Totally enclosed water-cooled motor
8. Totally enclosed air-to-air-cooled motor
9. Totally enclosed fan-cooled guarded motor

In special applications some of the open-class motors may use only one bearing. The purpose is to reduce the overall installation length by directly and solidly coupling one end of the induction-motor shaft directly to the driven-machine shaft. The load drive thus provides the bearing for both the induction motor and the load machine. Instead of motor and drive having a total of four bearings, only three are required.

11. Motor Characteristics. The motor characteristics indicate the properties associated with motor design and load requirements. The motor performance must be detailed in specifications to suit the requirements of the drive, and to assure economical and successful motor-drive operation. All performance data are based on nameplate voltage and frequency values unless otherwise specified.

a. The rating of a motor consists of the output together with any other characteristics such as speed, voltage, and current the manufacturer assigns (Ref. 1).

b. Nameplate marking contains the following minimum amount of information (Ref. 1). For squirrel-cage motors:

1. Manufacturer's name, serial number or date code, and suitable identification
2. Horsepower output
3. Time rating
4. Temperature rise
5. Speed at rated load
6. Frequency
7. Number of phases
8. Voltage
9. Rated-load current

For wound-rotor motors: In addition to items 1 to 9, inclusive, from above,

10. Secondary current at full load
11. Secondary voltage

c. Efficiency is the ratio of output to input. The electric power is measured to obtain the motor input. The mechanical-shaft power output may be measured by dynamometer or obtained by subtracting the motor losses from the electric-power input. Losses are obtained in accordance with the IEEE test procedure 112 (Ref. 3). Efficiency generally varies with motor rating, speed, and other quantities, as shown in Table 6. Motor-efficiency values are not standardized. Manufacturers, however, do publish listings of guaranteed values for many ratings.

TABLE 6. Efficiency of Polyphase Induction Motors

Motor type	Efficiency	Motor type
High speed	is greater than	low speed
Large hp	is greater than	low hp
Cage	is greater than	wound-rotor
Medium or low voltage	is greater than	high voltage
Normal slip	is greater than	high slip
Full load	is greater than	partial load

Rating is the same except for characteristic being compared.

d. Power factor is the ratio of the kilowatt input to the kilovolt-ampere (kVA) input. It is usually expressed as a percentage. The power factor for cage motors is slightly higher than that for wound-rotor motors of the same horsepower and speed ratings. The power factor is usually highest at the motor nameplate rating and drops off with load. The reduction with load decrease is slight for high-speed motors until light loads are reached. The reduction with load decrease may be quite noticeable in low-speed motors (see Table 7). Power-factor values are not standardized. Listings of guaranteed values for many ratings are published by the manufacturers.

TABLE 7. Power Factor of Polyphase Induction Motors

Motor type	Power factor	Motor type
High speed	is greater than	low speed
Large hp	is greater than	low hp
Cage	is greater than	wound-rotor
Medium or low voltage	is greater than	high voltage
Full load	is greater than	partial load

Rating is the same except for characteristic being compared.

e. Torque, with reference to polyphase induction motors, has several meanings, as indicated in Fig. 2. Minimum values (Ref. 1) are indicated in Table 8.

1. Full-load torque is the torque necessary to produce rated horsepower at full-load speed. In pounds at 1 ft radius, it is equal to the horsepower times 5252 divided by the full-load speed.
2. Locked-rotor torque (static torque) is the minimum torque the motor will develop at rest for all angular positions of the rotor, with rated voltage applied at rated frequency.

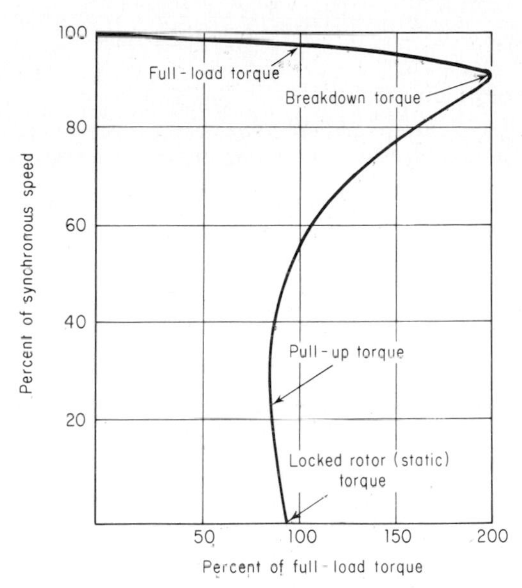

FIG. 2. Types of polyphase-induction-motor torques used in specifications and applications.

TABLE 8. Torque Standards*

Torques	*% of rated full-load torque*
Locked-rotor...........	60
Pull-up................	60
Breakdown............	175

* With rated voltage and frequency applied, torque should not be less than these values (Ref. 1).

3. Pull-up torque is the minimum torque developed by a motor during the acceleration period from rest to the speed at which breakdown torque occurs. For motors which do not have definite breakdown torque, the pull-up torque is the minimum torque developed up to rated speed.

4. Breakdown torque is the maximum torque a motor will develop at rated voltage and frequency without an abrupt drop in speed.

f. The service factor, when specified for a motor, is the multiplier applied to the normal horsepower rating to indicate the permissible loading that may be carried under the service-factor value specified.

g. Locked-rotor current is the steady-state motor current taken from the line with rotor locked at rated voltage and rated frequency.

12. Motor Tests. Motor tests may be (1) routine or (2) complete. Tests are made in accordance with the IEEE test procedure 112 (Ref. 3) and reported on forms suggested in the procedure.

a. Routine tests

1. Measurement of winding resistance
2. No-load readings of current and speed at normal voltage and frequency (This test is not taken on motors without complete shaft and bearings.)
3. Measurement of open-circuit voltage ratio on wound-rotor motors
4. High-potential test

b. Complete tests include all data taken during a routine test and, in addition, other data necessary to determine full-load temperature rise, efficiency, power factor, starting, pull-up and breakdown torque, and rated-load slip. Other tests which may be specified include noise, vibration, speed-torque, shaft currents, and insulation resistance (see Sec. 20).

APPLICATION

13. Motor-Application Data

a. Locked-rotor current and torque values for large polyphase squirrel-cage induction motors designated in specifications may vary widely. Locked current at full voltage should be kept low, consistent with required locked-rotor torque and breakdown torque, to minimize supply-voltage dip and large kVA demand. Locked-rotor torque can vary widely for a given locked current, but large breakdown torque requires high locked-rotor current.

The locked-rotor current may be specified in amperes, as a percentage of full-load current, or by means of a code letter.

b. Code-letter designations for large polyphase squirrel-cage induction motors having a locked-rotor kVA per horsepower as measured at full voltage and rated frequency are shown in Table 9. Code letters may be applied to the following types of squirrel-cage motors:

1. Single-speed
2. Multispeed
3. Dual-voltage
4. Part-winding-start

TABLE 9. Code-Letter Designations

Letter	kVA per hp	Letter	kVA per hp	Letter	kVA per hp	Letter	kVA per hp
A	0–3.15	F	5.0–5.6	L	9.0–10.0	S	16.0–18.0
B	3.15–3.55	G	5.6–6.3	M	10.0–11.2	T	18.0–20.0
C	3.55–4.0	H	6.3–7.1	N	11.2–12.5	U	20.0–22.4
D	4.0 –4.5	J	7.1–8.0	P	12.5–14.0	V	22.4 and up
E	4.5 –5.0	K	8.0–9.0	R	14.0–16.0		

For locked-rotor kVA per horsepower at rated voltage and frequency. For any letter the range spread includes the lower figure and up to but not including the higher figure.

For dual-voltage single-speed Y-start and delta-run, and part-winding-start motors, the connection giving the highest locked-rotor kVA per horsepower is used for the code-letter designation in applying polyphase squirrel-cage induction motors. For multispeed motors the code letter is used for the highest-speed winding or connection for which the motor can be started, and in the case of constant-horsepower types for the highest locked-rotor kVA per horsepower. For motors having 60- and 50-Hz ratings, the code letter for locked-rotor kVA per horsepower on 60 Hz is used.

c. Load Wk^2 for large polyphase squirrel-cage motors is shown in Table 10. Table 10, also generally applied to 60°C rise motors, provides a norm for standardization. Motors for applications having load Wk^2 in excess of values shown in the table or deviating therefrom in other respects can be designed. High-load Wk^2 or multiple successive starts may require special designs or larger motor frames and thus depart from standard motor design. Such motors are also more expensive to build than those for which standard Wk^2 applies.

d. The number of successive starts of squirrel-cage induction motors initially from ambient or rated-load temperature is indicated in Table 10. The same number of starts may also be specified for Wk^2 values greater than those in Table 10. However, a price penalty may apply over standard-motor designs.

Frequent motor starting, when required for an application, should definitely be made a part of the motor specification. During initial plant startup, etc., additional motor starts beyond the Table 10 standard may be desired. They should not be made without consulting the motor manufacturer. The life of the motor is affected by the number of motor starts.

Normal motor overload protective relays may not protect a motor against damage resulting from too frequent successive motor starts.

e. The overspeeds that squirrel-cage and wound-rotor polyphase induction motors will withstand without mechanical injury are 20 percent above synchronous speeds of 1801 r/min and over and 25 percent above synchronous speeds of 1800 r/min and below.

f. The direction of rotation for large polyphase induction motors is specified for the application. When two otherwise identical motors have opposite rotations in similar applications, specifications may require the motors to be suitable for either direction of rotation to provide motor interchangeability. Terminal boxes, etc., must be considered in the plan for interchangeability. Ventilation of large high-speed motors, two-pole motors, for example, may be more effective when designed with directional axial-flow or centrifugal-type fans. Hence, for some designs it may be difficult to provide otherwise identical motors which are suitable for operation in either direction of rotation.

g. Voltage and frequency variations from rated values have been standardized for running conditions (Ref. 1). The standards state that motors shall operate successfully (1) at rated frequency and load with voltage variation 10 percent above or below rated, or (2) at rated voltage and load with a frequency variation 5 percent above or below rated value, or (3) at rated load with a combined voltage and frequency variation 10 percent above or below rated values provided the frequency variation does not exceed 5 percent.

Operation under the variations indicated may not meet standards for normal rated conditions, such as efficiency or power factor. Deficiencies in motor torque during starting must be considered when voltage or frequency deviations are anticipated. Torque varies as the square of the voltage and inversely as the square of the frequency. Thus if the voltage is raised 10 percent, the torque increases 21 percent; or if the frequency is lowered 5 percent, the torque is raised 11 percent.

TABLE 10. Load Wk^2 (Exclusive of Motor Wk^2), lb·ft^2

hp	Speed, rpm						
	3,600	1,800	1,200	900	720	600	514
250					9,530	14,830	21,560
300				6,540	11,270	17,550	25,530
350				7,530	12,980	20,230	29,430
400				8,500	14,670	22,870	33,280
450				9,460	16,320	25,470	37,090
500	381	1,880	5,130	10,400	17,970	28,050	40,850
600	443	2,202	6,030	12,250	21,190	33,100	48,260
700	503	2,514	6,900	14,060	24,340	38,080	55,500
800	560	2,815	7,760	15,830	27,440	42,950	62,700
900	615	3,108	8,590	17,560	30,480	47,740	69,700
1,000	668	3,393	9,410	19,260	33,470	52,500	76,600
1,250	790	4,073	11,380	23,390	40,740	64,000	93,600
1,500	902	4,712	13,260	27,350	47,750	75,100	110,000
1,750	1,004	5,310	15,060	31,170	54,500	85,900	126,000
2,000	1,096	5,880	16,780	34,860	61,100	96,500	141,600
2,250	1,180	6,420	18,440	38,430	67,600	106,800	156,900
2,500	1,256	6,930	20,030	41,900	73,800	116,800	171,800
3,000	1,387	7,860	23,040	48,520	85,800	136,200	200,700
3,500	1,491	8,700	25,850	54,800	97,300	154,800	228,600
4,000	1,570	9,460	28,460	60,700	108,200	172,600	255,400
4,500	1,627	10,120	30,890	66,300	118,700	189,800	281,400
5,000	1,662	10,720	33,160	71,700	128,700	206,400	306,500
5,500	1,677	11,240	35,280	76,700	138,300	222,300	330,800
6,000	1,673	11,690	37,250	81,500	147,500	237,800	354,400
7,000	1,612	12,400	40,770	90,500	164,900	267,100	399,500
8,000	1,484	12,870	43,790	98,500	181,000	294,500	442,100
9,000	1,294	13,120	46,330	105,700	195,800	320,200	482,300
10,000	1,046	13,170	48,430	112,200	209,400	344,200	520,000

The table applies to standard polyphase squirrel-cage motors having locked-rotor torques equal to 60 percent of full-load torque. Motors can accelerate without injurious temperature rise under the following conditions:

1. Rated voltage and frequency applied.
2. During the accelerating period, the connected load torque should be equal to, or less than, a torque which varies as the square of the speed and is equal to 100 percent of full-load torque at rated speed.
3. Two starts in succession (coasting to rest between starts) with the motor initially at ambient temperature or one start with the motor initially at ambient temperature not exceeding its rated-load operating temperature.

Note. Data from Ref. 1. Also, for values of Wk^2 at lower speeds, see Ref. 1.

h. Service conditions should be included in motor specifications, especially conditions that may deviate from normal or standard, such as unusual voltage variations, ambient temperatures above 40°C or below 10°C, altitude operation exceeding 3300 ft (1000 m), or unusual overspeeding (pump overspeed, for example). Other service conditions relate to cooling-air quality resulting from contaminants, moisture or flammable mixtures which may require special motor enclosures, structure or insulation treatment, etc.

14. Starting Methods. Table 11 gives the starting methods together with the motor characteristics for large squirrel-cage induction motors. The motor torque and current from standstill to full speed are highest when full-voltage motor starting is used. The other motor-starting methods listed provide reduction of both motor current and torque during the starting period. To avoid kVA penalties, these alternate starting methods should be considered in applications. The method chosen, if evaluated from an economic viewpoint, must also be judged considering the motor torque. The motor torque must exceed the load torque at all times during starting, or acceleration to full speed will not be reached. This consideration is especially important when the voltage at the motor terminals or the type of starting method used produces a motor-torque curve inadequate to meet load-torque requirements during starting.

The following discussions apply to the starting methods listed in Table 11.

a. Full-voltage starting (Table 11) applies full line voltage to the motor terminals. Squirrel-cage motors are generally suitable for this duty. However, specifications should make this point clear.

Squirrel-cage motors may take from five to seven times full-load current when rated voltage is applied at locked rotor. In special applications even higher values of locked-rotor current may apply because a large motor breakdown torque is required.

Voltage-impedance drops in supply lines, transformers, etc., may result in less than full-voltage appearing at the motor terminals at start. The motor current is approximately proportional to the voltage, and the motor torque is approximately proportional to the voltage squared. Hence, in the selection and application of full-voltage starting of a squirrel-cage induction motor, consideration must be given to possible voltage drops; otherwise the expected motor torques may not be available to accelerate the load to full speed.

b. Reactor starting (Table 11) reduces the kVA the motor takes from the supply line during starting. Reactors can be placed in the motor lines or in the motor neutral (Fig. 3). In the latter case neutral leads must be brought out of the motor into the terminal box. The supply-line kVA at standstill is approximately proportional to the reactor tap used. If 50 percent tap provides 50 percent voltage at the motor at locked rotor, the line kVA is approximately 50 percent of the motor full-voltage locked-rotor kVA without reactor. Motor torque varies approximately as the tap value squared. See Table 12 for voltage taps available in standard reactor units.

TABLE 11. Starting Methods (Cage Motors)

Method	Per unit* volts	Per unit† kVA	Per unit‡ torque	Remarks
Full voltage	1.0	6.25	1.12	Assumed
Reactor	0.5	3.13	0.28	See Fig. 3
Autotransformer	0.5	1.56	0.28	See Fig. 6
Wye-delta	1.0	2.08	0.37	
Part winding	1.0	4.0	0.40	Approx

Different methods of reducing squirrel-cage motor-starting kVA. Taps assumed for reactor and autotransformer (magnetizing kVA neglected). See Art. 14.

*At motor terminals.

†Line value.

‡Rated full-load torque = 1.0 per unit.

As the motor speed increases, the motor impedance also increases. Because the reactor impedance remains constant, the voltage appearing at the motor terminals increases with the motor speed.

An example using a reactor to reduce the motor terminal voltage to 50 percent of rated value at zero speed is shown with detailed calculations in Table 13. Computations are identical for either line or neutral reactor location. The motor data needed for the computations are obtained from Fig. 4. All resistances, reactances, and impedances in Table 13 are expressed in per unit values. Per unit is percent divided by 100. Values in ohms per phase (terminal to neutral) may be obtained by multiplying per unit values by base ohms. For three-phase motors the value of base ohms is V/I, where V is the motor rated terminal-to-neutral voltage and I the rated motor full-load current in amperes. The terminal-to-neutral voltage for three phase motors is the rated (nameplate) voltage divided by $\sqrt{3}$.

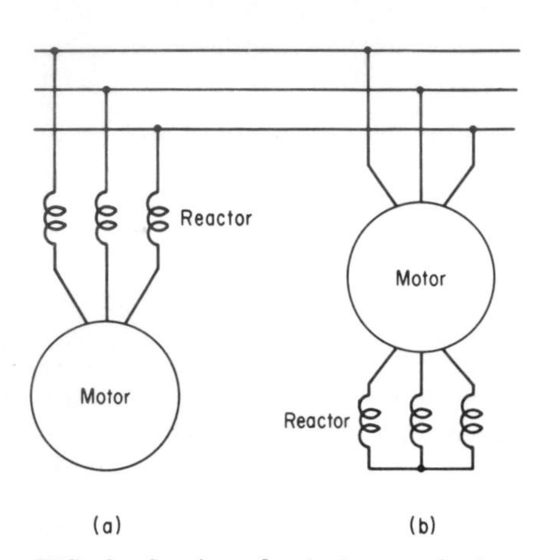

FIG. 3. Starting of polyphase squirrel-cage induction motors uses (*a*) line or (*b*) neutral located reactors which are short-circuited at full-speed operation.

Computations in Table 13 are made first for zero speed. All motor data are calculated, and reactor reactance X_r is assumed equal to motor reactance X as shown in the table. For most squirrel-cage induction motors, the assumption that $X_r = X$ provides a value of V_M very close to the required 50 percent. However, for some motors, especially those having high locked-rotor power factor, a value of X_r somewhat less than X gives the desired $V_M = 0.50$ at zero speed. Thus it may be necessary to make several estimates for X_r to make V_M equal to 0.50 at zero speed.

TABLE 12. Autotransformers and Reactors for Cage Motors (Ref. 4)

Duty	Motor hp rating	Approx % of line voltage at start
Heavy.	All	As required
Medium.	50 hp or less	65–80
Medium.	Above 50 hp	50–65–80
Light.	All	30–37.5–45

Once a zero-speed computation is determined, X_r may be used for other speeds. Torque, current, etc., may be plotted as shown in Fig. 5. Table 13 shows a calculation for motor characteristics at 85 percent speed.

c. Autotransformer starting of squirrel-cage motors is shown in the diagram of

TABLE 13. Computation for Reactor Starting

a. Per unit speed	0	0.85
b. Motor current I	6.25	4.53
c. Motor impedance $Z = 1/I$	0.16	0.22
d. Motor power factor/100	0.22	0.40
e. Motor $R + jX$	$0.035 + j0.156$	$0.088 + j0.203$
f. Reactor reactance jX_r	$j0.156$	$j0.156$
g. $R + j(X + X_r) = Z_t$	0.32	0.37
h. Line current $I_L = 1/Z_t$	3.12	2.70
i. Power factor $= R/Z_t$	0.10	0.24
j. Motor torque T	1.12	1.87
k. $T_t = (I_L/I)^2T$	0.28	0.66
m. $V_M = I_LZ$	0.50	0.597

For 50 percent voltage at start with 2500-hp 4000-V 304-A 1188-r/min three-phase 60-Hz squirrel-cage induction motor. See Fig. 3.

b and *d* from Fig. 4.

e. $R = Z \times d$; $X = Z[1 - d^2]^{1/2}$

f. Assume until V_M is desired value.

g. Z_t numerical $= [R^2 + (X + X_r)^2]^{1/2}$.

j. At full voltage, without reactor, Fig. 4.

k. Motor torque using reactor $= T_t$.

m. $V_M \times 100$ = percent of line voltage across motor.

Note. Base ohms $= 4000/(\sqrt{3} \times 304) = 7.6\ \Omega$.

Reactor ohms $= 0.156 \times 7.6 = 1.20\ \Omega$.

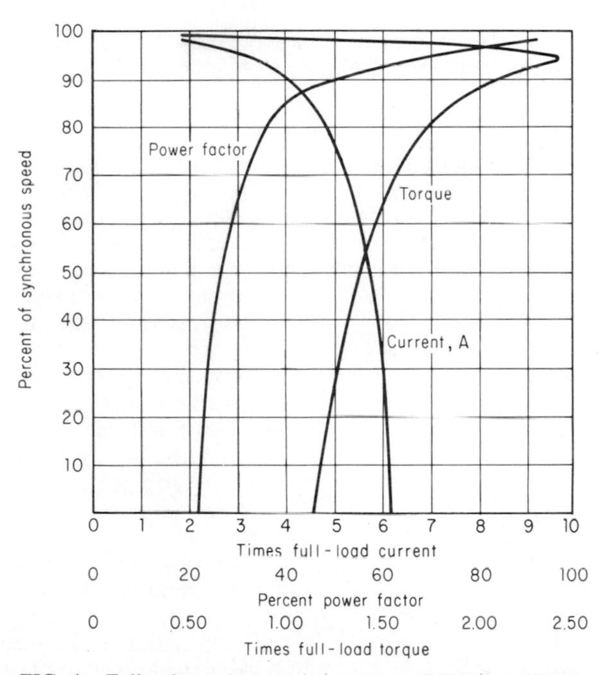

FIG. 4. Full-voltage characteristic curves of 2500-hp 4000-V 304-A full-load 1188-r/min three-phase 60-Hz squirrel-cage induction motor.

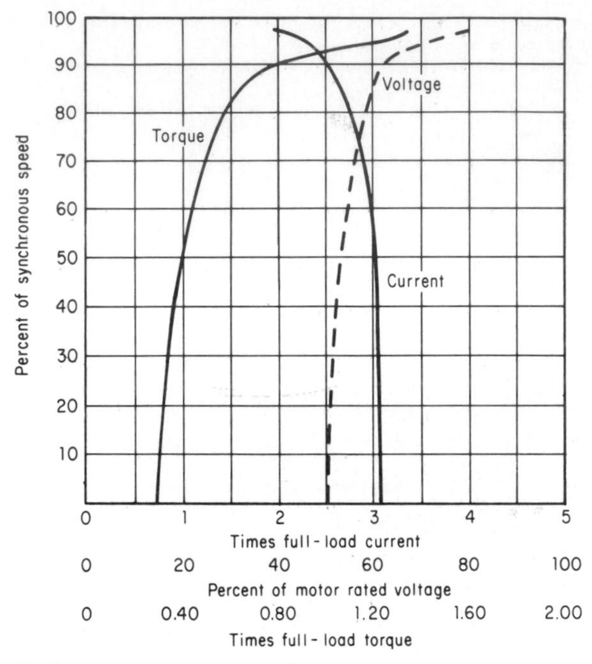

FIG. 5. Torque, current, and voltage of squirrel-cage induction motor with series reactor to give 50 percent motor voltage at standstill.

Fig. 6. The motor-starting method is on reduced voltage. On reaching full speed the motor is switched to full line voltage for normal continuous full-load operation. Because the starting device is a transformer, the motor terminal voltage is essentially constant at the transformer tap value during the complete acceleration period. The tap voltage chosen to start a motor must take into consideration that the motor torque varies as the square of its terminal voltage, because the motor must have adequate torque to accelerate its connected load to speed.

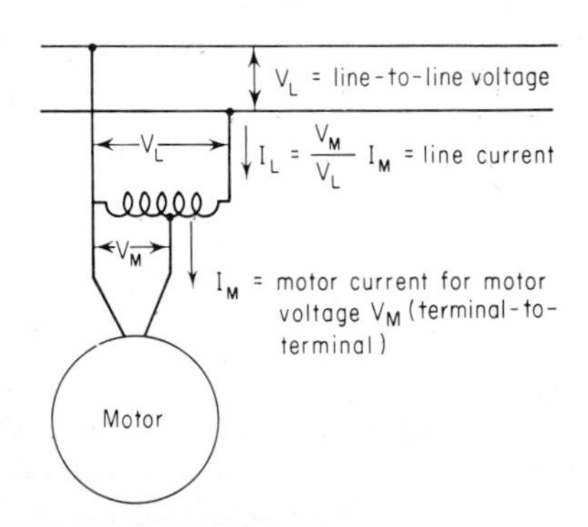

FIG. 6. Line diagram showing autotransformer starting arrangement.

Standard taps for autotransformers are shown in Table 12. Special taps can be obtained for nonstandard applications. A price penalty may apply, however, when deviating from standard designs.

Tap starting from an autotransformer reduces the starting kVA a motor takes from the supply lines. The kVA reduction from the full-voltage-starting method varies as the square of the tap used. Also the current taken from the supply line varies as the square of the motor full-voltage current. For example, using an autotransformer with appropriate tap,

$$\text{Motor current } I_M = I \text{ (tap used)}$$

$$\text{Motor voltage } V_M = V_L \text{ (tap used)}$$

$$\text{Motor kVA} = \sqrt{3}\, V_M I_M = \sqrt{3}\, V_L I \text{ (tap)}^2$$

Neglecting small kVA requirements (magnetizing, etc.) of the autotransformer, then

$$\text{Motor kVA} = \text{line kVA}$$

$$\sqrt{3}\, V_L I \text{ (tap)}^2 = \sqrt{3}\, V_L I_L$$

(Hence line current I_L varies as the square of the tap.)

where I = motor current on full voltage
I_M = motor current with tap voltage
I_L = line current with tap used for starting
V_M = motor tap voltage
V_L = line or full voltage

Example. A motor takes 2000 kVA at full voltage at locked rotor. Voltage V_L = 2300 V, terminal to terminal; I = 500 A. Assuming an 80 percent tap is used to start the motor,

$$\text{Motor kVA} = \text{line kVA} = 2000(0.80)^2 = 1280$$

$$\text{Line current} = 500(0.80)^2 = 320 \text{ A}$$

The formulas apply for any speed during the starting period. Full-voltage values of motor kVA or current must, of course, be known.

d. Wye-delta starting is not widely applied to ~~starting~~ large polyphase squirrel-cage induction motors in U.S. practice. Motors for this type of starting must be designed for normal full-load operation on the delta connection and must be started connected wye. Thus 73 percent more conductors per coil are required over designs which may be and frequently are designed for normal wye-operating connection at full load. The slot-space factor is reduced for the special wye-delta winding design, and more slot space is required for insulation than for the straight wye-designed winding. Started on the wye connection, the motor current and torque are one-third the values that would apply for starting on the delta connection.

e. Part-winding starting of squirrel-cage induction motors is sometimes specified for the smaller-horsepower higher-speed types. Sufficient winding leads are brought out to a terminal box so that a portion or all of the winding is externally accessible. When full voltage is applied to only part of the suitably designed stator winding, the motor accelerates its connected load. After accelerating on the part-winding connection for a short time, the remainder of the stator winding is connected and the motor acceleration continues on its full-winding connection to full speed. On the part-winding connection motor currents may be unbalanced but lower than full-winding starting values. Locked-rotor currents on the part-winding connection vary from 60 to 70 percent and locked-rotor torque is about 50 percent of full-winding values.

15. Starting of Large Polyphase Wound-Rotor Motors. The motors are started at full voltage. At startup, however, resistance must be inserted in the rotor or secondary to limit the starting current and torque. Access to the motor circuit is through shaft-mounted collector or slip rings to which the rotor winding leads are connected (Art. 9). Carbon-type brushes ride on the rings so that external electrical resistance connected to the brushes is effectively introduced into the rotor-winding circuit. Grid or similar resistors are used for smaller motor ratings and liquid-type slip regulators for the large-horsepower motors.

Wound-rotor motors should not be started with slip rings short-circuited (zero external resistance) without first consulting the motor manufacturer. With short-circuited rotor slip rings wound-rotor motors develop very little, if any, locked-rotor (static) torque. Also motor current under such conditions may be as high as or higher than squirrel-cage motors of similar rating. Rotor windings are not designed to carry large currents continuously.

Starting of a wound-rotor motor is illustrated by the curves in Fig. 7. Speed-torque curves are shown for three values of resistance R_3 (the largest value), R_2, and R_1 (the smallest value) placed in the rotor or secondary circuit through the slip rings and brushes. The curve R_0 is for short-circuited rings. To start the motor, resistance R_3 is used first and primary, or stator, voltage is applied to the motor. For 100 percent starting torque, the terminal-to-neutral value of R_3 in ohms is approximately equal to nameplate secondary, or rotor voltage divided by the product of $\sqrt{3}$ and full-load secondary nameplate current (see Art. 11*b*, nameplate marking). With this value of R_3, only full-load current is taken from the line at standstill, and the power factor and motor kilowatt input are approximately the same as for normal full-load running operation (100 percent torque) where slip rings are short-circuited. After the motor accelerates to, say, 55 percent speed

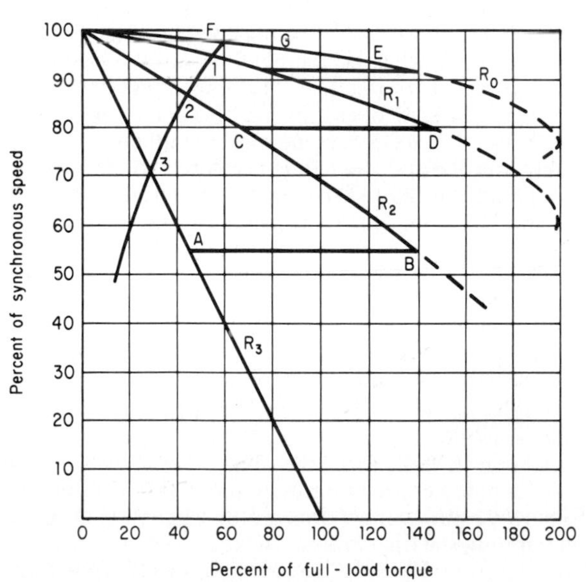

FIG. 7. Torque curves of wound-rotor induction motor with external resistance *R* across rotor slip rings.

along the torque of R_3, the resistance is removed from the rotor to obtain resistance R_2 in the rotor circuit. The motor torque changes from point A to point B in the figure and now operates on curve R_2. The motor accelerates to, say, 80 percent speed, where more rotor resistance is removed so that the motor torque changes from point C to point D on torque curve R_1. With a further increase in motor speed to, say, 92 percent rotor rings may then be short-circuited, and the speed increases from point E to point F along curve R_0 where the load torque is 60 percent (unloaded pump, fan, etc.). The load may be increased to full, or 100 percent, torque, and the speed drops slightly to point G. The motor then operates at normal full-load torque and speed.

Step-by-step removal of resistance from the rotor circuit was the procedure followed in the example just considered. Large motors use liquid slip regulators to vary the resistance across the slip rings smoothly. Theoretically, with suitable smooth resistance variations in the rotor circuit, the motor can develop 100 percent, or full-load, torque from zero to full speed with constant full-load current, power factor, and kilowatts during the speed change.

16. Speed Variation. A speed variation of induction motors (Ref. 5) may be obtained by:

1. Variation of stator voltage
2. Variation of rotor resistance
3. Slip-coupling device
4. Multispeed windings
5. Variable-frequency supply
6. Solid-state control

a. Variation of stator voltage weakens the motor air-gap magnetic flux and thereby reduces the induced rotor voltage and current at a given speed. The motor torque thus decreases. The load-torque demand now in excess of the motor-developed torque reduces the motor speed. The motor speed drops to a value such that the induced rotor voltage, current, and motor torque increase to meet the load demand at a lower speed. The method is rarely applied to large motors because of the variable-voltage-supply requirement and because of reduced breakdown torque with a decrease in motor voltage. The method applied to large motors provides only a small speed drop.

b. Rotor resistance variation in the secondary, rotor, circuit of wound-rotor motors is widely used. Resistance is inserted in the rotor circuit as described in Art. 15, with torque curves similar to those shown in Fig. 7. For example, a driven load may be operated at 70 percent synchronous speed (Fig. 7) continuously with resistance R_3 in the rotor circuit. The different speeds of operation are shown in the figure at points 1, 2, and 3, the intersections of the load-torque and motor-torque curves. Externally connected rotor resistances must be thermally suitable for continuous operation.

Wound-rotor motors using resistance speed control may be used in applications such as mine hoists, some types of pumps and fans where reduced output is desired, flywheel sets, starting motors for synchronous condensers and large waterwheel generators, and rolling-mill service. Speed reduction may be as much as 40 percent, provided the load torque decreases as the square of the speed. For special applications where speed reduction may be greater than 40 percent and torque conditions may not vary as the square of the speed, it may be necessary to supply external blowers to cool the motor. Speed regulation (change of speed with change in torque) may be a problem at low operating speeds. A minor change of load torque may result in a large speed change, that is, the speed regulation may be poor.

c. Slip-coupling devices, of the hydraulic or electric type, for example, are not a part of the motor or motor design but rather separate devices coupled between the motor and the driven load. The principle of operation is similar to that of the friction clutch. A primary member of the slip coupling is coupled to the motor and runs at the motor speed. The secondary member of the device, connected to the driven load, runs at a lower speed than the primary member, that is, there is slippage between members. The device is used with squirrel-cage motors for such drives as boiler-feed pumps and fans.

d. Multispeed-winding speed-reducing principles are discussed in Art. 9, Classification by Speed. Motors of this type are costly, especially if speeds other than the 2:1 ratio are required. Multispeed cage motors are applied to wire-drawing drives, fans, pumps, Banbury mixers, etc.

e. Variable-frequency supply causes motor speeds to vary with the frequency of the power-supply source [see Eqs. (1) and (2)]. The method of speed control is not extensively used for large motors because of the cost of frequency-conversion equipment. In some applications, however, the method may be economical and practical. An outstanding example of the application of this method is the (hydraulic) wave-making apparatus for the David Taylor Model Basin.

f. Solid-state developments have led to the common use of two methods of speed control of induction motors (Ref. 5):

1. Six-step adjustable frequency
2. PWM (pulse width modulation) adjustable frequency

17. Thermal and Electrical Protection of Motors. Such protection is recommended good practice in motor application. Many devices are available for continuous thermal monitoring of critical motor parts subject to possible damage from excessive temperatures. The most critical motor parts to be protected during continuous operation are stator windings and bearings.

Rotor windings of squirrel-cage or wound-rotor motors normally are not directly monitored for temperature because of a lack of suitable simple instrumentation for rotating members. Indirect thermal monitoring of rotating parts during the starting period may be obtained by special devices such as relays and timing devices.

a. Typical thermal-sensing detectors and their locations are described in Table 14. The physical location of the device should be as close to the heat-producing part as possible. Locating a thermal detector in the stator slots between the upper and lower coil sides or on coil ends provides quicker detector response to coil heating than a back-of-the-core sensor location. Time lags due to heat flow through intervening coil insulation or long heat-flow paths are not conducive to rapid thermal response and sensing. However, in some applications it is almost physically impossible or very costly to monitor in close proximity to stator coils. It may be necessary then to use back-of-core or stator-coil-end or even ventilating-air thermal monitoring. Monitoring ventilating air is a practical method of protection for machines having air-to-water heat exchangers.

b. Electrical protection of induction motors applies to the stator-winding insulation. Some of the methods indirectly provide a measure of protection for the rotor winding. Direct protection of the rotating winding is difficult and rarely used

TABLE 14. Motor Thermal Protection

Type	Winding		Core	Bearing	Motor* air	Water
	Slot	End				
a. Bimetal	. . .	x	x	. . .	x	
b. Dial thermometer	. . .	. . .	. . .	x	x	x
c. Switch	. . .	. . .	x	x		
d. Thermocouple	x	x	. . .	x		
e. RTD, 10-ohm	x	x	. . .	x		
f. RTD, 120-ohm	x	. . .	. . .	x		
g. Thermistor	x	x				

a. Bimetal snap switch.
b. Dial-type thermometer; may have electrical contacts.
c. Sensitive tip; bellows, etc., operates snap, mercury, etc., switch.
d. Thermocouples; copper or iron-constantan, etc.
e. Resistance temperature detector, three leads, for visual, recording, or switching operation.
f. Like *e*, but two leads.
g. Thermistor, positive or negative temperature coefficient, relay, etc., operation.
* Enclosed motors having air-water, air-air heat exchangers, etc.

except for very large motors. Stator-winding protection methods may use space heaters, overload relays, low-voltage devices, surge capacitors, lightning arresters, differential relaying, open-phase and phase-unbalance relays, phase-reversal relays, split-winding relays, etc. A discussion of the more commonly used methods follows.

1. Space heaters inside the motor enclosure keep the internal air above the dew point to prevent moisture condensation on the winding insulation. They may be switched by the motor control to be turned on when motor shutdown occurs.

2. Overload protection may be effected by a thermal-type relay. Heat, generated by motor current flowing through a resistance-type coil, is either conducted or radiated to a heat-sensitive element to cause relay-contact opening at a predetermined temperature. The relay characteristics are of the inverse-time type and can be compensated for variations in the ambient-air temperature (Ref. 5).

3. Induction- and magnetic-type overload relays usually have inverse-time characteristics like the thermal types. They are often used for instantaneous operation on large overcurrents. Both this type and the thermal-type relay depend on the motor current for operation. They thus protect the motor winding from overheating because of current overload. However, winding heating may also be caused by impairment of current overload. However, winding heating may also be caused by impairment of ventilation or by inadequate water in a heat exchanger used for cooling the motor-ventilating air. Winding overheating caused by ventilation impairment thus is not sensed by the thermal, induction, and magnetic overload relays.

4. Enclosed motors having air-water heat exchangers may, however, be protected by overload relays plus motor-ventilating-air temperature indicators. In larger motors a better method is to monitor winding temperatures directly by slot-embedded detectors or end-winding sensors. They may be made to operate alarms, etc.
5. Low-voltage motor protection may provide assurance that motors reach full speed on starting. Low supply voltage under full-load and full-speed operation causes the motor to lose speed and draw overcurrent. When the motor breakdown torque is reduced sufficiently by low voltage the motor and load may stall.
6. Surge capacitors and lightning arresters may be applied to motors of the 2300-V class and higher. High-voltage surges caused by lightning and high-speed line switching can break down stator-winding insulation to ground. Also the impact of the transient voltage entering the motor terminals is highest on the motor-terminal-coil conductor turns. Turn-to-turn voltage can be excessive, causing breakdown of insulation between turns of a motor-terminal coil. Connecting surge capacitors and lightning arresters in parallel with the motor phases provides stator-winding protection against voltage surges. The capacitors reduce the steepness of the oncoming voltage wavefront, and the arresters spill over at a given peak voltage.

18. Combined Characteristics. The combined characteristics of several induction motors may be needed in motor applications to evaluate bus overall current, power factor, etc. For this evaluation, efficiency and power-factor curves for the motors should be obtained from the motor manufacturer. Typical curves for a 100-hp motor are shown in Fig. 8. Many motor data can be computed from these

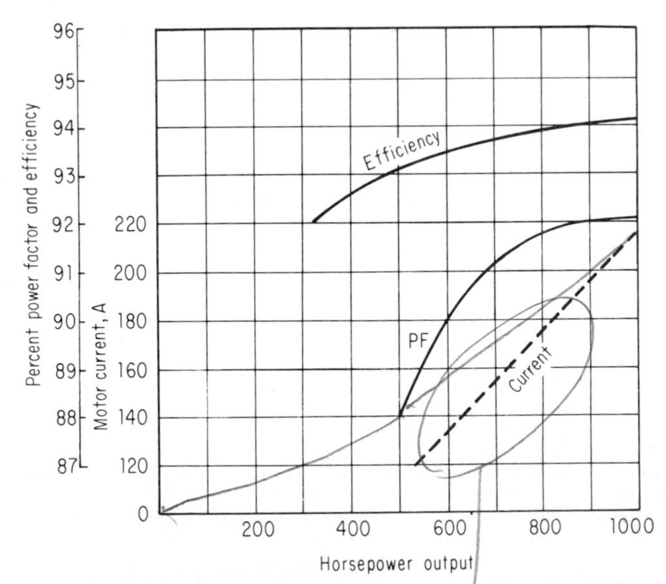

FIG. 8. Curves of efficiency, power factor, and current for 1000-hp 2300-V three-phase 60-Hz induction motor.

curves. For example, the motor current may be determined from the following equation for three-phase motors:

$$I = \frac{746 \times \text{hp}}{\sqrt{3}\, V \times \text{PF} \times \text{EFF}} \quad \text{amperes} \tag{4}$$

The motor current I in Fig. 8, shown as a dashed line, is calculated from Eq. (4).

The following motor data may also be obtained from curves such as those shown in Fig. 8:

$$\text{kVA} = \frac{\sqrt{3}\, VI}{1000} \tag{5}$$

$$\text{kW} = \text{kVA} \times \text{PF} \tag{6}$$

$$\text{kvar} = \text{kVA}(1 - \text{PF}^2)^{1/2} \tag{7}$$

$$\text{kVA} = (\text{kW}^2 + \text{kvar}^2)^{1/2} \tag{8}$$

where V = terminal-to-terminal voltage (three-phase motors)
PF = power factor, percent divided by 100
EFF = efficiency, percent divided by 100

An example problem of three motors is solved in Table 15 to obtain combined motor electrical characteristics. The manufacturer's curves applying to motor A are shown in Fig. 8, previously discussed. Similar curve data are assumed available, although not reproduced here, for the other two motors. Table 16 lists nameplate data for the motors being considered.

Computations for Table 15 require that the motor current be known. Switchboard or other metering instruments may be used to obtain this information. Then from the curves, such as those shown in Fig. 8, the motor horsepower, effi-

TABLE 15. Tabular Method to Obtain Bus Data (Combined Characteristics of Several Motors)

	Motor			Motor	
	A	B	C	sum	Remarks
Load hp	900	575	400	1,875	Manufacturer's curves
Efficiency	0.94	0.90	0.91		Manufacturer's curves
Power factor	0.92	0.64	0.6		Manufacturer's curves
I, current	195	187	135		Switchboard
kVA, Eq. (5)	775	745	538		
kW, Eq. (6)	713	477	328	1,518	
kvar, Eq. (7)	302	572	426	1,300	

Motor hp sum	1875
Bus kW = motor sum	1518
Bus kvar = motor sum	1300
Bus kVA = Eq. (8)	2000
Bus current, use Eq. (5)	503
Bus power factor = bus kW/bus kVA	0.76
Overall efficiency = 0.746 hp sum/bus kW	0.92

TABLE 16. Ratings of Three 2300-V Three-Phase 60-Hz Motors

Motor	hp	Synchronous speed, r/min	Curve Fig.	Full-load current, A*
A	1,000	1,200	8	216
B	700	200	*	211
C	500	300	*	153

*Curves like Fig. 8 assumed available from manufacturer. Amperes are switchboard values.

ciency, and power factor may be determined. Line-by-line computation follows as indicated in tabular form. Note that the bus, or total, current of all the motors is not simply the arithmetic sum of the individual motor currents. The deviation may be appreciable in some mixed motor systems.

19. Emergency Switching. To ensure the continuity of operation of important drives, emergency switching of induction motors from a faulted bus to an auxiliary bus may be required. This practice is warranted, especially in the power or utility industry, for critical drives, such as large pumps, fans, and pulverizers. It may also be considered in industry for critical processes requiring continuity of operation.

Precautions must be taken when switching from one power source to another. Reasons for the precautions will be discussed after the following analysis of motor-terminal voltage persistence.

Flux linked with the closed rotor-winding circuits does not immediately disappear after the stator winding is removed from the power source. Based on the theorem of constant-flux linkages, the magnetic-field flux linked with the rotor circuit is assumed the same immediately after power-line removal as it was before. However, as time increases, the flux decays. This flux, fixed to the rotor and rotating with it, generates a decreasing voltage in the stator winding. The rotor flux and also the generated stator open-circuit voltage both decay with time, in accordance with the following equation:

$$E_g = E_L \exp\left(\frac{-t}{T_0}\right) \tag{9}$$

where E_g = voltage generated in open-circuited stator winding
E_L = line voltage before disconnecting motor
exp = logarithm, base *e*, with indicated exponent
t = time, seconds
T_0 = motor open-circuit time constant, seconds
= $(x'_M + x_2)/2\pi f r_2$
x_2 = rotor reactance, ohms per phase, stator terms
x'_M = motor-magnetizing reactance, ohms per phase (stator)
f = line-supply frequency, usually 60 Hz
r_2 = rotor resistance, ohms per phase, stator terms

In Eq. (9) the open-circuit time constant T_0 is the time in seconds for the voltage to decay to 36.8 percent of its initial value after power removal from the motor. Another observation of Eq. (9) is that the voltage decreases to $E_g/2$ in $0.693T_0$, to $E_g/4$ in $2(0.693T_0)$, to $E_g/8$ in $3(0.693T_0)$, etc. (see Fig. 9).

Typical time constants of squirrel-cage motors vary from about 4 s for large

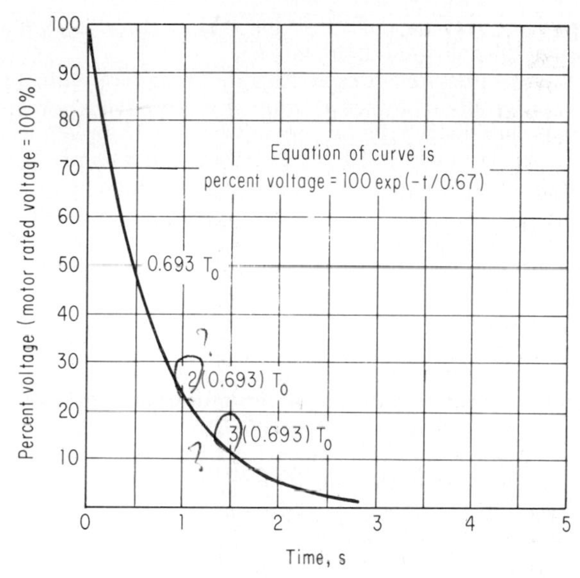

FIG. 9. Motor-terminal voltage decay on being disconnected from power source.

high-speed motors to ⅙ s (10 to 12 cycles on a 60-Hz basis) for small low-speed motors (see Fig. 10). It may be possible by special motor design to reduce the time constant T_0 of some motor ratings.

When switching occurs the motor open-circuit generated voltage may be substantial and 180° out of phase with the emergency bus voltage. The switching phenomenon then is similar to synchronizing a synchronous motor out of phase.

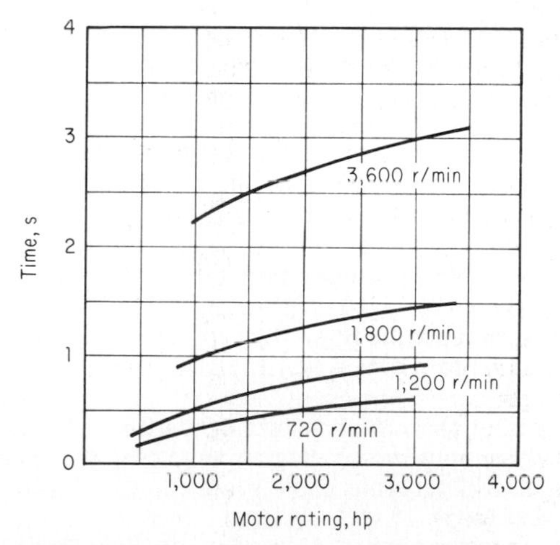

FIG. 10. Typical open-circuit time constant T_0 for 60°C 60-Hz squirrel-cage induction motors.

Coil-distorting forces, torsional stresses, etc., have values much higher than those occurring during normal running or starting.

To avoid undue forces and stresses because of high individual motor currents when switching a group of induction motors to an emergency or auxiliary bus, engineers may use a voltage relay to prevent transfer completion until the motor residual voltage has dropped to 25 percent of the rated nameplate value. Thus if a group of motors is switched to an emergency bus, which happens to be 180° out of phase with the motor open-circuit voltage, motor stresses, although above normal starting value, may be within acceptable limits. Because several motors of different characteristics on a common bus may be switched at the same time, the bus time constant of voltage decay on an open circuit is due to a mixture of the effects of all motors.

20. Capacitors. Applying capacitors at induction-motor terminals can compensate for lagging kvar of lightly loaded or low-speed motors. These types of machines may require considerable lagging kVA or kvar from the line. Capacitors compensate for this condition so that the combined motor-capacitor unit operates at a higher power factor than the motor alone.

Because capacitors are usually paralleled and switched with the motor, selection of the proper size of capacitor bank is important (1) to avoid excessive motor overvoltages and (2) to limit motor transient currents which can produce distorting winding forces and shaft torques. The problem is caused by the capacitors supplying excitation current to the motor, which then generates voltage even after removal from the supply lines. High transient motor currents and torques may result should the supply-line switch be reclosed when there is residual voltage on the motor. The following discussion explains the phenomenon.

Normally, without capacitors, opening the motor-line switch removes the motor excitation (magnetizing current) source so that the motor-terminal voltage decays to zero (Fig. 9). The magnetizing current is very closely equal to the motor idle running current. However, capacitors of the right amount placed at the motor terminals can supply motor full-voltage magnetizing kvar even after the motor-line switch is opened. Thus although disconnected from the power lines, the motor will continue to generate rated voltage provided the rotor speed is maintained by the load-connected inertia or another type of drive. A further increase in capacitance causes the motor to generate a voltage even greater than the previous line voltage. With enough excess capacitance at the motor terminals this generated voltage can become sufficiently raised to stress winding insulation. The proper value of capacitance to use to prevent greater than normal rated voltage generated after the line switch is opened and still provide power-factor correction may be illustrated in the following example. The effect of excess capacitance to produce greater than rated motor voltage is also treated.

A typical motor application driving a pump may serve as an example. The motor is rated 700 hp, 2300 V, three-phase, 60 Hz, 720 r/min, and has a no-load saturation curve as shown in Fig. 11. From the curve, the magnetizing current at rated voltage is 40 A. Because the power factor is low, the motor-magnetizing kvar are approximately equal to [Eq. (7)] $\sqrt{3}\,EI/1000$, or $\sqrt{3}\,(2300)(40)/1000$, or 159 kvar.

Capacitors, placed at the motor terminals, capable of supplying 159 kvar at 2300 V, will exactly supply motor-magnetizing kvar requirements. Under this condition the motor will generate 2300 V and no more after the line switch is opened and provided the rotor speed is maintained.

In microfarads the capacitance C is $1000 \times \text{kvar}/2\pi f(\text{kV})^2 = 1000(159)/$

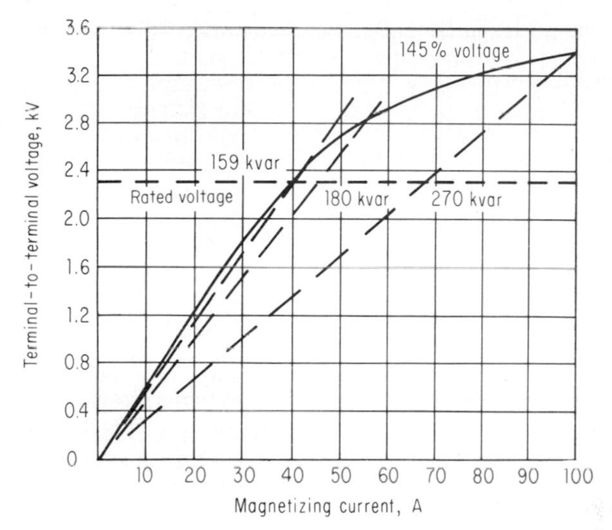

FIG. 11. No-load saturation curve of induction motor gives necessary data to determine critical value of capacitance at motor terminals.

$2\pi(60)(2.3)^2 = 80\ \mu$F. The capacitive reactance X_c, in ohms, is $1/2\pi fC$, or $10^6/2\pi(60)(80) = 33.2\ \Omega$. Thus three capacitors, each having $C = 80\ \mu$F, to obtain $X_c = 33.2\ \Omega$, may be connected in wye at the motor terminals.

Assuming motor and capacitors are still connected together after the line-supply switch is opened, the generated frequency decreases as the motor speed drops. Reactance X_c, being inversely proportional to frequency, will no longer supply sufficient kvar for motor requirements, and the terminal voltage will decay to zero.

Assume that capacitors of 180 kvar at 2300 V are connected to the motor instead of 159 kvar as discussed above. We may calculate as before: $C = 1000(180)/2\pi(60)(2.3)^2 = 90\ \mu$F and $X_c = 10^6/2\pi(60)(90) = 29.5\ \Omega$. At 2300 V the capacitor current is $2300/\sqrt{3}\,(29.5) = 45$ A. The straight line shown in Fig. 11 indicates the capacitor-voltage characteristics. The capacitor straight-line curve intersects the motor no-load magnetizing curve at 2800 V, where a stable condition of capacitor kvar supply equals motor kvar demand. Thus the motor now generates 2800 V at 60 Hz and will continue to do so as long as the rotor synchronous speed is maintained. As previously discussed, these conditions prevail after the motor-capacitor combination is disconnected from the power-supply source. Overvoltages may therefore result when capacitor values are too large.

The 700-hp motor of this example problem has a full-load efficiency and power factor of 94 and 90 percent, respectively. Using Eqs. (4) to (8), the motor kW input is calculated to be $0.746 \times$ hp/(EFF/100) = 555 kW. Also kVA drawn from the line equals kW/(PF/100) = 617 and kvar = 270. The latter is calculated by multiplying kVA by $[1.0 - \mathrm{PF}^2]^{1/2}$, where PF is percent power factor divided by 100. The vector diagram of Fig. 12*a* indicates the relationship of kW, kVA, and kvar.

Assume a 159-kvar capacitor bank is paralleled with the 700-hp motor. Under full load, the capacitor bank compensates for 159 of the 270 kvar of the induction motor. Thus in Fig. 12*b* the vector diagram indicates that only 111 kvar need be supplied by the power lines. The rotor-capacitor combination then requires 555 kW and 111 kvar from the supply source (capacitor losses are neglected). The power factor is calculated to be 97.9 percent from Eqs. (6) and (8).

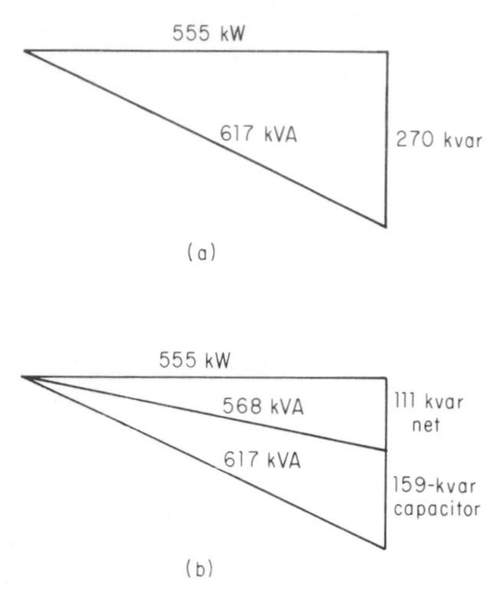

FIG. 12. (*a*) Full-load vector diagram of 700-hp motor shows kvar without capacitors. (*b*) Capacitor kvar is subtracted from motor kvar.

For 180 kvar of capacitors at 2300 V the motor-capacitor arrangement operates at approximately 99 percent power factor. As discussed previously, the voltage at the motor rises to 2800 V when the units are swiched from the supply lines.

To obtain 100 percent power-factor motor-capacitor operation, a capacitor bank of 270 kvar at 2300 V is required to compensate fully for the induction-motor 270 kvar. From Fig. 11 the voltage rises to 145 percent of normal when the combined motor-capacitor unit is switched from the supply lines.

Table 17 indicates the effect of capacitor kvar compensation (equal to motor no-load kvar) under motor full-load operation for several ratings of induction motors. The data were calculated according to procedures outlined above.

TABLE 17. Typical Motor Examples of Adding Capacitors

hp	Motor volts	Motor speed, r/min	I_M	Full load* % EFF	Full load* % PF	% PF†
200	2,300	450	36.0	92.0	70.0	99.2
200	440	300	187.0	90.0	67.0	98.6
1,250	4,160	1,800	31.5	95.2	92.3	98.3
700	2,300	720	40.0	94.0	90.0	97.9

I_M = no-load motor amperes.
*Without capacitors.
†With capacitors added at motor terminals.

21. Fault-Current Contributions. The contributions of fault currents of induction motors may be an important factor in relaying and circuit-breaker or fuse selection. Thus fault-current magnitudes and time constants influencing the rate of decay are of interest when studying machine short circuits.

At the instant following a short circuit at the terminals of a three-phase induction motor, the magnetic flux linking the stator and rotor windings becomes trapped or fixed to the windings. Stator magnetic-flux poles no longer rotate. Rotor flux is fixed to the rotor windings with which it rotates. The fluxes of both stator and rotor are maximum at the instant of a short circuit and start decaying immediately thereafter, in accordance with the winding time constants.

The stator-winding trapped flux decreases with time, but remains fixed with respect to the winding, as previously mentioned. The stator current associated with the decaying flux is approximately a unidirectional or direct current, decreasing with time as shown in Fig. 13. All phases of a three-phase motor may not have the same initial value of direct current because of the 120° relationship of the phases. Thus the initial value of direct current, following a short circuit, may vary from zero to a maximum in a phase winding.

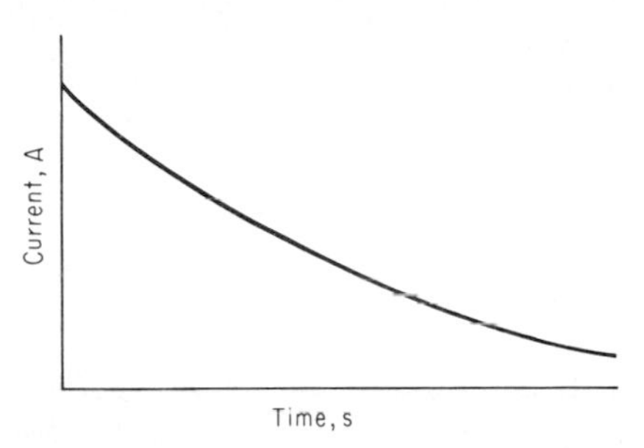

FIG. 13. Dc component i_{dc} of induction-motor short-circuit current decreases exponentially.

Magnetic flux fixed and revolving with the rotor generates normal frequency, 60 Hz, for example, in the stator windings if full rotor speed is maintained. The normal-frequency current generation is like a short-circuited synchronous machine having a dc-excited rotor winding. But in the case of the induction motor, the rotor flux, having no external excitation source, eventually dies out. Hence the alternating component of the stator short-circuit current induced by rotor flux decays with time, as shown in Fig. 14. The maximum value is shown at zero time.

The sum of the ac component i_{ac} and the dc component i_{dc} gives the total short-circuit current in a stator winding. Addition of the two components of Figs. 13 and 14 at a given time t is shown in Fig. 15.

Determination of the curves of Figs. 13 and 14 requires computing the instantaneous magnitudes and decay rates of the ac and dc components i_{ac} and i_{dc}. The following formulas may be used to compute numerical short-circuit values.

The ac component, assuming rotor-trapped flux does not decay after a terminal short circuit, would be a curve like that in Fig. 14, except that there would be no

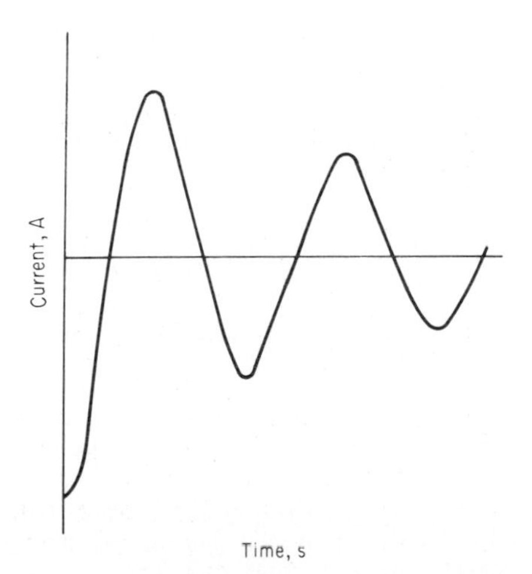

FIG. 14. Ac component i_{ac} of induction-motor short-circuit current.

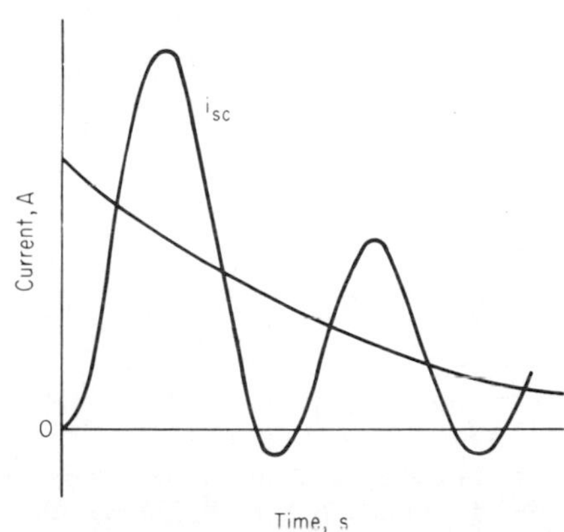

FIG. 15. Total short-circuit current i_{sc} of an induction motor is sum of direct and alternating components.

reduction in peak values with increasing time. The expression for such a curve is, for maximum current at zero time,

$$i'_{ac} = -\sqrt{2}\frac{E}{x'}\cos\omega t \qquad \text{amperes} \tag{10}$$

where E = terminal-to-neutral line voltage before short circuit
x' = transient reactance of induction motor, ohms per phase
ω = $2\pi f$ (f = supply frequency, Hz)
t = time, seconds

For x', the transient reactance of the motor, the following expression may be used:

$$x' = x_1 + \frac{x_2 x_M}{x_2 + x_M} = x_1 + x_2 \qquad \text{(approx)} \tag{11}$$

where x_1, x_2, and x_M are, respectively, the stator, rotor, and magnetizing reactances, in ohms per phase, as used in the induction-motor equivalent circuit.

Equation (10) is based on the assumption that rotor-trapped flux does not decay. Actually, however, the flux starts decreasing the instant after the short circuit and induces a decaying short-circuit alternating current in the motor stator windings. The expression may be modified by an exponential decrement factor to give the following expression for i_{ac}:

$$i_{ac} = -\sqrt{2}\frac{E}{x'}\left[\exp\left(-\frac{t}{T_{ac}}\right)\right]\cos\omega t \qquad \text{amperes} \tag{12}$$

The time constant T_{ac} may be calculated from

$$T_{ac} = \frac{x'(x_2 + x_M)}{(x_1 + x_M)r_2} \qquad \text{seconds} \tag{13}$$

$$T_{ac} = \frac{x'}{x_1 + x_M}T_0 \qquad \text{seconds} \tag{13a}$$

where r_2 is the induction-motor rotor resistance in stator terms, ohms per phase, and T_0 is the motor open-circuit time constant in seconds (Art. 19).

The dc component of short-circuit current results because at the instant of the short circuit the magnetic flux linked with the stator winding becomes trapped to the windings. As time passes following the short circuit, the flux decays and eventually dies out. The current induced in the stator windings by the decaying flux is approximately unidirectional or direct current, decaying exponentially as shown in Fig. 13. The maximum value of i_{dc} is equal to the peak value of i_{ac}, so that

$$i_{dc}\,(\text{max}) = i_{ac}\,(\text{peak}) = \sqrt{2}\frac{E}{x'} \qquad \text{amperes} \tag{14}$$

The decay or decrease in current with time is exponential in accordance with the dc time constant T_{dc}. Hence Eq. (14) multiplied by the decrement factor describes current decay, or

$$i_{dc} = \sqrt{2}\frac{E}{x'}\exp\left(-\frac{t}{T_{dc}}\right) \qquad \text{amperes} \tag{15}$$

The dc time constant T_{dc} may be computed from the following expression:

$$T_{dc} = \frac{x'}{r_1} \qquad \text{seconds} \tag{16}$$

where $\omega = 2\pi f$ (f = supply-line frequency, Hz)
r_1 = stator-winding resistance, ohms per phase (terminal-to-neutral value)

Total motor short-circuit current has a maximum phase value when the dc component i_{dc} expressed by Eq. (15) is added to the ac component i_{ac} [Eq. (12)]. The sum gives the fully asymmetrical or completely offset wave (Fig. 15). It may be expressed as

$$\begin{aligned} i_{sc} &= i_{dc} + i_{ac} \\ &= \sqrt{2}\frac{E}{x'}\left\{\exp\left(-\frac{t}{T_{dc}}\right) - \left[\exp\left(-\frac{t}{T_{ac}}\right)\right]\cos\omega t\right\} \end{aligned} \tag{17}$$

22. Frequency-Change Effects. The effects of frequency variations on polyphase-induction-motor characteristics were briefly discussed in Arts. 3 and 16 in relation to standards and speed control. Motors designed in accordance with standards operate successfully at rated load and at rated voltage with a variation in frequency up to 5 percent above or below rated value. Within this frequency variation the performance will not necessarily be in accordance with the standards established for operation at motor rated frequency. In reapplications of induction motors, approximate variations from rated characteristics may be desired for frequency changes wider than 5 percent, such as operation of a 60-Hz motor on a 50-Hz system.

a. The performance characteristics of interest when an induction motor is operated at a frequency differing from its nameplate value are motor speed, no-load current, horsepower, locked-rotor current, and torque and breakdown torque. The effects of a frequency change on power factor and efficiency are also important.

Frequency change varies the motor speed as indicated by Eqs. (1) and (2), that is, the no-load, or idle, running speed of a motor is proportional to frequency. For example, an unloaded 60-Hz motor running at approximately 600 r/min will run at about 500 r/min at 50 Hz.

Frequency change in the supply to a given motor may also be treated as a voltage change. When so considered, the motor magnetic densities remain the same, as indicated by

$$V = KBf \tag{18}$$

where V = motor-terminal voltage
K = a constant, includes winding turns, etc.
B = magnetic density
f = frequency, Hz

Examination of Eq. (18) indicates that a 440-V 60-Hz motor should be operated at 367 V, 50 Hz to maintain the same magnetic density B. Thus when the motor voltage is altered in proportion to frequency, the no-load, or idle, running current is roughly the same because, with magnetic density B unchanged, magnetizing ampere-turns are not changed. Motor operation at 440 V, 50 Hz would cause approximately a 20 percent rise in magnetic density. Such an increase may be excessive and accompanied by a high degree of iron saturation and high no-load running current.

Considering the previously mentioned 5 percent frequency deviation from the nameplate value as a standard allowance, the flux-density change from Eq. (18) is also 5 percent. It may be reasoned, therefore, that it should be permissible to increase the voltage (and thus the flux) from 367 to, say, 385 V for 50-Hz operation of a 60-Hz motor. The motor may then readily operate on 380 V, a common supply for 50-Hz systems. As discussed later, a gain in starting and breakdown torque results.

b. The horsepower loading at 50 Hz should be that permitted by the motor manufacturer. In emergencies it may be assumed that 50-Hz horsepower is five-sixths of the 60-Hz value. Because motor ventilation is reduced at the lower 50-Hz speed, the motor temperature should be monitored and the loading reduced if necessary so as not to exceed the 60-Hz nameplate value.

c. The locked-rotor current of a polyphase induction motor is limited by the motor impedance. Under locked-rotor conditions the motor power factor is low so that the impedance is almost entirely reactance. Reactance is equal to $2\pi fL$, where f is the supply frequency and L the inductance. Thus assuming L is constant (except as discussed later), the motor locked impedance varies approximately with the frequency only. Hence if the applied voltage varies in proportion to the supply frequency, the locked-rotor current is unchanged for frequency variation. For example, the locked-rotor current will be unchanged for a 440-V 60-Hz motor operated at 367 V 50 Hz.

Motors having eddy-current-type squirrel-cage-type rotor windings (deep, narrow bars, for example) may exhibit, on reduced frequency at locked rotor, a slight increase in the rotor-winding inductance L. Thus at a reduced supply frequency, the locked current may decrease slightly when frequency and applied voltage are reduced in the same proportion. The motor manufacturer can supply accurate figures.

From the previous discussion it may also be reasoned that a 60-Hz motor operated on 50 Hz without voltage change will have a locked-rotor current approximately 20 percent greater than on 60 Hz.

d. Motor torques may also vary with frequency, depending on the mode of operation and the applied voltage. For example, assuming 50-Hz load horsepower is five-sixths of the 60-Hz value, the full-load torque at 50 Hz will be the same as the 60-Hz value. This value may be verified from the familiar horsepower equation

$$\text{hp} = \frac{TN}{5252} \tag{19}$$

where T = torque, lb·ft
N = speed, r/min (50-Hz speed is approximately five-sixths of the 60-Hz speed)

e. The breakdown torque in pound-feet is unchanged (approximately) when the applied voltage is reduced in proportion to the frequency reduction. When the voltage remains unchanged with frequency change, the torque increases as the square of the inverse ratio of the frequency. Thus at a given voltage, the breakdown torque in pound-feet at 50 Hz is approximately $(6/5)^2$ of the 60-Hz value.

f. The locked-rotor torque varies like the breakdown torque for a change in frequency, except for squirrel-cage motors having eddy-current-type rotor bars. In the latter case the torque increase for a frequency reduction is not as great as the square of the inverse frequency ratio (voltage unchanged with frequency change). This ratio differs because the eddy-current effects of deep narrow squirrel-cage bars are less at lower frequencies.

g. Results of tests made at 50 and 60 Hz on a 200-hp 440-V 60-Hz motor are shown in Table 18. Tests were made at 440 V for each frequency. The no-load current increase is greater than the frequency ratio because of increased iron saturation and, accordingly, influenced the load power factor. The change in efficiency may be difficult to generalize because of the relationship between constant and variable losses. Constant losses are windage, friction, and core loss. Variable losses are due to winding copper losses and stray power losses. Breakdown and locked-rotor torques are influenced by the changes due to the deep-bar effects previously mentioned. These same effects also influence the locked-rotor current.

TABLE 18. Test Data at 60 and 50 Hz, 200-hp Motor

	60 Hz	50 Hz	% change*
No-load current, A	84	105	+25
% power factor	87	84	−3
% efficiency	92.6	92.7	+0.10
Full-load current, A	244	252	+3.3
Full-load torque, lb·ft	1,200	1,430	+20
Breakdown torque, lb·ft	3,100	4,400	+42
Locked-rotor current, A	1,450	1,700	+17

*50 Hz as related to 60 Hz.

23. Special Polyphase Induction Motors. Special polyphase-induction-motor types in the larger sizes do not normally appeal to the American market. The reason is one of economics. For example, the brush-shifting commutator polyphase induction motor is an excellent machine for speed control. It is widely applied abroad in some industries. U.S. practice prefers the squirrel-cage motor to attain moderate speed variation.

Another type of motor, less widely used than in the past in the larger sizes, is the double-squirrel-cage-type motor for high-starting-torque low-starting-current requirements. The double-cage motor, having two sets of rotor squirrel-cage bars, is costly to manufacture. Modern power systems provide capacity for the less expensive standard squirrel-cage motor with slightly higher starting current. This type of motor is the "workhorse" of industry and because of its simplicity requires the least maintenance. Maintenance, motor complexity as it affects downtime and production loss in the repair or replacement of parts, local repair facilities, availability of replacement or repair parts, etc., must be weighed in overall application economics.

REFERENCES

1. "Motors and Generators," ANSI/NEMA Standard MG 1-1978.
2. "Induction Motor Letter Symbols," ANSI/IEEE Standard 58-1978 (Reaff 1983).

3. "Test Procedure for Polyphase Induction Motors and Generators," ANSI/IEEE Standard 112-1983.
4. "Industrial Control and Systems," ANSI/NEMA Standard ICS 1-1978.
5. "Test Procedure for Airborne Sound Measurements on Rotating Electric Machinery," IEEE Standard 85-1973 (Reaff 1980).
6. "Alternating-Current Induction Motors, Induction Machines in General, and Universal Motors," ANSI Standard C50.2.

SECTION 5
SYNCHRONOUS MOTORS

Robert C. Moore*

FOREWORD 5-2

GENERAL CHARACTERISTICS 5-2

1. Synchronous Motor 5-2
2. Large Synchronous-Motor Types 5-2
3. Horsepower Ratings 5-5
4. Frequency of Supply 5-5
5. Number of Phases 5-5
6. Standard Voltage Ratings 5-5
7. Excitation Voltages 5-5
8. Power Factor 5-5
9. Insulation 5-6
10. Temperature Rise 5-6
11. Speed Ratings 5-6
12. Classification by Enclosure 5-6
13. Synchronous-Motor Characteristics 5-7
14. Tests for Synchronous Motors 5-8
15. V Curves for Synchronous Motors 5-9

APPLICATION 5-9

16. Motor-Application Data 5-9
17. Starting Characteristics 5-12
18. Methods of Starting Synchronous Motors 5-14
19. Reactor Starting 5-14
20. Overexcited Synchronous Motors 5-17
21. Kvar Compensation 5-17
22. Dynamic-Braking Requirements 5-18
23. Excitation 5-21
24. Protection 5-22

REFERENCES 5-25

*Senior Engineer, Motor-Generator Department, Allis-Chalmers Manufacturing Company, Milwaukee, Wis. (retired); Registered Professional Engineer (Wis.); Senior Member, IEEE; Distinguished Lecturer, Milwaukee School of Engineering.

FOREWORD

The large polyphase synchronous motors treated in this section are listed in Fig. 1. Types of special construction will also be discussed in such detail as is warranted by their usage; however, nonexcited-rotor-type motors, reluctance types, for example, will not be treated.

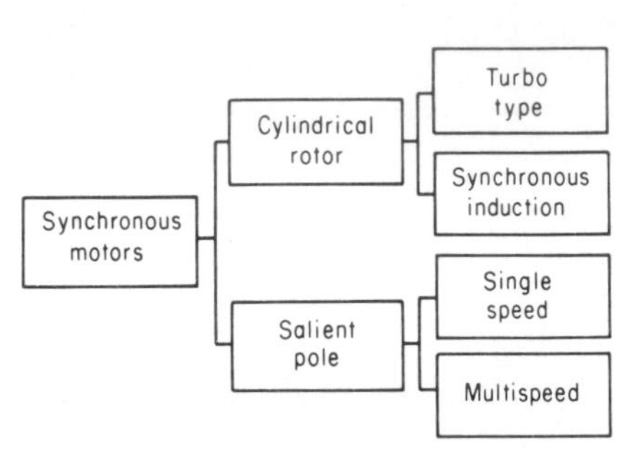

FIG. 1. Types of synchronous motors.

Motor ratings above 200 hp, approximately, are considered. Standardizing bodies have not assigned frame sizes or dimensions to these ratings. The motors do, however, follow standards relating to horsepower, voltage, frequency, Wk^2 capability, etc.

Extensive use of standards is made because misapplication of motors can result in an interpretation of specifications not related to standards. Standards provide for mutual understanding of motor mechanical and electrical technical terminology, motor characteristics, test procedures, and general motor specifications. They also promote motor-application economies in providing practical ranges of horsepower, voltage, speed, etc.

GENERAL CHARACTERISTICS

1. Synchronous Motor. The synchronous motor, like the induction motor, has two members: a stator and a rotor. A synchronous-motor stator with its polyphase winding closely resembles an induction motor and has the same function of receiving power from the supply lines to drive the connected load. The rotor, the rotating driving element, is supplied with dc excitation. The speed of rotation of the rotor is constant and does not vary with load as in the case of the induction motor. The motor speed in revolutions per minute is determined by the number of poles for which the stator is wound and by the frequency of the power supply. Hence Eqs. (1) and (2) in Sec. 4 may be used to predict synchronous-motor speed.

2. Large-Synchronous-Motor Types. The types of large synchronous motors used commercially are (1) salient-pole and (2) cylindrical-rotor (Fig. 1). Stator designs and windings are similar for each type, but rotor designs are quite different. Typical applications are shown in Table 1.

a. A salient-pole synchronous motor has a rotor of stacked thin laminated-steel prominences, or poles, on which insulated excitation coils are wound. These windings, supplied with direct current, make the steel poles strong magnets of alternate north and south polarity. The number of rotor poles is equal to the number of poles for which the stator winding is wound. During normal operation the rotor magnetic poles "lock in" with the magnetic field produced by the stator winding. Operation speed is in accordance with Eqs. (1) and (2) of Sec. 4.

b. A cylindrical-rotor synchronous motor may be one of two types. The high-speed type 3600 r/min, for example, has a high-strength solid forged magnetic steel rotor with electrical winding embedded axially in machined slots like a turbo-

TABLE 1. Applications of Synchronous Motors

Application	Torque, % of motor full-load torque: Locked-rotor	Pull-in	Pull-out	Ratio of Wk^2 of load to "normal Wk^2 of load" [see Eq. (1)]
Ball mills:				
For ore	150	110	175	1.5–4
For rock and coal	140	110	175	2–4
Banbury mixers	125	125	250	0.2–1
Blowers, positive-displacement, rotary-bypassed for starting	30	25	150	3–8
Blowers, centrifugal, starting with:				
Inlet or discharge valve closed	30	40–60*	150	3–30
Inlet and discharge valve open	30	100	150	3–30
Bowl mills (coal pulverizers), starting unloaded:				
Common motor for mill and exhaust fan	90	80	150	5–15
Individual motor for mill	140	50	150	4–10
Chippers, starting empty	60	50	250	10–100
Compressors, centrifugal, starting with:				
Inlet or discharge valve closed	30	40–60*	150	3–30
Inlet and discharge valve open	30	100	150	3–30
Compressors, reciprocating, starting unloaded:				
Air and gas	30	25	150	0.2–15
Ammonia (discharge pressure 100–250 lb/in^2)	30	25	150	0.2–15
Freon	30	40	150	0.2–15
Crushers, starting unloaded:				
Bradley-Hercules	100	100	250	2–4
Cone	100	100	250	1–2
Gyratory	100	100	250	1–2
Jaw	150	100	250	10–50
Roll	150	100	250	2–3
Fans, centrifugal (except sintering fans) starting with:				
Inlet or discharge valve closed	30	40–60*	150	5–60
Inlet and discharge valve open	30	100	150	5–60
Fans, centrifugal, sintering, starting with inlet gates closed	40	100	150	5–60
Fans, propeller type, starting with discharge valve open	30	100	150	5–60
Generators, ac	20	10	150	2–15
Generators, dc (except electroplating):				
150 kW and smaller	20	10	150	2–3
Over 150 kW	20	10	200	2–3
Hammer mills, starting unloaded	100	80	250	30–60
Hydrapulpers, continuous type	125	125	150	5–15
Plasticators	125	125	250	0.5–1
Pulverizers, B&W, starting unloaded:				
Common motor for mill and exhaust fan	105	100	175	20–60
Individual motor for mill	175	100	175	4–10

TABLE 1. Applications of Synchronous Motors (*Continued*)

Application	Torque, % of motor full-load torque			Ratio of Wk^2 of load to "normal Wk^2 of load" [see Eq. (1)]
	Locked-rotor	Pull-in	Pull-out	
Pumps, axial-flow, adjustable blade, starting with:				
Casing dry	5–40†	15	150	0.2–2
Casing filled, blades feathered	5–40†	40	150	0.2–2
Pumps, axial-flow, fixed-blade, starting with:				
Casing dry	5–40†	15	150	0.2–2
Casing filled, discharge closed	5–40†	175–250*	150	0.2–2
Casing filled, discharge open	5–40†	100	150	0.2–2
Pumps, centrifugal, Francis impeller, starting with:				
Casing dry	5–40†	15	150	0.2–2
Casing filled, discharge closed	5–40†	60–80*	150	0.2–2
Casing filled, discharge open	5–40†	100	150	0.2–2
Pumps, centrifugal, radial impeller, starting with:				
Casing dry	5–40*	15	150	0.2–2
Casing filled, discharge closed	5–40*	40–60*	150	0.2–2
Casing filled, discharge open	5–40*	100	150	0.2–2
Pumps, mixed-flow, starting with:				
Casing dry	5–40†	15	150	0.2–2
Casing filled, discharge closed	5–40*	80–125	150	0.2–2
Casing filled, discharge open	5–40*	100	150	0.2–2
Pumps, reciprocating, starting with:				
Cylinders dry	40	30	150	0.2–15
Bypass open	40	30	150	0.2–15
No bypass (three-cylinder)	150	100	150	0.2 15
Refiners, conical (Jordans, hydrafiners, Clafins, Mordens), starting with plug out	50	50–100‡	150	2–20
Refiners, disk-type, starting unloaded	50	50	150	1–20
Rod mills (for ore grinding)	160	120	175	1.5–4
Rubber mills, individual drive	125	125	250	0.5–1
Tube mills (see Ball mills)				
Vacuum pumps, reciprocating, starting unloaded	40	60	150	0.2–15

*The pull-in torque varies with the design and operating conditions. The machinery manufacturer should be consulted.

†For horizontal-shaft pumps having no thrust bearing (entire thrust laod carried by motor) the locked-rotor torque required is usually between 5 and 20 percent, while for vertical-shaft machines having their own thrust bearing a locked-rotor torque as high as 40 percent is sometimes required.

‡The pull-in torque varies with the design of the refiner. The machinery manufacturer should be consulted. Furthermore, even though 50 percent pull-in torque is adequate with the plug out, it is sometimes considered desirable to specify 100 percent to cover the possibility that a start will be attempted without complete retraction of the plug.

Note. See Ref. 1. Additional applications may be found there.

generator. The electrical winding in the rotor is of the turbogenerator rotor type, that is, it is not a polyphase winding. The rotor winding is distributed in slots and is brought out to two collector rings for dc-excitation supply. This type of motor is not used for accelerating large-inertia loads.

c. The low-speed-type cylindrical rotor construction differs from the high-speed type and closely resembles a wound-rotor induction-motor rotor. The rotor is built of laminated thin steel sheets in the magnetic-circuit portion and slotted on the periphery to receive a polyphase winding. Leads from the winding are brought out to (usually) five collector rings mounted on the rotor shaft. Three rings provide for wound-rotor induction-motor-type starting using external resistance (Sec. 4, Art. 15). The other two rings are used for supplying dc excitation for synchronous-motor operation. Thus the motor provides, like the wound-rotor induction motor, high starting torque. On reaching top speed the motor is synchronized to run as a synchronous motor. This type of motor requires a special low-voltage high-current dc-excitation source.

The motor of the laminated-steel rotor construction is also called asynchronous-synchronous motor or synchronous-induction motor.

3. Horsepower Ratings. The horsepower ratings have been standardized for practical applications, as shown in Table 2.

TABLE 2. Standard Horsepower Ratings of Synchronous Motors*

hp range	*hp steps*
250–500	50
500–1,000	100
1,000–2,500	250
2,500–6,000	500
6,000–20,000	1,000
20,000–40,000	2,500
40,000–80,000	5,000
80,000–100,000	10,000

* From Ref. 1. It is not practical to build motors of all horsepower ratings at all speeds. See Table 4.

4. Frequency of Supply. The supply frequency for motors is 50 to 60 Hz, the same as for induction motors (Ref. 1). Motors can be designed and supplied for other frequencies, but a price penalty may apply for such nonstandard machines.

5. Number of Phases. In general there are three phases, although two-phase motors can be obtained from manufacturers.

6. Standard Voltage Ratings. The standard voltage ratings are identical with those of induction motors (Sec. 4, Art. 5).

7. Excitation Voltages. The (direct current) excitation voltages for field windings are 62.5, 125, 250, 375, and 500 V. Voltages of 125 and 250 V are widely used for commercial applications. The voltages cited do not apply to motors of the brushless type. Also it is not practical to design all horsepower ratings of motors for all the foregoing excitation voltages. (Ref. 1).

8. Power Factor. The power factor is standardized at 1.0 (unity) or 0.80 leading (overexcited).

9. Insulation. For synchronous-motor stators the insulation systems are the same as for induction motors (Sec. 4, Art. 6). Field-winding insulations are Classes A, B, F, and H. Classes F and H insulations are not yet widely used in field-coil applications.

Rotor field-coil windings may be of round, square, or rectangular (strap) copper. The insulation of wire-wound copper (round or square) may be cotton, asbestos, glass, etc., consistent with the temperature class. Bare rectangular-strip copper-wound field coils may have turns separated from one another by insulating strips.

10. Temperature Rise. The temperature rise for stator and rotor windings and machine parts is shown in Table 3 for the different insulation classes.

11. Speed Ratings. The speed ratings for purposes of standardization are shown in Table 4. Other speed ratings, which can be obtained from manufacturers, may be determined from Eqs. (1) and (2) in Sec. 4.

By far the majority of synchronous motors are designed for single-speed operation. For special applications multispeed motors may be obtained. They are of special design and are infrequently used, however. Speed combinations are usually of a 2:1 ratio for multispeed motors, for example, 900 and 450 r/min.

12. Classification by Enclosure. The classification is the same as for induction motors (Sec. 4, Art. 10).

TABLE 3. Temperature Rise of Synchronous Motors*

Machine part	Method of temperature determination	Temperature rise, °C Class of insulation system			
		A	B	F	H
Armature winding:					
All hp ratings	Resistance	60	80	105	125
1500 hp and less	Embedded detector	70	90	115	140
Over 1500 hp:					
7000 V and less	Embedded detector	65	85	110	135
Over 7000 V	Embedded detector	60	80	105	125
Field winding:					
Salient-pole motors	Resistance	60	80	105	125
Cylindrical-rotor motors	Resistance		85	105	125
Cores, amortisseur windings, and mechanical parts such as collector rings, brushholders, and brushes may attain such temperatures as will not injure the machine in any respect.					

*Temperature rises under rated load shall not exceed table values and are based on 40°C external cooling air entering machine ventilating openings.

Note 1. For totally enclosed water-air-cooled machines, the temperature of the cooling air is the temperature of the air leaving the cooler. On machines designed for cooling water temperatures up to 30°C, the temperature of the air leaving the coolers shall not exceed 40°C. For higher cooling-water temperatures, the temperature of the air leaving the coolers may exceed 40°C provided the temperature rises of the machine parts are limited to values less than those given in the table by the number of degrees that the temperature of the air leaving the coolers exceeds 40°C.

Note 2. The temperature should be determined in accordance with the latest revision of IEEE test procedure 115 (Ref. 2). Rises apply for altitudes of 3300 ft (1000 m).

TABLE 4. Speed Ratings*

3,600	400	225	120
1,800	360	200	109
1,200	327	180	100
900	300	164	95
720	277	150	90
600	257	138	86
514	240	129	80
450			

*Speed at 60 Hz. At 50 Hz the speeds are five-sixths of the 60-Hz speeds.
Note. It is not practical to build motors of all horsepower ratings at all speeds (Ref. 1).

13. Synchronous-Motor Characteristics. These characteristics indicate the properties associated with motor design and application. Specifications must detail performance or characteristic requirements to ensure economical and successful motor-drive operation. All such data are based on nameplate voltage and frequency values.

a. The rating, efficiency, full-load and locked-rotor torque, and locked-rotor current as defined in Sec. 4, Art. 11, apply to synchronous motors. In calculating efficiency, losses are obtained in accordance with Ref. 2.

b. Nameplate data should contain the following minimum information (Ref. 1):

1. Manufacturer's name and serial number or other suitable information
2. Horsepower output
3. Time rating
4. Temperature rise
5. Speed at full load
6. Frequency
7. Number of phases
8. Voltage
9. Rated current per terminal
10. Rated field current
11. Rated exciter voltage
12. Rated power factor

c. Torque, in addition to full-load and locked-rotor torque mentioned above, has other meanings in synchronous-motor application. Definitions are:

1. *The pull-in torque* of a synchronous motor is the maximum constant torque under which the motor will pull its connected inertia load into synchronism, at rated voltage and frequency, when its field excitation is applied.

   The speed to which a motor will bring its load depends on the power required to drive it, and whether the motor can pull the load into step from this speed depends on the inertia of the revolving parts, so that the pull-in torque cannot be determined without knowing the WK^2 as well as the torque of the load (Ref. 1).

2. *The pull-out torque* of a synchronous motor is the maximum sustained torque that the motor will develop at synchronous speed, rated voltage, rated frequency, and with normal excitation (Ref. 1).
3. *The locked-rotor pull-in and pull-out torques* for salient-pole motors with rated voltage and frequency applied are shown in Table 5.

The locked-rotor torque of synchronous-induction motors varies, like that of the wound-rotor induction motor, with the resistance placed in the rotor circuit (Sec. 4, Art. 15). This type of motor of 100 percent power factor rating can provide as much as 150 percent starting torque with approximately 300 percent or less starting current, depending on motor design and speed. Pull-in torque of 100 percent or more is obtainable. Pull-out torque for 100 percent power-factor motors is approximately 150 percent.

d. Service factor, when specified for a motor, indicates a permissible overload the motor may carry. Conditions under which the service factor applies should be followed. For example, during service-factor operation of 0.8-power-factor motors, the field current is usually not increased over full-load excitation. Unity-power-factor motors have their field current increased on overloads to maintain 100 percent power factor.

14. Tests for Synchronous Motors. These tests may be (1) routine or (2) complete. All tests are made in accordance with IEEE test procedure 115 (Ref. 2).

a. Routine tests

1. Resistance of armature and field windings and field polarity
2. Measurement of no-load field current at rated voltage and frequency (omitted for engine-type motors which have no shaft or bearings)
3. High-potential test

b. Complete tests include all data necessary to determine full-load efficiency, temperature rise, locked-rotor torque and current, speed-torque, etc. Agreement between motor manufacturer and purchaser determines which motor character-

TABLE 5. Torque Standards

			Torques*		
Speed, r/min	hp	Power factor	Locked-rotor	Pull-in†	Pull-out
500–1,800	200 and below	1.0	100	100	150
	150 and below	0.8	100	100	175
	250–1,000	1.0	60	60	150
	200–1,000	0.8	60	60	175
	1,250 and larger	1.0	40	30	150
		0.8	40	30	175
450 and below	All ratings	1.0	40	30	150
		0.8	40	30	200

*Percent of rated full-load torque (minimum).

†Based on normal Wk^2 of load. Values of normal Wk^2 are given in Ref. 1 with rated excitation current applied.

istic tests shall be made. IEEE test procedure 115 supplies testing methods to follow to obtain the desired motor data.

c. Tests for locked-rotor torque and current and speed torque are not ordinarily made for synchronous-induction motors.

15. V Curves for Synchronous Motors. These are obtained from the manufacturer and provide useful application data. An example V curve is shown in Fig. 2. A study of the curve indicates that at constant load and rated voltage, the stator current varies with field or rotor excitation adjustment. The power factor may also be read from the power-factor curve shown when the field current is altered. Reducing the field current sufficiently can cause the motor to operate at unity power factor. Further reduction of field current may cause motor-torque instability or pullout as indicated by the rapidly changing stator-current curve. Thus field-current reduction tends to lower motor pull-out torque.

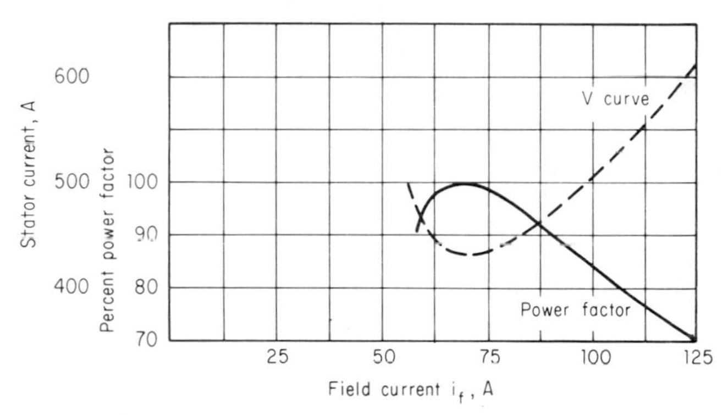

FIG. 2. Synchronous-motor V curve and power factor at rated load.

V curves for fractional loads may also be drawn, as shown in Fig. 3. This figure will be used in a later example to illustrate a valuable use of the V-curve system.

APPLICATION

16. Motor-Application Data

a. Locked-rotor torque values for large polyphase synchronous motors vary widely to suit the type of drive. Thus the locked-rotor torque for a generator drive or certain types of pump drives may be only 10 to 20 percent of full-load torque. For such drives as ball mills, crushers, rolling mills, or rubber mills, locked-rotor torque values of 140 to 200 percent of full-load torque may be specified. The motor torque listed in specifications is generally based on full or rated voltage at the motor terminals. Because the motor starting torque or locked-rotor torque decreases as the square of the applied voltage, appropriate allowance should be made for line voltage drop, etc., causing reduced voltage at the motor terminals.

Typical locked-rotor torque requirements for a large number of synchronous-

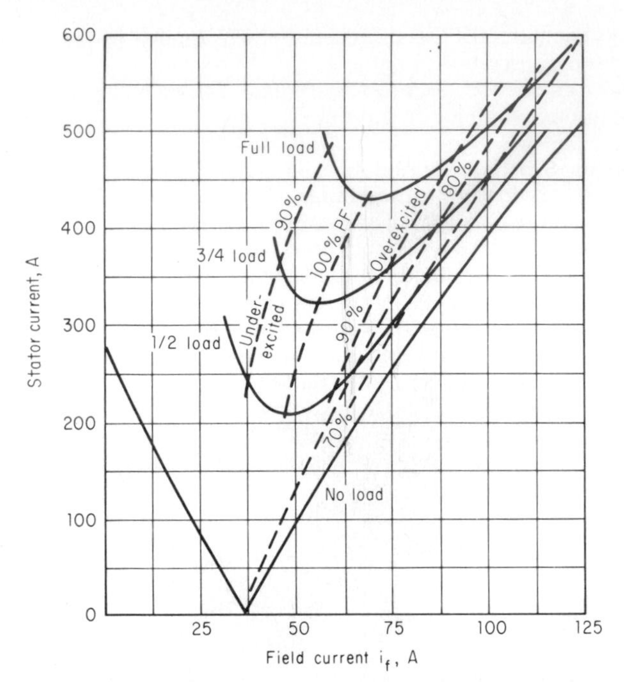

FIG. 3. Synchronous-motor V curves for motor of Fig. 2 at various loads.

motor applications may be found in Ref. 1 or in Table 1. It is suggested that these be referred to when applying motors. For locked-rotor torque requirements of application drives not listed, the manufacturer of the driven apparatus should be consulted.

b. Locked-rotor current varies from a low value of approximately 300 percent to a high value of 800 percent or more, depending on the torque required for starting the driven apparatus. Percentages are in terms of full-load or rated current.

Drives requiring high locked-rotor torque are likely to have the higher values of locked-rotor current. Locked-rotor currents are also a function of good motor design consistent with the torque required, and they also are influenced by the economic practicability of the manufacturer's standardization principles. It may not be economically feasible or desirable for the manufacturer to provide a large availability of frame sizes, dies, etc. For these and other reasons motor locked-rotor current values are not standardized but are suited to the manufacturer's offering or, as sometimes warrants, to the customer's requirements. The latter may require the manufacturer to deviate from standardizing principles for good reasons.

c. Pull-in torque requirements for large polyphase synchronous motors may be as low as 20 percent of rated full-load torque for generator drives to as high as 100 to 140 percent for certain types of pumps, crushers, rubber-mill drives, etc. Typical values for many types of loads are given in Ref. 1.

In general the supplier of the drive can provide information on the equipment so that pull-in torque requirements of the motor for various conditions of operation, such as for loaded or unloaded pull-in, can be specified.

Pull-in torque values are generally stated for rated voltage at the motor terminals. If the terminal voltage is less than rated, because of line drop, etc., the application engineer should be aware of the effect on pull-in torque. Of special importance during pull-in are the effects of inertia, or Wk^2, covered in Art. 17.

d. Pull-out torque requirements of some applications may exceed values given in Table 5. For example, a pull-out-torque requirement of 250 percent is not uncommon in the wood or rubber industries.

In any new or unusual application the torque which the driven machine can impose on the motor must be known. Knowing the maximum load torque during a cyclic varying load, for example, enables the application engineer and motor designer to provide a correct motor design. Effects of line-voltage dip, if any, must also be allowed for. The synchronous-motor pull-out torque is approximately proportional to the voltage for fixed excitation.

e. Normal Wk^2 values of the load, on which the pull-in-torque values of Table 5 are based, may be calculated from the following formula:

$$\text{Normal } Wk^2 \text{ of load} = \frac{0.375 \times \text{hp}^{1.15}}{(\text{r/min}/1000)^2} = \text{lb} \cdot \text{ft}^2 \qquad (1)$$

Numerical values calculated by Eq. (1) for a wide range of horsepower ratings and speeds are conveniently tabulated in Ref. 1. Load Wk^2 in excess of normal values requires special consideration in specifications and especially motor design. See Art. 17 for effects of Wk^2 during the motor- and load-starting period from standstill to full speed.

f. The number of starts considered in standards (Ref. 1) varies with the initial motor temperature. Initially at ambient temperature motors should be capable of two starts in succession, coasting to rest between starts. Only one start is permitted if the motor is at a temperature not exceeding its rated-load operating temperature. Conditions are that the load Wk^2, applied voltage, load torque during acceleration, and method of starting are those for which the motor was designed.

Starts in excess of the above must be carefully considered from the standpoint of motor heating. Hence users and application and design engineers must have complete mutual understanding when additional starts are desired or contemplated in a given application. This caution is especially important when Wk^2 exceeds normal values. In general the number of motor starts should be kept to a minimum. Motor stresses (torsional, coil end, etc.) are high during the starting period and affect the motor life. Special or oversize motor designs may also be necessary.

Normal motor overload protective relays will not protect a motor against damage resulting from too frequent successive motor starts.

g. The overspeed construction capability of salient-pole synchronous motors in an emergency is mentioned in standards (Ref. 1) as 25 percent without injury.

h. The direction of rotation should not be overlooked in specifying synchronous motors. For nonreversing motors it is counterclockwise when viewing the end of the motor opposite the drive (Ref. 1).

i. Voltage and frequency variations from rated values have been standardized for synchronous running conditions and rated field excitation. Motors shall operate successfully (1) at rated frequency and load with voltage variation 10 percent above or below rated, or (2) at rated voltage and load with frequency variation 5 percent above or below rated value, or (3) at rated load with a combined voltage and frequency variation 10 percent above or below rated values, provided the frequency variation does not exceed 5 percent (Ref. 1).

Operations under the variations indicated may not meet standards for normal rated conditions. Deficiencies in motor torque during starting must be considered when voltage or frequency deviations are anticipated. Locked-rotor and pull-in torques of a motor are approximately proportional to the square of the voltage and vary inversely as the square of the frequency.

j. Service conditions in application take account of altitude, ambient temperature, etc., as covered for induction motors in Sec. 4, Art. 13. The same problems and conditions apply to synchronous motors.

k. Pulsating armature current can occur when synchronous motors drive loads whose torque varies during each revolution. Typical drives are reciprocating loads such as compressors and pumps. The inertias of the rotating parts of the motor, load, and amortisseur winding design, when properly engineered, can limit the motor line-current pulsation. Standards suggest limiting values of armature-current pulsation. A top maximum value, for example, does not exceed 66 percent (Ref. 1).

Current variation must be obtained by oscillographic methods rather than with meter readings. A line drawn through the consecutive peaks of the current wave of the oscillogram traces out the envelope of the current wave. The variation is the difference between the maximum and minimum ordinates of the envelope. To comply with the preceding paragraph this variation must not exceed 66 percent of the maximum value of the rated full-load current of the motor. The latter is assumed as 1.41 times rated full-load current.

l. Noise levels of synchronous motors have not been standardized. The principles, covered in Sec. 20, may also be used for measurements of synchronous-motor noise. The test conditions may be to run the motor idle at its rated voltage and frequency. Field-excitation currents are set for minimum armature current (Rcf. 4).

m. Other conditions of application of synchronous motors may apply in special instances. These conditions may impose motor mechanical or electrical problems, or both.

17. Starting Characteristics. The starting characteristics of synchronous motors span the speed range from standstill to synchronism. Locked-rotor (static) torque, accelerating torques, and pull-in torque are involved.

a. The locked-rotor torque and the torque developed during acceleration (accelerating torque) depend on currents induced in rotor circuits by the stator-winding established air-gap magnetic field. The field induces torque-producing currents in the pole field-coil winding, which is usually short-circuited through a resistor, and in the pole-head starting amortisseur winding. The field-coil torque is small during most of the accelerating period because of the high reactance to resistance ratio of the winding. At high speeds, the field-coil reactance becomes small because of low induced frequency and the field-winding torque then becomes appreciable, especially at low values of rotor slip near synchronism.

The starting amortisseur winding, resembling an induction-motor squirrel cage, has copper or alloy bars or rods embedded in the outer periphery of the poles. Because the starting amortisseur is not used during synchronous operation, it can be designed primarily for torques to suit starting and acceleration requirements of the driven load. A typical synchronous-motor torque curve of a fan or pump drive is shown in Fig. 4.

b. The pull-in torque phenomenon encompasses the period of transition from induction-motor- to synchronous-motor-type operation when direct current is applied to the pole field winding. Thus after the motor has attained its highest speed operating like an induction motor, direct current may be applied to the pole winding, and pull-in starts and is completed as synchronism is successfully attained. In Fig. 4 the synchronous motor accelerates like an induction motor to an average speed of 96 percent of synchronous. However, the rotor speed will not be of uniform velocity because rotor pole pieces and interpole spaces do not present a uniform peripherally distributed magnetic circuit and starting winding to the stator rotating magnetic field. There are, therefore, variations in induced rotor currents causing torque variations. On application of dc field-winding excitation, additional torques develop. The rotor speed then alternately increases and decreases more widely than before. The rotor will pull into synchronism on one of the speed-increase excursions provided Wk^2 inertia is not excessive.

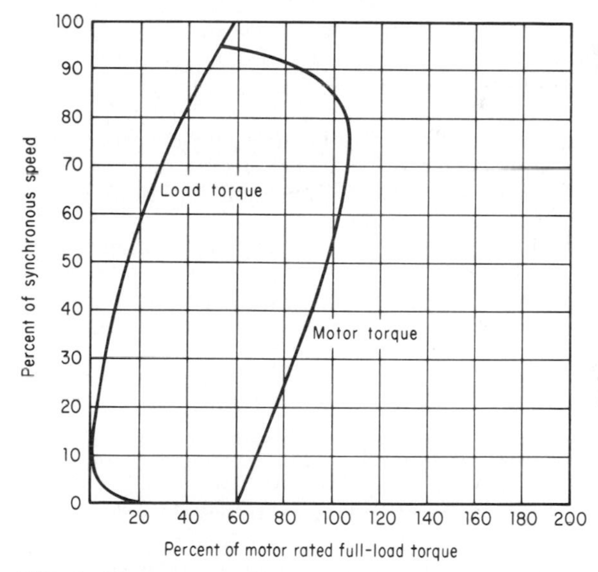

FIG. 4. Torque curves for synchronous motor of centrifugal pump or fan.

c. The effects of inertia Wk^2 of motor and load are important during pull-in. High-inertia loads, for example, may damp out the amplitude of rotor swings near synchronism when field excitation is applied, and swings then may not be sufficiently large to attain synchronous speed. If the driven load can be reduced, the motor may attain, say, 98 percent speed on its induction-type characteristic. Being closer to synchronous speed, the damped speed swings of the rotor may be adequate for the motor to pull into synchronism when dc excitation is applied to the field winding.

d. High Wk^2 values also require special design attention to the starting amortisseur windings in the pole faces. Sufficient material must be allowed for in the starting winding to absorb the winding heat loss incident to heavy inertia Wk_2 starting.

18. Methods of Starting Synchronous Motors. Synchronous motors are started at full or reduced voltage. Reduced voltage is obtained from autotransformers (Sec. 4, Art. 6) or from reactors in series with the motor, either in the supply line or in the motor neutral (Sec. 4, Fig. 3). Approximate starting voltages supplied by standard voltage-reducing devices are shown in Sec. 4, Table 12.

a. Autotransformer starting of synchronous motors operates in the same manner as for induction motors (Sec. 4, Art. 14). Reactor starting of syncchronous motors is discussed in the example of Art. 19.

b. Part-winding starting is sometimes specified for synchronous motors. Because it is not widely used for larger motors, a detailed discussion is not deemed warranted. In general, the comments in Sec. 4, Art. 14, relating to large polyphase induction motors also apply to large synchronous motors.

19. Reactor Starting. Reactor starting of synchronous motors may be used for such drives as centrifugal pumps, motor-generator sets, and sintering fans. The object of using reactors is to keep the motor-starting current and kVA as low as possible to avoid line disturbances or demand charges during the accelerating period of motor and load.

The motor design must be such that the motor-developed shaft torque with reactor is adequate to accelerate the load up to full speed. When load Wk^2 and torque requirements are known, the application and motor-design engineer can provide the right motor characteristics.

The curves of Fig. 5 show an actual reactor application to limit the motor-current increments to twice full-load current. An example based on these curves will illustrate a method of determining the reactor value to use in series with the motor. The computed reactor value is the same for either line or neutral location (Sec. 4, Fig. 3).

Calculations in the following example of reactor starting of synchronous motors use the per unit system. Computations using per unit values are simpler and generally more convenient than percentages or ohmic values. A short description of the per unit system as applied to the proposed example problem may be helpful.

For a three-phase motor, rated terminal-to-neutral voltage is rated terminal-to-terminal voltage divided by $\sqrt{3}$. This is the base voltage used in motor equivalent-circuit numerical computations. The voltage value so calculated is referred to as 100 percent in the percent system or as 1.0 per unit in the per unit system.

The motor rated full-load current is the base current and has a value of 100 percent in the percent system or 1.0 in the per unit system.

The base ohms impedance of a motor is the base voltage divided by the base current. In the per unit system the motor base impedance is 1.0, that is, the base voltage of 1.0 divided by the base current of 1.0.

In Fig. 5 the motor current at rated voltage and zero speed (locked rotor) without reactor is 4.7 per unit at 0.25 (25 percent) power factor. Hence the motor impedance is

$$Z_M = \text{motor impedance} = \frac{1.0}{4.7} = 0.213 \text{ per unit}$$

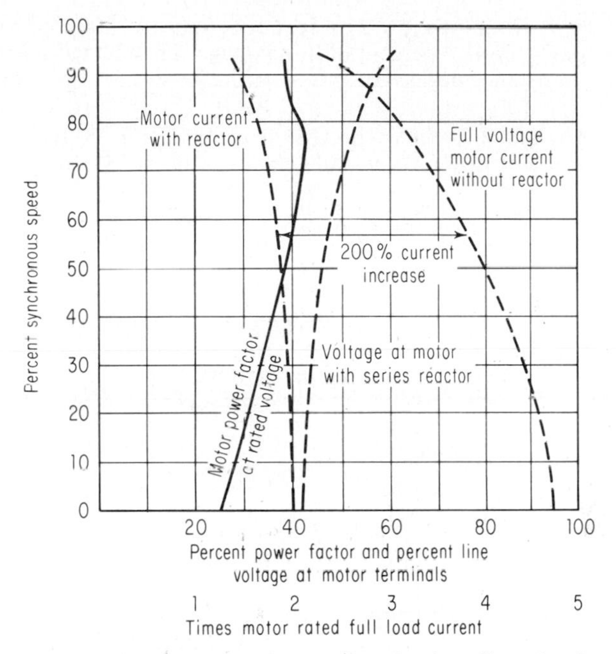

FIG. 5. Synchronous-motor starting showing effect of series reactor to limit starting-current increments to 200 percent of full-load current.

$$= \text{motor resistance} + j(\text{motor reactance})$$
$$= 0.213[0.25 + j(1.00 - 0.25^2)^{1/2}]$$
$$= 0.053 + j0.206 \text{ per unit}$$

To limit the motor current, and hence the line current, to twice the full-load value at zero speed, a reactor is connected in series with the motor (Sec. 4, Fig. 3), so that

$$2.0 = \frac{1.0}{Z_M + Z_R} \qquad \text{per unit current}$$

where Z_R is the per unit reactor impedance.

Neglecting reactor resistance $Z_R = X_R$, or reactor reactance, then

$$2.0 = \frac{1.0}{0.053 + j(0.206 + X_R)}$$
$$= \frac{1.0}{[0.053^2 + (0.206 + X_R)^2]^{1/2}}$$

The motor is rated 1000 hp, 100 percent power factor, 2300 V, three-phase, 60 Hz, 900 r/min, 197 A at full load. Base ohms = $2300/(\sqrt{3} \times 197) = 6.75\ \Omega$. Thus reactor ohms = $0.290 \times 6.75 = 1.96\ \Omega$ in each line of Fig. 3, Sec. 4.

Voltage across the motor at zero speed = IZ_M = 2.0(0.053 + j0.206) = 2.0(0.212) = 0.424, or 42.4 percent of line voltage. The voltage across the reactor is $IZ_R = IX_R$ (resistance neglected) = 2.0 × 0.290 = 0.58, or 58 percent of line voltage.

Computations for 50 percent speed follow. From Fig. 5, the motor current without reactor = 4.0, or four times full-load current, at 38 percent power factor.

$$Z_M = \frac{0.38}{4.0} + \frac{j(1.00 - 0.38^2)^{1/2}}{4.0} = 0.095 + j0.231 + 0.250$$

$$jX_R \text{ is unchanged} = j0.290 = Z_R$$

$$Z_M + Z_R = 0.095 + j0.521 = 0.529$$

$$\text{Motor current with reactor} = \frac{1.00}{0.529} = 1.89 \text{ per unit}$$

$$\text{Motor current} = 1.89 \times 197 = 373 \text{ A}$$

$$\text{Motor-terminal voltage} = IZ_M = 1.89 \times 0.250 = 0.47 \text{ per unit}$$

$$= 0.47 \times 2300 = 1080 \text{ V line to line}$$

As the motor speed increases, the motor impedance also increases, and the voltage across the motor increases.

The motor torque without reactor at full voltage and at 50 percent speed = 122 percent (Fig. 6). In the per unit system this torque is expressed as 1.22. Using the above, the computed reactor torque (varies as the square of motor current) equals $(1.89/4.00)^2 \times 1.22 = 0.27$ per unit, or 27 percent of full-load torque. Full-load torque in pound-feet may be calculated from Eq. (19) in Sec. 4.

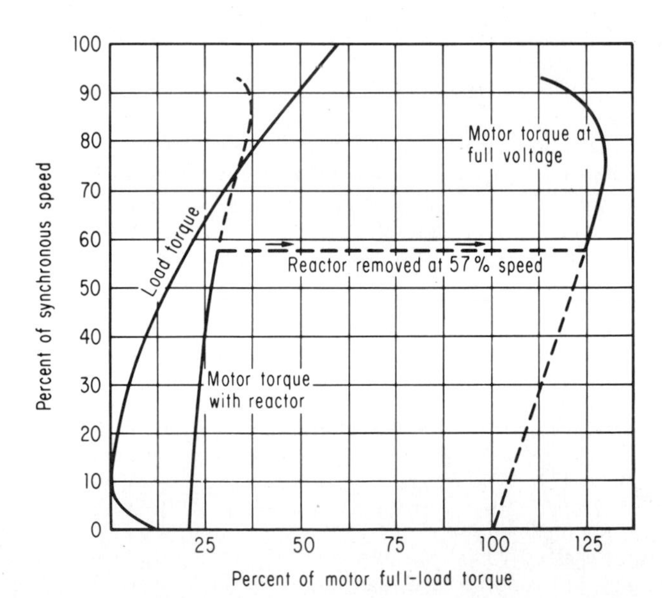

FIG. 6. Synchronous-motor torque curves with series starting reactor.

As shown in Fig. 5, the reactor may be short-circuited at 57 percent speed so that the motor is connected to full voltage with a current increment not exceeding two times the rated value as shown.

An important consideration in the above discussion may be pointed out in Fig. 6. If 57 percent speed must be attained for a current increment not exceeding twice full-load value when the reactor is cut out, then the motor must have adequate torque with the reactor to attain 57 percent speed. In Fig. 6 the motor torque with the reactor could accelerate to 75 percent speed, where the motor and load torques just balance. Thus 57 percent speed is readily attained. In fact, had the motor accelerated to, say, 65 percent speed, the current increment, when the reactor was removed, would have been 3.60 − 1.75 = 1.85 times full-load current.

20. Overexcited Synchronous Motors. Overexcited motors operate to provide reactive kilovolt-amperes (kvar) like capacitors. Underexcited, the motors operate like induction motors as regards the kvar effect on a system. Induction motors tend to depress or reduce system voltage by requiring lagging kVA (kvar). Overexcited synchronous motors thus benefit system performance by compensating for induction-motor kvar.

The amount of kvar compensation a synchronous motor can provide may be predicted from a type of motor characteristic called V curves because of their distinctive shape. Such a curve for full-load operation is shown in Fig. 2. Note that as the field current i_f is adjusted during full-load operation, stator (armature) current and motor power factor vary.

For example, from Fig. 2, at 100-A field current i_f, the motor power factor is 84 percent, the stator current 508 A. Hence when the rated voltage = 4160 V,

$$\text{kVA} = \sqrt{3} \times 4160 \times \frac{508}{1000} = 3660 \qquad \text{[from Eq. (5), Sec. 4]}$$

$$\text{kW} = 3660 \times 0.84 = 3070 \qquad \text{[from Eq. (6), Sec. 4]}$$

$$\text{kvar} = 3660(1.00 - 0.84^2)^{1/2} = 1985 \qquad \text{[from Eq. (7), Sec. 4]}$$

V curves for 50, 75, and 100 percent load are shown in Fig. 3. Power-factor lines may be drawn in as shown. This method of drawing the curves is of more general utility because motor-load variations are taken into account. Interpolation between curves can be made to obtain approximate data. In applying the curves, the manufacturer's maximum values of stator and field current should not be exceeded.

21. Kvar Compensation. As mentioned in the previous article, kvar compensation can be supplied by synchronous motors for lagging kVA (kvar) of induction motors. Example problems will illustrate methods of computation.

A simple problem assumes that a generator in Fig. 7*a* supplies two identical induction motors. In Fig. 7*b* an 80 percent power-factor (overexcited) synchronous motor of the same horsepower replaces one of the induction motors of *a*. Computations in the figure show that the generator current is reduced when the synchronous motor replaces the induction motor. Computations for each case are shown in the respective vertical columns of the figure. Note that in Fig 7*b* net kvar is zero since the synchronous motor compensates for the kvar of the replaced induction motor.

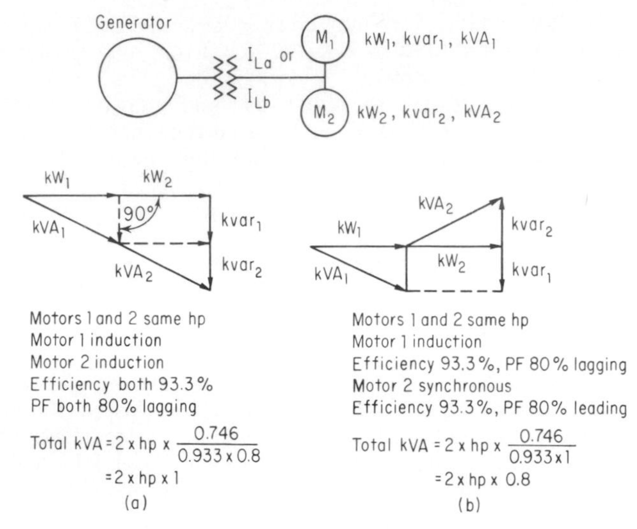

FIG. 7. (*a*) Generator supplies two induction motors. (*b*) One induction motor is replaced by synchronous motor.

A general method of kvar computation for mixed synchronous and induction motors is illustrated in the example problem of Fig. 8. Characteristic data of Fig. 8 are used to calculate the bus information shown in Table 6. Computations in the table are detailed and self-explanatory. Line losses are neglected. Note that the 80 percent power factor (overexcited) synchronous motor compensates for a large portion of the lagging kVA (kvar) of two induction motors. The power factor at the bus is thus raised considerably over that of the two induction motors acting alone.

Should more motors of either type be involved, the computation form in Table 6 may be readily extended. Calculations may be made according to the indicated routine. In col. 4, kvar should be made + (positive) for induction motors and − (negative) for synchronous motors (overexcited). For a 100 percent power-factor synchronous motor enter a zero in col. 4.

22. Dynamic-Braking Requirements. Dynamic braking requirements may arise in the application of synchronous motors. For example, in emergencies, rubber mill rolls must be stopped rapidly to protect the operator. Laws and safety codes prescribe the stopping of individual new mills in no greater than 1½ percent of roll travel as measured in feet per minute on their surface. For new mills driven in groups of two or more the figure of 1¾ percent of roll

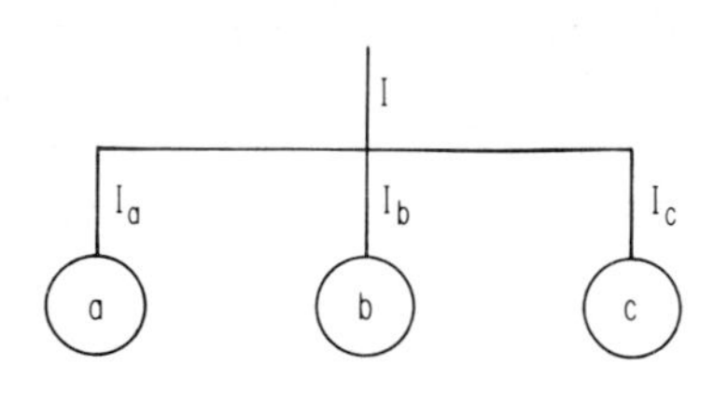

Motor	hp	Type	A	% eff	% PF
a	200	Induction	53.5	91	77
b	250	Induction	59.5	91	74
c	500	Synchronous	128.5	92	80

FIG. 8. Feeder supplies bus for mixed 2300-V three-phase 60-Hz induction and synchronous motors.

TABLE 6. Synchronous-Motor kvar Compensation for Induction Motor

Motor	hp (1)	Output kW (2)	Input kW (3)	kvar (4)	kVA (5)	I, amperes, (6)
A	200	149.2	164	−136	213	53.5
B	250	186.5	205	−186	277	59.5
C	500	373	406	+305	510	128.5
Sum	950	708.7	775	−17		

For each motor:
col. 2 = 0.746 × col. 1
col. 3 = col. 2/efficiency from Fig. 8
col. 4 = Eq. (7), Sec. 4
col. 5 = Eq. (5), Sec. 4; volts = 2300
col. 6 = Eq. (4), Sec. 4
Bus computation:
Bus kW = 775; col. 3
Bus kvar = −17; col. 4
Bus kVA = 775 (approx); Eq. (8), Sec. 4
Bus amperes = 196; Eq. (5), Sec. 4
Bus power factor = 1.0 (approx); col. 3/bus kVA
Bus efficiency = 0.914; col. 2/col. 3
Note that bus kVA and I are not the arithmetic sums of the individual motor values but must be calculated according to the indicated equations.

travel is used. However, a check with the machine manufacturer should be made for possible updating of the data.

Systems of rotating inertias may possess considerable kinetic energy in some applications. In the stopping process this kinetic energy is removed from the revolving system and appears in the form of heat somewhere in the braking apparatus. With mechanical braking heat appears at the brake lining or shoes. With electrical braking conditions pervail as described in the following discussion.

Electrical rather than mechanical braking may be used in synchronous-motor applications. Two methods are (1) plugging or (2) short-circuiting the stator terminals across resistors. Method 1, plugging, is little used because of high damper-bar losses and heating possibilities as well as large motor line-supply currents. In plugging, two stator-terminal leads are reversed to reverse the direction of the synchronously rotating air-gap field of the motor. Braking currents are induced in the pole-face damper bars (the field winding is short-circuited) to produce the torque to stop the rotating parts. If the supply line is not removed when standstill speed is reached, the motor will accelerate back up to speed in the reverse direction. To prevent reverse rotation, a speed switch may be used to initiate circuits to remove the primary or stator electric power automatically at the proper speed close to standstill speed. In this method damper-bar current losses may cause excessive bar heating where the application Wk^2 is large. Braking by method 2 is therefore more commonly used.

Braking method 2 is illustrated in Fig. 9, where the motor armature (stator usually) winding is shown suddenly switched from the supply-line power source to the resistor bank. The field current is usually not changed, although in some applications the exciter voltage is suddenly increased (field forcing) to increase induced resistor currents and loss. The braking torque thereby increases.

In Fig. 9 the switched synchronous motor becomes a generator to supply power to the resistor bank. The load and motor-rotor inertias become the generator prime mover or driving force to supply generator electrical-energy output and losses in the short-circuiting resistors. As resistor losses develop, kinetic energy is drawn from rotating inertia parts and the system of load and motor slows down and stops. Reverse speed does not occur.

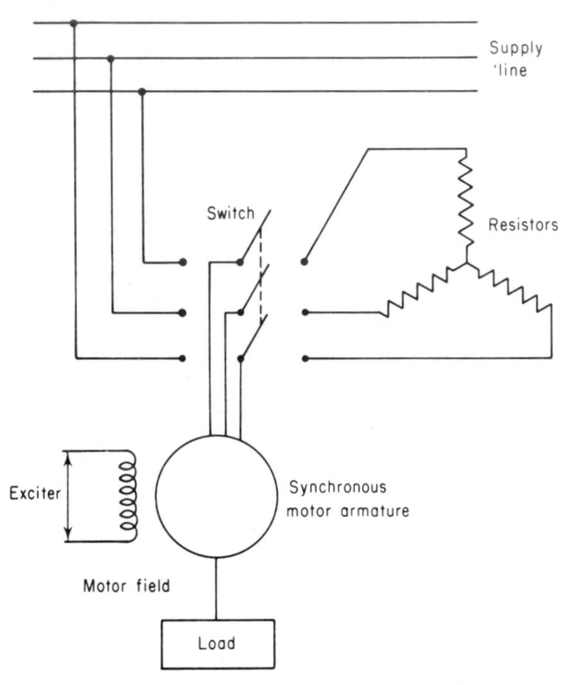

FIG. 9. Elementary diagram showing dynamic braking of synchronous motor suddenly connected across resistors.

An approximate value for the resistor in Fig. 9 required to provide the minimum number of revolutions to stop is given by

$$r = \frac{x'_d}{\sqrt{3}} \tag{2}$$

where r = terminal-to-neutral resistor. The resistance includes both stator winding and external values in ohms.

x'_d = motor transient reactance, direct axis, in ohms, terminal-to-neutral value. It may be obtained from the manufacturer.

The value calculated is the single-valued resistor for the fewest stopping revolutions. Actually the resistor would be required to vary continuously for theoretically best results. Such a procedure is impractical in motor application.

a. Minimum revolutions to stop using the resistance from Eq. (2) may be shown to be approximately

$$\text{Revolutions to stop (minimum)} = \frac{N_0 H x'_d}{26(e'_d)^2} \tag{3}$$

where N_0 = synchronous speed, r/min
H = kWs/kVA
= $0.231\ Wk^2(N_0/1000)^2$/kVA
kVA = rated full-load value
Wk^2 = lb · ft² on N_0 basis
x'_d = per unit transient reactance, direct axis
e'_d = voltage behind transient reactance, per unit (see Fig. 10)

If x'_d is in ohms from Eq. (2), it may be divided by base ohms to obtain x'_d in per unit. Base ohms is the rated terminal-to-terminal voltage divided by $\sqrt{3}I$, where I is the rated full-load current in amperes.

b. The braking resistor must have sufficient thermal capacity to dissipate electrically generated losses during the braking period. Because the time to brake to rest is usually very short, all kinetic energy removed from the rotating-system inertias may be assumed to appear as heat in the braking resistor and generator armature. The slight loss due to windage, friction, and iron losses may be ignored. Thus computations for braking resistors usually specify them to have thermal-storage capacity for the number of successive stops desired. The value of H [Eq. (3)] times kVA can be used to compute the kW-seconds energy they must absorb per stop without undue heating.

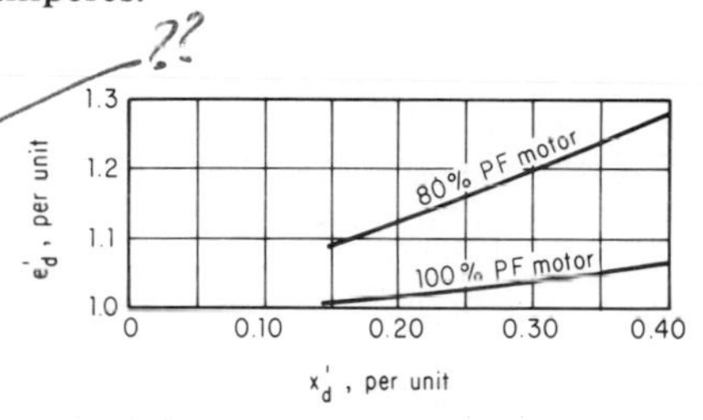

FIG. 10. Per unit values of e'_d for typical values of per unit x'_d to calculate dynamic-braking revolutions to stop [Eq. (3)].

23. Excitation. Excitation systems provide a dc supply for the rotor-field winding of synchronous motors. Dc field supply is necessary not only for normal synchronous-speed operation, but also for pulling into synchronism at the end of the starting period. Several methods of obtaining a dc supply are available, such as direct shaft connected, belted or chain-driven exciters driven by the motor itself, motor-generator sets, and local dc bus. Static and brushless exciters are more recently developed sources which are becoming popular because of certain advantages discussed below.

a. Rotating dc exciters of the commutator type are familiar because of their wide application since the earliest days of synchronous-motor drives. In smaller motors the exciter may be directly connected to the motor shaft extension. The exciter must then be capable of building up its own voltage as the motor speed increases. Variation of excitation (exciter) voltage during normal operation of the motor may be obtained by exciter-field rheostatic control. Exciter stability is essential when the exciter-field current is reduced to provide the lowest voltage required for the lowest motor-field current. In the reapplication of rewinding of synchronous motors it may be found that the exciter used is unstable at the low voltage required for reduced motor-field currents. In such cases the exciter can be operated at a higher stable voltage and a series resistor or rheostat can be placed in series with the motor field for low collector-ring voltage requirements. Rheostat or series-resistor losses are, or course, undesirable but may have to be tolerated.

Low-speed synchronous motors usually do not have direct-connected exciters. Low-speed exciters are larger and more costly than higher-speed types. For low-speed synchronous motors an exciter belted or chain-driven by the motor to run at a higher speed provides an economic solution.

b. A motor-generator set driven by an induction motor provides an economical excitation supply for low-speed synchronous motors where space in the vicinity of the motor is at a premium. The motor-generator set can be placed in a convenient location. These sets are designed for speeds to provide economical motor and exciter physical size, weights, etc.

c. A local bus, where available, can also supply excitation for synchronous motors. Motor-field current adjustments can be made using a series rheostat or a tapped resistor to obtain motor collector-ring voltage variations. Rheostat or resistor losses may not be large if the motor-field current is maintained at or near nameplate value (Art. 20).

d. Static exciters using silicon rectifiers provide a factory-wired and -assembled nonrotating system of excitation for synchronous motors. Such units, connected to an ac supply, develop conventional 125- or 250-V dc output. Static exciters provide high efficiency, 90 percent or more, at a high power factor of 93 percent or more.

A typical static-excitation system may use a transformer with primary voltages of 208, 230, 460 V, etc. Taps of plus or minus 5 percent may be used on the primary winding for close line-voltage matching. The secondary winding may have a number of taps to provide, say, 4 percent steps in dc voltage down to about 70 percent of rated value. Various types of protection are incorporated in static-exciter units. The primary winding may be fused for short-circuit protection, diodes protected against surges of pull-out or pull-in origin, etc.

e. Brushless excitation is, as the name signifies, a method of supplying synchronous-motor field excitation without requiring exciter or motor brushes. Hence motor collector rings are not needed. The system may employ an alternator, the output of which is rectified to obtain the dc motor-field supply. The alternator armature is mounted on the motor shaft either inside or outside the motor bearing opposite the motor-load drive end. The alternator field is mounted stationary with a small air gap between it and the rotating armature. By adjusting the alternator-field excitation, the alternator-armature voltage may be varied. The alternator armature leads connect directly to rectifier elements, switches, surge-protection devices, etc., mechanically fastened to and rotating with the synchronous-motor rotor. Leads from the rectified output are connected to the synchronous-motor field winding to provide motor-field excitation.

When the rotating armature of the exciter is mounted outside a motor bearing, its leads are ducted through a shaft drilled hole exiting on the inner side of the bearing. The leads are then connected to rectifier elements, etc., as described above (see Sec. 10, Part 2).

f. The rotating transformer principle has a stationary primary supplied from an ac source. The secondary is mounted opposite the primary on the shaft with which it rotates. Rectifying elements and other instrumentation for field, pull-in, etc., requirements are rotor-mounted.

24. Protection. Protection methods for synchronous motors may utilize devices listed for induction motors in Table 14, Sec. 4. These devices are, for example, space heaters in the motor enclosure, surge capacitors and lightning arresters in the motor-terminal box, and slot-embedded or end-winding thermal sensors. Other protective devices are primarily external to the motor, such as overload relays and differential relays, mentioned in Sec. 4, Art. 17. These are applicable to synchronous as well as induction motors.

There are types of protection designed primarily for synchronous motors, for example, protection at pullout, for loss of field, starting-winding protection, and incomplete sequence. Another important type of protection relates to ground faults and is important for and applicable to either synchronous or induction motors.

a. Pull-out protection prevents prolonged out-of-synchronism operation on the starting winding of salient-pole-type synchronous motors. Pullout may be due to overloading, insufficient field current, low line voltage, line disturbances, etc. From a thermal standpoint the starting winding is not adequate for prolonged operation at full motor load. Besides, because of the incomplete nature of amortisseur windings, out-of-step (asynchronous) operation is accompanied by motor-torque variations and motor-line-current disturbances.

The control engineer can design suitable relaying and circuitry to permit resynchronizing should the situation causing the pullout be corrected. Automatic load unloading, for example, may provide a suitable correction. However, if such load correction is not obtained, motor shutdown occurs. Some motors cannot resynchronize under full load.

b. Loss-of-field protection should be considered for synchronous motors. Loss of field or insufficient field is sensed by a current relay in series with the motor field. Relays used in pull-out protection are not operative should no transient field currents flow, as would occur with an open field, field-line break, collector-brush breakage, etc. The loss-of-field relay offers important protection. Control engineers can offer valuable suggestions for loss-of-field protection.

c. Starting-winding protection for synchronous motors can be provided by special relays and circuitry. Power for the protective devices may be obtained from the field-discharge circuit during the starting period. Thus the field-source signal, of varying magnitude and frequency when the motor speed varies, acts on circuit components having thermal characteristics approximating motor starting-winding thermal-storage and time constants.

d. Incomplete sequence protection has as its purpose the protection of important intermittently rated controller components. Such components may be reactors, autotransformers, etc., used for starting duty. Operated on the basis of timing out, the protective device is inoperative should the starting of the motor proceed to successful completion to synchronous speed. In case completion is not attained, the protective device stops the motor.

e. Ground-fault protection for either synchronous or induction motors has as its purpose the detection and initiation of switching to interrupt and clear a grounded phase conductor. Thus in the event a stator-winding conductor is grounded, the ground-fault protective equipment acts to initiate tripping of the power-supply circuit. Power flow and damage at or in the vicinity of the fault are minimized by rapid relaying. The power system itself is also protected.

With the neutral of the power-transformer ground connected through a resistor or reactor, ground-fault currents may be limited to a safe value. The current-interrupting device (line switch, circuit breaker, etc.) must be able to handle safely the maximum available ground-fault current.

1. A common method of ground-fault protection uses a doughnut or toroid-type current transformer through the window of which all three motor leads are threaded (Fig. 11). Normally all three motor leads carry equal currents so that transformer flux and hence secondary current are zero. During ground-fault conditions unbalanced motor currents produce current-transformer flux and hence secondary currents. The relay in the transformer secondary operates an instantaneous overcurrent relay to remove the motor from the line.

   This method has the advantage of simplicity because only one transformer

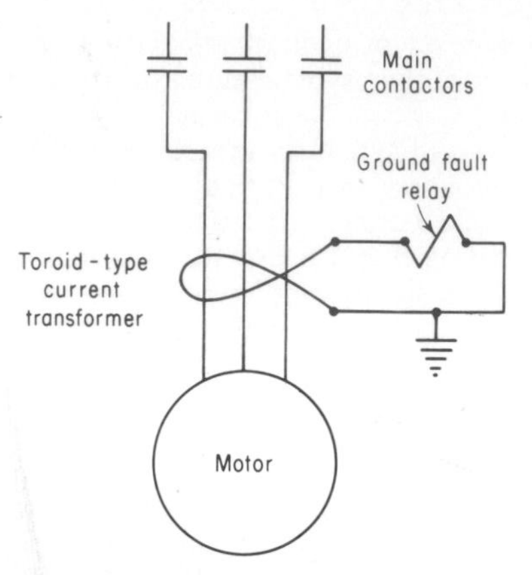

FIG. 11. Toroid-type current transformer ground-fault protection.

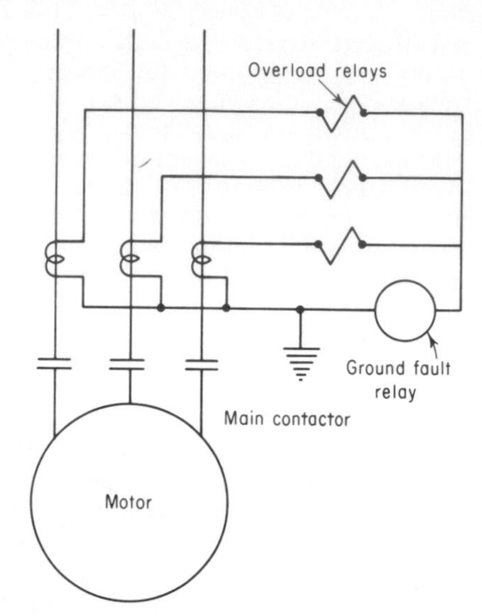

FIG. 12. Ground-fault protection using overcurrent relays.

and one overcurrent relay are needed. The method may be difficult to apply to larger low-voltage motors because of fitting large-sized conductors or cables of all three motor-line leads through the transformer window opening.

2. Another method of ground-fault protection has been labeled residual tripping. The method uses a sensitive relay placed in the grounded neutral of the secondarics of three current transformers (Fig. 12). During three-phase balanced conditions neutral zero-sequence currents do not flow and the relay does not operate. Under motor phase-current unbalance during a ground fault, the current-transformer neutral current operates the relay.
3. Differential current-relay protection is the most sensitive and most expensive ground-fault protection. Six current transformers are required, one at each end of the three motor windings of a three-phase motor. In this method equal current in and out of a motor-winding phase is monitored during normal operation. Under motor ground-fault conditions, the current at the ends of a motor winding is unequal, that is, the current entering is not equal to the current leaving the faulted phase. The difference in current causes relay operation to remove the motor from the line.

f. Ungrounded systems require a different treatment for motor-fault protection. When a motor conductor grounds, no fault current flows. Thus tripping the motor off the line would serve no useful purpose. A hazard results, however, because motor windings may be subjected to much higher voltages than normal when a ground occurs. For example, if the ground appears near a motor-terminal conductor, the voltage between the terminal coils of the other two phases and the grounded conductor is approximately 73 percent greater than normal in a wye-connected motor winding. Chances for coil or conductor faulting of the two unfaulted phase are increased. In fact, a line-to-line fault results should another ground occur near a terminal coil of a second phase. Control engineers should be consulted for protective means for motors operating on an ungrounded system.

REFERENCES

1. "Motors and Generators," ANSI/NEMA Standard MG 1-1978.
2. "Test Procedures for Synchronous Machines," ANSI/IEEE Standard 115-1983.
3. "General Requirements for Synchronous Machines," ANSI Standard C50.10-1977.
4. "Test Procedure for Airborne Sound Measurements on Rotating Machinery," IEEE Standard 85-1973 (Reaff 1980).
5. "Guide for ac Motor Protection," ANSI/IEEE Standard C37.96-1976 (Reaff 1983).
6. M. S. Sarma, *Synchronous Machines: Their Theory, Stability, and Excitation Systems* (Gordon & Breach, New York, 1979).

SECTION 6

AC THREE-PHASE MOTORS (680 Frame and Smaller)

Paul J. Dobbins*

FOREWORD 6-2

POLYPHASE-INDUCTION-MOTOR TYPES 6-3
1. Design A 6-3
2. Design B 6-3
3. Design C 6-4
4. Design D 6-4
5. Design F 6-4
6. NEMA Rerate 6-5
7. Multispeed Induction Motors 6-7
8. Wound-Rotor Induction Motors 6-8

SERVICE CONDITIONS 6-10
9. Service Factor of General-Purpose ac Motors 6-14
10. Proper Selection of Apparatus 6-14
11. Usual Service Conditions 6-14
12. Unusual Service Conditions 6-16
13. Operation at Altitudes above 3300 ft (1000 m) 6-17
14. Variation from Rated Voltage 6-18
15. Variation from Rated Frequency 6-18
16. Combined Variation of Voltage and Frequency 6-18
17. Service Factors of General-Purpose and Other 40°C Motors 6-18
18. Overspeeds 6-18
19. Short-Time-Rated Electric Machines 6-18
20. Noise Quality 6-19
21. Noise Measurement 6-19

MECHANICAL ARRANGEMENTS 6-19
22. Flat-Belt Pulleys 6-19
23. V-Belt Sheaves 6-20
24. Chain Sprockets 6-20
25. Gears (Spur and Helical) 6-22

*Electrical Design and Development Engineer, Large AC Motor and Generator Department, General Electric Company, Schenectady, N.Y. (retired); Member, IEEE.

NEMA DESIGNATIONS 6-22
26. NEMA Standard Code Letters for Locked-Rotor kVA 6-22
27. Frame Assignments 6-22
28. Lettering Dimensions and Outline Tables 6-22
29. Hazardous Atmospheres 6-22

HELPFUL INFORMATION 6-24
30. Application Data 6-24
31. Secondary Data 6-40
32. Motor-Selection Factors 6-42
33. Basic Steps 6-42
34. Duty Classification 6-43
35. Horsepower Required 6-46
36. Duty Cycle and Inertia 6-46
37. Moment of Inertia 6-47
38. Brake Selection 6-47
39. Service Factor 6-47
40. Axial Thrust on Bearings 6-47
41. Thermal Protection 6-47
42. Handy Formulas 6-48
43. Rules of Thumb 6-48

ENERGY CONSIDERATIONS 6-49
44. Motor-Selection Factors 6-49
45. Efficiency 6-50
46. Motor Losses 6-52
47. Efficiency Testing Methods 6-53
48. Manufacturing Variations 6-53

EVALUATION OF EFFICIENCY ECONOMICS 6-53
49. Simple-Payback Analysis 6-54
50. Present-Worth Life-Cycle Analysis 6-54

OPERATING CONSIDERATIONS 6-55
51. Power Factor 6-55
52. Application Analysis 6-57
53. Applications Involving Load Cycling 6-57
54. Applications Involving Extended Periods of Light-Load Operation 6-60

FOREWORD

Ac motors in frame 680 and smaller sizes have the widest application of any group of motors. They are used in all industries and drive the greatest variety of machines. These motors are produced in large volume, and because of their degree of standardization they can be stocked and readily substituted for motors of like rating of other manufacturers.

This section provides considerable data on these motors and should be helpful in selecting the proper motor for any load.

POLYPHASE-INDUCTION-MOTOR TYPES

The polyphase induction motor may be of either the squirrel-cage or the wound-rotor type (Figs. 1 and 2). The squirrel-cage induction motor has been classified by the National Electrical Manufacturers Association (NEMA) according to the following designs. The tables in this section are all taken from NEMA standards.

1. Design A. A Design A motor is a squirrel-cage motor designed to withstand full-voltage starting and to develop locked-rotor torque as shown in Table 1. It has a breakdown torque as shown in Table 3. It has a locked-rotor current higher than the value shown in Table 4 and a slip at rated load of less than 5 percent.

Design A motors are usually used for applications where extremely high efficiency and extremely high full-load speed are required. Therefore Design A motors tend to be special motors.

2. Design B. A Design B motor is a squirrel-cage induction motor designed to withstand full-voltage starting, developing locked-rotor and breakdown torques adequate for general application as specified in Tables 1 and 3, drawing locked-

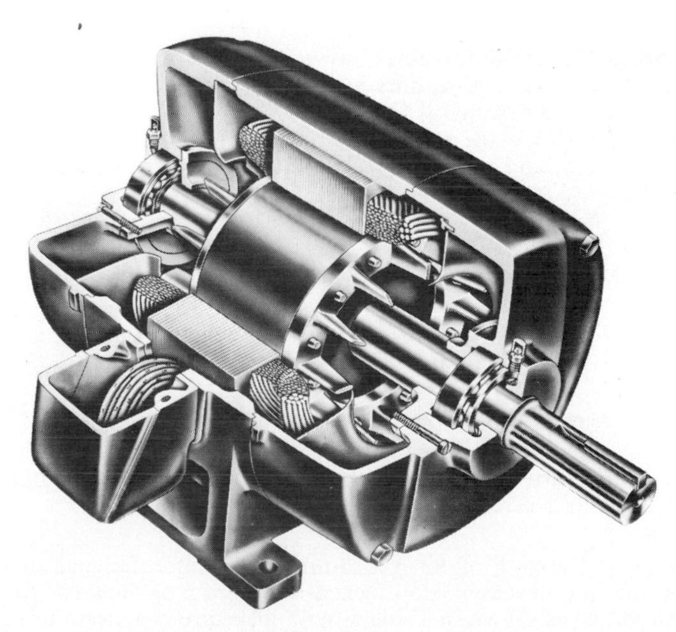

FIG. 1. Cutaway view of dripproof motor in frames 254U through 326U.

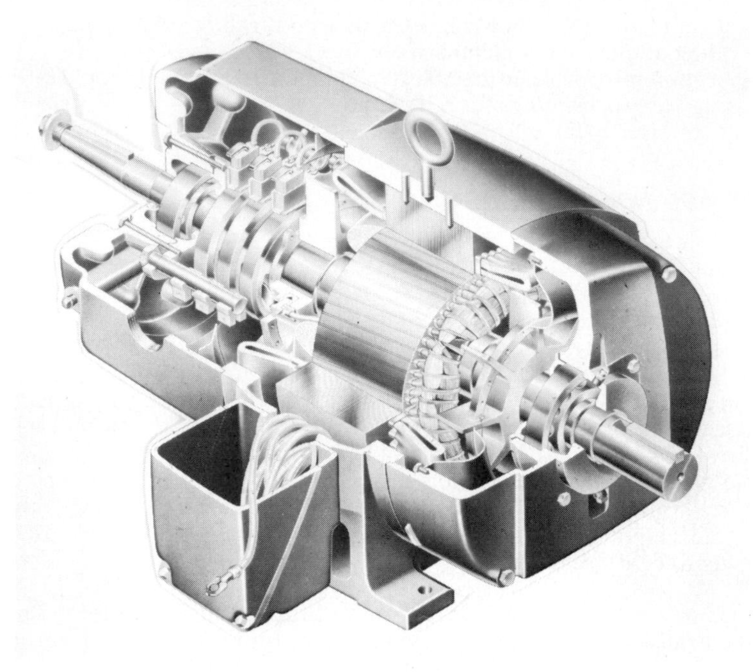

FIG. 2. Cutaway view of wound-rotor induction motor.

rotor current not to exceed the values shown in Table 4, and having a slip at rated load of less than 5 percent. Motors with 10 and more poles may have a slip slightly greater than 5 percent.

Design B motors are the standard general-purpose motors used where low locked-rotor current and moderate locked-rotor torque are required along with high full-load speed and efficiency.

3. Design C. A Design C motor is a squirrel-cage motor designed to withstand full-voltage starting, developing locked-rotor torque for special high-torque applications up to the values shown in Table 2, breakdown torque up to the values shown in Table 5, with locked-rotor current not to exceed the values shown in Table 4, and having a slip at rated load of less than 5 percent.

4. Design D. A Design D motor is a squirrel-cage motor designed to withstand full-voltage starting, developing high locked-rotor torque as shown in Table 2, with locked-rotor current not greater than that shown in Table 4, and having a slip at rated load of 5 percent or more.

5. Design F. A Design F motor is a squirrel-cage motor designed to withstand full-voltage starting, developing low locked-rotor torque as shown in Table 1 with breakdown torque as shown in Table 5, with locked-rotor current not to exceed the values shown in Table 4, and having a slip at rated load of less than 5 percent.

TABLE 1. Locked-Rotor Torque in Percent of Full-Load Torque of Single-Speed Polyphase Squirrel-Cage Integral-Horsepower Motors with Continuous Ratings, Designs A and B

hp	Synchronous speed, r/min			
60 Hz	3600	1800	1200	900
50 Hz	3000	1500	1000	750
½	...	...	...	140
¾	...	...	175	135
1	...	275	170	135
1½	175	250	165	130
2	170	235	160	130
3	160	215	155	130
5	150	185	150	130
7½	140	175	150	125
10	135	165	150	125
15	130	160	140	125
20	130	150	135	125
25	130	150	135	125
30	130	150	135	125
40	125	140	135	125
50	120	140	135	125
60	120	140	135	125
75	105	140	135	125
100	105	125	125	125
125	100	110	125	120
150	100	110	120	120
200	100	100	120	120
250	70	80	100	100
300	70	80	100	
350	70	80	100	
400	70	80		
450	70	80		
500	70	80		

For other speed ratings, see NEMA Standard MG 1-12.37.

Figure 3 shows typical speed-torque curves for Design A, B, C, and D motors. Figure 4 shows typical rotor punchings and rotor slot shapes used on Design A, B, C, D, and wound-rotor motors.

6. NEMA Rerate. NEMA has changed the suggested horsepower-frame assignments when the state of the art has advanced to a point where more horsepower can be taken from a given frame by use of new materials, design innovations, of both electrical and mechanical nature, etc. These changes are usually accomplished by a realignment of torque and current values. Since the latest rerate took

TABLE 2. Locked-Rotor Torque in Percent of Full-Load Torque of Single-Speed Polyphase Squirrel-Cage Integral-Horsepower Motors with Continuous Ratings, Design C

hp	Synchronous speed, r/min		
	60 Hz 1800	1200	900
	50 Hz 1500	1000	750
3		250	225
5	250	250	225
7½	250	225	200
10	250	225	200
15	225	200	200
20	200	200	200
25 and larger	200	200	200

The locked-rotor torque of Design D, 60- and 50-Hz, four-, six-, and eight-pole, single-speed, polyphase squirrel-cage motors, with rated voltage and frequency applied, should be 275 percent, expressed in percentage of full-load torque, which represents the upper limit of application for these motors.

The locked-rotor torque of Design F, 60- and 50-Hz, four- and six-pole, single-speed, polyphase squirrel-cage motors, rated 30 hp and larger, with rated voltage and frequency applied, should be 125 percent, expressed in percentage of full-load torque, which represents the upper limit of application for these motors.

For motors having nameplate voltages of 208 to 220/440 V or 208 to 220 V, the locked-rotor torques given above apply to operation on 220 V (or 440 V); on 208 V the locked-rotor torques are somewhat lower, as given in NEMA Standard MG 1-14.31.

From NEMA Standard MG 1-12.37.

TABLE 3. Breakdown Torque in Percent of Full-Load Torque of Single-Speed Polyphase Squirrel-Cage Integral-Horsepower Motors with Continuous Ratings, Designs A and B

hp	Synchronous speed, r/min				
	60 Hz	3600	1800	1200	900
	50 Hz	3000	1500	1000	750
½		...	...	...	225
¾		...	...	275	220
1		...	300	265	215
1½		250	280	250	210
2		240	270	240	210
3		230	250	230	205
5		215	225	215	205
7½		200	215	205	200
10–125, inclusive		200	200	200	200
150		200	200	200	200
200		200	200	200	200
250		175	175	175	175
300–350		175	175	175	
400–500, inclusive		175	175		

Design A values are in excess of values shown above.
NEMA Standard MG 1-12.38.

TABLE 4. Locked-Rotor Current of Three-Phase 60-Hz Integral-Horsepower Squirrel-Cage Induction Motors Rated at 220 or 230 V

hp	Locked-rotor current, A	Design letters
½	20	B, D
¾	25	B, D
1	30	B, D
1½	40	B, D
2	50	B, D
3	64	B, C, D
5	92	B, C, D
7½	127	B, C, D
10	162	B, C, D
15	232	B, C, D
20	290	B, C, D
25	365	B, C, D
30	435	B, C, D
40	580	B, C, D
50	725	B, C, D
60	870	B, C, D
75	1,085	B, C, D
100	1,450	B, C, D
125	1,815	B, C, D
150	2,170	B, C, D
200	2,900	B, C
250	3,650	B
300	4,400	B
350	5,100	B
400	5,800	B
450	6,500	B
500	7,250	B

Note. The locked-rotor current of motors designed for voltages other than 230 volts shall be inversely proportional to the voltages.
From NEMA Standard MG 1-12.34.

place in 1965, the tables for torques and current for both prererated motors and rerated motors are given in this section.

7. Multispeed Induction Motors. Multispeed induction motors usually have one or two primary windings. The one-winding motor can be connected to give either of two speeds, which have a ratio of 2:1, such as 1800 to 900 r/min, or it can be reconnected to have eight poles and give 900 r/min, assuming a 60-Hz power source. The two windings can be wound to give any synchronous speed and reconnected to give one other speed which is half the higher synchronous speed. This arrangement makes it possible for the motor to have four speeds, such as 3600, 1800, 1200, and 600 r/min. One winding is used for the 3600- and 1800-r/min speeds, and the second is used for the 1200- and 600-r/min speeds.

TABLE 5. Breakdown Torque of Single-Speed Polyphase Squirrel-Cage Integral-Horsepower Motors with Continuous Ratings, Designs C and F

hp	Synchronous speed, r/min (60 and 50 Hz)	Breakdown torque, % of full-load torque	
		Design C	Design F
3	3,600–3,000		
	1,800–1,500		
	1,200–1,000	225	
	900–750	200	
	Lower than 750		
5	3,600–3,000		
	1,800–1,500	200	
	1,200–1,000	200	
	900–750	200	
	Lower than 750		
7½	3,600–3,000		
	1,800–1,500	190	
	1,200–1,000	190	
	900–750	190	
	Lower than 750		
10	3,600–3,000		
	1,800–1,500	190	
	1,200–1,000	190	
	900–750	190	
	Lower than 750		
15–25	All speeds	190	
30 and larger	All speeds	190	135

For motors having nameplate voltages of 208 to 220/440 volts or 208 to 220 volts, the breakdown torques given apply to operation on 220 volts (or 440 volts); on 208 volts, the breakdown torques are somewhat lower.

From NEMA Standard MG 1-12.38.

These motors may be classified as constant-horsepower, constant-torque, and variable-torque. Constant-horsepower motors have the same horsepower at all speeds and are used on applications such as machine-tool drills. Constant-torque motors have horsepower ratings which are directly proportional to the speed and are used on such applications as conveyors. Variable-torque motors have horsepower ratings which are proportional to the square of the speed and are used on fans where the load decreases at least as the square of the speed.

Figure 5 shows connections used for connecting two-, three-, and four-speed polyphase induction motors.

8. Wound-Rotor Induction Motors. Wound-rotor induction motors have the secondary, or rotor, winding insulated from the magnetic structure in much the same

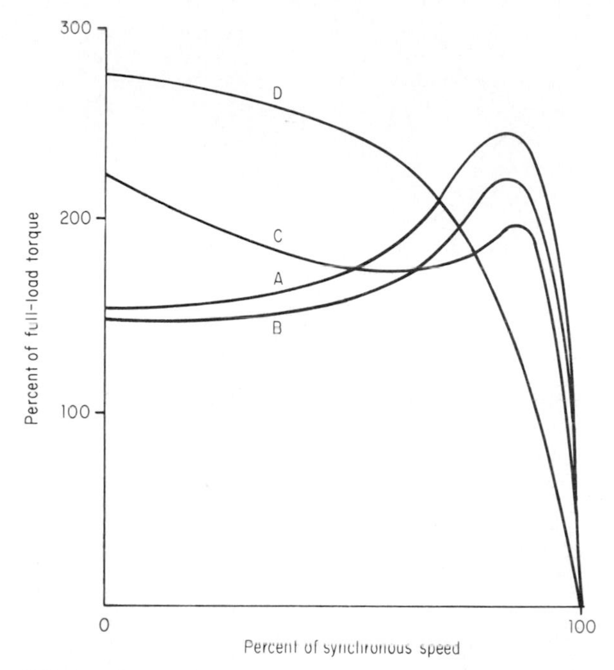

FIG. 3. General shape of speed-torque curves for motor with NEMA Designs A, B, C, and D.

manner as the primary, or stator, winding (Fig. 2). The winding, in the case of a three-phase winding, which is most common, has the three line leads brought out each to a separate collector ring. Brushes, which are held stationary by brushholders, are in contact with the rings, and these brushes are connected through variable resistors. The varying of the resistance value of these resistors gives the motor the corresponding variations in operating characteristics.

Wound-rotor motors are normally started with secondary resistance in the circuit, and the resistance is reduced periodically to allow the motor to come up to speed. This procedure allows the motor to develop substantial torque while limiting the locked-rotor current. Figure 6 shows speed-torque and speed-current curves for a wound-rotor motor with different values of external resistance.

The secondary resistance can be designed for continuous service so that it will dissipate the heat produced from continuous operation at reduced speed, frequent accelerations, or acceleration with a large inertia load. The external resistor in the rotor circuit will give the motor a characteristic which results in a large drop in speed for a rather small change in load. This method for obtaining reduced speed is generally used down to only 50 percent speed. The motor efficiency under these conditions is, needless to say, rather low.

The breakdown torque of polyphase wound-rotor integral-horsepower motors with continuous ratings is shown in Table 6. Secondary data for wound-rotor motors are given in Table 7.

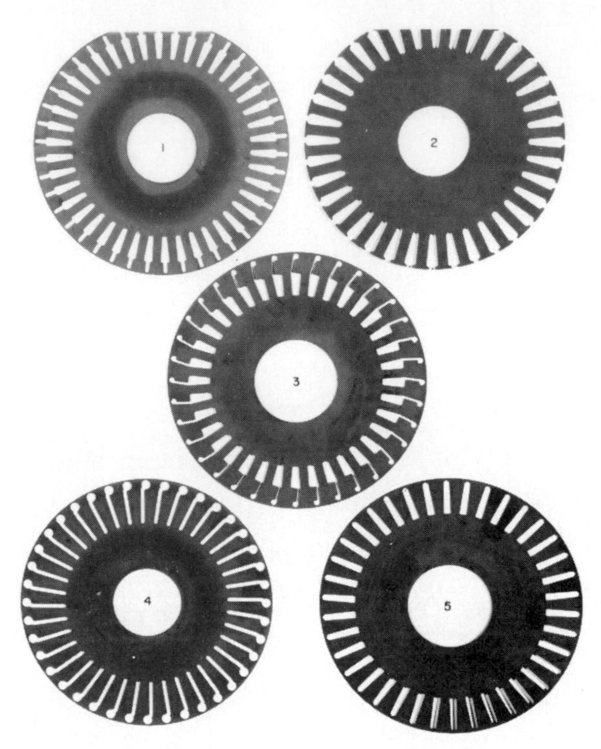

FIG. 4. Rotor punchings for NEMA Design A, B, C, and D induction motors. 1—Normal torque, NEMA Design B; 2—wound rotor; 3—high torque, low slip, NEMA Design C; 4—high torque, high slip, NEMA Design D; 5—normal torque, NEMA Design A or B.

SERVICE CONDITIONS

A general purpose ac motor having a service factor in accordance with Table 8 is suitable for continuous operation at rated load under the usual service conditions given later in this section. When the voltage and frequency are maintained at the value specified on the nameplate, the motor may be overloaded up to the horsepower obtained by multiplying the rated horsepower by the service factor shown on the nameplate.

When operated at the service-factor load, the motor will have a temperature rise as specified in Table 9. If the service factor is greater than 1, the motor may have efficiency, power factor, and speed different from those at rated load, but the locked-rotor torque and current and the breakdown torque will remain unchanged.

FIG. 5. Connection arrangements for two-, three-, and four-speed polyphase integral-horsepower motors.

1

Three-phase two-speed one-winding constant hp

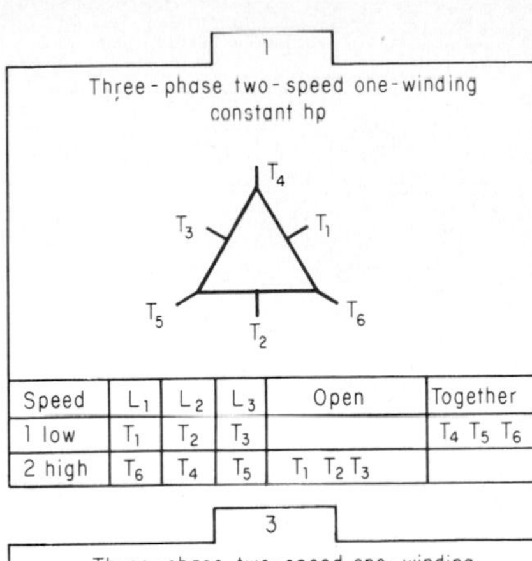

Speed	L_1	L_2	L_3	Open	Together
1 low	T_1	T_2	T_3		T_4 T_5 T_6
2 high	T_6	T_4	T_5	T_1 T_2 T_3	

2

Three-phase two speed one-winding constant torque

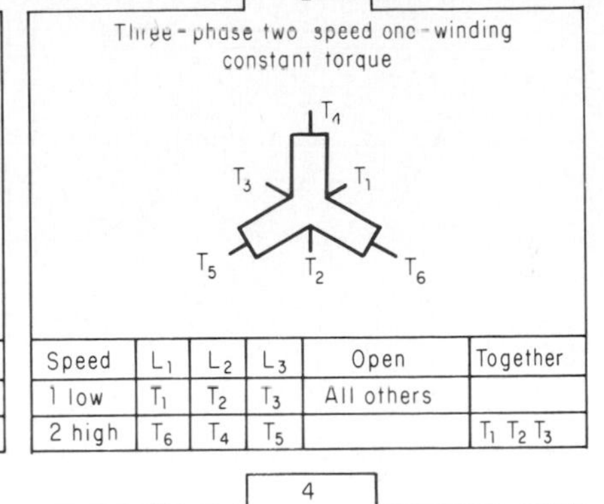

Speed	L_1	L_2	L_3	Open	Together
1 low	T_1	T_2	T_3	All others	
2 high	T_6	T_4	T_5		T_1 T_2 T_3

3

Three-phase two-speed one-winding variable torque

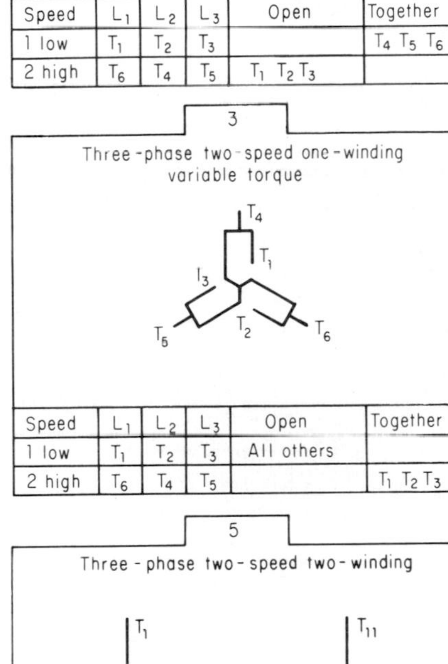

Speed	L_1	L_2	L_3	Open	Together
1 low	T_1	T_2	T_3	All others	
2 high	T_6	T_4	T_5		T_1 T_2 T_3

4

Two-phase two-speed one-winding variable torque

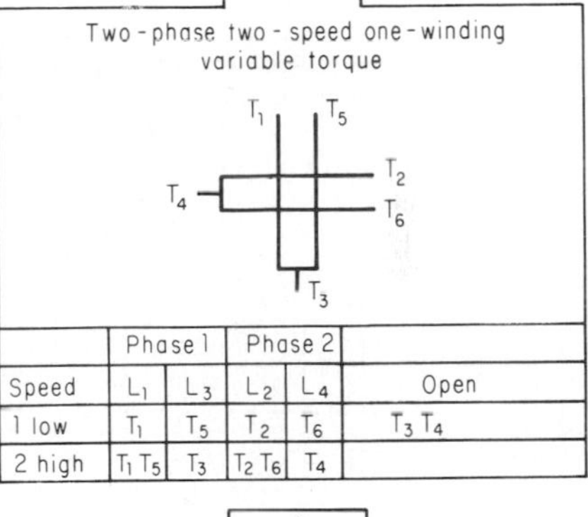

	Phase 1		Phase 2		
Speed	L_1	L_3	L_2	L_4	Open
1 low	T_1	T_5	T_2	T_6	T_3 T_4
2 high	T_1 T_5	T_3	T_2 T_6	T_4	

5

Three-phase two-speed two-winding

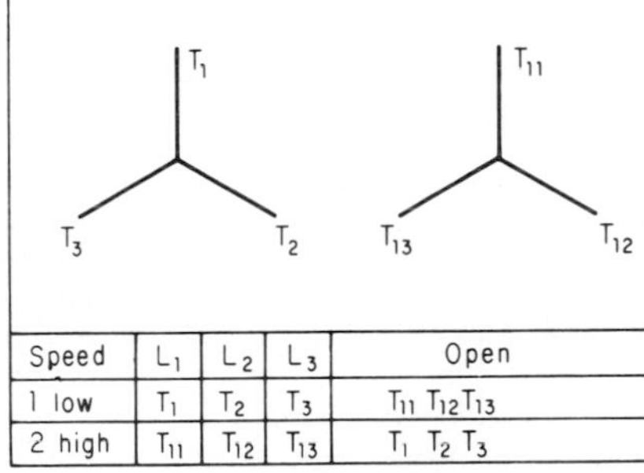

Speed	L_1	L_2	L_3	Open
1 low	T_1	T_2	T_3	T_{11} T_{12} T_{13}
2 high	T_{11}	T_{12}	T_{13}	T_1 T_2 T_3

6

Three-phase two-speed two-winding

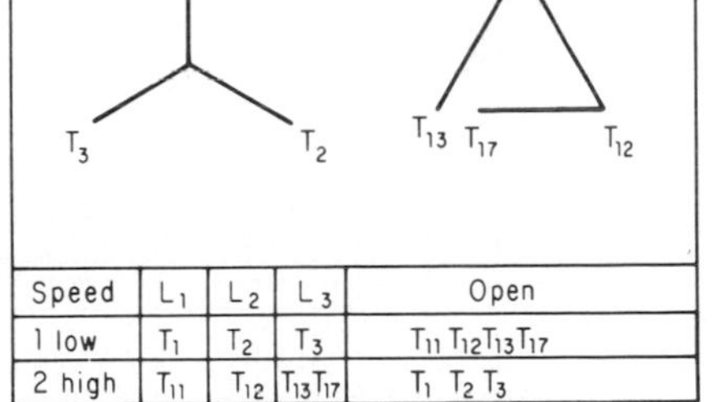

Speed	L_1	L_2	L_3	Open
1 low	T_1	T_2	T_3	T_{11} T_{12} T_{13} T_{17}
2 high	T_{11}	T_{12}	T_{13} T_{17}	T_1 T_2 T_3

7

Three-phase two-speed two-winding

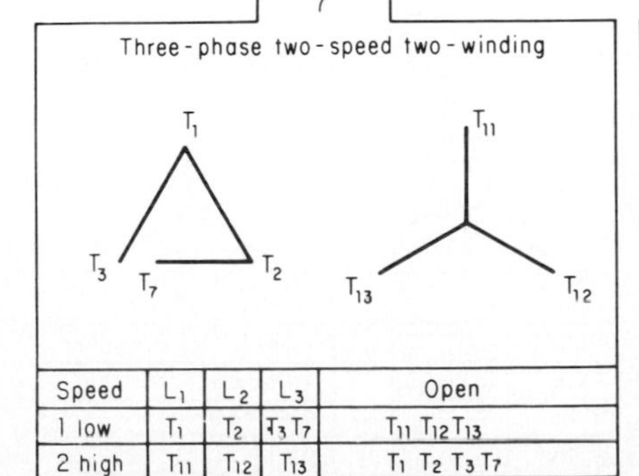

Speed	L_1	L_2	L_3	Open
1 low	T_1	T_2	T_3 T_7	T_{11} T_{12} T_{13}
2 high	T_{11}	T_{12}	T_{13}	T_1 T_2 T_3 T_7

8

Two-phase two-speed two-winding

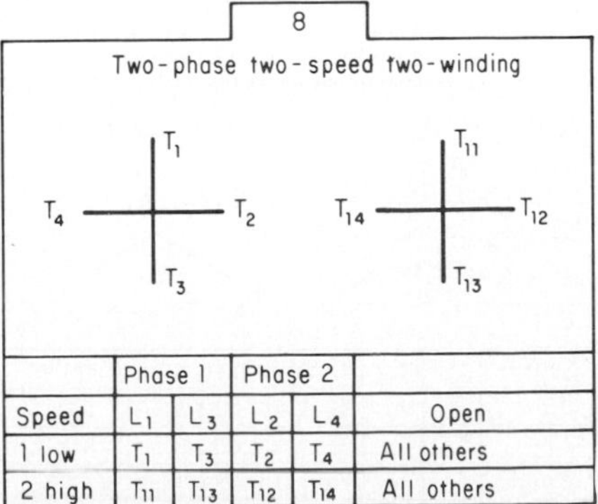

	Phase 1		Phase 2		
Speed	L_1	L_3	L_2	L_4	Open
1 low	T_1	T_3	T_2	T_4	All others
2 high	T_{11}	T_{13}	T_{12}	T_{14}	All others

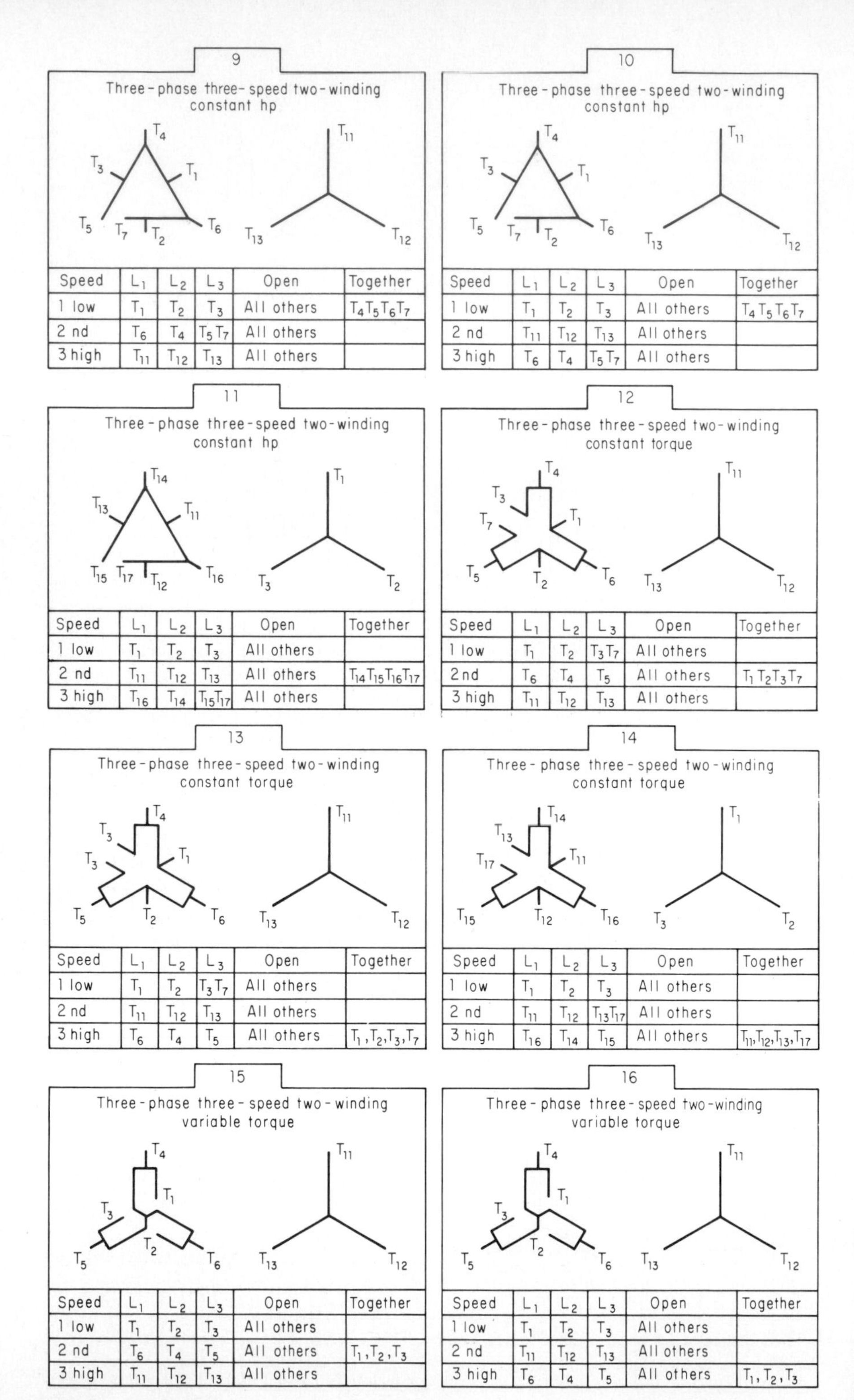

FIG. 5. (*Continued*)

17

Three-phase three-speed two-winding variable torque

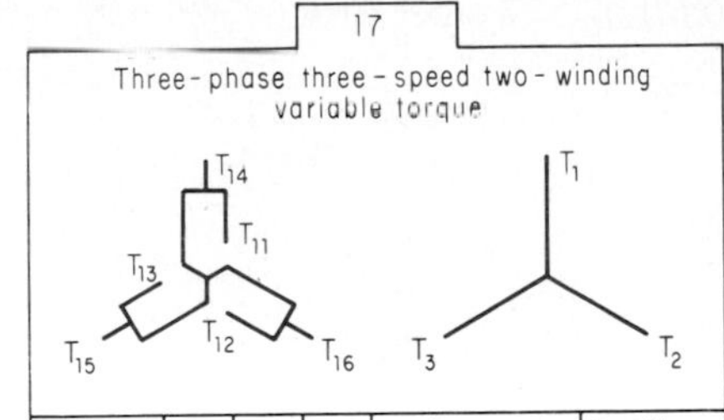

Speed	L_1	L_2	L_3	Open	Together
1 low	T_1	T_2	T_3	All others	
2 nd	T_{11}	T_{12}	T_{13}	All others	
3 high	T_{16}	T_{14}	T_{15}	All others	T_{11}, T_{12}, T_{13}

18

Three-phase four speed two-winding constant hp

Speed	L_1	L_2	L_3	Open	Together
1 low	T_1	T_2	T_3	All others	T_4, T_5, T_6, T_7
2 nd	T_{11}	T_{12}	T_{13}	All others	$T_{14}, T_{15}, T_{16}, T_{17}$
3 rd	T_6	T_4	$T_5 T_7$	All others	
4 high	T_{16}	T_{14}	$T_{15} T_{17}$	All others	

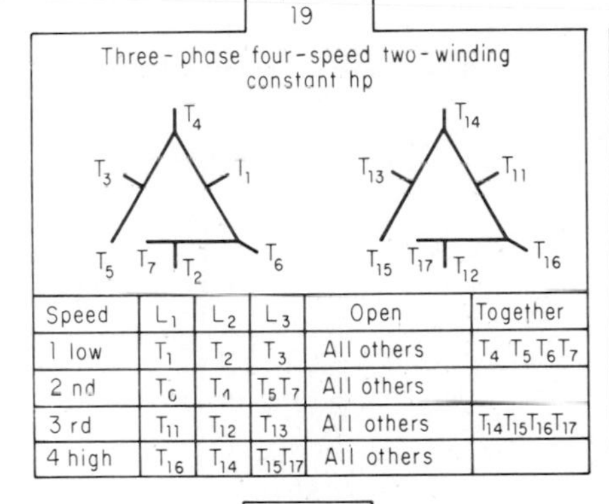

19

Three-phase four-speed two-winding constant hp

Speed	L_1	L_2	L_3	Open	Together
1 low	T_1	T_2	T_3	All others	$T_4 T_5 T_6 T_7$
2 nd	T_6	T_4	$T_5 T_7$	All others	
3 rd	T_{11}	T_{12}	T_{13}	All others	$T_{14} T_{15} T_{16} T_{17}$
4 high	T_{16}	T_{14}	$T_{15} T_{17}$	All others	

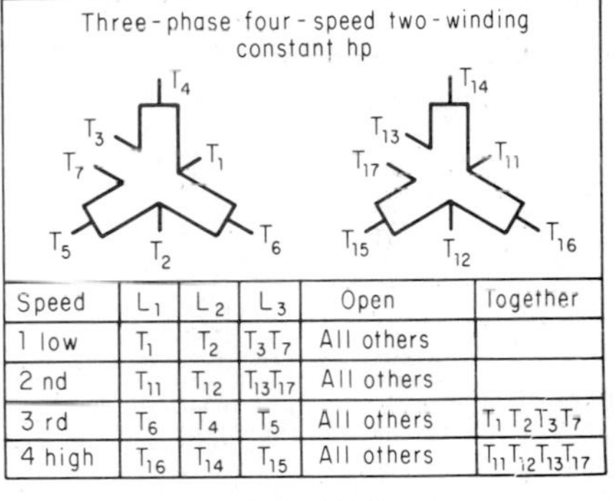

20

Three-phase four-speed two-winding constant hp

Speed	L_1	L_2	L_3	Open	Together
1 low	T_1	T_2	$T_3 T_7$	All others	
2 nd	T_{11}	T_{12}	$T_{13} T_{17}$	All others	
3 rd	T_6	T_4	T_5	All others	$T_1 T_2 T_3 T_7$
4 high	T_{16}	T_{14}	T_{15}	All others	$T_{11} T_{12} T_{13} T_{17}$

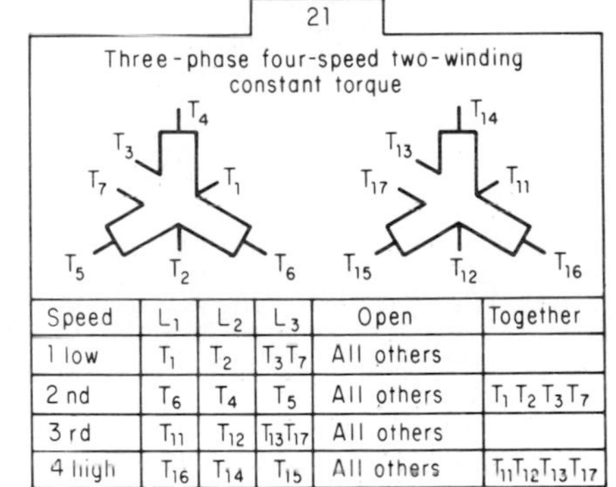

21

Three-phase four-speed two-winding constant torque

Speed	L_1	L_2	L_3	Open	Together
1 low	T_1	T_2	$T_3 T_7$	All others	
2 nd	T_6	T_4	T_5	All others	$T_1 T_2 T_3 T_7$
3 rd	T_{11}	T_{12}	$T_{13} T_{17}$	All others	
4 high	T_{16}	T_{14}	T_{15}	All others	$T_{11} T_{12} T_{13} T_{17}$

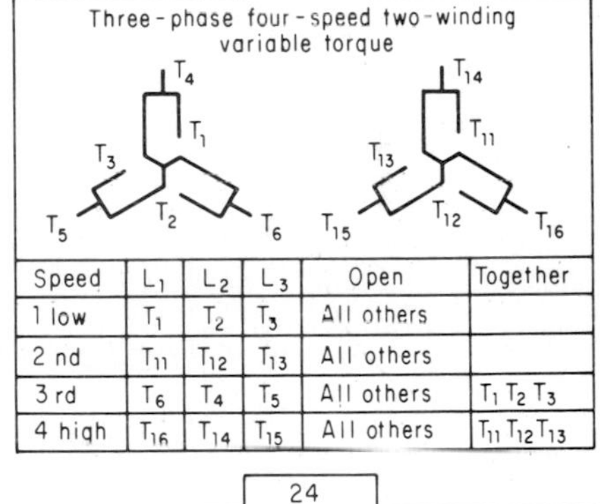

22

Three-phase four-speed two-winding variable torque

Speed	L_1	L_2	L_3	Open	Together
1 low	T_1	T_2	T_3	All others	
2 nd	T_{11}	T_{12}	T_{13}	All others	
3 rd	T_6	T_4	T_5	All others	$T_1 T_2 T_3$
4 high	T_{16}	T_{14}	T_{15}	All others	$T_{11} T_{12} T_{13}$

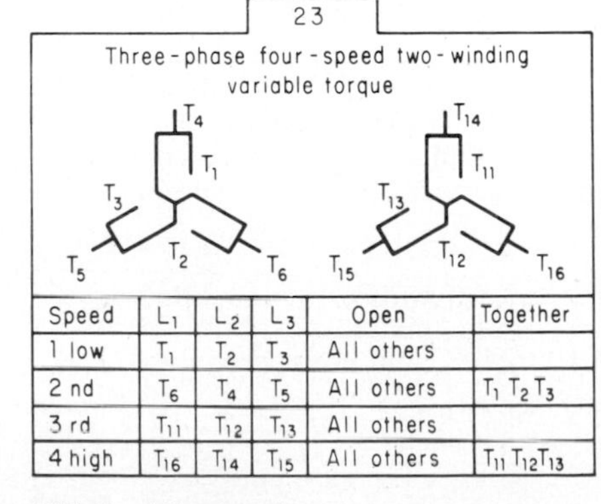

23

Three-phase four-speed two-winding variable torque

Speed	L_1	L_2	L_3	Open	Together
1 low	T_1	T_2	T_3	All others	
2 nd	T_6	T_4	T_5	All others	$T_1 T_2 T_3$
3 rd	T_{11}	T_{12}	T_{13}	All others	
4 high	T_{16}	T_{14}	T_{15}	All others	$T_{11} T_{12} T_{13}$

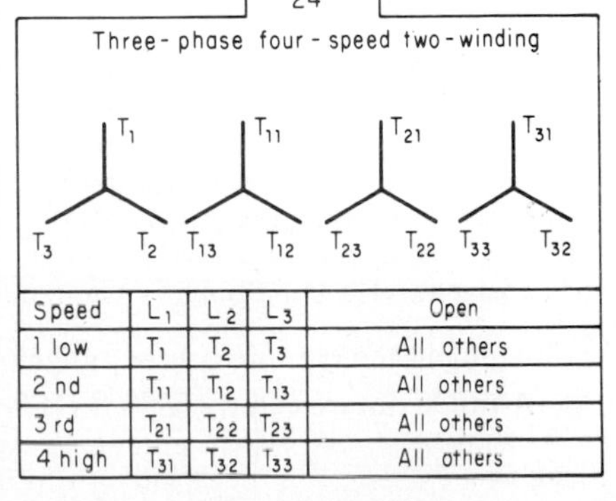

24

Three-phase four-speed two-winding

Speed	L_1	L_2	L_3	Open
1 low	T_1	T_2	T_3	All others
2 nd	T_{11}	T_{12}	T_{13}	All others
3 rd	T_{21}	T_{22}	T_{23}	All others
4 high	T_{31}	T_{32}	T_{33}	All others

FIG. 5. *(Continued)*

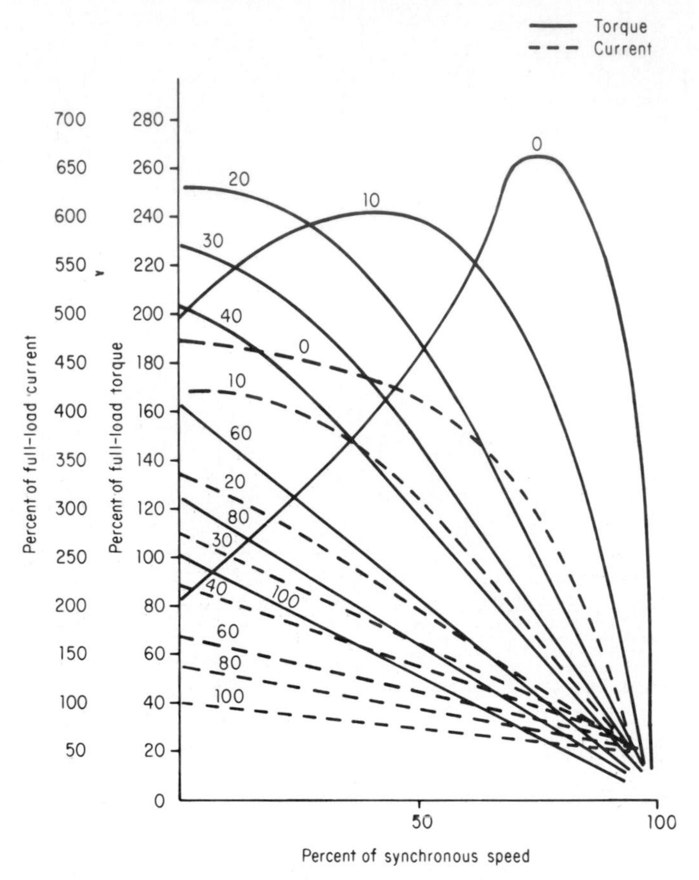

FIG. 6. Current and torque characteristics of wound-rotor motor with changing external resistance in rotor circuit. Numbers on curves are percent of total rated resistance in rotor circuit.

9. Service Factor of General-Purpose ac Motors. The service factor for general-purpose ac motors having a rated temperature rise in accordance with Table 9 and operated at rated voltage and frequency should be as shown in Table 8.

10. Proper Selection of Apparatus. Motors should be properly selected to ensure their satisfactory service. Machines conforming to NEMA standards are suitable for operation in accordance with their ratings under usual service conditions. Where the machines are subjected to unusual service conditions, the manufacturer should be consulted.

11. Usual Service Conditions. The usual service conditions are as follows:

1. Ambient temperature in the range of 10 to 40°C, but not exceeding 40°C
2. Altitude not exceeding 3300 ft (100 m) (see Art. 13)

TABLE 6. Breakdown Torque of Polyphase Wound-Rotor Integral-Horsepower Motors with Continuous Ratings

hp	Breakdown torque, % of full-load torque		
	Speed, r/min		
	1800	1200	900
1	...	...	250
1½	...	...	250
2	275	275	250
3	275	275	250
5	275	275	250
7½	275	250	225
10	275	250	225
15	250	225	225
20–200, inclusive	225	225	225

From NEMA Standard MG 1-12.40.

TABLE 7. Secondary Data for Wound-Rotor Motors

hp	Secondary volts*	Max secondary A
1	90	6
1½	110	7.3
2	120	8.4
3	145	10
5	140	19
7½	165	23
10	195	26.5
15	240	32.5
20	265	38
25	220	60
30	240	65
40	250	75
50	280	84
60	300	92
75	235	146
100	275	170
125	305	190
150	340	207

* Tolerance plus or minus 10 percent.

3. Voltage variation in accordance with Art. 14 for ac motors
4. Frequency variation as specified in Art. 15
5. Combined variation in voltage and frequency as specified in Art. 17
6. V-belt drive in accordance with Art. 23
7. Flat-belt chain and gear drives in accordance with Arts. 22, 24, and 25
8. Installed on a rigid mounting surface

TABLE 8. Service-Factor Synchronous Speed, r/min

hp	3,600	1,800	1,200	900
½	1.25	1.25	1.25	1.15*
¾	1.25	1.25	1.15*	1.15*
1	1.25	1.15*	1.15*	1.15*
1½–200	1.15*	1.15*	1.15*	1.15*
250	1.0	1.0	1.0	1.0
300 and 350	1.0	1.0	1.0	
400 and 500	1.0	1.0		

* In the case of polyphase squirrel-cage integral-horsepower motors, these service factors apply only to Design A, B, and C motors.

From NEMA Standard MG 1-12.47.

TABLE 9. Temperature Rise, °C, for Single-Phase and Polyphase Induction Motors

Class of insulation system	B	F	H
Time rating (may be continuous or any short-time rating given in MG1-10.37)			
Temp rise (based on a max ambient temp of 40°C), windings, integral-horsepower motors:			
Motors other than those given below resistance	80	105	125
All motors with 1.15 or higher service factor, resistance	90	115	
Totally enclosed fan-cooled motors, including variations thereof, resistance	80	105	125
Totally enclosed nonventilated motors, including variations thereof, resistance	85	110	135
Encapsulated motors with 1.0 service factor, all enclosures, resistance	85	110	

The temperature rise, above the temperature of the cooling medium, for each of the various parts of the motor should not exceed the values given when tested in accordance with the rating, except that, for motors having a service factor greater than 1.0, the temperature rise should not exceed the values given when tested at the service-factor load.

The temperatures attained by squirrel-cage windings, cores, and mechanical parts (such as brush-holders, pole tips, uninsulated shading coils, commutators, and collector rings) should not injure the machine in any respect.

Temperatures should be determined in accordance with the latest revision of the following: (1) for single-phase motors, IEEE test code 114; (2) for polyphase induction motors, IEEE test code 112.

From NEMA Standard MG 1-12.42.

9. Location or supplementary enclosures which do not seriously interfere with the ventilation of the machine

12. Unusual Service Conditions. In all cases where the service conditions are different from those specified as usual, the manufacturer should be consulted. Typical unusual service conditions are:

1. Exposure to chemical fumes
2. Exposure to combustible dust
3. Exposure to dusts of explosives
4. Exposure to abrasive or conducting dust
5. Exposure to lint
6. Exposure to steam
7. Exposure to flammable or explosive gases
8. Exposure to ambient temperatures above 40°C or below 10°C
9. Exposure to oil vapor
10. Exposure to salt air
11. Exposure to abnormal shock or vibration from external sources
12. Exposure to external mechanical loads involving thrust or overhang
13. Exposure to radiant heat
14. Exposure to nuclear radiation
15. Exposure to vermin infestation or atmospheres conducive to growth of fungus

16. Operation in damp or very dry places
17. Operation in poorly ventilated rooms
18. Operation at speeds above the highest rated speed
19. Operation in pits or enclosed in boxes without an adequate ventilation system
20. Operation where there is other than usual departure from rated voltage and/or frequency
21. Operation where low noise levels are required
22. Operation in an inclined position
23. Operation where dynamic balance better than that specified is required
24. Operation when subjected to torsional impact loads
25. Operation of machine at standstill with any winding continuously energized

13. Operation at Altitudes above 3300 ft (1000 m). The standard temperature rises given for motors in Table 9 are based upon operation at altitudes of 3300 ft (1000 m) or less and a maximum ambient temperature of 40°C. It is also recognized as good practice to use motors at altitudes greater than 3,300 ft as indicated below.

a. Class A or B insulated motors having a standard temperature rise will operate satisfactorily at altitudes above 3300 ft (1000 m) in those locations where the decrease in ambient temperature compensates for the increase in temperature rise, as shown in Table 10.

TABLE 10. Temperature versus Altitude

Ambient temp, °C	Max altitude, ft (m)
40	3,300 (1000)
30	6,600 (2000)
20	10,000 (3000)

b. Motors having a 1.15 or higher service factor will operate satisfactorily at unity service factor at an ambient temperature of 40°C at altitudes above 3300 ft (1000 m) up to 10,000 ft (3000 m).

c. Motors intended for use at altitudes above 3300 ft (1000 m) at an ambient temperature of 40°C should have temperature rises at sea level not exceeding the values calculated from the following formula:

$$T_{RSL} = T_{RA} \frac{1 - (\text{alt} - 3300)}{33{,}000}$$

where T_{RSL} = test temperature rise, °C, at sea level
T_{RA} = temperature rise, °C, from Table 9
alt = altitude above sea level, feet at which machine is to be operated

d. Preferred values of altitude are 3300, 6600, 10,000, 13,300 and 16,600 ft (1000, 2000, 3000, 4000, and 5000 m)

14. Variation from Rated Voltage. Induction motors should operate successfully at rated load with a variation in the voltage up to plus or minus 10 percent with rated frequency.

Performance within these voltage variations will not necessarily be in accordance with the standards established for operation at rated voltage.

15. Variation from Rated Frequency. Ac motors should operate successfully at rated load and at rated voltage with a variation in the frequency up to 5 percent above or below the rated frequency. Performance within this frequency variation will not necessarily be in accordance with the standards established for operation at rated frequency.

16. Combined Variation of Voltage and Frequency. Ac motors should operate successfully at rated load with a combined variation in the voltage and frequency up to 10 percent above or below the rated voltage and the rated frequency, provided that the frequency variation does not exceed 5 percent. Performance within this combined variation will not necessarily be in accordance with the standards established for operation at rated voltage and rated frequency. Table 11 shows the effects of variation of voltage and frequency on motor performance.

TABLE 11. Effects of Voltage and Frequency Variation

Characteristics	Voltage		Frequency	
	110%	90%	105%	95%
Torque, starting and breakdown	Increase 21%	Decrease 19%	Decrease 10%	Increase 11%
Speed:				
Synchronous	No change	No change	Increase 5%	Decrease 5%
Full load	Increase 1%	Decrease 1.5%	Increase 5%	Decrease 5%
Full-load efficiency	Increase 4–6 points	Decrease 2 points	Slight increase	Slight decrease
Full-load power factor	Decrease 4 points	Increase 1 point	Slight increase	Slight decrease
Currents:				
Starting	Increase 10–12%	Decrease 10–12%	Decrease 5–6%	Increase 5–6%
Full-load	Decrease 6%	Increase 10%	Slight decrease	Slight increase

17. Service Factors of General-Purpose and Other 40°C Motors. Service factors for general-purpose motors and, when applicable, for other continuous-duty motors having a rated temperature rise of 40°C, when operated at rated voltage and frequency, should be in accordance with Table 12.

18. Overspeeds. Ac motors, except synchronous and crane motors, should be so constructed that, in an emergency, they will withstand without mechanical injury overspeeds above synchronous speed in accordance with Table 13.

19. Short-Time-Rated Electric Machines. Short-time-rated electric machines should be applied so as to ensure performance without injury. They should not

TABLE 12. Service Factors for General-Purpose Motors

hp	*Service factor*
½	1.25
¾	1.35
1	1.25
1½	1.20
2	1.20
3 and larger	1.15

TABLE 13. Allowable Overspeed for ac Motors

Synchronous speed, r/min	Overspeed, % of synchronous speed
1801 and over	25
1201–1800	25
1200 and below	50

be used (except on the recommendation of the manufacturer) on any application where the driven machine may be left running continuously.

20. Noise Quality. Noise quality, that is, the distribution of effective sound intensities as a function of frequency, affects the objectionability of the sound (see Sec. 20).

A measurement of total noise does not completely define noise acceptability because machines with the same overall decibel noise level may have different noise qualities. For these reasons, machine noise is specified in terms of (1) total noise and (2) octave bands.

21. Noise Measurement. Noise tests should be taken with the machine unloaded because of the impracticability of isolating the noise of the load. It should be recognized that the decibel readings are not exact and are subject to many external influences (see Sec. 20).

MECHANICAL ARRANGEMENTS

In general, the closer the pulleys, sheaves, sprockets, or gears are mounted to the bearing on the motor shaft, the less will be the load on the bearing. This will give greater assurance of trouble-free service.

The following arrangements are recommended for the mounting of pulleys, sheaves, sprockets, and gears on motor and generator shafts.

22. Flat-Belt Pulleys. The centerline of the belt must not be beyond the end of the motor and generator shaft.

The following is recommended for pulley design:

1. The inner edge of the pulley rim should be in line with the shoulder on the shaft. Placing this edge closer to the bearing may create interference with lubricating fittings.

2. The inner edge of the pulley hub should be recessed from the inner edge of the rim as follows:
   ¼ in for pulleys of 2½ to 5 in diameter
   ⅜ in for pulleys of 6 to 10 in diameter
   ⅝ in for pulleys of 11 to 15 in diameter
3. The outer edge of the pulley hub should be in line with the end of the motor shaft.

23. V-Belt Sheaves. The center point of the system of V belts must not be beyond the end of the motor shaft.

The inner edge of the sheave rim should not be closer to the bearing than the shoulder on the shaft but should be as close to this point as possible. Tables 14 and 15 show recommended dimensions.

24. Chain Sprockets. The outer edge of a chain sprocket must not extend beyond the end of the motor shaft.

TABLE 14. Application of V-Belt Sheave Dimensions to General-Purpose Motors, Integral-Horsepower Motors, Polyphase Induction

Frame No.	hp at						V-belt sheave	
	Synchronous speed, r/min						Min pitch diam, in.	Max. width, in.
	3600	1800	1200	900	720	600		
182	1½	1	¾	½	....	....	2¼	3
184	3–2	2–1½	1½–1	¾	....	....	2½	3
213	5	3	2	1½–1	1	....	2½	3½
215	7½	5	3	2	1½	1	3	3½
254U	10	7½	5	3	2	1½	3	5½
256U	15	10	7½	5	3	2	3¾	6¾
284U	20	15	10	7½	....	....	4½	7¾
286U	25	....	......	......	....	....	4½	7¾
286U	....	20	......	10	5	3	4½	9¾
324U	....	25	15	......	7½	5	4½	9¾
326U	....	30	20	15	10	7½	5¼	11
364U	....	40	25	20	15	10	6	11
365U	....	50	30	25	....	....	6¾	12
404U	....	60	......	......	....	....	7½	12
404U	....	....	40	30	20	15	6¾	13
405U	....	75	......	......	....	....	9	13
405U	....	....	50	40	25	20	8¼	13
444U	....	100	......	......	....	....	10	17
444U	....	....	60	50	30	25	9	17
445U	....	125	......	......	....	....	11	17
445U	....	....	75	60	40	30	10	17

From NEMA Standard MG 1-14.43.

TABLE 15. Application of V-Belt Sheave Dimensions to ac General-Purpose Motors, Integral-Horsepower Motors, Polyphase Induction

Frame no.	hp at				V-belt sheave			
	Synchronous speed, r/min				Conventional A, B, C, D, and E		Narrow 3V, 5V, and 8V	
	3600	1800	1200	900	Min. pitch diam, in.	Max width*	Min OD, in.	Max width†
143T	1½	1	¾	½	2.2	...	2.2	
145T	2–3	1½–2	1	¾	2.4	...	2.4	
182T	3	3	1½	1	2.4	...	2.4	
182T	5	....	.....	.....	2.6	...	2.4	
184T	....	....	2	1½	2.4	...	2.4	
184T	5	....	.....	.....	2.6	...	2.4	
184T	7½	5	.....	.....	3.0	...	3.0	
213T	7½–10	7½	3	2	3.0	...	3.0	
215T	10	....	5	3	3.0	...	3.0	
215T	15	10	.....	.....	3.8	...	3.8	
254T	15	....	7½	5	3.8	...	3.8	
254T	20	15	.....	.....	4.4	...	4.4	
256T	20–25	....	10	7½	4.4	...	4.4	
256T	....	20	.....	.....	4.6	...	4.4	
284T	....	....	15	10	4.6	...	4.4	
284T	....	25	.....	.....	5.0	...	4.4	
286T	....	30	20	15	5.4	...	5.2	
324T	....	40	25	20	6.0	...	6.0	
326T	....	50	30	25	6.8	...	6.8	
364T	....	....	40	30	6.8	...	6.8	
364T	....	60	.....	.....	7.4	...	7.4	
365T	....	....	50	40	8.2	...	8.2	
365T	....	75	.....	.....	9.0	...	8.6	
404T	....	....	60	.....	9.0	...	8.0	
404T	....	....	.....	50	9.0	...	8.4	
404T	....	100	.....	.....	10.0	...	8.6	
405T	....	....	75	60	10.0	...	10.0	
405T	....	100	.....	.....	10.0	...	8.6	
405T	....	125	.....	.....	11.5	...	10.5	
444T	....	....	100	.....	11.0	...	10.0	
444T	....	....	.....	75	10.5	...	9.5	
444T	....	125	.....	.....	11.0	...	9.5	
444T	....	150	.....	.....	....	...	10.5	
445T	....	....	125	.....	12.5	...	12.0	
445T	....	....	.....	100	12.5	...	12.0	
445T	....	150	.....	.....	....	...	10.5	
445T	....	200	.....	.....	....	...	13.2	

Ac general-purpose motors having continuous time rating with the frame sizes, horsepower, and speed ratings listed are designed to operate with V-belt sheaves within the limited dimensions listed. Selection of V-belt sheave dimensions is made by the V-belt drive vendor and the motor purchaser, but to assure satisfactory motor operation, the selected diameter should be not smaller than, nor should the selected width be greater than, the dimensions listed. For the assignment of horsepower and speed ratings to frames, see NEMA Standards MG 1-13.02.a and MG 1-13.06.a.

*The width of the sheave should be no greater than that required to transmit the indicated horsepower but in no case should it be wider than 2(N-W)-¼.

†The width of the sheave should be no greater than that required to transmit the indicated horsepower but in no case should it be wider than (N-W).

From NEMA Standard MG 1-14.43.

25. Gears (Spur and Helical). The outer edge of the gear must not extend beyond the end of the motor shaft.

NEMA DESIGNATIONS

26. NEMA Standard Code Letters for Locked-Rotor kVA. The nameplate of every ac motor, except polyphase wound rotor rated 1/20 hp and larger, should be marked with the caption "Code" followed by a letter selected from Table 9, Sec. 4, to show locked-rotor kVA per horsepower.

Multispeed motors should be marked with a code letter designating the locked-rotor kVA per horsepower for the highest speed, except constant-horsepower motors, which should be marked with a code letter for the speed giving the highest locked-rotor kVA per horsepower.

Single-speed motors starting on wye connection and running on delta connection should be marked with a code letter corresponding to the locked-rotor kVA per horsepower for the wye connection.

Dual-voltage motors which have different locked-rotor kVA on the two voltages should be marked with the code letter for the voltage giving the highest locked-rotor kVA per horsepower.

27. Frame Assignments. For the prererated and rerated NEMA motors frame sizes are shown in Tables 16 to 20.

28. Lettering Dimensions and Outline Tables. For three-phase induction motors built to NEMA standards, these data are shown in Figs. 7 and 8 and in Tables 21 through 30.

29. Hazardous Atmospheres. Motors are used in hazardous locations at times, and the Underwriters Laboratories has published the following data to classify locations. Hazardous atmospheres of both gaseous and dusty nature are classified by the National Electrical Code as follows:

Class I, Group A. Atmospheres containing acetylene

Class I, Group B. Atmospheres containing hydrogen or vapors of equivalent hazard, such as manufactured gas

Class I, Group C. Atmospheres containing ethyl ether vapors

Class I, Group D. Atmospheres containing gasoline, petroleum, naphtha, alcohols, acetone, lacquer-solvent vapors, and natural gas

Class II, Group E. Atmospheres containing metal dust

Class II, Group F. Atmospheres containing carbon black, coal, or coke dust

Class II, Group G. Atmospheres containing grain dust

Note. Each of the above classes is further divided into Division I, which may be considered as continuously hazardous, and Division II, occasionally but not normally hazardous.

TABLE 16. Frame Sizes for Polyphase, Squirrel-Cage, General-Purpose, Design A, B, and C, Horizontal and Vertical Motors, Open Type, 40°C, 60 Hz

hp	Speed, r/min					
	3600	1800	1200	900	720	600
	550 volts and less					
½	....	....	.....	182*		
¾	....	....	182*	184*		
1	....	182*	184*	213*	213*	215*
1½	182*	184*	184*	213*	215*	254U*
2	184*	184*	213*	215*	254U*	256U*
3	184*	213*	215†	254U†	256U*	286U*
5	213*	215*	254†	254U†	286U*	324U*
7½	215*	254U†	256U†	284U†	324U*	326U*
10	254U*	256U†	284U†	286U†	326U*	364U*
15	256U*	284U†	324U†	236U†	364U*	404U*
20	284U*	286U†	326U†	364U†	404U*	405U*
25	286U*	324U†	364U†	365U†	405U*	444U*
30	324S*	326U†	365U†	404U†	444U*	445U*
40	326S*	364U†	404U†	405U†	445U*	
50	364US*	365US†‡	405U†	444U†		
60	365US*	404US†‡	444U†	445U†		
75	404US*	405US†‡	445U†			
100	405US*	444US†‡				
125	444US*	445US†‡				
150	445US*					
	2,300 volts					
60	....	444US*†	445U*			
75	444US*	444US*†				
100	444US*	445US*†				
125	445US*					

* These frame sizes apply to Design A and B motors only.
† These frame sizes apply to Design A, B, and C motors.
‡ When motors are to be used with V-belt or chain drives, the correct frame size is the frame size shown but with the suffix letter S omitted. For the corresponding shaft dimensions, see MG 1-11.31.

Local inspection authorities will usually determine the degree of hazard involved for a given installation, Group A being the most hazardous of the first class, and Group E the most hazardous of the second class.

Explosionproof motors and control, utilizing the wide-flange principle, are available in generally used ratings for Class I, Group D locations, and in a few ratings for Class I, Group C. Apparatus for Groups A and B require very special designs and are not generally available.

TABLE 17. Frame Sizes for Polyphase, Squirrel-Cage, Design A and B, Horizontal and Vertical Motors, 60 Hz, Class B Insulation System, Open Type, 1.15 Service Factor

hp	Speed, r/min			
	3600	1800	1200	900
½	......	......	......	143T
¾	......	......	143T	145T
1	......	143T	145T	182T
1½	143T	145T	182T	184T
2	145T	145T	184T	213T
3	145T	182T	213T	215T
5	182T	184T	215T	254T
7½	184T	213T	254T	256T
10	213T	215T	256T	284T
15	215T	254T	284T	286T
20	254T	256T	286T	324T
25	256T	284T	324T	326T
30	284TS	286T	326T	364T
40	286TS	324T	364T	365T
50	324TS	326T	365T	404T
60	326TS	364TS†	404T	405T
75	364TS	365TS†	405T	444T
100	365TS	404TS†	444T	445T
125	404TS	405TS†	445T	
150	405TS	444TS†		
200	444TS	445TS†		
250*	445TS			

*The 250-hp rating at the 3600-r/min speed has a 1.0 service factor.

†When motors are to be used with V-belt or chain drives, the correct frame size is the frame size shown but with the suffix letter S omitted. For the corresponding shaft extension dimensions, see NEMA Standard MG 1-11.31.a.

See NEMA Standard MG 1-11.31.a for the dimensions of the frame numbers.

Dust-explosion-proof motors and control are available in most ratings for Class II, Groups F and G dust locations, and in a few ratings for Class II, Group E.

HELPFUL INFORMATION

30. Application Data. The following facts, which can be helpful to anyone who is working with three-phase induction motors, are included for reference:

TABLE 18. Frame Sizes for Polyphase, Squirrel-Cage, Design A and B, Horizontal and Vertical Motors, 60 Hz, Class B Insulation System, Totally Enclosed Fan-Cooled Type, 1.00 Service Factor

hp	Speed, r/min			
	3600	1800	1200	900
½	......	......	....	143T
¾	......	......	143T	145T
1	......	143T	145T	182T
1½	143T	145T	182T	184T
2	145T	145T	184T	213T
3	182T	182T	213T	215T
5	184T	184T	215T	254T
7½	213T	213T	254T	256T
10	215T	215T	256T	284T
15	254T	254T	284T	286T
20	256T	256T	286T	324T
25	284TS	284T	324T	326T
30	286TS	286T	326T	364T
40	324TS	324T	364T	365T
50	326TS	326T	365T	404T
60	364TS	364TS*	404T	405T
75	365TS	365TS*	405T	444T
100	405TS	405TS*	444T	445T
125	444TS	444TS*	445T	
150	445TS	445TS*		
200				
250				

*When motors are to be used with V-belt or chain drives, the correct frame size is the frame size shown but with the suffix letter S omitted. For the corresponding shaft extension dimensions, see NEMA Standard MG 1-11.31.a.

See NEMA Standard MG 1-11.31.a for the dimensions of the frame numbers.

1. Torque is proportional to the square of voltage.
2. Rotor losses are proportional to slip (squirrel cage).
3. Secondary losses are proportional to slip (slip ring).
4. Secondary voltage is proportional to slip.
5. Secondary voltage is inversely proportional to speed.
6. Slip is proportional to percent secondary resistance.
7. Starting torque is inversely proportional to secondary resistance within certain limits beyond which proportionality ceases.

TABLE 19. Typical Rerate NEMA Motor Characteristics, 230 V, Three-Phase, 60 Hz, Dripproof

hp	Synchronous speed, r/min	Full-load speed, r/min	Frame	Full load		
				Amperes*	Eff*	pf*
½	900	850	143T	2.7	62.5	56.0
¾	1,200	1,140	143T	3.1	72.5	63.5
	900	850	145T	3.8	66.5	55.0
1	1,800	1,735	143T	3.5	77.0	70.5
	1,200	1,140	145T	3.8	74.0	66.0
	900	870	182T	5.0	70.0	54.0
1½	3,600	3,475	143T	4.4	76.0	84.0
	1,800	1,715	145T	4.6	79.0	77.5
	1,200	1,160	182T	5.4	76.5	68.5
	900	865	184T	6.5	75.0	57.5
2	3,600	3,495	145T	5.6	79.0	85.5
	1,800	1,725	145T	6.1	81.0	75.0
	1,200	1,155	184T	7.1	78.5	67.5
	900	865	213T	8.6	77.0	56.5
3	3,600	3,475	145T	8.2	80.0	85.5
	1,800	1,745	182T	8.8	81.0	79.0
	1,200	1,160	213T	10.8	81.0	64.5
	900	865	215T	13.0	78.0	55.5
5	3,600	3,510	182T	13.2	84.0	84.5
	1,800	1,745	184T	14.1	83.5	79.5
	1,200	1,150	215T	16.2	83.0	69.5
	900	860	254T	17.6	80.0	67.0
7½	3,600	3,510	184T	19.1	85.0	86.5
	1,800	1,755	213T	21.8	84.0	76.5
	1,200	1,155	254T	22.5	83.0	75.5
	900	860	256T	23.6	81.5	73.5
10	3,600	3,520	213T	26.0	84.0	85.5
	1,800	1,750	215T	27.2	85.5	80.5
	1,200	1,155	256T	29.2	84.0	76.0
	900	865	284T	32.0	82.0	71.5
15	3,600	3,520	215T	37.4	86.0	87.0
	1,800	1,750	254T	38.6	89.5	81.0
	1,200	1,165	284T	42.5	85.0	78.5
	900	865	286T			
20	3,600	3,530	254T	49.0	88.0	87.0
	1,800	1,750	256T	50.0	90.0	83.0
	1,200	1,165	286T	55.0	86.0	79.0
	900	870	324T	67.3	87.0	75.0
25	3,600	3,530	256T	61.0	88.0	87.0
	1,800	1,750	284T	64.0	88.5	82.5
	1,200	1,165	324T	68.0	88.5	78.0
	900	875	326T	71.5	87.0	75.0

TABLE 19. Typical Rerate NEMA Motor Characteristics, 230 V, Three-Phase, 60 Hz, Dripproof (*Continued*)

hp	Synchronous speed, r/min	Full-load speed, r/min	Frame	Full load		
				Amperes*	Eff*	pf*
30	3,600	3,530	284TS	73.2	87.5	87.5
	1,800	1,750	286T	76.5	89.0	82.0
	1,200	1,165	326T	81.2	88.5	78.0
	900	875	364T	86.0	87.5	75.0
40	3,600	3,530	286TS	94.5	88.0	90.0
	1,800	1,755	324T	104	90.0	80.0
	1,200	1,170	364T	104	89.0	81.0
	900	875	365T	114	87.5	75.0
50	3,600	3,545	324TS	120	89.0	87.5
	1,800	1,760	326T	128	91.5	80.0
	1,200	1,170	365T	128	89.5	82.0
	900	875	404T			
60	3,600	3,545	326TS	141	89.5	88.5
	1,800	1,770	364T	150	90.5	82.5
	1,200	1,175	404T	158	90.5	79.5
	900	875	405T	174	89.0	73.0
75	3,600	3,555	364TS	173	90.5	90.0
	1,800	1,770	365T	179	91.0	86.0
	1,200	1,175	405T	187	91.0	82.0
	900	880	444T	192	90.0	81.0
100	3,600	3,550	365TS	225	91.0	90.5
	1,800	1,775	404T	243	91.5	82.0
	1,200	1,175	444T	237	91.5	85.5
	900	880	445T	248	92.0	82.0
125	3,600	3,560	404TS	147†	90.5	87.5
	1,800	1,780	405T	149†	92.0	84.5
	1,200	1,175	445T	149†	91.5	85.0
150	3,600	3,560	405TS	171†	91.5	89.0
	1,800	1,770	444TS	170†	92.0	88.5
200	3,600	3,560	444TS	230†	91.5	89.0
	1,800	1,770	445TS	226†	93.5	86.5
250	3,600	3,560	445TS	278†	92.5	91.0

For any other voltage the current varies inversely as the voltage.
* Estimated values.
† Currents are at 460 volts.

TABLE 20. Standard Frame Assignments, 230 V, Three-Phase, 60 Hz

hp	Synchronous speed, r/min	Full-load speed, r/min	Frame	Full load		
				Amperes*	Eff*	pf*
½	900	850	143T	2.7	62.5	56.0
¾	1,200	1,140	143T	3.1	72.5	63.5
	900	850	145T	3.8	66.5	55.0
1	1,800	1,735	143T	3.4	77.0	70.5
	1,200	1,140	145T	3.8	74.0	66.0
	900	870	182T	5.0	70.0	54.0
1½	3,600	3,475	143T	4.4	76.0	84.0
	1,800	1,715	145T	4.6	79.0	77.5
	1,200	1,160	182T	5.4	76.5	68.5
	900	865	184T	6.5	75.0	57.5
2	3,600	3,495	145T	5.6	79.0	85.5
	1,800	1,725	145T	6.1	81.0	75.5
	1,200	1,155	184T	7.1	78.5	67.5
	900	865	213T	8.6	77 0	56.5
3	3,600	3,510	182T	8.6	80.0	81.5
	1,800	1,745	182T	8.8	81.0	79.0
	1,200	1,160	213T	10.7	81.0	64.5
	900	865	215T	13.0	78.0	55.5
5	3,600	3,510	184T	13.2	84.0	84.5
	1,800	1,745	184T	14.1	83.5	79.0
	1,200	1,150	215T	16.2	83.0	69.5
	900	860	254T	23.6	81.5	73.5
7½	3,600	3,530	213T	20.6	83.0	82.0
	1,800	1,755	213T	21.8	84.0	76.5
	1,200	1,155	254T	22.5	82.0	76.0
	900	860	256T	23.5	81.5	73.5
10	3,600	3,520	215T	26.0	84.0	85.5
	1,800	1,750	215T	26.7	86.0	81.5
	1,200	1,155	256T	29.6	82.5	76.5
	900	870	284T	32.0	85.0	69.0
15	3,600	3,520	254T	37.6	84.0	89.0
	1,800	1,750	254T	37.6	88.0	85.0
	1,200	1,170	284T	44.0	87.0	73.5
	900	870	286T	47.4	86.5	68.5
20	3,600	3,520	256T	50.0	85.0	88.0
	1,800	1,750	256T	48.6	89.0	87.0
	1,200	1,170	286T	58.5	88.0	73.5
	900	875	324T	57.4	88.5	74.0
25	3,600	3,525	284T	61.0	85.5	89.0
	1,800	1,765	284T	67.0	89.0	79.0
	1,200	1,170	324T	67.0	89.0	79.0
	900	875	326T	71.5	88.5	74.0

TABLE 20. Standard Frame Assignments, 230 V, Three-Phase, 60 Hz (*Continued*)

hp	Synchronous speed, r/min	Full-load speed, r/min	Frame	Full load		
				Amperes*	Eff*	pf*
30	3,600	3,525	286TS	72.0	87.0	89.5
	1,800	1,765	286T	76.0	90.0	82.0
	1,200	1,170	326T	79.4	89.5	79.5
	900	885	364T	84.0	90.0	73.0
40	3,600	3,550	324TS	102	86.0	85.0
	1,800	1,770	324T	102	91.0	80.5
	1,200	1,180	364T	107	89.5	78.0
	900	885	365T	115	90.0	72.0
50	3,600	3,555	326TS	122	87.5	87.5
	1,800	1,770	326T	126	91.5	81.0
	1,200	1,180	365T	132	90.5	78.0
	900	880	404T	134	89.5	77.5
60	3,600	3,555	364TS	143	87.5	89.0
	1,800	1,775	364T	152	91.0	81.5
	1,200	1,180	404T	155	90.0	79.0
	900	880	405T	158	91.5	78.0
5	3,600	3,560	365TS	169	89.5	91.5
	1,800	1,775	365T	184	91.5	83.0
	1,200	1,180	405T	187	90.5	82.0
	900	885	444T	199	89.0	81.0
100	3,600	3,560	405TS	231	90.5	89.5
	1,800	1,780	405T	243	91.5	84.0
	1,200	1,180	444T	234	91.5	85.5
	900	885	445T	264	92.5	76.0
125	3,600	3,565	444TS	152†	89.5	86.0
	1,800	1,775	444T	150†	92.0	85.0
	1,200	1,185	445T	147†	91.5	86.5
150	3,600	3,570	445TS	169†	91.0	91.5
	1,800	1,775	445TS	174†	93.0	87.0

* Estimated values.
† Currents are at 460 volts.

8. Torque is inversely proportional to the square of the number of turns per coil in the stator winding.
9. Starting torque depends on rotor reactance as well as rotor resistance.
10. Low rotor resistance is used for continuous duty only. High rotor resistance is used for duty such as elevators, cranes, and tapping machines.
11. Medium rotor resistance is used for punch press and long acceleration of heavy inertia loads like extractors and centrifuges. The best rotor resistance combination for extractors is obtained with rotor and stator losses equal.

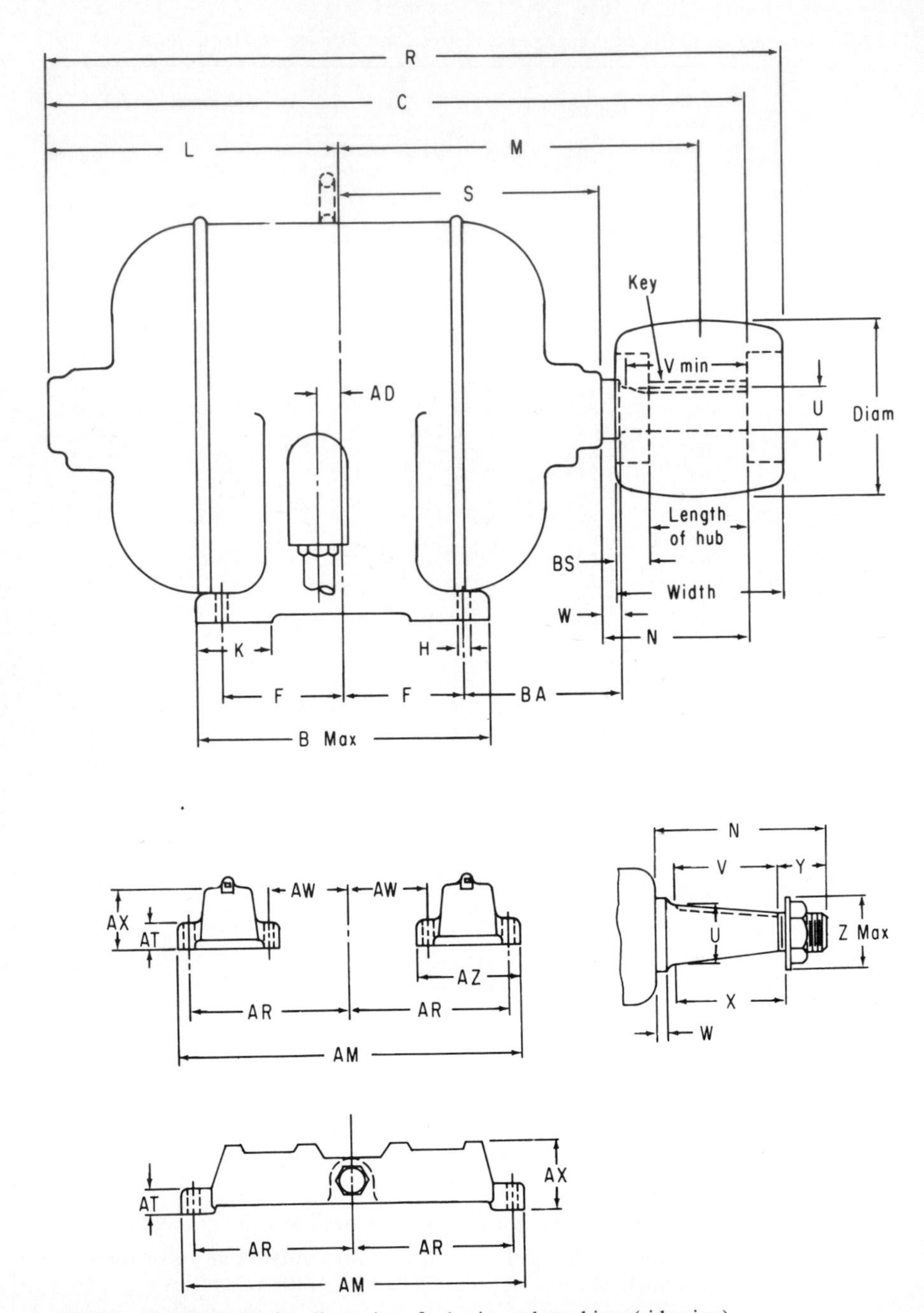

FIG. 7. Standard lettering dimensions for horizontal machines (side view).

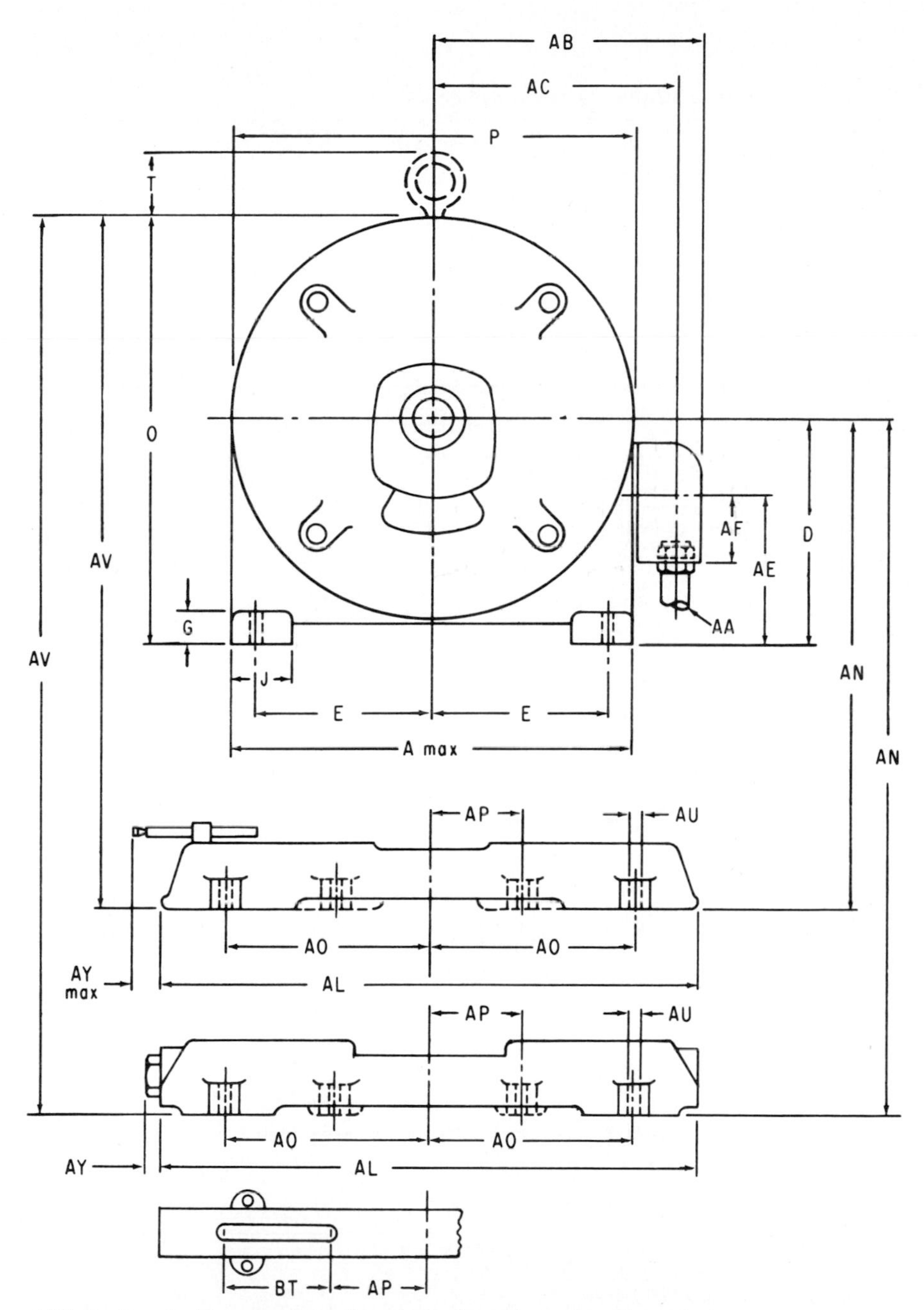

FIG. 8. Standard lettering dimensions for horizontal machines (front view).

TABLE 21. Motor Mounting Assembly Symbols

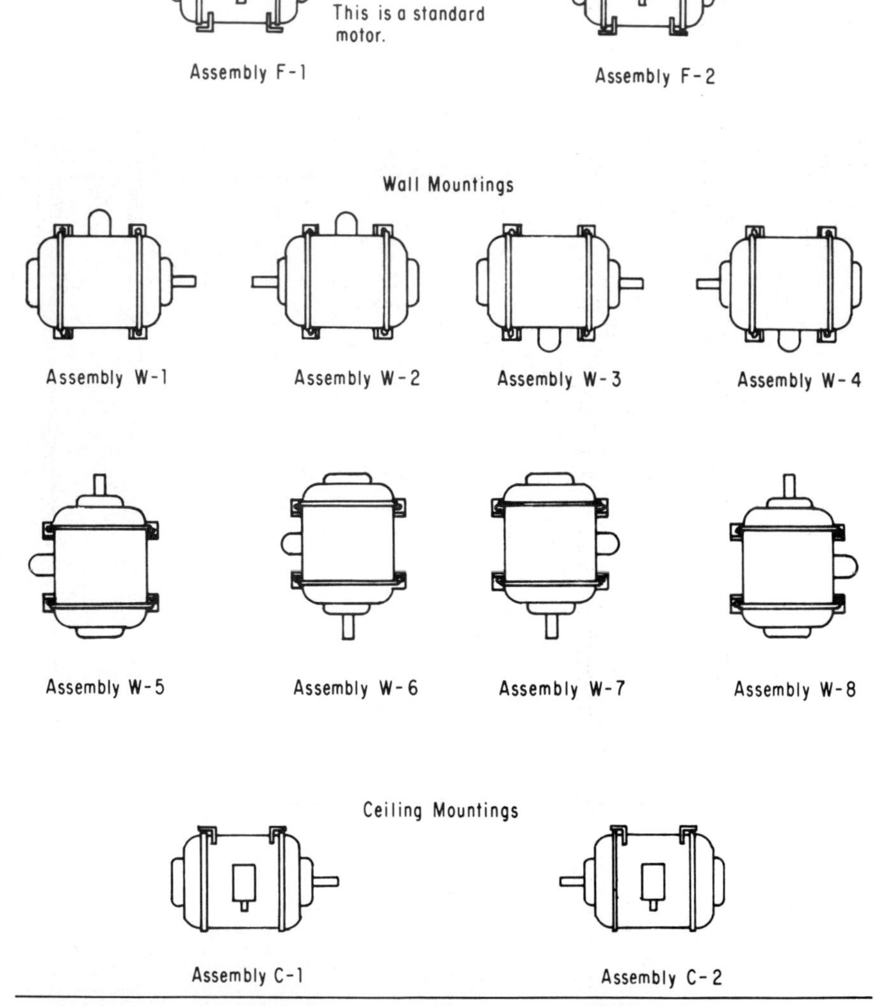

Standard lead location—F-1, W-2, W-3, W-6, W-8, C-2.
Lead location opposite standard—F-2, W-1, W-4, W-5, W-7, C-1.
See NEMA Standard MG 1-1978.

TABLE 22. Lettering of Dimensions

Letters	*Dimensions indicated*
A (max)	Overall dimension across feet of horizontal machine (end view)
B (max)	Overall dimension across feet of horizontal machine (side view)
C	Overall length of machine less pulley (for single-shaft extension)
D	Centerline of shaft to bottom of feet
E	Centerline of machine to centerline of mounting holes in feet (end view)
F*	Centerline of base or feet of machine to centerline of mounting holes in feet (side view)
G	Thickness of feet of machine
H	Diameter of holes or width of slot in feet of machine
J	Width of end of feet of horizontal or vertical machine
K	Length of feet of horizontal or vertical machine (side view)
L	Centerline of base or feet of machine to end of housing (end opposite the shaft)
M	Centerline of base or feet of machine to end of housing (shaft end)
N	Length of shaft from end of housing to end of shaft, drive end
O	Top of horizontal machine to bottom of feet
P	Diameter of vertical or horizontal machine
R	Overall length of horizontal machine including pulley (side view) (for single-shaft extension)
S	Centerline of base or feet of machine to centerline of pulley
T	Height of eye bolt above top of machine
U	Diameter of shaft extension (for tapered shaft this is a diameter at a distance *V* from the threaded portion of the shaft)
V	Length of shaft available for coupling, pinion or pulley hub, drive end (on a straight shaft extension this is a minimum value)
W	For straight and tapered shaft, end of housing to shoulder. For shaft extensions without shoulders, it is a clearance to allow for all manufacturing variations in parts and assembly
X	Length of hub of pinion when using full length of taper, drive end
Y	Distance from end of shaft to outer end of taper, drive end
Z	Width across corners of nut or diameter of washer, on tapered shaft, drive end
AA	Diameter of conduit (pipe size)
AB	Centerline of machine to extreme outside part of conduit box (end view)
AC	Centerline of machine to centerline of hole AA for conduit entrance (end view)
AD	Centerline of base or feet to centerline of conduit box inlet (side view)
AE	Centerline of conduit box inlet to bottom of feet (horizontal)
AF	Centerline of conduit box inlet to conduit entrance
AG	Mounting surface of face, flange or base of vertical machine to opposite end of housing (side view)
AH	Distance from mounting surface of face or flange to end of shaft
AJ	Diameter of bolt circle in face or flange or in base of vertical machine
AK	Diameter of male or female pilot on face or flange or on base of vertical machine
AL	Overall length of sliding base or rail
AM	Overall width of sliding base or outside dimensions of rails
AN	Distance from centerline of machine to bottom of sliding base or rails
AO	Centerline of sliding base or rail to centerline of mounting boltholes (end view)
AP	Centerline of sliding base or rails to centerline of inner-mounting boltholes (motor end view)
AR	Centerline of base or feet to centerline of sliding-base or rail-mounting boltholes (side view)
AT	Thickness of sliding base or rail foot
AU	Size of mounting holes in sliding base or rail
AV	Bottom of sliding base or rail to top of horizontal machine
AW	Distance from centerline of machine to centerline of inner mounting holes in rails (side view)
AX	Height of sliding base or rail
AY (max)	Maximum extension of sliding base (or rail) adjusting screw
AZ	Width of slide rails

TABLE 22. Lettering of Dimensions (*Continued*)

Letters	*Dimensions indicated*
BA	Centerline of mounting hole in the nearest foot to the shoulder on drive end shaft. Equal to M + W − F. For motors without a shaft shoulder, centerline of mounting hole in the nearest foot to limit line for mounting pulley, coupling, etc.
BB	Depth of male or female pilot of face or flange or base of vertical machine
BC	Distance from mounting surface of face or flange or base of vertical machine to shoulder on shaft
BD (max)	Outside diameter of face or flange or base of vertical machine
BE	Thickness of flange or base of vertical machine
BF	Diameter of mounting holes in flange or base of vertical machine
BG	Centerline of base or feet of horizontal machine to mounting surface of face or flange
BH	Outside diameter of core or shell (side view)
BJ (max)	Overall length of coils (side view). Actual dimensions may be less depending on the number of poles and winding construction
BK (max)	Distance from centerline of stator to lead end of coils
BL (max)	Diameter over coils, both ends (BL = 2 times maximum radius)
BM	Overall length of stator shell
BN	Diameter of stator bore
BO	Length of rotor at bore
BP (max)	Length of rotor over fans
BR	Diameter of finished surface or collar at ends of rotor
BS	Recess in pulley
BT	Movement of horizontal motor on base or rail
BU	Angle between centerline of conduit box inlet and centerline of motor (end view)
BV	Centerline of conduit box inlet to mounting surface of face or flange
BW (min)	Inside diameter of rotor fan or end ring for shell-type and hermetic motors
BX	Diameter of bore in top-drive coupling for hollow-shaft vertical motor
BY	Diameter of mounting holes in top-drive coupling for hollow-shaft vertical motor
BZ	Diameter of bolt circle for mounting holes in top-drive coupling for hollow-shaft vertical motor
CA	Rotor bore diameter
CB	Rotor counterbore diameter
CC	Depth of rotor counterbore
CD	Distance from the top of coupling to the bottom of the base on type P vertical motors
CE	Overall diameter of mounting lugs
CF	Distance from the end of the stator shell to the end of the rotor quill at compressor end. Where either the shell or quill is omitted, the dimension refers to the driven load end of the core
CG (max)	Distance from the end of the stator shell to the end of the stator coil at compressor end
CH (max)	Distance from the end of the stator shell to the end of the stator coil at end opposite the compressor
CJ (max)	Distance from end of stator core, commutator end, to end of coil at end opposite commutator for universal motors (side view)
CK	Distance from end of stator core, commutator end, to end of shaft insulation at end opposite the commutator for universal motors (side view)
CL	Distance between clamp-bolt centers for two-hole clamping of universal-motor stator cores
CM	Distance between horizontal clamp-bolt centers for four-hole clamping of universal-motor stator cores
CN	Distance between vertical clamp-bolt centers for four-hole clamping of universal-motor stator cores
CO	Clearance hole for maximum size of clamp bolts for clamping universal-motor stator cores
CP	Distance from end of stator core, commutator end, to end of commutator sleeve for universal motors (side view)

TABLE 22. Lettering of Dimensions (*Continued*)

Letters	*Dimensions indicated*
CR	Distance from end of stator core, commutator end, to centerline of brush for universal motors
CS	Brush width for universal motors
CT	Diameter of commutator for universal motors
CU	Diameter of bearing journal, commutator end, for universal motors
CV	Diameter of shaft under winding, end opposite commutator, for universal motors
CW (min)	Distance from centerline of shaft to outside end of brushholder supporting area for universal motors
CX (min)	Axial distance from centerline of brushholder to inside of housing for universal motors
CY (max)	Distance from centerline of shaft to centerline of brushholder setscrew for universal motors
CZ	Diameter of brushholder supporting hole for universal motors
DA	Distance from centerline of shaft to inside of brushholder boss for universal motors
DB	Outside diameter of rotor core
DC	Distance from the end of stator shell (driven load end) to the end of rotor fan or end ring (opposite driven-load end). Where the shell is omitted, the dimension is to the driven-load end of the stator core
DD	Distance from the end of stator shell (driven-load end) to the end of rotor fan or end ring (driven-load end). Where the shell is omitted, the dimension is to the driven-load end of the stator core
DE (min)	Diameter inside coils, both ends (DE = 2 times minimum radius)
DF	Distance from driven-load end of stator core or shell to centerline of mounting hole in lead clip or end of lead if no clip is used
DG (max)	Distance from driven-load end of stator core or shell to end of stator coil (opposite driven-load end)
DH	Centerline of base or feet to centerline of secondary-lead conduit-box inlet (side view)
DJ	Centerline of secondary-lead conduit-box inlet to bottom of feet (horizontal)
DK	Centerline of machine to centerline of hole DM for secondary-lead conduit entrance (end view)
DL	Centerline of secondary-lead conduit-box inlet to entrance for conduit
DM	Diameter of conduit (pipe size) for secondary-lead conduit box
DN	Distance from the end of stator shell to the bottom of rotor counterbore (driven-load end). Where the shell is omitted, the dimension is to the driven-load end of the stator core
DO	Dimension between centerlines of base-mounting grooves for resilient ring-mounted motors or, on base drawings, the dimension of the base which fits the groove
DP	Radial distance from center of type C face at end opposite drive to center of circle defining the available area for disk brake lead opening(s)
DQ to EJ, incl.	(No longer in use)
EK	Overall length of motor without feet from mounting face to opposite end of motor

Dimensions added by individual manufacturers for their own use will be designated by symbols with the prefix X.

* Any letter dimension normally applying to the drive end of the machine will, when prefixed with the letter F, apply to the end opposite the drive.

TABLE 23. Dimensions for ac Foot-Mounted Motors and Generators with Single Straight Shaft Extension

Frame No.[a]	Key			A max	B max	D[d]	E	F	BA	H	N-W[e,f]	U	V min	AA min size of conduit	AL	AM	AO	AR	AU	AW	AX	AY max bases	AY max rails	BT
	Width	Thickness	Length[b]																					
42	...	3/64 flat	....	....	....	2 5/64	1 3/4	27/32	2 1/16	9/32 slot	1 1/8	3/8												
48	...	3/64 flat	....	....	....	3	2 1/8	1 3/8	2 1/2	11/32 slot	1 1/2	1/2												
48H	...	3/64 flat	....	....	....	3	2 1/8	2 3/8	2 1/2	11/32 slot	1 1/2	1/2												
56	3/16	3/16	1 3/8[c]	....	....	3 1/2	2 7/16	1 1/2	2 3/4	11/32 slot	1 7/8	5/8												
56H	3/16	3/16	1 3/8[c]	....	....	3 1/2	2 7/16	2 1/2	2 3/4	11/32 slot	1 7/8	5/8												
66	3/16	3/16	1 7/8[c]	....	....	4 1/8	2 15/16	2 1/2	3 1/8	13/32 slot	2 1/4	3/4												
182	3/16	3/16	1 3/8	9	6 1/2	4 1/2	3 3/4	2 1/4	2 3/4	13/32	2 1/4	7/8	2	3/4	12 3/4	9 1/2	4 1/2	4 1/4	1/2	...	1 1/2	1/2	...	3
184	3/16	3/16	1 3/8	9	7 1/2	4 1/2	3 3/4	2 3/4	2 3/4	13/32	2 1/4	7/8	2	3/4	12 3/4	10 1/2	4 1/2	4 3/4	1/2	...	1 1/2	1/2	...	3
213	1/4	1/4	2	10 1/2	7 1/2	5 1/4	4 1/4	2 3/4	3 1/2	13/32	3	1 1/8	2 3/4	3/4	15	11	5 1/4	4 3/4	1/2	...	1 3/4	1/2	...	3 1/2
215	1/4	1/4	2	10 1/2	9	5 1/4	4 1/4	3 1/2	3 1/2	13/32	3	1 1/8	2 3/4	3/4	15	12 1/2	5 1/4	5 1/2	1/2	...	1 3/4	1/2	...	3 1/2
254U	5/16	5/16	2 3/4	12 1/2	10 3/4	6 1/4	5	4 1/8	4 1/4	17/32	3 3/4	1 3/8	3 1/2	1	17 3/4	15 1/8	6 1/4	6 5/8	5/8	1 5/8	2	5/8	5	4
256U	5/16	5/16	2 3/4	12 1/2	12 1/2	6 1/4	5	5	4 1/4	17/32	3 3/4	1 3/8	3 1/2	1	17 3/4	16 7/8	6 1/4	7 1/2	5/8	2 1/2	2	5/8	5	4
284U	3/8	3/8	3 3/4	14	12 1/2	7	5 1/2	4 3/4	4 3/4	17/32	4 7/8	1 5/8	4 5/8	1 1/4	19 3/4	16 7/8	7	7 1/2	5/8	2	2	5/8	5 3/4	4 1/2
286U	3/8	3/8	3 3/4	14	14	7	5 1/2	5 1/2	4 3/4	17/32	4 7/8	1 5/8	4 5/8	1 1/4	19 3/4	18 3/8	7	8 1/4	5/8	2 3/4	2	5/8	5 3/4	4 1/2
324U	1/2	1/2	4 1/4	16	14	8	6 1/4	5 1/4	5 1/4	21/32	5 5/8	1 7/8	5 3/8	1 1/2	22 3/4	19 1/4	8	8 1/2	3/4	2	2 1/2	3/4	6 1/2	5 1/4
324S	3/8	3/8	1 7/8	16	14	8	6 1/4	5 1/4	5 1/4	21/32	3 1/4	1 5/8	3	1 1/2										
326U	1/2	1/2	4 1/4	16	15 1/2	8	6 1/4	6	5 1/4	21/32	5 5/8	1 7/8	5 3/8	1 1/2	22 3/4	20 3/4	8	9 1/4	3/4	2 3/4	2 1/2	3/4	6 1/2	5 1/4
326S	3/8	3/8	1 7/8	16	15 1/2	8	6 1/4	6	5 1/4	21/32	3 1/4	1 5/8	3	1 1/2										
364U	1/2	1/2	5	18	15 1/4	9	7	5 5/8	5 7/8	21/32	6 3/8	2 1/8	6 1/8	2	25 1/2	20 1/2	9	9 1/8	3/4	2 1/8	2 1/2	3/4	7 1/2	6
364US	1/2	1/2	2	18	15 1/4	9	7	5 5/8	5 7/8	21/32	3 3/4	1 7/8	3 1/2	2										
365U	1/2	1/2	5	18	16 1/4	9	7	6 1/8	5 7/8	21/32	6 3/8	2 1/8	6 1/8	2	25 1/2	21 1/2	9	9 5/8	3/4	2 5/8	2 1/2	3/4	7 1/2	6
365US	1/2	1/2	2	18	16 1/4	9	7	6 1/8	5 7/8	21/32	3 3/4	1 7/8	3 1/2	2										

404U	5/8	3/8	5 1/2	20	16 1/4	10	8	6 1/8	6 5/8	13/16	7 1/8	2 3/8	6 7/8	2	28 3/4	22 3/8	10	9 7/8	7/8	2 3/8	3	7/8	7 7/8	7
404US	1/2	1/2	2 3/4	20	16 1/4	10	8	6 1/8	6 5/8	13/16	4 1/4	2 1/8	4	2										
405U	5/8	3/8	5 1/2	20	17 3/4	10	8	6 7/8	6 5/8	13/16	7 1/8	2 3/8	6 7/8	2	28 3/4	23 7/8	10	10 5/8	7/8	3 1/8	3	7/8	7 7/8	7
405US	1/2	1/2	2 3/4	20	17 3/4	10	8	6 7/8	6 5/8	13/16	4 1/4	2 1/8	4	2										
444U	3/4	3/4	7	22	18 1/2	11	9	7 1/4	7 1/2	13/16	8 5/8	2 7/8	8 3/8	2 1/2	31 1/4	24 5/8	11	11	7/8	3 1/2	3	7/8	8 3/8	7 1/2
444US	1/2	1/2	2 3/4	22	18 1/2	11	9	7 1/4	7 1/2	13/16	4 1/4	2 1/8	4	2 1/2										
445U	3/4	3/4	7	22	20 1/2	11	9	8 1/4	7 1/2	13/16	8 5/8	2 7/8	8 3/8	2 1/2	31 1/4	26 5/8	11	12	7/8	4 1/2	3	7/8	8 3/8	7 1/2
445US	1/2	1/2	2 3/4	22	20 1/2	11	9	8 1/4	7 1/2	13/16	4 1/4	2 1/8	4	2 1/2										
504U	3/4	3/4	7 1/4	25	21	12 1/2	10	8	8 1/2	15/16	8 5/8	2 7/8	8 3/8	2 1/2	35	28	12 1/2	12 1/2	1	3 1/2	3 1/2	1	9	8
504S	1/2	1/2	2 3/4	25	21	12 1/2	10	8	8 1/2	15/16	4 1/4	2 1/8	4	2 1/2										
505	3/4	3/4	7 1/4	25	23	12 1/2	10	9	8 1/2	15/16	8 5/8	2 7/8	8 3/8	2 1/2	35	30	12 1/2	13 1/2	1	4 1/2	3 1/2	1	9	8
505S	1/2	1/2	2 3/4	25	23	12 1/2	10	9	8 1/2	15/16	4 1/4	2 1/8	4	2 1/2										

All dimensions in inches.

[a] The system used for obtaining frame numbers is given in MG 1-11.01.

[b] Tolerance on length of key, plus or minus 1/32 in.

[c] Effective length of keyway.

[d] Frames 42, 48, 56, and 66. Dimension D will never be greater than the above values for rigid-base motors, but it may be less, so that shims are usually required for coupled or geared machines. When the exact dimension is required, shims up to 1/32 in. may be necessary. No tolerance for the D dimension of resilient-mounted motors has been established.

Frames 182 to 505S, inclusive. Dimension D will never be greater than the above values, but it may be less, so that shims are usually required for coupled or geared machines. When the exact dimension is required, shims up to 1/32 in. may be necessary on frame sizes whose dimension D is 8 in. and less: on larger frames, shims up to 1/16 in. may be necessary.

[e] N-W is the length of the shaft extension from the shoulder to the end of the shaft.

[f] Frames 42, 48, 56, and 66. If the shaft-extension length of the motor is not suitable for the application, it is recommended that deviations from this length be in 1/4-in. increments.

For the meaning of the letter dimensions, see MG 1-11.02.

For the tolerances on shaft extension diameters and keyways, see MG 1-11.06.

It is recommended that all machines with keyways cut in the shaft extension for pulley, coupling, pinion, etc., should be furnished with a key unless otherwise specified by the purchaser.

Frames 42, 48, 56, and 66. The tolerance for the 2F dimension should be plus or minus 1/32 in. and for the H (width of slot) dimension shall be plus 3/64 in. minus 0 in. Frames 182 to 505S, inclusive: The tolerance for the 2E and 2F dimensions should be plus or minus 1/32 in. and for the H dimension plus 3/64 in. minus 0 in.

TABLE 24. Dimensions for ac Foot-Mounted Motors with Single Straight Shaft Extension

Frame No.[a]	Key			A max	B max[c]	D[d]	E	F	BA	H	N-W[e]	U	V min	AA min size of conduit
	Width	Thickness	Length[b]											
143	3/16	3/16	1 3/8	7	6	3 1/2	2 3/4	2	2 1/4	11/32	2	3/4	1 3/4	3/4
145	3/16	3/16	1 3/8	7	6	3 1/2	2 3/4	2 1/2	2 1/4	11/32	2	3/4	1 3/4	3/4
143T	3/16	3/16	1 3/8	7	6	3 1/2	2 3/4	2	2 1/4	11/32	2 1/4	7/8	2	3/4
145T	3/16	3/16	1 3/8	7	6	3 1/2	2 3/4	2 1/2	2 1/4	11/32	2 1/4	7/8	2	3/4
182T	1/4	1/4	1 3/4	9	6 1/2	4 1/2	3 3/4	2 1/4	2 3/4	13/32	2 3/4	1 1/8	2 1/2	
184T	1/4	1/4	1 3/4	9	7 1/2	4 1/2	3 3/4	2 3/4	2 3/4	13/32	2 3/4	1 1/8	2 1/2	
213T	5/16	5/16	2 3/8	10 1/2	7 1/2	5 1/4	4 1/4	2 3/4	3 1/2	13/32	3 3/8	1 3/8	3 1/8	
215T	5/16	5/16	2 3/8	10 1/2	9	5 1/4	4 1/4	3 1/2	3 1/2	13/32	3 3/8	1 3/8	3 1/8	
254T	3/8	3/8	2 7/8	12 1/2	10 3/4	6 1/4	5	4 1/8	4 1/4	17/32	4	1 5/8	3 3/4	
256T	3/8	3/8	2 7/8	12 1/2	12 1/2	6 1/4	5	5	4 1/4	17/32	4	1 5/8	3 3/4	
284T	1/2	1/2	3 1/4	14	12 1/2	7	5 1/2	4 3/4	4 3/4	17/32	4 5/8	1 7/8	4 3/8	
284TS	3/8	3/8	1 7/8	14	12 1/2	7	5 1/2	4 3/4	4 3/4	17/32	3 1/4	1 5/8	3	
286T	1/2	1/2	3 1/4	14	14	7	5 1/2	5 1/2	4 3/4	17/32	4 5/8	1 7/8	4 3/8	
286TS	3/8	3/8	1 7/8	14	14	7	5 1/2	5 1/2	4 3/4	17/32	3 1/4	1 5/8	3	
324T	1/2	1/2	3 7/8	16	14	8	6 1/4	5 1/4	5 1/4	21/32	5 1/4	2 1/8	5	
324TS	1/2	1/2	2	16	14	8	6 1/4	5 1/4	5 1/4	21/32	3 3/4	1 7/8	3 1/2	
326T	1/2	1/2	3 7/8	16	15 1/2	8	6 1/4	6	5 1/4	21/32	5 1/4	2 1/8	5	
326TS	1/2	1/2	2	16	15 1/2	8	6 1/4	6	5 1/4	21/32	3 3/4	1 7/8	3 1/2	
364T	5/8	5/8	4 1/4	18	15 1/4	9	7	5 5/8	5 7/8	21/32	5 7/8	2 3/8	5 5/8	
364TS	1/2	1/2	2	18	15 1/4	9	7	5 5/8	5 7/8	21/32	3 3/4	1 7/8	3 1/2	
365T	5/8	5/8	4 1/4	18	16 1/4	9	7	6 1/8	5 7/8	21/32	5 7/8	2 3/8	5 5/8	
365TS	1/2	1/2	2	18	16 1/4	9	7	6 1/8	5 7/8	21/32	3 3/4	1 7/8	3 1/2	

404T	3/4	3/4	5 5/8	20	16 1/4	10	8	6 1/8	6 5/8	1 3/16	7 1/4	2 7/8	7
404TS	1/2	1/2	2 3/4	20	16 1/4	10	8	6 1/8	6 5/8	1 3/16	4 1/4	2 1/8	4
405T	3/4	3/4	5 5/8	20	17 3/4	10	8	6 7/8	6 5/8	1 3/16	7 1/4	2 7/8	7
405TS	1/2	1/2	2 3/4	20	17 3/4	10	8	6 7/8	6 5/8	1 3/16	4 1/4	2 1/8	4
444T	7/8	7/8	6 7/8	22	18 1/2	11	9	7 1/4	7 1/2	1 3/16	8 1/2	3 3/8	8 1/4
444TS	5/8	5/8	3	22	18 1/2	11	9	7 1/4	7 1/2	1 3/16	4 3/4	2 3/8	4 1/2
445T	7/8	7/8	6 7/8	22	20 1/2	11	9	8 1/4	7 1/2	1 3/16	8 1/2	3 3/8	8 1/4
445TS	5/8	5/8	3	22	20 1/2	11	9	8 1/4	7 1/2	1 3/16	4 3/4	2 3/8	4 1/2

All dimensions in inches.

For the meaning of the letter dimensions, see MG 1-11.02.

For the tolerances on shaft extension diameters and keyways, see MG 1-11.06.

It is recommended that all machines with keyways cut in the shaft extension for pulley, coupling, pinion, etc., should be furnished with a key unless otherwise specified by the purchaser.

The tolerance for the 2E and 2F dimensions should be $\pm 1/32$ in.; for the H dimension it shall be $+2/64$ in. -0 in.

[a] The system used for obtaining frame numbers is given in MG 1-11.01.

[b] Tolerance on length of key: plus or minus 1/32 in.

[c] The B dimension does not necessarily have the same centerline as the F dimension.

[d] Dimension D will never be greater than the above values, but it may be less, so that shims are usually required for coupled or geared machines. When the exact dimension is required, shims up to 1/32 in. may be necessary on frame sizes whose dimension D is 8 in. and less; on larger frames shims up to 1/16 in. may be necessary.

[e] N-W is the length of the shaft extension from the shoulder to the end of the shaft.

TABLE 25. Shaft Extension and Key Dimensions for ac Foot-Mounted Motors and Generators with Tapered or Double Shaft Extension

Frame No.	Drive-end tapered shaft extension										
									Key		
	BA	N-W	U	V	X	Y	Z max	Shaft thread	Width	Thickness	Length
182	2¾	2⅝	⅞	1¾	1⅞	¾	1⅜	⅝–18	3/16	3/16	1½
184	2¾	2⅝	⅞	1¾	1⅞	¾	1⅜	⅝–18	3/16	3/16	1½
213	3½	3⅜	1⅛	2¼	2⅜	⅞	1½	¾–16	¼	¼	2
215	3½	3⅜	1⅛	2¼	2⅜	⅞	1½	¾–16	¼	¼	2
254U	4¼	4⅛	1⅜	2⅝	2¾	1¼	2	1 –12	5/16	5/16	2¼
256U	4¼	4⅛	1⅜	2⅝	2¾	1¼	2	1 –12	5/16	5/16	2¼
284U	4¾	4½	1⅝	2⅞	3	1¼	2	1 –12	⅜	⅜	2½
286U	4¾	4½	1⅝	2⅞	3	1¼	2	1 –12	⅜	⅜	2½
324U 324S	5¼	4¾	1⅞	3⅛	3¼	1¼	2⅜	1½– 8	½	½	2⅞
326U 326S	5¼	4¾	1⅞	3⅛	3¼	1¼	2⅜	1¼– 8	½	½	2⅞
364U 364US	5⅞	5¼	2⅛	3½	3⅝	1⅜	2¾	1½– 8	½	½	3¼
365U 365US	5⅞	5¼	2⅛	3½	3⅝	1⅜	2¾	1½– 8	½	½	3¼
404U 404US	6⅝	5¾	2⅜	3¾	3⅞	1½	3¼	1¾– 8	⅝	⅝	3½
405U 405US	6⅝	5¾	2⅜	3¾	3⅞	1½	3¼	1¾– 8	⅝	⅝	3½
444U 444US	7½	6⅝	2⅞	4⅜	4½	1¾	3⅝	2 – 8	¾	¾	4⅛
445U 445US	7½	6⅝	2⅞	4⅜	4½	1¾	3⅝	2 – 8	¾	¾	4⅛
504U 504S	8½	6⅝	2⅞	4⅜	4½	1¾	3⅝	2 – 8	¾	¾	4⅛
505 505S	8½	6⅝	2⅞	4⅜	4½	1¾	3⅝	2 – 8	¾	¾	4⅛

All dimensions in inches.

31. Secondary Data

$$100\%\ \text{ohms} = \frac{249 \times \text{hp}}{I^2}$$

where hp = horsepower rating
I = secondary current per phase, amperes

$$\text{Rated hp} = \frac{I^2 \times 100\%\ \text{ohms} \times 3}{746}$$

TABLE 26. Shaft Extension and Key Dimensions for ac Foot-Mounted Motors and Generators with Double Shaft Extension

Frame No.	Opposite drive end straight shaft extension					
	FN-FW*	FU	FV min	Key†		
				Width	Thickness	Length‡
143T	1⅝	⅝	1⅜	3/16	3/16	⅞
145T	1⅝	⅝	1⅜	3/16	3/16	⅞
182T	2¼	⅞	2	3/16	3/16	1⅜
184T	2¼	⅞	2	3/16	3/16	1⅜
213T	2¾	1⅛	2½	¼	¼	1¾
215T	2¾	1⅛	2½	¼	¼	1¾
254T	3⅜	1⅜	3⅛	5/16	5/16	2⅜
256T	3⅜	1⅜	3⅜	5/16	5/16	2⅜
284T	4	1⅝	3¾	⅜	⅜	2⅞
284TS	3¼	1⅝	3	⅜	⅜	1⅞
286T	4	1⅝	3¾	⅜	⅜	2⅞
286TS	3¼	1⅝	3	⅜	⅜	1⅞
324T	4⅝	1⅞	4⅜	½	½	3¼
324TS	3¾	1⅞	3½	½	½	2
326T	4⅝	1⅞	4⅜	½	½	3¼
326TS	3¾	1⅞	3½	½	½	2
364T	4⅝	1⅞	4⅜	½	½	3¼
364TS	3¾	1⅞	3½	½	½	2
365T	4⅝	1⅞	4⅜	½	½	3¼
365TS	3¾	1⅞	3½	½	½	2
404T	5¼	2⅛	5	½	½	3⅞
404TS	4¼	2⅛	4	½	½	2¾
405T	5¼	2⅛	5	½	½	3⅞
405TS	4¼	2⅛	4	½	½	2¾
444T	5⅞	2⅜	5⅝	⅝	⅝	4¼
444TS	4¾	2⅜	4½	⅝	⅝	3
445T	5⅞	2⅜	5⅝	⅝	⅝	4¼
445TS	4¾	2⅜	4½	⅝	⅝	3

All dimensions in inches.
For tolerances on standard shaft extension diameters and keyways, see MG 1-11.06.
For the meaning of letter dimensions, see MG 1-11.02.
* FN-FW is the length of the shaft extension from the shoulder to the end of the shaft, if shoulder is used.
† It is recommended that all machines with keyways cut in the shaft extension for pulley, coupling, pinion, etc., should be furnished with a key.
‡ Tolerance on length of key is plus or minus 1/32 in.

$$\text{Secondary voltage} = \frac{\text{hp} \times 746}{I \times 3}$$

(usually gives 5 to 10% less than the value given the design engineer).

TABLE 27. Shaft Extension and Key Dimensions for ac Motors Built in Frames Larger than the 445U Frame

U	Key* Width	Key* Thickness	U	Key* Width	Key* Thickness
2	$\frac{1}{2}$	$\frac{1}{2}$	$3\frac{3}{8}$	$\frac{7}{8}$	$\frac{7}{8}$
$2\frac{1}{8}$	$\frac{1}{2}$	$\frac{1}{2}$	$3\frac{1}{2}$	$\frac{7}{8}$	$\frac{7}{8}$
$2\frac{1}{4}$	$\frac{1}{2}$	$\frac{1}{2}$	$3\frac{5}{8}$	$\frac{7}{8}$	$\frac{7}{8}$
$2\frac{3}{8}$	$\frac{5}{8}$	$\frac{5}{8}$	$3\frac{3}{4}$	$\frac{7}{8}$	$\frac{7}{8}$
$2\frac{1}{2}$	$\frac{5}{8}$	$\frac{5}{8}$	$3\frac{7}{8}$	1	1
$2\frac{5}{8}$	$\frac{5}{8}$	$\frac{5}{8}$	4	1	1
$2\frac{3}{4}$	$\frac{5}{8}$	$\frac{5}{8}$	$4\frac{3}{8}$	1	1
$2\frac{7}{8}$	$\frac{3}{4}$	$\frac{3}{4}$	$4\frac{1}{2}$	1	1
3	$\frac{3}{4}$	$\frac{3}{4}$	$4\frac{7}{8}$	$1\frac{1}{4}$	$1\frac{1}{4}$
$3\frac{1}{8}$	$\frac{3}{4}$	$\frac{3}{4}$	5	$1\frac{1}{4}$	$1\frac{1}{4}$
$3\frac{1}{4}$	$\frac{3}{4}$	$\frac{3}{4}$			

Horsepower and speed ratings

Synchronous speed, r/min	hp: Squirrel-cage and wound-rotor motors	hp: Synchronous motors power factor 1.0	hp: Synchronous motors power factor 0.8
3,600	500	200	150
1,800	500	200	150
1,200	350	200	150
900	250	150	125
720	200	125	100
600	150	100	75
514	125	75	60

All dimensions in inches.
*For belt-connected and direct-connected drives.

32. Motor-Selection Factors. The correct selection and application of a motor involves a great many factors affecting the installation, operation, and subsequent servicing of the motor. The selection is determined in whole, or in part, by the user or certain intermediaries, such as the manufacturer of motor-driven devices, based on known facts and calculations, field tests, and close study of the processes and driven machines.

While this section cannot include all the special problems and conditions, it is presented as a general guide to the conditions most commonly encountered in motor application.

33. Basic Steps. The basic steps in the selection of a motor consist of determining the power supply, horsepower rating, speed, motor type, and enclosure (see Sec. 2). In addition, the surrounding conditions, mounting, connection of the motor to the load, and mechanical accessories or modifications must be considered.

TABLE 28. Dimensions for Type C Face-Mounted ac Motors

Frame No.	U	AH	AJ	AK	BB min	BG	BD max	BG	BF hole No.	BF hole Tap size	BF hole Bolt-penetration allowance	Key Width	Key Thickness
42C	3/8	1 5/16	3 3/4	3	5/32	3/16	5	2 23/32	4	1/4–20			
48C	1/2	1 11/16	3 3/4	3	5/32	3/16	5 5/8	3 11/16	4	1/4–20			
56C	5/8	2 1/16	5 7/8	4 1/2	5/32	3/16	6 1/2	4 1/16	4	3/8–16			
66C	3/4	2 7/16	5 7/8	4 1/2	5/32	3/16	6 1/2	5 7/16	4	3/8–16			
182C	7/8	2 1/8	5 7/8	4 1/2	5/32	−1/8	6 1/2	.....	4	3/8–16	9/16	3/16	3/16
184C	7/8	2 1/8	5 7/8	4 1/2	5/32	−1/8	6 1/2	.....	4	3/8–16	9/16	3/16	3/16
213C	1 1/8	2 3/4	7 1/4	8 1/2	1/4	−1/4	9	.....	4	1/2–13	3/4	1/4	1/4
215C	1 1/8	2 3/4	7 1/4	8 1/2	1/4	−1/4	9	.....	4	1/2–13	3/4	1/4	1/4
254UC	1 3/8	3 1/2	7 1/4	8 1/2	1/4	−1/4	10	.....	4	1/2–13	3/4	5/16	5/16
256UC	1 3/8	3 1/2	7 1/4	8 1/2	1/4	−1/4	10	.....	4	1/2–13	3/4	5/16	5/16
284UC	1 5/8	4 5/8	9	10 1/2	1/4	−1/4	11 1/4	.....	4	1/2–13	3/4	3/8	3/8
286UC	1 5/8	4 5/8	9	10 1/2	1/4	−1/4	11 1/4	.....	4	1/2–13	3/4	3/8	3/8
324UC	1 7/8	5 3/8	11	12 1/2	1/4	−1/4	14	.....	4	5/8–11	15/16	1/2	1/2
324SC	1 5/8	3	11	12 1/2	1/4	−1/4	14	.....	4	5/8–11	15/16	3/8	3/8
326UC	1 7/8	5 3/8	11	12 1/2	1/4	−1/4	14	.....	4	5/8–11	15/16	1/2	1/2
326SC	1 5/8	3	11	12 1/2	1/4	−1/4	14	.....	4	5/8–11	15/16	3/8	3/8
364UC	2 1/8	6 1/8	11	12 1/2	1/4	−1/4	14	.....	8	5/8–11	15/16	1/2	1/2
364USC	1 7/8	3 1/2	11	12 1/2	1/4	−1/4	14	.....	8	5/8–11	15/16	1/2	1/2
365UC	2 1/8	6 1/8	11	12 1/2	1/4	−1/4	14	.....	8	5/8–11	15/16	1/2	1/2
365USC	1 7/8	3 1/2	11	12 1/2	1/4	−1/4	14	.....	8	5/8–11	15/16	1/2	1/2
404UC	2 3/8	6 7/8	11	12 1/2	1/4	−1/4	15 1/2	.....	8	5/8–11	15/16	5/8	5/8
404USC	2 1/8	4	11	12 1/2	1/4	−1/4	15 1/2	.....	8	5/8–11	15/16	1/2	1/2
405UC	2 3/8	6 7/8	11	12 1/2	1/4	−1/4	15 1/2	.....	8	5/8–11	15/16	5/8	5/8
405USC	2 1/8	4	11	12 1/2	1/4	−1/4	15 1/2	.....	8	5/8–11	15/16	1/2	1/2
444UC	2 7/8	8 3/8	14	16	1/4	−1/4	18	.....	8	5/8–11	15/16	3/4	3/4
444USC	2 1/8	4	14	16	1/4	−1/4	18	.....	8	5/8–11	15/16	1/2	1/2
445UC	2 7/8	8 3/8	14	16	1/4	−1/4	18	.....	8	5/8–11	15/16	3/4	3/4
445USC	2 1/8	4	14	16	1/4	−1/4	18	.....	8	5/8–11	15/16	1/2	1/2
504UC	2 7/8	8 3/8	14 1/2	16 1/2	1/4	−1/4	18	.....	4	5/8–11	15/16	3/4	3/4
505C	2 7/8	8 3/8	14 1/2	16 1/2	1/4	−1/4	18	.....	4	5/8–11	15/16	3/4	3/4
504SC	2 1/8	4	14 1/2	16 1/2	1/4	−1/4	18	.....	4	5/8–11	15/16	1/2	1/2
505SC	2 1/8	4	14 1/2	16 1/2	1/4	−1/4	18	.....	4	5/8–11	15/16	1/2	1/2

See NEMA Standard MG 1.11-34.

34. Duty Classification. Motors drive a load which may be classified as continuous, intermittent, or varying duty.

a. Continuous duty is a requirement of service that demands operation at a substantially constant load for an indefinitely long time. This duty is the most common duty classification and accounts for 90 percent of motor applications.

b. Intermittent duty is a requirement of service that demands operation for

TABLE 29. Opposite Drive End Straight and Tapered Shaft Extension

Frame No.	Opposite drive end straight shaft extension						Opposite drive end tapered shaft extension*									
	FN-FW†	FU	FV min	Key§			FN-FW†	FU	FV	FX	FY	FZ max	Shaft thread‡	Key§		
				Width	Thickness	Length¶								Width	Thickness	Length¶
182	2 1/4	7/8	2	3/16	3/16	1 3/8	2 5/8	7/8	1 3/4	1 7/8	3/4	1 3/8	5/8–18	3/16	3/16	1 1/2
184	2 1/4	7/8	2	3/16	3/16	1 3/8	2 5/8	7/8	1 3/4	1 7/8	3/4	1 3/8	5/8–18	3/16	3/16	1 1/2
213	2 1/4	7/8	2	3/16	3/16	1 3/8	2 5/8	7/8	1 3/4	1 7/8	3/4	1 3/8	5/8–18	3/16	3/16	1 1/2
215	2 1/4	7/8	2	3/16	3/16	1 3/8	2 5/8	7/8	1 3/4	1 7/8	3/4	1 3/8	5/8–18	3/16	3/16	1 1/2
254U	3	1 1/8	2 3/4	1/4	1/4	2	3 3/8	1 1/8	2 1/4	2 3/8	7/8	1 1/2	3/4–16	1/4	1/4	2
256U	3	1 1/8	2 3/4	1/4	1/4	2	3 3/8	1 1/8	2 1/4	2 3/8	7/8	1 1/2	3/4–16	1/4	1/4	2
284U	3 3/4	1 3/8	3 1/2	5/16	5/16	2 3/4	4 1/8	1 3/8	2 5/8	2 3/4	1 1/4	2	1 –12	5/16	5/16	2 1/4
286U	3 3/4	1 3/8	3 1/2	5/16	5/16	2 3/4	4 1/8	1 3/8	2 5/8	2 3/4	1 1/4	2	1 –12	5/16	5/16	2 1/4
324U	4 7/8	1 5/8	4 5/8	3/8	3/8	3 3/4	4 1/2	1 5/8	2 7/8	3	1 1/4	2	1 –12	3/8	3/8	2 1/2
324S	3 1/4	1 5/8	3	3/8	3/8	1 7/8										
326U	4 7/8	1 5/8	4 5/8	3/8	3/8	3 3/4	4 1/2	1 5/8	2 7/8	3	1 1/4	2	1 –12	3/8	3/8	2 1/2
326S	3 1/4	1 5/8	3	3/8	3/8	1 7/8										
364U	5 5/8	1 7/8	5 3/8	1/2	1/2	4 1/4	4 3/4	1 7/8	3 1/8	3 1/4	1 1/4	2 3/8	1 1/4– 8	1/2	1/2	2 7/8
364US	3 3/4	1 7/8	3 1/2	1/2	1/2	2										
365U	5 5/8	1 7/8	5 3/8	1/2	1/2	4 1/4	4 3/4	1 7/8	3 1/8	3 1/4	1 1/4	2 3/8	1 1/4– 8	1/2	1/2	2 7/8
365US	3 3/4	1 7/8	3 1/2	1/2	1/2	2										
404U	6 3/8	2 1/8	6 1/8	1/2	1/2	5	5 1/4	2 1/8	3 1/2	3 5/8	1 3/8	2 3/4	1 1/2– 8	1/2	1/2	3 1/4
404US	4 1/4	2 1/8	4	1/2	1/2	2 3/4										
405U	6 3/8	2 1/8	6 1/8	1/2	1/2	5	5 1/4	2 1/8	3 1/2	3 5/8	1 3/8	2 3/4	1 1/2– 8	1/2	1/2	3 1/4
405US	4 1/4	2 1/8	4	1/2	1/2	2 3/4										

444U	7⅛	2⅜	6⅞	⅝	⅝	5½	5¾	2⅜	3¾	3⅞	1½	3¼	1¾– 8	⅝	⅝	3½
444US	4¼	2⅛	4	½	½	2¾										
445U	7⅛	2⅜	6⅞	⅝	⅝	5½	5¾	2⅜	3¾	3⅞	1½	3¼	1¾– 8	⅝	⅝	3½
445US	4¼	2⅛	4	½	½	2¾										
504U	4¼	2⅛	4	½	½	2¾	5¾	2⅜	3¾	3⅞	1½	3¼	1¾– 8	⅝	⅝	3½
504S	4¼	2⅛	4	½	½	2¾										
505	4¼	2⅛	4	½	½	2¾	5¾	2⅜	3¾	3⅞	1½	3¼	1¾– 8	⅝	⅝	3½
505S	4¼	2⅛	4	½	½	2¾										

All dimensions in inches.

For tolerances on standard shaft extension diameters and keyways, see MG 1-11.06.

For the meaning of the letter dimensions, see MG 1-11.02.

* The taper of shaft should be at the rate of 1¼ in. in diameter per foot of length.

† N-W and FN-FW is the length of the shaft extension from the shoulder to the end of the shaft, if shoulder is used.

‡ The threaded end of the tapered shaft should be provided with a nut and suitable locking device.

§ It is recommended that all machines with keyways cut in the shaft extension for pulley, coupling, pinion, etc., shall be furnished with a key.

¶ Tolerance on length of key is plus or minus $\frac{1}{32}$ in.

TABLE 30. Dimensions for Frame Series 580 and 680 for Foot-Mounted Motors and Generators

Frame series*	D†	E	F (for third digit in frame No.)‡										BA
			0	1	2	3	4	5	6	7	8	9	
580	14½	11½	5½	6¼	7	8	9	10	11	12½	14	16	10
680	17	13½	7	8	9	10	11	12½	14	16	18	20	11½

All dimensions in inches.

For the meaning of the letter dimensions, see MG 1-11.02.

For suggested shaft extension dimensions see MG 1-11.61 for dc motors and MG 1-11.33 for ac motors.

The tolerance for the 2E and 2F dimensions should be ±1/32 in.

* The system for obtaining frame numbers is given in MG 1-11.01.

† Dimension D will never be greater than the above values, but it may be less, so that shims are usually required for coupled or geared machines. When the exact dimension is required, shims up to 1/16 in. may be necessary.

‡ In place of a last digit in the frame number, the successive integers following 9 will be used when F dimensions greater than those shown for the 9 frames are required. In such cases, the recommended values of F are, in succession, the 20 series preferred numbers rounded to the next largest inch or ½ in. For example, for frame 5811, F equals 20 in.

alternate intervals of load and no load; or load and rest; or load, no load, and rest, each interval of which is definitely specified.

c. Varying duty is a requirement of service that demands operation at loads and for intervals of time both of which may be subject to wide variation.

35. Horsepower Required. The horsepower required by the driven machine determines the motor rating (see Sec. 1). Where the load varies with time, a horsepower versus time curve will permit determining the peak horsepower required and the calculation of the root-mean-square (rms) horsepower, indicating the proper motor rating from a heating standpoint. In case of extremely large variations in load, or where shutdown, accelerating, or decelerating periods constitute a large portion of the cycle, the rms horsepower may not give a true indication of the equivalent continuous load, and the motor manufacturer should be consulted.

Where the load is maintained at a constant value for an extended period (varying from 15 min to 2 h, depending upon the size), the horsepower rating will not usually be less than this constant value, regardless of other parts of the cycle.

If the driven machine is to operate at more than one speed, the horsepower required at each speed must be determined.

36. Duty Cycle and Inertia. In any type of duty-cycle operation, it is necessary to determine not only the horsepower requirements, but the number of times the motor will be started, the inertia of the driven machine, and the method of stopping the motor. The inertia or flywheel effect (Wk^2) of the rotating parts of the driven machine affects the accelerating time and therefore the heating of the motor, particularly where reversing duty or frequent starting is involved.

37. Moment of Inertia. The moment of inertia is expressed as Wk^2 in terms of pound-feet squared. It is the product of the weight of the object in pounds and the square of the radius of gyration in feet.

The radius of gyration may be defined as the average of the radii from the axis of rotation of each infinitesimal part of the object.

If the application is such that the motor is driving through a pulley or gear so that the driven equipment is operating at a higher or lower speed than the motor, it is necessary to calculate the inertia referred to the motor shaft, that is, an equivalent Wk^2 based on the speed of the motor,

$$Wk^2 \text{ (referred to motor shaft)} = Wk^2 \text{ (driven equipment)} \times \frac{(\text{driven equipment speed})^2}{(\text{motor speed})^2}$$

The method of stopping the motor is important. For example, if the motor is plug-reversed, this operation increases the heating to a much larger extent than if the motor is allowed to coast to a stop, or if a mechanical brake is used.

38. Brake Selection. If a motor-mounted brake is required, it is necessary to determine the proper brake size. The majority of applications are satisfied by selecting a brake with a torque rating equal to the full-load motor torque since this rating will stop the load in approximately the same time required by the motor to accelerate the load.

In brake selection it is necessary not only to determine the torque required, but to be assured that the thermal capacity of the brake is sufficient. This rating is the ability of the brake to dissipate the heat generated during the stopping cycle. If the required stops per minute exceed four, or if the inertia is large, the motor or brake manufacturer should be consulted to determine if the brake selected has sufficient thermal capacity.

39. Service Factor. When determining motor-horsepower requirements, it should be kept in mind that open, general-purpose motors do have a service factor, depending upon the particular rating of the motor.

40. Axial Thrust on Bearings. Following the standard recommendations on the use of minimum sheave sizes will prevent exceeding the allowable radial bearing load. On certain applications such as fan and close-coupled pumps, it is necessary to assure that the maximum thrust capacity of the bearing is not exceeded.

Table 31 lists the recommended maximum axial thrust loading for standard horizontal ball-bearing motors. The bearing-loading data given are for reference only. Any special applications which involve high radial or axial loading should be referred to the manufacturer.

41. Thermal Protection. Built-in thermal protection is available.

a. Thermostats. These devices, which can be furnished on polyphase machines, are mounted next to the stator winding. They are temperature-sensing only and are available with normally open or normally closed contacts. However, they are only available with automatic reset.

Two leads are provided for connection to a magnetic control or indicating cir-

TABLE 31. Recommended Maximum Axial Thrust, in Pounds

	r/min		
Frame*	3600	1800	1200
140	55	80	110
180	55	80	110
210	100	150	195
250	150	225	250
280	300	350	475
320	375	500	625
360	475	675	800
400	500	700	825
440	600	825	950

*Frames with prefix T or TS may have higher capacity. Refer to manufacturer.

cuit. Thermostats will protect the motor from high ambient temperatures, exceedingly long accelerating cycles (where the design is not rotor-limited), repeated or excessive overloads, and loss of ventilation. The devices are rugged, easy to apply, and relatively inexpensive. However, they do not give protection under stalled or locked-rotor conditions. They must be used with some external current-sensitive motor control for complete motor protection.

b. Inherent protection system. This system consists of two or more normally closed heat-sensing miniature switches connected in series and embedded within the motor windings. It offers the same built-in inherent protection as the current-sensing thermal protector since it is sensitive to both the total temperature and the rate of temperature rise. This protection system is connected in series with the coil of the motor's magnetic contactor and can be arranged to restart either automatically or manually. The only limitation is that it must be used with magnetic control. This protection would normally be recommended above 10 hp.

42. Handy Formulas

$$\text{Torque (lb}\cdot\text{ft)} = \frac{\text{hp} \times 5250}{\text{r/min}}$$

$$\text{hp} = \frac{\text{torque} \times \text{r/min}}{5250}$$

$$\text{Speed (r/min)} = \frac{120 \times \text{frequency}}{\text{no. of poles}}$$

$$°\text{C} = (°\text{F} - 32) \times 5/9$$

$$°\text{F} = (°\text{C} \times 9/5) + 32$$

43. Rules of Thumb

At 3600 r/min a motor develops 1.5 lb·ft/hp.

At 1800 r/min a motor develops 3 lb·ft/hp.

At 1200 r/min a motor develops 4.5 lb·ft/hp.
At 550 V a three-phase motor draws 1 A/hp.
At 440 V a three-phase motor draws 1.25 A/hp.
At 220 V a three-phase motor draws 2.5 A/hp.

*ENERGY CONSIDERATIONS**

The shortage and the large cost increases of vital national energy resources have demonstrated the need to conserve such resources. It is important that motor users and specifiers understand the selection, application, and maintenance of electric motors in order to improve the management of electric energy consumption. Energy management as related to electric motors is the consideration of the factors that contribute to reducing the energy consumption of a total electric-motor-drive system. Among the factors to be considered are the motor design and application.

An electric motor is an energy converter, converting electric energy to mechanical energy. For this reason, an electric motor should be considered as always being connected to a driven machine or apparatus, with specific operating characteristics, which dictate the starting- and running-load characteristics of the motor (see Sec. 1). Consequently, the selection of the motor most suitable for a particular application is based on many factors, including the requirements of the driven equipment (such as starting and acceleration, speed, load, duty cycle), service conditions, motor efficiency, motor power factor, and initial motor cost. These application factors often conflict with one another. The driven-system efficiency is the combination of the efficiencies of all of the components in the system. In addition to the motor, these components include the driven equipment (such as fans, pumps, and compressors) and the power transmission components (such as belts, pulleys, gears, and clutches). Other components which are not part of the driven system will affect the overall system efficiency. Some of these are refrigerator and air-conditioning evaporator and condenser coils, piping associated with pumps, ducts and baffles associated with fans and blowers, and motor controllers (ac variable-speed drives and power-factor controllers).

Good energy management is the successful application of the motor controller, the motor, and the driven components that results in the least consumption of energy. Since all motors do not have the same efficiency, careful consideration must be given to their selection and application.

44. Motor-Selection Factors. The proper selection and application of polyphase induction and synchronous motors involves the consideration of many factors affecting installation, operation, and maintenance. The basic steps in selecting a motor consist of determining the power supply, horsepower rating, speed, duty cycle, motor type, and enclosure. In addition, environmental conditions, mounting, connections of the motor to the load, and mechanical accessories or modifications must be considered. Motors must also be properly selected with respect to the known service conditions, often referred to as usual and unusual, as defined

*Information in these paragraphs is from NEMA Standard MG 10-1983.

in NEMA Standard MG 1-1978, "Motors and Generators." Usual service conditions are considered to be:

1. Exposure to an ambient temperature within the range of 0 to 40°C
2. Installation in areas or supplementary enclosures which do not seriously interfere with the ventilation of the machine
3. Operation within a tolerance of ±10 percent of rated voltage
4. Operation from a sine wave voltage source (not to exceed 10 percent deviation factor)
5. Operation within a tolerance of ±5 percent of rated frequency
6. Operation within a voltage unbalance of 1 percent or less

Operation at other than usual service conditions may result in the consumption of additional energy.

45. Efficiency. Motor efficiency is a measure of the effectiveness with which electric energy is converted to mechanical energy. It is expressed as the ratio of power output to power input:

$$\text{Efficiency} = \frac{\text{output}}{\text{input}} = \frac{\text{output}}{\text{output} + \text{losses}}$$

Motor efficiencies are usually given for rated load, ¾ load, and ½ load.

The efficiency of a motor is primarily a function of load, horsepower rating, and speed, as indicated below:

1. A change in efficiency as a function of load is an inherent characteristic of motors (Fig. 9). Operation of the motor at loads substantially different from rated load may result in a change in motor efficiency.
2. In general, the full-load efficiency of motors increases as the motor horsepower rating increases (Fig. 9).
3. For the same horsepower rating, motors with higher speeds generally, but not necessarily, have a higher efficiency at rated load than motors with lower rated speeds. This does not imply, however, that all apparatus should be driven by high-speed motors. Where speed-changing mechanisms, such as pulleys or gears, are required to obtain the necessary lower speed, the additional power losses could reduce the efficiency of the system to a value lower than that provided by a direct-drive lower-speed motor.

A definite relationship exists between the slip and the efficiency of a polyphase induction motor, that is, the higher the slip, the lower is the efficiency, for slip is a measure of the losses in the rotor winding. Slip of an induction motor is the difference between synchronous speed and full-load speed. Slip, expressed in percent, is the difference in speeds divided by the synchronous speed and multiplied by 100. Therefore, under steady load conditions, Design A, B, and C squirrel-cage induction motors having a slip of less than 5 percent are more efficient than Design D motors having a higher slip, and should be used when permitted by the application. However, for applications involving pulsating-inertia loads, such as a punch press or oil-well pumping, the overall efficiency of motors with high slip may be higher than that of motors having a slip of less than 5 percent. In addition,

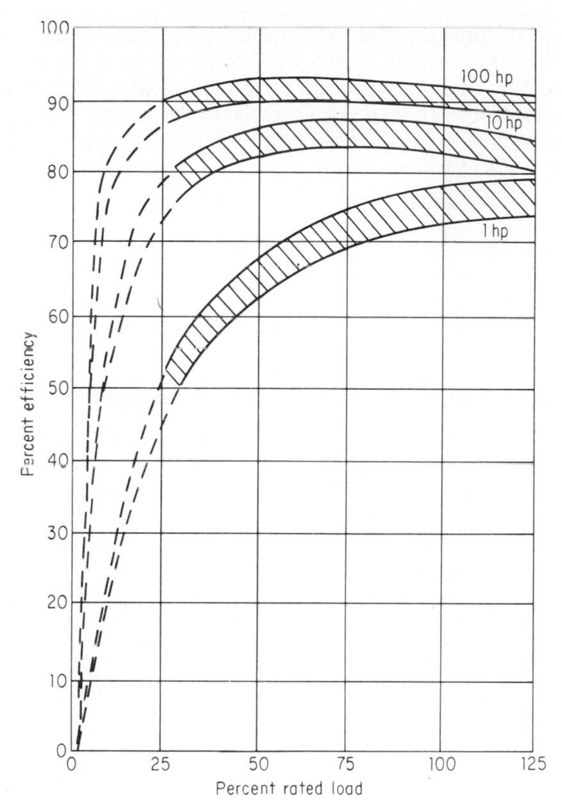

FIG. 9. Typical efficiency versus load curves for 1800-r/min three-phase 60-Hz Design B squirrel-case induction motors.

high-slip motors may be necessary for applications requiring high starting (locked-rotor) torque.

If load permits, it may be possible to make a significant saving in energy by utilizing a multispeed motor operating at low speed whenever possible and using high speed only when necessary. However, it should be noted that the efficiency of a multispeed motor at each operating speed is somewhat lower than that of a single-speed motor having a comparable rating. Single-winding multispeed motors are generally more efficient than two-winding multispeed motors.

Motors which operate continuously or for long periods of time provide a significant opportunity for reducing energy consumption. Examples of such applications are processing machinery, air-moving equipment, pumps, and many types of industrial equipment. A small change in motor efficiency will make a significant change in the total energy cost due to the lengthy operating time.

While many motors are operated continuously, some motors are used for very short periods of time and for a very low total number of hours per year. Examples of such applications are valve motors, dam-gate operators, and industrial door openers. In these instances, a change in motor efficiency would not substantially change the total energy cost since very little total energy is involved.

A modest increase of a few percentage points in motor efficiency can represent a rather significant decrease in percentage of motor losses. For example, for the same output, an increase in efficiency from 75 to 78.9 percent, from 85 to 87.6 percent, or from 90 to 91.8 percent represents a 20 percent decrease in losses in each case.

46. Motor Losses. An electric motor converts electric energy into mechanical energy and, in so doing, incurs losses which are generally described as follows.

a. Electrical (stator and rotor) losses (vary with load). Current flowing through the motor windings produces losses which are proportional to the current squared times the winding resistance (I^2R).

b. Iron (core) losses (essentially independent of load). These losses are confined mainly to the laminated core of the stator and rotor. The magnetic field, essential to the production of torque in the motor, causes hysteresis and eddy-current losses.

c. Mechanical (friction and windage) losses (independent of load). Mechanical losses occur in the bearings, fans, and brushes (when used) of the motor. These losses are generally small in open low-speed motors, but may be appreciable in large high-speed or totally enclosed fan-cooled motors.

d. Stray load losses (dependent on load). These are made up of several minor losses which result from such factors as leakage flux induced by motor currents, nonuniform current distribution in stator and rotor conductors, air gap, and so forth. These losses combined make up 10 to 15 percent of the total motor losses and tend to increase with load.

Table 32 gives the motor loss components together with the typical percentage of total motor losses they represent, and the design and construction factors which influence their magnitude.

In general, by increasing the active material in the motor, that is, the type and quantity of conductors and magnetic materials, the losses can be reduced.

In the manufacture of motors there are periodic variations in the motor characteristics even in motors of duplicate design and repetitive manufacture. These are caused by normal variations in purchased raw materials such as copper and steel. In addition, there are variations caused by manufacturing processes and testing-accuracy deviations. Thus in forecasting the efficiency of a given motor, one can speak of the nominal efficiency (the average efficiency of a population of motors of duplicate design) or the minimum efficiency (the level reached when both raw materials and manufacturing processes are at the least favorable end of

TABLE 32. Motor Loss Components

	Typical percent of losses, 4-pole motors	Factors affecting these losses
Stator losses	35 to 40	Stator conductor size
Rotor losses	15 to 20	Rotor conductor size
Core losses	15 to 20	Type and quantity of magnetic material
Stray-load losses	10 to 15	Primarily manufacturing and design methods
Friction and winding	5 to 10	Selection and design of fans and bearings

their specified tolerances). Both of these values are of use to the prospective purchaser of a motor. The nominal efficiency should be used in estimating the power required to supply a number of motors. The minimum efficiency permits the motor user the assurance of having received the specified level of performance. Thus the NEMA system of specifying efficiency is required to include both values.

47. Efficiency Testing Methods. There are a number of test methods for determining motor efficiency. The standard method for testing induction machines in the United States is ANSI/IEEE Standard 112-1983, "Test Procedure for Polyphase Induction Motors and Generators," which recognizes five methods for determining motor efficiency, each of which has certain advantages as to accuracy, cost, and ease of testing, depending primarily on the motor rating. The common practice for 1 to 125 hp is to test the motor with a load absorption device called dynamometer, and to measure carefully the power input and output to determine loss components and thus efficiency. This method is given in IEEE test code 112 as Method B; it forms the basis of determining the NEMA efficiency rating.

48. Manufacturing Variations. All manufactured products are subject to variances associated with materials and manufacturing methods. No two products will perform exactly the same, even though they are of the same design and produced on the same assembly line.

This is also true for electric motors. Product variances in materials, such as steel used for laminations in the stator and rotor cores, will lead to variances in electromagnetic properties and ultimately affect losses and motor efficiency. Using a 10-hp motor as an example, a 10 percent increase in iron loss (300 to 330 W), which is within the tolerance offered by steel suppliers, would increase the total motor losses from 1167 to 1197 W and reduce efficiency from 86.5 to 86.2 percent.

Variances also occur as the result of manufacturing process limitations. There is an economic limit to the practical dimensional tolerances on motor parts. Combinations of mating parts contribute to dimensional variations, such as the size of the air gap, which cause variations in stray-load loss and hence motor efficiency.

Due to reasonable manufacturing variations in materials, components, and processes, as illustrated above, efficiencies of specific type and horsepower motors produced by the same manufacturer will vary.

EVALUATION OF EFFICIENCY ECONOMICS

To obtain accurate results during the economic evaluation process, it is extremely important that only efficiencies determined by the same method be compared. There are a number of ways to evaluate the economic impact of motor efficiency. Two of these methods are simple-payback analysis and present-worth life-cycle analysis. A third method, not detailed here, is the cash-flow and payback analysis. This method considers motor-cost premium, motor-depreciation life, energy cost and energy-cost-inflation rate, corporate tax rate, tax credit, and the motor-operating parameters covered in the simple-payback analysis. Details of this method are explained in current financial management texts.

49. Simple-Payback Analysis. The simple-payback method gives the number of years required to recover the differential investment for higher efficiency motors. To determine the payback period, the premium for the higher-efficiency motor is divided by the annual savings. First the annual savings must be determined using the following formula:

$$S = 0.746 \times \text{hp} \times L \times C \times N\left(\frac{100}{E_B} - \frac{100}{E_A}\right)$$

where S = annual savings, dollars
hp = horsepower rating
L = percentage load divided by 100
C = energy cost, \$/kWh
N = annual hours of operation
E_B = lower motor efficiency
E_A = higher motor efficiency

The efficiencies for the percent load at which the motor will be operated must be used, since motor efficiency varies with load.

Then the motor-cost premium is divided by the annual savings to compute the payback period for the higher-efficiency motor. If motor A costs \$300 more than motor B and yields annual savings of \$100, the simple-payback period is \$300/\$100, or 3 years. The payback method is easy to apply but ignores savings occurring after the end of the payback period, and it does not consider the time value of money or changes in energy costs.

50. Present-Worth Life-Cycle Analysis. For greater precision, the present-worth method of life-cycle savings may be employed. This method considers both the time value of money and the energy-cost inflation. First the user will determine the required internal rate of return and the expected energy-cost inflation. Then the effective interest rate can be determined using the following formula:

$$i = \frac{1 + R_2}{1 + R_1} - 1$$

where i = effective interest rate
R_1 = expected annual rate of energy-cost inflation
R_2 = required internal rate of return on investments

The next step is to determine the present worth by inserting the effective interest rate into the basic formula for the present value of an annuity:

$$\text{PW} = \frac{(1 + i)^n - 1}{i(1 + i)^n}$$

where PW = present worth
n = expected operating lifetime of the motor

Once PW is solved, the present-worth evaluation factor (PWEF) can be calculated:

$$\text{PWEF (\$/kW)} = C \times N \times \text{PW}$$

where C = energy cost, \$/kWh
N = annual hours of operation
PW = present worth

Finally the present worth of the life-cycle savings resulting from the higher-efficiency motor can be computed:

$$\text{PW savings} = 0.746 \times \text{hp} \times \text{PWEF}\left(\frac{100}{E_B} - \frac{100}{E_A}\right)$$

The present worth of savings due to the investment is the basis on which the investment decision is made.

Example. Given:

100-hp motor A, 95% efficient
100-hp motor B, 92% efficient

Assumed:

5.5¢/kWh
12% inflation of energy costs
25% return on investment
8-yr expected life
4160 h/yr (2 shifts, 5 days per week)

Thus,

$$i = \left(\frac{1.25}{1.12} - 1\right) = 0.116$$

$$\text{PW} = \frac{1.116^8 - 1}{0.116 \times 1.116^8} = 5.037$$

$$\text{PWEF} = 0.055 \times 4160 \times 5.037 = 1152$$

$$\text{PW savings} = 0.746 \times 100 \times 1152 \times \left(\frac{100}{92} - \frac{100}{95}\right) = \$2950$$

OPERATING CONSIDERATIONS

51. Power Factor. The connected motor load in a facility is usually a major factor in determining the system power factor. Low system power factor results in increased losses in the distribution system. Induction motors inherently cause a lagging system power factor.

The power factor of an induction motor decreases as the load decreases, as shown in Fig. 10. Figure 11 indicates that the rated-load power factor increases with an increase in the horsepower rating of the motor. A number of induction motors, all operating at light load, can cause the electric system to have a low power factor. The power factor of induction motors at rated load is less for low-

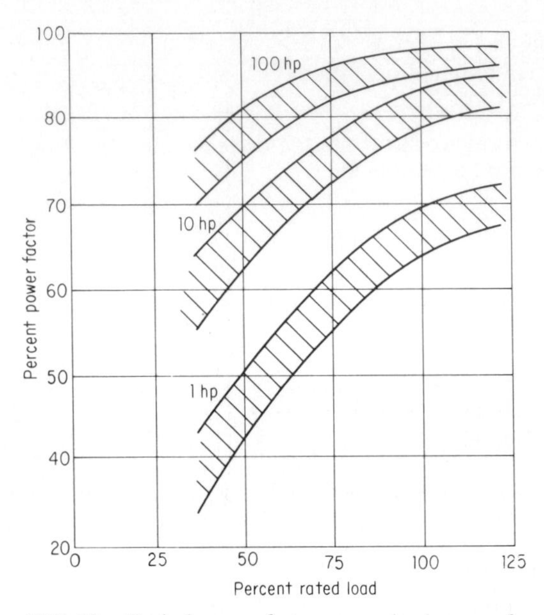

FIG. 10. Typical power factor versus load curves for 1800-r/min three-phase 60-Hz Design B squirrel-cage induction motors.

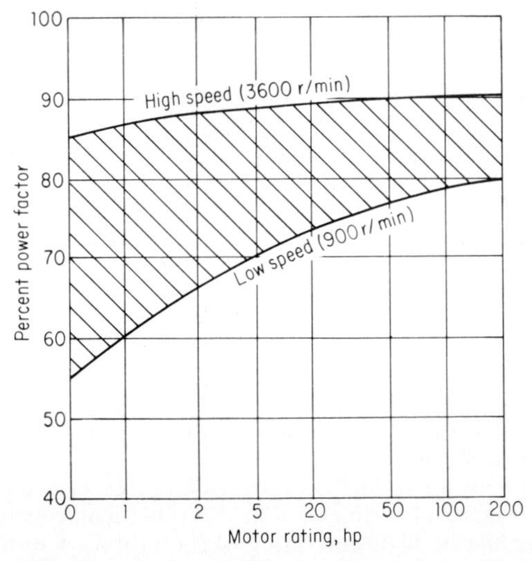

FIG. 11. Typical full-load power factor versus horsepower rating curves for three-phase 60-Hz Design B squirrel-cage induction motors.

speed motors than for high-speed motors as shown in Fig. 11. A small increase in voltage (less than or equal to 10 percent) above rated voltage will decrease the power factor, and a small decrease in voltage (less than or equal to 10 percent) below rated voltage will improve the power factor of an induction motor. However, other performance characteristics may be adversely affected by such a change in voltage, and operation as close as possible to the nameplate voltage and horsepower rating is recommended.

Power-factor-correction capacitors can be used to improve the power factor of the electric system. However, if they are used, they should be carefully selected and applied to avoid unsafe operating conditions. It is recommended that the motor manufacturer be consulted for the proper value of the corrective capacitance.

The power factor of synchronous motors can range from unity to approximately zero power factor leading, depending on the rated power factor, the field excitation, and the load. Standard designs are usually rated for either unity power factor or 0.8 leading power factor. As previously stated, synchronous motors have the capability of improving the power factor of the electric system.

An analysis of the electric system will indicate whether improvement in the power factor is needed and whether capacitors, synchronous motors, or other corrective measures should be used (see Secs. 2 and 5).

52. Application Analysis. When the device being driven by an electric motor is producing a relatively constant and continuous level of useful work, the primary motor selection concern is its rated-load efficiency. However, many applications are cyclic in nature. In these cases specific application techniques can be used to obtain substantial energy savings.

Other applications require intermittent or continuous absorption of energy. Again there are application techniques that will recover a significant percentage of the otherwise wasted energy.

A few of these cases are cited here to illustrate the technology that is available to the user. The motor manufacturer should be consulted to determine the most effective solution.

53. Applications Involving Load Cycling. Some applications require running at load for a period of time, followed by a period during which no useful work is being done by the driven machine. In this case energy may be saved by stopping the motor and restarting it at the beginning of the next load period.

The following example illustrates the savings to be realized by start-stop cycling. This example assumes that the friction and winding loss of the driven machine is 2.5 hp.

Example. Assumptions:

Motor rating, 10 hp

Motor full-load efficiency, 89.2%

Motor output with drive machine running idle, 2.5 hp

Motor efficiency at 2.5 hp, 86%

Motor acceleration loss plus system stored energy ($Wk^2 = 1.4\ \text{lb}\cdot\text{ft}^2$), 3750 Ws

Driven machine is loaded for 15 min followed by a 15-min idle period

Utilization, 2000 h/yr

Cost of energy, \$0.06/kWh

Motor inertia, 7 lb·ft²
Driven-machine inertia, 7 lb·ft²

Annual cost of energy if motor runs continuously:

1. Cost to run full load:

$$\text{Cost} = \frac{\text{run hours} \times \text{hp} \times 0.746 \times \$/\text{kWh}}{\text{FL EFF}}$$

$$= \frac{1000 \times 10 \times 0.746 \times 0.06}{0.892} = \$501$$

2. Cost to run driven machine idle:

$$\text{Cost} = \frac{1000 \times 2.5 \times 0.746 \times 0.06}{0.86} = \$130$$

Then, $\text{Total cost} = 501 + 130 = \$631$

Annual cost of energy if motor is shut down after each load period:

1. Cost to run full load: same as above, = $501
2. Cost to restore system energy and supply motor acceleration losses:

$$\text{Cost} = \frac{N_s \times J \times \$/\text{kWh}}{1000 \times 60 \times 60}$$

where N_s = number of starts per year
J = motor acceleration loss plus system stored energy, watt-seconds

$$\text{Cost} = \frac{4000 \times 3750 \times 0.06}{1000 \times 60 \times 60} = \$0.25$$

Then, $\text{Total cost} = 501 + 0.25 = \$501.25$

Hence,

$$\text{Energy savings per year} = 631 - 501.25 = \$129.75$$

When making a decision whether to stop a motor or to run at no load, a number of factors must be considered. These include motor type, horsepower rating, speed, starting frequency, restriction on inrush current, power demand charges, and the extra winding stress imposed by repeated accelerations and associated reduction in life expectancy.

NEMA Standard MG 1-12.50 provides guidance on the number of successive starts (i.e., two starts from ambient or one start from rated-load operating temperature). However, the information in MG 1-12.50 is not applicable to repetitive start-run-stop-rest cycles resulting from energy-management programs. Accordingly, Table 33 has been prepared as a guide to (1) the minimum off-time required to allow the motor to cool sufficiently to permit another start, (2) the maximum number of starts per hour (irrespective of load Wk^2) to minimize the effect of winding stress imposed by repeated starts, and (3) a means of adjusting the number of starts per hour as a function of the load inertia. It should be recognized that

TABLE 33. Allowable Number of Starts and Minimum Time between Starts for Design A and B Motors

	Two-pole			Four-pole			Six-pole		
hp	*A*	*B*	*C*	*A*	*B*	*C*	*A*	*B*	*C*
1	15	1.2	75	30	5.8	38	34	15	33
1.5	12.9	1.8	76	25.7	8.6	38	29.1	23	34
2	11.5	2.4	77	23	11	39	26.1	30	35
3	9.9	3.5	80	19.8	17	40	22.4	44	36
5	8.1	5.7	83	16.3	27	42	18.4	71	37
7.5	7.0	8.3	88	13.9	39	44	15.8	104	39
10	6.2	11	92	12.5	51	46	14.2	137	41
15	5.4	16	100	10.7	75	50	12.1	200	44
20	4.8	21	110	9.6	99	55	10.9	262	48
25	4.4	26	115	8.8	122	58	10.0	324	51
30	4.1	31	120	8.2	144	60	9.3	384	53
40	3.7	40	130	7.4	189	65	8.4	503	57
50	3.4	49	145	6.8	232	72	7.7	620	64
60	3.2	58	170	6.3	275	85	7.2	735	75
75	2.9	71	180	5.8	338	90	6.6	904	79
100	2.6	92	220	5.2	441	110	5.9	1181	97
125	2.4	113	275	4.8	542	140	5.4	1452	120
150	2.2	133	320	4.5	640	160	5.1	1719	140
200	2.0	172	600	4.0	831	300	4.5	2238	265
250	1.8	210	1000	3.7	1017	500	4.2	2744	440

A—maximum number of starts per hour; *B*—maximum product of starts per hour times load Wk^2; *C*—minimum rest or off-time in seconds.

Allowable starts per hour is the lesser of (1) *A* or (2) *B* divided by the load Wk^2, that is,

$$\text{Starts per hour} \leq A \leq \frac{B}{\text{load } Wk^2}$$

Note: The table is based on the following conditions:

1. Applied voltage and frequency in accordance with NEMA Standard MG 1-12.43.
2. During the accelerating period, the connected load torque is equal to or less than a torque which varies as the square of the speed and is equal to 100 percent of rated torque at rated speed.
3. The external load Wk^2 is equal to or less than the values listed in NEMA Standard MG 1-12.50.

For other conditions, the manufacturer should be consulted.

each start is one factor in the life expectancy and reliability of the motor, and as a result, some reduction in life expectancy and reliability must be accepted when a motor is applied at the upper range of the starting duty determined by Table 33.

Example 1. 50-hp four-pole Design B motor, direct-connected to a pump with a Wk^2 of 20 lb·ft². From Table 33,

$$A = 6.8$$

$$B/\text{load } Wk^2 = 232/20 = 11.6$$

$$\text{Minimum off-time } C = 72 \text{ s}$$

The value of B/load Wk^2 exceeds the maximum number of starts per hour. Therefore, the motor must be limited to a maximum of 6.8 starts per hour, with a minimum off-time between starts of 72 s.

Example 2. 25-hp two-pole 3550-r/min Design B motor, belt connected to a 5000-r/min blower with a Wk^2 of 3.7 lb·ft^2.

$$\text{Load } Wk^2 \text{ referred to motor shaft} = \left(\frac{5000}{3550}\right)^2 \times 3.7 = 7.34 \text{ lb·ft}^2$$

From Table 33,

$$A = 4.4$$

$$B/\text{load } Wk^2 = 26/7.34 = 3.5$$

$$\text{Minimum off-time } C = 115 \text{ s}$$

The value of B/load Wk^2 is less than the maximum number of starts per hour, and therefore the motor must be limited to 3.5 starts per hour, with a minimum off-time between starts of 115 s.

54. Applications Involving Extended Periods of Light-Load Operation. A number of methods have been proposed to reduce the voltage applied to the motor in response to the applied load, the purpose of this being to reduce the magnetizing losses during periods when the full torque capability of the motor is not required. Typical of these devices is the power-factor controller. It is a device that adjusts the voltage applied to the motor to approximate a preset power factor.

These power-factor controllers may, for example, be beneficial for use with small motors operating for extended periods of light loads where the magnetization losses are a relatively high percentage of the total loss. Care must be exercised in the application of these controllers. Savings are achieved only when the controlled motor is operated for extended periods at no load or light load.

Particular care must be taken when considering their use with other than small motors. A typical 10-hp motor will have idle losses on the order of 4 or 5 percent of the rated output. In this size range the magnetization losses that can be saved may not be equal to the losses added by the controller plus the additional motor losses caused by the distorted voltage waveform introduced by the controller.

SECTION 7

FRACTIONAL-HORSEPOWER AC MOTORS

Cyril G. Veinott, D. Eng.*

Walter G. Stiffler†

FOREWORD . . . 7-3
1. What Is a Fractional-Horsepower Motor? . . . 7-3

CLASSIFICATION BY APPLICATION . . . 7-3
2. General-Purpose Motor . . . 7-3
3. Definite-Purpose Motor . . . 7-3
4. Special-Purpose Motor . . . 7-3

CLASSIFICATION BY ELECTRICAL TYPE . . . 7-3
5. Polyphase Induction Motors . . . 7-3
6. Single-Phase Induction Motors . . . 7-6
7. Universal Motors . . . 7-7
8. Synchronous Motors . . . 7-7
9. Split-Phase Motors . . . 7-7
10. Capacitor-Start Motors . . . 7-9
11. Dual-Voltage Capacitor-Start Motors . . . 7-10
12. Permanent-Split Capacitor Motors . . . 7-10
13. Two-Value Capacitor Motors . . . 7-11
14. Repulsion Motors . . . 7-11
15. Repulsion-Start Induction-Run Motors . . . 7-11
16. Repulsion-Induction Motors . . . 7-12
17. Shaded-Pole Motors . . . 7-12

CLASSIFICATION BY METHOD OF ENCLOSURE . . . 7-13
18. Open Machine . . . 7-13
19. Totally Enclosed Machine . . . 7-13

*Machinery Consultant and Chief Engineering Analyst, Reliance Electric Company, Cleveland, Ohio (retired); Registered Professional Engineer (Ohio); Fellow, IEEE; Member, ACM.

†Manager of Engineering, Rotating Machinery Group, Reliance Electric Company, Cleveland, Ohio; Member, IEEE and NEMA.

NOTE: This section was written for the first edition by Cyril G. Veinott, D. Eng., and was reviewed and revised for this edition by Walter G. Stiffler

CLASSIFICATION BY VARIABILITY OF SPEED 7-13
20. Constant-Speed Motor 7-13
21. Varying-Speed Motor 7-13
22. Adjustable-Speed Motor 7-13
23. Adjustable Varying-Speed Motor 7-13
24. Pole-Changing Polyphase Motor 7-13
25. Pole-Changing Single-Phase Motor 7-14
26. Multispeed Motor 7-14

MISCELLANEOUS USEFUL DEFINITIONS 7-14
27. Efficiency 7-14
28. Power Factor 7-14
29. Service Factor 7-14
30. Torque 7-14

GENERAL-PURPOSE AND STANDARD MOTORS . . . 7-15
31. Voltages 7-16
32. Frequencies 7-16
33. Ratings and Performance Characteristics 7-16
34. Time and Temperature Ratings 7-17
35. Dimensions and Frame Assignments 7-17

APPLICATION DATA 7-17
36. Proper Selection of Apparatus 7-17
37. Usual Service Conditions 7-17
38. Direction of Rotation 7-19
39. Effects of Variation of Voltage and Frequency 7-19
40. Application of Motors with Service Factor 7-19

DEFINITE-PURPOSE FRACTIONAL-HORSEPOWER AC MOTORS 7-20
41. General 7-20
42. Permanent-Split Capacitor Industrial-Instrument Motors and Gear Motors 7-20
43. Low-Inertia Servo Industrial-Instrument Motors and Gear Motors 7-21
44. Universal-Motor Parts 7-21
45. Motors for Hermetic Refrigeration Compressors . . . 7-23
46. Motors for Shaft-Mounted Fans and Blowers 7-23
47. Shaded-Pole Motors for Shaft-Mounted Fans and Blowers 7-24
48. Motors for Belted Fans and Blowers 7-24
49. Motors for Air-Conditioning Condensers and Evaporator Fans 7-25
50. Motors for Cellar Drainers and Sump Pumps 7-26
51. Motors for Gasoline-Dispensing Pumps 7-26
52. Motors for Oil Burners 7-26
53. Motors for Home-Laundry Equipment 7-28
54. Motors for Jet Pumps 7-29
55. Motors for Coolant Pumps 7-30
56. Submersible Motors for Deep-Well Pumps, 4 in . . . 7-31

REFERENCES 7-31

FOREWORD

In the preparation of this section, the author has drawn, in considerable measure, upon previously published material, especially upon the first three works listed in the References. Because NEMA standards reflect the collective thinking of an industry, rather than the opinions of one individual, extensive use has been made of them in the preparation of this section. So much interpreting, editing, abridging, and expanding of the NEMA material has been done that it seemed hardly practical to put all NEMA words in quotes—or set them off in smaller type—and the net effect of so doing probably would have been to make this section harder to read and less useful to the user. If it becomes important to know the exact wording of the NEMA standard, reference should be made directly to the NEMA publication (Ref. 1).

1. What Is a Fractional-Horsepower Motor? A fractional-horsepower motor is generally understood to be any motor built in a fractional-horsepower frame, such as NEMA frames 56, 48, 42, or smaller, regardless of the actual horsepower rating of the motor. Generally, the ratings involved are those given in Tables 1 and 2. Of late, however, even larger ratings have often been built in these frame sizes. The physical sizes of these frames are shown in Fig. 12.

Fractional-horsepower motors may be classified in many ways.

CLASSIFICATION BY APPLICATION

2. General-Purpose Motor. An open motor, of continuous rating, built in standard ratings, with standard operating characteristics and standard mechanical constructions, intended for operation under usual service conditions (Art. 37) without restriction to a specific application or type of application.

3. Definite-Purpose Motor. One built in standard ratings, with standard operating characteristics for use under service conditions other than usual, and designed for use on a particular type of application (see Arts. 41 through 56).

4. Special-Purpose Motor. One built with special operating characteristics or special mechanical construction, or both, designed for a specific application.

CLASSIFICATION BY ELECTRICAL TYPE

5. Polyphase Induction Motors. In fractional-horsepower sizes these motors generally have characteristics comparable to those of Design B motors in their larger counterparts (see Sec. 6). Stator windings, energized by a three-phase supply, set up in the air gap a rotating magnetic field of an even number of poles. The stator does not actually have physical "salient" poles, like a dc motor, for example, but the winding arrangement is such as to produce "poles" in the revolving magnetic field. Rotors of such machines are invariably of squirrel-cage construction and generally die-cast. The speed of the rotating magnetic field is known as synchronous speed and is given by the formula

$$\text{Synchronous speed} = \frac{120 \times \text{frequency}}{\text{no. of poles}} \tag{1}$$

TABLE 1. Performance Characteristics of Fractional-Horsepower ac Motors

hp rating	Poles	Single- and three-phase		Single-phase				Three-phase			Permanent-split capacitor				
		Full-load		Torques		Locked-rotor current, A		Torques		Locked-rotor current, A	Full-load		Torques		Locked-rotor current, A
		Speed, r/min	Torque	Breakdown	Lock.	115 volts	230 volts	Breakdown	Lock.		Speed, r/min	Torque	Breakdown	Lock.	
1⁄20	2	3,450	1.22	3.7	...	20	12	5.18	....	....	3,250	1.29			
	4	1,725	2.44	7.1	...	20	12	9.94	....	....	1,625	2.58	4.13	2.3	2.5
	6	1,140	3.69	10.4	...	20	12	14.56	....	....	1,075	3.91	6.09	3.5	2.5
	8	850	4.89	13.5	...	20	12	18.90	....	....	825	5.09	8.00	4.6	2.5
1⁄12	2	3,450	2.03	6.0	...	20	12	8.40	....	....	3,250	2.15			
	4	1,725	4.06	11.5	...	20	12	16.10	....	....	1,625	4.31	6.39	3.1	4.0
	6	1,140	6.14	16.5	...	20	12	23.1	....	....	1,075	6.51	9.42	4.7	4.0
	8	850	8.14	21.5	...	20	12	30.1	....	....	825	8.50	12.4	6.0	4.0
1⁄8	2	3,450	3.04	8.7	...	20	12	12.18	....	....	3,250	3.23			
	4	1,725	6.09	16.5	24	20	12	23.1	....	....	1,625	6.46	10.4	4.0	5.8
	6	1,140	9.21	24.1	32	20	12	33.7	....	....	1,075	9.77	15.3	6.1	5.8
	8	850	12.21	31.5	...	20	12	44.1	....	....	825	12.73	20.1	7.9	5.8
1⁄6	2	3,450	4.06	11.5	15	20	12	16.10	9.7	4.0	3,250	4.31			
	4	1,725	8.12	21.5	33	20	12	30.1	18.0	4.0	1,625	8.62	12.7	5.2	7.5
	6	1,140	12.29	31.5	43	20	12	44.1	26.5	4.0	1,075	13.02	18.8	7.9	7.5
	8	850	16.29	40.5	...	20	12	56.7	34.0	4.0	825	17.0	24.6	10.0	7.5
1⁄4	2	3,450	6.09	16.5	21	26	15	23.1	13.9	6.0	3,250	6.46			
	4	1,725	12.18	31.5	46	26	15	44.1	26.5	6.0	1,625	12.92	21.0	7.4	11.0
	6	1,140	18.43	44.0	59	26	15	61.6	37.0	6.0	1,075	19.54	31.5	11.2	11.0
	8	850	24.42	58.0	...	26	15	81.2	49.0	6.0	825	25.45	41.0	14.5	11.0

1/3	2	3,450	8.12	21.5	26	31	18	30.1	18.0	8.0	3,250	8.62			
	4	1,725	16.24	40.5	57	31	18	56.7	34.0	8.0	1,625	17.35	31.5	19.6	14.5
	6	1,140	24.6	58.0	73	31	18	81.2	49.0	8.0	1,075	26.0	47.0	14.5	14.5
	8	850	32.6	77.0	...	31	18	108	65.0	8.0	825	34.0	61.0	18.7	14.5
1/2	2	3,450	12.19	31.5	37	45	25	44.1	26.5	12.0	3,250	12.92			
	4	1,725	24.4	58.0	85	45	25	81.2	49.0	12.0	1,625	25.8	47.5	13.5	21.0
	6	1,140	36.9	82.5	100	45	25	115.5	70.0	12.0	1,075	39.1	70.8	20.5	21.0
3/4	2	3,450	18.28	44.0	50	61	35	61.6	37.0	18.0	3,250	19.38			
	4	1,725	36.6	82.5	119	61	35	115.5	70.0	18.0	1,625	38.8	63.5	19.4	30.0
1	2	3,450	24.4	58.0	61	...	...	81.2	49.0	24.0	3,250	25.8			

All torques are in ounce-feet.

Single-phase. Torques are NEMA minimum values. Locked-rotor currents are NEMA maximum values for Design N (general-purpose) motors. For Design O motors, rated ½ hp or less, NEMA maximum values of locked-rotor current are 50 A for 115-V motors and 25 A for 230-V motors.

Three-phase. Breakdown torques are NEMA minimum values. Locked-rotor torques and currents are approximate values, taken from Ref. 3. Currents given are for 220 V; for 110 V, multiply by 2; for 440 V, divide by 2.

Permanent-split capacitor. Breakdown torques are NEMA minimum values. Locked-rotor torques and currents are approximate values, taken from Ref. 3. Currents given are for 115 V; for 230 V, divide by 2. In addition to the horsepower ratings shown, the following additional horsepower ratings are standard: 1/15, 1/10, and 1/5 hp.

TABLE 2. Torques of Shaded-Pole and Permanent-Split Capacitor Motors (Ref. 1)

Rating, mhp	4-pole		6-pole		8-pole	
	1,550 r/min		1,050 r/min		800 r/min	
	Full-load	Breakdown	Full-load	Breakdown	Full-load	Breakdown
	Torques, oz-in.					
1	0.65	0.89– 1.1	0.96	1.3 – 1.6	1.26	1.7 – 2.1
1.25	0.81	1.1 – 1.4	1.20	1.6 – 2.1	1.58	2.1 – 2.7
1.5	0.98	1.4 – 1.7	1.44	2.1 – 2.5	1.89	2.7 – 3.3
2	1.30	1.7 – 2.1	1.92	2.5 – 3.1	2.52	3.3 – 4.1
2.5	1.63	2.1 – 2.6	2.40	3.1 – 3.8	3.15	4.1 – 5.0
3	1.95	2.6 – 3.2	2.88	3.8 – 4.7	3.78	5.0 – 6.2
4	2.60	3.2 – 4.0	3.84	4.7 – 5.9	5.04	6.2 – 7.8
5	3.25	4.0 – 4.9	4.80	5.9 – 7.2	6.30	7.8 – 9.5
6	3.90	4.9 – 6.2	5.76	7.2 – 9.2	7.56	9.5 –12.0
8	5.20	6.2 – 7.7	7.68	9.2 –11.4	10.1	12.0 –14.9
10	6.50	7.7 – 9.6	9.60	11.4 –14.2	12.6	14.9 –18.6
12.5	8.13	9.6 –12.3	12.0	14.2 –18.2	15.8	18.6 –23.8
16	10.41	12.3 –15.3	15.4	18.2 –22.6	20.2	23.8 –29.6
20	13.00	15.3 –19.1	19.2	22.6 –28.2	25.2	29.6 –37.0
25	16.26	19.1 –23.9	24.0	28.2 –35.3	31.5	37.0 –46.3
30	19.51	23.9 –30.4	28.8	35.3 –44.9	37.8	46.3 –58.9
40	26.01	30.4 –38.3	38.4	44.9 –56.4	50.4	58.9 –74.4
Rating, hp	Torques, oz-ft					
1/20	2.71	3.20– 4.13	4.00	4.70– 6.09	5.25	6.20– 8.00
1/15	3.61	4.13– 5.23	5.33	6.09– 7.72	7.00	8.00–10.1
1/12	4.51	5.23– 6.39	6.67	7.72– 9.42	8.75	10.1 –12.4
1/10	5.42	6.39– 8.00	8.00	9.42–11.8	10.50	12.4 –15.5
1/8	6.77	8.00–10.4	10.00	11.8 –15.3	13.13	15.5 –20.1
1/6	9.03	10.4 –12.7	13.33	15.3 –18.8	17.50	20.1 –24.6
1/5	10.84	12.7 –16.0	16.00	18.8 –23.6	21.00	24.6 –31.0
1/4	13.55	16.0 –19.8	20.00	23.6 –29.2	26.25	31.0 –38.4

When acting as a motor, the actual speed of the rotor is always less than that of the rotating field. The difference is known as slip, usually expressed as a percentage of synchronous speed:

$$\text{Slip (\%)} = 100 \times \frac{\text{synchronous speed} - \text{actual speed}}{\text{synchronous speed}} \tag{2}$$

A more detailed explanation of the above is given in Ref. 2, Chap. 11.

6. Single-Phase Induction Motors. Most fractional-horsepower ac motors are used where only single-phase power is available. A single-phase induction motor is not inherently self-starting because a single winding, excited by a single-phase supply, produces a stationary, rather than a revolving, magnetic field in the air gap. Once up to speed, however, the motor will develop a useful torque. Various means for starting single-phase motors have been devised, giving rise to a number of different types of single-phase motors. Important types are split-phase, capacitor-start, two-value capacitor, permanent-split capacitor, repulsion, repulsion-start, repulsion-induction, and shaded-pole.

7. Universal Motors. These are essentially series-wound dc motors so designed and constructed as to be suitable for operation on alternating current of any frequency of 60 Hz or less or on direct current. For more information, see Art. 44.

8. Synchronous Motors. Synchronous motors are characterized by exact speed of operation—so exact, in fact, that they are used for driving electric clocks. Like induction motors, they are provided with stator windings that produce a rotating magnetic field, but unlike an induction motor, the rotor locks into step with the field and rotates at the same speed, as given by Eq. (1). Unlike large synchronous motors, the small ones are not provided with dc excitation, but the equivalent is sometimes supplied by means of permanent magnets. The basic types follow.

a. Reluctance motors. These are essentially induction motors with a special squirrel-cage rotor with one cutout per pole, providing as many salient projections as there are poles. These cutouts cause the magnetic reluctance to be greater between poles than it is along the pole axis; hence the name reluctance motor, because the motor depends for its operation upon a difference in reluctance. Polyphase reluctance motors have stator windings like those of polyphase induction motors. Single-phase reluctance motors may be provided with any of the following types of stator windings, found also in induction motors: split-phase, capacitor-start, permanent-split capacitor, two-value capacitor.

b. Hysteresis motors. These motors, like reluctance motors, use induction-motor stators to provide a rotating field, but they depend upon the hysteresis effect of a permanent-magnet material for their operation. One form of rotor construction consists of rings of chrome or cobalt steel pressed on over a nonmagnetic material, without any squirrel-cage winding. The rotating magnetic field brings the rotor up to speed, where it locks into synchronism because of the remanence effect of the permanent-magnet steel. The stator winding may be polyphase, split-phase, permanent-split capacitor, capacitor-start, or shaded-pole.

c. Permanent-magnet motors. These are essentially similar to reluctance motors except that the reluctance effect is augmented by a permanent-magnet material in the rotor.

d. Synchronous-inductor motors. These motors are essentially inductor alternators, operated as motors, except that the dc excitation, used in large inductor alternators, is provided by built-in permanent magnets. These are made in small sizes for light-duty applications.

9. Split-Phase Motors. A split-phase motor is represented schematically in Fig. 1. Note that there are two windings, displaced 90 electrical degrees. The auxiliary winding uses relatively fine wire, giving a high ratio of resistance to reactance, bringing the current in this winding more in phase with the supply voltage than is the current in the main winding. The effect is shown in Fig. 2. Observe that there is only a small time-phase displacement between the currents; hence the locked-rotor torque of this motor is comparatively low, and the locked-rotor current relatively high, because the line current is nearly equal to the sum of the currents in the two windings.

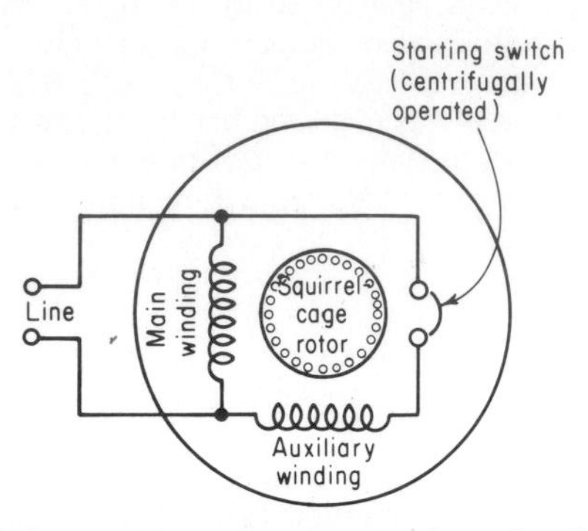

FIG. 1. Schematic representation of split-phase motor. *(Ref. 3, p. 24.)*

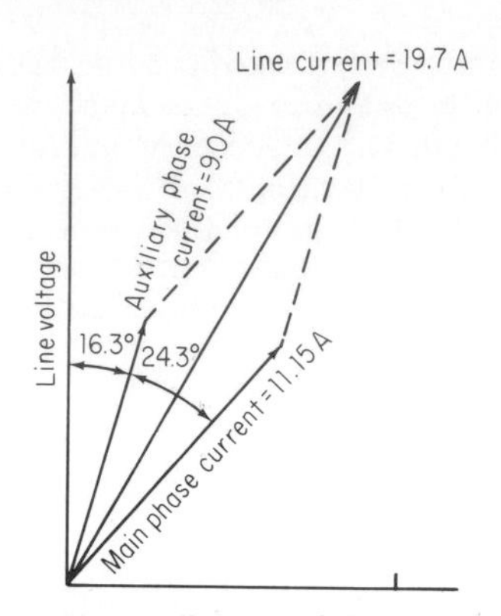

FIG. 2. Vector diagram of the locked-rotor currents of Design N split-phase motor. *(Ref. 3, p. 26.)*

Design N motors are offered in a wide variety of ratings and mechanical constructions for numerous applications. A typical speed-torque characteristic curve is shown in Fig. 3. A centrifugal starting switch (or magnetic relay) cuts out the auxiliary winding when the motor is nearly up to speed. Otherwise the auxiliary winding would quickly burn out. Breakdown torques are 10 to 15 percent lower than for general-purpose single-phase motors. Design O motors are made for a few special applications, generally involving infrequent starting, because of the higher locked-rotor currents.

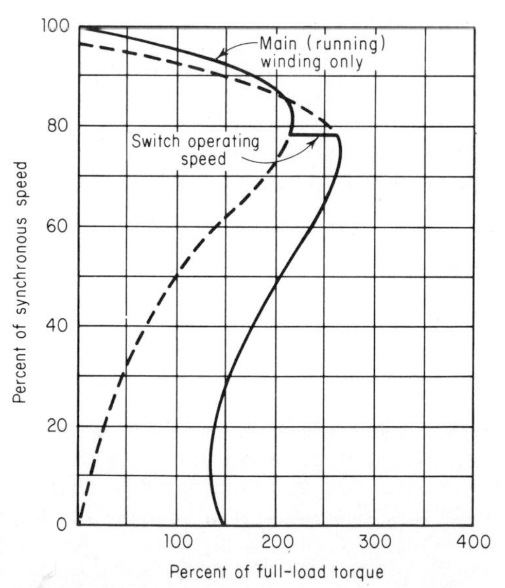

FIG. 3. Speed-torque characteristics of Design N split-phase motor. *(Ref. 3, p. 26.)*

10. Capacitor-Start Motors. This is the most popular type for general-purpose applications. It is shown schematically in Fig. 4, which shows two windings, displaced 90 electrical degrees in space, a capacitor in series with the auxiliary winding, and a centrifugal switch (or a magnetic relay) to cut out the auxiliary phase after the motor is nearly up to speed. Note in Fig. 5 a large phase displacement, which gives high locked-rotor torque and low locked-rotor current, because the currents add out of phase. Speed-torque characteristics of this type are shown in Fig. 6.

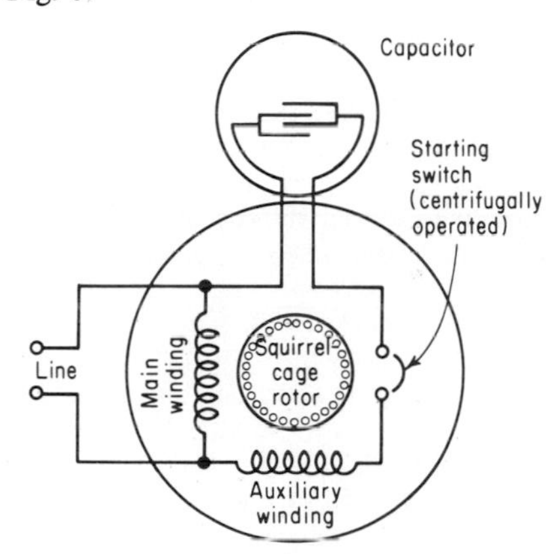

FIG. 4. Schematic representation of capacitor-start motor. *(Ref. 3, p. 53.)*

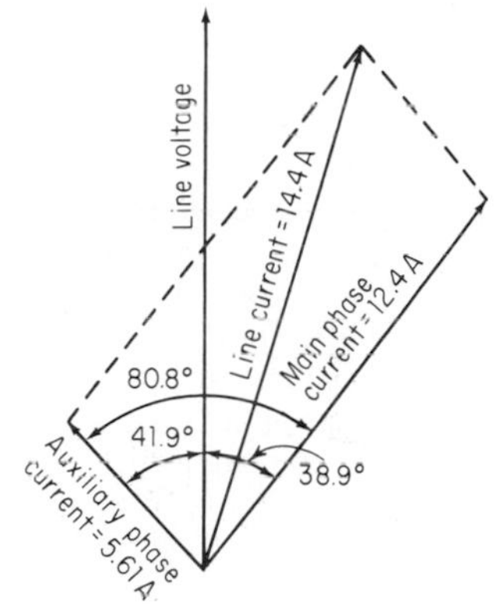

FIG. 5. Vector diagram of the locked-rotor currents of capacitor-start motor. *(Ref. 3, p. 55.)*

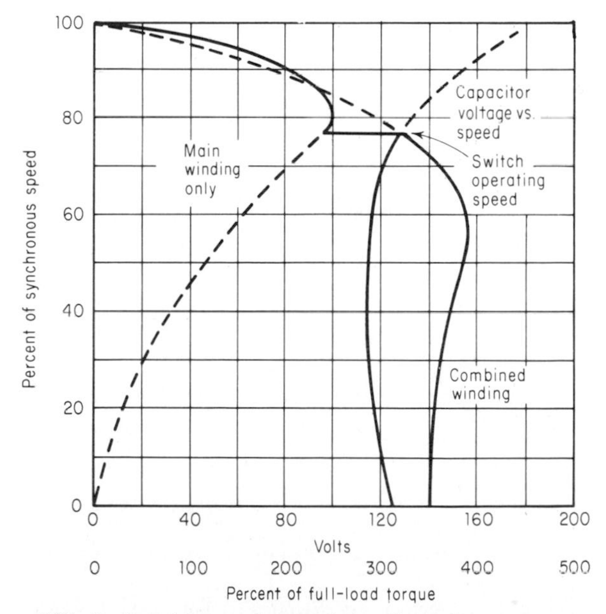

FIG. 6. Speed-torque characteristics of capacitor-start motor. *(Ref. 3, p. 56.)*

11. Dual-Voltage Capacitor-Start Motors. Most capacitor-start motors are built for dual voltage, that is, they can be connected in the field for either 115 or 230 V. Figure 7 shows schematically how this is accomplished. The main winding is split into two sections, but there is only one auxiliary winding, one starting switch, and one capacitor. For 115 V the two main-winding sections are connected in parallel, and the auxiliary phase is connected across the line. For 230 V the main-winding sections are connected in series, and the auxiliary winding is connected in parallel with one of the main-winding sections, so that it "sees" only 115 V, for the main winding serves as an autotransformer.

12. Permanent-Split Capacitor Motors. Schematically, a permanent-split capacitor motor would be represented by Fig. 4 or Fig. 7 if the starting switch were omitted, leaving the auxiliary phase (including capacitor) permanently in the circuit. Such a motor uses an oil-type capacitor of few microfarads, as compared with the electrolytic capacitor with many microfarads, used in a comparable capacitor-start motor. The capacitance for the permanent-split motor is only on the order of 5 to 10 percent of that for the capacitor-start motor. Hence the locked-rotor torque is quite low, much below that of a general-purpose motor. The principal advantages of this type over general-purpose motors are the absence of a starting switch, the capability of speed adjustment (see Arts. 46 and 49), and the quiet operation under load conditions, if not at no load. It is generally furnished with dual-voltage windings, similar to the windings shown in Fig. 7.

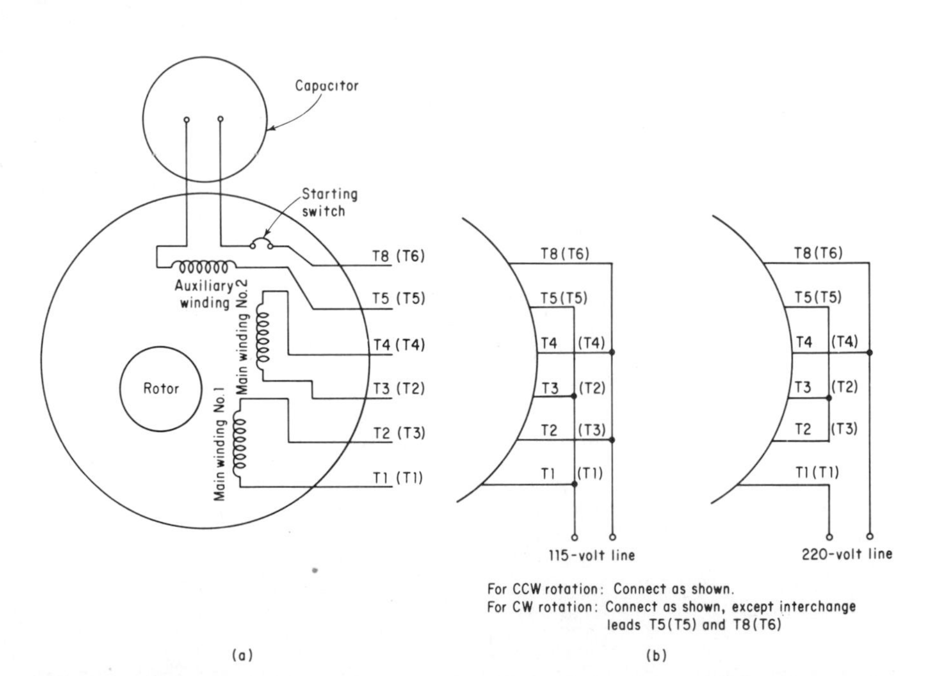

FIG. 7. (*a*) Wiring and (*b*) line-connection diagrams for four-lead dual-voltage capacitor-start motor. Terminal markings conform, in principle, to ANSI standards. Markings in parentheses show another system that has been used. *(Ref. 2, p, 131.)*

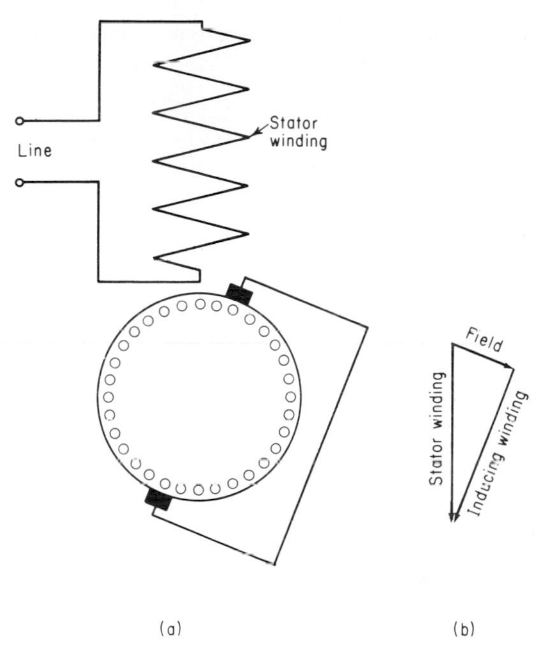

FIG. 8. Schematic representation of repulsion motor. *(Ref. 3, p. 114.)*

13. Two-Value Capacitor Motors. A two-value capacitor motor uses a starting capacitor, as in Fig. 4, and a running capacitor permanently in the circuit. Considering any given motor, the effect of the running capacitor is generally to increase the breakdown torque 5 to 30 percent, to reduce the full-load current and the full-load noise, to increase the locked-rotor torque some 5 to 20 percent, and to increase the efficiency.

14. Repulsion Motors. A straight repulsion motor is represented schematically in Fig. 8. There is a single stator winding; the rotor has an armature winding, commutator, and brushes, like a dc machine. The brushes are short-circuited and shifted. The shifting of the brushes gives the same effect as if there were two stator windings, (1) a field winding, at right angles to the brush axis, and (2) an inducing winding along the brush axis. By transformer effect, the inducing winding induces a current in the short-circuited armature, which reacts with the field set up by the field winding, thereby producing torque. Since the armature of this machine, being the secondary of the inducing winding, is effectively in series with the field winding, this motor has the speed-torque characteristics of a dc series motor. Very few of these straight repulsion motors are built today, although many repulsion-start motors are still manufactured.

15. Repulsion-Start Induction-Run Motors. A repulsion-start induction-run motor starts as a repulsion motor, as described above, but at about 75 percent of synchronous speed a centrifugal device short-circuits the bars of the commutator to one another, making the rotor winding become, in effect, a squirrel cage. Thus the machine runs as an induction motor. A repulsion-start motor develops more starting torque per ampere than other types of single-phase motors. However, for economic reasons primarily, its popularity has dwindled, and it is now built by only a few manufacturers, and generally in the larger ratings. For convenience, this motor is frequently called repulsion-start motor for short. Unfortunately, it is sometimes erroneously referred to as a repulsion-induction motor, which is quite a different motor.

16. Repulsion-Induction Motors. Schematically, this type of motor is similar to the repulsion motor of Art. 14, except that the armature has an embedded squirrel cage under the armature winding. The squirrel cage improves the speed regulation and eliminates the short-circuiter. It is little used today.

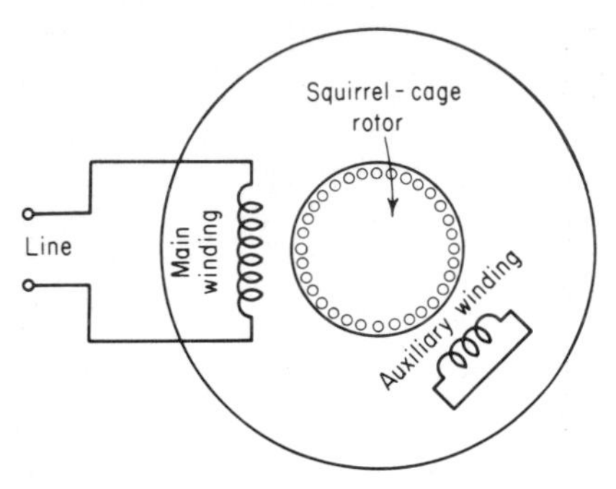

FIG. 9. Schematic representation of shaded-pole motor. *(Ref. 3, p. 99.)*

17. Shaded-Pole Motors. In ratings below 1/12 hp, the shaded-pole motor is the most popular and most widely used type of all. It is represented schematically in Fig. 9. Essentially it has one stator winding connected to the line and a second stator winding displaced in space and short-circuited upon itself. Usually the stator is constructed with salient poles. One form of construction is illustrated in Fig. 10; here there are four salient poles, with a coil on each pole, and a short-circuiting coil over a portion of each pole. Direction of rota-

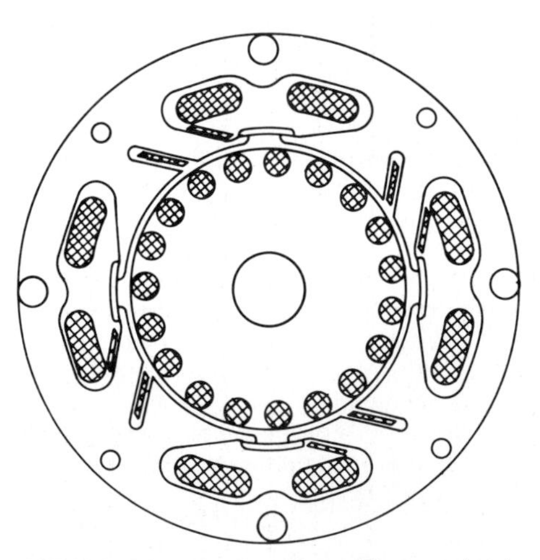

FIG. 10. Shaded-pole motor with tapered poles and magnetic wedges. *(Ref. 3, p. 99.)*

tion is toward the shading coil (clockwise in the figure). In addition, most motors today cut away or chamfer a portion of the leading tip of each pole to improve performance. In the smaller ranges, a skeleton-type construction is usually employed. A single stator coil is arranged on a C-type core, much like that of a transformer, with the rotor positioned in the opening of the C.

The shaded-pole motor of Fig. 10 is not reversible, but reversibility is often provided by use of two main windings or two shading windings. In either case, only one of the pair of windings is used at a time. A big advantage of the shaded-pole motor is its rugged simplicity and freedom from use of a starting switch. The biggest disadvantage is its low efficiency, a feature which inhibits its use in larger horsepower ratings.

CLASSIFICATION BY METHOD OF ENCLOSURE

18. Open Machine. Open-type enclosures used in fractional-horsepower motors include dripproof, splashproof, semiguarded, and guarded types. These enclosures are described in Sec. 3, Art. 18. The semiguarded and splashproof enclosures are not generally offered in fractional-horsepower sizes. In these sizes totally enclosed motors are commonly used in lieu of splashproof motors.

19. Totally Enclosed Machine. Totally enclosed motors in the fractional-horsepower sizes include totally enclosed nonventilated, totally enclosed fan-cooled, and explosionproof machines as covered in Sec. 3, Art. 18.

CLASSIFICATION BY VARIABILITY OF SPEED

20. Constant-Speed Motor. Such a motor changes speed only slightly from no load to full load.

21. Varying-Speed Motor. This is a motor which changes speed substantially from no load to full load; generally the speed decreases as load is applied. Examples are a universal motor, a straight repulsion motor, and a dc series motor.

22. Adjustable-Speed Motor. The speed of an adjustable-speed motor can be adjusted gradually over a considerable range, but once adjusted, it remains practically unaffected by the load. A dc shunt motor with field control is an example.

23. Adjustable Varying-Speed Motor. The speed of such a motor can be adjusted gradually, but once adjusted for a given load, it will vary substantially with changes in load. A dc series motor with armature resistance control is an example.

24. Pole-Changing Polyphase Motor. This polyphase induction motor is so arranged that by changing the connections of the external leads to the power source the number of poles of the rotating field can be changed, thereby producing two or more operating speeds, each one of which remains substantially unaffected by changes in load. This may be done by using two or more completely independent sets of windings with different numbers of poles. In such a case, any of the windings can have any number of poles—within limits—so that considerable flexibility of choice of speeds is possible.

Another arrangement is to use a single stator winding so designed that by reversing the direction of current flow in half of the winding, twice as many poles are set up by the winding. Such a connection is often referred to as a consequent-pole connection. It gives a 2:1 speed ratio and uses all the copper in the stator winding on both speeds.

25. Pole-Changing Single-Phase Motor. In this single-phase induction motor the main winding is so arranged that by changing the connections of the external leads to the power source the number of poles produced by the main winding can be changed to produce two or more substantially constant operating speeds. One way to accomplish this change is to use two independent main windings; another way is to arrange the main winding so that it is reconnectable to produce consequent poles, giving thereby a 2:1 change in speed. In either case, two auxiliary windings can be used, one for each number of poles set up by the main winding. A single auxiliary winding is often used and the external control is so arranged that, no matter which speed is selected, the motor always starts on the number of poles for which the auxiliary winding is connected, and then transfers to the other speed if required. In general, it is not practical to reconnect the auxiliary winding for consequent poles because this doubles the electrical space angle between the main and the auxiliary windings.

26. Multispeed Motor. Initially a multispeed motor was a motor having two or more speeds that could be set by changing winding connections (pole-changing), and which would hold the speed for which it was set, substantially independent of the load. By common usage, however, the meaning of this term has been extended to include permanent-split capacitor and shaded-pole motors for fan applications when so arranged as to provide speed adjustment. Such motors are further discussed in Arts. 46, 47, and 49. Strictly speaking, such motors are adjustable varying-speed motors, even though they are commonly called multispeed motors. In these cases, speed adjustment is obtained by adjusting the slip by reconnecting or by tapping the windings.

MISCELLANEOUS USEFUL DEFINITIONS

27. Efficiency. (Power output)/(power input), usually in percent.

28. Power Factor. (Watts input)/(volt-amperes input), usually in percent.

29. Service Factor. A factor by which the rated horsepower can be multiplied to obtain a permissible loading (see Art. 40 relative to use of service factor).

30. Torque. Torques of fractional-horsepower motors down to 1/20 hp are usually expressed in ounce-feet; for motors smaller than that, torques are usually expressed in ounce-inches, and output ratings in millihorsepower (1 mhp = 0.001 hp). There are many kinds of torque. Some of them are shown in Fig. 11.

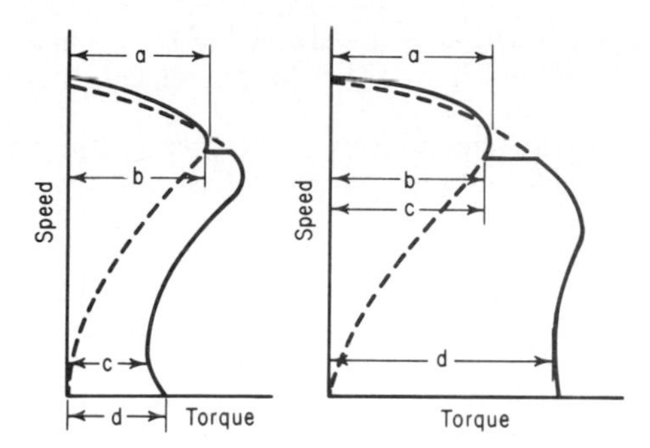

FIG. 11. Typical speed-torque curves of single-phase motors illustrating definitions of various kinds of torque. *a*—breakdown torque; *b*—switching torque; *c*—pull-up torque; *d*—locked-rotor torque. *(Ref. 2, p. 470.)*

a. Full-load torque. The torque necessary to produce rated horsepower output at full-load speed. For fractional-horsepower motors,

$$\text{Torque (oz}\cdot\text{ft)} = \frac{84{,}000 \times \text{hp}}{\text{r/min}} \tag{3}$$

$$\text{Torque (oz}\cdot\text{in)} = \frac{1008 \times \text{mhp}}{\text{r/min}} \tag{4}$$

b. Locked-rotor torque (static torque, breakaway torque). The minimum torque a motor will develop at rest for all angular positions of the rotor, with rated voltage applied at rated frequency (see Fig. 11).

c. Pull-up torque. The minimum torque developed during a period of acceleration from rest to breakdown-torque speed (for two examples, refer to Fig. 11). For motors which have no definite breakdown torque, the pull-up torque is the minimum torque developed up to rated speed.

d. Breakdown torque. The maximum torque a motor will develop with rated voltage applied at rated frequency without an abrupt drop in speed.

e. Switching torque. The minimum torque developed by a single-phase motor as it accelerates through switch-operating speed.

f. Pull-out torque. The maximum sustained torque a synchronous motor can carry with rated voltage applied at rated frequency without falling out of synchronism.

g. Pull-in torque. The maximum constant torque under which a synchronous motor will pull its connected inertia load into synchronism with rated voltage applied at rated frequency. This torque is markedly affected by the inertia of the load connected to the motor.

GENERAL-PURPOSE AND STANDARD MOTORS

The discussion that follows through Art. 35 pertains primarily to general-purpose and standard off-the-shelf motors; similar information for various definite-purpose motors is given in Arts. 41 through 56.

31. Voltages. Standard voltages are 115 and 230 V for single-phase and universal motors; 115, 200, 230, 460, and 575 V for polyphase motors.

32. Frequencies. Standard frequencies are 60 and 50 Hz. The latter frequency has all but disappeared from this country and is not even considered in this section.

33. Ratings and Performance Characteristics. Ratings and performance characteristics of fractional-horsepower ac motors are given in Tables 1 to 3.

a. Polyphase motors. These are shown under the heading Three-phase in Table 1. In general, they are seldom used in ratings below ⅙ hp. They meet general-purpose requirement.

b. Split-phase motors. Standard split-phase motors are built for a wide variety of applications, but they do not have enough torque to be classified as general-purpose. Torques are given in Table 3; full-load speeds and torques are given in Table 1; and locked-rotor currents are given in Table 1 under Single-phase.

c. Capacitor-start motors. General-purpose capacitor-start motors have the torques and locked-rotor torques shown under Single-phase in Table 1; horsepower and speed ratings are as shown, from ⅙ hp up.

d. Two-value capacitor motors. These motors have about the same torques and locked-rotor currents as capacitor-start motors but are more expensive and are available on a more limited basis than capacitor-start motors.

e. Permanent-split capacitor motors. These motors are used primarily for shaft-mounted fans and blowers, and they are built in the full gamut of fractional-horsepower ratings. Performance characteristics for the fractional range are given in Table 1. Torques, horsepower, and speed ratings in the subfractional range are given in Table 2. They are often used as adjustable-speed motors for driving fans (see Arts. 46 and 49).

f. Repulsion-start motors. These motors are in limited supply and generally more costly than capacitor-start motors. Generally they are available only in the upper horsepower ranges. Breakdown torques are about the same as those of

TABLE 3. Torques of 60-Hz Standard Split-Phase Motors (Ref. 3)

hp rating	Min breakdown torque				Min locked-rotor torque			
	No. of poles				No. of poles			
	2	4	6	8	2	4	6	8
1/20	3.1	6.0	8.8	11.5	3.0	6.8	10.0	13.0
1/12	5.1	9.8	14.0	18.3	4.4	9.4	12.7	15.0
1/8	7.4	14.0	20.5	26.8	6.5	12.7	16.2	18.0
1/6	9.8	18.3	26.8	34.4	7.5	16.0	20.0	21.0
1/4	14.0	26.8	37.4	49.3	9.0	17.8	22.5	23.5
1/3	18.3	34.4	49.3	65.5	10.8	19.8	25	26

All torques are in ounce-feet.

capacitor-start motors, but locked-rotor torques are much higher; locked-rotor currents are lower than those of capacitor-start motors.

g. Shaded-pole motors. Horsepower and speed ratings are given in Table 2. Note that, in the subfractional-horsepower range, ratings are given in millihorsepower and torques in ounce-inches. A wide variety of ratings are available. Torques and efficiencies are both low. A major advantage of this type is that no starting switch is needed or used.

34. Time and Temperature Ratings. Most fractional-horsepower ac motors use Class A insulation, and the temperature-rise ratings for continuous-duty motors, based upon thermometer observations, are as follows:

1. Open, general-purpose 40°C
2. Dripproof 40°C
3. Splashproof 50°C
4. Totally enclosed 55°C

The time rating is usually continuous for general use. Time ratings for intermittent-duty motors are 5, 10, 15, 30, and 60 min.

35. Dimensions and Frame Assignments. Dimensions for fractional-horsepower frames, as established by NEMA, are given in Fig. 12. There are no NEMA standards on what ratings are built in these frames, and there may be some variation between manufacturers, but the various ratings of fractional-horsepower motors are commonly built in the frames shown in Table 4.

APPLICATION DATA

36. Proper Selection of Apparatus. While general-purpose motors are designed for a wide variety of applications, there are a number of conditions which require special consideration and consultation with the manufacturer. Some of these special environmental conditions are exposure to chemical fumes; exposure to combustible, gritty, or conducting dust; exposure to steam, lint, flammable or explosive gases, oil vapor, or salt air; operation in damp places, poorly ventilated rooms, pits, or enclosures; subjection to ambient temperatures below 10°C or above 40°C; exposure to abnormal shock and vibration; operation on a line voltage abnormally high or low or, in the case of polyphase motors, unbalanced.

37. Usual Service Conditions. NEMA gives the following:

1. An ambient temperature not exceeding 40°C
2. A plus or minus variation in voltage not exceeding, for induction motors, 10 percent; for universal motors, 6 percent
3. A variation in frequency of not over 5 percent
4. A combined variation in frequency and voltage not exceeding 10 percent, so long as the frequency does not vary more than 5 percent
5. An altitude not exceeding 3300 ft (1000 m)

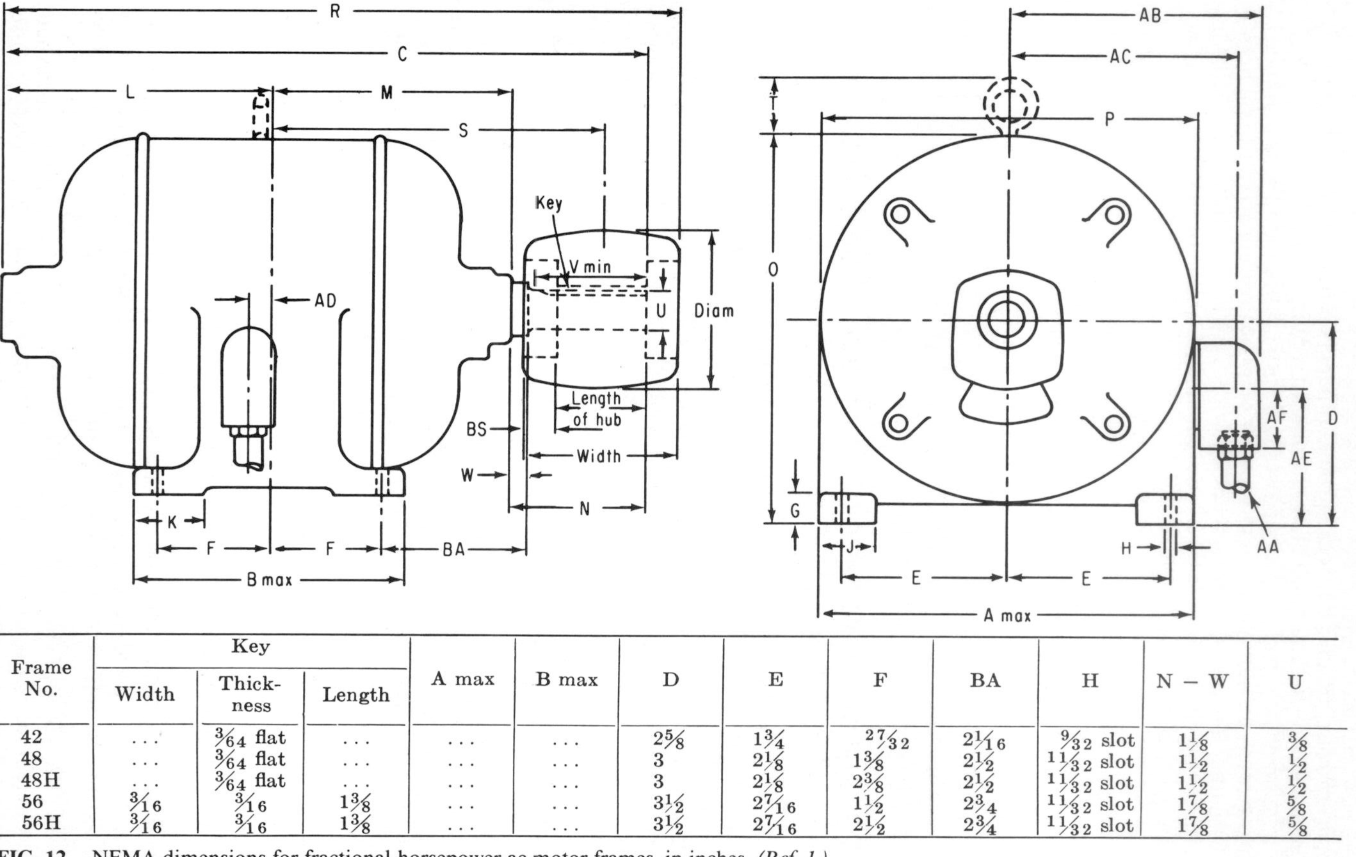

Frame No.	Key Width	Key Thickness	Key Length	A max	B max	D	E	F	BA	H	N − W	U
42	...	3/64 flat	...	...	...	2 5/8	1 3/4	27/32	2 1/16	9/32 slot	1 1/8	3/8
48	...	3/64 flat	...	...	...	3	2 1/8	1 3/8	2 1/2	11/32 slot	1 1/2	1/2
48H	...	3/64 flat	...	...	...	3	2 1/8	2 3/8	2 1/2	11/32 slot	1 1/2	1/2
56	3/16	3/16	1 3/8	...	...	3 1/2	2 7/16	1 1/2	2 3/4	11/32 slot	1 7/8	5/8
56H	3/16	3/16	1 3/8	...	...	3 1/2	2 7/16	2 1/2	2 3/4	11/32 slot	1 7/8	5/8

FIG. 12. NEMA dimensions for fractional-horsepower ac motor frames, in inches. *(Ref. 1.)*

TABLE 4. Frame Sizes for General-Purpose Fractional-Horsepower Motors

hp rating	No. of poles			
	2	4	6	8
1/8	48	48	48	56
1/6	48	48	48	56
1/4	48	48	56	56
1/3	48	48	56	56
1/2	48	56	56	
3/4	56	56		
1	56			

6. Location or supplementary enclosures which do not seriously interfere with the ventilation of the motor
7. Solid mounting and properly applied belt or chain drives or gearing

When the service conditions are less favorable than the above, the motor manufacturer should be consulted.

38. Direction of Rotation. The standard direction of rotation is counterclockwise, as viewed from the end opposite the shaft extension (front end). Single-phase motors are normally preconnected at the factory for this rotation, but polyphase motors may start in either direction. A number of definite-purpose motors normally specify clockwise rotation, as viewed from the drive end.

39. Effects of Variation of Voltage and Frequency. A given change in the voltage applied to an induction motor changes its torques in proportion to the square of the voltage, and from a torque standpoint, the true horsepower rating is changed by the same amount, though the heating might be exceeded. A 10 percent increase in voltage may often result in a slight improvement in efficiency, with a decided lowering of the power factor. Such an increase in voltage will decrease the full-load slip inversely as the square of the voltage. A reduction of 10 percent in voltage will probably increase the power factor, decrease the efficiency, and increase the temperature rise and slip. A change in frequency will cause a change in speed nearly proportional to the change in frequency; if the motor is driving a fan, a 5 percent increase in frequency will cause the motor to be considerably overloaded since the fan load will increase 15 percent.

40. Application of Motors with Service Factor. Motors having a service factor stamped on the nameplate may be safely operated at service-factor load without injurious overheating provided rated voltage is applied at rated frequency. However, if a motor is operated continuously at service-factor load, the insulation life may be only as much as one-third of what it would have been if the motor had been operated continuously at rated load. Where really long life is important, use of the service factor on a continuous basis is not recommended. Usual service factors are given in Table 5.

TABLE 5. Service Factors for General-Purpose Motors*

hp	*Service factor*
1/20	1.4
1/12	1.4
1/8	1.4
1/6	1.35
1/4	1.35
1/3	1.35
1/2	1.25
3/4	1.25
1	1.25

* Except for jet-pump motors; see Art. 54*f*.

DEFINITE-PURPOSE FRACTIONAL-HORSEPOWER AC MOTORS

41. General. Definite-purpose motors are defined in Art. 3. NEMA standards for definite-purpose motors fill 100 pages in Ref. 1. The following is abstracted from NEMA. To conserve space, only 60-Hz motors are considered below, although NEMA still recognizes 50 Hz as a standard frequency.

42. Permanent-Split Capacitor Industrial-Instrument Motors and Gear Motors. These are subfractional motors intended for application in such instruments as recorders and timing devices.

a. Types. Synchronous capacitor; nonsynchronous capacitor, normal slip; nonsynchronous capacitor, high slip.

b. Voltage ratings. 115 V.

TABLE 6. Characteristics of Permanent-Split Capacitor Motors for Industrial Instruments

	Axial length of motor body, in		
	1.75	2.00	2.50
Synchronous capacitor motors:			
Full-load speed, r/min	1,800	1,800	1,800
Full-load and pull-in torque, oz·in	0.25	0.33	0.60
Locked-rotor torque, oz·in	0.25	0.35	0.60
Nonsynchronous capacitor motors, normal slip:			
Approx full-load speed, r/min	1,550	1,550	1,550
Full-load torque, oz·in	1.0	1.4	2.4
Breakdown torque, oz·in	1.7	2.4	4.2
Locked-rotor torque, oz·in	1.0	1.4	2.4
Nonsynchronous capacitor motors, high slip:			
Approx full-load speed, r/min	1,200	1,200	1,200
Full-load torque, oz·in	0.75	1.10	1.85
Breakdown torque, oz·in	1.5	2.2	3.7
Locked-rotor torque, oz·in	1.5	2.2	3.7

c. Torque and speed ratings. See Table 6 for nongear motors. Comparable gear motors are built with gear ratios from 6:1 to 1800:1.

d. Temperature rise. 55°C by thermometer, 65°C by resistance.

e. General mechanical features. Totally enclosed, ball-bearing, operable in any position; mounting is face-type, by means of two 8-32 tapped holes, 2 11/16 in apart; shaft diameter 0.1875 in, with flat.

43. Low-Inertia Servo Industrial-Instrument Motors and Gear Motors. These are subfractional motors intended for applications in such instruments as self-balancing recorders and remote-positioning devices.

a. Types. Permanent-split capacitor; two-phase.

b. Voltage ratings and control-phase impedances. 115 V on the fixed phase; on the control phase, as in the following table.

Control-phase impedance, ohms	Rated control-phase voltage, volts	Min control-phase voltage to start (breakaway), volts
5,000–5,500	165	16.5
2,500–2,750	115	11.5
40–44	15	1.5

c. Speed ratings. Approximately 1500 r/min at no load, 1200 r/min at full load; gear ratios of 6:1 to 1800:1 for gear motors.

d. Rated torques (nongear). Full-load, 1.1 oz·in; locked-rotor and breakdown, 2.5 oz·in.

e. Rotor inertia. 0.14 oz·in^2 maximum.

f. Temperature rise. 55°C by thermometer, 65° by resistance.

g. Control voltage to start. Must start without load when voltages of Par. *b* are applied; control-phase voltage is applied gradually, starting from zero.

h. General mechanical features. Totally enclosed, ball-bearing, operable in any position; mounting is face-type, by means of two 8-32 tapped holes, 2 11/16 in apart; shaft diameter 0.1875 in, with flat.

44. Universal-Motor Parts. A set of universal-motor parts consists of stator (field), rotor (armature), and brushholder mechanism without end shields, bearings, or conventional frame.

a. Types. Salient-pole (as in Fig. 13); distributed, in which the stator punching resembles an induction-motor punching. The former type is by far the more popular, because it is less expensive; the latter type has better universal characteristics below 5000 r/min.

b. Voltage ratings. 115 and 230 V.

c. Frequencies. Direct current, or alternating current of any frequency not over 60 Hz.

d. Speeds. In the thousands of revolutions per minute. At speeds below 5000 r/min there will be a marked difference in performance characteristics between operations on alternating and on direct current, especially motors with salient-pole stators. For a more complete discussion see Ref. 2, Chap. 12.

e. Temperature rise. Depends upon the cooling system provided by the user.

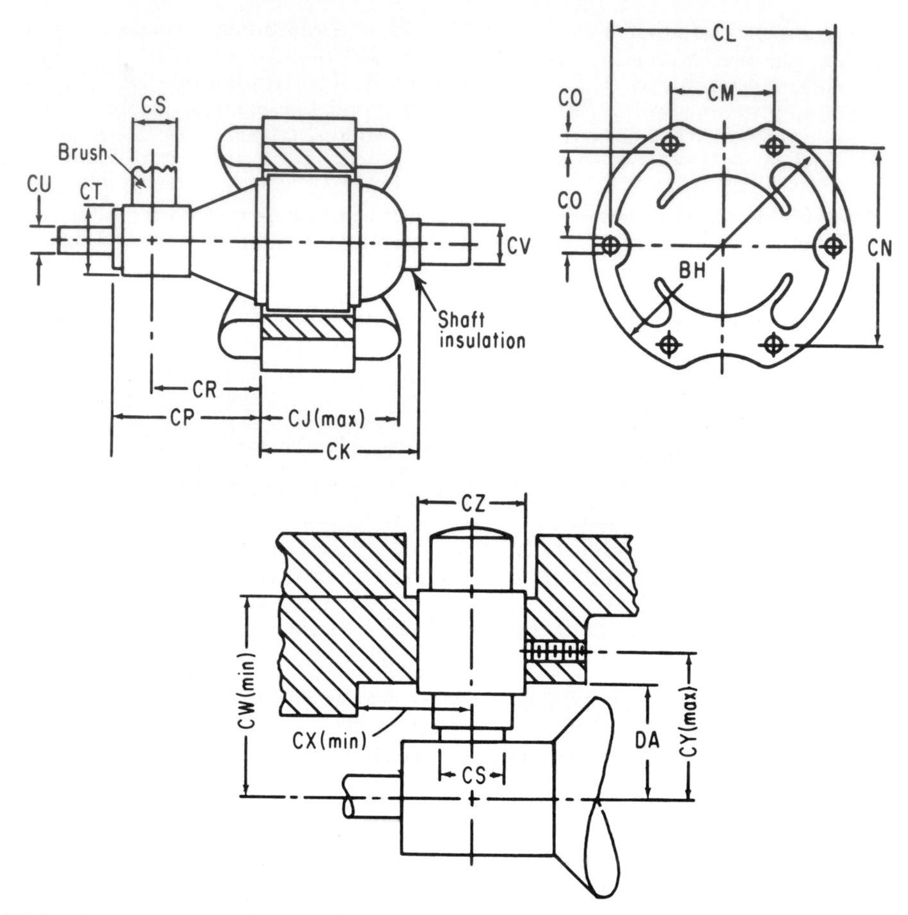

FIG. 13. Outline dimensions for universal motors. *(Ref. 1.)*

f. Physical sizes. Figure 13 is a standard NEMA dimension drawing. Tabulated dimensions range over the following:

BH (OD)	2.125–4.375
CJ (max)	$1\frac{7}{16}$–$3\frac{15}{16}$
CP	$\frac{29}{33}$–$2\frac{7}{16}$
CT (commutator OD)	$\frac{7}{8}$–$1\frac{7}{8}$
CU	0.2363–0.4724
CV	$\frac{5}{16}$–$\frac{7}{8}$

g. Common practices in motor-frame design. The frame usually consists of two aluminum castings, held together by screws entering tapped holes in one of the castings. Alignment is obtained by machined male and female rabbet fits. The front-end (commutator-end) casting is deep enough to hold the stator, which is positioned by lands or cast ribs; this casting also contains the front-bearing housing and brushholder assembly. Cartridge-type brushholders are held in bosses in the front-end casting. The drive-end casting is relatively shallow and often is a part of the driven device as well. Ventilation is provided by a centrifugal fan, mounted on the drive end of the shaft, and so arranged as to draw cool air in over the commutator, over the stator between the lands, and out the drive end.

45. Motors for Hermetic Refrigeration Compressors. A hermetic motor consists of a stator and rotor without shaft, end shields, or bearings for installation in refrigeration compressors of the hermetically sealed type.

a. Types. Split-phase; capacitor-start; two-value capacitor; permanent-split capacitor.

b. Voltage ratings. Single-phase, 115 and 230 V.

c. Horsepower and speed ratings. Unlike most motors, these carry no horsepower rating; instead, they are rated on the basis of breakdown torque. Standard breakdown-torque ratings for both four-pole and two-pole motors are given in Table 7. Normal operating speeds are on the order of 1725 and 3450 r/min, respectively.

d. Locked-rotor current. See Table 7.

e. Terminal-lead markings. Main winding, white with red tracer; auxiliary, white; common, white with black tracer.

TABLE 7. Breakdown Torques and Locked-Rotor Currents of Single-Phase Motors for Hermetic Refrigeration Compressors

Four-pole motors			Two-pole motors	
Breakdown torque, oz·ft	Locked-rotor current at 115 volts, A		Breakdown torque, oz·ft	Locked-rotor current at 115 volts, A
10.5	20		5.25	20
12.5	20		6.25	20
15	20		7.5	20
18	20		9.0	20
21.5	20		10.75	21
26	21.5		13.0	23
31	23		15.5	26
37	28	23*	18.5	29
44.5	34	23*	22.0	33
53.5	40	...	27.0	38
64.5	48	46*	32.0	43
77	57	46*	38.5	49
92.5	68	46*	46.0	56

*Motors having these locked-rotor currents usually have lower locked-rotor torques than those with the higher locked-rotor currents.

46. Motors for Shaft-Mounted Fans and Blowers. These motors are totally enclosed and are designed for propeller fans or centrifugal blowers mounted on the motor shaft, with or without air drawn over the motors; they are generally not suitable for belted loads.

a. Types. Split-phase; permanent-split capacitor; polyphase induction for ⅛ hp and larger.

b. Voltage ratings. Split-phase, 115 and 230 V; permanent-split capacitor, 115/230 V; polyphase, 200, 230, 460, and 575 V.

c. Horsepower and speed ratings. For single-speed motors, all the horsepower and speed ratings in Table 1, except that there are no two-pole ratings.

d. Two-speed motors. Permanent-split capacitor motor with suitable switch or control means to give a second speed on the order of two-thirds of synchronous speed, when driving a fan load. (For details as to how this is accomplished, see Chap. 7 of Ref. 2, or Chap. 4 of Ref. 3.)

e. Adjustable varying-speed (multispeed) motors. Permanent-split capacitor motor with suitable controller which permits adjusting the speed in many steps down to about half synchronous speed. (For details, see the reference cited above.)

f. Temperature rise. 55°C by thermometer, or 65°C by resistance.

g. General mechanical features. Totally enclosed; horizontal motors, sleeve bearings with provision for taking axial thrust of a fan in the front bearing; vertical motors, ball bearings; end-shield clamp bolts extend at least ⅜ in beyond the nut on the back end of the motor to provide means for attaching a fan guard or for mounting the motor itself; shaft extensions as in Fig. 12; permanent-split capacitor motors, rated ¼ hp or less, may sometimes be provided with blade terminals.

47. Shaded-Pole Motors for Shaft-Mounted Fans and Blowers

a. Voltage ratings. 115 and 230 V.

b. Horsepower and speed ratings. See Table 2.

c. Breakdown torques. See Table 2.

d. Temperature rise. 65°C by resistance.

e. Speed classifications. Single-speed, two-speed, three-speed (usually obtained by tapped windings).

f. Terminal-lead markings. Two-speed motors, common is white or 1, low speed is red or 5, and high speed is black or 3; three-speed motors, common is white or 1, low speed is red or 5, medium speed is blue or 4, and high speed is black or 3. When the motors are provided with blade terminals, the latter are identified by the numbers given.

g. General mechanical features. Open or totally enclosed, sleeve bearings, resilient mounting.

48. Motors for Belted Fans and Blowers. These motors are built for operating belt-driven fans or blowers such as are commonly used in conjunction with hot-air heating installations and attic ventilators.

a. Types. Split-phase, capacitor-start, or repulsion-start for single-speed applications; split-phase or capacitor-start for two-speed applications.

b. Voltage ratings. 115 and 230 V.

c. Speed ratings. Single-speed motors, 1725 r/min; two-speed motors, 1725/1140 r/min.

d. Horsepower ratings for single-speed motors. Split-phase, ⅙, ¼, and ⅓ hp; capacitor-start and repulsion-start, ⅓, ½, and ¾ hp.

e. Horsepower ratings for two-speed motors. Split-phase, ⅙ and ¼ hp at higher speed; capacitor-start, ⅓, ½, and ¾ hp at higher speed.

f. Temperature rise. 40°C by thermometer, 50°C by resistance.

g. Breakdown torques. See Tables 1 and 3.

h. Locked-rotor current. See Table 1.

i. General mechanical features. Open, dripproof sleeve bearings for horizontal operation, resilient mounting; automatic-reset thermal-overload protector; mounting dimensions and shaft extensions as in Fig. 12.

49. Motors for Air-Conditioning Condensers and Evaporator Fans

a. Types. Shaded-pole; permanent-split capacitor.

b. Voltage ratings. 115 and 230 V.

c. Horsepower and speed ratings. See Table 2.

d. Breakdown torques. See Table 2.

e. Temperature rise. By resistance, 65°C for Class A or 85°C for Class B insulation.

f. Variations from rated speed. As can be seen in Table 2, these motors operate at high slips; hence variations from rated speed are greater than they are for general-purpose motors. As single-speed motors, the slip may vary as much as 20 percent (i.e., if the full-load nominal slip is 20 percent, the actual slip could vary from 16 to 24 percent).

g. Terminal-lead markings, multispeed shaded-pole motors. Same as in Art. 47*f*.

h. Terminal-lead markings, permanent-split capacitor motors. See Ref. 1, 2, or 3.

i. Variations in speed due to voltage and/or manufacturing variations. Permanent-split capacitor motors, driving a fan, may be subject to wide variations in speed due to manufacturing variations or variations in line voltage, or both; this effect is greatest at low operating speeds (Fig. 14). The solid curve intersecting

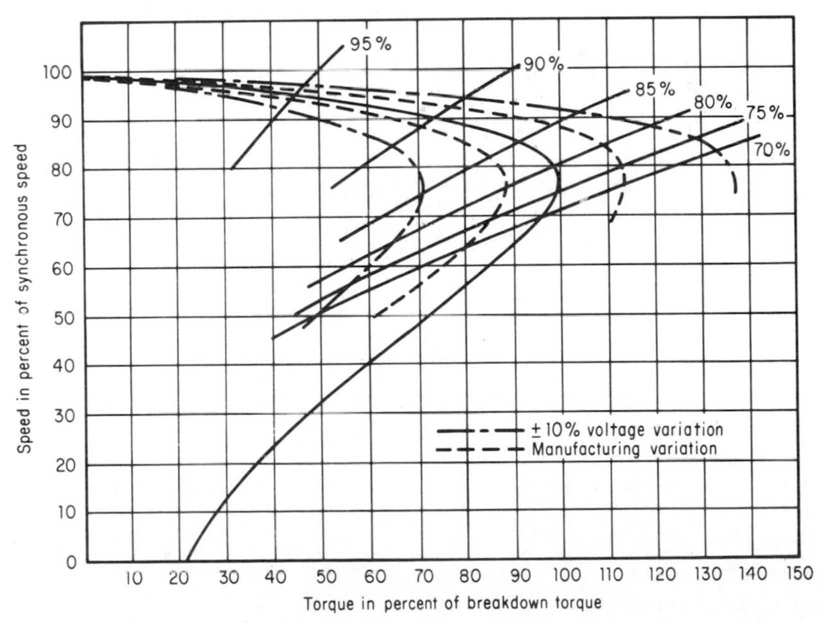

FIG. 14. Effect of manufacturing and line-voltage variations on operating speed of a permanent-split capacitor motor driving a fan. *(Ref. 1.)*

the vertical axis near 100 percent synchronous speed illustrates the speed-torque characteristic of an average motor of a typical design. The dashed curves on either side of this curve illustrate changes in the characteristic that can occur because of manufacturing variations. The dot-dash curves on either side of these illustrate the variation in the speed-torque characteristic that would be caused by a plus or minus 10 percent variation in line voltage. The solid curves, marked 95%, 90%, etc., represent typical fan speed-torque curves; the corresponding fans would cause the average motor, at rated voltage, to operate at speeds of 95%, 90%, etc., as these are the points where the respective fan curves cross the motor curve. Now a study of the figure shows that if the operating speed of the fan is 95 percent, the manufacturing and line-voltage variations cause only a small change in speed—in this case, only about 5 percent of synchronous speed; but if the nominal operating speed is 70 percent, normal manufacturing and normal line-voltage fluctuations may cause the operating speed to vary from 50 to 83 percent of synchronous speed. Shaded-pole motors are even more sensitive to these conditions.

50. Motors for Cellar Drainers and Sump Pumps. A cellar-drainer motor furnishes power for operating a pump for draining cellars, pits, or sumps.

a. Types. Split-phase.

b. Voltage ratings. 115 or 230 V.

c. Horsepower and speed ratings. ⅓ hp at 1725 r/min.

d. Temperature rise. 50°C by thermometer, 60°C by resistance.

e. Torque characteristics. Breakdown, 32 oz·ft; locked-rotor, 20 oz·ft.

f. Locked-rotor current. 50 A or less, at 115 V.

g. General mechanical features. Designed for vertical operation; open construction top end shield totally enclosed or ventilating openings protected by louvers, or the equivalent; bottom end shield has hub machined for direct mounting on support pipe; may have automatic thermal-overload protector.

51. Motors for Gasoline-Dispensing Pumps. These motors are of Class 1, Group D explosionproof construction as approved by Underwriters Laboratories for belt or direct-coupled drive of gasoline-dispensing pumps of the size commonly used in automobile service stations.

a. Types. Capacitor-start; repulsion-start induction; polyphase, squirrel-cage.

b. Voltage ratings. Capacitor-start or repulsion-start, 115/230 V; polyphase, 200 and 230 V.

c. Horsepower and speed ratings. ½ hp at 1725 r/min.

d. Temperature rise. 55°C by thermometer, 65°C by resistance.

e. Torque characteristics. Breakdown, 46 oz·ft; locked-rotor, 48.8 oz·ft.

f. Locked-rotor current. Not over 45 A at 115 V.

g. Direction of rotation. Clockwise, facing end opposite drive.

h. General mechanical features. Totally enclosed, explosionproof, Class 1, Group D; sleeve bearings; rigid base mounting; built-in line switch and operating lever; for single-phase motors, built-in thermal-overload protector optional; for single-phase motors, voltage-selector switch built in on same end as line switch; line leads 36 in long, brought out through swivel connector on line switch end.

i. Dimensions. See Fig. 15.

j. Frame number. Motors built as above carry the frame designation 61G.

52. Motors for Oil Burners. These motors are for operating mechanical-draft oil burners for domestic applications.

a. Types. Split-phase.

b. Voltage ratings. 115 and 230 V.

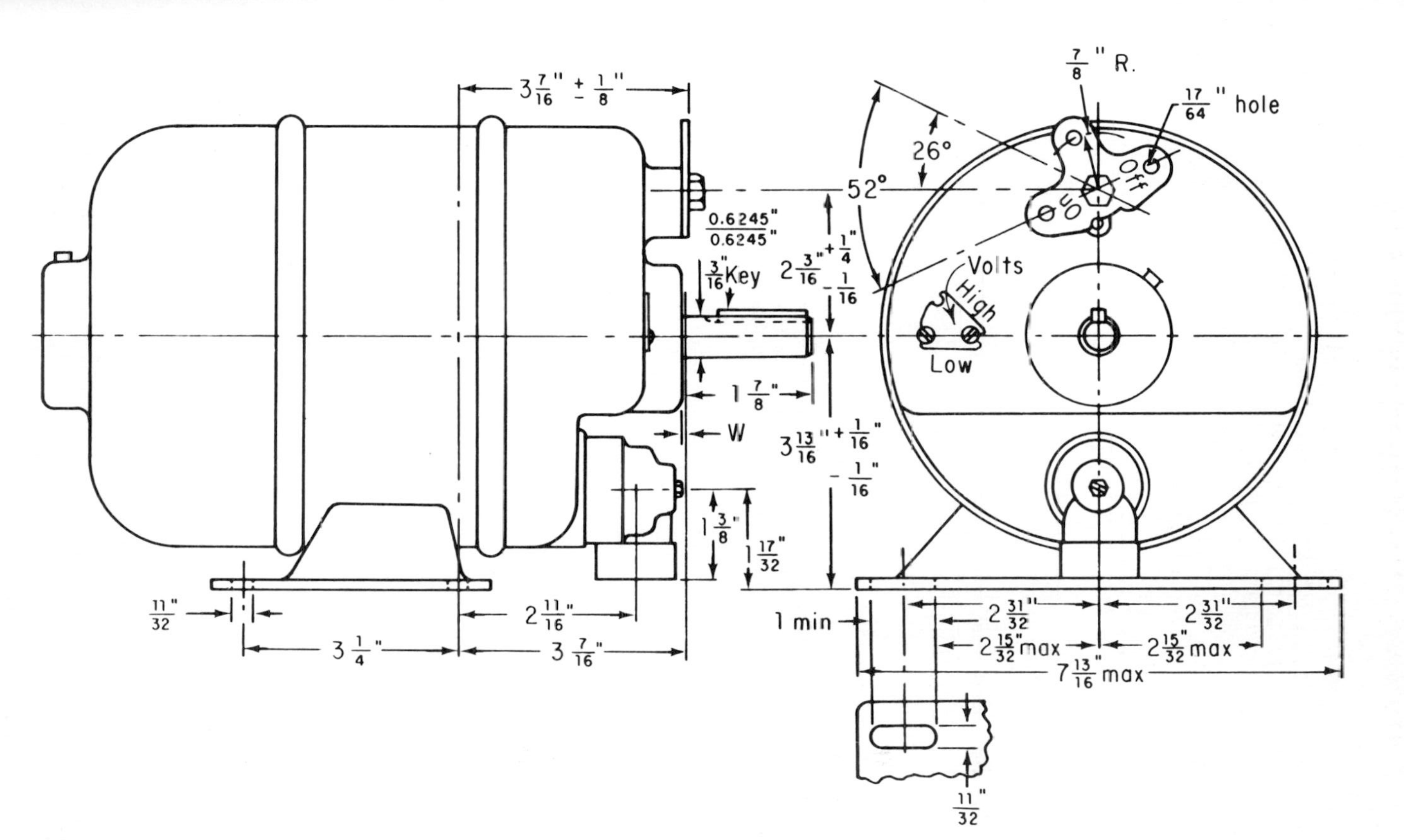

FIG. 15. Outline dimensions for gasoline-dispensing-pump motors, type G. All dimensions in inches. *(Ref. 1.)*

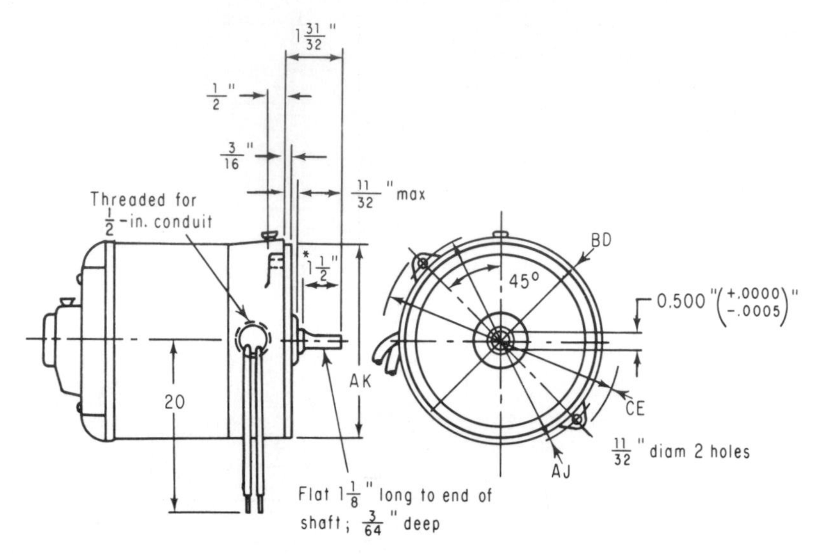

FIG. 16. Outline dimensions for face-mounted oil-burner motors, types M and N. All dimensions in inches. If the motor shaft extension length is not suitable for application, it is recommended that deviations from this length be in ¼-in increments. *(Ref. 1.)*

c. Horsepower and speed ratings. 1/12, ⅛, and ⅙ hp, all at 1725 r/min.

d. Temperature rise. Guarded motors, 50°C by thermometer or 60°C by resistance; totally enclosed motors, 55°C by thermometer or 65°C by resistance.

e. Breakdown torques. Approximately as given in Table 3.

f. Locked-rotor current. See Table 1 under Single-phase.

g. Direction of rotation. Clockwise, facing end opposite drive.

h. General mechanical features. Guarded or totally enclosed housing; nameplate carries words "oil burner motor"; manual-reset thermal-overload protector, with directions for resetting displayed; two terminal leads, consisting of two 20-in lengths of flexible single-conductor wire, brought out from the motor through a hole tapped for a ½-in conduit.

i. Dimensions and frame numbers. Figure 16 is an outline covering two sets of face mounting. The dimensions in inches for Fig. 16 are as follows:

Frame suffix	AK	AJ	CE	BD
M	5½	6¾	7¾ max	6¼ max
N	6⅜	7¼	8¼ max	7 max

53. Motors for Home-Laundry Equipment. These motors furnish power for driving a home washing machine, dryer, or a combination washer-dryer.

a. Types. Split-phase, capacitor-start.
b. Voltage ratings. 115 and 230 V.
c. Horsepower ratings. ⅙, ¼, ⅓, and ½ hp.
d. Speed ratings. Single-speed, 1725 r/min; two-speed, 1725/1140 r/min.
e. Temperature rise. By resistance, 60°C for Class A insulation or 80°C for Class B insulation.
f. Locked-rotor current. Not over 50 A, 3 s after application of power.
g. Breakdown torques. Minimum values are as follows, in ounce-feet:

	Type I	Type II
Single-speed:		
½ hp	45.0	
⅓ hp	35.0	40.5
¼ hp	23.0	31.5
⅙ hp	18.0	
Two-speed:	1,725 r/min	1,140 r/min
½ hp	45.0	40.0
⅓ hp	40.5	35.0

h. Locked-rotor torques. Minimum values are as follows, in ounce-feet:

	Type I	Type II
⅓-hp (single and two-speed).....	27.0	33.0
¼ hp..........................	19.0	24.0
⅙ hp..........................	16.0	

i. General mechanical features. Open; sleeve bearing; mounting may be either ungrounded mounting rings for resilient mounting, or extended studs; ½-in shaft extension with flat; built-in connection box in front end; sometimes, blade terminals.
j. Frame suffix letter. Motors built in a frame size shown in Fig. 12, and according to the description above, use the suffix letter L after the frame number to indicate "laundry motor."
k. Schematic connection diagrams. See NEMA Standard MG 1-18.317, Ref. 1.

54. Motors for Jet Pumps. These are open ball-bearing motors for horizontal or vertical operation for direct connection to direct-driven centrifugal ejector pumps.
a. Types. Split-phase; capacitor-start; repulsion-start; polyphase, squirrel-cage.
b. Voltage ratings. Split-phase, 115 and 230 V; capacitor-start, 115/230 V; repulsion-start, 115/230 V; polyphase, 200, 230, and 460 V.
c. Horsepower and speed ratings. 2, 1, ¾, ½, ⅓, and ¼ hp, all at 3450 r/min.

d. Temperature rise. 40°C by thermometer, 50° by resistance.

e. Breakdown torques. See Table 1 for single-phase and three-phase.

f. Service factors. Motors with service factors substantially in excess of those shown in Table 5 are generally available for this application.

g. Maximum locked-rotor current. See Table 1.

h. Direction of rotation. Clockwise, facing end opposite the drive.

i. General mechanical features. Open construction; grease-lubricated ball bearings suitable for horizontal or vertical operation and provision for taking axial thrust; back end shield machined as in Fig. 17; front end shield provided with a ⅝-18 tapped hole in the center of the bearing hub to accommodate a drip cover when used vertically; standard shaft extension as in Fig. 17 (frame 56C); an alternate shaft extension, not shown, has the end of the shaft threaded for a 7⁄16-20 nut; terminals or leads located in, or near, front end shield; automatic-reset thermal-overload protector on single-phase motors.

j. Dimensions. For motors with standard shaft see Fig. 17.

k. Frame suffix letter. Motors built as above, and to the dimensions of Fig. 12, use the suffix letter C if the shaft extension is as in Fig. 17, or the suffix letter J if the shaft extension is threaded.

55. Motors for Coolant Pumps. These are enclosed ball-bearing motors built for horizontal or vertical operation for direct connection to direct-driven centrifugal coolant pumps.

a. Types. Split-phase; capacitor-start; repulsion-start.

b. Voltage ratings. Split-phase, 115 and 230 V; capacitor-start, 115 and 230 V for ¼ hp or less, 115/230 V for ⅓ hp or larger; repulsion-start, 115/230 V; polyphase, 200, 230, and 460 V.

c. Horsepower and speed ratings. All the two-pole and four-pole ratings shown in Table 1.

d. Temperature rise. 55°C by thermometer, 65°C by resistance.

e. Breakdown torques. See Table 1 for single-phase and three-phase.

f. Maximum locked-rotor current. See Table 1.

g. Direction of rotation. Clockwise, facing end opposite drive.

h. General mechanical features. Generally the same as for jet-pump motors; see Art. 54*i*.

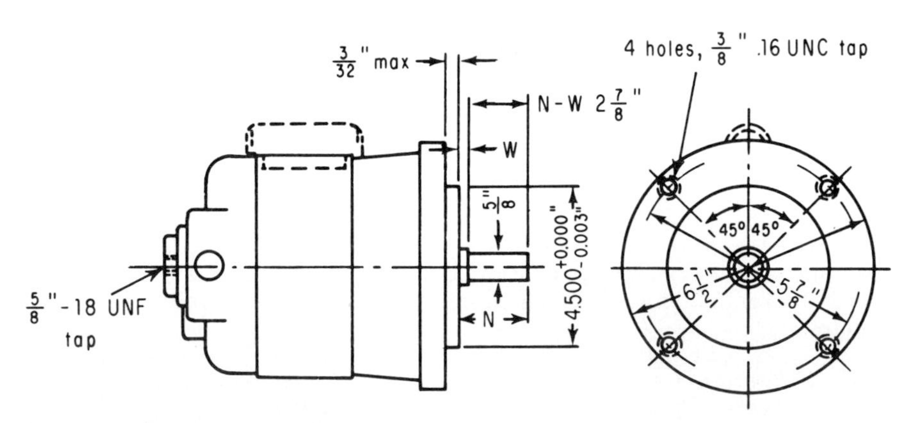

FIG. 17. Outline dimensions for jet-pump motors. *(Ref. 1.)*

56. Submersible Motors for Deep-Well Pumps, 4 in. A submersible motor for deep-well pumps is a motor designed for operation while totally submerged in water having a temperature not exceeding 25°C (77°F).

a. Types. Split-phase; capacitor; polyphase, squirrel-cage.

b. Voltage ratings. Single-phase, 115 and 230 V; polyphase, 200, 230, and 460 V.

c. Horsepower ratings. Single-phase 115 V, ¼, ⅓, and ½ hp; single-phase 230 V, ¼, ⅓, ½, ¾, 1, 1½, 2, and 3 hp; polyphase induction, ¼, ⅓, ½, ¾, 1, 1½, 2, 3, and 5 hp.

d. Speed rating. 3450 r/min.

e. Breakdown torques. Approximately as shown in Table 1.

f. Maximum locked-rotor current. For single-phase fractional-horsepower motors see Table 1.

g. Direction of rotation. Clockwise, facing end opposite drive.

h. Thrust capacity. When operated in a vertical position with the shaft up, they are capable of withstanding the following thrust: ¼–1½ hp inclusive, 300 lb; 2–5 hp inclusive, 900 lb.

i. Terminal-lead markings. Main winding, black; auxiliary winding, red; common, yellow.

REFERENCES

1. "Motors and Generators," ANSI/NEMA Standard MG 1-1978.
2. C. G. Veinott, *Fractional and Subfractional Horsepower Electric Motors* (McGraw-Hill, New York, 1970).
3. C. G. Veinott, *Theory and Design of Small Induction Motors* (McGraw-Hill, New York, 1959).
4. "National Electrical Code," ANSI/NFPA No. 70-1984.
5. "Alternating-Current Induction Motors, Induction Machines in General, and Universal Motors," ANSI Standard C50.2.
6. "Standard Dictionary of Electrical and Electronics Terms," ANSI/IEEE Standard 100-1984.
7. "Test Procedure for Polyphase Induction Motors and Generators," ANSI/IEEE Standard 112-1978.
8. "Test Procedure for Single-Phase Induction Motors," ANSI/IEEE Standard 114-1969.
9. "Safety Standard for Motor-Operated Appliances," ANSI/UL Standard 73-1980.
10. "Safety Standard for Electric Fans," ANSI/UL Standard 507-1076.
11. "Safety Standard for Thermal Protectors for Motors," ANSI/UL Standard 547-1979.

SECTION 8
DIRECT-CURRENT MOTORS

John F. Sellers*

FOREWORD 8-2

CHARACTERISTICS AND SELECTION FACTORS OF DC MOTORS IN INTEGRAL-HORSEPOWER SIZES . . 8-2

1. Typical Applications 8-2
2. Power Source 8-2
3. Classification of dc Motors 8-4
4. Factors Determining Speed Range 8-16
5. Motor Horsepower and Torque Capabilities 8-18
6. Temperature Rise and Ventilation 8-18
7. Overload Capability 8-20
8. Inertia Acceleration and Deceleration 8-21
9. Transient-or Impact-Speed Drop 8-23
10. Output-Factor Application 8-25
11. Mechanical Modifications and Accessories 8-28

REFERENCES 8-29

*Chief Engineer, DC Machines, Allis-Chalmers Manufacturing Company, Milwaukee, Wis. (retired); Fellow, IEEE.

FOREWORD

A closer coordination between the power source, the motor design, and the load characteristics is required for dc motor application than for ac motors. This coordination requires a thorough knowledge of dc motor characteristics to assemble an optimum drive system. This section provides the basic considerations needed to apply dc motors.

CHARACTERISTICS AND SELECTION FACTORS OF DC MOTORS IN INTEGRAL-HORSEPOWER SIZES

1. Typical Applications. The application of dc motors occurs most often where:

1. Wide speed range with essentially stepless variation in speed setting is required.
2. Either variable or constant output torque is needed, or the combination of both, as required in most processes (Fig. 1).
3. Fast acceleration, deceleration, or reversal of rotation is required, such as for hoists, traction, propulsion, and metal rolling or processing.
4. Fine accuracy of speed control is needed, such as for tension reels.
5. Accurate speed correlation between two or more parts of a process line has to be maintained.
6. High overload-torque requirements are needed at the lower part of a wide-speed-range process.
7. Variable regenerative braking torque is needed.

2. Power Source. Since there is now very little use of dc power-distribution systems, each dc motor usually has its own power supply in the form of an ac-to-dc motor-generator set or one of several available types of static rectifiers. All these systems permit stable speed control of a dc motor over its total speed range, by variation of the applied armature voltage while keeping the motor field constant. The motor has a constant horsepower per r/min output capability (constant torque and armature current) when using this means of speed adjustment. The type of ventilation provided will determine the time duration of available torque.

a. Rectifiers. Ac-to-dc statically rectified power is now most commonly used because of high efficiency, economy, and elimination of the bearing, brush, and commutator maintenance required on a motor-generator set.

Both the dc motor and the power supply must, however, be carefully matched. Commutation trouble and overheating may occur in the motor at low speeds if a sufficient number of phases are not used, and the rectifier output may become too highly pulsating. In many cases it will be necessary to use a series reactor to smooth out the power pulses to the motor. The dc motor itself performs best, and reduces the need for a series reactor, when the largest possible number of commutator bars is used in the motor. This design observance results in the lowest possible commutating voltage across the brush face.

b. Overhauling loads. When an overhauling load requires energy to be returned to the power line (called regenerative braking), the cost of a supporting rectifier unit is appreciably increased. In many cases a motor-generator set may become more attractive. Each case of overhauling load must be considered sep-

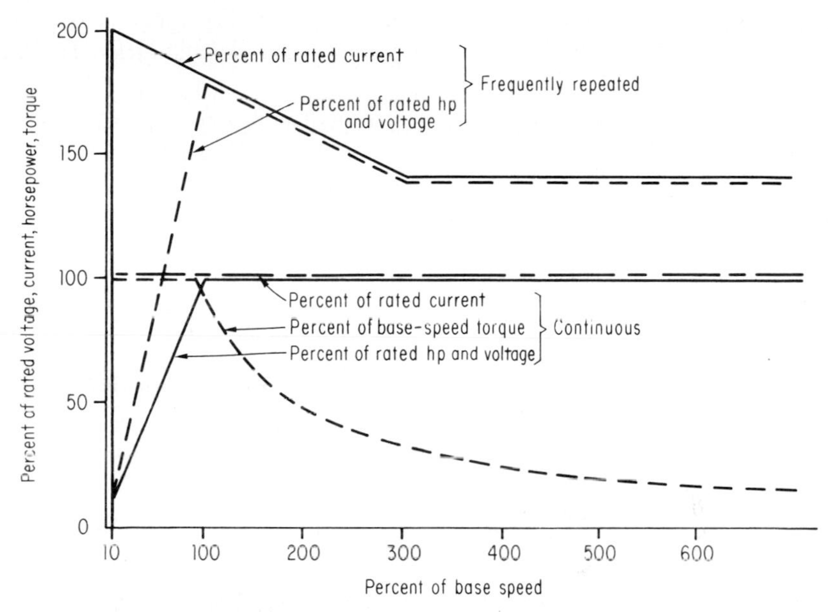

FIG. 1. Typical characteristics of standard-shunt or shunt-stabilized 60°C rise industrial dc motors, separately ventilated.

arately, since a dynamic-braking resistor may be satisfactory or even a mechanical brake rather than regenerative braking, which increases the rectifier cost.

Dynamic braking discards the energy of deceleration as heat from a resistor, increasing the overall losses of a load cycle when compared with the use of regenerative braking. However, dynamic braking is inexpensively and easily applied and can be made very effective by the use of one or two resistor-short-circuiting contactors (Fig. 2). The additional contactors can operate by either timing, current, or voltage signal.

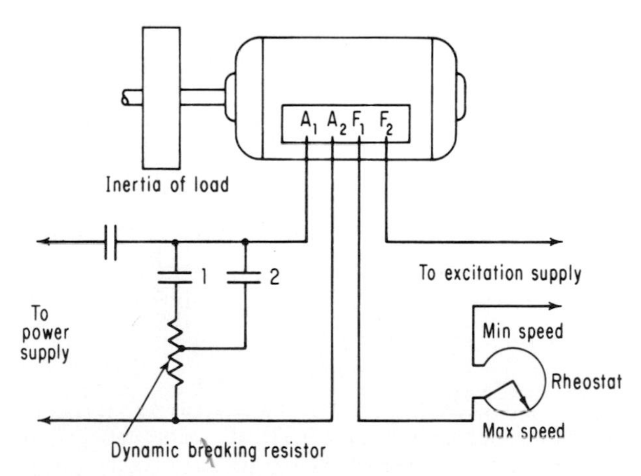

FIG. 2. Dynamic braking with two-step resistor.

3. Classification of dc Motors. Dc motors are defined in six different categories.

a. The physical size of dc motors is divided into fractional integral and large industrial.

1. Fractional-horsepower sizes are the frames that are smaller than those required for 1 hp at 1700 r/min, open, continuous, 60°C rise (Table 1).
2. Integral-horsepower sizes are those frames from 1 hp at 1700 r/min up to 0.75 hp per r/min.
3. Large general industrial sizes are above 0.75 hp per r/min.

b. The application of dc motors can be divided into five groups.

1. Industrial dc motors (Table 2) are of the integral-horsepower size (1 hp at 1700 r/min to 0.75 hp per r/min) and have the following features: (a) They are continuously rated. (b) They have Class B insulation, with 60°C rise. (c) They have a dripproof enclosure. (d) They are shunt-wound or shunt-stabilized-wound. (e) They are mounted by feet attached to the frame and have a single straight shaft extension.
2. Large general industrial motors are all motors larger than 0.75 hp per r/min except those which are used for metal rolling.
3. Metal-rolling-mill motors are of the same torque classification as the large industrial motors, but generally they are made of heavier construction for higher momentary overloads. They come in two classes, Class N and Class S. Class N (Tables 3 and 4) are normally liberally designed, with moderate field-weakening range. Class S are designed for still wider speed ranges (Tables 5 and 6) and require heavier coil bands, more elaborate commutator design, and more testing at the manufacturer's plant.
4. Reversing hot-mill motors are designed to withstand rapid reversals and shock loads and are limited to the base speeds shown in Table 7. When the product of hp × base speed is not over 250,000 and the speed range by field control is not over 2:1, these motors will reverse at no load in 1.5 s from base speed to base speed. When the product of horsepower and base speed is greater than 250,000, reversing time is 2 s at base speed.

TABLE 1. Dc Fractional- and Integral-Horsepower Motor Speed and Voltage Ratings*

hp	Approx full-load speed, r/min			
1/20	3,450	1,725	1,140	850
1/12	3,450	1,725	1,140	850
1/8	3,450	1,725	1,140	850
1/6	3,450	1,725	1,140	850
1/4	3,450	1,725	1,140	850
1/3	3,450	1,725	1,140	850
1/2	3,450	1,725	1,140	
3/4	3,450	1,725		
1	3,450			

Motors rated 0.75 hp per r/min, open type, and smaller.
*From NEMA Standard MG 1-10.61.

TABLE 2. Industrial dc Integral-Horsepower Motors with Field Weakening

	hp at base speed*	hp at 300% of base speed and higher speeds (dripproof motors only)	Base speed, r/min								
			3,500	2,500	1,750	1,150	850	650	500	400	300
			Speed by field control, r/min								
120 and 240 volts	½	0.65	.....	.....	.....	.....	3,000	2,600	2,000	1,600	
	¾	1	.....	.....	.....	3,200	3,000	2,600	2,000	1,600	
	1	1.3	.....	.....	3,500	3,200	2,800	2,600	2,000	1,600	
	1½	2	4,000	4,000	3,500	3,000	2,800	2,600	2,000	1,600	
	2	2.6	4,000	4,000	3,300	3,000	2,600	2,600	2,000	1,600	1,200
	3	4	4,000	3,700	3,300	2,800	2,600	2,600	2,000	1,600	1,200
	5	6.5	3,700	3,700	3,000	2,800	2,600	2,400	2,000	1,600	1,200
	7½	10	3,500	3,500	3,000	2,800	2,600	2,400	2,000	1,600	1,200
	10	13	3,500	3,500	3,000	2,800	2,500	2,200	2,000	1,600	1,200
240 volts	15	20	3,500	3,300	3,000	2,600	2,500	2,200	2,000	1,600	1,200
	20	26	3,500	3,300	3,000	2,600	2,400	2,200	1,800	1,600	1,200
	25	33	.....	3,100	3,000	2,600	2,400	2,000	1,800	1,600	1,200
	30	40	.....	3,100	3,000	2,600	2,400	2,000	1,800	1,600	1,200
	40	52	.....	3,100	2,700	2,400	2,200	2,000	1,800	1,600	1,200
	50	65	.....	.....	2,700	2,400	2,200	1,800	1,800	1,600	1,200
	60	80	.....	.....	2,400	2,200	2,000	1,800	1,600	1,600	1,200
	75	100	.....	.....	2,400	2,200	2,000	1,800	1,600	1,600	1,200
	100	130	.....	.....	2,200	2,000	1,800	1,600	1,600	1,600	1,200
	125	165	.....	.....	2,000	2,000	1,800	1,600	1,600	1,600	1,200
	150	200	.....	.....	2,000	2,000	1,800	1,600	1,600	1,600	1,200
	200	260	.....	.....	1,900	1,800	1,700	1,600	1,600	1,200	1,200
250 and 500 volts	250	......	.....	.....	.....	1,700	1,600	1,600	1,400	1,200	
	300	......	.....	.....	.....	1,600	1,500	1,500	1,300	1,200	
	400	......	.....	.....	.....	1,500	1,500	1,400			
	500	......	.....	.....	.....	1,500	1,400				
500 and 700 volts	600	......	.....	.....	.....	1,500	1,300				
	700	......	.....	.....	.....	1,300					
	800	......	.....	.....	.....	1,250					

*When dripproof motors are rated in accordance with the table:

1. They should be capable of carrying continuously a load equal to 1.15 times the base-speed horsepower rating when operating at base speed.

2. For all ratings through 0.25 hp per r/min (ratings above the line in the table), they shall be capable of carrying continuously a load equal to 1.3 times the base-speed horsepower rating when operating at 150 to 300 percent of base speed but within the speed range listed in the table.

3. For all ratings through 0.25 hp per r/min (ratings above the line in the table), they should be capable of being rated as shown in column 2 of the table only when the motor is operating at speeds by field control within the range of 300 percent of base speed and up to the highest speed listed in the table for the given horsepower and base-speed combination.

When a dripproof motor is operated at any of the loadings described, its operating characteristics, such as efficiency and temperature rise, may differ from those specified when the motor is operated at the base-speed rating.

TABLE 3. Speed Ratings, General Industrial and Metal-Rolling 250-V dc Motors, Class N*

hp	Base speed, r/min														
	850	650	500	450	400	350	300	250	225	200	175	150	125	110	100
	Speed by field control, r/min—nonreversing service†														
250	.....	.....	.....	.....	.....	.....	1,025	910	855	790	725	660	585	540	510
300	.....	.....	.....	.....	.....	1,085	985	875	820	760	700	630	560	520	490
400	.....	.....	1,200	1,140	1,080	1,000	910	810	765	710	650	595	525	490	460
500	.....	1,275	1,110	1,055	1,000	930	855	765	720	670	615	565	500	465	440
600	.....	1,190	1,040	990	940	880	805	725	690	640	590	540	480	445	425
700	1,230	1,120	975	935	890	830	765	690	655	615	560	515	460	430	410
800	1,145	1,050	925	890	840	795	735	660	630	590	540	500	450	420	395
900	.....	1,000	875	845	800	760	700	630	605	570	525	480	435	405	385
1,000	.....	965	840	800	770	725	675	615	585	550	505	460	420	395	370

*From NEMA Standard MG 1-23.11. Speed ratings by field control may vary between the base speed and the speeds listed.

†Speed ratings by field control of motors designed for reversing service (operation with either direction of rotation) may vary between the base speed and a speed equal to 90 percent of the value listed.

Note. The speeds indicated take into consideration both electrical and mechanical limitations. Operation at speeds above those indicated by increasing the armature voltage is not recommended.

TABLE 4. Speed Ratings, General Industrial and Metal-Rolling 500- or 700-V dc Motors, Class N*

hp	Base speed, r/min																					
	850	650	500	450	400	350	300	250	225	200	175	150	125	110	100	90	80	70	65	60	55	50
	Speed by field control, r/min—nonreversing service†																					
250	.....	.....	.....	.....	.....	.....	1,140	1,030	960	890	820	730	645	590	550							
300	.....	.....	.....	.....	.....	1,190	1,090	990	920	855	790	700	620	570	530							
400	.....	.....	1,290	1,250	1,200	1,110	1,020	920	860	800	735	655	580	540	500							
500	.....	1,400	1,220	1,170	1,110	1,040	960	870	810	750	700	625	550	510	480	450						
600	.....	1,330	1,160	1,120	1,060	980	910	830	775	720	660	590	530	490	455	430	400					
700	1,370	1,270	1,110	1,065	1,010	940	870	790	740	690	640	570	510	470	440	415	385	350				
800	1,320	1,220	1,070	1,020	970	900	830	760	710	660	610	550	490	450	425	400	370	340	315	300		
900	1,270	1,170	1,030	980	930	870	805	730	690	640	590	530	475	440	410	385	360	330	305	285	260	
1,000	1,220	1,130	990	950	900	840	780	710	660	620	570	515	460	425	400	375	350	320	295	275	255	240
1,250	1,115	1,030	920	870	830	770	720	660	620	575	530	480	430	400	385	350	330	300	280	260	240	225
1,500	1,030	960	850	810	770	720	670	610	580	540	500	450	410	380	355	330	310	285	265	250	230	215
1,750	960	900	800	760	720	670	630	575	540	510	475	430	385	360	340	315	300	270	250	235	220	205
2,000	.....	840	750	720	675	630	590	540	515	485	450	410	370	340	320	300	285	260	240	225	210	200
2,250	.....	795	710	680	640	600	560	515	490	460	430	390	350	330	310	290	275	250	230	220	205	195
2,500	.....	750	675	650	600	570	535	490	470	440	410	370	340	315	300	280	260	240	225	210	200	190
3,000	.....	.....	610	585	540	510	490	450	430	405	380	340	315	295	280	260	245	225	210	200	190	180
3,500	.....	.....	.....	530	490	470	445	410	395	380	350	320	295	275	260	245	230	210	200	190	180	170
4,000	.....	.....	.....	.....	450	430	410	380	365	350	330	300	275	260	250	235	220	200	190	180	170	160
4,500	.....	.....	.....	.....	.....	390	380	355	340	340	330	310	285	260	235	220	205	190	180	170	165	155
5,000	.....	.....	.....	.....	.....	.....	350	330	320	310	290	270	250	235	225	210	195	180	170	165	160	150
6,000	.....	.....	.....	.....	.....	.....	.....	...	...	...	260	240	225	210	205	190	180	165	155	150	145	140
7,000	.....	.....	.....	.....	.....	.....	.....	...	...	...	...	220	205	195	190	175	165	155	145	140	135	130
8,000	.....	.....	.....	.....	.....	.....	.....	...	...	...	...	...	190	180	170	160	150	140	135	130	128	125

*From NEMA Standard MG 1-23.12. Speed ratings by field control may vary between the base speed and the speeds listed.

†Speed ratings by field control of motors designed for reversing service (operation with either direction of rotation) may vary between the base speed and a speed equal to 90 percent of the value listed.

Note. The speeds indicated take into consideration both electrical and mechanical limitations. Operation at speeds above those indicated by increasing the armature voltage is not recommended.

TABLE 5. Speed Ratings, 250-V dc Metal-Rolling Motors, Class S*

hp	Base speed, r/min														
	850	650	500	450	400	350	300	250	225	200	175	150	125	110	100
	Speed by field control, r/min—nonreversing service†														
250	.....	.....	.....	.....	.....	.....	1,200	1,070	1,000	930	850	775	690	640	600
300	.....	.....	.....	.....	.....	1,260	1,150	1,025	960	890	820	745	660	615	580
400	.....	.....	1,380	1,320	1,240	1,160	1,060	950	890	830	765	695	620	575	545
500	.....	1,440	1,270	1,220	1,150	1,075	990	890	840	785	720	660	590	550	520
600	.....	1,340	1,180	1,135	1,080	1,010	930	840	800	745	690	630	565	525	500
700	1,350	1,250	1,105	1,070	1,020	960	885	800	760	715	660	600	540	505	480
800	1,265	1,175	1,040	1,010	960	910	840	765	730	685	630	580	525	490	465
900	.....	1,110	985	950	915	870	805	735	700	660	610	560	510	475	450
1,000	.....	1,050	935	905	870	830	775	710	675	640	590	540	490	460	435

*From NEMA Standard MG 1-23.11. Speed ratings by field control may vary between the base speed and the speeds listed.

†Speed ratings by field control of motors designed for reversing service (operation with either direction of rotation) may vary between the base speed and a speed equal to 90 percent of the value listed.

Note. The speeds indicated take into consideration both electrical and mechanical limitations. Operation at speeds above those indicated by increasing the armature voltage is not recommended.

TABLE 6. Speed Ratings, 500- to 700-V dc Metal-Rolling Motors, Class S

hp	Base speed, r/min																				
	850	650	500	450	400	350	300	250	225	200	175	150	125	110	100	90	80	70	65	60	55	50
	Speed by field control, r/min—nonreversing service*																					
250	.....	.....	.....	.....	.....	.....	1,340	1,200	1,130	1,050	965	860	760	700	650							
300	.....	.....	.....	.....	.....	1,390	1,280	1,160	1,085	1,010	930	825	730	675	625							
400	.....	.....	1,480	1,440	1,390	1,290	1,200	1,075	1,010	940	860	775	680	635	590							
500	.....	1,590	1,400	1,350	1,310	1,210	1,120	1,020	950	880	820	735	650	600	565	525						
600	.....	1,500	1,320	1,280	1,220	1,130	1,060	965	905	840	775	695	620	575	535	505	470					
700	1,510	1,420	1,260	1,210	1,160	1,085	1,010	915	860	805	740	670	600	550	515	490	455	410				
800	1,460	1,330	1,215	1,160	1,110	1,030	960	880	825	770	710	645	575	530	500	470	435	400	370	350		
900	1,400	1,310	1,165	1,110	1,060	1,000	930	850	800	745	685	620	555	515	480	450	420	390	360	335	310	
1,000	1,330	1,260	1,120	1,080	1,025	960	900	820	765	720	660	600	540	500	470	435	400	375	350	325	300	275
1,250	1,210	1,140	1,040	1,020	1,000	875	825	765	715	670	615	560	500	470	450	410	385	350	325	300	285	265
1,500	1,100	1,050	950	920	870	820	765	700	670	625	580	525	475	440	415	385	365	335	310	295	270	255
1,750	1,030	980	885	850	810	755	720	660	620	585	550	500	445	415	395	365	350	315	295	275	260	240
2,000	.....	910	830	800	755	720	670	625	590	555	520	475	430	395	370	350	330	300	285	265	245	235
2,250	.....	855	785	750	715	670	635	585	560	525	490	450	405	380	360	335	320	290	270	255	240	230
2,500	.....	800	740	710	665	635	600	550	530	500	465	425	390	365	345	325	300	280	260	245	235	220
3,000	.....	.....	660	635	590	565	540	510	485	460	430	390	360	340	325	300	280	260	245	230	220	210
3,500	.....	.....	.....	570	530	515	490	460	445	430	395	365	335	315	300	285	265	240	230	220	210	200
4,000	.....	.....	.....	.....	480	465	450	420	405	395	370	335	310	295	285	270	250	230	220	210	195	185
4,500	.....	.....	.....	.....	.....	420	410	390	375	355	340	320	295	280	270	250	235	220	205	195	190	180
5,000	.....	.....	.....	.....	.....	.....	370	360	350	340	320	300	280	265	255	240	225	210	195	190	185	175
6,000	.....	.....	.....	.....	.....	.....	.....	.....	.....	.....	285	265	250	240	230	215	205	190	180	170	165	160
7,000	.....	.....	.....	.....	.....	.....	.....	.....	.....	.....	...	240	230	220	215	195	185	175	165	160	155	150
8,000	.....	.....	.....	.....	.....	.....	.....	.....	.....	.....	...	...	210	200	190	180	170	160	155	150	145	140

*Speed ratings by field control of motors designed for reversing service (operation with either direction of rotation) may vary between the base speed and a speed equal to 90 percent of the value listed.

Note. The speeds indicated take into consideration both electrical and mechanical limitations. Operation at speeds above those indicated by increasing the armature voltage is not recommended.

TABLE 7. Reversing Hot-Mill Motors

Base speed, r/min																
200	175	150	125	110	100	90	80	70	65	60	55	50	45	40	35	30
Speed by field control, r/min																
400	350	300	250	220	200	180	160	140	130	120	110	100	90	80	70	60

Note. The speeds indicated take into consideration both electrical and mechanical limitations. Operation at speeds above those indicated by increasing the armature voltage is not recommended.

5. Traction motors are very ruggedly built to withstand heat, moisture, vibration, and high operating speeds. They usually have series-type field coils, with characteristics as shown in Fig. 3 (see Sec. 10, Part 3).

c. Insulation systems. Four types of insulation are available, for gradually increasing levels of temperature rise, for the purpose of obtaining smaller frame sizes for a given motor torque rating. These systems are Class A, B, F, and H, and the limiting temperature rises are shown in Table 8. In many cases it is not safe to use the highest-temperature-rise motor because of hazard to personnel or location. The use of silicone varnish in Class H insulation must be carefully considered because of the harmful effect on carbon brushes when a concentration of silicone is present in the atmosphere. The life of each type of insulation, when operating at the respective temperature rises shown, will be the same on an average basis. The relative capability in horsepower per r/min of any given frame size when using the different insulations is covered in Art. 6*d*.

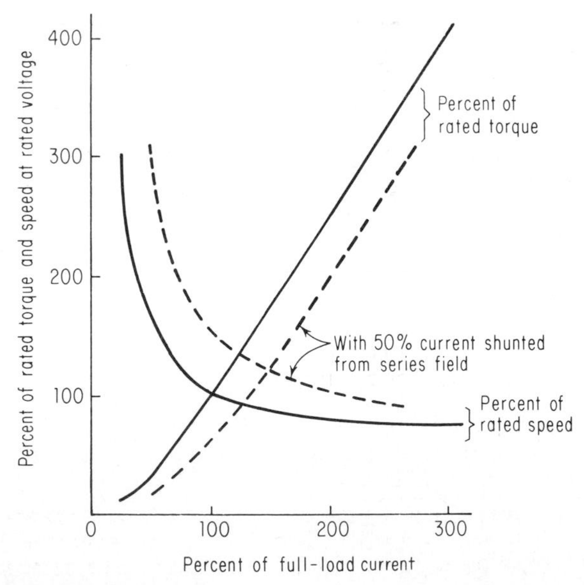

FIG. 3. Typical series-motor characteristics with constant applied voltage.

TABLE 8. Continuous-Time-Rated Industrial dc Motors

	Dripproof and forced-ventilated motors	Totally enclosed nonventilated and totally enclosed fan-cooled motors, including variations thereof				Motors with all other enclosures†			
Class of insulation system	B	A	B	F	H	A	B	F	H
Time rating	Continuous	Continuous				Continuous			
Temp rise, °C:*									
Armature windings and all other windings other than those given below:									
Thermometer	60	55	75	95	115	50	70	90	110
Resistance	90	70	100	130	155	70	100	130	155
Multilayer field windings:									
Thermometer	60	...	...	...	...	50	70	90	110
Resistance	90	70	100	130	155	70	100	130	155
Single-layer field windings with exposed uninsulated surfaces and bare copper windings:									
Thermometer	70	65	85	110	135	60	80	105	130
Resistance	90	70	100	130	155	70	100	130	155
Cores and mechanical parts in contact with or adjacent to the insulation:									
Thermometer	60	...	...	...	...	50	70	90	110
Commutators, thermometer	75	65	85	105	125	65	85	105	125
Miscellaneous parts (such as brushholders, brushes, and pole tips) may attain such temperatures as will not injure the machine in any respect									

*Where two methods of temperature measurement are listed, a temperature rise within the values listed, measured by either method, demonstrates conformity with the standard.

†Includes dripproof motors not covered by first column.

All temperature rises are based on a maximum ambient temperature of 40°C. Abnormal deterioration of insulation may be expected if this ambient temperature is exceeded in regular operation.

Temperatures measured by either the thermometer or the resistance method shall be determined in accordance with the latest revision of the IEEE test code 113.

The foregoing values of temperature rise are based upon operation at altitudes of 3300 ft (1000 m) or less. For temperature rises for motors intended for operation at altitudes above 3300 ft (1000 m), see NEMA Standard MG 1-14.08.

d. Protective covers and method of cooling. A range of protective covers and cooling methods is available for dc motors.

1. A motor is most accessible for service and inspection if it is completely open and self-ventilated. However, there are not many applications for dc motors in which open motors are practical.
2. Because of the vulnerability of the commutator to damage from falling objects, it is customary to use drip covers over the top half-openings in the motor. These covers increase the temperature rise of the motor at rated load by about 20 percent.
3. Further protection is obtained by a screen over all enclosures, with a mesh

such that a ½-in-diameter bar cannot enter the screen. The motor is then classed as guarded.

4. When motors have to operate for long periods at rated armature current at low speeds, self-ventilation is insufficient and it becomes necessary to provide separate ventilation. This cooling can be obtained by either a top-mounted motor-driven blower or a duct connection to a central ventilating system. In either case from 100 to 120 ft^3/min of air per kilowatt loss at rated continuous load is required. This amount of cooling will give 15°C rise in temperature of the air leaving the motor. Even with separate ventilation, a stall current of rated magnitude should not be held longer than 5 s, since the copper segments will heat up rapidly at the brush contact and cause local expansion of the copper. The segments seldom return exactly to their original position; so a bar-to-bar roughness starts. This condition is detrimental to commutation. However, 50 percent of rated continuous current is permissible for unlimited stall time if separate ventilation is maintained.
5. When a motor must operate in an extremely dusty or corrosive atmosphere, it is necessary to enclose the motor totally. Such an enclosure reduces the effective continuous rating of the motor, but intermittent and occasional overload capability is retained, provided the effective heating of the variable load does not exceed the temperature rise allowed for continuous load.

   To maintain normal output (open rating) from a given motor frame when totally enclosed, it is necessary to provide a heat exchanger, mounted on the motor frame. This unit is usually of the air-to-water type, but air-to-air heat exchangers are available for motors as large as 1000-hp rating. The internal air is circulated by a motor-driven blower at a volume of 100 ft^3/min per kilowatt loss, and a face velocity of 600 to 1000 ft/min through the heat exchanger. When an air filter is also provided in the internal air circuit, a face velocity of 600 to 800 ft/min is usually used. However, the air-to-air type may cost five to ten times that of a common air-to-water heat exchanger, and have a volume almost equal to that of the motor.

   An external shaft-driven fan is sometimes provided to produce a forced draft of air over the motor frame. This enables almost full rating to be obtained from motors in the 1-hp and smaller sizes, but its effectiveness decreases gradually with increasing horsepower. This decrease is because the ratio of (watts loss) to (total area of the outside surface of the frame) gradually increases, causing a rise in frame temperature in direct proportion to the value of this ratio.

e. Time and service conditions. The short-time temperature limits vary considerably with motor size, and the ranges for 5- to 60-min loading are shown in Tables 9 and 10. The short-time overload capability in terms of continuous rating is usually as follows:

60 min	135 percent
30 min	140 percent
15 min	145 percent
5 min	150 percent

A service factor varying from 1.15 to 1.3 is also available in limited ranges of rating.

All motors will operate at rated load when the applied armature voltage is within plus or minus 10 percent of rated, but the performance may vary from that at rated voltage.

TABLE 9. Short-Time-Rated Industrial dc Motors (5 and 15 min)

	Dripproof, forced-ventilated, and other enclosures				Totally enclosed nonventilated and totally enclosed fan-cooled motors, including variations thereof			
Class of insulation system	A	B	F	H	A	B	F	H
Temp rise, °C:*								
Armature windings and all other windings other than those given below:								
Thermometer	50	70	90	110	55	75	95	115
Resistance	80	115	145	175	90	125	155	185
Multilayer field windings:								
Thermometer	50	70	90	110	...	...	...	...
Resistance	80	115	145	175	90	125	155	185
Single-layer field windings with exposed uninsulated surfaces and bare copper windings:								
Thermometer	60	80	105	130	65	85	110	135
Resistance	80	115	145	175	90	125	155	185
Cores and mechanical parts in contact with or adjacent to the insulation:								
Thermometer	50	70	90	110				
Commutators, thermometer	65	85	105	125				
Miscellaneous parts (such as brushholders, brushes, and pole tips) may attain such temperatures as will not injure the machine in any respect								

*Where two methods of temperature measurement are listed, a temperature rise within the values listed, measured by either method, demonstrates conformity with the standard.

All temperature rises are based on a maximum ambient temperature of 40°C. Abnormal deterioration of insulation may be expected if this ambient temperature is exceeded in regular operation.

Temperatures measured by either the thermometer or the resistance method shall be determined in accordance with the latest revision of the IEEE test code 113.

The foregoing values of temperature rise are based upon operation at altitudes of 3300 ft (1000 m) or less. For temperature rises for motors intended for operation at altitudes above 3300 ft (1000 m), see NEMA Standard MG 1-14.08.

Motors operating in an ambient temperature above 40°C will have the continuous rating reduced to obtain a temperature rise which is lower than the standard value by the number of degrees the ambient temperature is above 40°C.

A similar condition exists when a motor must operate in a 40°C ambient at elevations above 3300 ft (1000 m). A motor in this application must have a temperature rise at sea level equal to

$$T_{RSL} = T_{RA}\left(1 - \frac{\text{alt} - 3300}{33{,}000}\right)$$

where T_{RSL} = temperature rise, °C, at sea level
T_{RA} = temperature rise, °C, for the appropriate motor classification
alt = altitude above sea level, feet, at which the motor will operate

See NEMA Standard MG 1-14.08.

f. Type of field winding influences the operating characteristics of the motor. The field excitation for a motor can be obtained from either of two sources or a combination of both.

TABLE 10. Short-Time-Rated Industrial dc Motors (30 and 60 min)

	Dripproof, forced-ventilated, and other enclosures				Totally enclosed nonventilated and totally enclosed fan-cooled motors, including variations thereof			
Class of insulation system	A	B	F	H	A	B	F	H
Temp rise, °C:*								
Armature windings and all other windings other than those given below:								
Thermometer	50	70	90	110	55	75	95	115
Resistance	70	100	130	155	80	110	140	165
Multilayer field windings:								
Thermometer	50	70	90	110				
Resistance	70	100	130	155	80	110	140	165
Single-layer field windings with exposed uninsulated surfaces and bare copper windings:								
Thermometer	60	80	105	130	65	85	110	135
Resistance	70	100	130	155	80	110	140	165
Cores and mechanical parts in contact with or adjacent to the insulation:								
Thermometer	50	70	90	110				
Commutators, thermometer	65	85	105	125				
Miscellaneous parts (such as brushholders, brushes, and pole tips) may attain such temperatures as will not injure the machine in any respect								

*Where two methods of temperature measurement are listed, a temperature rise within the values listed, measured by either method, demonstrates conformity with the standard.

All temperature rises are based on a maximum ambient temperature of 40°C. Abnormal deterioration of insulation may be expected if this ambient temperature is exceeded in regular operation.

Temperatures measured by either the thermometer or the resistance method shall be determined in accordance with the latest revision of the IEEE test code 113.

The foregoing values of temperature rise are based upon operation at altitudes of 3300 ft (1000 m) or less. For temperature rises for motors intended for operation at altitudes above 3300 ft (1000 m), see NEMA Standard MG 1-14.08.

One is from a constant-voltage source in which a rheostat is inserted in a field circuit to obtain the desired motor speed above that at full field (Fig. 2). The field in this circuit is termed shunt-field winding.

The other source is variable and proportional to the armature current, consisting of turns of heavy wire around the main field poles, which carry the armature current (Fig. 4). This winding is termed series winding, and typical series-motor characteristics are shown in Fig. 3. With either source of excitation, to obtain reverse rotation it is necessary to reverse the polarity on either the armature circuit or the main field circuit, but never both. If the armature circuit is reversed, the commutating poles, if used, must be reversed along with the armature for proper commutation. It is also necessary to have the brushes located on electrical neutral for symmetrical speed characteristics in both rotations. Since reversing the field of a series motor requires the use of four large contactors for rated load current, the control is more expensive than that for a shunt-wound motor. The series motor rises in speed above rated value for light loads; so this type of motor can be used only where the load does not decrease to low values, since otherwise dangerous speeds will be attained. On the other hand, some speed adjustment is available by use of one or two steps of series-field shunting (Fig. 4). The usual practice is to reduce the series-field strength to one-half by a shunting contactor in series with a resistor which is equal to the value of the series-field hot resistance. When the motor design characteristics permit, as much as 75 percent of the series-field current may be shunted. A series motor is applied almost exclusively to the traction type of work, since speed is always under manual control. The torque per ampere increases with overload, because the magnetic-field strength in the main pole gap increases until a value of high saturation in the steel is reached.

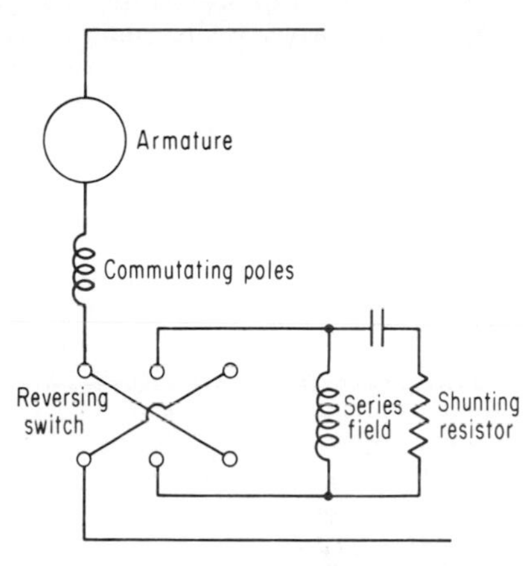

FIG. 4. Series motor with reversing switch.

Since a shunt-wound motor, which does not have a pole-face winding, has a tendency to rise in speed from ½ to 1½ load because of a loss of flux from distortion, it is desirable to provide a stabilizing effect on the main poles to maintain the total air-gap flux. This effect is accomplished with a light series field in addition to the shunt field. This series field may have about 5 to 10 percent of the full value of the shunt-field strength (Fig. 5). This amount of excitation will produce

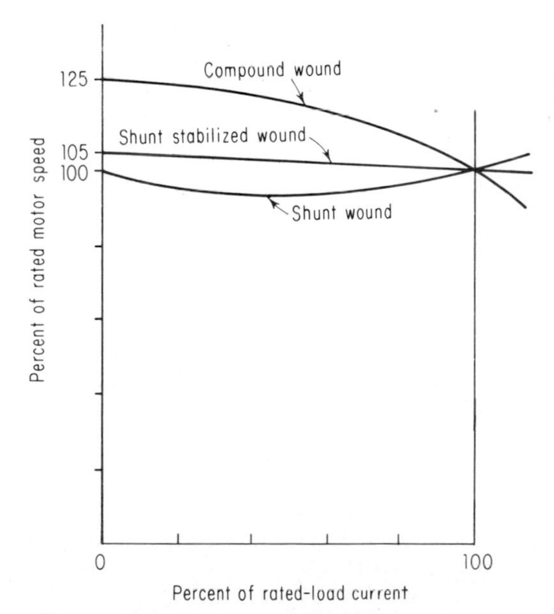

FIG. 5. Typical speed regulation, shunt and compound motors.

a constantly drooping speed characteristic with increasing motor load. When such a motor is operated at weak shunt-field strength, the drop in speed from no load to full load remains almost the same as at base speed, especially if the motor has pole-face windings. If the motor does not operate with shunt-field variation, the series part of the field may have as much as one-third to one-half of the total field strength. This arrangement is termed compound-wound motor, and its speed-load characteristic is indicated in Fig. 5. A stabilized, or compound, motor is usually used only for the one-rotation type of load, because reversing requires contactors for both the shunt and the series fields, or alternately for the armature and commutating poles.

Instead of using a light series field for stabilizing, it is often more convenient to use a small field on the main poles, which is connected in parallel with the commutating winding with an adjustable resistor in series for the desired speed droop. The resistor is usually set for 2 to 5 percent speed droop at top rated motor speed, and shunts out from ½ to 1 percent of the load current from the commutating winding (Fig. 6). The air gap of the commutating poles is adjusted on test to compensate for the reduction in commutating current.

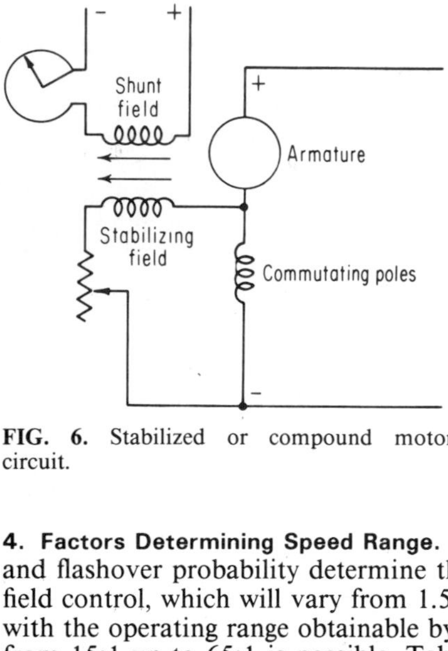

Stabilizing may also be done external to the motor, by use of a field power supply which incorporates, as needed, a drooping effect with load, speed setting, a constant-speed control, and paralleling load with other motors. This arrangement is economically practical only in the larger sizes of industrial and steel-mill motors. Motors in the fractional and smaller integral-horsepower sizes often use a resistor in series with the armature circuit to obtain load speeds below base motor speed. At fractional loads the speed will still be near base value since the series resistor has little effect on reducing the voltage applied to the armature circuit.

FIG. 6. Stabilized or compound motor circuit.

4. Factors Determining Speed Range. Commutating ability, mechanical stress, and flashover probability determine the maximum permissible motor speed by field control, which will vary from 1.5 to 6.5 times rated base speed. Combined with the operating range obtainable by armature voltage variation, a total range from 15:1 up to 65:1 is possible. Tables 3 through 7 show the standard horsepower, base speed, and maximum speeds available for the various classifications of motors at standard voltages. These standards incorporate all the above-listed factors which determine the respective speed ranges by shunt-field control. These three factors are considered individually in the order of their most common occurrence:

a. The commutation ability might be thought to be unlimited if the motor has commutating poles which do not exceed 50,000 lines/in^2 density, and the current density in the brushes does not exceed 75 A/in^2 of contact area on the commutator, at nameplate continuous rating.

However, a factor called divergence causes the commutating-field strength to be too great at high speed, or else too weak at base speed if the strength is opti-

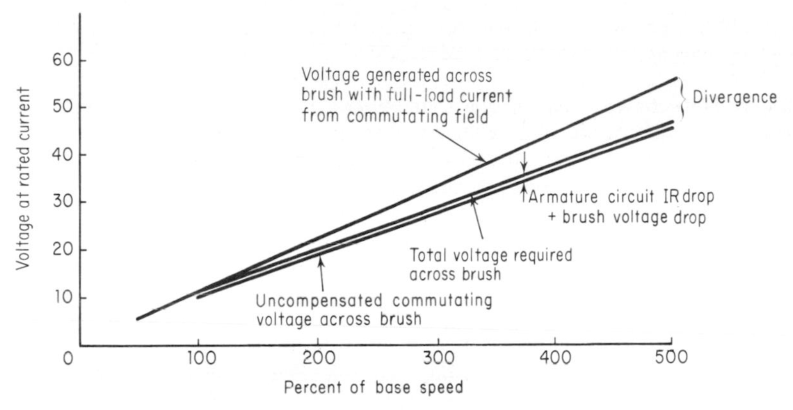

FIG. 7. Typical divergence of actual and required commutating voltage.

mum at high speed. Figure 7 shows this relation, and one of the major factors is the brush and armature-winding *IR* voltage. For this reason the motor designer must know the maximum commutating voltage permissible across the brush face at rated load and standard overloads, and from this value the maximum field-weakening speed of any particular motor design can be accurately established. The maximum commutating voltage is the major limitation in the tabulations of Tables 3 through 7.

b. Mechanical stress caused by centrifugal force affects the commutator design and the bands over the ends of the armature coils. On smaller-sized machines these two considerations can be easily taken care of in the design, with the limitation occurring in the bearings, balance, or natural critical speed of the mass elastic system. On medium and large machines, the commutator design is limited to about 8500 ft/min peripheral speed, although traction motors go as high as 10,000 ft/min. Stress in coil bands limits the armature-core peripheral speed to about 12,000 ft/min. With the application requirement for most dc motors to have low inertia, the mechanical limits are seldom reached, but rather commutation sets the limits.

c. Flashover limitation is the heavy arcing which may occur around a commutator, caused either by poor commutation or more usually by zones of high voltage between commutator segments which connect to armature coils passing under the leading section of the main poles. This condition is caused by the local summation of armature ampere-turns with main-field ampere-turns consumed in the armature teeth and the air gap. To reduce this hazard it is customary to use a "flared," or nonconcentric, pole face, where the gap at the pole tips is two to four times the gap at the center of the main pole. If this method is not enough, it is always possible (but expensive) to use a pole-face winding to neutralize the armature ampere-turns.

If the percent *IR* drop in the armature circuit is 5 percent or more of rated voltage at continuous-current rating, it is probably not necessary to use the stabilizing series field. The divergence effect causes a rising speed characteristic from no load to full load, at weak field speeds; so this effect must be compensated for either by the percent *IR* drop or by the stabilizing field. Another method which is used to obtain a stable (drooping) motor-speed characteristic, when an individual power source is furnished, is to build a 5 percent or more voltage droop into the power pack or generator. This droop eliminates the need for the motor's stabilizing series field and its reversing contactors.

5. Motor Horsepower and Torque Capabilities. A dc motor has constant rated horsepower capability from base speed up to its maximum field-weakening speed. This capability is obtained because the applied armature voltage and current capability are constant, giving a constant product for power input. The basic power equation is

$$\text{hp} = \frac{\text{torque (ft}\cdot\text{lb)} \times \text{r/min}}{5250} \tag{1}$$

With constant horsepower output in the field-weakening range, the torque is variable and is equal to (5250 × hp)/r/min. This relation is produced by the reaction of the magnetic fields from the armature winding and the main-pole windings, whose value in rotational terms is

$$T_a = K\phi i_a \tag{2}$$

where T_a = air-gap torque, ft·lb
K = constant for a specific armature winding
ϕ = total lines of main-pole flux
i_a = armature current

$$T_a = \frac{7.04(1 - d)(e_a - i_a r_a)}{\text{r/min}} i_a \tag{3}$$

where d is the per unit loss in main-pole flux due to the cross-magnetizing field of the armature current. For noncompensated (without pole-face windings) motors with commutating poles, $(1 - d) \approx 0.95$. For compensated motors, $(1 - d) = 1.00$ since the armature field is practically all neutralized. e_a is the applied armature-circuit voltage, and $i_a r_a$ the internal voltage used in the brushes and winding resistance of the total armature circuit, which varies from 2 to 5 percent of rated e_a.

Thus the quantity $(1 - d)(e_a - i_a r_a)$ is the effective countervoltage generated by the motor armature winding. This quantity multiplied by the armature current equals the total power in watts developed in the motor-air gap. This power will be less at the motor coupling, by the amount of winding, brush friction, bearing friction, and core loss. Those losses are usually on the order of 2 to 5 percent of full-load motor rating. The torque T_s at the motor shaft is, within plus or minus 2 percent,

$$T_s = \frac{6.82(1 - d)(e_a - i_a r_a)}{\text{r/min}} i_a \tag{4}$$

6. Temperature Rise and Ventilation. Since dc motors are applied to continuously varying speed requirements, low armature inertia is most desirable.

The lowest values are obtained by use of 60 or 75°C rise motors, rather than 40°C, since the armatures are smaller in diameter or length. These motors have Class B or F insulation, which permits hot spots of 130 and 155°C, respectively, with normal life expectancy.

a. Separately ventilated motor. The effective heating of the armature current in a separately ventilated motor is calculated approximately as follows. Starting with basic equation (2) ($T = K\phi i_a$), various load torques require armature currents in direct ratio, assuming constant main-field flux. The armature current also is double if the main-pole flux is changed from 100 to 50 percent. For a sample load, suppose a 25-hp motor, 600/1800 r/min by field control, 60°C rise, separately ventilated, has the following pattern of load torque repeated continuously at 600-r/min field flux:

1. 400 ft·lb for 1 min at 200 r/min
2. Stopped for 2 min
3. 300 ft·lb for 2 min at 400 r/min
4. Stopped for 2 min
5. 200 ft·lb for 3 min at 600 r/min
6. Stopped for 2 min

Since a separately ventilated motor has cooled independent of motor speed, the effective cycle time is the simple sum of all the intervals 1 to 6, namely, 12 min. By Eq. (1), a 25-hp 600-r/min motor has 219 ft·lb torque at rated armature current and full-field flux. Torque = hp × 5250/r/min. Since the motor torque per ampere is essentially constant at any speed of 600 r/min and less, then

$$400 \text{ ft·lb requires } \frac{400}{219} \times 100 \text{ percent} = 182 \text{ percent of rated current}$$

$$300 \text{ ft·lb requires } \frac{300}{219} \times 100 \text{ percent} = 137 \text{ percent of rated current}$$

$$200 \text{ ft·lb requires } \frac{200}{219} \times 100 \text{ percent} = 91 \text{ percent of rated current}$$

The resultant heating is

$$T\,(^\circ\text{C rise}) \cong \frac{\Sigma i^2 t}{T}\,60^\circ\text{C} \tag{5}$$

where $\Sigma i^2 t$ is the summation of the heating effect of all the components of the load cycle and T is the total load-cycle time.

Then for the above example,

$$T = \frac{(1.82^2 \times 1) + (1.37^2 \times 2) + (0.91^2 \times 3)}{1 + 2 + 3 + 6}\,60^\circ\text{C}$$

$$= 0.80 \times 60 = 48^\circ\text{C}$$

The equivalent current is $0.80^{1/2}$ = 89.5 percent of rated current. In comparison consider the same load cycle of the separately ventilated motor applied to a self-ventilated 25-hp 600/1200-r/min motor. The following considerations must first be applied to the effectiveness of the motor's self-ventilation.

1. Volume flow of air through the motor varies directly with the motor speed below base speed, so that the effective time of each load component should be increased inversely as the actual motor speed is increased to 600 r/min.
2. At standstill, only radiation of heat loss is present. So the effective standstill

time is taken at one-tenth of actual. The temperature rise of the sample cycle is

$$T = \frac{\left(1.82^2 \times \frac{600}{200} \times 1\right) + \left(1.37^2 \times \frac{600}{300} \times 2\right) + \left(0.91^2 \times \frac{600}{600} \times 3\right)}{1 + 2 + 3 + \frac{3 \times 2}{10}} \, 60°\text{C}$$

$$= 3.02 \times 60 = 181°\text{C rise}$$

The equivalent current is $3.02^{1/2} = 174$ percent of rated current, which overheats the motor. Try a 30-hp 600/1200-r/min 60°C motor, self-ventilated. The rise becomes 127°C and the equivalent current is 145 percent of rated. A motor with this value would still operate too hot for a reasonable insulation life. It will be found by trial that a 40-hp 600/1200-r/min motor would be needed to keep the rms current within the 15 percent service factor which is inherent in this class of self-ventilated motor.

b. An alternate choice would be to use a 25-hp 500/1200-r/min self-ventilated motor. This arrangement gives a temperature rise of 84°C, or an equivalent current of 118 percent of rated. Thus the larger-horsepower motor is a slightly better choice than a lower-base-speed motor for this particular example of load cycle.

c. Uses for self-ventilated motors. As can be seen from the examples, self-ventilated motors are most suited to loads which are chiefly in the field control range of the motor speed, with only short duration and infrequent or low loads below base motor speed. This kind of machine becomes almost impossible for low inertia requirements, where the peripheral speed of the armature seldom gets to 2500 ft/min at base speed.

d. Insulation life. In all cases of continuous or rms loading beyond a machine's nameplate rating, each 12°C additional temperature rise reduces the life of the insulation by one-half. In some cases where the process to which the motor is applied may become obsolete or planned to be replaced in 5 to 10 years, it would be most economical to apply a motor which will have an rms load that gives approximately 15°C higher temperature rise than its nameplate rating.

7. Overload Capability. NEMA standard industrial rated motors are capable of 150 percent of full-load current for 1 min at all speeds. However, at less than base (full-field) motor speed, all motors are capable of much higher momentary loads, up to 200 or even 400 percent. This capability is an extremely variable factor between different motor designs; so there is no hard-and-fast rule to use for overloads beyond standard. Commutating capability becomes the actual overload limitation at speeds above base. It is usually necessary to use a larger motor frame when repeated loads, above 300 percent of base speed, exceed 150 percent of rated. The larger frame allows the use of more shallow armature slots and lower commutating voltage, even though the armature core length is increased. All motors will operate safely continuously within a service factor of 115 percent of full-load ratings.

Some motors which have an average voltage between commutator segments of 10 V or less at rated voltage are capable of operating with full-load current and from 150 to 200 percent of rated voltage continuously. For instance, most 240-V motors can be operated up to at least 360 V, giving 150 percent of rated base speed and horsepower with full main field. This versatility often makes it possible to apply a standard motor to special load and speed requirements. However, the power supply must be furnished with the same continuous overvoltage capability, which increases cost very little when a rectifier is used.

8. Inertia Acceleration and Deceleration. Since many dc motor applications require a fast change in speed or a small error in maintaining a set speed, the total Wk^2, or the combined inertia of the motor armature and load, is of extreme importance.

a. The power used to increase speed can have a large effect on the rms power of a load cycle. The basic relations used to calculate torque and the power required for acceleration are:

$$\text{Total inertia at motor shaft} = \text{motor inertia} + \frac{\text{load inertia}}{\text{gear ratio}} \qquad (6)$$

where a step-down gear is used from motor to load. This type of drive is the most common since it permits the lowest cost. Usually the cost of a direct drive would require a prohibitive outlay for the motor.

Stored energy in a rotating mass

$$= Wk^2\ (\text{lb}\cdot\text{ft}^2) \times (\text{r/min})^2 \times 3.1 \times 10^{-7} \quad \text{hp}\cdot\text{s} \qquad (7)$$

The horsepower at $(\text{r/min})_t$ required to accelerate at a constant rate by increasing the armature voltage with constant field and constant armature current is

$$\text{hp} = \frac{(\text{r/min})_t^2}{1.615 \times 10^{-6} \times t} \qquad (8)$$

from zero speed to $(\text{r/min})_t$ in t seconds.

The torque to accelerate at a constant rate of (r/min)/s from $(\text{r/min})_1$ to $(\text{r/min})_2$ is

$$\text{Torque} = \frac{Wk^2 \times [(\text{r/min})_2 - (\text{r/min})_1]}{307.5 \times t} \quad \text{ft}\cdot\text{lb} \qquad (9)$$

using constant armature current and field strength.

The horsepower to accelerate from $(\text{r/min})_1$ to $(\text{r/min})_2$ by motor field weakening, with armature voltage and current constant, is

$$\text{hp} = \frac{Wk^2 \times [(\text{r/min})_2^2 - (\text{r/min})_1^2]}{3.23 \times 10^6 \times t} \qquad (10)$$

b. The winding-reel drive with a separately ventilated motor is taken as an example to calculate the horsepower load, the horsepower for acceleration, and the rms heating of the complete load cycle. Assume coil data as follows:

Full strip speed 6000 ft/min
Tension in strip 5500 lb
Coil diameters 16.5 in ID and 84 in OD
Accelerate from 0 to 6000 ft/min in 12 s
Decelerate from 6000 to 0 ft/min in 8 s
Wk^2 of an empty reel is 4500 lb·ft^2

1. The reel speed at the end of a 12-s accelerating period to 6000 ft/min, assuming no buildup of strip, is

$$\text{Speed} = \frac{(\text{ft/min}) \times \text{seconds}}{\text{reel circumference}} = \frac{6000 \times 12}{16.5\pi} = 1390 \text{ r/min}$$

2. The reel speed with full coil at 6000 ft/min is

$$\text{Speed} = \frac{6000 \times 12}{84\pi} = 273 \text{ r/min}$$

3. The continuous load with 5500-lb pull at 6000 ft/min is

$$\text{Load} = \frac{5500 \times 6000}{33{,}000} = 1000 \text{ hp}$$

Referring to the NEMA motor list (Table 6), a 1000-hp motor at 250 r/min base speed is limited to 820 r/min top speed. Since a total speed range of 1390/273 = 5.1:1.0 by field range is required, Table 5 shows that an 80-r/min base-speed motor with 273/80 = 3.42:1.0 step-up gear would be required. This arrangement is expensive; so the obvious solution comes to a triple-armature motor, 60°C rise, 270/1390 r/min by field control. This range is somewhat greater than shown in Table 6 but should be available with 60°C rise.

4. Wk^2 of this motor is 3 × 2130 = 6390 lb·ft². Because of the critical design parameters required for this motor rating, it is necessary to obtain the value of motor Wk^2 from a motor supplier or to apply the principles outlined in Art. 10. The total Wk^2 of motor and empty reel is

$$Wk^2 = 4500 + 6390 = 10{,}890 \text{ lb}\cdot\text{ft}^2$$

The horsepower required for acceleration is, by Eq. (8),

$$\frac{10{,}890 \times 1390^2}{1.615 \times 10^6 \times 12} = 1087 \text{ hp}$$

for a total of 1087 + 1000 = 2087 hp during the 12-s acceleration period.

The horsepower on the motor during 8-s deceleration from 6000 ft/min will have to be calculated with a full 84-in coil. To find the total Wk^2 involved, a strip width will have to be assumed. This width will be taken as 46 in. The Wk^2 of the strip is

$$0.283 \times 46 \times \frac{\pi}{4}(84^2 - 16.5^2) \times \frac{1}{2}\left[\left(\frac{84}{2 \times 12}\right)^2 + \left(\frac{16.5}{2 \times 12}\right)^2\right] = 440{,}000 \text{ lb}\cdot\text{ft}^2$$

$$\text{hp to decelerate} = \frac{(440{,}000 + 4500 + 6390)273^2}{1.615 \times 10^6 \times 8} = 2600$$

$$\text{Net hp during deceleration} = 1000 - 2600 = -1600 \text{ hp (regeneration)}$$

5. Time to fill the reel at 6000 ft/min, with a strip 0.012 in thick, is

$$\text{Wraps} = \frac{84 - 16.5}{2 \times 0.012} = 2812$$

$$\text{Length of average wrap} = \frac{84 + 16.5}{2 \times 12}\pi = 13.16 \text{ ft}$$

$$\text{Time to wind} = \frac{2812 \times 13.16}{6000} = 6.167 \text{ min}$$

6. The rms load of one coil, with 1 min down between coils, is

$$\text{rms load} = \left[\frac{\left(2087^2 \times \frac{12}{60}\right) + (1000^2 \times 6.167) + \left(1600^2 \times \frac{8}{60}\right)}{0.20 + 6.167 + 0.133 + 1.0}\right]^{1/2}$$

$$= 992 \text{ hp}$$

c. A triple-armature motor would be required for the above example, with total rating 1000 hp, 270/1390 r/min, 60°C rise, with frequently repeated load capability of 220 percent at base speed and 175 percent at 1390 r/min. This overload capability is sufficient to take care of the total load during the acceleration part of the load cycle. The three armatures should be connected in series across an 800-kW 750-V generator or rectifier, for 250 V per armature.

9. Transient or Impact-Speed Drop. When two or more motors do work on a process simultaneously, it is necessary to keep the speeds very closely matched.

a. Speed matching is especially important when a thin metal strip or rod is in two or more tandem stands of a rolling mill simultaneously, with tension on the strip to produce reduction in thickness or to keep it from folding over while being rolled.

Before rectifier power packages were available, with fast-acting speed regulators, it was the practice to use motors having high inertia, with low armature-circuit resistance and inductance. The low armature-circuit resistance permitted obtaining steady-state no-load to full-load speed regulation of 0.5 to 1 percent, compared with a normal 2 to 4 percent with constant armature voltage and field strength. A fast buildup of armature current occurs with low circuit inductance when a load torque is suddenly applied to the motor shaft. The high inertia is a source of stored energy to supply a gradually decreasing amount of load torque while the armature current increases to the required amount needed for the load torque.

If only one operating speed is required from the motor, it is possible to obtain an ideal relation between inertia, inductance, and resistance. But a dc motor is usually chosen because of its versatility in a wide operating-speed range, with simple control.

With the gradual improvement in accuracy of control, especially with rectifier power supply, the high-inertia armature becomes less important in continuous rolling, such as multiple-strand rod mills; in fact it is a hindrance for fast response by the motor to control signals.

b. Impact-speed drop. To set the value of the control components, it is necessary to have accurate motor data to supply computer programs. A preliminary calculation of the inherent transient-speed drop of the motor can be obtained by use of the NEMA formula for impact-speed drop, at constant voltage and current,

$$\text{ISD} = \frac{88 \times IR}{E}\left(\frac{1000 \times L \times E^2}{Wk^2 \times R^2 \times S^2} + 1\right) \tag{11}$$

where ISD = percent impact-speed drop at base speed
E = rated voltage, volts
I = rated current, amperes
L = inductance of armature circuit, henrys
R = resistance of armature circuit at 75°C, including the equivalent resistance of 3-V brush drop, ohms
S = rated basic (full-field) speed, r/min
Wk^2 = total inertia, armature + load referred to motor shaft, lb·ft²

The maximum impact-speed drop varies from 1.5 to 4 times the steady-state motor-speed regulation. It occurs in 0.25 to 0.75 s after load is applied at base speed. At speeds above base, the impact-speed drop is less in percent, but the maximum drop occurs in two to five times the number of seconds required to reach this point at base speed.

c. Armature-circuit inductance is calculated with good accuracy by

$$L_a = \frac{2.85 \times E}{I \times S \times \text{no. of main poles}} \quad \text{henrys} \tag{12}$$

for motors having pole-face windings. For motors without pole-face windings use a constant of 11.4 instead of 2.85. In other words, an uncompensated motor has four times the inductance of a compensated motor of the same rating. A small motor without commutating poles has probably five to ten times the inductance it would have with commutating poles, depending on the strength of the series field which is usually added to these motors.

When extremely accurate or fast response is needed, only the compensated motor is suitable. However, the cost of this is usually prohibitive on motors whose armature diameter is 12 in or less. Also, when the average voltage per bar is 12 V or more, a motor with field-weakening speeds should be compensated to prevent flashovers at weak field speeds.

d. Using ISD and L_a. With the above values for ISD and L_a established, let us apply them to the 1000-hp reel motor of Art. 8*b,* item 3, using only one armature alone, of the three in tandem for 1000 hp total.

From Eq. (12),

$$L_a = \frac{2.85 \times 250}{(333 \times 800)/250 \times 270 \times 4 \text{ poles}} = 6.2 \times 10^{-4} \text{ henry}$$

$$I = \frac{333 \times 800}{250} = 1070 \text{ A}$$

$$Wk^2 = 2130 + \frac{4500}{3} = 3630 \text{ lb}\cdot\text{ft}^2$$

for 1 armature + ⅓ of empty reel.

$$R = \frac{0.05 \times 250}{1070} = 0.0117\ \Omega$$

for a 5 percent *IR* drop. From Eq. (11),

$$\text{ISD} = \frac{88 \times 1070 \times 0.0117}{250}\left(\frac{1000 \times 6.2 \times 10^{-4} \times 250^2}{3630 \times 0.0117^2 \times 270^2} + 1\right)$$

$$= 4.4\left(\frac{38{,}800}{30{,}300} + 1\right) = 4.4 \times 2.28 = 10.0 \text{ percent}$$

maximum speed drop when full-load torque is suddenly applied at base speed.

e. Effect of power supply on ISD should be considered. If the generator is very large compared with the motor rating, its L_a and R_a do not noticeably affect the ISD. However, with the 800-kW generator chosen, its L_a and R_a should be added to those of the motor for a more accurate analysis of motor speed drop.

10. Output-Factor Application. The horsepower output of any motor frame size is limited basically by temperature rise and by commutation at the overload or maximum speed needed.

a. Three factors are to some extent, interdependent. For a given frame size and horsepower per r/min at base speed, a 60°C rise design will permit a higher field-weakening speed than a 40°C design. This difference arises because a more shallow armature slot, using less weight of copper, can be used in the 60°C rise design. A more shallow slot reduces the slot inductance and the commutating voltage across the brush face, allowing a higher motor speed before the definite limit of this criterion is reached. Higher than 175 percent repeated loads will also limit the top speed of the motor because of the brush commutating voltage.

b. Horsepower per inch of armature diameter. For 60°C rise and 175 percent frequently applied load, the output factor of an armature becomes approximately constant, regardless of the core length. For armatures of 36-in diameter and larger, about 40 hp per inch of diameter can be used. This factor approaches 50 hp for the largest diameters (12 ft). At 25-in diameter it is 35 hp, and it decreases rapidly below 18-in diameter because the effective diameter for current and flux loading becomes more of a problem, and the space factor in the armature slot for copper decreases (Fig. 8).

c. A mechanical limitation in armature output due to shaft strength or critical speed becomes noticeable when the armature core length exceeds its diameter. Although armatures are designed with core lengths as great as 150 percent of diameter, their use is rather infrequent, so that for safe approximation a 1:1 ratio between core length and armature diameter should be observed.

d. Output-factor formula. Limitations in the armature core length can be determined by the parameter termed output factor (OF):

$$\text{OF} = \frac{746 \times \text{hp}}{S \times D^2 \times C} \tag{13}$$

where S = base speed, r/min
D = armature diameter, inches
C = core length, inches

For armature diameters of 36 in and larger, with 60°C rise, this factor varies from 0.08 to 0.105, depending on the ratio of maximum speed to base motor speed. For 16-in-diameter armatures, this factor is in the range of 0.045 to 0.55 and decreases rapidly below 16-in diameter for the reasons cited in Par. *b*. It can be seen that the two design parameters—output factor and horsepower per inch

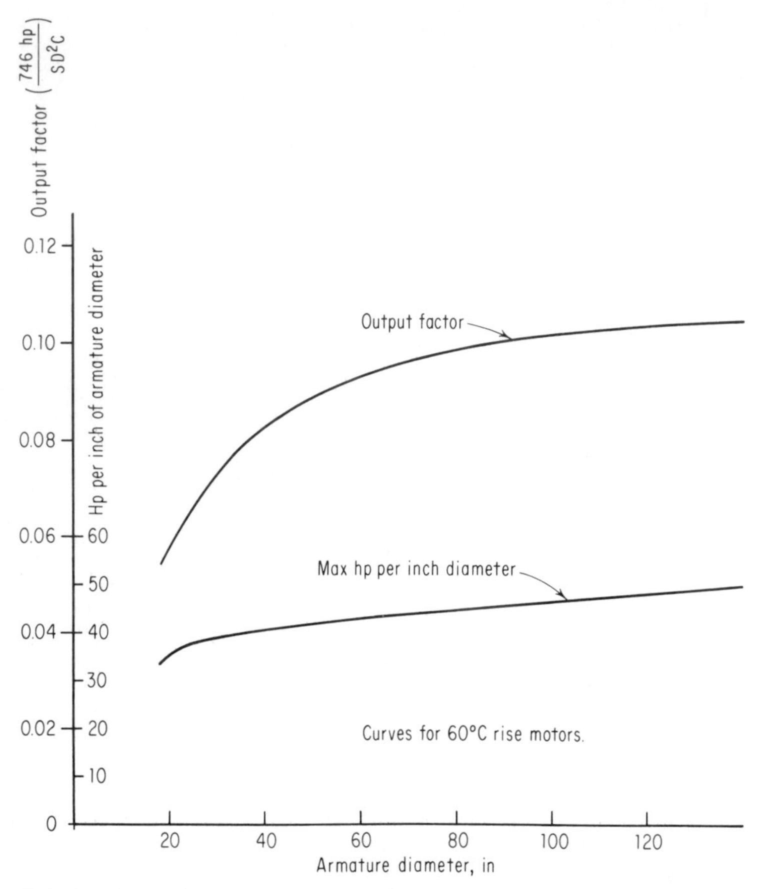

FIG. 8. Output factor related to armature diameter.

of armature diameter—are closely related to one another, depending on current loading, magnetic loading, and commutating voltage (Fig. 8).

e. The output-factor formula is applied to the triple-armature reel motor (Art. 8*b*, item 3). At 20 hp per inch of diameter, the minimum diameter would be 333/20 ≈ 17 in, with an output factor of 0.05:

$$0.05 = \frac{746 \times 333}{270 \times 17^2 \times C}$$

or

$$C = 64 \text{ in}$$

This ratio of 64:17 is so far from the practical 1:1 ratio that it is apparent that the commutating voltage is not the important factor in this rating, but rather a mechanical consideration. Now assume that $D = C$:

$$0.065 = \frac{746 \times 333}{270 \times D^2 \times D}$$

or

$$D^2 = 14{,}200$$
$$D = 24.2 \text{ in} = C$$

Figures 9 and 10 are plots of armature Wk^2 relating diameter to core length for representative diameters, and indicate the number of main poles.

The example of this paragraph can now be based on a 25-in armature diameter, which is the next larger value available than the 24.2 in required for $D = C$. The

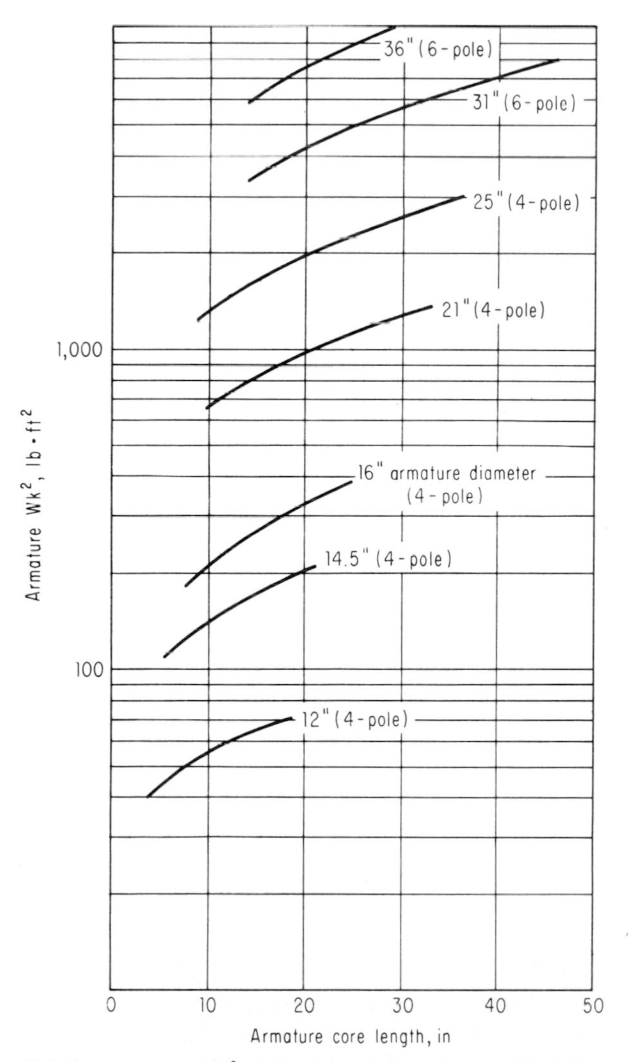

FIG. 9. Armature Wk^2 with average commutator, medium-diameter cores.

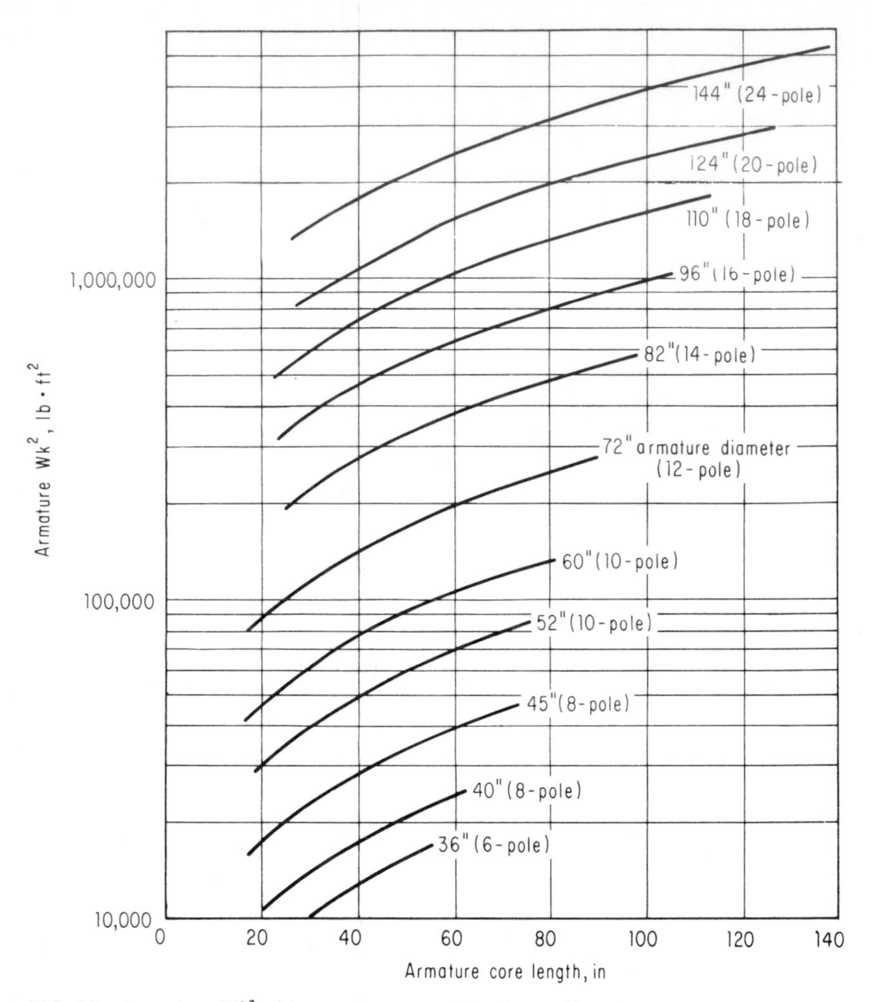

FIG. 10. Armature Wk^2 with average commutator, larger-diameter cores.

required core length C, for 25-in diameter, is

$$C = \frac{746 \times 333}{0.065 \times 25^2 \times 270} = 23 \text{ in}$$

From Fig. 9, $Wk^2 = 2130 \text{ lb} \cdot \text{ft}^2$, which is the value used in previous calculations.

11. Mechanical Modifications and Accessories. These depend on the application of the motor.

a. Bearings. Dc motors are often applied to drive in a vertical or inclined position. Antifriction bearings are usually employed to support the weight of the motor armature. When an added weight or thrust of large magnitude must be taken, a tilting-pad type thrust bearing is used for vertical installations.

b. Armature skew. When the motor torque must be kept very smooth at low speeds, the pulsations caused by the armature slots are eliminated by spiraling or skewing the armature punchings by one slot pitch. This design produces practically uniform reluctance at any armature position, with no torque variation at constant armature current.

c. Forced-oil lubrication may be a necessity for sleeve-type bearings which must operate for long periods at low peripheral speeds. The oil film may become very thin at low speed with oil-ring lubrication.

d. Constant-pressure brushholder springs are available which greatly reduce maintenance costs. This type of spring allows the use of brushes having about twice the length of wear as compared to those used with holders having clock springs or coil springs. This reduces the cost of brush maintenance and keeps the brush always at the ideal pressure on the commutator. The total deviation in pressure is only 10 percent, from unwound to wound-up spring.

e. Accessories normally applied include the following.

1. Overspeed trip devices are used as a safety precaution should the shunt-field circuit open or should speed occur beyond the normal 115 percent of top rated speed. Overspeed is especially apt to occur with an overhauling load, when a large percentage of series field is used.
2. One or more tachometers are often supplied for either speed indication or speed regulation, or both. Pulse generators may also be used for control functions where extreme accuracy is required.
3. Temperature detectors are used on main- or commutating-field coils for either alarm or indicating and recording instruments. On special applications detectors are embedded in the armature winding, and the leads are brought out to brass collector rings, on which ride silver or brass composition brushes to pick up the small currents for recorders.
4. Magnetically operated shoe brakes are generally used only for holding the load when at rest. The rating is usually 25 percent of the motor torque at top motor speed. They are mounted on a shaft extension opposite the motor drive end.
5. Space heaters are provided on the larger types of dc machines when these are out of service for more than 8 h. Quick rises in ambient temperature would otherwise lead to excessive sweating. They are also used on much of the larger marine equipment. Heaters have sufficient capacity to raise the machine's metal parts from 5 to 10°F above room temperature. They are usually installed in a location where convection-air flow will transmit the heat over the greatest possible internal surface area of the machine.
6. Bearing-temperature detectors or relays are usually furnished as optional equipment on medium and large dc machines.

REFERENCES

1. M. G. Say and E. O. Taylor, *Direct-Current Machines* (Halsted Press, New York, 1980).
2. Peter Walker, *Direct-Current Motors: Characteristics and Applications* (Tab Books, Blue Ridge Summit, Pa., 1978).

SECTION 9

SMALL SPECIALTY MOTORS

Richard R. Annis*

FOREWORD 9-2

MOTOR TYPES AND CHARACTERISTICS 9-3

1. Specialty-Motor Types 9-3
2. Universal-Type Motor 9-3
3. Small Shaded-Pole Motor 9-19
4. Magnetic Vibrating Motor 9-23
5. Permanent-Magnet Motor 9-27
6. Impulse Motor 9-29
7. Stepping Motor 9-31
8. Synchronous Motor 9-34
9. Printed-Circuit Motor 9-37
10. Other Miniature Motors 9-39
11. Small Specialty Motors, Selection Parameters 9-42

REFERENCES 9-43

*Vice President, Electrical Engineering, John Oster Manufacturing Company, Milwaukee, Wis.; Registered Professional Engineer (Wis.); Member, IEEE.

FOREWORD

Small specialty motors—universal, shaded-pole, permanent-magnet, small synchronous, impulse—are in the fractional- and subfractional-horsepower sizes. In most cases they are designed for a specific business machine, military unit, toy, appliance, portable tool, or automation component of a larger machine. These motors are commonly designed into the product. In many present-day consumer products the motor frame, brushholder assembly, and bearings are an integral part of the housing. In most applications the duty cycle is intermittent, and large amounts of power may be required for short periods of time. Consumer-product use of these motors necessitates a relatively low cost motor. They are designed for mass production. Manufacturing operations such as stamping of laminations, machining of motor shaft and commutator, insulating the windings of armatures, fields, or stators, connecting of armature leads, casting of rotors, varnish impregnation, balancing, and testing are done on automated equipment. The portability of many consumer products dictates the need for very high horsepower and torque requirements for a given size and weight. Since many of these consumer products find their end use in the home, particular attention must be paid to motor noise, vibration, and radio and television interference. With the present trend toward suburban living and great interest in television, these problems become more pressing every day.

As is the case with all motors, large and small, the continuous-duty horsepower output of these motors depends on the degree of cooling. Many times the appearance and mechanical design of a consumer product will not allow the use of an optimum cooling system, and motor efficiency must be sacrificed.

NEMA has made some attempt to standardize the universal and shaded-pole motor types by motor frame sizes, but many frame sizes exist which are not included in these designations. It is common practice for motor designers to design a new motor with an entirely different lamination for a consumer-product application. Since the anticipated volume of most consumer products is expected to be hundreds of thousands per year, is not difficult to write off the cost of tooling if a few cents in savings per unit will result by changing the motor-lamination design.

Most local governments require consumer products to meet the necessary safety requirements of the various safety-approval laboratories that examine products, such as Underwriters Laboratories in the United States, the Canadian Standards Association in Canada, Semko in Sweden, Demko in Denmark, Nemko in Norway, and other organizations throughout the world. In purchasing or designing a small specialty motor for use in a consumer product, one must be sure to require that the motor meet the necessary approval-laboratory requirements.

A look at all the wonderful electric-motor-operated products in the home such as can openers, meat grinders, blenders, food mixers, knife sharpeners, hair dryers, vacuum cleaners, tooth brushes, shavers, clocks, manicure sets, ice crushers; electric tools such as drills, sanders, and saws; outdoor gardening equipment such as edger trimmers, hedge trimmers, and lawn mowers, having selling prices under $50, makes one realize that small specialty motors are doing a great job and have a very important place in the motor field.

MOTOR TYPES AND CHARACTERISTICS

1. Specialty-Motor Types. All the major types of small specialty motors are summarized in chart form in Table 1. The product designer can select one or two motors from the motor-selection chart for a particular application and obtain more detailed information in the paragraph covering the various motor types.

2. Universal-Type Motor. The universal motor is so named because it will run on both alternating and direct current. This motor has almost identical performance characteristics from direct current to 60 Hz alternating current. The universal motor has a slightly improved speed-torque curve and efficiency on direct current. These motors are probably the most widely used small specialty motors for portable electric appliances. They are found in many consumer products such as sewing machines, food mixers, business machines, liquefier blenders, meat grinders, knife sharpeners, can openers, and in many portable tools such as saws, grinders, sanders, and drills. The main reason this motor is used so extensively is its high horsepower and torque per pound relationship, as shown in Table 1. The amount of horsepower per pound that can be obtained from this motor is greater than that for any other dc or 25- to 60-Hz motor. This motor is relatively simple in construction, consisting of a wound armature and wound field as shown in Fig. 1. Universal motors are normally designed with either of two shapes of laminations. The conventional salient-pole type is shown in Fig. 2, and the C-shaped lamination where the field winding is bobbin-wound in place on the field lamination is shown in Fig. 3.

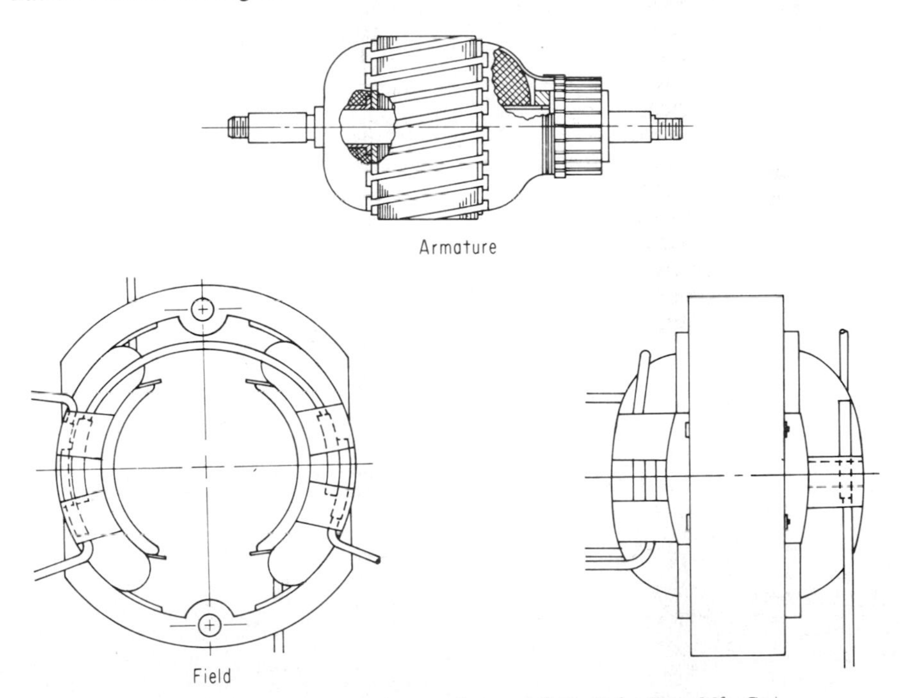

FIG. 1. Universal-motor wound armature and wound field. *(John Oster Mfg. Co.)*

TABLE 1. Selection of Small Specialty Motors

Parameters	Universal		Shaded-pole	Synchronous		
				Two-pole		
	Low-speed	High-speed	Two- and four-pole	Reluctance	Hysteresis	Polarized*
Continuous duty, hp/lb	0.057	0.091	0.004–0.007	0.000068	0.011	0.0021
Continuous duty, hp/in^3	0.014	0.019	0.001–0.004	0.000011	0.0017	0.0003
Max developed hp/lb	0.071	0.12	0.004–0.009		0.011	
Max developed hp/in^3	0.016	0.024	0.001–0.005		0.0017	
Continuous-duty torque, oz·in/lb	8.5	12	1.30–4.7	0.019	1.9	28.44†
Continuous-duty torque, oz·in/in^3	0.74	1.1	0.46–2.80	0.003	0.28	3.9†
Stalled torque, oz·in/lb	26	41	1.3–2.30	0.038	2.2	
Stalled torque, oz·in/in^3	2.5	3.7	0.44–1.40	0.006	0.33	
Starting torque, oz·in/lb	26	41	0.92–1.45	0.038	2.14	
Starting torque, oz·in/in^3	2.5	3.7	0.29–0.91	0.006	0.32	
% efficiency at max developed hp	41	35	21.6–20		37	
% efficiency at continuous-duty hp	46	40	22–20	0.31	37	17
Input at continuous-duty hp, W/lb	118	130	12–23	16	24	9.2
Input at continuous-duty hp, W/in^3	12	11	4.2–14	2.5	3.4	1.3
Input at max developed hp, W/lb	157	194	12–24		24	9.2
Input at max developed hp, W/in^3	14	16	4.3–13		3.4	1.3
Input, W/lb	250	261	4–25	16	31	
Input, stalled rotor, W/in^3	30	22	4.9–15	2.5	4.50	
Power supply (frequency), Hz	dc to 60		50–60	50–60 and 400	50–60 and 400	50–60 and 400
Radio and TV interference	Extreme		None	None	None	None
Range of stator lamination size, OD	1–4¾ in diam		2.25 in^2 5.0 in diam	1.5–4.5 in diam		
Noise	Extreme		Low	Low		
Maintenance	Brush replacement lubrication		Lubrication	Lubrication		
Speed control	Easy		Moderately easy	Difficult		
Speed range, r/min	4000–10,000		3000–1,500	1200–24,000; depends on frequency		
Recommended duty cycle	Intermittent		Intermittent or continuous	Continuous		

*These data are for a 100-pole motor.

†The torque per pound and torque per cubic inch for this type of motor are high compared with the other types of synchronous motors because of a low synchronous speed of 72 r/min.

‡A range is specified for this type of motor since a small increase in rotor diameter results in a substantial increase in output.

§A range is specified for this type of motor since output can be increased by increasing the stepping rate by addition of an external resistor, with basic motor size remaining unchanged.

Note The values shown represent average data obtained from motor manufacturers' catalogs. Continuous-duty horsepower is defined as the maximum horsepower output that can be taken from a motor,

Magnetic vibrating		Permanent-magnet motor		Stepping				
					Pulsed			
					Permanent-magnet rotor			
Permanent-magnet type	Non-permanent magnet type	Printed-circuit armature‡	Wound armature	Free-running permanent-magnet stator§	Cyclical	Sequential	Variable reluctance	Synchronous motor dc stepping
0.006	0.0042	0.013–0.029	0.023	0.00026–0.010	0.00059	0.0034	0.0053	0.0060
0.0022	0.0016	0.0015–0.0040	0.0029	0.000036–0.00015	0.000071	0.00069	0.0011	0.000074
0.018	0.012	0.016–0.057	0.033			0.0035	0.0053	0.006
0.0065	0.0049	0.0021–0.0078	0.0042			0.00069	0.0011	0.000074
123	98	1.0–11	4.2	1.8–6.8	2.9	3.2	4.3	12
47	37	0.13–1.5	0.57	0.24–0.98	0.33	0.63	0.84	1.4
264	208	3.5–75	17		4.7	6.4	8.5	22
101	79	0.46–10	2.3		0.52	1.3	1.7	2.7
264	208	3.5–75	17	0.88–3.4				
101	79	0.46–10	2.3	0.12–0.49				
65	43	33–48	34			9.6	6.7	11
40	19	33–62	46	3.4–5.3	8	10	7	11
11	17	18–35	41	6–142	6.6	28	60	4.2
4.2	6.2	2.4–4.8	5.8	0.83–21	0.74	5.5	12	0.52
11	21	30–89	77			28	60	4.2
6.5	8	3.9–12	9.4			6	12	0.52
		55–188	129			28	60	
		7.3–26	17			5.5	12	
50–60	50–60	dc	dc	50–60	Pulsed dc	Pulsed dc	Pulsed dc and 50–60 ac	dc
None	None	Moderate	Moderate	None	None	None	None	None
1¾ in long × ⅞ in wide, 2½ in long × 1½ in wide		Not applicable	0.75–1.5 in diam	0.5–4.5 in diam				
Moderate		Moderate		Low				
None		Brush replacement lubrication		Lubrication				
Difficult		Moderately easy		Moderately easy				
3600–7200, depends on frequency		0–3000	6000–10,000	3–2500				
Intermittent or continuous		Intermittent		Intermittent				

designed with Class A insulation, continuously without exceeding the 105° C maximum temperature limitation. The data assume average motor cooling. The output per unit weight can be increased by improved motor cooling or use of higher-temperature insulations.

The motor weight and volume for the shaded-pole, permanent-magnet, synchronous, and stepping motors include the motor housing, fan, and bearings. Since universal motors—armatures and fields—are often sold without bearings and the motor frame, and the vibrating motors do not have bearings and are mounted in the product, the weight and volume of these types exclude the motor frame and bearings.

The author was unable to obtain any data on the impulse-motor type.

TABLE 1. Selection of Small Specialty Motors

Parameters	Universal		Shaded-pole	Synchronous		
				Two-pole		Polarized*
	Low-speed	High-speed	Two- and four-pole	Reluctance	Hysteresis	
Connection diagram	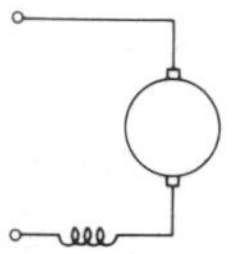			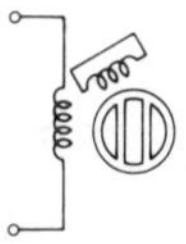	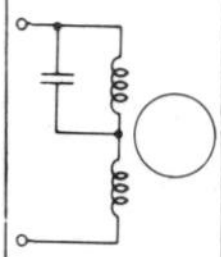	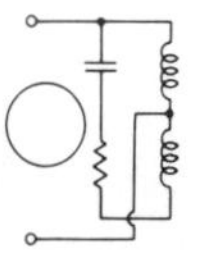
Speed-torque curve	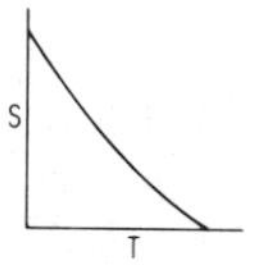		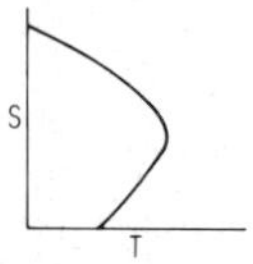	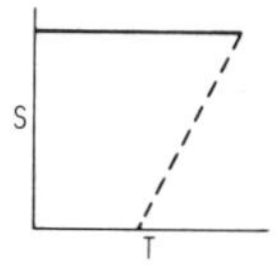		
Sources of data	Bodine Electric, Dayton Electric Mfg., General Electric, Howard Industries, Lamb Electric, John Oster Mfg. Co., Robbins Myers, Globe Industries, Inc.		Alliance Mfg. Co., Barber Colman, Emerson Electric, General Electric, General Industries, Howard Industries, John Oster Mfg. Co.	Barber Colman, Borg Instruments, Bristol Motors, Cramer Controls Corp., General Electric, Globe Industries, Inc., A. W. Haydon Company, Superior Electric		
Typical applications	Electrical appliances, blenders, food mixers, can openers, massage devices, hair clippers, animal clippers, vacuum cleaners, etc., portable tools, drills, saws, grinders, sanders, etc., typewriters, calculators, sewing machines, dental		Electrical appliances, knife sharpeners, can openers, fans, humidifiers, vending machines, pencil sharpeners, heat exchangers, tools, sanders, hair dryers, rotisseries, space heaters, refrigerators, circulators, electric organs, business machines, vending machines	Electric clocks, timing controls, appliances, activators, fans and blowers, machine tools, business machines, recorders, tape drives, radio controls, timers, drive motors		

Magnetic vibrating		Permanent-magnet motor		Stepping				
					Pulsed			
					Permanent-magnet rotor			
Per-manent-magnet type	Non-per-manent magnet type	Printed-circuit arma-ture‡	Wound arma-ture	Free-running permanent-magnet stator §	Cyclical	Sequential	Variable reluctance	Synchronous motor dc stepping
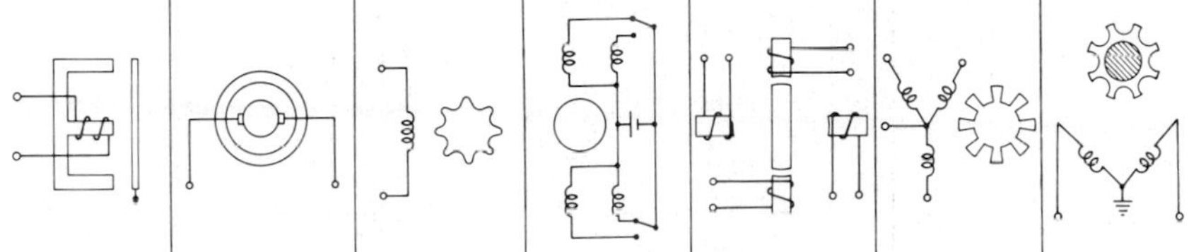								
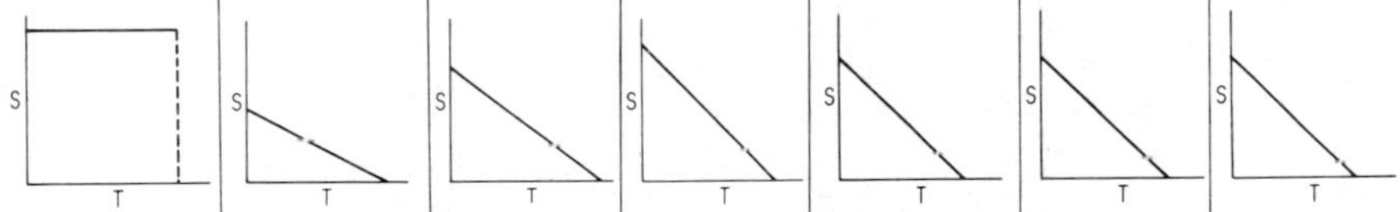								
		Barber Colman Co., Controls Co. of America, Globe Industries, Inc., A. W Haydon Company, Indiana General, Printed Motors, Inc., Rowe Industries, Inc., Superior Electric Co.		American Electronics, Inc., Electric Tachometer Corporation, Enercon Manufacturing Co., General Electric, General Precision, Inc., Globe Industries, Inc., Gap Instrument Corporation, A. W. Haydon Company, Ledex, Inc., Muirhead Instrument, Inc., Powerton Div., Giannini Controls Corporation, J. G. Ruckelshaus Lab., Inc., Sigma Instruments, Inc., Superior Electric Co., Viking Tool & Machine Corp.				
Hair clippers, massage devices, pumps, doorbells, shavers, sanders, jig saws, toys		Battery-powered products, toothbrush, electric knife, ice crusher, shavers, lawn mowers, hair clippers, drink mixers, toys, military		Servosystems, timing controls, appliances				

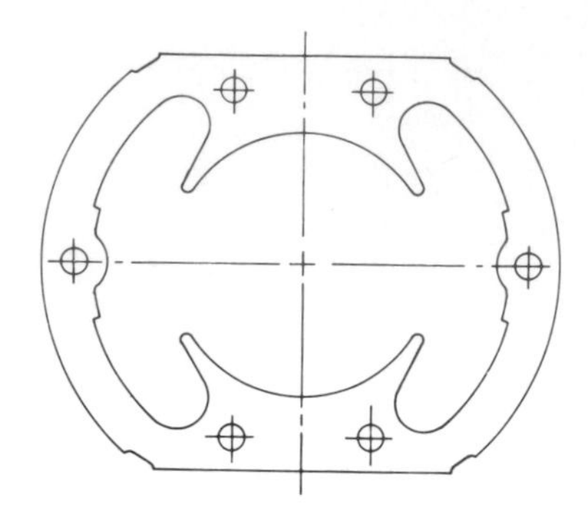

FIG. 2. Universal-motor salient-pole-type lamination. *(John Oster Mfg. Co.)*

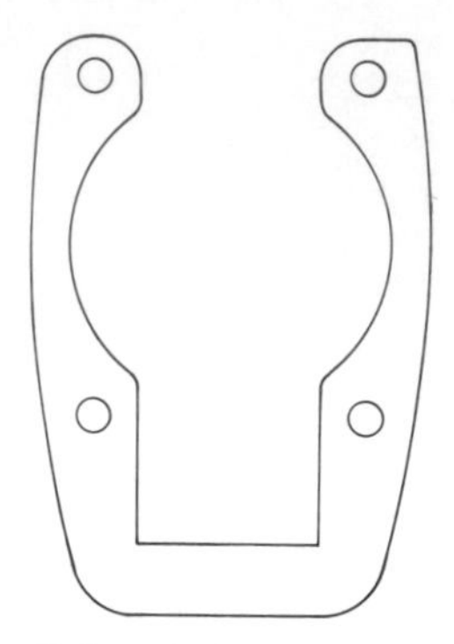

FIG. 3. C-shaped lamination. *(Sunbeam Corp.)*

The salient-pole lamination type allows a symmetrical design and provides a larger surface area for cooling. Where the design does not require a symmetrical shape and appearance, the C-shaped lamination can be used. This design is somewhat simpler since only one coil is required. Some economies are possible if the air flow is sufficient and directed over the coil, since the amount of copper in the field coil will be less.

a. Performance. The performance curves of the universal motor are shown in Fig. 4. The speed-torque curve is very steep, and the speed varies inversely as the load. This motor has a very high free speed. It is basically a high-speed motor and operates from speeds at the light loads of 25,000 to 35,000 r/min to normal operating speeds of 8100 to 10,000 r/min. Because of the very high operating speeds, it has a considerable amount of rotational air noise. Particular attention must be paid to balancing to prevent vibration. It will develop motor torque up to the point of stalling, as shown in the curves, and it has tremendous overload capacity. The designer tries to select the motor 25 to 30 percent below the maximum horsepower point so that reserve power is available. Since this motor has tremendous overload capabilities, care has to be taken in the application and use of the universal motor because it can be self-destructive. As the load is increased, the current increases and the speed decreases, and the motor cooling is reduced. The motor heating varies approximately as the third power of the load.

b. Speed control. There are several methods of controlling the speed of the universal motor. One method is the use of a variable series resistor. The series resistor varies the voltage of the motor, which in turn affects its speed. The basic equation for the speed of the universal motor is

$$N = K\frac{E_a - I_L Z}{BlAD}$$

where K = constant
N = speed, r/min
I_L = line current
E_a = line voltage
A = total number of inductors
Z = motor impedance
D = diameter of armature
B = motor field flux density
l = length of armature conductor

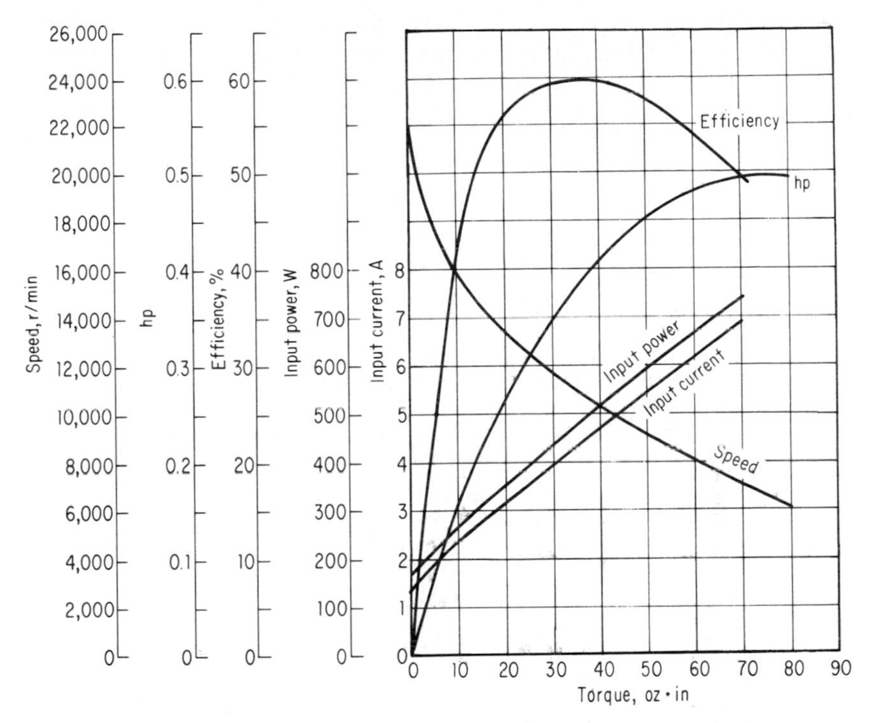

FIG. 4. Performance curves of universal salient-pole motor. Normal operating horsepower is at 40 oz·in. Speed decreases in approximately a straight line to zero as torque increases beyond curve. *(John Oster Mfg. Co.)*

Another method of controlling speed is by means of tapped field windings, varying the impedance of the motor which results in a change in speed.

The speed of the universal motor can also be varied by use of an electromechanical governor (Figs. 5 and 6). With low-speed operation the governor is not in the circuit in this motor design. A centrifugal governor is used to cause the opening and closing of contacts, which place a series resistance in the motor circuit. The addition of the series resistance to the motor impedance creates a drop in voltage and results in a decrease in motor speed. As the motor speed drops, the governor contacts close, taking the resistance out of the circuit and causing the motor to speed up again. This method results in a slight motor hunting at a very high rate, which is not noticeable to the user.

The speed is varied in a governor-controlled universal motor by varying the contact pressure between the two governor contacts. In the design in Fig. 5, maximum speed is obtained by adjusting the control knob in a maximum clockwise position. The cammed interlock pushes the contacts together and requires maximum centrifugal force by the flyballs to cause the riding contact to move inward and the governor contacts to open. Figure 7 shows the curves of a universal motor utilizing an electromechanical type of governor. The speed can be varied with this type of control and also can be held constant over a range of load.

Recently considerable work has been done with solid-state-type controls. Currents through the motor can be varied by use of a silicon-controlled rectifier (thy-

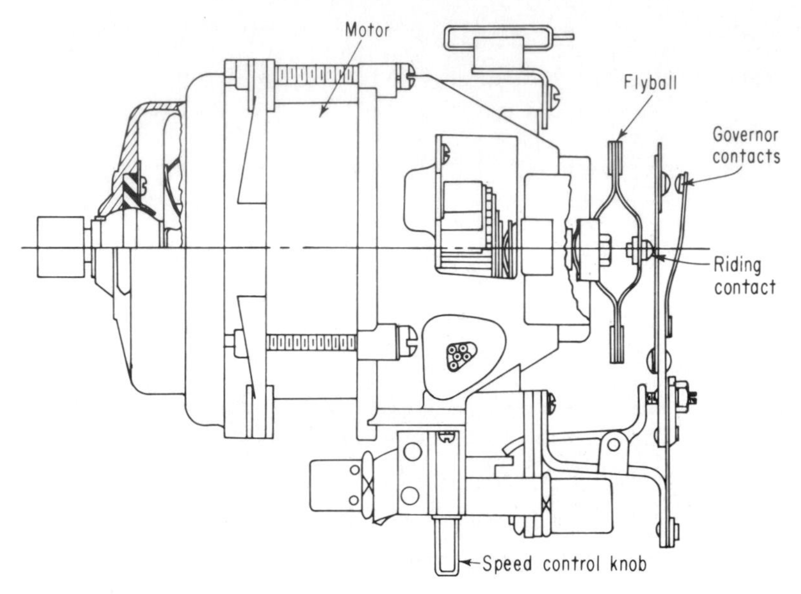

FIG. 5. Universal motor with flyball governor. *(John Oster Mfg. Co.)*

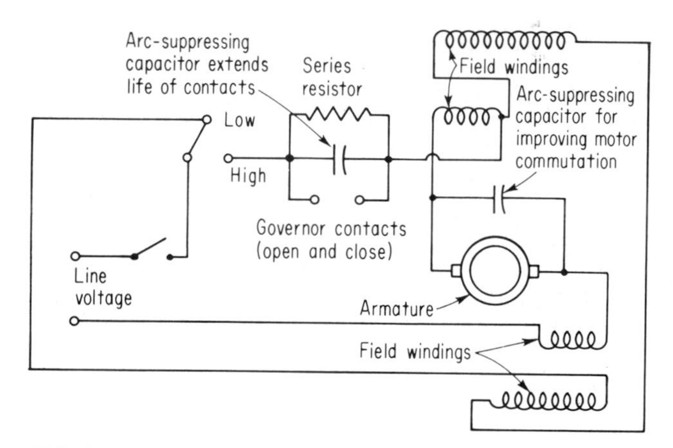

FIG. 6. Electrical schematic for flyball-governor-controlled universal motor. *(John Oster Mfg. Co.)*

ristor). A silicon-controlled rectifier is a silicon diode in which the conduction of current is controlled by a third element, the gate, and the rectifier can be made to conduct at different points in the cycle. Full-wave conduction results in maximum current flow and maximum speed. Low speed is produced by a minimum of conduction (Fig. 8). The silicon-controlled rectifier is capable of handling large amounts of current. A typical example of this circuit is illustrated in Fig. 9. Figure 10 shows one of these controls. Figure 11 presents the typical motor-performance curve utilizing a solid-state silicon-controlled-rectifier speed control without feedback. Figure 12 shows a schematic of a solid-state timer and speed control for a universal motor. The motor can be adjusted for zero to maximum speed and run at any speed setting over a range of zero time to 60 s. The timing function is obtained by charging and discharging a capacitor in the control (Fig. 13).

c. Brushholders. Several types of brushholder designs are commonly used in universal motors.

1. The phenolic brass-insert type shown in Fig. 14 consists of a machined brass insert pressed into a phenolic insulator designed to support a square or rectangular brush. Provision is made for connecting the brush leads to the brass brush tube by use of a hairpin clip or some other fastener. The other end of the brush tube is threaded to allow for insulated or uninsulated brush caps.
2. Formed brass brushholder. Where the spacing is adequate and there is no need to replace brushes externally, a simple rectangular stamped brass brushholder can be used (Fig. 15). Because this holder has no machined parts, it is low in cost.

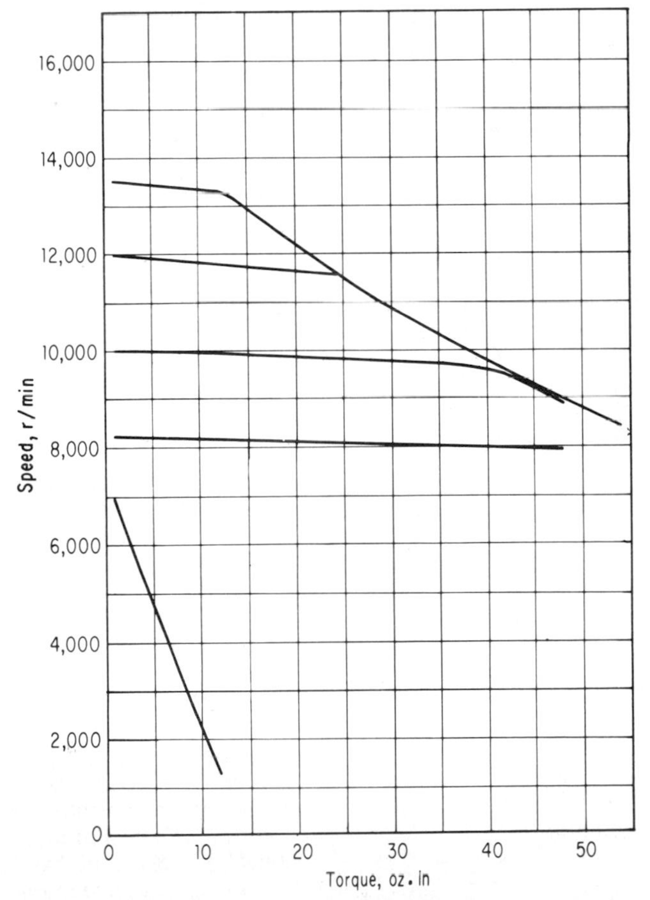

FIG. 7. Curves of universal motor utilizing flyball electromechanical-type governor. Upper curves are for governed speeds; lowest-speed curve ungoverned. *(John Oster Mfg. Co.)*

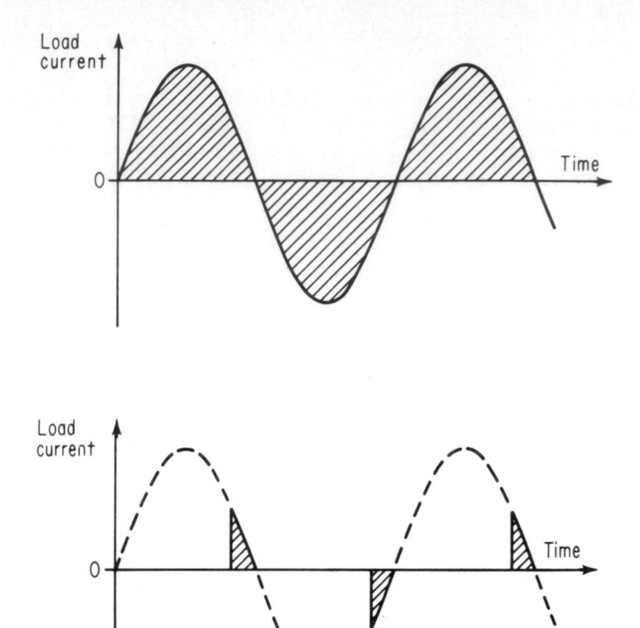

FIG. 8. Full speed has maximum current through motor. Silicon-controlled rectifiers are triggered to cause conduction through complete cycles (upper curves). Low speed has minimum current through motor. Silicon-controlled rectifiers are triggered to cause conduction for only small portion of cycles (lower curve).

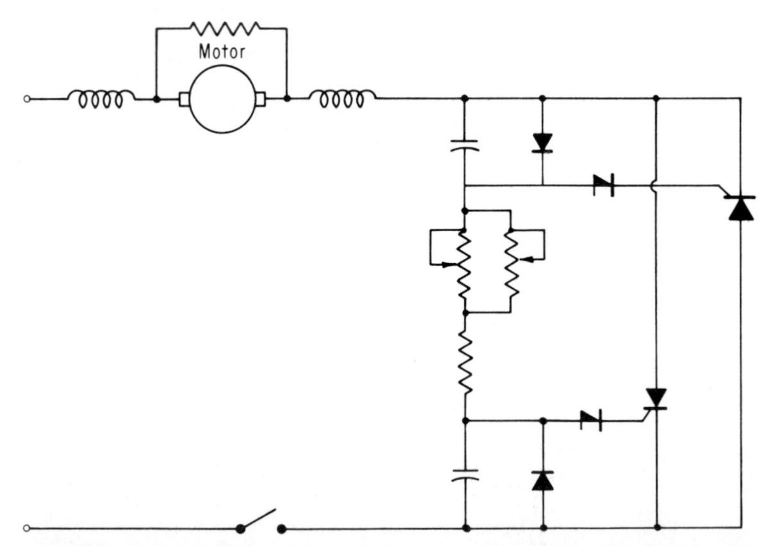

FIG. 9. Schematic of silicon-controlled-rectifier solid-state speed control for universal motor. *(John Oster Mfg. Co.)*

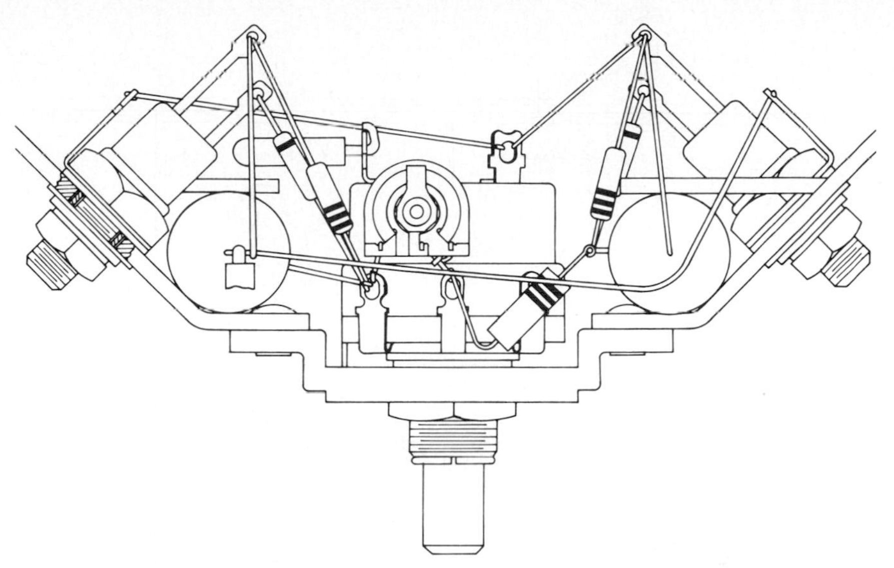

FIG. 10. Assembly drawing of silicon-controlled-rectifier solid-state speed control for universal motor. *(John Oster Mfg. Co.)*

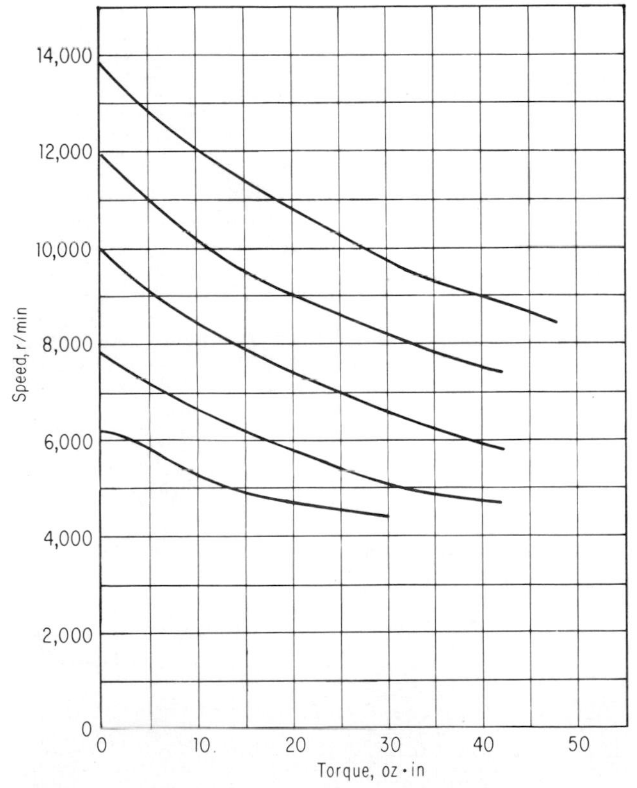

FIG. 11. Motor-performance curve using silicon-controlled-rectifier solid-state control. *(John Oster Mfg. Co.)*

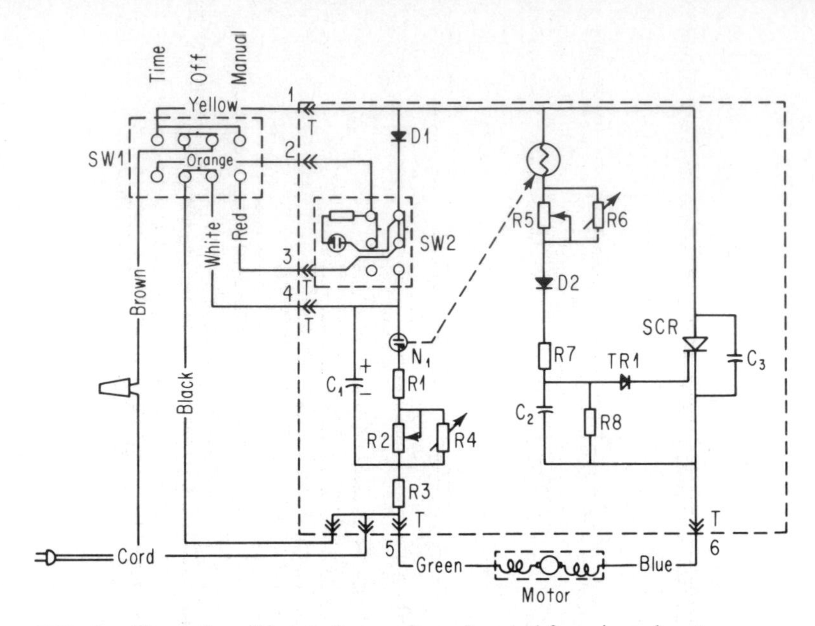

FIG. 12. Circuit for solid-state timer and speed control for universal motor.

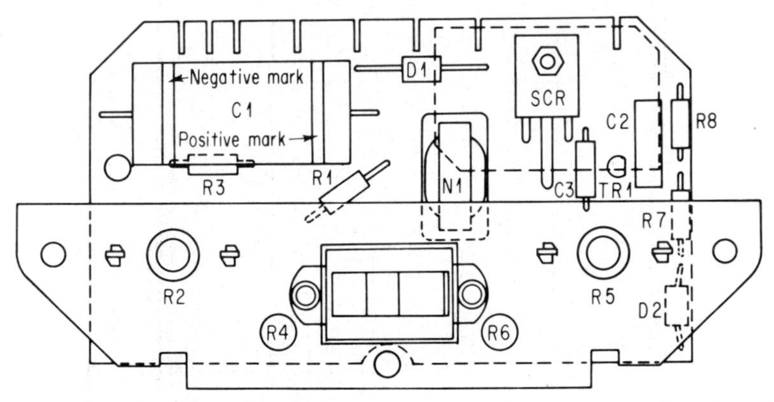

FIG. 13. Assembly arrangement of solid-state timer and speed control for universal motor.

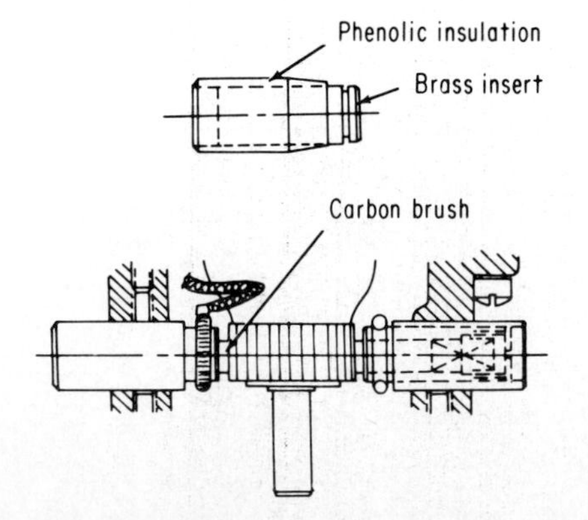

FIG. 14. Phenolic brass-insert-type brush-holder. *(John Oster Mfg. Co.)*

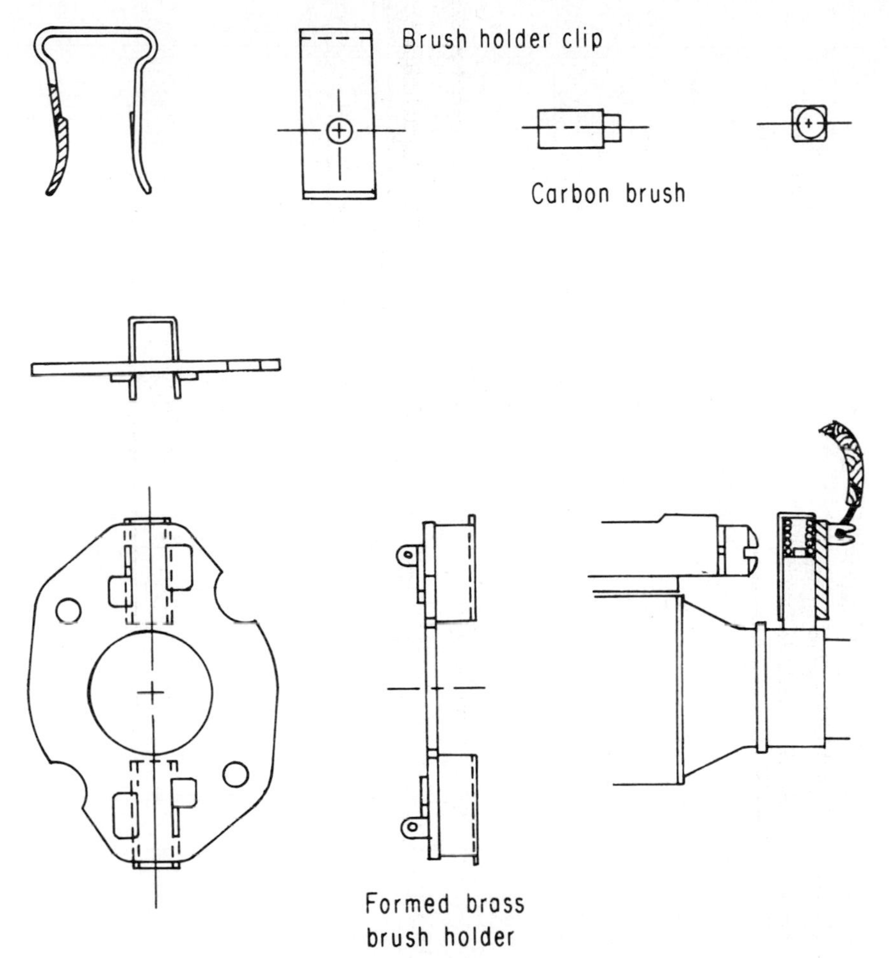

FIG. 15. Formed brass brushholder. Complete assembly is at lower right. *(John Oster Mfg. Co.)*

3. Plastic holder with special brush terminal for use with an unshunted brush. Figure 16 shows a molded phenolic holder designed to support a brush used in conjunction with a brush terminal designed to fit around the neck of the brush and provide the electric contact. This brushholder type is well suited for designs where the housing is split in half, as shown in Fig. 17.
4. Plastic brushholder without special brush terminal for use with a shunted brush. Figure 18 shows a plastic brushholder without a special terminal. In this type, current is carried through the brush shunt to the terminal. The addition of the shunt increases the brush assembly cost, and this method is used only where the spacings do not allow the use of a terminal around the brush.

d. Availability of universal motors at different voltages and frequencies (other than 120 V, 60 Hz). The universal motor will operate on dc voltage and frequencies up to 60 Hz alternating current with very little change in the motor-performance characteristics. It can be readily designed for other voltages such as 150-,

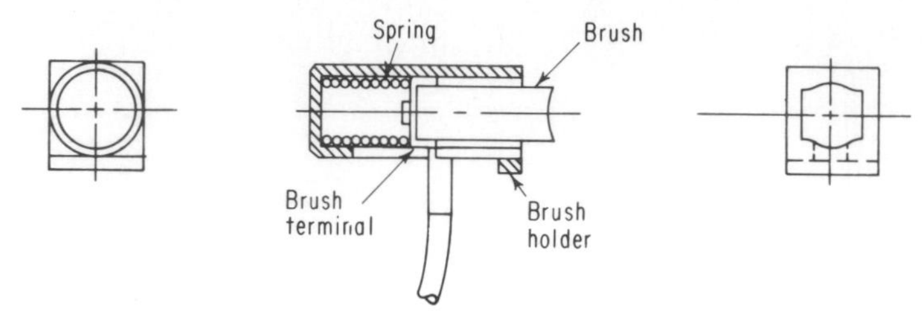

FIG. 16. Plastic brushholder with special brush terminal for use with unshunted brush. *(John Oster Mfg. Co.)*

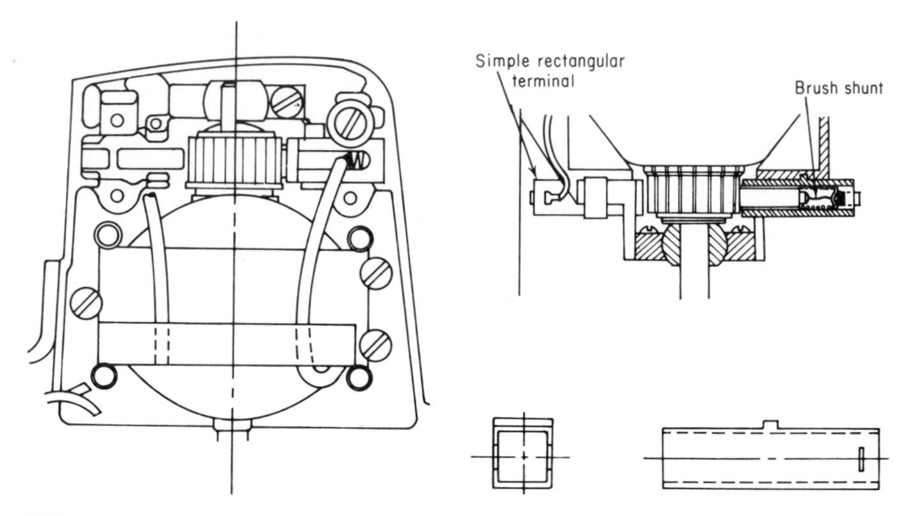

FIG. 17. Split-housing design showing use of plastic brush terminal for use with unshunted brush. *(John Oster Mfg. Co.)*

FIG. 18. Plastic brushholder without special brush terminal for use with shunted brush. *(John Oster Mfg. Co.)*

230-, and 250-V applications by a change in the motor windings. On heavy-duty applications it is necessary to change the number of commutator bars to obtain satisfactory commutation.

e. Motor cooling. The cooling of the universal motor, as with all other motors, is dependent on the type of fan and the resulting volume and pattern or direction of the air flow through the motor. Because of the high speeds of these motors and the resulting air noise, the axial-type fan, as shown in Fig. 19, is best suited for the universal motor. In those applications where cooling becomes a big factor and where higher air-noise levels can be tolerated, a vane-axial or centrifugal-type fan, as shown in Fig. 20, may be used to obtain better cooling. A continuous-air-flow-type cooling system is preferred in the universal motor. Increased cooling efficiency can be obtained if the cool-air-inlet side of the system allows the air to pass over the brush assembly, because the commutator is generally the hottest part of the motor.

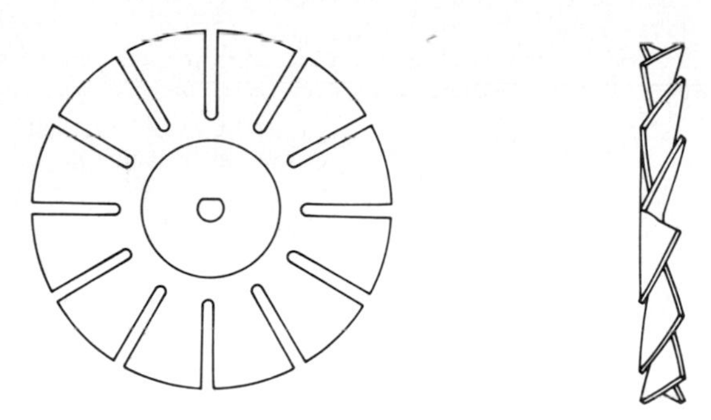

FIG. 19. Axial-type fan. *(John Oster Mfg. Co.)*

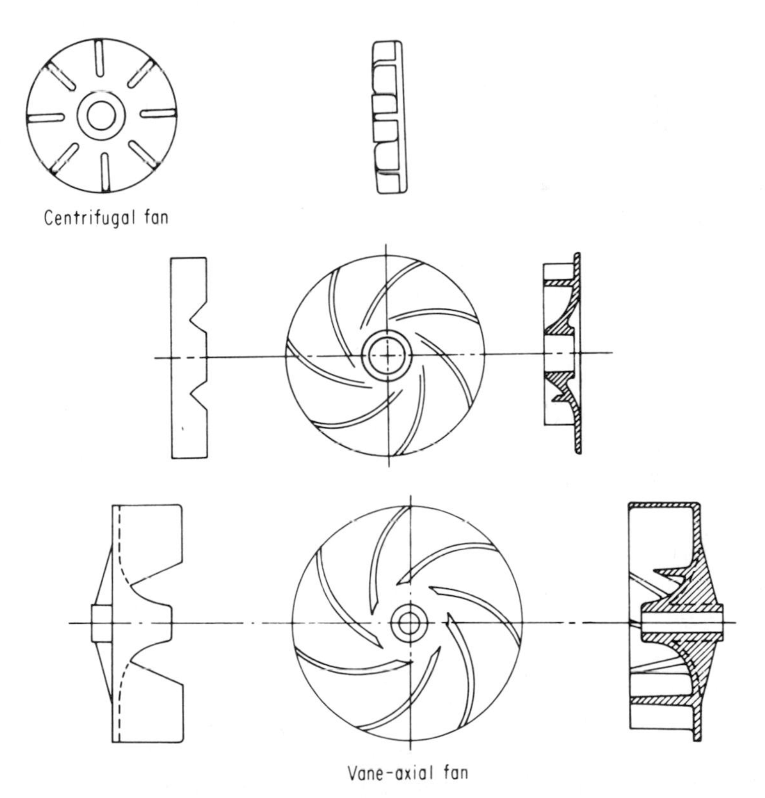

FIG. 20. Centrifugal and vane-axial fans. *(John Oster Mfg. Co.)*

f. Lubrication and bearings. In most universal-motor applications self-aligning oilite-type bearings are adequate and have proved to be quite satisfactory. Frictionless bearings are used in extra-heavy-duty motor applications, such as portable tools. Felt-type oil reservoirs can be used to provide long-life lubrication. Figure 21 shows some of these self-aligning oilite-type bearings. High-speed ball bearings may be used, but in most cases ball bearings are noisier than the oilite type and considerably more expensive.

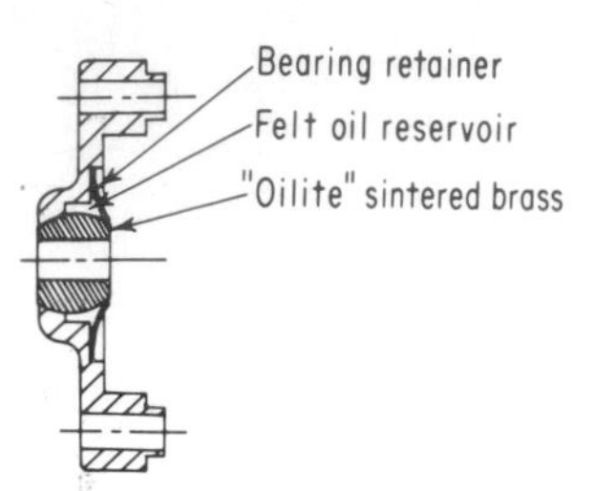

FIG. 21. Self-aligning oilite bearings. *(John Oster Mfg. Co.)*

Lubrication consists of saturating copper-alloy-type sleeve bearings with a high grade of light turbine oil. Felt-type oil reservoirs provide adequate oiling over long periods of time. In applications that receive heavy duty, provisions for external oiling are provided.

g. Maintenance. The only points of maintenance required in universal motors are (1) replacement of carbon brushes and resurfacing of the commutator after several years of use; and (2) in heavy-duty applications the adding of oil to the bearings. In most intermittent-duty universal-motor applications, the brush is designed to provide a minimum of 3 to 5 years of life. It is normally recommended by the manufacturer of universal motors that the commutator be refinished after two sets of brushes have been replaced.

h. Advantages of universal motors

1. Size and weight. As shown in the motor tables, the universal motor develops more horsepower per pound and per unit volume than any other 60 Hz ac motor. This characteristic is extremely important in applications requiring portability.
2. High starting and stalled torque. These motors will develop torque output up to the point of stalling as compared with shaded-pole ac motors which have a breakdown-torque point.
3. Universal motors will operate on direct current and ac frequencies up to 60 Hz.
4. Tapped fields provide an easy method of controlling speed.

i. Disadvantages of universal motors

1. Since these motors operate at very high speeds, up to 40,000 r/min, considerable air noise is present.
2. Very little stalled-rotor protection. These motors can be self-destructive and can be overloaded up to the point of stalling. Because of the large increase in the power input under stalled conditions and the loss of motor cooling, they can burn out within 30 s.
3. Intermittent-duty application only. Because these motors are commutator-type motors where there is brush and commutator wear, they are limited in most cases to several hundred hours of use and are thus most suitable for intermittent application.
4. They produce radio and television interference. The process of commutating, which consists of shorting and opening a circuit across the motor coils with the brushes, generates power frequencies from audio up to the television-frequency range. In many cases, it is necessary to add filters to minimize this interference problem.

3. Small Shaded-Pole Motor. The shaded-pole motor is one of several types of single-phase motors and is called a shaded-pole type because its stator poles contain shading coils which are usually pieces of copper placed around a portion of the pole. The shading coils have the effect of delaying the flux buildup in that portion of the pole. Figure 22 shows a typical two-pole shaded-pole motor.

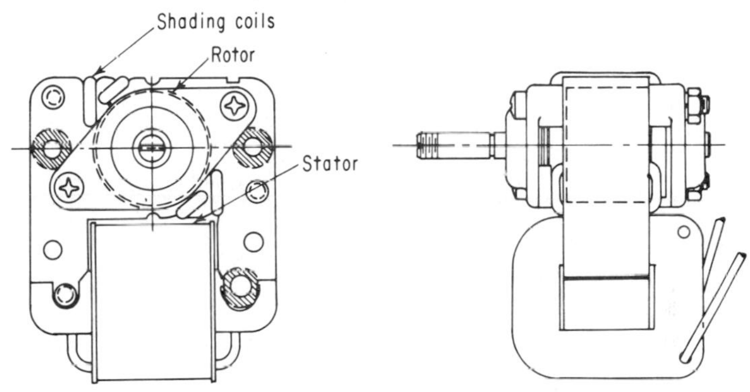

FIG. 22. Two-pole shaded-pole motor. *(John Oster Mfg. Co.)*

As is the case with all ac motors, the speed of these motors is dependent on the number of poles and the line frequency:

$$\text{Synchronous speed} = \frac{120f}{P}$$

where f = line frequency
P = number of poles

Free-running motor speeds are always 5 to 10 percent less than the synchronous speed. These motors are cataloged in two, four, six, and eight poles with a 3500- to 800-r/min free-speed range.

The shaded-pole motors are available in two general shapes, salient-pole and bobbin-wound C-shape types. Figure 22 shows the C-shaped lamination. Figure 23 shows a two-pole salient-pole type. The stator coils are bobbin-wound or wound in place by the use of automatic coil winders. The rotors are of the squirrel-cage type and may have cast-aluminum or copper-riveted construction. These motors are very simple in construction and lend themselves to fully automatic production techniques.

a. Performance characteristics. Figure 24 shows a typical performance curve for a two-pole shaded-pole motor. The efficiency of shaded-pole motors is quite low and usually ranges between 20 and 30 percent. The speed-torque curve for a shaded-pole motor is not as steep as that of the universal motor. The shaded-pole motor does not develop torque uniformly down to zero but reaches a breakdown point, noted in Fig. 24 by point X, at which the motor stalls.

The input current on shaded-pole motors is quite high under free and light-load conditions, and their power factor is quite poor. Their starting torque can be increased by designing their rotors with higher resistance. High rotor resistance can also be used to lower their speed by introducing additional slip. The speed-

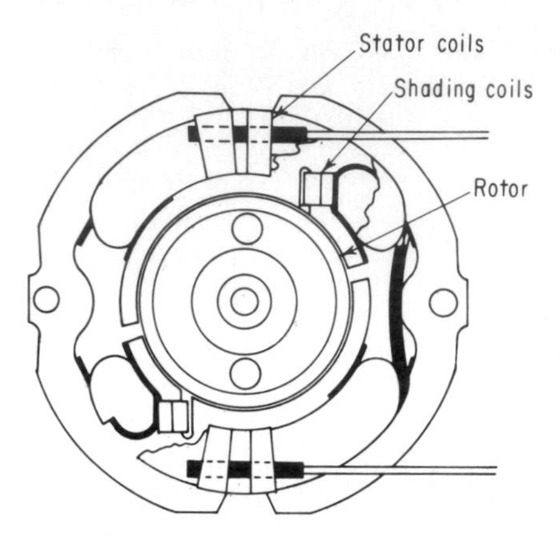

FIG. 23. Two-pole salient-pole motor. *(John Oster Mfg. Co.)*

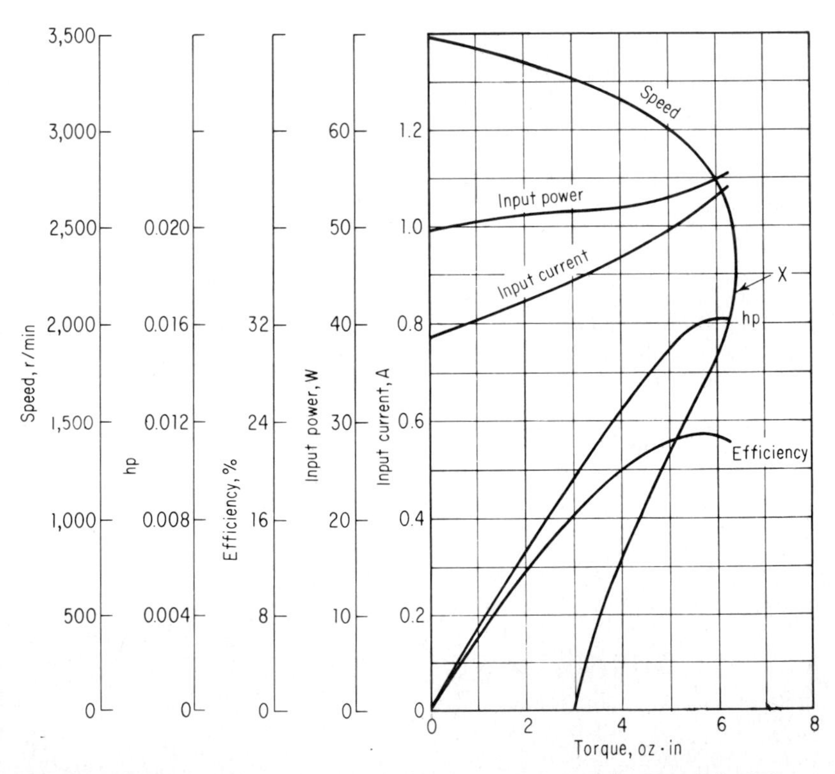

FIG. 24. Performance curve for two-pole shaded-pole motor. *(John Oster Mfg. Co.)*

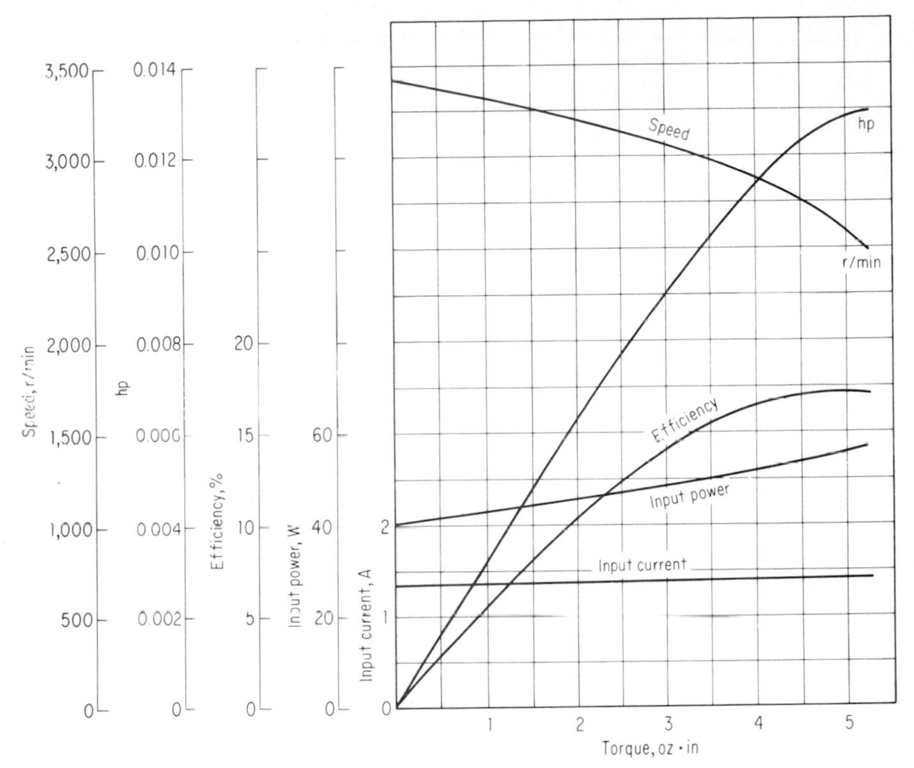

FIG. 25. Performance curve for two-pole shaded-pole motor with high-resistance rotor. *(John Oster Mfg. Co.)*

torque curve is thereby straightened out. Figure 25 shows the performance curves for a two-pole motor with a high-resistance rotor.

b. Control of speed. The speed of shaded-pole motors depends primarily on the line frequency and the number of poles. The speed of the shaded-pole motor can be varied by changing the voltage impressed across the stator winding by inserting a resistor or inductance in series with the stator winding, or tapping the stator winding. Another method of changing the speed of a shaded-pole motor is accomplished by putting a rectifier in the line and shunting the rectifier by a variable resistance (Ref. 1). By increasing the resistance (Fig. 26*c*), less current is bypassed around the rectifier and thus the direct current is increased. The direct current in the fields results in braking action to slow down the motor. In all these methods, the output of the motor is reduced and care must be taken not to overload the motor and burn it out. Figure 26 shows these three methods.

c. Availability of shaded-pole motors at different voltages and frequencies (other than 120 V, 60 Hz). Shaded-pole motors can be designed for higher voltages by changing the stator winding. For the ultimate efficiency it is desirable to redesign the rotor.

For the operation of shaded-pole motors on lower frequencies, such as 50 Hz, the motor speed will be 16 percent lower and the current higher, resulting in higher motor temperatures. Many shaded-pole motors are suitable for 50- and 60-

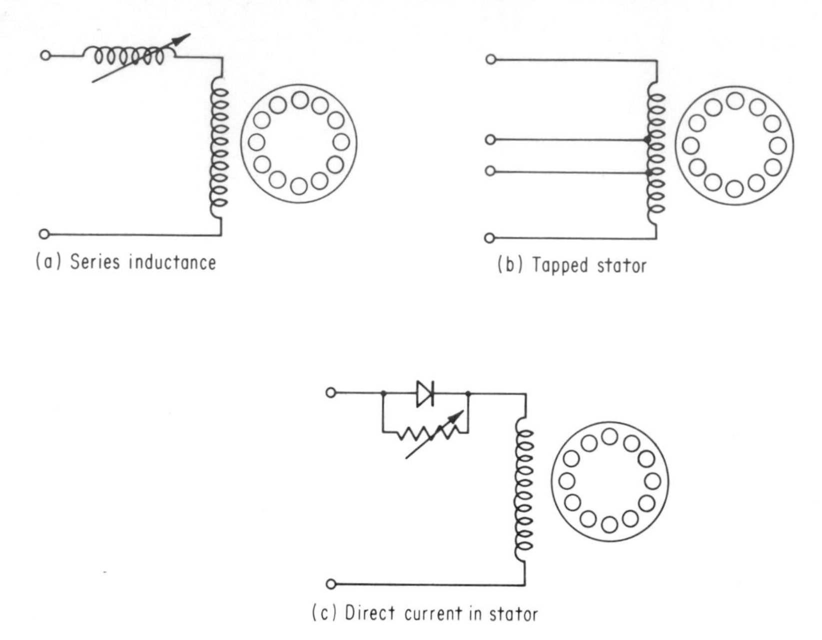

FIG. 26. Three methods of controlling the speed of a shaded-pole motor.

Hz operation. In some applications, however, a motor larger than the 60-Hz version will be required.

d. Lubrication and bearings. Most shaded-pole motors use oilite self-aligning bearings with felt oil reservoirs. See Fig. 21 for an example of the common lubrication system. Since the duty cycle in most shaded-pole motors is quite long, it is necessary to have large oil-reservoir felts and, in some cases, provision for adding oil externally.

e. Motor cooling. Since the speed of shaded-pole motors is relatively low (varying from 3550 r/min for a two-pole motor to 850 r/min for an eight-pole motor) compared with universal motors, the cooling-fan types are usually the centrifugal or vane types shown in Figs. 19 and 20.

The shaded-pole motor can be designed to have stalled-rotor protection by providing the stator winding with enough impedance (inductance and resistance) that the motor has the ability to withstand stalled-rotor currents for long periods of time. Because of the inherent high impedance, the shaded-pole motor has very good overload protection, and most shaded-pole motors can be stalled for several hours without damaging the motor.

f. Maintenance. The shaded-pole motor is almost maintenance-free. Since this is a commutatorless motor, there is no need for brush replacement or commutator refinishing. Since these motors have excellent inherent overload and stalled-rotor protection, they require very little replacement. The only serious maintenance requirement is the need for oiling the reservoir felts every 5 to 6 years.

g. Advantages of shaded-pole motors

1. The horsepower per dollar cost is very high. This motor lends itself to automated production.

2. Long life on a continuous duty cycle.
3. Very little maintenance.
4. Good overload and stalled-rotor endurance.
5. No radio and television interference.
6. Quiet operation. This motor is basically a low-speed motor.

h. Disadvantages of shaded-pole motors

1. Low starting and stalled torque.
2. Low motor efficiencies, 25 to 35 percent for two-pole motors.
3. Low horsepower per pound.
4. Low horsepower per cubic foot.
5. Speed control is expensive and difficult to achieve.
6. Availability of motors at different speeds is primarily dependent on the number of poles and the line frequency.

4. Magnetic Vibrating Motor. The magnetic vibrating motor is a very simple impulse-type motor. Figure 27 shows a typical magnetic vibrating motor used in a hair clipper. The solenoid coil produces an instantaneous field which results in poles that attract the movable armature. When the applied alternating current decreases to zero and reverses its polarity, the resulting solenoid poles decrease to zero and the armature moves back because of the energy stored in its spring. The cycle then starts all over again. Figure 28 shows curves of armature displacement and coil current. With a 60-Hz line frequency, the magnetic vibrating motor

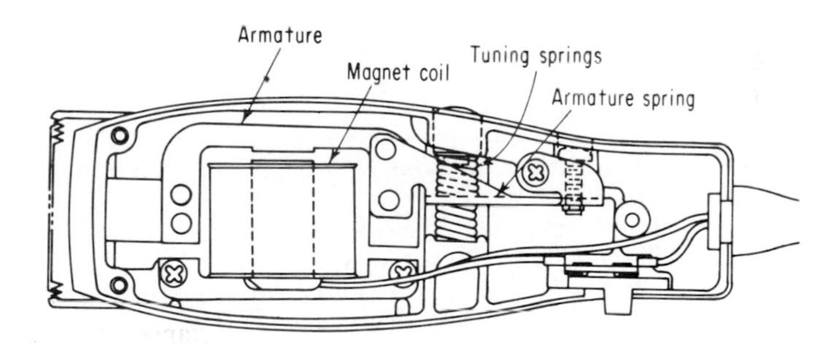

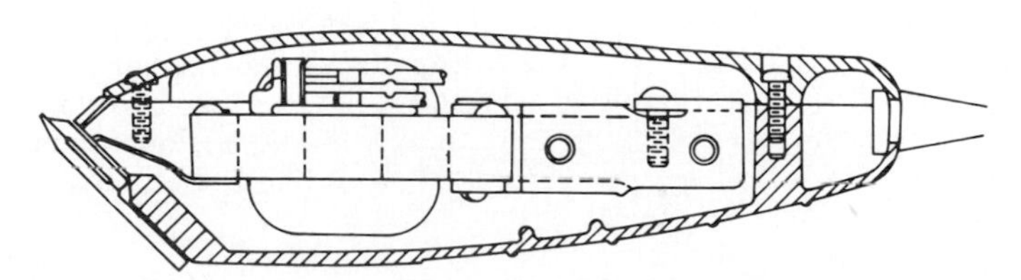

FIG. 27. Magnetic vibrating motor. *(John Oster Mfg. Co.)*

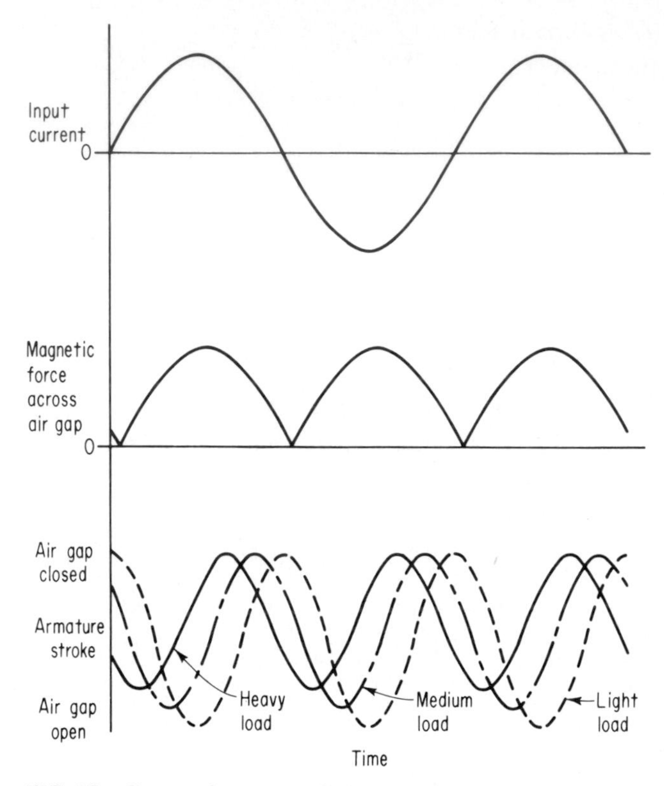

FIG. 28. Curves of armature displacement and coil current versus time for loaded magnetic vibrating motor.

moves at a frequency of 7200 cycles per minute, or 14,400 armature displacements per minute. With a rectifier in series with the circuit, the resulting speed will be 3600 cycles per minute.

There are magnetic-vibrating-type motors that employ permanent-magnet fields in the solenoid design. Figure 29 shows an example of this type of motor. The poles of the permanent magnet react with the coil on the right side to produce

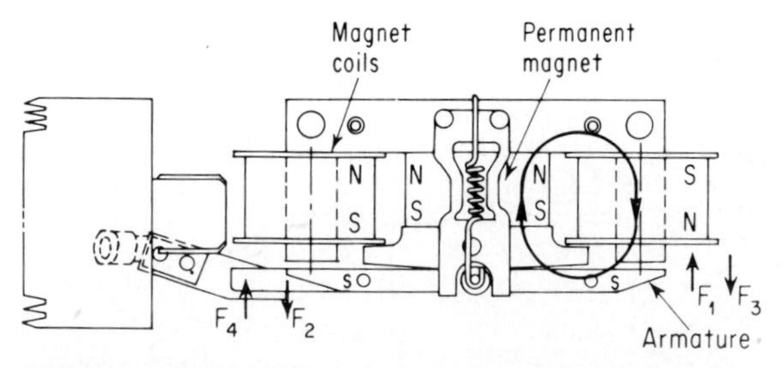

FIG. 29. Vibrating-type motors that employ permanent-magnet fields in solenoid design. *(Eisemann GmbH.)*

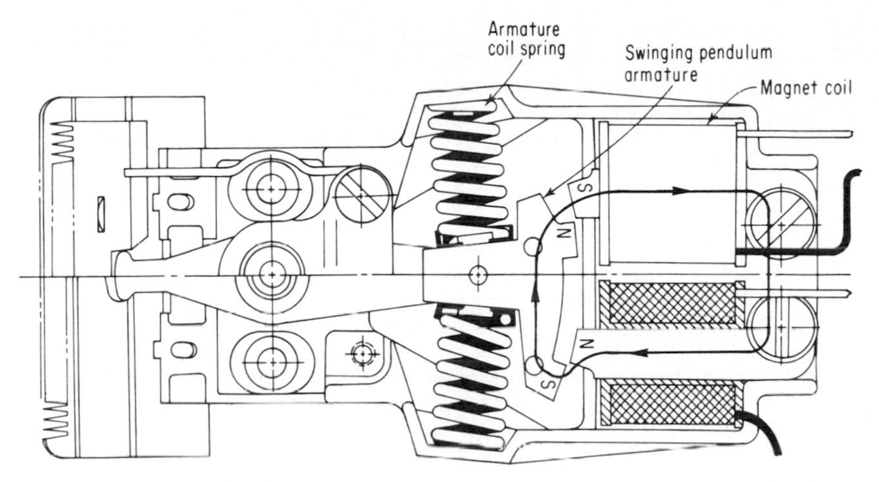

FIG. 30. Magnetic vibrating motor, pivoted swinging-pendulum type. *(Sunbeam Corp.)*

an attractive force F_1, and the permanent magnet reacts with the coil on the left to produce a repulsive force F_2 during the first half-cycle. This action is reversed during the next half-cycle, resulting in a rocking motion of the armature producing 3600 power cycles or strokes per minute.

These motors have a rectangular shape and produce translated motion. Magnetic vibrating motors are also designed as a pivoted swinging-pendulum type. Figure 30 shows one of these motors. This motor can also be designed to run at 3600 cycles per minute by adding a permanent magnet, as shown in Fig. 31.

a. Characteristic curves. The magnetic vibration motor produces oscillating translational motion such as that described in the hair clipper or magnetic sander. It is designed to produce a given force required to overcome the load and to oscillate back and forth with a required stroke. As the load is increased, the movement is reduced. The speed of the magnetic motor will remain constant. These motors are tailored for the application. The efficiency depends considerably on the configuration of the design and how well the mechanical spring system is tuned to the applied electrical frequency. The magnetic motors with permanent magnets are much more efficient.

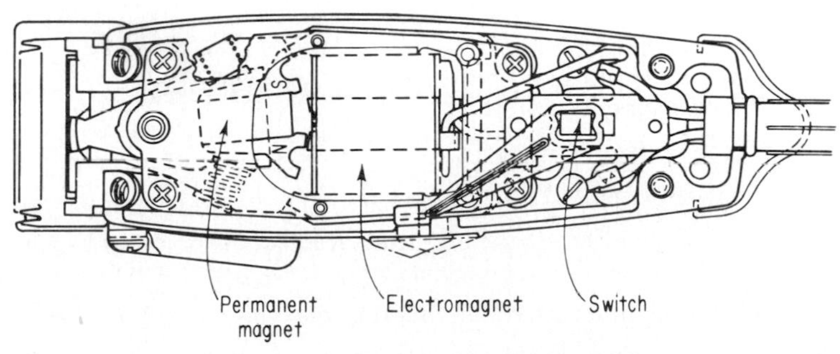

FIG. 31. Pivoted swinging-pendulum-type motor with permanent magnet. *(John Oster Mfg. Co.)*

b. Variation and control of speed. As mentioned in the description of this motor, the speed, or vibration frequency, can be varied by varying the applied frequency. It is common in this type of motor to obtain half-speed by use of a rectifier or a permanent magnet so that the motor runs at line frequency—3600 cycles per minute, or 3600 complete strokes per minute (over and back).

The amplitude of vibration or power output can be varied by tapping the magnet coil winding and varying the ampere-turns and the resulting magnet strength.

c. Availability of magnetic vibrating motors at different voltages and frequencies (other than 120 V, 60 Hz). The magnetic motor can be designed for different voltages by varying the magnet coil winding. If a rectifier is used, then the rectifier must be suitable for the higher voltages.

In operating a given magnetic motor designed for 50 to 60 Hz, the design must have sufficient cooling to compensate for the increased heating, and the spring system must be satisfactory. In most cases it is necessary to change the magnet coil winding and the spring system.

d. Lubrication and bearings. As can be seen in the sketches, these motors are very simple with very few moving parts. The simple magnetic-motor type has no bearings and requires no lubrication. The oscillating type has a pivot-type bearing made of sintered bronze saturated with oil that requires very little lubrication.

These motors are cooled by radiation and convection, much like other solenoid and relay-type structures.

e. Maintenance. These motors are virtually maintenance-free. There are very few moving parts and with the exception of the pivot-oscillating type, there are no bearings that require lubricating and few wear points in a complete motor.

f. Advantages of magnetic vibrating motors

1. Low cost. Because of the simplicity of design of the magnetic motor, it lends itself to mass-production techniques and the resulting low costs.
2. Maintenance-free. The simplicity of design and lack of moving parts result in an extremely long-life trouble-free motor.
3. Constant-speed motor. The speed of this motor depends upon the line frequency—only the amplitude varies with load.
4. No radio and television interference. Since there are no brushes and commutators, this motor does not generate any radio or television interference.
5. This motor is ideal for vibratory translated motion.

g. Disadvantages of magnetic vibrating motors

1. Limited speed control. Since the speed of these motors depends on the line frequency, the control of speed, without utilizing expensive electronic variable-frequency power supplies, is limited.
2. Efficiency is moderate.
3. The horsepower per pound and horsepower per cubic foot are quite low.
4. These motors depend to a large extent on free convection and radiation for cooling.
5. Magnetic vibrating motors have an inherent magnetic hum and vibration.
6. They are not readily available on the open market and must be custom-designed for the application.

5. Permanent-Magnet Motor. A permanent-magnet motor basically has a shunt-type dc motor characteristic with a permanent-magnet field. The most common construction has a salient-pole shape made of a permanent-magnet material such as alnico steel (nickel-cobalt material) or one of the newer ceramic magnets. Figures 32 and 33 show these types. Some special permanent-magnet motors designed by the German firm Dunker Motors have the permanent magnet located in the armature region, and a special cylindrical armature winding molded into an armature structure rotates about the magnet. This motor has a very small mechanical air gap and produces high efficiencies (Fig. 34).

FIG. 32. Alnico magnet. *(Indiana Steel Products Co.)*

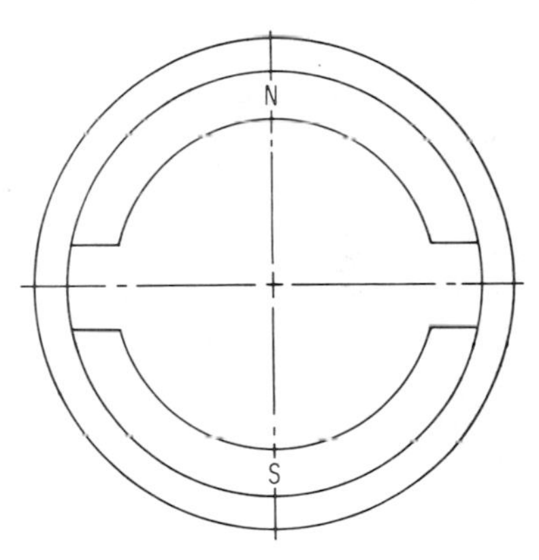

FIG. 33. Oriented ceramic magnet. *(John Oster Mfg. Co.)*

The permanent-magnet motor requires dc power for its operation, and thus this motor is operated from a dc power supply consisting of batteries or rectified alternating current.

With the advent of many improved large ampere-hour per pound rechargeable

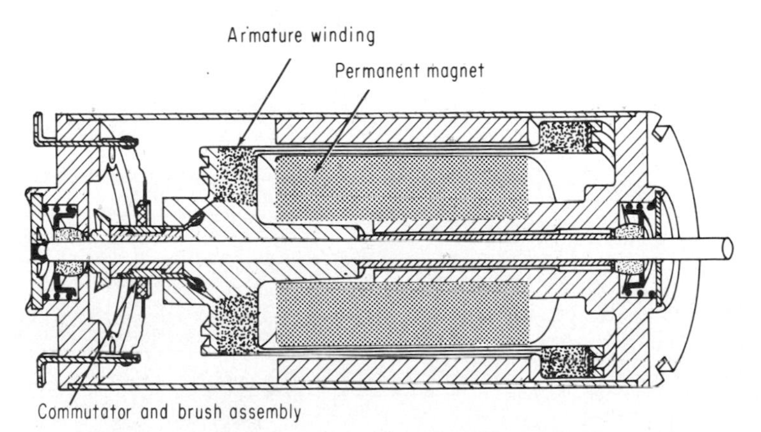

FIG. 34. Permanent-magnet motor with winding molded in thin shell. *(Christian Dunker Praezisions-Kleinstmotoren-Gesellschaft.)*

batteries, such as the nickel-cadmium or silver-zinc ores, more cordless appliances and tools are appearing on the market. Low-voltage permanent-magnet motors have many advantages such as ease in manufacture, no shock hazard, good commutation, little radio and television interference, low speed, and good horsepower per pound.

a. Performance. Figure 35 shows some typical performance curves for a permanent-magnet motor. These motors have high efficiency. The speed-torque curve has less slope than that of the universal-type motor and is not subject to as much speed variation as the universal motor. The permanent-magnet motor is a medium-speed type with loaded speeds of around 1500 to 8000 r/min.

b. Availability of permanent-magnet motors at different voltages and frequencies (other than 120 V, 60 Hz). As is true with all wound-motor types, this motor can be designed for a wide range of voltages. If the motor is designed for voltages greater than 24 V, the problems of motor insulation and commutation must be considered.

If a transformer is not used with a full-wave rectifier for the dc power, then the motor will operate satisfactorily over a wide range of frequencies. If a transformer is used, then the transformer must be suitable for the range of frequencies.

c. Speed control. The speed of these motors can be varied by varying the voltage applied to the armature through transformer taps, using a variable impedance in series with the armature or a solid-state pulsed silicon-controlled rectifier (thyristor) control. Another method of controlling the speed of permanent-magnet

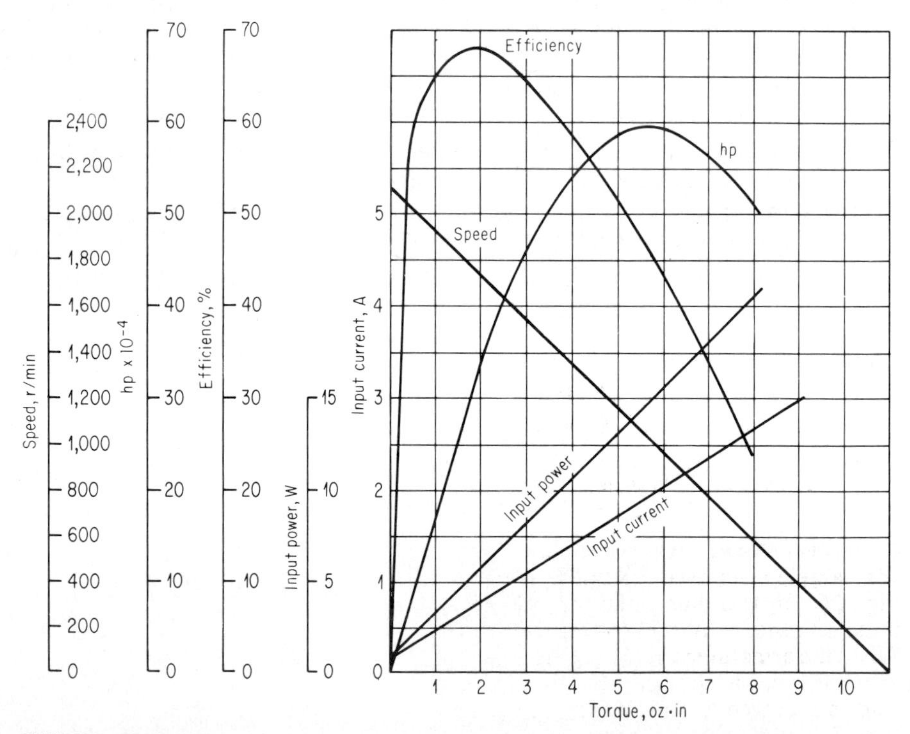

FIG. 35. Performance curves for permanent-magnet motor.

motors consists of varying the reluctance in the permanent magnet field, which in turn varies the field strength and results in a change in motor speed.

d. Lubrication and bearings. In the design of permanent-magnet motors for portable battery-powered appliances and tools, the designer must strive for the ultimate in efficiency. The shaft journals and bearings must be as small as possible to minimize the bearing friction. In many applications ball bearings are used. In battery-powered permanent-magnet motor applications adequate felt oil reservoirs must be provided and in many cases provision for external oiling.

e. Maintenance. Since this is a brush-type motor, it will be necessary to examine the brushes periodically and replace them. After several brush replacements, it will be necessary to refinish the commutator. Since these motors are normally a low-voltage high-current design, the brushes are copper or silver-graphite alloy designed for high conductivity. Since the commutating voltages are small and the commutator surface speed is low, the electrical and mechanical wear of the motor brushes is low, and it is not unusual to obtain a brush life of several thousand hours. Sintered-bronze oil-impregnated bearings are used in these motors, and provision for adding oil externally may be required.

f. Advantages of permanent-magnet motors

1. These motors find their major application in battery-powered products. The toy industry, each year, uses millions of the low-cost three-coil motors with inexpensive alnico magnets. In the past few years the more highly refined permanent-magnet motors are finding their place in many appliances such as the toothbrush, food mixer, ice crusher, portable vacuum cleaner, and shoe polisher, and in portable electric tools such as drills, saber saws, and hedge trimmers. Since the life of the battery is a big factor in battery-powered products, a very high-efficiency motor such as the permanent-magnet motor is needed.
2. The permanent-magnet motor is basically a medium-speed motor and produces very little air noise.
3. It produces very little radio and television interference when designed for low voltage (12 V and lower).
4. It has longer brush life than the universal motor and is suitable for longer-duty-cycle applications when designed for low power and speed.
5. Permanent-magnet motors have flatter speed-torque curves than the universal motor.

g. Disadvantages of permanent-magnet motors

1. The demagnetizing effect on the permanent-magnet field because of the armature ampere-turns may limit the size of this type of motor.
2. The motor runs on direct current and will not operate on alternating current unless rectifiers are used.

6. Impulse Motor. An impulse motor is a very simple ac-dc-type motor. As shown in Fig. 36, this motor runs on electrical impulses and must be started by hand. When the armature is in the position shown in the figure, the contacts are closed and a field is established that attracts the armature. The momentum carries the armature

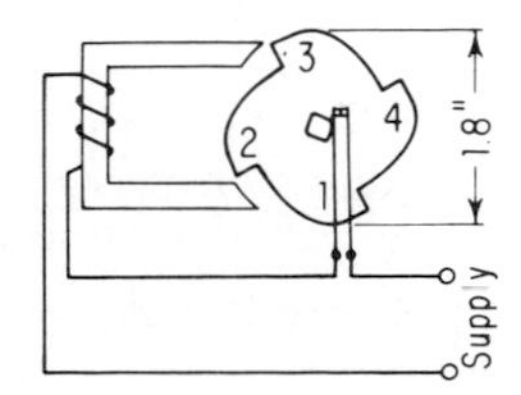

FIG. 36. Impulse motor. *(Spreadbury.)*

around until it advances to the next pole and the contacts close, producing another impulse. This motor is very inefficient and has only limited use. Many of these motors were used in the first electric shavers. The impulse motor is a very special type of miniature motor and is used only in special applications.

a. Performance. Figure 37 shows some typical operating curves for impulse motors. These motors have operating characteristics that are similar to those of universal motors. The need for contacts to interrupt the current limits the horsepower output of this motor, and thus it finds very limited application and has been used mainly for electric shavers.

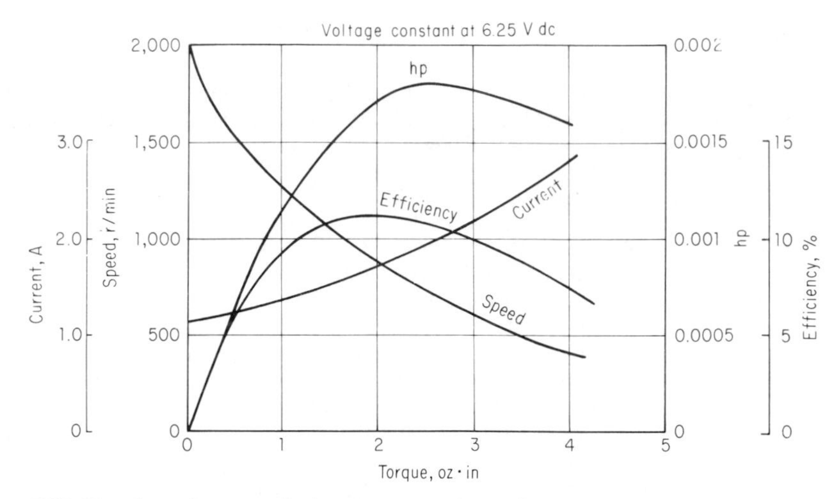

FIG. 37. Operating curve for impulse motor. *(Spreadbury.)*

b. Speed control. The speed of this motor can be changed by modifying the shape of the segmented rotor. The speed can be controlled by tapping the stator winding or inserting a variable inductance or resistance in series with the stator.

c. Availability of impulse motors at different voltages and frequencies (other than 120 V, 60 Hz). The impulse motor will operate on alternating and direct current. Because of the need for interrupting the current through contacts and obtaining good contact life, this motor is limited to the lower voltages and is not very practical to operate over 115 V.

d. Lubrication and bearings. The problems of lubrication for these motors are similar to those for the universal-type motor. Since the contact life limits its application to intermittent operation, the bearings and bearing oil reservoirs (felt) will be small. The short-duty-cycle use of these motors eliminates the need for fan cooling in most applications.

e. Maintenance. Since these motors are brushless, they require no brush or commutator maintenance. The contacts in the interrupter become pitted from use and do require replacement.

f. Advantages of impulse motors

1. Low cost. These motors have a very simple one-coil winding and no armature winding.
2. Satisfactory operation on alternating and direct current.

g. Disadvantages of impulse motors

1. Limited horsepower. The need for interrupting the line current with contacts limits the power that can be obtained from these motors.
2. Requires manual starting. The contacts must be closed before the motor will start.
3. Low starting and stalled-torque characteristics and poor motor efficiencies.
4. Interrupting the current with the contacts results in radio and television interference.

7. Stepping Motor. Stepping motors (Ref. 2) can be divided into two distinct types: the mechanical type, which consists of a solenoid supplying electromagnetic pull to drive a ratchet and pawl or inclined cylindrical cams to translate axial pull into shaft rotation, and the magnetic-electric-motor types. Figure 38 shows a mechanical-type stepping motor. There are numerous electric-motor types: the rotary transmitter follower, permanent-magnet rotor, permanent-magnet stator, variable reluctance, ac low-speed synchronous motor adaptable to stepper operation through control circuitry, and ac stepping motor.

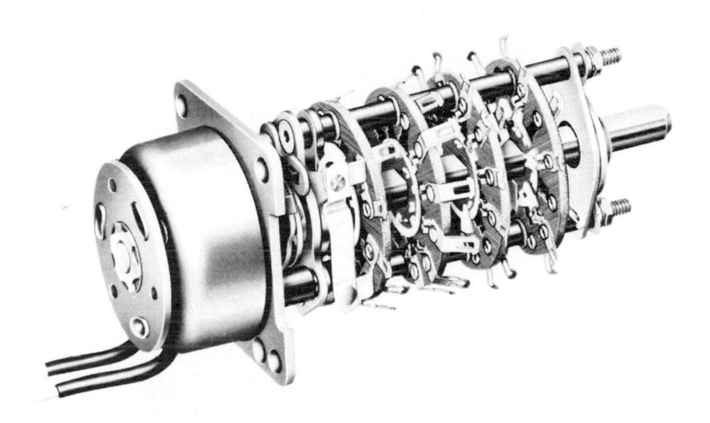

FIG. 38. Mechanical-type stepping motor. *(Oak Mfg. Co.)*

In all the electric-motor-type stepping motors the stator winding consists of one or more windings out of phase with each other and the rotors are of the salient-pole type. Permanent magnets can be used in the stator or the rotor.

Many stepping motors require dc input pulses to drive the motor, and in some designs it is necessary to switch the applied pulse from one coil to another. There are two or three types of stepping motors that are free-running and will operate on ac line voltage. Figure 39 shows a cross-sectional view of a free-running stepping motor which consists of a magnetic-vibrating-type motor with a permanent-magnet pole, which drives a shaft through a spring-loaded roller-type friction clutch. Figure 40 shows a low-speed synchronous-inductor-type motor that will operate on sinusoidal ac power or pulsed direct current. This motor is basically a four-pole split-phase motor with a toothed permanent-magnet rotor that runs at 75 r/min.

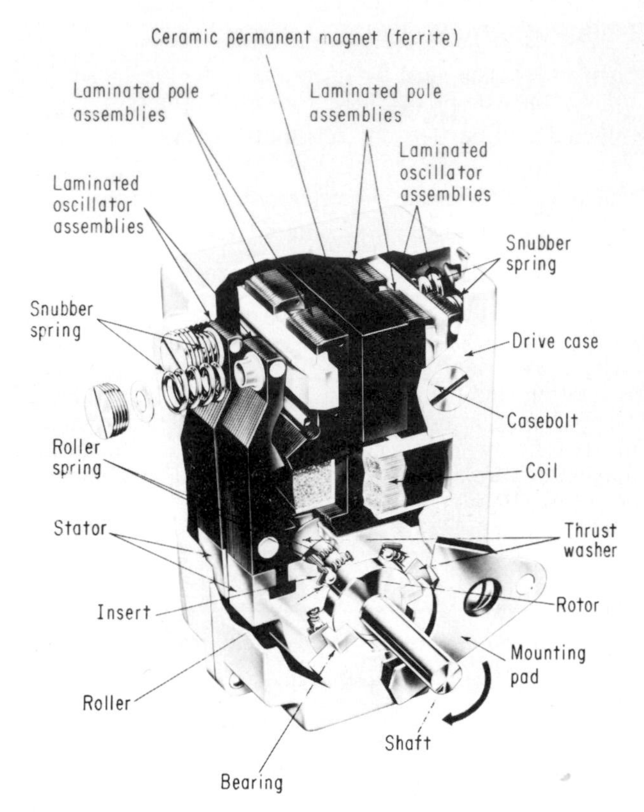

FIG. 39. Free-running stepping motor with permanent magnet. *(Enercon Conversion Systems Corp.)*

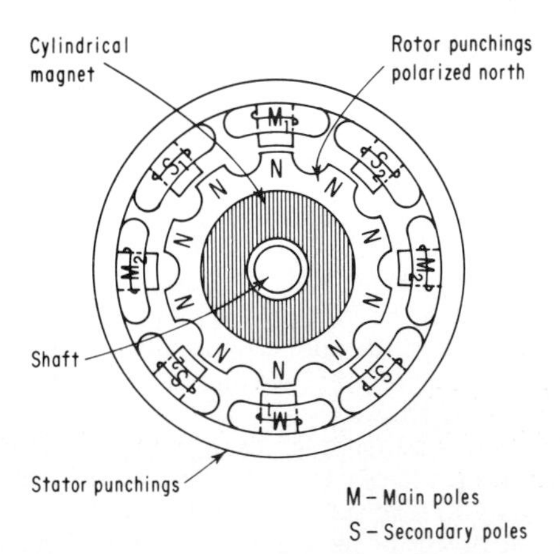

FIG. 40. Low-speed synchronous inductor-type stepping motor. *(General Electric Co.)*

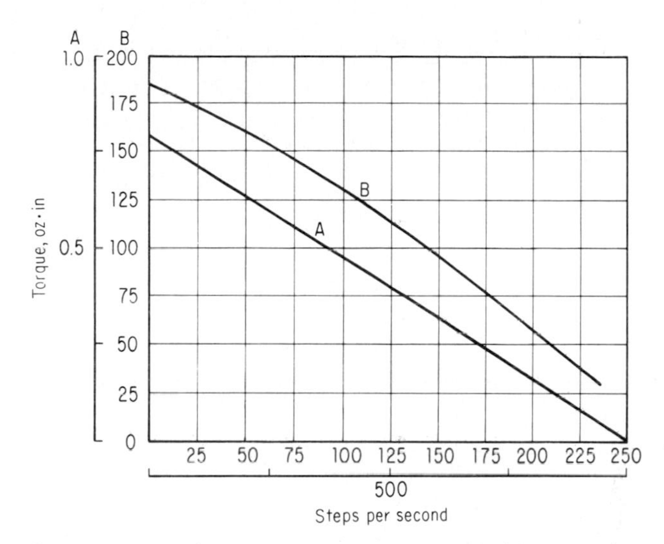

FIG. 41. Plots of torque versus steps per second. *(American Electronics Inc. and Superior Electric Co.)*

a. Performance. Stepping motors are rated in terms of the number of steps per second, the stepping angle and load capacity in ounce-inches, and the pound-inches of torque that the motor can overcome. Figure 41 shows two plots of pull-in torque versus pulses per second for variable reluctance and permanent-magnet-type stepping motors.

b. Speed control. The stepping speeds can be changed by varying the applied pulse rate or frequency of the power supply. A system of gearing can also be incorporated in the output of the motor to change the torque and speed characteristics.

c. Availability of stepping motors for different voltages and frequencies (other than 120 V, 60 Hz). Stepping motors are available for a range of dc and ac voltages and frequencies. In the free-running types a change in windings and the spring system is necessary for operation on different frequencies.

d. Lubrication and bearings. Since these motors operate by a stepping action, starting friction is very important. Since the output speed of these motors is very low, they cannot be cooled by putting a fan on the drive motor shaft. Stepping motors depend upon natural radiation and convention for cooling.

e. Maintenance. These motors are brushless and contain very few moving parts. Periodic relubrication of parts will be required. In motors that have clutches, it may be necessary to adjust the clutch periodically and make replacement after long periods of operation.

f. Advantages of stepping motors

1. Rapid acceleration, stopping, and reversing.
2. Certain types are simple low-speed drive devices without gearing.
3. Low rotor moment of inertia.
4. No radio or television interference.
5. External pulse control.
6. Alternating current or direct current.

g. Disadvantages of stepping motors

1. Mechanical types require pulse control and are not free-running.
2. Mechanical types have mechanical parts such as springs, cams, and latches, which can wear and require maintenance.
3. Some of the free-running types have inherent hum and vibration.
4. To control speed, a variable frequency or pulse rate is required.
5. The cost in dollars per horsepower is greater than for shaded-pole or magnetic-vibrating-type ac motors.
6. Free-running types operate on alternating current only.

8. Synchronous Motor. A synchronous motor runs at the synchronous speed of the rotating field:

$$\text{Synchronous speed} = \frac{120f}{P}$$

where f = line frequency
P = number of poles

Conventional-type synchronous motors employ a three-phase, two-phase, or split-phase type electrically rotating stator field and a dc-excited mechanically rotating field. This type of synchronous motor generally has a rotor which contains a squirrel-cage winding and salient poles which are excited with direct current. It is started as an induction motor, and then the dc power is applied to the rotating field and the rotor locks into synchronous speed.

In the unexcited hysteresis-type synchronous motor the rotor field is produced by the hysteresis or reluctance effect in the rotor. The rotor poles lag behind the corresponding polar position of the exciting field and produce a resulting torque. This lag is due to hysteresis, and the greater the hysteresis effect, the larger the output torque.

Figure 42 shows a cross-sectional view of the rotor of a hysteresis-type motor. Materials that have large hysteresis losses such as alnico and other nickel-cobalt materials commonly used in permanent magnets are used for the active material in the rotor.

There are two other types of nonexcited-type synchronous motors: the reluctance and the polarized-cylindrical-rotor types. The reluctance type has a spider-web-shaped rotor with two or more poles. The flux becomes concentrated in the crossarms of the rotor and permanently magnetizes the poles, which lock into synchronism with the ac rotating field (Fig. 43).

The polarized-cylindrical-rotor synchronous motor consists of a conventional squirrel-cage induction rotor in tandem with a permanent magnet having as many poles as the stator. The motor starts as a squirrel-cage induction motor, accelerating to a speed near synchronism, and then the permanent magnet locks into synchronism with the rotating field (Ref. 3) (Fig. 44).

a. Performance. Figure 45 shows typical motor-performance curves for these three types of unexcited synchronous motors. The polarized-cylindrical-rotor type that employs a permanent-magnet rotor has improved efficiency and operating performance.

b. Speed control. Since this motor operates at synchronous speed, the speed can be varied by changing the line frequency or the poles.

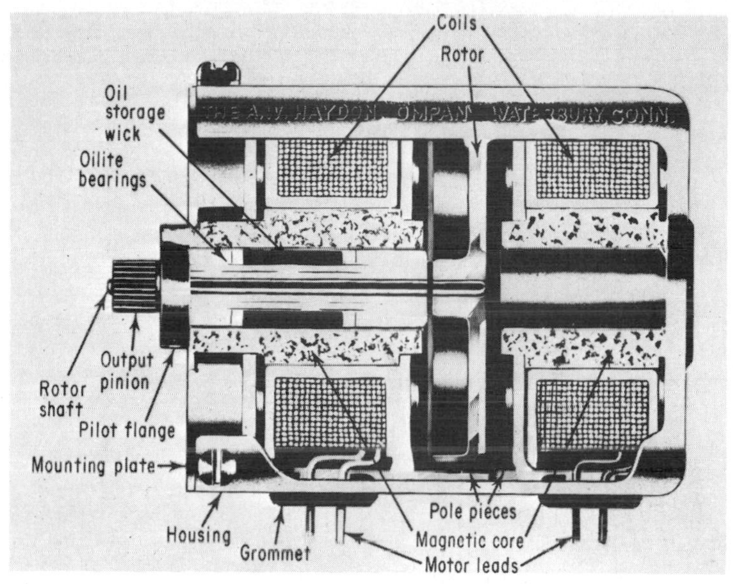

FIG. 42. Cross-sectional view of rotor of a hysteresis-type synchronous motor. *(A. W. Haydon Co.)*

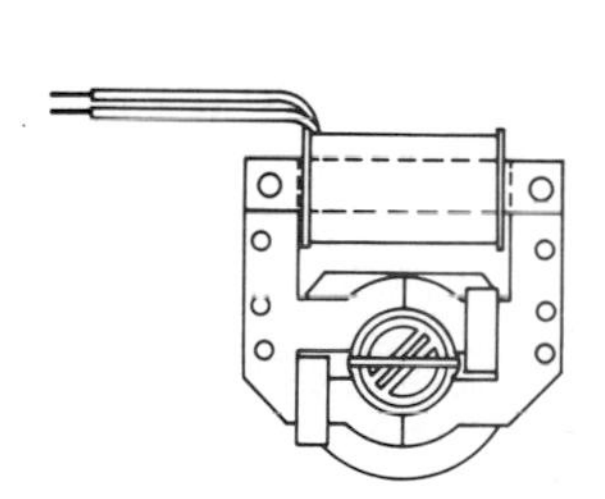

FIG. 43. Reluctance-type synchronous motor. *(Sunbeam Corp.)*

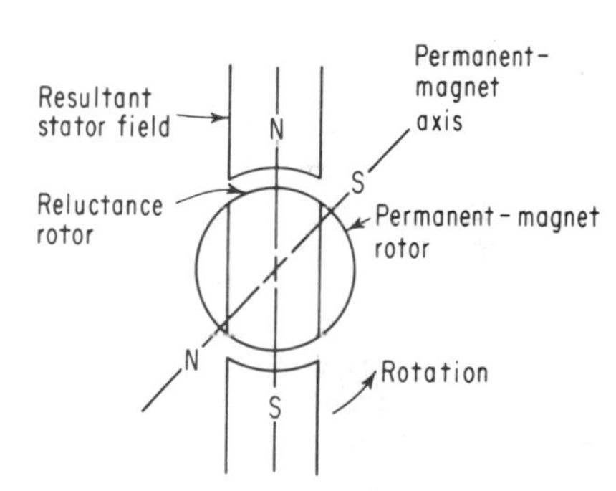

FIG. 44. Polarized-cylindrical-rotor synchronous motor. *(Rotodyne Corp.)*

c. Availability of synchronous motors at different voltages and frequencies (other than 120 V, 60 Hz). The motor can be designed for different voltages by changing the stator winding. Operating the motor on other frequencies, such as 50 Hz, will affect the speed and cooling of the motor and may require an increase in size at the lower frequencies.

d. Lubrication and bearings. This motor requires the same type of bearing used in shaped-pole motors, the oilite-type large felt oil reservoirs, and provision for external oiling. Most small clock motors use no fan cooling and depend on free radiation and convection for cooling.

e. Maintenance. The small synchronous motor requires very little maintenance except the need for adding oil to the bearings periodically.

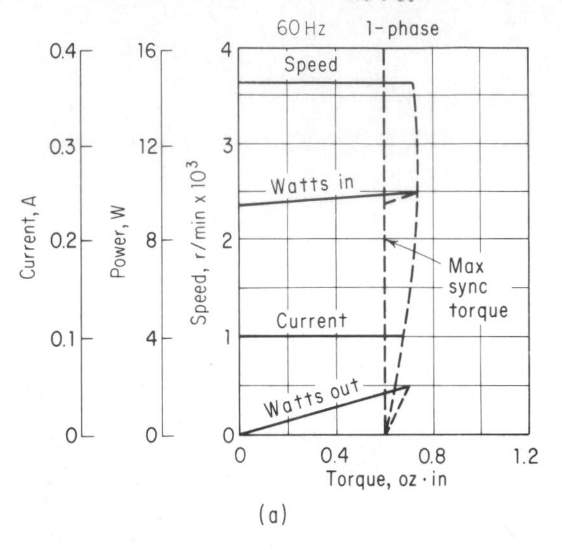
115 V ac
60 Hz 1-phase
Current, A
Power, W
Speed, r/min x 10^3
Speed
Watts in
Max sync torque
Current
Watts out
Torque, oz · in

(a)

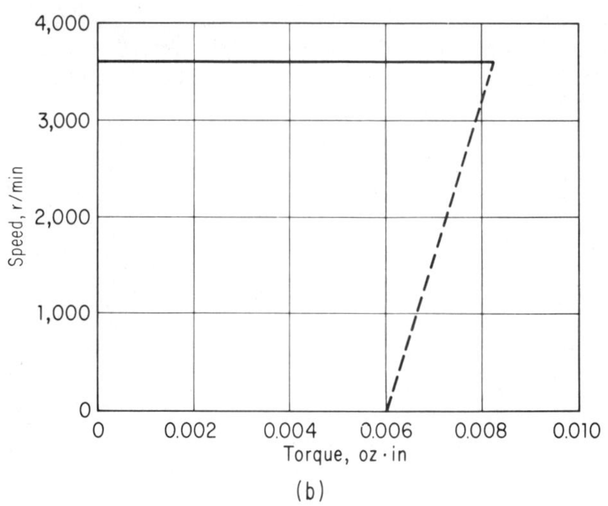
Speed, r/min
Torque, oz · in

(b)

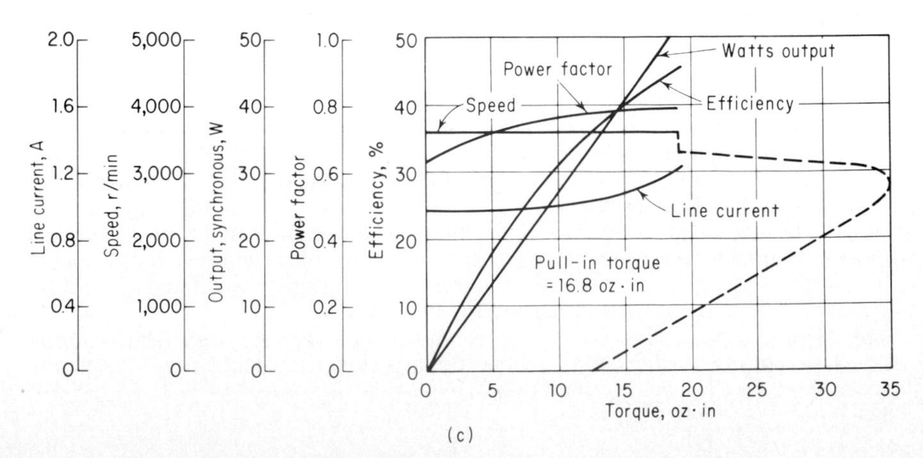
Line current, A
Speed, r/min
Output, synchronous, W
Power factor
Efficiency, %
Watts output
Power factor
Speed
Efficiency
Line current
Pull-in torque
= 16.8 oz · in
Torque, oz · in

(c)

f. Advantages of synchronous motors

1. Constant-speed motor

g. Disadvantages of synchronous motors

1. Low starting torque
2. High cost for medium- and high-power-output motors.
3. Ac operation only
4. Difficult to vary motor speed

9. Printed-Circuit Motor. These low-voltage dc motors have their armature (conductors) windings and commutators printed on a thin disk of insulating material such as laminated phenolic. In some designs the etched-copper conductors are printed on both sides of the laminated disk, and in other designs the winding is printed on one side only. At present, printed-circuit motors are of the permanent-magnet-field type (Fig. 46).

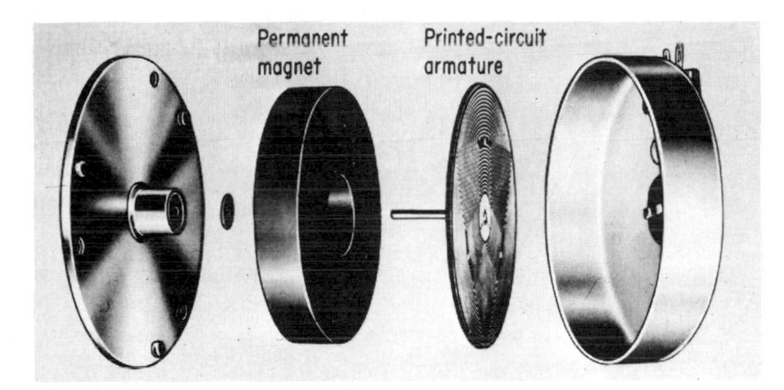

FIG. 46. Printed-circuit permanent-magnet motor. *(Haydon Switch and Instrument, Inc.)*

Since the number of conductors that can be printed in a given diameter is limited by the required current capacity and spacing between conductors required for insulation, these motors are limited to lower-voltage application. A typical motor has an inherently large diameter and is a short stacked (pancake) type of motor.

The maximum horsepower that can be designed into these motors is dependent on the degree of cooling. Excessive currents can cause thermal stresses and resulting distortion of the disk-type armature.

a. Performance. Since the armature has no electrical steel, the mass is low and inertia forces are small. They can accelerate and decelerate rapidly.

Since this motor is a special type of permanent-magnet motor, the characteristics are quite similar to those of the conventional permanent-magnet motor (Fig. 47). The speed-torque curve is flatter than that of the universal motor. The line current increases with increased load.

FIG. 45. Performance curves for (*a*) hysteresis, (*b*) reluctance, and (*c*) polarized-cylindrical-rotor motors. (Hysteresis-type synchronous motor, *Globe Industries Inc.*; reluctance-type synchronous motor, *Sunbeam Corp.*; polarized synchronous motor with rotor flats shifted 45 electrical degrees in direction of rotation, *Rotodyne Corp.*)

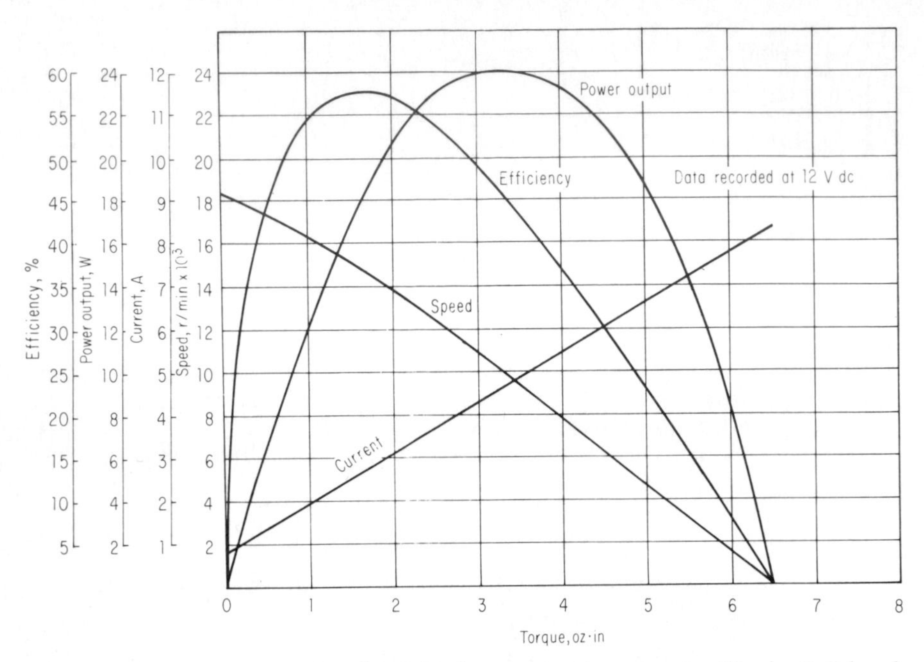

FIG. 47. Performance curves for printed-circuit permanent-magnet motor. *(Haydon Switch and Instrument, Inc.)*

b. Control. Since this motor is designed with a permanent-magnet field, the speed must be controlled by varying the applied armature voltage or controlling the amount of current passing through the armature.

c. Availability of printed-circuit motors at different voltages and frequencies (other than 120 V, 60 Hz). This motor presently is being offered as a low-voltage dc motor. Because of the large diameter of the armature and the limitation of the mechanical strength of the glass or cloth phenolic disk material, it is limited to lower voltages, 75 V direct current and lower. Limited winding space dictates low back emfs.

d. Lubrication and bearings. Since the weight of the rotating armature is very small, the load on bearings is quite low. The lower operating speeds (15,000 r/min and lower) permit the use of ball bearings. Provision for relubricating the ball bearings or adding oil to a phosphor-bronze sleeve or oilite bearing depends on the duty cycle.

Since these motors have high efficiency, fan cooling is not provided in many applications.

e. Maintenance. The motor has brushes that require periodic inspection and replacing. The commutator is also printed on the armature disk. Since the thickness of the printed armature and commutator section is limited in thickness to several thousandths of an inch, the amount of wear and life of the armature is limited and it will require replacing. For a long-duty-cycle application, oil or grease must be added to the bearings.

f. Advantages of printed-circuit motors

1. High motor efficiency.
2. Simplified armature construction.

3. Low profile, pancake shape. Very attractive for large-diameter short-height shapes.
4. Small inertia and low acceleration and deceleration times make application for controls very attractive.
5. Minimum of radio and television interference because of low-voltage dc design.
6. Important motor consideration for battery-powered products where a large-diameter low silhouette is desirable.

g. Disadvantages of printed-circuit motors

1. Low horsepower per pound and per cubic inch.
2. Limitation on voltage (low voltage, 75 V and below, direct current).
3. Low horsepower per dollar, high-cost motor.
4. Limitation on horsepower as a result of difficulty in cooling the armature disk.
5. Short armature life.
6. Best suited for intermittent-duty cycles.

10. Other Miniature Motors. Some other miniature types for special applications follow.

a. Eddy-current motor found in watthour meters. This motor consists of a shaded-pole stator and a flat cylindrical disk-type rotor. Its principle of operation is similar to that of the shaded-pole motor. Figure 48 shows an eddy-current motor.

b. Solid-state oscillator drive for a shaded-pole motor. One manufacturer makes a motor which they call brushless electronic variable-speed shaded-pole motor (Fig. 49).

c. Commutatorless transistorized motors. Several attempts have been made to eliminate the commutator and brushes in universal motors by using transistor switching circuits to obtain the proper current flow in the armature, thus elimi-

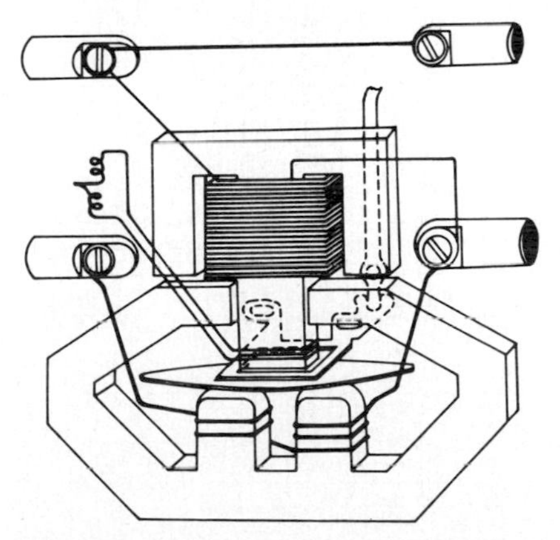

FIG. 48. Eddy-current motor. *(Chester L. Dawes,* A Course in Electrical Engineering, *4th ed., McGraw-Hill, New York, 1952.)*

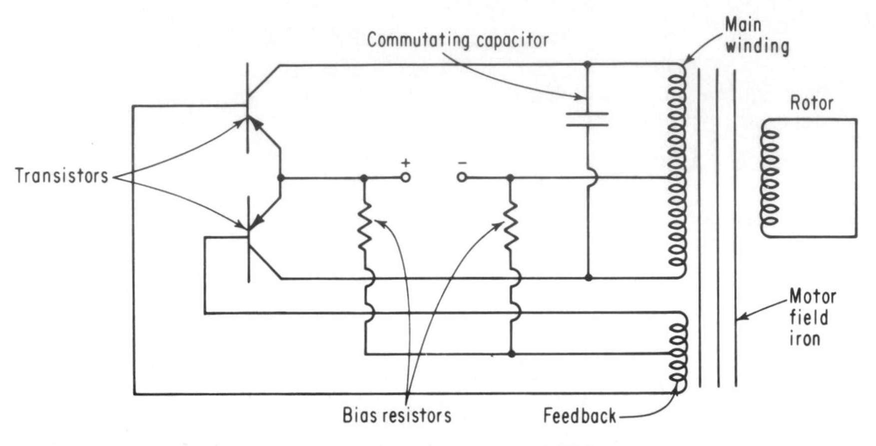

FIG. 49. Brushless electronic variable-speed motor. *(Lamb Electric.)*

nating the brushes and commutator, which are a big source of trouble. To date, the cost of the electronic circuitry limits the application of this motor (Fig. 50).

d. High-frequency (180-Hz induction motors). High-frequency motors have been designed for use in naval and other military applications where communication interference cannot be tolerated. These motors are of the three-phase induction-motor type. Large amounts of power can be obtained in small sizes. They have replaced the universal motor in military portable-tool applications (Ref. 4) (Figs. 51 and 52). These motors require a high-frequency generator. The

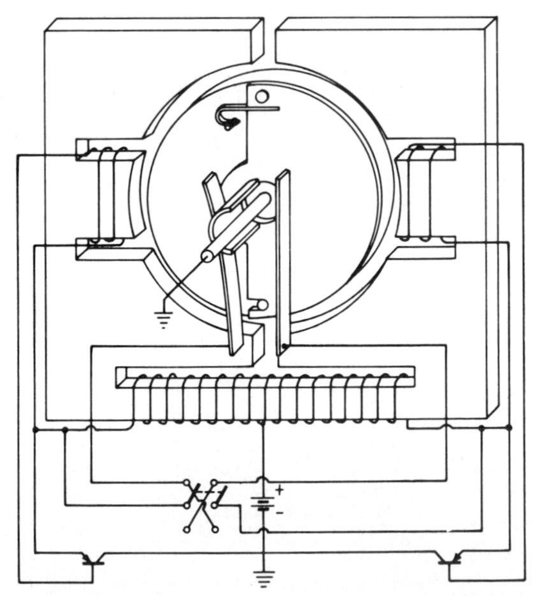

FIG. 50. Commutatorless transistor motor. *(H. D. Brailsford, U.S. Patent 2,753,501.)*

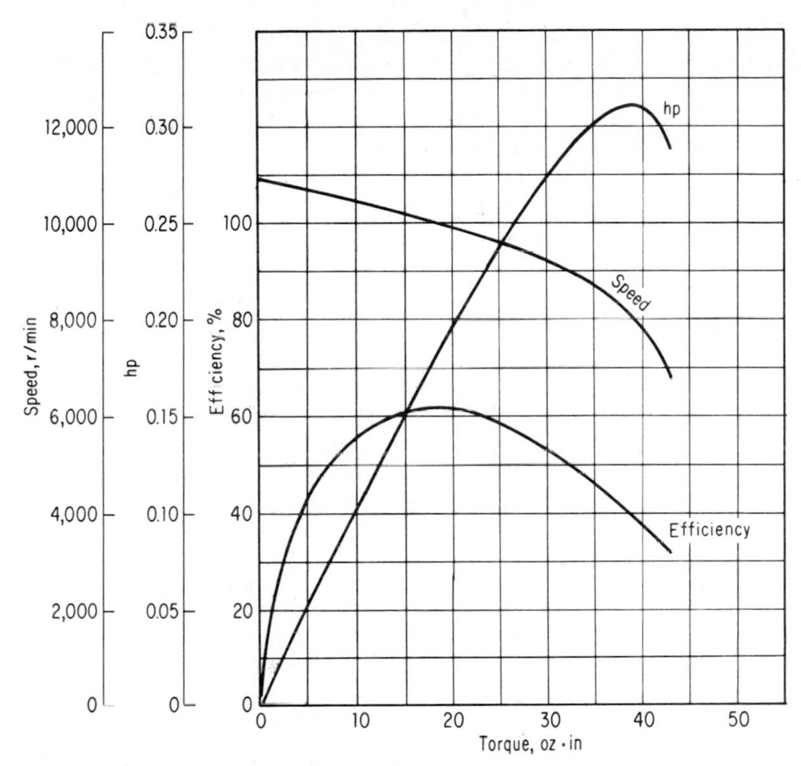

FIG. 51. Performance curves for 180-Hz induction motor. *(Robbins & Myers, Inc.)*

FIG. 52. 180-Hz induction motor. *(Robbins & Myers, Inc.)*

extra cost of the generating equipment and the cost of this special high-frequency induction motor limit the application.

e. Display motors. Hundreds of thousands of battery-powered dc impulse-type motors are used every year in the advertising business.

Figure 53 shows an oscillating type. When the motor magnet coil is energized, a dc field is established (Fig. 54). A magnetic-repulsion force is produced between the poles of the electromagnet and the permanent-magnet armature, causing the armature to swing outward in an arc. At the same time the circuit of the electromagnet is broken, stopping the drain on the battery. Momentum carries the swinging armature to the top of its arc. As the rocker arm moves downward and passes through the coil, the circuit is closed and the electromagnet is reenergized and repels the armature in the opposite direction. The swinging armature reaches to the starting position and the cycle starts all over again. Figure 55 shows another style of dc impulse motor that drives a set of gears to produce rotating motion. Table 2 shows performance data for this type of motor.

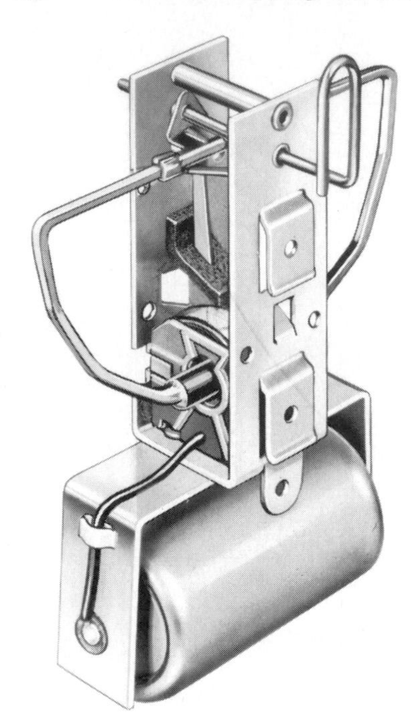

FIG. 53. Oscillating-type dc impulse motor. *(Hankscraft Company.)*

11. Small Specialty Motors, Selection Parameters. These motors are of special types and in some cases, such as the magnetic vibrating types, are not readily available on the open market. In most applications these motors are designed to do a specific job and become an integral part of the design.

Some of the common pitfalls to be avoided in selecting a small specialty motor are as follows:

1. Do not select an intermittent-duty motor to perform a continuous-duty job.
2. Become familiar with the safety laboratories' requirements for motors used in product design such as Underwriters Laboratories, the Canadian Standards Association, and Semko (Swedish).
3. Provide for adequate motor cooling.
4. Give serious consideration to starting requirements under all types of weather conditions, and select a motor that will handle the requirements.
5. Do not try to use a shaded-pole-type motor for heavy-starting-torque application.

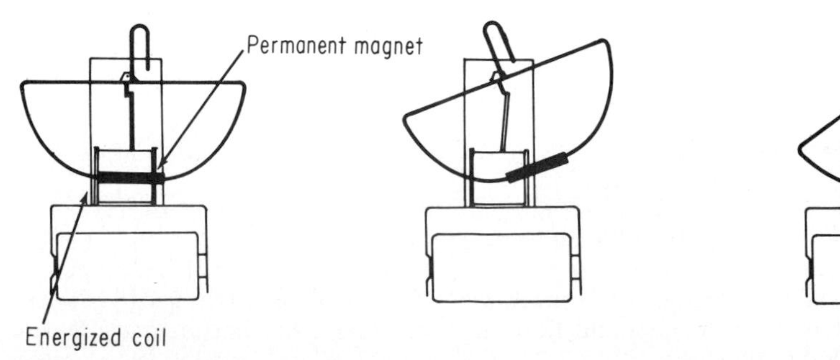

FIG. 54. Sketches showing operation of oscillating dc impulse motor. *(Hankscraft Company.)*

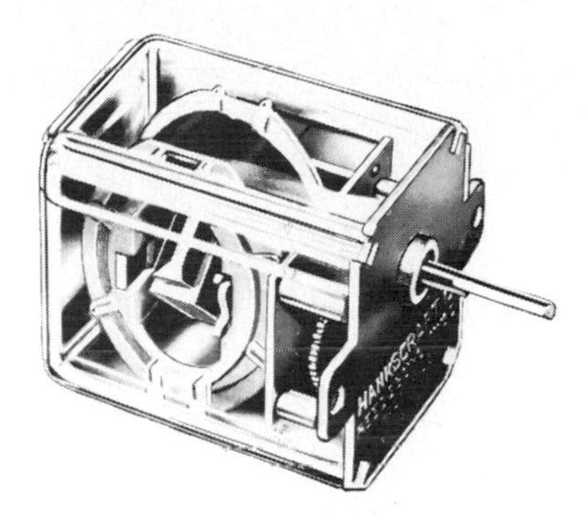

FIG. 55. Oscillating dc impulse motor with gears to produce rotary motion. *(Hankscraft Company.)*

TABLE 2. Typical Display-Type-Motor Characteristics

Volts, dc	Current, mA	Watts input	Output-shaft speed, r/min	Armature speed, r/min	Output-shaft torque, oz · in	Hp output $\times 10^{-3}$	Watts output	Efficiency, %	Gear ratio
3	11.2	0.03	1.94	1,810	0	0	0	0	932.24:1
3	17.0	0.051	1.66	1,555	5	0.0083	0.006	11.7	
3	23.0	0.069	1.42	1,325	10	0.014	0.0104	15.1	
3	27.5	0.083	1.20	1,110	15	0.018	0.0134	16.2	
3	34.5	0.104	0.99	910	20	0.020	0.0149	14.3	
3	41.0	0.123	0.78	725	25	0.0195	0.0145	11.8	
3	76.0	0.228	Stalled	Stalled	32	0	0	0	

6. Give careful thought to the method of cooling the motor. Designing the product with the best possible cooling fan and directing the air through the motor windings will result in the lowest overall motor cost.
7. Be sure to check the voltage and frequency of the country in which the product is sold, and use the motor designed for this system voltage.
8. Where size and weight are not factors and the power requirements can be fulfilled by using commutatorless motors, such as the shaded-pole, synchronous, magnetic vibrating, stepping, and impulse types, the motor cost and maintenance problems will be less with these ac-type motors.

REFERENCES

1. F. G. Spreadbury, *Fractional Horsepower Electric Motors, Their Principles, Characteristics and Design* (Pitman, Toronto), pp. 231–233.
2. John Proctor, "Stepping Motors Move In," *Prod. Eng.*, pp. 74–88 (Feb. 4, 1963).
3. Joseph Winston, "The Polarized Synchronous Motor," *Electro-Technol.*, pp. 92–95 (July 1963).
4. T. C. Lloyd, "Motor Parts Will Be Important to Postwar Design," *Elec. Mfg.* (1943).
5. Cyril G. Veinott, *Fractional and Subfractional Horsepower Electric Motors* (McGraw-Hill, New York, 1970).

SECTION 10
SPECIAL-APPLICATION MOTORS

FOREWORD 10-3

PART 1. INTEGRAL-HORSEPOWER SYNCHRONOUS RELUCTANCE MOTORS 10-4

1. Synchronous Reluctance Integral-Horsepower Motors 10-4
2. Performance of Synchronous Reluctance Motors . . . 10-5
3. Factors to Consider 10-7
4. Constant-Speed Applications 10-8
5. Adjustable-Speed Applications 10-9

PART 2. BRUSHLESS SYNCHRONOUS MOTORS 10-10

6. Basic Equations 10-10
7. Construction 10-10
8. Brush-Type dc-Excited Synchronous Motors 10-10
9. Brushless dc-Excited Synchronous Motors 10-10
10. Advantages of Brushless Synchronous-Motor Systems 10-14
11. Brushless Synchronous-Motor Control 10-14

PART 3. BATTERY-POWERED MOTORS FOR INDUSTRIAL APPLICATIONS 10-15

12. Power Source 10-15
13. Basic Motor Requirements 10-16
14. Motor Enclosures 10-16
15. Motor Ratings 10-19
16. Motor Insulation and Temperatures 10-19
17. Motor Voltages 10-20
18. Motor Characteristics 10-20
19. Motor Design 10-22
20. Dual Series Field 10-25

PART 4. ENGINE-CRANKING MOTORS 10-26

21. Motor Sizes 10-26
22. Engine-Cranking Requirements 10-26
23. Cranking-Motor Performance Characteristics 10-27
24. Motor Construction 10-30
25. Engagement Methods and Mechanisms 10-32

PART 5. MINING MOTORS 10-34

26. Auxiliary Motors 10-34
27. Face Motors 10-34

28. Power Supply 10-34
29. Nameplate Rating 10-35
30. Dc Motor Characteristics 10-35
31. Ac Motor Characteristics 10-36
32. General-Mine-Duty Requirements 10-36
33. Explosionproof Requirements 10-37
34. Mine-Service Requirements 10-37
35. Enclosures 10-37
36. Mounting and Coupling Arrangement 10-38
37. Bearings 10-38
38. Shaft Seals 10-38
39. Lubricant 10-39
40. Rotating Assembly 10-39
41. Stator Assembly 10-40
42. Brush-Support Assembly 10-40
43. Insulation and Winding Techniques 10-40

PART 6. TEXTILE MOTORS 10-42
44. Code 10-42
45. Motor Temperature 10-43
46. Enclosures 10-43
47. Torque 10-44
48. Speed Control 10-48
49. Operating Conditions 10-49

PART 7. NAVY AND MARINE MOTORS 10-51
50. Shipboard Service 10-51
51. Philosophy of Specifications 10-51
52. Considerations for All Shipboard Motors 10-52
53. Commercial Marine Motors 10-53
54. Navy Motors 10-53
55. Service C Motors 10-61

PART 8. CRANE MOTORS 10-62
56. Motor Requirements 10-62
57. Applications 10-62
58. Operating Conditions 10-63
59. Rating Crane Motors 10-63
60. Safety and Reliability 10-63

PART 9. ELECTRONICALLY COMMUTATED MOTORS 10-65
61. Shaft-Position Sensing 10-65
62. Electronics 10-65
63. Applications 10-66

PART 10. PERMANENT-MAGNET-EXCITED SYNCHRONOUS MOTORS 10-74
64. Rare-Earth Magnets 10-74
65. Performance 10-75
66. Applications 10-75

FOREWORD

These motors, not fully covered in other sections, are in the integral-horsepower and larger sizes and are built for specific applications. While these motors are generally built under special specifications, they are nevertheless important and are applied in considerable quantity. This section gives background information on these motors and the special considerations that make them different from motors covered in other sections.

PART 1

INTEGRAL-HORSEPOWER SYNCHRONOUS RELUCTANCE MOTORS

Paul D. Wagner*

1. Synchronous Reluctance Integral-Horsepower Motors. Automated processes require a large number of simple and efficient motors which operate at a fixed speed between no load and full load, even under adverse conditions such as voltage fluctuations or temperature variations. Reluctance integral-horsepower motors were developed to fill this requirement. Most of these applications, 80 percent or more, are provided with NEMA frames 140, 180, and 210 in the two-, four-, and six-pole configurations.

The enclosure parts and the three-phase stators of these motors are similar to those of the standard squirrel-cage motors. The rotor has a simple die-cast aluminum cage winding and requires no brushes. The rotor laminations (Fig. 1) have special slot arrangements which serve as flux guides as well as flux barriers in the interior of the rotor. Near the periphery of the rotor the usual induction-motor rotor bars are located. The lamination stack is die-cast in the manner of squirrel-cage induction motors. Except for the shaft hole, all rotor interior openings are filled with aluminum; so are the rotor "saliencies," which results in a cylindrical body with end rings, balancing lugs, and ventilating fan blades forming a homogeneous structure, in a manner similar to induction motors.

The conventionally excited synchronous motor is characterized by its constant speed when operating on a fixed-frequency power system, regardless of load and voltage. Unlike a reluctance motor, it uses windings on the rotating field poles and requires dc excitation through brushes and slip rings to these windings. Usually this means that an exciter or dc power supply is required to produce the dc excitation. Exciters and power supplies add to the complexity and the maintenance needs of this device. It is this complicated mechanical construction of the excited synchronous motor that has limited its general use to applications above 100 hp.

On the other hand, the industrial polyphase induction motor is characterized by rugged construction, particularly in the rotor. The rotor windings usually are die-cast, and the absence of rotor-conductor insulation precludes a limited life because of excessive temperatures. No brushes, slip rings, or rotating rectifiers can cause trouble; and the rotor does not require dc excitation as in the case of the synchronous motor.

However, the induction motor does not operate at constant speed, but changes speed with load and voltage. In fact, it is difficult to find two induction motors that will operate at exactly the same speed, even when operated on the same power line, that is, on identical voltage and frequency, and driving identical loads.

*Manager, Advanced Technology, Industrial Motor Division, Siemens Energy and Automation, Inc., Little Rock, Ark.; Senior Member, IEEE.

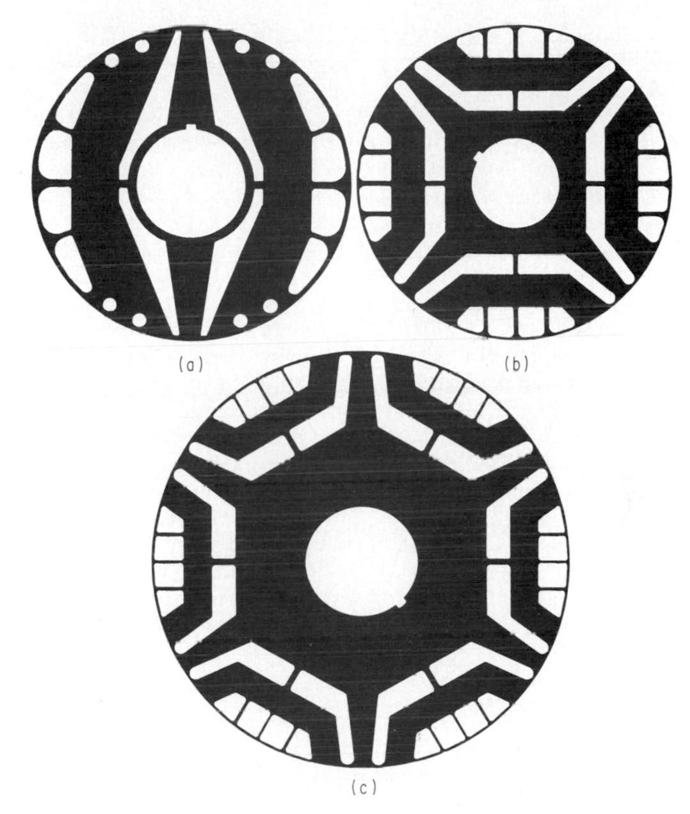

FIG. 1. Rotor laminations for (*a*) two-pole, (*b*) four-pole, and (*c*) six-pole synchronous reluctance motors.

A large number of applications have required the fixed-speed characteristics of the synchronous machine combined with low first cost, rugged mechanical construction, and simplicity of maintenance. Developments in recent years in some of the process industries have made this need more apparent than ever.

2. Performance of Synchronous Reluctance Motors. These motors start and accelerate in the manner of an induction motor. Although they have a tendency toward cogging, this characteristic does not produce any undue difficulties with a skewed rotor. With the aid of induction-motor torque produced in the rotor bars, the rotor is accelerated to a speed just below synchronism. If the combination of load inertia and load torque is within design limits, the motor speed quickly rises to synchronism.

Having attained synchronous speed, the motor runs at constant speed regardless of load or voltage, provided the pull-out point is not exceeded and the supply frequency is held constant. Beyond pullout, the motor runs as a squirrel-cage induction motor; slip is present and the speed decreases with load. The motor, however, is designed not to operate beyond the pull-out point, except in a transient manner.

Speed-torque curves for one motor rating at different frequencies are shown in

Fig. 2. The load inertia is 2.2 times that of the motor rotor. The pull-in torque varies with applied frequency. At high frequencies the motor windage is considerable and detracts from the available synchronizing torque. Curve (*a*) shows a pull-in torque of approximately 95 percent rated torque; curve (*c*) indicates the low-frequency operating speed-torque relationship. Windage and bearing friction produced by the motors is significantly lower, permitting a larger load torque, 105 percent of rated torque, during the synchronizing process. Curve (*b*) falls between the two extreme operating frequencies and shows that the motor is capable of synchronizing the inertia against 100 percent load torque. Larger external inertias can be synchronized by this motor, but at reduced load torques. Conversely, larger load torques may be applied during the synchronization process if the external inertia is reduced.

By design the motor pull-out torque is held to 150 percent of rated torque. However, if higher synchronous torques are required to cover torque peaks, or if the user has voltage fluctuations on the line which reduce the pull-out torque (as a square of the applied voltage), then higher pull-out torques should be specified. The values of pull-out torque can be adjusted by design. Table 1 shows typical efficiencies and power factors of synchronous reluctance motors.

Figure 3 shows curves for efficiency, power factor, speed, and line current of a typical 10-hp four-pole motor. The curves show performance both in the synchronous-speed operating range up to pullout, and beyond the pull-out point when the motor operates on induction torque. The transition from synchronous to induction-torque operation at the pull-out point is abrupt, as illustrated by the

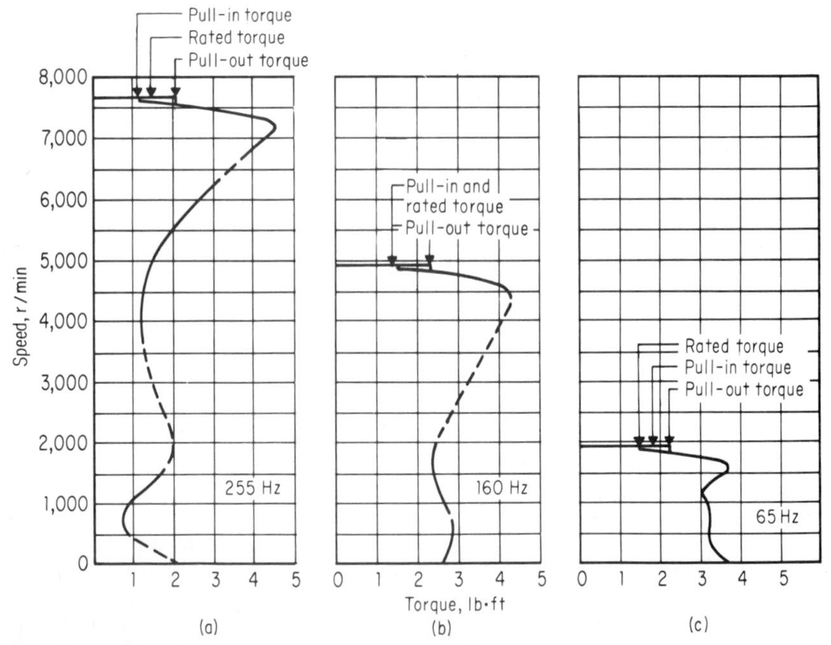

FIG. 2. Speed-torque curves for 2.07/0.526-hp 230/61.4-V 255/65-Hz 7650/1950-r/min four-pole motor with rotor inertia of 0.09 lb·ft^2 and load inertia of 0.2 lb·ft^2. Pull-in torque Wk^2 is 0.2 lb·ft.

TABLE 1. Typical Synchronous-Reluctance-Motor Performance*

Rating, hp	Number of poles	Full-load efficiency, %	Full-load power factor, %
1	4	75	65
1½	2	80	70
2	4	80	66
7½	4	87	68
10	4	90	69
30	6	90	65
50	6	91	68

*All ratings are given for 60 Hz.

curves. When the pull-out point is passed, the motor abruptly becomes noisy, and meter readings fluctuate.

The starting torque is usually lower and the starting current of these motors generally is greater than those of induction motors. Table 2 compares the starting performance of typical 20-hp electric motors for 60-Hz operation. It should be noted that these relationships vary for other horsepower ratings.

3. Factors to Consider. Synchronous reluctance motors can be braked by plugging, regenerative, capacitive, dc, and mechanical methods in the same way as induction motors. With an overhauling load these motors generate power back

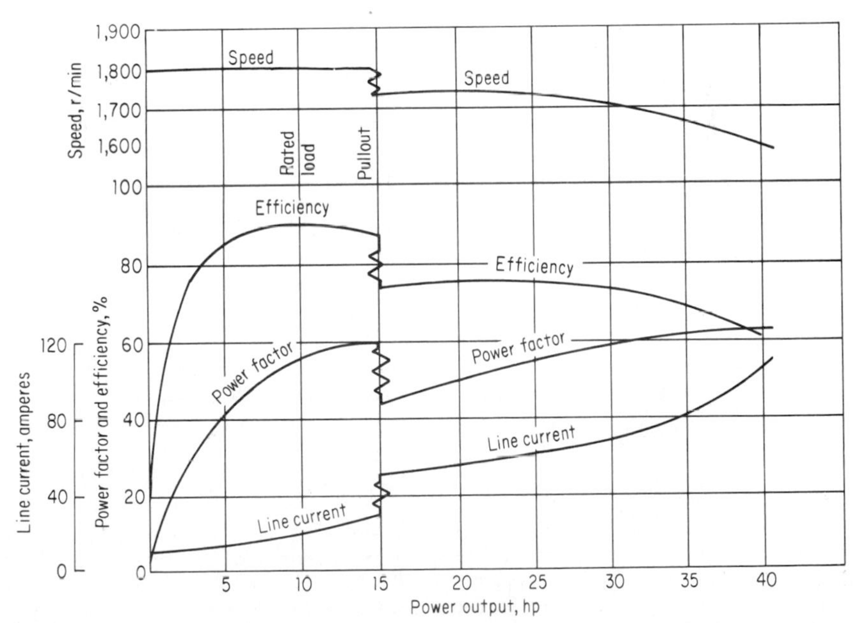

FIG. 3. Output characteristics of 10-hp four-pole 1800-r/min 220/440-V synchronous reluctance motor.

TABLE 2. Comparison of Typical Performance Characteristics for 20-hp 1800-r/min 60-Hz Three-Phase Motors

	Excited synchronous motor		Synchronous reluctance motor	Induction motor
	100% power factor	80% power factor		
Efficiency, %	88	87	88	88
Power factor, %	100	80 (lead)	60 (lag)	88 (lag)
Pull-out torque, % full-load torque	150	200	175	
Pull-in torque,* % full-load torque	110	125	120	
Locked-rotor torque, % full-load torque	110	125	130	150
Locked-rotor current, % full-load current	675	600	675	575

*Based on 8-lb·ft^2 load inertia.

into the system like a synchronous generator up to the pull-out point and like an induction generator beyond that point.

4. Constant-Speed Applications. The synchronous-reluctance-motor speed is calculated by the formula

$$\text{Speed (r/min)} = \frac{120 \times \text{frequency}}{\text{number of poles}}$$

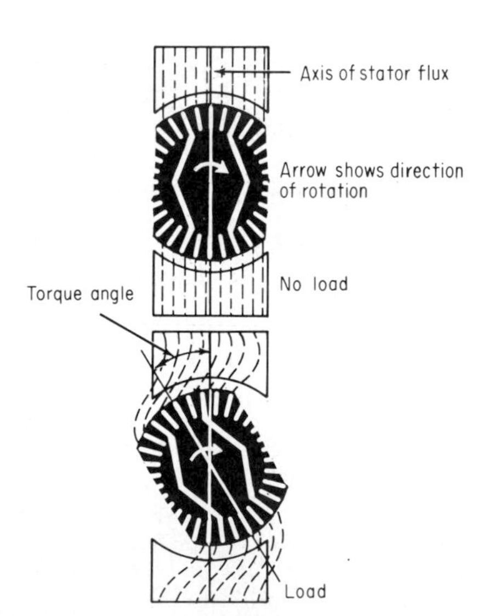

FIG. 4. Torque angle, formed by stator flux and rotor-pole axis, is minimum at no load and maximum at pullout.

The motor speed is constant to pull-out if the frequency is constant. The reluctance motor will run at this speed despite temperature changes, manufacturing variances, adverse environments, or voltage and load changes or fluctuations. It will, however, pull out of synchronism if its pull-out torque is exceeded (Figs. 4 and 5).

Because of this constant-speed characteristic, these motors are especially suitable for applications requiring precisely constant flow of gases or liquids, precisely constant travel or displacement, and for automated processes requiring precise synchronization with other motors or timed operations in the system.

A typical application uses a synchronous reluctance motor to drive one of many winches on a marine elevator or ship hoist. Each winch will

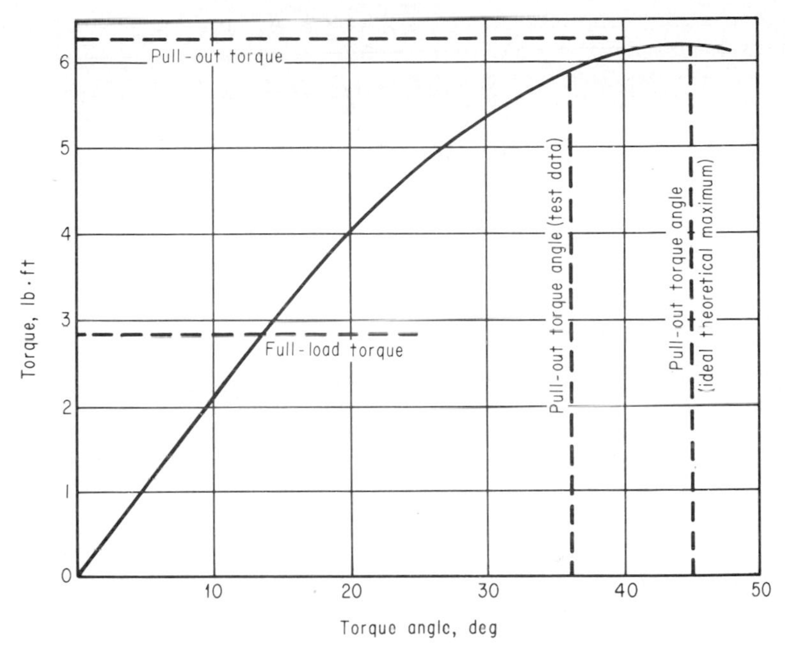

FIG. 5. As load increases, torque angle increases until pullout occurs; speed remains constant up to pullout.

operate precisely at the same speed, and together they operate to provide the elevator with a smooth, level lift so that tilting is not experienced by the ship being hoisted. In larger installations as many as 64 winches operate at the same speed, that is, synchronously. Other applications include drives for proportioning pumps in chemical processes and for conveyors where the speed of part travel is synchronized with another activity, such as putting bottle caps on bottles.

5. Adjustable-Speed Applications. Synchronous-reluctance-motor speeds can be varied over a wide range by varying the frequency. A chemical process may require that 90 to 100 machines be operated in synchronism over a range of preset or regulated speeds. The frequency of a single power source, a solid-state inverter, for example, is changed, and the speed of all connected motors will change by a corresponding amount.

In these applications the frequency of the power supply may be changed by changing the speed of the generator, by means of frequency-changing motor-generator sets, or by precisely controlled static frequency changers. In the plastic, paper, pulp, rubber, glass, and metal industries these motors are used in processing continuous sheets or films. These applications require precise, adjustable speed control. Synchronous reluctance motors are also applied as drives for printing, machine-tool-transfer lines, and packaging and folding machines.

These motors are ideal for high-speed operation or low-speed drives. Special designs can attain operating speeds of 12,000 r/min at 200 Hz with a two-pole motor, down to 200 r/min at frequencies as low as 10 Hz with a six-pole motor.

PART 2

BRUSHLESS SYNCHRONOUS MOTORS

Raymond F. Horrell*

A brushless synchronous motor is a dc-excited motor as described in Sec. 3. The term "brushless" refers to the combination of a synchronous motor and its field-excitation system, which eliminates the need for collector rings and brushes. Brushless motors are available in sizes from 50 to 20,000 hp at conventional speeds and voltages. They provide performance characteristics suitable for the same applications for which brush-type synchronous motors are utilized.

6. Basic Equations. The basic equations for brushless-synchronous-motor operation are the same as those given for synchronous-motor operation in Sec. 3, Table 3.

7. Construction. The construction of brushless-synchronous-motor parts is the same as that given for synchronous motors in Sec. 3, except that the collector-ring and brush assembly is omitted.

8. Brush-Type dc-Excited Synchronous Motors. A brush-type dc-excited synchronous motor with a commonly used dc-exciter excitation system is illustrated in Fig. 6*a*. The dc exciter may be direct-connected, overhung, belted, or separately driven. Other sources of direct current may also be used instead of the dc exciter. Basically, direct current is supplied through the field contactors and the synchronous-motor collector-ring and brush assembly to the motor field winding. The collector rings are mounted on the rotor shaft and the brush assembly is generally bolted to the motor frame parts. The motor primary ac line contactors, field contactors, field-discharge resistor, polarized-field frequency relay, and the exciter-field rheostat are usually mounted in a motor-control enclosure. Because carbon brushes are required to carry the current to the collector rings and from the commutator of the dc exciter when one is used, the motor is known as a brush-type synchronous motor, although the term "brush-type" is seldom used.

9. Brushless dc-Excited Synchronous Motors. A brushless dc-excited synchronous motor with a commonly used three-phase-exciter excitation system is illustrated in Fig. 6*b*. The ac-exciter-field assembly may be mounted on the motor

*Chief Engineer, Induction Machines, Electric Machinery Manufacturing Company, Minneapolis, Minn. (retired); Registered Professional Engineer (Minn.); Senior Member, IEEE.

FIG. 6. Dc-excited synchronous motor and control systems. (*a*) Brush-type motor uses dc-exciter current supplied through collector-ring and carbon-brush assemblies to motor field winding. Control components are shown inside dashed lines. (*b*) Brushless motor uses rectified ac-exciter current supplied directly to motor field windings. Many control-circuit components are also mounted on rotor shaft, as shown within lower dashed-line enclosure.

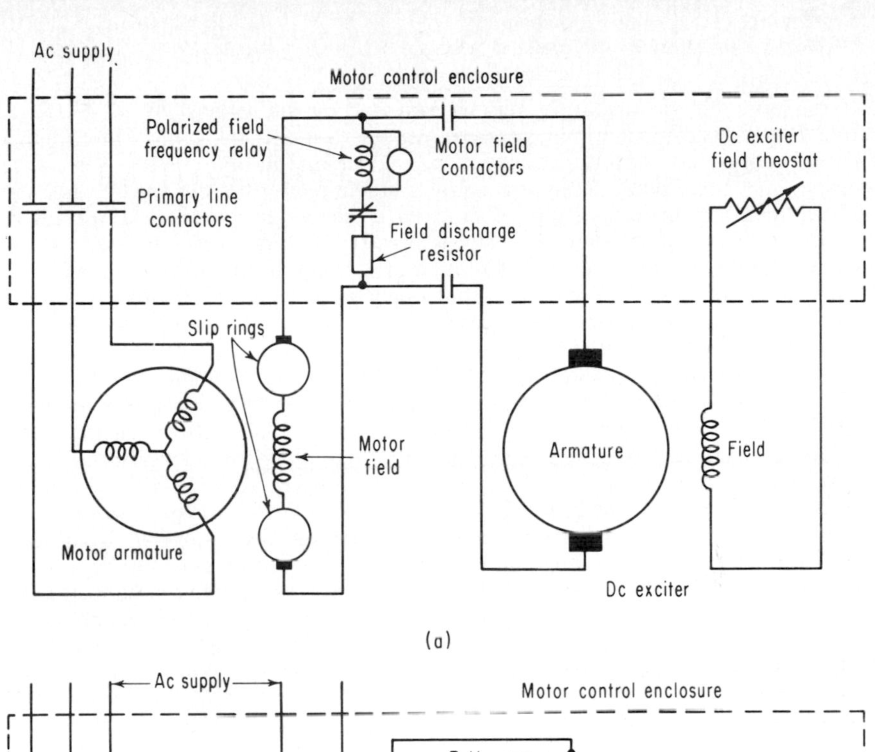
Ac supply
Motor control enclosure
Polarized field frequency relay
Motor field contactors
Dc exciter field rheostat
Primary line contactors
Field discharge resistor
Slip rings
Motor field
Armature
Field
Motor armature
Dc exciter

(a)

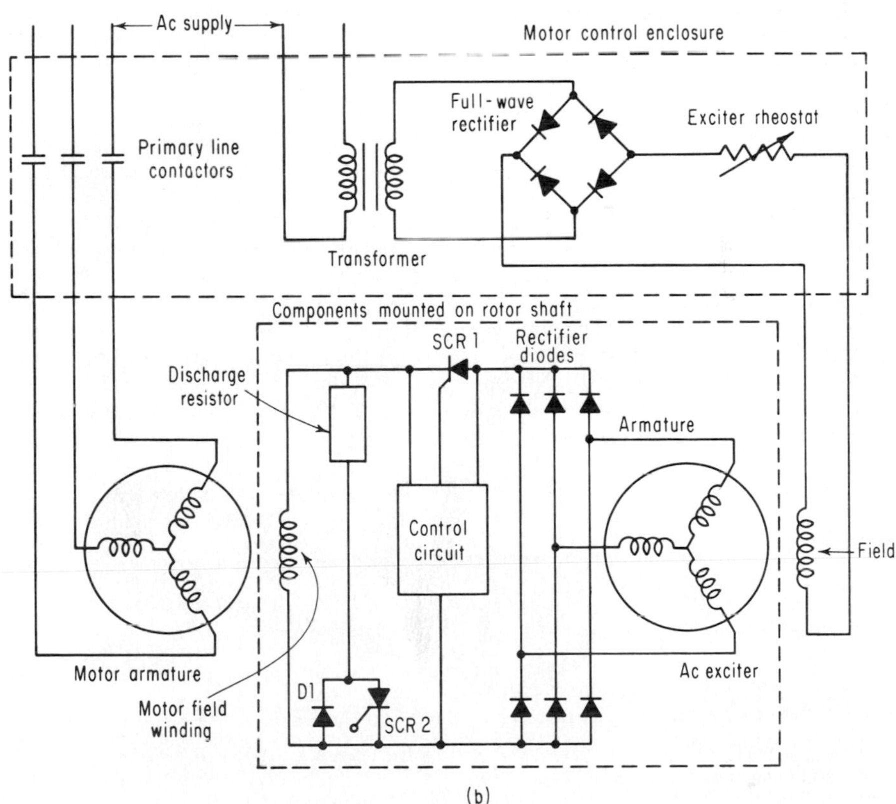
Ac supply
Motor control enclosure
Full-wave rectifier
Exciter rheostat
Primary line contactors
Transformer
Components mounted on rotor shaft
SCR 1
Rectifier diodes
Discharge resistor
Armature
Control circuit
Field
Motor armature
Ac exciter
D1
SCR 2
Motor field winding

(b)

frame parts and the armature may be mounted on the rotor shaft, as shown in Fig. 7 for bracket-type or pedestal-type motors, or as shown in Fig. 8 for engine-type synchronous motors. Basically, the exciter alternating current is rectified either by a three-phase full-wave or by a single-phase rectifier bridge, and the resulting direct current is supplied directly to the motor field without the use of collector rings and brushes. The motor primary ac line contactors and the ac-excited-field rheostat are usually mounted in a motor-control enclosure. The rectifier-bridge diodes and the silicon-controlled rectifiers, the field-discharge resistor, and certain control-circuit components are mounted on the rotor shaft, as illustrated in Figs. 7 and 8*b*. The control-circuit components comprise the frequency-sensitive static-field application system that replaces the polarized-field frequency relay and the field contactor used by the brush-type synchronous-motor system. This system applies positive field excitation at the proper time and phase to obtain maximum pull-in torque. The point of applying field excitation may be adjusted very easily as indicated in Fig. 7 and requires no attention after the initial adjustment. Excitation is removed automatically if the motor pulls out of step. Instantaneous shutdown or unloading upon pullout, accomplished by a field-monitoring relay which senses the power factor, provides out-of-step cage-winding protection. The ac-exciter field may be supplied by a rectifier and tapped resistor unit, which is connected to an ac power supply as illustrated in Fig. 8*b*. Other dc sources may also be used. Since the collector-ring and brush assembly is eliminated, the term "brushless" is used.

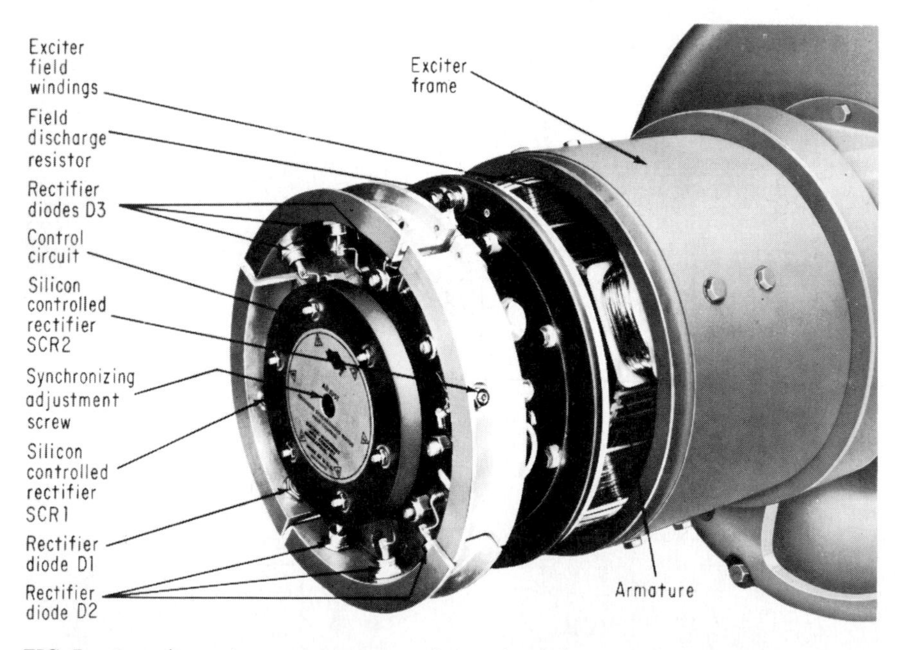

FIG. 7. Ac-exciter and control components for overhung direct-coupled mounting on synchronous-motor parts. Exciter parts, field-discharge resistor, diode and silicon-controlled rectifier ring, and control-circuit unit with convenient synchronizing-point adjustment are shown. *(Electric Machinery Mfg. Company, Minneapolis, Minn.)*

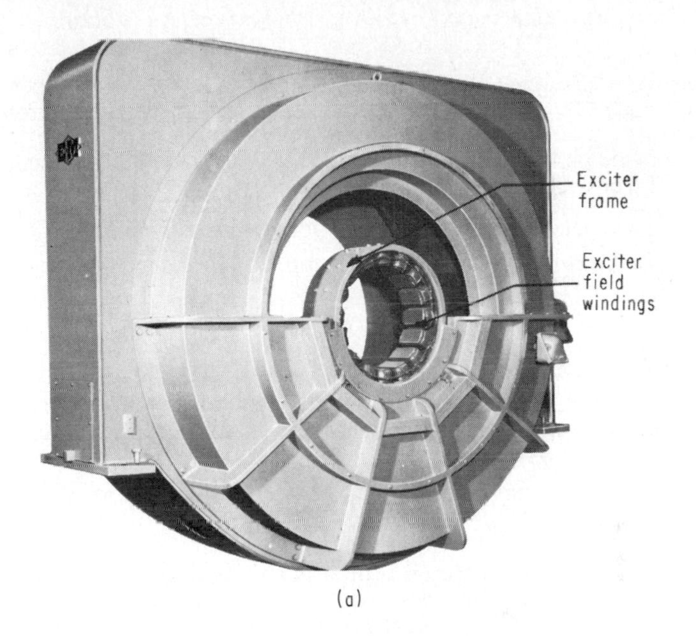

(a)

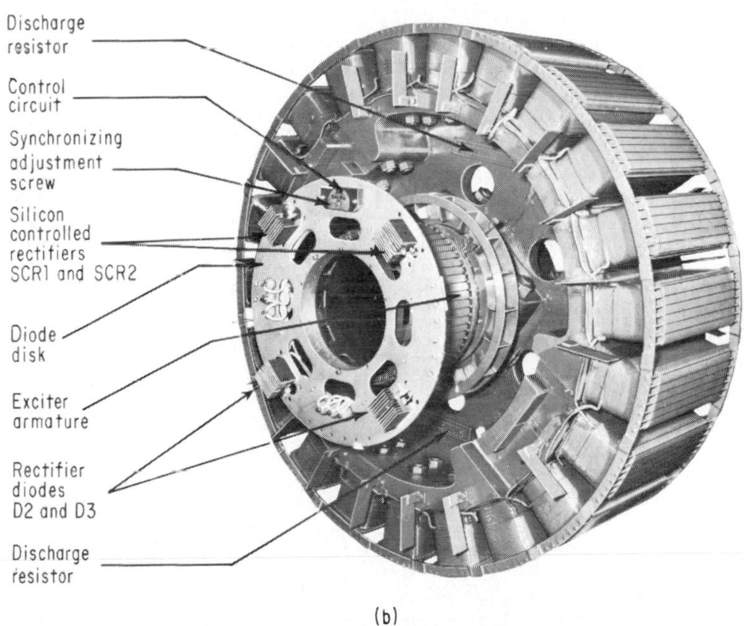

(b)

FIG. 8. Large engine-type brushless synchronous motor. The method of mounting ac exciter and control-system components is shown. (*a*) Ac-exciter-field ring is centrally supported by brackets which are bolted to motor frame. (*b*) Ac-exciter armature is mounted to spider hub, which is mounted on large shaft of engine-type motor. Field-discharge resistors are mounted on rotor spider. Rectifier-bridge diodes and silicon-controlled rectifiers are mounted on heat sinks which, together with control-circuit components, are all mounted on a disk, which is generally bolted to armature spider. *(Electric Machinery Mfg. Company, Minneapolis, Minn.)*

10. Advantages of Brushless Synchronous-Motor Systems. The advantages of brushless systems over brush-type systems are gained by the elimination of:

1. A separate dc power supply for the motor-field excitation
2. Maintenance of commutator and brushes of a dc exciter
3. Maintenance of collector rings and brushes of a brush-type synchronous motor
4. Routine carbon-brush selection, inspection, and replacement
5. Hazardous carbon-brush sparking
6. Hazardous carbon-dust accumulations
7. Routine slip-ring inspection and maintenance
8. Need for special enclosures and/or forced-ventilation blowers and ductwork for collector-ring and brush assemblies when used in hazardous locations common to chemical, gas, and petroleum industries
9. Need for separately mounted field contactors and conventional field application and removal relays

11. Brushless Synchronous-Motor Control. This is basically the same as that described for synchronous-motor control in Sec. 3, except that certain control functions are provided in the control-circuit components (Fig. 6*b*). The availability of controller functions such as speed changing, inching, reversing, and braking should be checked with the motor manufacturer. Almost any type of protection shown in Sec. 3, Table 7, is available.

PART 3

BATTERY-POWERED MOTORS FOR INDUSTRIAL APPLICATIONS

C. F. Cobb*

12. Power Source. The use of low-voltage direct current has, in general, been made practical by the advent of a means to store electric energy. Improvements made in storage batteries over the years have made it practical to apply this stored electric energy to motors which transform electric energy into mechanical energy (torque).

There are two general classes of storage batteries: (1) the primary type and (2) the secondary, or storage, type. Both types generate electric energy by chemical means, but in different ways.

a. Primary batteries generate electricity by chemical action with the consumption of material such as zinc. This material must be replaced when the battery is exhausted. The electrolyte also becomes exhausted and must be replaced. Primary-battery cells generate from 0.9 to 1.5 V at minimum discharge rates and lower, depending upon the current draw. These primary cells may be connected in series to obtain a higher voltage and in parallel to obtain higher currents. While it is possible to recharge these primary batteries several times, they are not generally used for an industrial power source. They are at present used in great quantities as a power source for flashlights and driving miniature and subfractional motors such as cordless electric clocks and shavers where very small power output is required.

b. Secondary, or storage, batteries must be charged before they can generate electricity. Energy is put into the battery in the form of direct current and discharged from it as direct current. This energy is stored as chemical energy. There are many types of storage batteries, but perhaps the two most familiar are the lead-acid and the nickel-iron-alkaline cells. Although both operate on the same general principle, they differ in the materials used in their construction and have different characteristics. These batteries may be recharged many times after being discharged by removing electric energy from them. The voltage output per cell does not exceed 2 V.

The surface area of the battery plates determines their size in current capacities. By series and paralleling these cells, various voltages and current capacities may be obtained to suit various applications. The most usual voltages for power requirements are 6, 12, 18, 24, 28, 30, 36, 48, 60, and 72 V. The 6- and 12-V batteries are used to power many automotive devices, such as cranking motors, fan motors, air-conditioning motors, power-window motors, and lights.

*Chief Engineer, Electrical Engineering Section, Allis-Chalmers Manufacturing Company, Norwood, Ohio (retired); Member, IEEE.

Many of the small fractional and integral dc motors are powered from 12- to 36-V batteries. The motors power such industrial equipment as electric-operated platform trucks, both propulsion and driving pumps for hoisting loads, and power steering by hydraulic action. Many thousands of battery-powered motors are used in such applications as golf carts, personnel carriers, and emergency equipment in many industrial plants and hospitals.

For moderate- to heavy-duty output for powering such vehicles as electric material-handling trucks, 36 V is a popular and economical choice. On large heavy-duty requirements, 48, 60, or 72 V is being applied to reduce the size of the current-carrying parts of the electric system.

13. Basic Motor Requirements. Battery motors must be designed and built to satisfy a special-purpose application with special duty cycles. They differ in many respects from general-purpose motors as built in standard frame sizes for normal motor applications.

In general, special-duty motors have been developed for propulsion service, pump drives for hoisting, power steering, and other devices using hydraulic power to perform these functions on electric material-handling equipment and related equipment.

Propulsion motors are normally developed to be mounted to the housing of the truck transmission with the drive pinion generated or mounted on the motor shaft, engaging the ring gear to drive the wheels. For electric hoists, the motor is usually connected through a gearbox to the load.

14. Motor Enclosures. Hoists are usually totally enclosed nonventilated. They have a short duty cycle of from 1 to 3 min.

Propulsion motors are supplied with several types of enclosures, open protected (Fig. 9), dripproof protected, totally enclosed nonventilated (Fig. 10), explosionproof (Fig. 11), and totally enclosed fan-cooled. The enclosure is predicated on the environment in which the truck is to be operated. These motors are usually rated for 1-h duty at nameplate current and voltage rating.

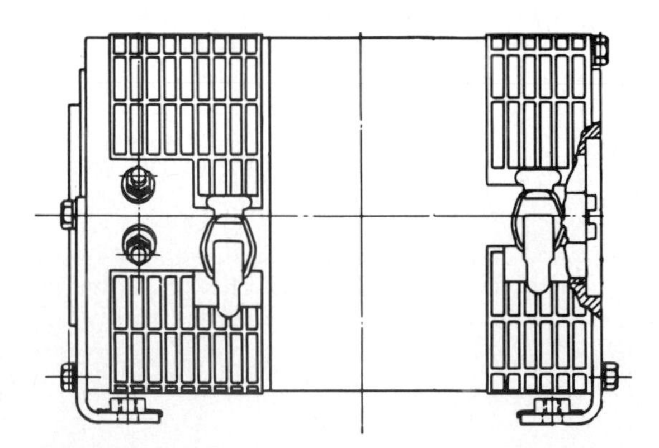

FIG. 9. Open protected propulsion-motor enclosure.

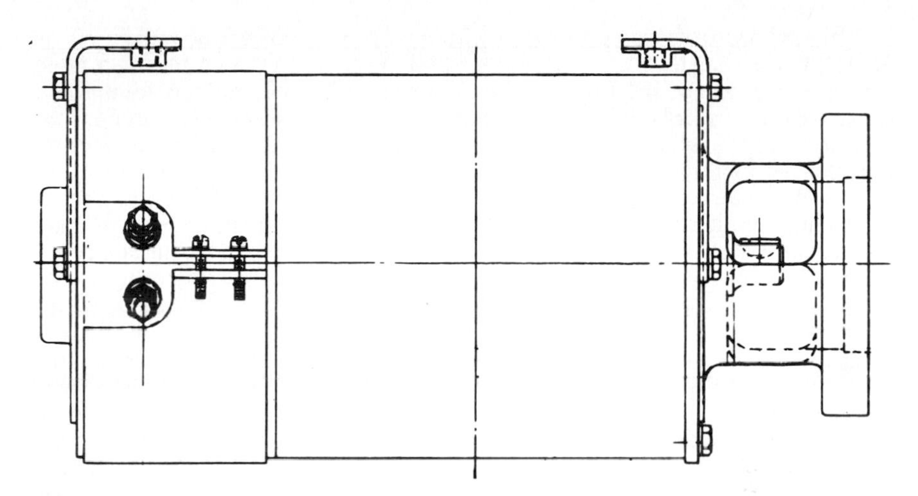

FIG. 10. Totally enclosed propulsion-motor enclosure.

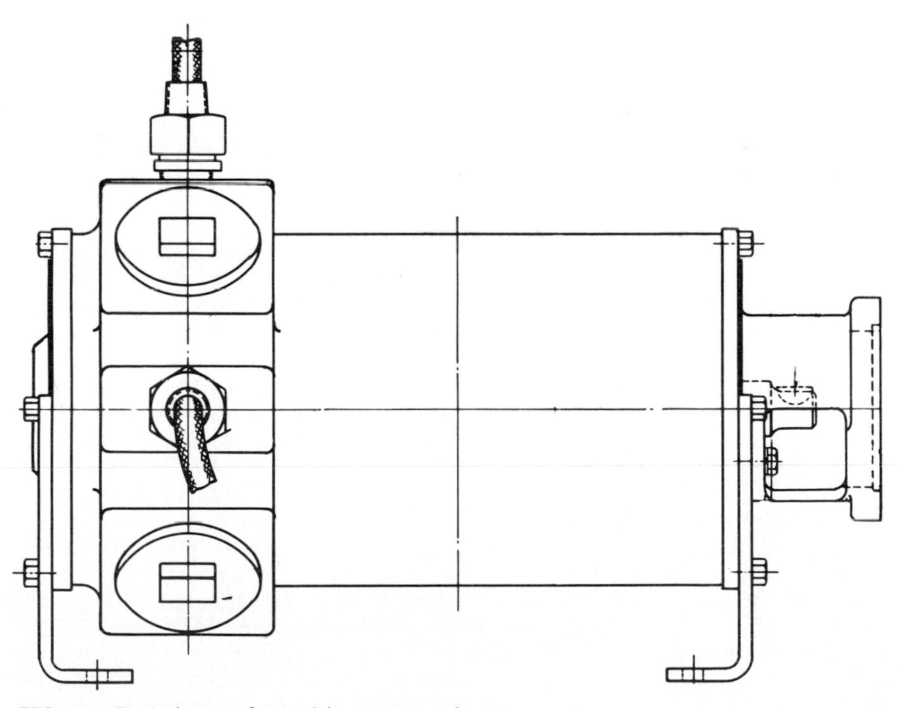

FIG. 11. Explosionproof propulsion-motor enclosure.

Motors used to drive pumps for hydraulic operation of various functions on electric trucks usually have an enclosure similar to that of the propulsion motor. These motors are of short duty cycle except for the power-steering motor. The power-steering motor is continuous rated and runs whenever the operator has the truck control turned on. The other motors are usually rated 3- or 5-min duty with a cooling period between operations. The torque per r/min is high on these short-duty-cycle motors.

An open protected machine is one having ventilating openings which permit passage of external cooling air over and around the windings of the machine. A protective cover of screen, expanded metal, or airline-type material is used over the air openings as a protection against foreign material entering the machine.

A dripproof protected machine is one in which the ventilating openings and protective covers are designed so that successful operation is not interfered with when drops of liquid or solid particles strike or enter the enclosure at any angle from 0 to 15° downward from the vertical or so that foreign material will not enter the machine.

A totally enclosed nonventilated machine is one that is not equipped for cooling by means external to the enclosing parts.

A great percentage of motors in present electric-operated material-handling trucks are of the open protected type in contrast to the totally enclosed nonventilated type thought necessary several years ago. It was thought that an open protected motor would become contaminated because of its location in the truck and from dust, dirt, and foreign material in areas where the trucks are operated. Experience gained from several years of service has confirmed the fact that this type of enclosure is satisfactory in a large percentage of applications. These enclosures are not suitable, however, where air is contaminated by gases of flammable or explosive materials of any kind. One major influence in using open protected motors is the additional torque which can be obtained in a given size of motor because of the free exchange of outside and inside air to cool the electrical and mechanical parts. The higher duty cycle expected from modern trucks would require a larger motor or make the change to open protected motors necessary to obtain this increased power requirement.

The dripproof protected motor is a modification of the open protected motor. Instead of being of open material over all openings, the protecting covers are made of solid material in the upper half of the enclosure for protection from drops of liquid or solid particles, and of expanded metal or screen in the lower half. This type of enclosure is used extensively where protection is needed from leaking battery acid or where outdoor operation may allow water or other liquid to drop on the motor frame. These motors usually are equipped with an armature fan to provide additional movement of air through the motor windings to cool the motor during high torque demands.

The totally enclosed nonventilated motor is indicated where there is excessive contamination of ventilating air and materials which can enter the protective openings of other types of enclosures. These motors have a reduced rating over the open protected and dripproof protected motors since the heat generated in the motor windings and other internal parts must be dissipated through the frame of the motor to the outside air surrounding the motor. The transfer of heat through a stationary frame, by conduction, is much slower than where it can be picked up by air passing through the motor windings. It therefore requires a reduction in motor losses (heat), which means a lower-rated motor. To service the motor brushes, enclosing covers should be easily removable. This accessibility can be accomplished by means of screws and catches to clamp enclosures to the motor frame. Special gaskets are used under the solid enclosing covers of totally enclosed nonventilated motors to seal out moisture and other contaminants which may be deposited on the frame of the motor.

Where the motor is to be used in areas of explosive gases and liquids, a totally enclosed explosionproof motor is used. The enclosure is designed and constructed to withstand an explosion of a specified gas or vapor which may occur within it and to prevent the ignition of the specified gas or vapor surrounding the machine by sparks, flashes, or explosions of the specified gas or vapor, which may occur within the machine casing. Motors of this type are usually built to meet Underwriters Laboratories specifications for Class I, Group D, or Class II, Group G, for use in hazardous locations. This type of motor must not only provide adequate seals over brush openings but have special provisions to seal any leakage areas around shaft and joints between and frames and yoke. Electric leads must be sealed by pressure devices or similar means to meet the conditions imposed by exploding gases or vapors within the machine.

While it is not the usual practice, some totally enclosed motors are ventilated by a separate blower. The blower forces air through the motor during operation or when it is at standstill, providing additional cooling and thereby permitting the motor to be uprated. A thermostat used in some applications provides automatic temperature protection for the motor under various torque settings. It measures the temperature of the field-coil winding and turns the blower motor on or off. The blower starts when the winding reaches a precalibrated thermostat setting and is disconnected from the line when the temperature drops to a predetermined minimum setting.

The open protected motors are best suited for the majority of electric material-handling trucks used in industry. Trucks with these motors produce more power than trucks with totally enclosed motors of the same size. Because they are satisfactory for most operating conditions, the open protected motors have become the number one choice for high-speed operations where fast response is essential. Trucks with motors having other enclosures are used only for work in environments unsuited for open-type-motor operation.

15. Motor Ratings. Motor ratings cover a wide range to suit applications, from subfractional to integral sizes. A ¼-hp motor may be suitable to propel a single personnel carrier, while a 15- to 20-hp motor may be required for a 10,000-lb industrial truck. The short-time rating of the pump-drive motors on a truck is many times of higher horsepower than that of the propulsion motor, to provide the high torque and fast hoisting speeds demanded by industry. These motors normally cover a range from 3 to 20 hp for trucks rated up to 10,000-lb pay load. Propulsion motors range from ¼ to 15 hp in trucks up to 10,000-lb pay load. In some drives two or more motors are used to obtain sufficient torque to perform satisfactorily.

16. Motor Insulation and Temperatures. The full range of insulation types is used in low-voltage motors. Class A is used in some of the smaller ratings, but this low-temperature insulation is being rapidly superseded by the higher-temperature insulation systems of the B, F, and H types. Total allowable temperatures for short-time-rated motors are 95°C for A, 115°C for B, 135°C for F, and 155°C for H. Industrial truck motors are considered to operate in 25°C ambient and industrial dc motors in 40°C ambient. This means that industrial truck motors can have a basic temperature rise of 70°C for Class A, 90°C for Class B, 110°C for

Class F, and 130°C for Class H, while industrial short-time-duty motors would have temperature rises of 55, 75, 95, and 115°C for Classes A, B, F, and H, respectively.

New materials and new or improved high-temperature varnishes are now available which greatly increase the service life of motors over those available a few years ago. By upgrading insulation, the motors will withstand higher temperatures without shortening the insulation life, as would happen with a lower-temperature system. Class A insulation was used quite generally 50 and more years ago, but is used much less at present. Typical Class A insulation systems include materials such as cotton, paper, cellulose acetate films, enamel-coated wire, and similar organic materials impregnated with suitable substances. Class B insulation systems include such materials as mica, glass fiber, asbestos, and other material, not necessarily inorganic, with compatible bonding substances having suitable thermal stability. Class F insulation systems include such materials as mica, glass fiber, asbestos, and other materials, not necessarily inorganic, with compatible bonding substances having suitable thermal stability. Class H insulation systems include such materials as mica, glass fiber, asbestos, silicone elastomer, and other materials, not necessarily inorganic, with compatible bonding substances such as silicone resins having suitable thermal stability.

17. Motor Voltages. In this discussion we are considering low-voltage motors, 72 V or less. There is no clear-cut division between motor ratings and voltages. One horsepower rating may be designed for one of several voltages depending upon the type of battery or the size of other motors which may be used in the power system. In general, motors up to 3 hp are operated from a 12-V system, although 18- and 24-V systems are also in use. Motors above 3 hp are usually 30 and 36 V, with some larger ones built for 48, 60, and 72 V. At the present time 12- and 36-V systems appear to be the most popular in small and medium-heavy-duty applications. Perhaps the 36-V system is most widely used for industrial trucks of the 2000- through 10,000-lb pay-load range since this voltage permits a good balanced electrical design and still keeps within the current limits that can be handled by the control devices in the electric system. The lower voltages, within limits, tend to improve motor commutation since the reactance voltage and the voltage between bars, etc., are reduced. This consideration is very important in these noninterpole-type motors where there are no commutating poles to help reduce the voltage in the armature coils undergoing commutation as they pass through the commutation zone.

18. Motor Characteristics. Low-voltage motors, such as higher-rated industrial motors, are designed with various output characteristics to suit the torque requirements of the drive. Propulsion motors are usually of the straight series-field type. Power-steering motors are of the compound wound-field type. Motors for powering hydraulic pumps, which operate the hoist and power attachments, have been designed as straight series-wound, stabilized series-wound, compound-wound, and stabilized shunt-wound, depending upon the service required and also the size of the motor.

a. Straight series-wound motors have field coils of heavy copper section capable of carrying the line current. This field is connected in series with the armature, through the brushes and commutator. It produces a magnetomotive force in proportion to the current in the line. Therefore the lighter the load current, the smaller the magnetomotive force, or field strength. Field strength can be equated to machine flux ϕ per pole. Since the speed equation for a designed motor is

$$\text{Speed (r/min)} = \frac{K \times E_{\text{cemf}}}{\phi}$$

it can be readily seen that the smaller the flux ϕ, the faster the speed. Here K is the motor constant and E_{cemf} is the counter-electromotive force in volts available to produce power.

It is therefore the characteristic of a straight series motor to "run away" under light- or no-load conditions. Unless the direct-coupled load is sufficient to keep the motor within safe speed limits at light loads, a stabilized series field must be used. The very small motors can sometimes withstand this high no-load or light-load speed since their internal losses, winding and friction losses, and coupled load are sufficient to limit the speed to within the safe centrifugal forces of this high speed. These conditions may not be found in larger-sized frames since the high no-load or light-load speeds might damage the rotating parts. Most propulsion motors are designed as straight series-wound since they are permanently connected to a load through gears and cannot be entirely unloaded (Fig. 12).

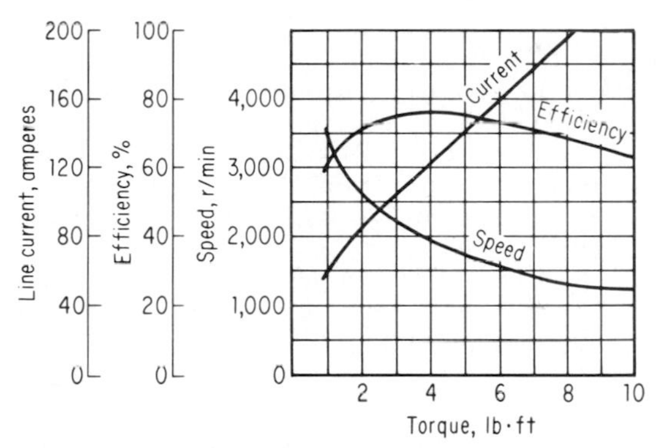

FIG. 12. Calculated characteristic curves for series, open-type, Class B insulated, 90°C 1-h rated 1.10-hp 12-V 2300-r/min 12-V drive motor.

b. Stabilized series-wound motors are essentially series motors but have sufficient shunt-field winding on each main pole to limit the no-load speed of the motor to a safe value. This shunt-field winding is on the same pole as the series-field winding but is connected in parallel with the battery or power source. It is not influenced by motor current but by battery voltage and is, therefore, essentially a constant value supplying approximately constant flux ϕ, which limits the motor speed to within safe mechanical limits. Motors having this characteristic are used on such applications as hoists, winch drives, and small propeller drives.

c. Compound-wound motors vary in characteristics from the series and stabilized series motors in that they have a series-field and a shunt-field winding on each pole, with the shunt winding being the major control winding.

This shunt winding may produce 75 to 90 percent of the total full-load ampere-turns required, while the series field supplies 10 to 25 percent as required by application. The larger-sized motors driving hydraulic pumps for hoisting and motors for power steering are made compound-wound. This motor has much less speed change from full load to no load than the motors with stronger series fields (Fig. 13).

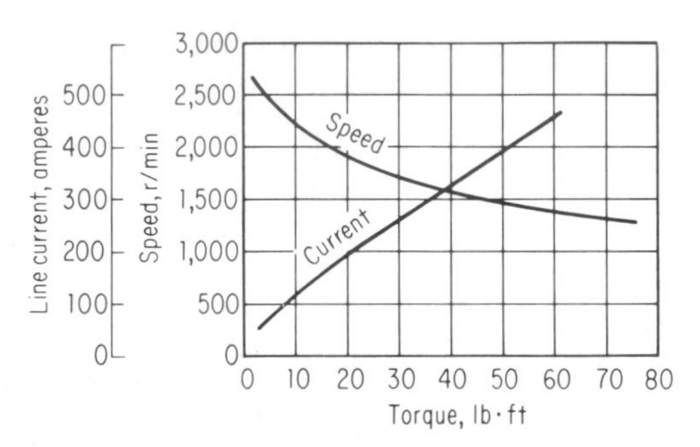

FIG. 13. Calculated characteristic curves for open-protected-type fan-cooled compound-wound, Class F insulated, 90°C ½-h rated 10.4-hp 36-V 1660-r/min 275-A hoist-pump motor.

d. Stabilized shunt-wound motors are quite similar to the compound motor, with the difference being in the percentage of shunt ampere-turns compared to series ampere-turns at rated load and voltage. The series ampere-turns may be from 3 to 7½ percent of the total required. The purpose of the stabilized shunt winding is to provide a slightly drooping speed characteristic from no load to full load and also to provide more starting torque than the straight shunt motor. This type of motor is used very generally in industry. The drooping speed characteristic with load application allows a slight shock-absorber action, which allows wide load changes with minimized shock to the driven equipment. This motor characteristic is desirable in many low-voltage applications as well as for higher-rated types.

e. Straight shunt-wound motors have the field in parallel with the power supply or separately excited from some outside constant-potential source. The field strength is therefore not dependent on load current as for the previously discussed motors. It is considered an essentially constant-speed motor and is applied where the starting torque is small and it is desired to maintain good speed regulation over a load range. Shunt motors are applied to such devices as centrifugal pumps and fan drives.

19. Motor Design. The design of low-voltage motors is comparable with the design of the higher-voltage type. Because of low-voltage operation, the armature winding is of heavy section and usually designated as series, or wave, type. It is a two-circuit winding irrespective of the number of main poles. Many high-voltage designs require the use of the multiple, or lap, type of armature winding, which has as many circuits as there are main poles, that is, each circuit must carry $1/N$ percent of load current, where N equals the number of main poles. The commutator must be of suitable dimensions to permit mounting of sufficient brushes to keep the brush current density within specified limits and have a surface area such that heat losses can be dissipated to keep the temperature rise within the guaranteed limit. The watts per square inch which can be tolerated will depend upon the peripheral speed and the cooling air over the commutator. There are normally as many brush studs as main poles. Each brush stud may have one or more brushes per stud as required to transmit the current. The brushes on each stud must carry $2/P$ times load current, where P is the number of main poles; that is, in a four-pole motor each brush stud must carry one-half the line current, in a six-pole motor one-third line current, etc. On a 1-h-rated propulsion motor for material-handling trucks it is usual to provide sufficient brush area to obtain a brush density at rated load current of 70 to 80 percent of the brush manufacturer's recommended current density. This is necessary to have the capacity to carry the heavy peak currents imposed during truck acceleration, ramp climbing, and dynamic braking. Overloaded brushes cause excessive heating, poor commutation, and rapid brush wear. The short-duty-rated motors (rated 3 to 5 min) may use brushes which operate at 150 percent recommended brush density at rated motor current. The short load cycle tends to provide time for brushes to cool before the next load cycle is applied. It is usual practice to apply a special grade of electrographitic brush developed primarily for this variable-load service. Brushes are retained in a close-fitting holder to assure good face contact with the commutator. Brush pressure is maintained by spring pressure on the top of the brush. The springs are designed to maintain essentially constant pressure over the wearing length of the brush. These constant-pressure springs change position as brushes wear, to maintain the desired pressure automatically. Brush pressures vary with different grades of brushes and with the peripheral speeds of commutators. The pressure range is normally from 2½ to 4½ lb/in^2 of brush contact with the commutator surface.

The armature design is predicated upon the voltage, horsepower, and speed desired. The output of a machine is proportional to the armature volume. The well-known output equation for 40°C rise machines is

$$D_a^2 L_c = \frac{\text{watts} \times 60.8 \times 10^7}{\text{r/min} \times B_g \times q \times \Psi} \quad \text{in}^3$$

where D_a = armature diameter, inches
L_c = core length, inches
B_g = apparent gap density, lines per square inch
q = ampere-turns per inch of armature circumference, $= ZI_c/\pi D_a$
Ψ = percent enclosure, = pole arc divided by pole pitch

This output equation may be adapted to other temperature-rise values by changing the constant in the numerator.

The emf equation for dc machines is

$$E = Z\phi_a \frac{\text{r/min}}{60} \frac{\text{poles}}{\text{paths}} \times 10^{-8} \quad \text{volts}$$

where E = voltage at machine terminals, volts
Z = total number of active conductors
ϕ_a = flux per pole which crosses the air gap and is cut by the armature conductors
r/min = armature revolutions per minute
poles = number of main poles and paths, = number of parallel circuits (electric) in armature

A series, or wave, wound armature has only two circuits regardless of the number of main poles, while a multiple, or lap, wound armature has as many paths as there are main poles. Other types of windings have been employed in armatures, but for the purpose presented here, the simple series- and multiple-wound armatures are referenced.

Where certain machine characteristics are known, another output equation may be derived from the emf equation as follows:

$$\frac{\text{watts}}{\text{r/min}} = \frac{ZI_a}{\text{paths}} \phi_a \times \text{poles} \frac{1}{60 \times 10^{-8}}$$

The factor ZI_a/paths, called electric loading, is the total number of ampere-turns on the armature periphery. The larger this factor, the greater the ratio of copper to steel in the machine. The value of $\phi_a \times$ poles is the magnetic loading and is the total flux in the armature. The larger this magnetic loading, the greater the ratio of steel to copper in the machine. The value of B_g is limited by the density at the bottom of the teeth where the highest density occurs. The value of ampere-turns per inch q is limited by heating and by commutation. The higher the value of q, the larger the copper section required for heat dissipation. A larger value of q means deeper slots and higher reactance voltage and tends to cause sparking at the brushes, poor commutation, and rapid brush wear. Since many low-voltage machines are built without commutating poles to assist commutation, care must be exercised in keeping to shallow slots and low reactance voltages.

The field coils are mounted on the main poles. Because of the small confined space in many small low-voltage motors, the field coils are wound on a flat mold and then bent on a press to obtain a curvature which allows them to fit against the field frame and the supporting pole horns. This contact with the frame allows good heat transfer from coil to frame and to the surrounding atmosphere. The design of the field coil depends upon the characteristics expected from the machine. As stated previously, they may be shunt, stabilized shunt, compound wound, stabilized series, and series coils. Each coil must supply sufficient ampere-turns at rated load to force the full-load flux through the magnetic circuit of the machine. This value is determined by the following steps: (1) The density B in each part of the magnetic circuit is obtained by dividing the net area of each part into the rated full-load flux ϕ. (2) By referring to the permeability curves (B per inch length of path) of the various materials, the ampere-turns per inch, at this density, are determined except for the air gap. The length of the flux path in each magnetic path is determined from a scale layout of the parts.

By multiplying the inch length of each path by the required ampere-turns per inch from the permeability curves, the total ampere-turns for the steel sections are obtained. The air-gap ampere-turns are determined by the formula $0.313 \times B_g \times$ effective length of gap (actual length + Carter's coefficient). The summation of the ampere-turns required for each flux path provides the apparent total ampere-turns to force this full-load flux through the magnetic circuit. To obtain the actual full-load ampere-turns of excitation (magnetomotive force) for each main-field coil, additional ampere-turns must be allowed because of armature distortion. On small low-voltage machines the value of the distortion ampere-turns varies within an approximate range of 5 to 10 percent of the full-load armature ampere-turns per pole, (ZI_a/poles) $\times$ paths. By adding this distortion ampere-turn value to the apparent total ampere-turns noted above, the final total is obtained. A field design that will provide this total value of excitation must be designed to fit in the coil space available on the field pole. In the event of a series-type motor the turns per pole are determined by dividing the total ampere-turns per pole by the rated full-load current of the motor. This number of turns per pole of sufficient copper section to carry this current, and keep within the heating limits assigned to the motor, must be wound to fit in the allowable coil space. In the case of a straight shunt machine, the wire size may be determined by using the following formula to determine the approximate circular mils and then finding the wire size from a wire table:

$$\text{Circular mils} = \frac{1.035 \times \text{number of poles} \times \text{ampere-turns} \times \text{length of mean turn in winding}}{\text{field voltage available}}$$

For the combination of series and shunt fields, to obtain various characteristics, the field design may be arrived at by first deciding upon the percent of series-field full-load ampere-turns required. The remaining ampere-turns must be supplied by the shunt field. The series-field coil may be placed on top or at either side of the shunt coil on the field pole. The coils must be thoroughly insulated from each other and the pole. High voltages may be induced by the shunt winding in the event the field circuit is quickly opened without suitable discharge resistor or reduction of excitation voltage before the circuit is opened.

The design of the series and shunt fields is made by following the procedure given above for individual series and shunt designs.

20. Dual Series Field. To simplify the controls used in some material-handling trucks, a dual series field is employed. This reduces the number of control contactors and resistors in the accelerating circuit. A control with four accelerating steps and a single series field would have three steps of resistance, and the fourth step would connect the motor across the power source. With dual field coils, the fields are connected in series through the third point of power, so that only two steps of resistance are required in the control, the third point has both fields in series, and the fourth point connects the motor to the power source with only the high-speed series field in the circuit. This arrangement saves power by eliminating the I^2R loss in resistors and in addition provides greater accelerating torque during acceleration by producing greater magnetomotive force across the air gap.

The division of series-field turns of low- and high-speed windings varies greatly. This division depends upon the desired motor characteristics and may be from 50 percent in each winding to perhaps 65 percent in high-speed windings and 35 percent in low-speed windings. This characteristic is also obtained by using a single series field and shunting it to provide increased motor speeds. This latter method is used with the pulse-width-moderated control which provides practically stepless acceleration and motor-current limitation without the necessity of dissipating power in stepping resistors.

A refinement of the stepped control is the carbon pile where the pressure on carbon disks varies the resistance value, thereby changing the speed of the motor. A silicon-controlled-rectifier-type control (thyristor) is also used to provide varying voltage to the motor from an essentially constant voltage source such as a battery.

PART 4

ENGINE-CRANKING MOTORS

A. N. Kaiser* and E. J. Szabo†

Engine-cranking motors are often called starting motors, or starters, but the correct terminology is cranking motor, since the function of the cranking motor is to provide rotation of the engine for starting. It cannot provide the many other things, such as fuel and ignition, which are required for the engine to start and run under its own power.

Although this section is confined to electric cranking motors for engines, it should be mentioned that there are other types of motors or methods also used for this purpose. These include air motors, of both the rotary-vane type and the piston type; hydraulic motors, explosive cartridges; auxiliary gasoline engines; and also combination units known as starter generators. Engine-generator sets are often cranked by energizing the generator windings if the main generator is a dc unit.

The purpose of this section is to provide information to engine designers and engine-application engineers regarding the sizes and types of electric cranking motors used to crank various engines. It also provides information on the complete electric system involving the electric cranking motor under normal operating conditions and also unusual conditions, including low-temperature cranking.

21. Motor Sizes. The physical sizes of electric cranking motors vary from a 3-in-diameter unit used for cranking marine outboard engines to 6⅝-in-diameter motors weighing 125 lb. Sizes include 4-in-diameter units for small engines, 4½-in-diameter units commonly used on passenger automobiles and gasoline-engine-powered light trucks, 5⅛-in-diameter units for larger gasoline engines and also for the majority of truck and bus diesel engines, and 5⁹⁄₁₆-in- and 6⅝-in-diameter motors for the larger diesel engines used in engine-generator sets, stationary engines, marine-propulsion engines, and army-tank engines.

These cranking motors range in horsepower output from about ⅛ hp for the smallest to about 30 hp in the largest units. The operating voltages cover a range from 6 V dc to 110 V dc and include 12, 24, 30, 32, 48, and 64 V, and also 110 V ac. The selection of the size and horsepower output for the cranking motor depends upon the engine requirements. The operating voltage, however, is determined by other factors such as system voltages already established to operate various electrical loads, and thus the electric cranking motor must be designed and operated at a system voltage which may not necessarily be the optimum for the particular cranking system.

22. Engine-Cranking Requirements. Much experimental work has been done by engine manufacturers and cranking-motor manufacturers to determine mathe-

*Manager of Product Engineering, Prestolite Division of Eltra Corporation, Toledo, Ohio (retired); Member, SAE.

†Manager of Engineering, Regular Products Division, Leece-Neville Company, Cleveland, Ohio (retired); Registered Professional Engineer (Ohio); Member, IEEE, SAE, NSPE.

matical procedures for applying properly rated cranking motors to various engines under various operating conditions. This type of information is proprietary and has not been published for general usage. Excellent papers (Refs. 1 and 2) have been written on the general approach and analysis of electric cranking systems and applications, as well as troubleshooting of systems that do not operate properly. None of these papers, however, provides a mathematical procedure for determining the torque, speed, and horsepower required to crank and start an engine. Cranking-motor applications are based on either (1) previous experience with similar applications, (2) actual cranking tests at room ambient or in a cold room, or (3) a calculation procedure which at best can give only about plus or minus 25 percent accuracy, because of the many variables involved.

Actual engine tests are widely used to obtain the cranking torque and speed for a particular engine under all the various operating conditions in which it may be expected to start successfully.

The engine lubricating-oil viscosity is one of the most important factors for consideration in the determination of the torque required to crank an engine. Friction load and compression load are the two components of the total load to be overcome in cranking an engine over the cranking-speed range. Friction load is the larger of these two, and engine oil viscosity represents the largest part of friction load and is the major variable in friction load, since it is necessary to shear the oil film on the piston walls during cranking.

It has been shown (Ref. 2) that for equal displacement, the horsepower required to crank a gasoline engine is about one-half that required for a two-stroke-cycle diesel engine. A four-stroke-cycle diesel engine of the same displacement required 1½ times the cranking torque of the two-stroke-cycle engine. The number of cylinders is a factor, because for the same total engine displacement, with a greater number of cylinders, less torque is required to rotate the engine through the first compression stroke.

To establish engine-cranking requirements, tests should be conducted to obtain the torque to rotate the engine at the desired cranking speed established by the engine manufacturer, and the torque required to rotate the engine through the first compression stroke. These data can be obtained by rotating the engine with a torque-producing means, measuring the torque directly with strain gages, or using a dc motor and variable-voltage source, recording the current, voltage, and speed. The motor characteristics can then be reproduced on a dynamometer to determine the cranking-torque requirements of the engine.

The engine used for these tests should be in a normal operating condition with the lubricating oil well dispersed through the engine and with a short period of operation. In the case of four-cycle gasoline engines, caution should be used not to dilute the oil through excessive choking. Oils should be of the type recommended by the engine manufacturer for the lowest temperatures expected in normal usage, and soak periods of a minimum of 24 h at this temperature should be employed before engine-cranking-torque tests are conducted. Compression-relief devices, when incorporated in the engine design, should be actuated to provide realistic cranking-torque values.

In actual engine use there are many occasions when a load is applied without a means for disconnecting. Examples are hydraulic pumps and automatic transmissions. These parasitic loads must be taken into account when establishing engine-cranking torques, and if possible, the device producing the load should be included during cold tests.

23. Cranking-Motor Performance Characteristics. The selection of a suitable motor must be based upon the drive ratio available between the cranking motor and the engine, with due consideration being given to the power available from the battery at temperatures that the system will be expected to withstand.

The normal means used to couple the motors to the engine while cranking is through the use of a pinion on the cranking-motor output shaft, which is moved into engagement with a ring gear mounted on the engine flywheel. The diameter of this ring gear is usually limited in the case of passenger cars by the flywheel-housing clearance to ground, and in the case of outboard engines, the size of the engine shroud or clearance to accessories.

Knowing the limits of the ring-gear size and the smallest practical pinion size while maintaining sufficient tooth strength, a tentative ratio can be established. Using this ratio and the cranking speeds and torques previously obtained by engine tests, required performance of the cranking motor can be established.

Cranking-motor characteristics require high torque at low speeds, and by necessity series-connected motors are used. Most motors are straight-line in that the armature shaft is directly coupled through the drive mechanism and gears during cranking to the engine ring gear. However, some electric cranking motors have a gear-reduction integral with the motor itself to provide additional gear-ratio multiplication. The optimum gear ratios depend on the engine requirements. Small engines permit the use of small high-speed cranking motors, whereas larger diesel engines require considerably more torque at cranking speed.

Motor manufacturers, given an engine-cranking requirement, will probably offer one or more of their standard motor types equipped with standard drives to minimize tooling and to provide the lowest-cost unit to the engine manufacturer. A careful review of the performance curves for these motors will reveal the voltage and temperature conditions under which the curves were taken. Typical performance curves are shown in Figs. 14 to 17.

When the engine-application temperature or voltage conditions differ from those of the curves, corrections should be applied to the curves to achieve a more accurate definition of motor performance. Since the torque-current relationship

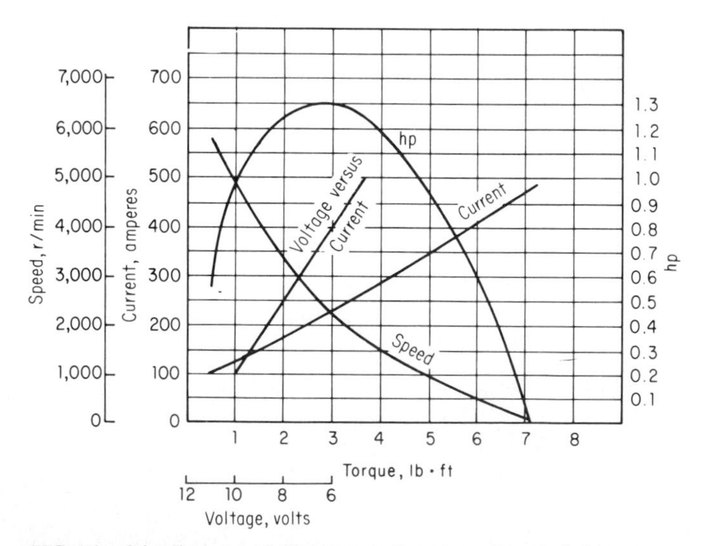

FIG. 14. 3-in-diameter 12-V motor performance characteristics.

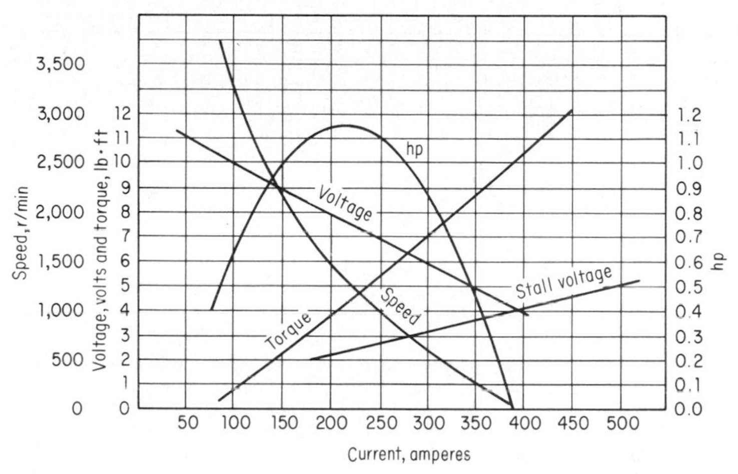

FIG. 15. 4½-in-diameter 12-V motor performance characteristics.

of the motor stays fairly constant through a wide range of voltage and temperature conditions, the speed will be the one characteristic that must be modified.

The cranking system consists of the battery, connecting cables, contactors, and cranking motor. Batteries are important to successful cranking performance since they must deliver the power required to the cranking motor. The battery capacity to produce the required power is dependent on the plate area, the state of charge,

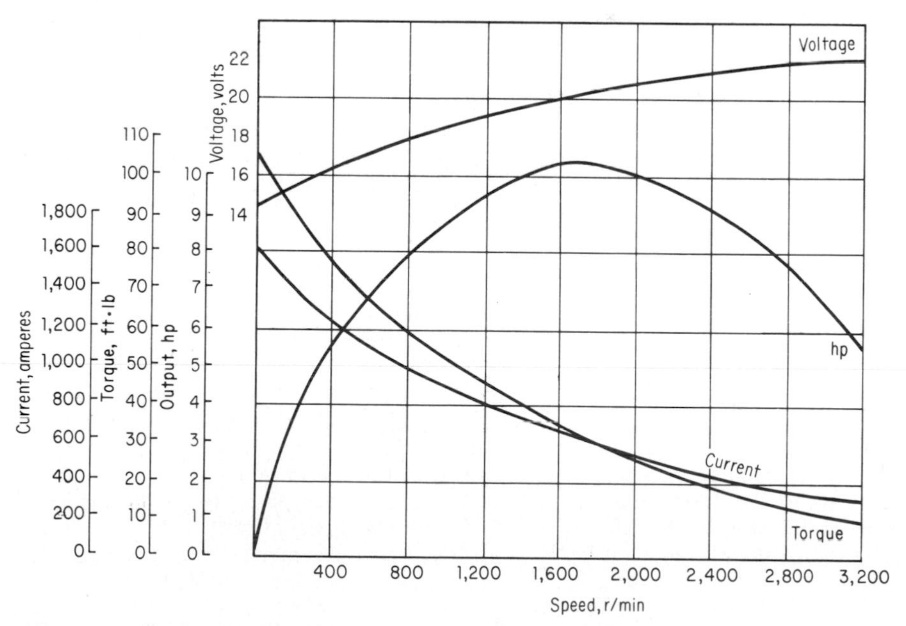

FIG. 16. 5⅛-in-diameter 24-V motor performance characteristics.

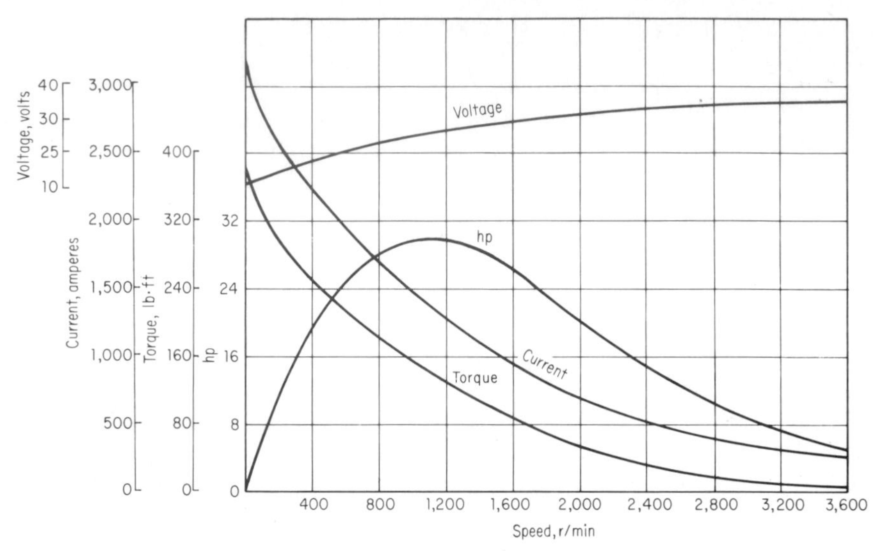

FIG. 17. 6⅝-in-diameter 36-V motor performance characteristics.

and the temperature of the battery (Ref. 3). The first two of these are evident, but it is not generally recognized that the battery temperature has a large effect on the power available. Based on 100 percent at 27°C (80°F), the available battery power is only 60 percent at 0°C (32°F) and is further reduced to 40 percent at −18°C (0°F). Unfortunately, when the power requirement is greater at lower temperature, the battery has the lowest available power. It is therefore important to establish the proper size and characteristics of the battery to do an effective cranking job. The connecting cables must be designed and maintained to provide voltage drops within reasonable limits. This factor is covered in Ref. 4.

The battery voltages available for various temperature and load conditions can be taken from curves supplied by the battery manufacturers. A typical curve is shown in Fig. 18. Further data on battery characteristics are covered in Ref. 5. It must be noted, however, that these voltages must be reduced by the voltage drop in connecting cables and solenoid switches to arrive at the voltage available at the motor terminals.

The selection of a motor should allow a safety factor of at least 20 percent over the torque required to move the engine over the first compression stroke and at least 10 percent over the speed required for proper cranking.

24. Motor Construction. The construction of various types of cranking motors is shown in Figs. 19 and 20. Figure 19 is a typical passenger-car-type cranking motor; Fig. 20 shows a 5⅛-in motor, solenoid shift. A typical solenoid-shift cranking motor consists of three major parts, the electric-motor assembly, the solenoid assembly, and the drive mechanism. The electric motor consists of an armature assembly, a field frame and coils, and a brush-rigging assembly. All circuits are designed to carry high currents, up to 1000 A, on an intermittent-duty basis. The commutator connections and field-coil connections are welded, brazed, or soldered with high-temperature solder to withstand momentary high temperatures, even though the grade of electrical insulation in the motor is Class A, since the

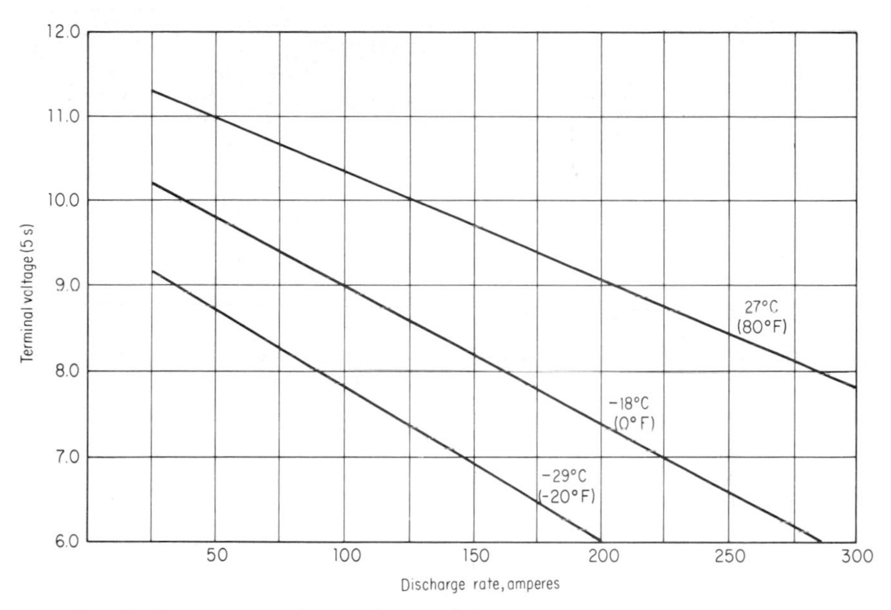

FIG. 18. 40-Ah passenger-car battery characteristics.

FIG. 19. 4½-in-diameter motor construction.

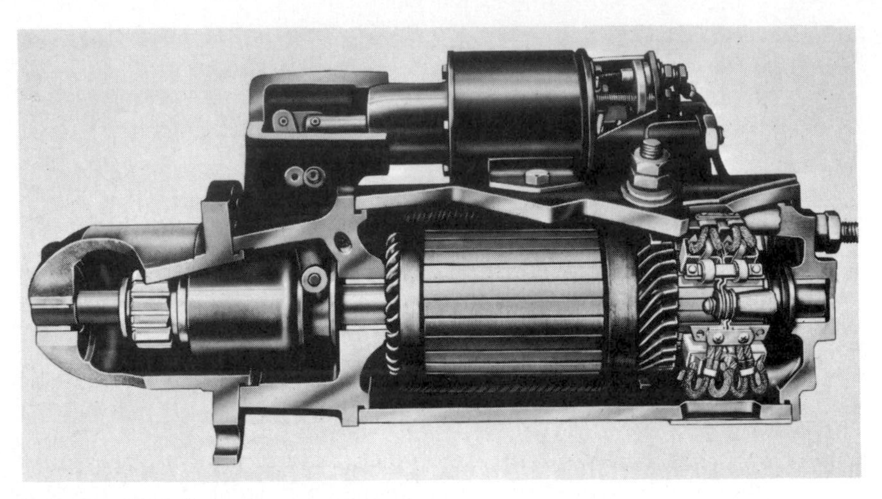

FIG. 20. 5⅛-in-diameter motor construction.

total running-life requirement is quite short. The insulation-life requirement of an electric cranking motor is the opposite of most electric motors, in that it must be able to withstand high momentary operating temperatures in the range of 260°C (500°F) on various parts, but Class A insulation is generally adequate for the life requirement, which is based on 10,000 to 50,000 starts of 1- to 5-s duration each. The brush and commutator construction is of extreme importance because of the high currents, which can reach 3000 A on some of the largest motors.

The solenoid mechanism includes an axial solenoid serving two purposes: (1) to engage the pinion gear into mesh with the engine ring gear through a system of leverages and linkages, and (2) to close a set of electric contacts in series with the motor to complete the electric circuit through the motor after the pinion gear is in mesh. Two coils are used in the solenoid assembly. The first, the pull-in coil, is energized only until the pinion is in mesh and then is shunted by the solenoid-switch contacts during cranking. The second, termed hold-in coil, keeps the drive in the engaged position and the solenoid contacts closed with sufficient pressure.

25. Engagement Methods and Mechanisms. Cranking motors must have some means of engagement, or mechanically coupling to the engine during the cranking period, and of disengaging from the engine after it has started. The requirement for disengagement from the engine is based primarily on the relative running speeds of the units, which are such that the gear ratio between them would result in excessive and destructive speeds to the cranking motor if it were not disengaged after the engine has started. For example, typical gear ratios are in the range of 15:1. Gasoline engines may have maximum running speeds as high as 5000 r/min, which would result in cranking-motor speeds of 75,000 r/min if the cranking motor were not disengaged.

Two basic types of drives are generally employed, the inertia type and the positive-shift type.

The inertia-type drive is used quite extensively on outboard engines and small air-cooled engines. In operation the drive tends to resist rotation because of its inertia, but is forced toward the ring gear by the rotation of the threaded portion of the motor shaft assembly. Upon engine startup the pinion is driven by the ring gear at a speed greater than the armature and the drive is driven out of engagement. A spring, rubber shock-absorbing device, or friction clutch is used in the drive assembly to prevent mechanical damage. An additional spring is used in the drive to prevent the pinion from vibrating into the ring gear during engine operation.

The positive-shift type drive employs a solenoid actuator which acts through a mechanical linkage to force the drive pinion into engagement with the ring gear. When the drive is fully engaged, a set of electric contacts contained in the solenoid are connected by the motion of the solenoid plunger, allowing current to flow to the motor, which starts the cranking operation. A motor using this type of drive is shown in Fig. 19.

After a motor and a drive have been selected, it would be well to review the ratios and other mechanical and electrical conditions to arrive at an expected performance. A final test of the complete engine and cranking-motor assembly under actual operating conditions should then be made to check the design.

Engine testing should include cranking tests for the required number of starts, and a continuous endurance run of the engine should be completed to check the cranking motor for the effects of vibration. Cranking-motor tests at the highest expected voltage using a locked ring gear will aid in checking the tooth strength in the pinion and the ring gear, the rigidity of the motor mounting, and in determining whether the shaft strength is adequate.

REFERENCES

1. Cameron, Pettit, and Rowls, "Cold Cranking Team," SAE paper 894-B, Sept. 1964.
2. Pettit and Cameron, "Cranking Motor Requirements," SAE paper 660012, Jan. 1966.
3. Robert W. Smeaton, Ed., *Switchgear and Control Handbook* (McGraw-Hill, New York, 1987).
4. *SAE Handbook* (1966).
5. Kruger and Barrick, "Battery Ratings," SAE paper 660029, Jan. 1966.

PART 5

MINING MOTORS

Peter J. Tsivite, Ph.D.,* and Walter G. Stiffler†

Electric motors are the prime source of mechanical power in the mining industry. They are used extensively in all areas of mining, and installations include the mining of coal, potash, salt, iron, copper, and other minerals.

26. Auxiliary Motors. The application of mine motors is divided into two basic groups. Motors in the first group are used as auxiliaries to the basic mining effort. These motors drive compressors, fans, pumps, conveyors, and hoists.

27. Face Motors. The second group includes the use of motors at the mine face that are directly involved in the cutting and removal of minerals. These motors are used on drills, continuous mine loaders, cutting machines, shuttle cars, long-wall shearers, and conveyors, where they drive pumps, drills, cutting chains, ripper bars, boring arms, loading arms, drum cutters, conveyors, and traction drives.

There is a considerable difference between auxiliary motors and face motors. Auxiliary motors are generally modifications of general-purpose industrial motors, while face motors are specially designed for specific functions. Auxiliary motors are normally fixed to the floor, while face motors are mounted on mining machines. The duty of the auxiliary motors is generally well defined and steady, while the face-motor duty consists of random loading and contains a number of high-shock loads.

The overall electrical and mechanical demands placed on face-type motors are considerably greater than those placed on auxiliary motors. These demands have evolved the face motors into a highly specialized design that is tailored for specific mining machines. The remaining portion of Part 5 will be devoted to the special features of the design and construction of face-type mine motors.

28. Power Supply. Both ac and dc mine motors are used. Dc mine motors operate on constant-potential 250- or 500-V power from variable-voltage on-board rectified alternating current for traction service.

Ac mine motors normally operate on 440, 550, and 950 V 60 Hz, and on 380, 415, and 865 V 50 Hz. The 950-V motor is usually dual voltage 950/865 V 60/50 Hz. The 60-Hz power systems that provide this power are 480-, 600-, or 1000-V supplies.

Since 1964 NEMA has standardized on 460 and 575 V instead of 440 and 550 V. There is a trend in mining toward the higher voltages in new installations. This results in 990 V in the 1000-V classification, rather than 950 V.

*Chief Engineer, Rotating Machinery Group, Reliance Electric Company, Cleveland, Ohio; Registered Professional Engineer (Ohio); Member, IEEE.

†Manager of Engineering, Rotating Machinery Group, Reliance Electric Company, Cleveland, Ohio; Member, IEEE, NRMA.

29. Nameplate Rating. Mine motors have traditionally been rated in terms of the horsepower to meet a given winding temperature rise on the basis of continuous or intermittent duty. Continuous duty is definite, but intermittent duty is often defined only as mine duty, although in other cases it is defined in terms of a definite short-time duty, such as 15-, 30-, or 60-min duty.

One rating scheme that is being successfully used defines the horsepower rating in terms of both a continuous rating and a 60-min rating. The motors are then applied to loads such that the rms of the load is matched to the continuous horsepower rating and the peak horsepower loads are limited to the 60-min rating.

A variety of temperature-rise standards are in current use. They include standards such as B rise, H rise, or a numerical rise that depends on the type of insulation used. Although the majority of mine motors use Class F or Class H insulation materials, in an effort to obtain uniformity and liberality of motor application, a temperature-rise standard has been developed using a given numerical rise that is related to the temperature-rise standards of Class B insulated industrial motors. Enclosed industrial Class B rise motors are usually based on a 40°C ambient temperature and an 80°C rise measured by resistance. The mine-motor temperature standard is based on a 25°C ambient, the average normally encountered in mines, and uses 95°C rise by resistance to give the same total temperature as used in industrial motors.

Although the nameplate temperature rise is based on a thermometer rise, the actual motor rating is normally determined by measuring the temperature rise with embedded thermocouples and by the shutdown resistance method per IEEE Standard 1-1969 to take into account temperature hot spots that cannot be reached with a thermometer.

30. Dc Motor Characteristics. Dc mine motors normally have base speeds of 1175 or 1750 r/min. These speeds have been selected to match the speeds of four- and six-pole ac motors and give a compromise between having enough speed to provide the required horsepower and yet a speed low enough to provide reliability.

The design characteristics depend on the actual use of the motor (Fig. 21). Pump motors are generally stabilized series or compound motors with no-load to full-load speed regulation ranging from 10 to 15 percent. Loading-arm and cutting motors are normally heavily compounded with 30 to 35 percent speed regulation. Traction motors are series-wound and the armature assembly is constructed to withstand speeds as high as 6000 r/min, which can occur during overhauling operations of shuttle cars.

In cases where motors are mechanically paralleled, the motors are often matched in speed to within a ±3 percent speed variation to obtain the correct load-sharing characteristic. In other cases proper load sharing with mechanically paralleled motors is obtained by using cumulatively com-

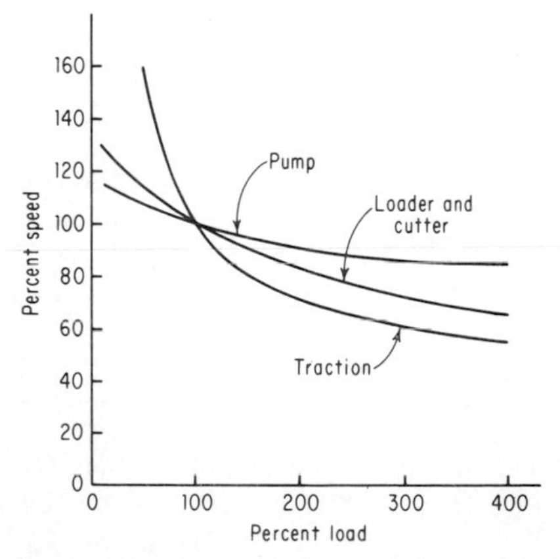

FIG. 21. Typical dc mine-motor characteristics for pump, loader and cutter, and traction.

pounded motors with differential series fields that are cross-connected in the armature circuits of the other motor.

31. Ac Motor Characteristics. The common ac induction-motor synchronous speeds are 1200 and 1800 r/min on 60 Hz, which are a compromise between horsepower ability and reliability.

Pump motors normally have speed-torque characteristics similar to NEMA Design B or A (Fig. 22). Loading-arm and cutting motors normally have special locked and maximum torques that are higher than NEMA Design A torques. In addition, higher slip characteristics are sometimes used for load sharing and for softness during sudden load application. Direct-connected traction motors have a high-slip NEMA type D torque characteristic. Single-winding two-speed and double-winding three-speed traction motors are used to give the proper torques and speeds during jogging and reversing operations. The traction-motor locked torque is normally selected so that the wheels or crawler tractor treads will slip before the motor stalls. Traction motors must be capable of frequent cycling occurring in the form of starting, stopping, plug reversing, or regenerative braking during an overhauling operation. Conveyor motors frequently have speed-torque characteristics similar to NEMA Design C, but sometimes they have higher slips for load sharing.

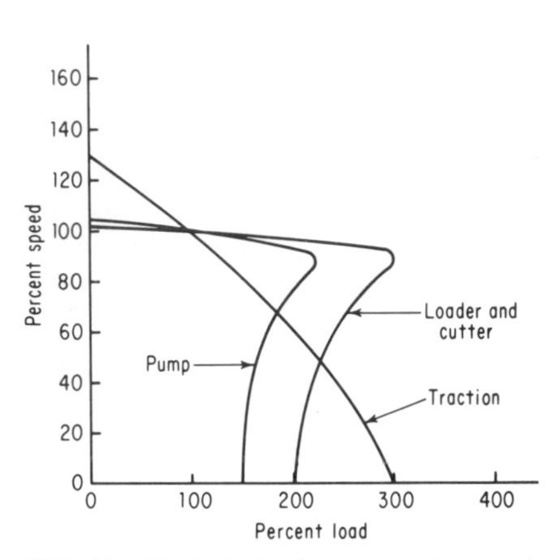

FIG. 22. Typical ac mine-motor characteristics for pump, loader and cutter, and traction.

32. General-Mine-Duty Requirements. Mine motors are generally supplied by a trailing cable that is often 500 ft long. Because of the line drop, the current requirement of mine motors is important in determining the starting and maximum torques available. The torque per ampere becomes an important figure of merit in evaluating the torque ability of a mine motor.

Direct-connected traction motors and head-mounted loading and cutting motors in general are subjected to the most severe duty that electric motors encounter. The motors are normally operated at their full rating and are subjected to severe overloads and stalls of unpredictable frequency and duration as well as frequent cycling. External ventilation may be lacking and high temperatures exist. In addition, extreme mechanical stress from heavy shock loads may occur when the mining machines are bumped during loading operations or when the loading and cutting heads are suddenly loaded or stalled.

It is well known that when the line voltage is suddenly applied to a motor, transient torques, currents, and voltages are developed that can be many times the steady-state values. These electric and mechanical forces can be particularly severe if the voltage is applied when the motor is not at standstill or during speed changing of a multispeed motor. The normal operation of head and traction motors includes the above operations to a much higher degree than generally found with industrial motors. Consequently mine motors must be constructed more heavily than industrial motors to stand the transient voltages, currents, torques, and external mechanical shocks that severely stress the windings and mechanical structure.

33. Explosionproof Requirements. Because of the ever-present danger of concentrations of explosive methane gas and combustible coal dust at the working faces of the mine, the motors used in these areas are specially designed to be explosionproof, that is, they are so constructed that they can be safely operated in the hazardous gassy and dusty atmosphere of the mine-face areas without causing the ignition of fires and explosions or contributing to the propagation of fires resulting from other sources of ignition.

The explosionproof qualities of a motor result from a construction that will safely contain within the motor enclosure any fire or explosion which may occur there, without the emission of incendiary sparks or flame which would, in turn, result in the ignition of the gas or dust surrounding the motor.

This characteristic is accomplished by provision for a strong and rugged motor enclosure to withstand the force of an internal explosion, and by long, tight-fitting joints between the parts of the enclosure to cool and extinguish the flames and sparks of an internal explosion before they can reach the surrounding atmosphere.

The construction features which make a motor explosionproof are tailored to the rigid standards of the Mining Safety and Health Association (MSHA). MSHA inspects and tests all new designs of motors under explosion conditions and must be satisfied as to their explosionproof adequacy before they will certify them as suitable for use in the face areas of a mine.

34. Mine-Service Requirements. Because of the limited space at the mine face, the mining machine and consequently the mine motor is limited by space; so the motor external geometry is tailored to fit into available space on the mining machine.

The mine-motor construction must also include features of reliability and serviceability. Although the mechanical and electrical stresses on mine motors are severe, the adverse working conditions at the mine face require that the motor perform reliably and that maintenance and service be readily accomplished. Servicing at the mine face is complicated by the lack of space, light, cleanliness, tools and equipment, technical information, and skilled personnel. When applying motors to mine equipment, the motor construction must be considered with the above limitations in mind.

35. Enclosures. The enclosures are of a rugged, compact construction and are provided with generous handhole openings and removable brackets to give access to parts that require adjustment, inspection, or maintenance. The component parts of the enclosure such as the frame, covers, brackets, and bearing cartridges are made of cast or fabricated steel or cast ductile iron with adequate sections to withstand the extreme stresses associated with mine service (Fig. 23).

Because of the explosionproof requirement all mine motors are totally enclosed. The most common types are nonventilated or fan-cooled, although motors with liquid-cooled frames or shafts are also used. Some special motors are liquid-filled and contain an internal liquid-circulating system which uses the liquid to transfer the motor losses to a water- or liquid-cooled frame jacket.

Mine-motor enclosures are commonly designated as follows:

Totally enclosed, nonventilated, explosionproof	TENVXP or TEXP
Totally enclosed, fan-cooled, explosionproof	TEFCXP or FCXP

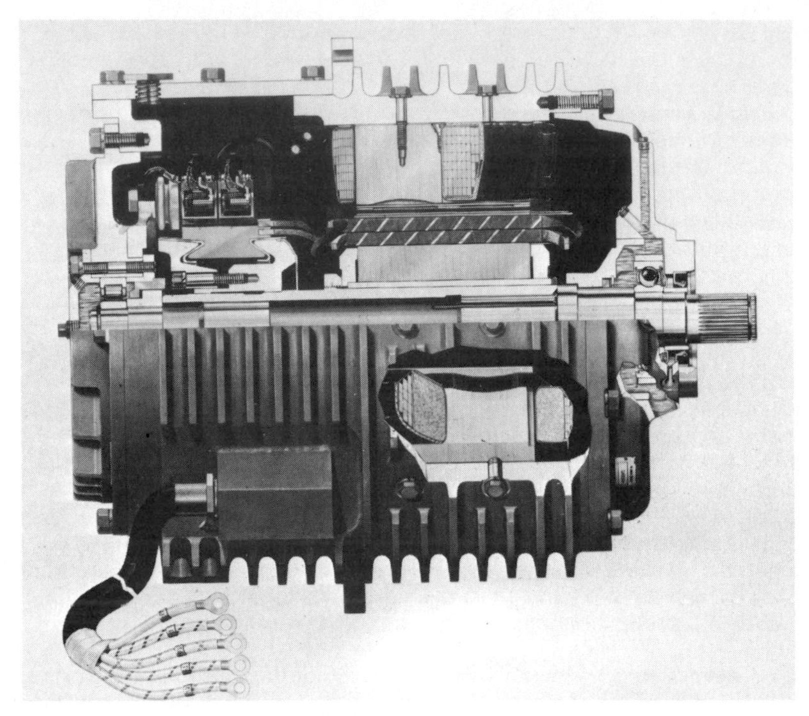

FIG. 23. Typical explosionproof mine motor rated 20 hp at 250 V dc.

Totally enclosed, liquid-cooled, explosionproof	TEWCXP or WCXP
Totally enclosed, liquid-filled, explosionproof	TEOCXP or OCXP

36. Mounting and Coupling Arrangement. Mine motors are normally flange-mounted directly to a gearbox with additional support provided by "belly bands" or by "feet" or "pads" on the motor frame. The shaft extension is normally splined or tapered to accept gears or couplings that are connected to the driven load through pinion gears, couplings, or universal joints, or they are direct-connected. The shaft is normally made of alloy steel and machined free of grooves to decrease stress risers. Splined extensions are hardened to minimize wear.

37. Bearings. Antifriction bearings are almost universally used, and depending on the load requirements, they are used as ball bearings at each end or a ball and roller combination. The bearing size is generally predicted by the available space, but large, liberally rated bearings are preferred. To prevent radial preload, the minimum internal clearance for ball bearings is AFBMA no. 4 and the equivalent for roller bearings. The fits between inner race and shaft as well as between outer race and housing are designed tight to minimize "pound-out." To permit the lubricant to enter the bearing readily, open bearings are commonly used.

38. Shaft Seals. Mine motors are commonly provided with shaft seals inboard of the bearings. A typical seal is a stationary brass sleeve mounted in the bracket or cartridge. The seal runs with a small clearance to the shaft bore and thus provides a labyrinth path that retards flame or grease leakage. In addition the sleeve prevents galling of the shaft in the event of a bearing failure.

In cases where the motor is close connected to an oil-filled gearbox, a dry well or equivalent open space between motor and gearbox is preferred to prevent oil leakage into the motor. However, because of space limitations it is usually necessary for the shaft to extend into the oil-filled gearbox and oil seals are required. These seals are of the rubbing-lip type and are made of metal or high-temperature elastomers (see Sec. 13).

Because of the high-temperature operation encountered by the normal mine motor, frequent lubrication is desired; so grease entrances are made readily accessible. Oil and grease drain passages are made as large as possible to minimize plugging.

39. Lubricant. Drive-end bearings are sometimes lubricated directly by the gearcase oil. However, the most common methods of lubrication use grease. High-quality polyurea- and lithium-soap-based greases are commonly used, but some high-temperature motors require synthetic-base and oil greases (see Sec. 14).

40. Rotating Assembly. Ac induction mine motors normally use cast-aluminum or Dow metal squirrel-cage rotors, but high-torque severe-duty traction motors have rotor cages fabricated with silver-brazed or electron-beam-welded high-resistance copper alloys, nickel alloys, or stainless steel. The rotors can be mounted directly on the shaft or on a spider or sleeve. The rotor is secured against axial movement by means of welds, or by the use of locknuts or snap rings if ease of rotor replacement is desired.

The dc armature assembly is also constructed for the rugged mine duty and for serviceability. The armature core is normally mounted on a sleeve or spider, but bolted cores mounted directly on the shaft are also used. It is preferred to have the armature core and commutator mounted on a common quill to permit shaft removal without disturbing the armature winding.

The commutator is normally of the bolt or nut type with V-ring and core materials made from cast, forged, or fabricated steel or cast ductile iron. The commutator bars use silver-bearing copper to resist loss of strength at high temperatures. The commutator bars have sufficient depth to allow for wear and repeated machining of the face. The commutator riser is integral with the bar and has a diameter close to the armature diameter to facilitate banding and insulation of the coil heads. To facilitate repair, the commutators are designed to be removable from the armature assembly by pressing, after disassembly of the mechanical locking means, without destruction of welds. A centerline relationship between the armature slot and the commutator bar is held to a close tolerance to ensure interchangeability of armatures without adjustment of the brush location.

Armature coils are normally retained in their slots, with slot wedges cut from a woven-glass laminate bonded with a high-temperature resin. The armature-coil heads are banded against insulated coil supports that are mounted to the armature cores. Steel banding wire is used, but impregnated glass roving is popular because of the limited destruction caused during a banding failure. The armature coils are welded to the commutator risers by tungsten inert gas (TIG) or metal inert gas (MIG) processes.

Both internal and external fans are often mounted on a rotating assembly to provide internal or external ventilation. Nonsparking materials such as brass or aluminum are required for external fans.

41. Stator Assembly. Ac motors use laminated-silicon sheet-steel stator cores with either open or semienclosed slots. The cores are secured against movement by keys, welds, or dowel pins. End laminations or fingers are used to avoid core flare.

The dc-motor stator assembly contains field poles and interpoles constructed with low-carbon-steel laminations to minimize eddy currents and residual magnetism so that commutation, field response, and pole losses will be favorable. Field poles are normally designed for use without shims, but if shims are required to adjust to a small speed tolerance, the shims are permanently fastened to the pole to avoid possible discrepancies during future repair. Interpoles often use brass and steel shims that are permanently fastened to the interpole.

42. Brush-Support Assembly. The dc motor brush-support assembly is of rugged construction, providing mechanical stability and resistance to damage from commutator flashing. The brushholders are for radial brushes and designed to give constant brush pressure over the full wear depth. The brushholder support mounts and insulates the brushholder from the rocker. Adequate creepage distances to ground and nontracking insulating materials that can resist coal dust, carbon-brush dust, oil vapor, chemical products, and heat resulting from commutator flashing are used.

The rocker is made of steel, cast ductile iron, brass, or bronze, although when the rocker is made of ferrous metal it is nickel-plated to ensure ease of rotation after exposure to the corrosive action of a mine atmosphere. The rocker is normally marked at the bracket fit to provide for accurate positioning of the rocker after disassembly or rotation for maintenance.

Normally brushes of the hammer-plate type with the brush shunt riveted to the hammer plate are used. The brushes are made large to keep the peak brush current densities under 80 A/in^2.

43. Insulation and Winding Techniques. The majority of mine motors utilize Class F or Class H insulation systems. The stator windings are dipped or vacuum-pressure impregnated in epoxy resins, which are highly resistant to moisture, oil, grease, and acids. Their high bond strength is retained at elevated temperatures to improve the mechanical strength of the winding under the severe mechanical loads of mine duty. While some polyester and silicone resins have slightly higher temperature ability, the epoxies excel in resistance to environments and bond strength at elevated temperatures.

The wire, coil-to-coil, and ground insulations are also selected for their high-temperature and mechanical abilities. Glass-covered wire and high-quality India mica composites are used for these purposes.

Where added temperature, mechanical, and chemical abilities are required, polyamide insulations are being used.

The preferred armature and stator coil construction is an open-slot rectangular-wire preinsulated drum-formed coil that is hot-and-cold pressed to the slot size.

In cases where semienclosed-slot mush coils are used, coils with formed and taped knuckles are preferred. Insulating end laminations are used to prevent cutting of the slot cell at the slot edges.

Field coils and intercoils are commonly made to fit the pole body snugly and are secured against movement with metal coil tighteners at the armature side so that the coils bear against the frame. The pole-fitting frame-bearing coils have excellent mechanical and heat-transfer characteristics. Glass-reinforced plastic or metal bobbins that are formed to fit the pole are also used to contain the field and intercoil windings.

In general, the insulation materials and winding techniques used on mine motors are of the highest quality to produce motors that can reliably withstand the severe mine duty.

REFERENCES

1. "General Principles for Temperature Limits in the Rating of Electric Equipment," IEEE Standard 1-1969.

PART 6

TEXTILE MOTORS

L.P. Gregory* and Paul D. Wagner†

The same basic considerations apply to the selection and application of motors for the textile industry as for any industrial application. Each industrial application, however, has certain characteristics or peculiarities which distinguish it from any other application. In the textile industry two outstanding characteristics that must be considered are environment and accelerating torque.

44. Code. The National Electrical Code defines areas where combustible fibers or flyings are present as a Class III environment. This classification applies to areas in a textile mill where fibers are being handled or processed. The NEC imposes a limit to the surface temperature of motors applied in a Class III environment of 165°C (329°F) for equipment not subject to overload, and of 125°C (257°F) for equipment that may be overloaded.

In the past few years the use of so-called hot motors, those rated 75°C rise or with Class B or higher insulation and service factor, has been at the root of a controversy as to whether these represent an increased fire hazard over the previously common 40°C motors. Table 3 lists some common textile materials along with common materials for comparison and states the self-ignition temperature determined by test. The samples used were pure and clean. Actually, the self-ignition temperature is affected by the length of time of exposure, the presence of

TABLE 3. Self-Ignition Temperatures for Pure, Clean Samples

Material	°C	°F
Cotton, absorbent, rolls	266	511
Cotton, plating, rolls	230	446
Woolen blanket, roll	205	401
Viscose rayon, roll	280	536
Nylon roll	475	887
Silk roll	570	1058
Pure scoured wool	525	977
Pine shavings	228	442
Newsprint	230	446
Wood fiber board	216	421

*Senior Application Engineer, Allis-Chalmers Manufacturing Company, Norwood, Ohio; Member, IEEE.

†Manager, Advanced Technology, Industrial Motor Division, Siemens Energy and Automation, Inc., Little Rock, Ark; Senior Member, IEEE.

impurities, and the technique used to prepare and expose samples. For example, wool-blanket material containing less than 1 percent cotton and some wool fat was found to have a self-ignition temperature of 215°C (420°F) while pure scoured wool demonstrated a temperature of 525°C (977°F). Cotton-mill floor sweepings including dirty, oily lint exposed to 130°C (266°F) for 24 h showed no sign of impending ignition.

45. Motor Temperature. The total motor temperature is governed by the total temperature that its insulation system can withstand continuously with acceptable insulation life. The motor designer takes this factor into consideration when designing a given machine for a given set of conditions, and selects an insulation system accordingly. The total insulation temperature is set by standards at 105°C for Class A, 130°C for Class B, 155°C for Class F, and 180°C for Class H. These temperatures represent continuous hot-spot conditions in the motor winding and are well below the self-ignition temperatures listed in Table 3.

Since these temperatures are internal, and cooling of the motor takes place by migration of heat from this area to a cooler area, that of the surface of the motor, it is obvious that the surface temperature of the motor is well below the self-ignition temperatures of Table 3, and below the NEC limit of 125°C. The only consideration even remotely resembling an exception would be a totally enclosed nonventilated motor where the surface temperature would be closer to that of the internal hot spot. Such motors are normally 75°C winding-temperature rise, and again this temperature would place the surface temperature within acceptable limits.

This margin does not imply that, once installed, a motor needs no further attention. Good housekeeping should prevail at all times, particularly in textile mills. Excess lubricant should be wiped from motor surfaces at the time of relubrication, and any bleeding of lubricant should be investigated at once for cause. Lint buildup on motor surfaces should be prevented, or minimized, by frequent brushing or blowing. A blanket of lint, whether caused by adherence to excess lubricant or to rough surfaces or projections, represents also a blanket of insulation which, while it may not be the direct cause of self-ignition, will cause an increase in motor temperature and premature failure.

46. Enclosures. Motor enclosures for textile application are often described by reference to "lint-free" motors, a carryover from the day of the special open-motor design. The totally enclosed fan-cooled (TEFC) and the totally enclosed nonventilated (TENV) motors are generally used in areas where lint and combustible fibers may be loose in the atmosphere. The TENV construction is available into the 2-hp higher-speed ratings, and TEFC motors are available in all sizes. It should be known that the enclosures used in TEFC textile motors are not identical to those used in TEFC motors for other industrial applications. The motor manufacturer makes a special effort to provide smooth surfaces and contours to minimize the accumulation of lint and to omit grids and screens upon which lint might accumulate. Again, good housekeeping must prevail. It is not possible to call out specifically what type of motor enclosure must be employed. Generally motors in areas that may involve a linting condition should be of the "textile" type, TENV or TEFC, while those for auxiliary equipment, pumps, fans, compressors, etc., located out of the production area can be of conventional designs. The matter of interchangeability should be given some thought as a matter of basic economics.

47. Torque. Accelerating torque is a matter of prime importance in selecting motors for textile applications. Most motor applications in the textile industry can use a standard NEMA Design B motor. However, where a departure from this does occur, it is sufficient to require a special design, the performance of which has come to be described with the machine name, such as card-drive motor, draw-frame motor, roving-frame motor, twister motor, or loom motor. This nomenclature may also be said to define at least the basic requirements of the motor.

a. Card drive. The outstanding characteristic of a card, as far as the drive motor is concerned, is high inertia as compared with that of the motor. In the old line-shaft–group-drive concept compensation for this inertia was obtained in the tight-loose pulley-drive scheme. This slippage has been carried into the individual drive in a variety of ways. One method employs belt drives which permit belt slippage on startup; another employs some type of clutch, either mechanical-manually controlled or centrifugally governed. In any event, the motor should be permitted to accelerate well toward full speed before the card load is imposed on the motor. This precaution reduces the need for a highly special motor, although the motor should have higher accelerating torques and thermal capacity because of the driven inertia. Some drives have employed gear motors, eliminating the need for belt drives. However, in such cases a motor with very high starting torque, in the area of 400 percent, and with high thermal capacity is required.

For multicylinder cards, the drive is usually composed of a motor, a fluid coupling, and a gear reducer. This equipment is floor-mounted, with final transmission to the cylinders by means of belts. Here again, slip is introduced by means of the fluid coupling, and in most cases a standard NEMA Design B motor is applicable.

The important factor to remember in a card drive is the inertia reflected to the motor shaft, and this inertia needs to be reviewed each time an increase in speed is considered; the card is basically a constant-torque machine, and as speed is increased, the required horsepower increases proportionately. A drive that may have been satisfactory at a given speed may not be capable of satisfactory performance at a higher speed. Figure 24 shows a typical single-cylinder card speed-torque curve, with motor curve, centrifugal-clutch curve, and the resultant torque curve. A similar construction may be employed to analyze the acceleration performance of other types of card drives.

b. Draw-frame motors usually are designed with a starting torque of 100 percent and a breakdown torque of 180 percent. This design is based quite often on a rating of 4 hp rather than the usual standard of 3 or 5 hp.

c. Roving-frame motors are often referred to as soft-start motors. The problem here results from the lack of strength in the yarn in process. The starting torque cannot be applied with impact, and acceleration must be smooth and uniform if yarn breakage is to be avoided. This characteristic is accomplished by providing a dual- or triple-torque motor. The motor is provided with a special winding and a multiplicity of leads. It is reconnectable in the motor conduit box to permit selection of the torque characteristic that will no more than start and accelerate the frame without damage to the material in process. In this application it is the usual practice to refer to average accelerating torque rather than starting torque. Extreme care must be used when specifying these motors. The dual-torque motor is connected two-circuit wye for high torque and series delta for low torque. The high torque is approximately 35 percent greater than the low torque. The triple-torque motor has a delta winding that is provided with taps which usually result in torques of 100, 84, and 69 percent.

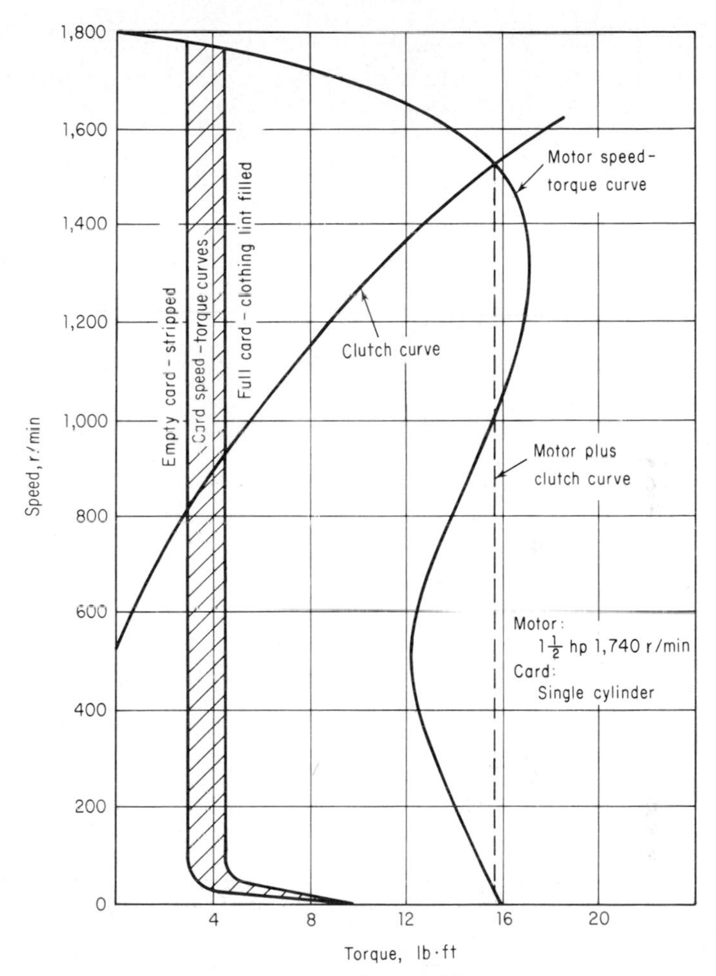

FIG. 24. Speed-torque relationship for card drive.

In practice the proper torque connection is determined by trial. Usually on new frames or those recently rebuilt, the high-torque connection is used initially until the frame has had time to run in; then reconnnection to a lower torque is made. There is no rule of thumb or empirical equation by which to determine the torques required or the connection to be employed. Design torques and percent steps evolve from the design and performance data that the machinery manufacturer has accumulated over years of experience with the machine design in question. Figure 25 shows the soft-start principle, and Fig. 26 the typical motor connection.

d. Twister applications require consideration of the average accelerating torque. Normally the motor has less than the NEMA Design B torque characteristic. The starting torque is usually 140 to 160 percent that of a NEMA Design B motor.

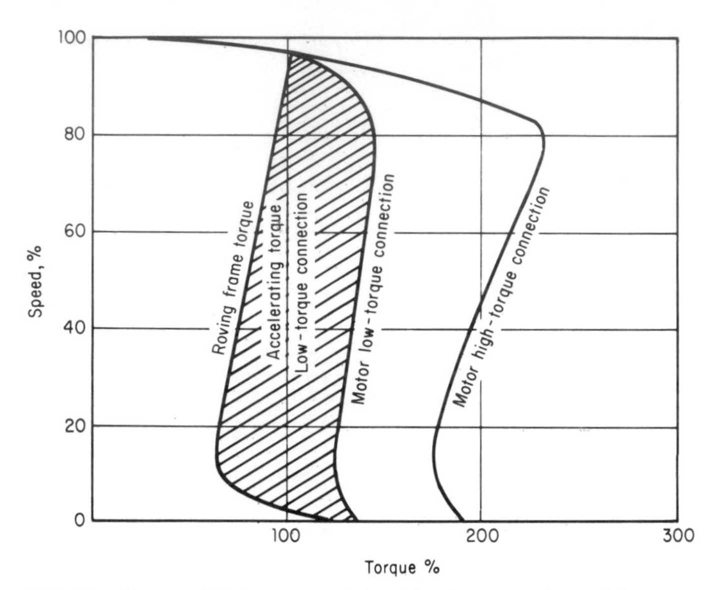

FIG. 25. Low- and high-torque relationships for roving-frame drive.

e. Loom motors deserve more care in application than perhaps any other than the roving-frame drive. These motors also are categorized as to whether they are used for a conventional shuttle-type loom or for a shuttleless loom. The more difficult and exacting applications are those for the conventional loom. The motor requirements must be defined by stating loom speed (picks per minute), loom size (width in inches), fabric being produced, and auxiliaries present on the loom. The horsepower demand has been determined by experience and an accumulation of a vast amount of data by the loom manufacturer. The loom manufacturer should be consulted as to the horsepower required for a given set of conditions. Loom motors have a high starting torque and high thermal capacity and are subject to regeneration during certain portions of the loom cycle. There are periods in which the loom is actually driving the motor. Figure 27 shows a typical power swing. Smoothing of the cycle can be accomplished by adding inertia to the motor in the form of flywheels. Matching a motor to a given loom actually becomes a delicate balancing proposition. It is desirable to drive the loom at as high an average speed as possible, and at as constant a speed as possible. The drive must, however, be consistent with the particular application; yet some softness in the drive is required. Addition of too much flywheel effect in the motor results in a stiff drive; and although the regeneration effect may be reduced and higher average speed attained, the wear will be directed to other parts of the system, including drive pinion and loom clutch. Some slippage will occur, and it is best and least expensive to take it in the motor. The conventional loom drive employs a clutch and brake as part of the loom mechanism. The power-transmitter type drive incorporates a clutch and brake inside the motor housing. These parts then are operating at a higher speed and are smaller than their counterparts in the loom. The advantage here is a smaller overall package, with an overall economic advantage and a power package that can be replaced quickly in the event of transmission problems. Again, close coordination with the loom manufacturer is required if a

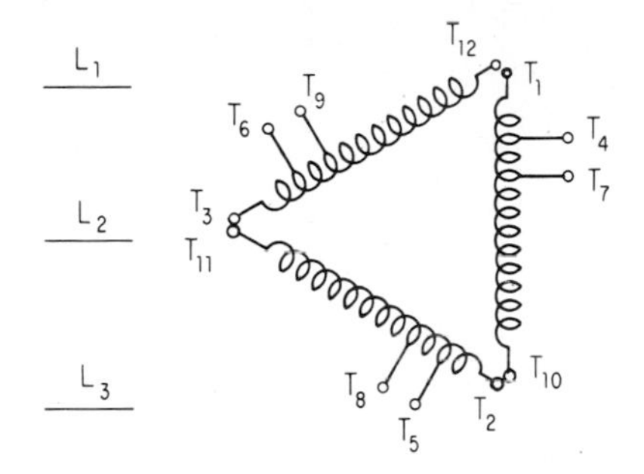

	To line			Together
	L_1	L_2	L_3	
High torque	T_{1-12}	T_{2-10}	T_{3-11}	
Medium torque	T_1	T_2	T_3	T_{4-12} T_{5-10} T_{6-11}
Low torque	T_1	T_2	T_3	T_{7-12} T_{8-10} T_{9-11}

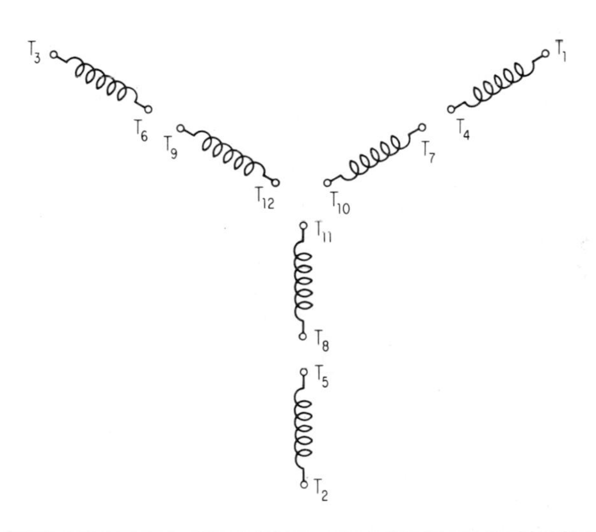

	To line			Together
	L_1	L_2	L_3	
High torque (YY)	T_1 T_7	T_2 T_8	T_3 T_9	T_{4-5-6} – $T_{10-11-12}$
Low torque (△)	T_1 T_{12}	T_2 T_{10}	T_3 T_{11}	T_{4-7} – T_{5-8} – T_{6-9}

FIG. 26. Typical triple- and dual-torque connections for roving-frame motor.

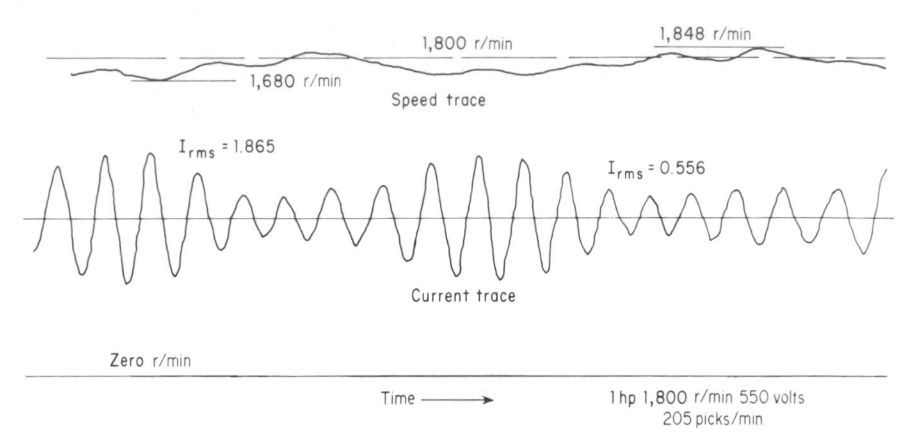

FIG. 27. Typical loom-motor characteristics as a function of time.

satisfactory application is to result. Shuttleless looms operate at a considerably higher speed than shuttle types, primarily because many of the inertia-causing elements have been eliminated. In most cases, normal-torque motors are applicable.

f. Warpers, slashers, and range drives provide an area where variable speed is required and introduce a fine degree of economics into the drive selection. The speed range may be desired merely for flexibility in use of the machine for inching, tension control, or setup, or for combinations of all these factors.

48. Speed Control. Table 4 lists representative methods of obtaining the speed range and compares these various systems with respect to the features available. The table lists the drives roughly in the order of increasing cost, although it is difficult to state that one system is more expensive than the other as there is no common basis for cost comparison. The machine designer must determine what work is to be done, how it is to be done, what is required to do it, and then note the systems available which are capable of meeting these requirements. Then as a final step, if there is an overlap between systems, the designer must determine which is best for the present and future use of the machine. Variable-frequency ac inverter drives have come of age, particularly in the field of man-made fiber processing. This system consists basically of providing a source of power in which the frequency and voltage can be varied over a desired range, and a motor, or motors, can be designed to operate over this frequency range. The characteristic of an induction, or reluctance, motor, operated over a frequency range, is that of constant torque, and the system operates at constant voltage per cycle. To clarify, at 30 Hz the voltage and speed are half what they would be at 60 Hz. The normal frequency range is 25 to 130 Hz. Below 25 Hz some means must be provided to raise the voltage of the system to produce torque, for at very low frequencies the machine would lack sufficient torque to be practical. Motors for variable-frequency systems usually employ a higher-grade steel in the core structure than standard machines. Again, in the design of a variable-frequency system, the electrical designer must work closely with the machine designer to assure adequate results. A specific bench mark is usually given as a starting point, such as a specific horsepower at a specific speed, and a speed range desired. Starting conditions

TABLE 4. Typical Adjustable-Speed Drives

	Ac, mechanical speed changers	Dc, SCR source	Dc, MG source	Ac, magnetic coupling	Ac, inverter source
Hp range of sizes	1–60	1–300	5–200	1–500, up	1–200
Speed range, Std	2:1, 10:1	20:1	8:1, 16:1	16:1	8:1, 16:1
% speed regulation, Std	3	5	5	2	1
Speed control by	Adjustable sheaves	Armature and field	Armature and field	Excitation voltage	Internal oscillator
Optional features:					
Preset speed	A	Std	Std	Std	Std
Timed aceleration and deceleration	NA	Std	Std	A	Std
Preset jog	NA	A	A	A	A
Threading	NA	A	A	A	A
Braking, regenerating	A	NA	Std	NA	A
Braking, other	Mechanical	Dynamic	Dynamic	Eddy current	Dynamic
Speed regulation	0.5	0.1	0.1	0.25	0.1

Std = standard. A = available. NA = not available. SCR = silicon-controlled rectifier. MG = Motor-generator.

must also be defined so that a power source of adequate capacity can be provided. The worst condition possible must be anticipated and the power unit sized accordingly.

The advantages of such a system include those of stepless speed range, inching, threading, controlled acceleration, the simplicity and economics of the induction, or reluctance, motor, and for multimotor drives, the assurance of synchronized operation when synchronous reluctance motors are used.

49. Operating Conditions. Cotton-gin motors should also be of the TEFC textile design. Here the matter of foreign solids in the atmosphere is basically one of lint. However, there is much other organic trash freed in the operating area in the way of stems, twigs, etc. In older gins it becomes a matter of keeping this material out of the motor. In more modern gins, with effective machinery enclosure and more sophisticated material-handing equipment leading to greater cleanliness, drip-proof motors have been used with success. This does not condone the use of drip-proof motors, for the Class III hazard is potentially present, and TEFC or TENV motors are recommended where available. Normal-torque Design B motors are generally satisfactory except for mechanical conveyors (Design C) or baling presses (Design D). Here again the machine design and method of operation must be reviewed for accurate motor selection.

Problem areas are not peculiar to this particular industry; however, they do differ somewhat from the problems experienced in other industries. The preferred enclosure for production areas is the TEFC textile type or TENV. The fan location, whether shaft end and/or opposite shaft end on TEFC machines, should be determined by the amount of lint anticipated in the atmosphere and the proximity of other equipment and mechanisms to the airflow from the motor. Good housekeeping should be promoted to keep the motor and its locations as clean and free of lint accumulation as possible. With greater utilization of floor space of paramount importance, special mechanical designs of motors to fit under the frame, particularly on spinning frames, give rise to special problems. Compact high-speed belt drives for such applications lead the motor designer to a more specific type of design. Any consideration of increased frame speed or load should be coordinated with the motor and frame manufacturer, to assure that the equipment is capable of that for which it is intended.

Poor starting characteristics and ends down on startup can result from too little or too much starting torque. Investigate the motor characteristics, and make certain that the motor has the exact torque characteristic for the application, and that it is supplied with nameplate voltage and frequency. Close coordination with the machinery manufacturer is necessary for a successful application, particularly when existing equipment is to be modified for speed-up, increased load, or other factors not originally included in the specification.

REFERENCES

1. "National Electrical Code," ANSI/NFPA No. 70-1984, chap. 5, arts. 500-6, 503-1, 503-6, 503-7.
2. "Recommended Practice for Electric Installations on Textile Machinery," IEEE Standard 77-1965 (Reaff 1972).
3. "Ignition Temperature of Solids," Oregon State Agricultural College Engineering Experiment Station Bull. 26.
4. "Allowable Surface Temperatures," AIEE Textile Industry Committee Rept., 1962.
5. "Storage of Combustible Fibers," NFPA No. 44.
6. "Dust Explosion in Industrial Plants," NFPA No. 63-1971.

PART 7

NAVY AND MARINE MOTORS

Alfred P. Nickley*

Motors for shipboard service are a special type, and the application engineer must be aware of the peculiar requirements of these motors that set them apart from the general-purpose industrial motor. Industrial motors are often mounted on rigid, vibration-free foundations in a relatively low humidity atmosphere. When a motor is installed on board a ship, the foundation is no longer a thick concrete slab but a steel deck subject to vibration, shock, pitch, and roll, and the air is usually very humid and salt-laden. The environment in which the motor is to operate as well as the function it is to perform point up the important differences between industrial and shipboard motors.

50. Shipboard Service. To meet the demands of shipboard service, two general classifications of motors have evolved: commercial marine-type motors designed for ships in commercial-type service and military-type motors designed for certain naval ships. Each type of motor is covered by a different set of specifications.

a. Commercial regulatory bodies. Motors for commercial-ship services are usually governed by three major specifications: the American Bureau of Shipping Rules for Building and Classing Steel Vessels; the U.S. Coast Guard Electrical Engineering Regulations; and IEEE Standard 45 (Ref. 1).

b. U.S. Navy specifications. The requirements for motors intended for naval shipboard service are contained in four general navy motor specifications: ac integral horsepower, ac fractional horsepower, dc integral horsepower, and dc fractional horsepower. The motors within these four specifications are divided into distinct groups. Those motors essential to the military effectiveness of the ship are classed as Service A, while those not essential to the military effectiveness of the ship are classed Service C. Service A motors for submarine duty are often considered as a third class. There is no Service B motor.

51. Philosophy of Specifications. The governing specification for all equipment on board ship and for the construction of the ship itself is the specification for the construction of the ship, more commonly referred to as the ship's spec. Contained within this specification are the applicable equipment specifications, contract plans, and contract guidance plans with which the completed ship must comply.

The requirements for each item installed on board ship either are spelled out in the ship's spec in detail or reference is made to a commercial or military specification for that particular item. Whenever a conflict exists between the ship's spec and an equipment specification, the ship's spec will take precedence. A similar hierarchy exists between an equipment specification and the component specifications referenced within it. The equipment specification will always override the individual-component specification.

*Supervisor, Motor and Motor-Generator Branch, Electrical Division, Naval Sea Systems Command, Washington, D.C.

Since the individual equipment specifications as well as the ship's spec are modified from time to time to take advantage of the latest engineering developments, the application engineer is well advised to ascertain with which issue of a specification he or she must comply. As a general rule, the specification issue in effect at the date of the request for bid applies. An exception to this rule may occur when, for the sake of standardization within a class of ships, the specification in effect at the date of the first ship contract is specified in a later bid request.

52. Considerations for All Shipboard Motors. The following are general considerations which apply to all motors designed for shipboard service. The application engineer must refer to the ship's spec to determine the specific motor requirements.

a. Ambient temperature. Motors located in machinery spaces or boiler rooms are generally rated on the basis of a 50°C ambient. Unless otherwise specified, motors for other locations may be rated on the basis of a 40°C ambient.

b. Motor size. All motors should be as small and lightweight as is practicable.

c. Squirrel-cage motors preferred. To keep the number of brush-type motors at a minimum, squirrel-cage motors are recommended wherever suitable and as permitted by the applicable specifications.

d. Corrosion resistance. The metal frame, end shields, terminal-box assembly, and all hardware must be treated for corrosion resistance or made from a corrosion-resistant material. Squirrel-cage rotors should be protected by coating with an insulating varnish. Shafts of corrosive material should be similarly varnished; however, extremely corrosive environments may dictate the use of a noncorrosive material such as monel or stainless steel.

Consideration should also be given to the union of dissimilar metals, particularly in a salt-laden atmosphere. Since galvanic corrosion takes place when two dissimilar metals are in contact in the presence of an electrolyte, a strong electrolyte such as saltwater will increase the rate of corrosion. To minimize this electrolytic corrosion, reference to the galvanic series of dissimilar metals should be made to avoid the contact of two metals far apart in the series. Metal-to-metal contact should be considered to exist between parts that depend only upon paint for corrosion resistance.

e. Inclination. Motors should be designed to operate satisfactorily when the shaft is permanently inclined or the motor base is tilted at an angle as specified in the applicable specifications.

f. Lubrication. Because oil-lubrication systems are more sensitive to roll and pitch than grease-lubrication systems, the use of oil lubrication should be limited to those applications where the speed of the ball bearing requires oil lubrication or sleeve bearings must be used to satisfy an application requirement. Although motors are generally installed with their shafts in the fore and aft direction of the ship, an application may require athwartship mounting. In this case the oil-lubrication system will require special consideration to prevent oil spills during pitch and roll.

g. Silicone materials. The use of silicone varnishes, tapes, laminates, compounds, rubber, greases, or silicone materials of any type should be avoided in enclosed recirculating-air brush-type motors. The silicone vapor evolved during operation of the motor causes excessive brush wear and premature failure of the motor.

h. Accessibility. The motors should be designed so that all parts that may require servicing, repair, or replacement during the life of the motor are readily accessible for repair or replacement. Motors employing brushes should have access openings of sufficient size to permit a view of the brushes while the motor is running and to allow ease of brush replacement.

i. Lifting means. Motors weighing 150 lb or more should be provided with removable lifting eyes.

j. Enclosure. The degree of enclosure is dependent upon the environment in which the motor is required to operate. Watertight, waterproof, or spraytight motors are required where the motor is subjected to weather exposure. Motors with external cooling fans must not be used in applications exposed to the weather. As ships may operate in frigid climates, the fans will become coated with ice and prevent the motor from starting. Totally enclosed motors are suitable for nonhazardous locations where protection against dirt, dust, water vapors, etc., is required. Explosionproof motors are necessary when the motors must operate in an explosive atmosphere. Motors for submerged operation may be of the free-flooding type, or the motor enclosure may be so designed as to prevent the entrance of water into the motor. Dripproof motor construction is the most often used for internal shipboard applications and is commonly found in machinery spaces.

Considerable attention should be given to the selection of an enclosure since the wrong choice may result in a motor larger than necessary or a motor totally unsuited for the environment in which it must operate.

k. Motor drawings. Drawings supplied for commercial marine motors are generally the vendor's standard outline drawings. Service C drawings closely parallel the commercial marine drawings with the additional requirements of winding data and insulation materials.

Drawings delineating Service A motors are usually more detailed than those for commercial marine or Service C motors. Construction details, winding data, insulation materials, test data, and a complete list of materials are required to indicate compliance with the specifications and to provide information for repair and maintenance by the Navy.

53. Commercial Marine Motors. In general a commercial marine motor is a high-grade commercial motor, meeting the requirements of the commercial regulatory bodies and the general considerations listed above plus the special requirements of the applicable specifications.

54. Navy Motors. Since Service A motors are essential to the military effectiveness of the ship, the requirements for this type of motor are considerably more rigorous than those for commercial marine or Service C. The following considerations are presented to acquaint the application engineer with the salient characteristics of Navy Service A motors and should not be construed to be the only considerations, as the applicable ship's spec and motor specification must be consulted for detail requirements.

a. Material restrictions. Minimum requirements for material are included in each motor specification. The majority of material used in Navy motors is required to be in accordance with a military specification, whether it is a ball bearing, an eyebolt, or a magnet wire. Commercial materials are permitted only when specifically allowed by the specifications or where it has been demonstrated that a particular commercial material is equal to or better than the military.

Certain materials are prohibited for use in Service A motors because of inherent characteristics which make them undesirable. For example:

1. Materials capable of producing dangerous volumes of toxic gases or other effects when subjected to elevated temperatures.

2. Any material in a form that will ignite or explode from an electric spark, flame, or from heating and will independently support combustion in the presence of air.
3. Because of the toxicity of mercury and the ease with which it contaminates nonferrous metals, mercury in any form should not be used in the manufacture or test of motors for shipboard service. Possible sources of contamination are thermometers, switches, and manometers or the handling of mercury in the vicinity of the motor.
4. Magnesium or magnesium-base alloys should not be used. Magnesium is considered to be undesirable for two basic reasons: (1) magnesium and its alloys have definite fire-hazard characteristics and may be ignited by an adjacent or surrounding fire; and (2) once burning has been established, extinguishment is exceedingly difficult.
5. Aluminum alloys should not be used for motor frames and end shields except as specifically permitted for nonmagnetic applications. When used, steel inserts must be provided for all threaded positions and bearing housings.

b. Standardization. Service A motors are required to meet the standardization requirements included in each motor specification. The Navy has standardized on horsepower ratings and on frame size and dimensions for each horsepower-speed combination. Maximum values of starting current are indicated as well as minimum values of starting and breakdown torques. The foregoing requirements closely parallel those established by the National Electrical Manufacturers Association, with specific modifications for Service A applications.

c. High shock resistance. All motors that are essential to the military effectiveness of the ship are required to be shockproof in accordance with the Navy specifications for high-impact shipboard machinery. Whether the shock transmitted to the motor is generated by a torpedo or missile exploding nearby, in direct contact with the ship's hull, or by the firing of the ship's own armament, these motors must continue to operate satisfactorily. Accordingly, the first design of each motor type is required to be subjected to a high-shock test in accordance with the high-impact shock specification.

To ensure that a motor will be shockproof, certain shock-resistant features must be designed into the motor. For example:

1. Brittle materials such as cast iron, semisteel, or similar materials should not be used for any motor part.
2. The mounting boltholes should provide for a minimum diametrical clearance. Slotted holes should be avoided wherever possible.
3. Special high-tensile-strength bolts should not be used to assure a shockproof connection since such bolts may not be readily available for replacement in the event of loss.
4. The relative deflection of the components may be quite large during a high-impact shock, so that sufficient clearances must be provided to prevent collision damage or overstrained connections.

d. Bearings. Ball bearings are used on the majority of Navy motors with sleeve and roller bearings utilized only when dictated by the application.

Navy ball bearings are required to meet either the specification for general-purpose annular ball bearings or the specification for annular ball bearings for quiet operations. Each specification indicates the metrology and tolerances for each size of bearing and the recommended shaft and housing tolerances. Table 5 shows typical fits for general-purpose and low-noise bearings. Unless otherwise modified by the motor specification, these tolerances should be used in applying the bearings.

The motor specifications have standardized on two types of medium series bearings: radial-single-row and cartridge-type. Special bearings such as the angular contact type may be used, however, when required by the application.

TABLE 5. Typical Navy Service A Bearing Fits for General-Purpose and Low-Noise Bearings

Bearing size	Shaft diam		Housing bore	
	Min	Max	Min	Max
305:				
General-purpose	0.9844	0.9847	2.4409	2.4416
Low-noise	0.98430	0.98450	2.44090	2.44110
310:				
General-purpose	1.9686	1.9690	4.3307	4.3316
Low-noise	1.9685	1.96870	4.33070	4.33090
315:				
General-purpose	2.9529	2.9534	6.2992	6.3005
Low-noise	2.95280	2.95300	6.29920	6.29940
320:				
General-purpose	3.9371	3.9377	8.4646	8.4660
Low-noise	3.93700	3.93730	8.46460	8.46490

e. Bearing lubrication. Grease lubrication is used in the majority of Navy motors. However, oil lubrication must be used for ball bearings when the nd_m value of 350,000 is exceeded, n being the rotating speed (r/min) and d_m the bearing mean diameter. Compared with oil, greases have relatively poor transfer characteristics and tend to deteriorate more rapidly at the higher operating speeds.

Lubrication of grease-lubricated ball bearings should be achieved by the use of compression-type grease cups (with pipe extensions where necessary) in lieu of fittings designed for pressure grease guns. To prevent excessive lubrication, the grease cups are not permanently mounted on the motor but should be attached to the motor only when relubrication is necessary. The grease inlet is normally fitted with a pipe plug.

f. Radio noise. The motors should be designed to ensure a minimum amount of radio interference.

g. Ambient temperature. Motors must be designed to operate in an ambient temperature of 50°C unless otherwise specified.

h. Sealed insulation system. Where severe environmental conditions exist, such as exposure to condensate, steam, flooding, salt spray, splashing, or temporary submergence, a squirrel-cage induction motor may be provided with a sealed insulation system.

By Navy definition, a sealed insulation system is one in which the varnish treatment of wound components is by the vacuum pressure impregnation process. Using solventless (100 percent solids) epoxy varnish, the sealed insulation system provides maximum protection against moisture. The following features have been incorporated into the Navy sealed insulation system.

1. Random-wound motors
   a. Each coil is taped iron to iron for resin retention.
   b. Slot cells are lapped and folded under the wedges.
   c. Slot cells and wedges are extended beyond the laminations for additional protection.
   d. No exposed magnet wire is permitted.
   e. Lead connections are sealed to maintain flexibility.
2. Form-wound motors
   a. Preimpregnated coils may be used.
   b. Taped construction is used for all coils.
   c. Slot cells are lapped and folded under the wedges.
   d. Slot cells and wedges are extended beyond the laminations.
   e. Lead connections are sealed.

All motor stators are tested submerged in conductive water for 24 h. Stators with an insulation resistance less than 100 MΩ at 25°C are unacceptable. Motors using the sealed insulation system should not experience insulation failure for the life of the motor (20 to 30 years).

Because so much is expected of a motor with a sealed insulation system, the Navy exercises rigorous control over the materials, procedures, and tests. Before a commercial or Navy activity can produce a Navy motor with a sealed insulation system, a detailed step-by-step procedure must be prepared and accepted by the Navy. A suitability test is then conducted on a sample stator. Among other tests, the motor is submerged and operated for 200 h. An insulation resistance less than 2 MΩ at 25°C is unacceptable.

i. High efficiency. In order to conserve energy (thereby reducing the amount of fuel a ship must carry or increasing its range due to reduced fuel consumption) the Navy introduced the requirement for high-efficiency motors in mid-1981.

Single-speed NEMA Design B motors driving centrifugal auxiliaries with synchronous speeds of 1200 to 3600 r/min are required to have a minimum full-load efficiency as shown in Table 6.

To produce a high-efficiency motor, the motor designer must pay close attention to such electrical and mechanical design considerations as low-loss steel, low copper losses, optimized air gap, low-stray-load losses, improved cooling fans, selection of bearings, and low-viscosity grease. When these design features are all brought together in a high-efficiency motor, many side benefits accrue which result not only in energy savings but in low airborne and structureborne noise, low cold-to-hot dimensional changes, improved power factor, and high reliability. Figure 28 illustrates the efficiencies obtained in Navy motors.

j. Low-noise requirements. The generation of noise by shipboard equipment is a major concern on combatant-type ships in general and on submarines in particular, where noise levels must be reduced to a minimum (see Sec. 20).

Motors generate two types of noise: (1) airborne noise that is radiated through

TABLE 6. Motor Efficiency (Minimum)

	Efficiency, %*			Efficiency, %*	
Motor hp	Dripproof enclosures	All other enclosures	Motor hp	Dripproof enclosures	All other enclosures
200 & over	97	97	25	92	94
150	97	97	20	92	93
125	96	97	15	91	93
100	95	97	10	90	92
75	95	96	7½	90	91
60	94	96	5	89	89
50	94	95	3	88	88
40	93	95	2	87	87
30	92	94	1½	86	86
			1	86	86

*All single-speed Design B motors rated 900 r/min and less shall have efficiencies not less than 2 percent less than shown. All other single-speed Design B motors shall have efficiencies not less than 1 percent less than shown. All multispeed Design B motors, for the highest speed only, shall have efficiencies not less than 3 percent less than shown.

the air and (2) structureborne noise that is conducted through the motor structure to the ship's hull and ultimately to the surrounding water.

1. Airborne noise. Motors installed in spaces where conversation must be carried on should not create noise of such an intensity as to interfere with the intelligibility of conversation. Motors mounted in unmanned spaces or areas where speech interference is not a consideration may have higher noise levels.

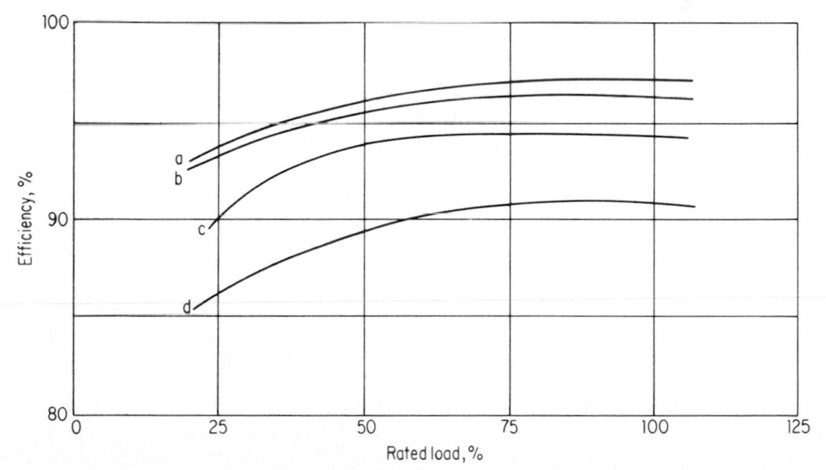

FIG. 28. Typical high-efficiency motors. Curve *a*—150 hp, 3600 r/min, TEFC; curve *b*—75 hp, 1800 r/min, TEFC; curve *c*—30 hp, 1800 r/min, dripproof; curve *d*—7½ hp, 3600 r/min, TEFC.

No airborne noise may be generated that will cause discomfort to the ship's crew when in their normal rest and recreation areas.

In an effort to avoid the creation of a wide variety of motors, each meeting a different airborne-noise level, the motor specifications contain only one set of sound-pressure airborne-noise limits. These limits will usually meet the speech-interference levels spelled out in the ship's spec as well as the machinery-space-noise levels. Of course, a ship's spec may require an airborne-noise level more stringent than those in the motor specifications.

Since the motor-ventilating system is the main cause of airborne noise, considerable care should be taken in the design of the fan and the ventilating passages. The number of fan blades should be selected to avoid exciting frequencies of noise which are also produced by other sources. Sharp edges and burrs should be eliminated from the parts subject to contact with the airstream, and air velocities should be kept as low as is consistent with the limiting temperature of the insulation (Fig. 29).

2. Structureborne noise. When structureborne-noise limits are imposed by the motor specification or the ship's spec, the motors are generally required to meet the limits specified in the military standard for airborne and structureborne noise measurements and acceptance criteria of shipboard equipment. The limits applied to a specific motor will depend upon the manner in which it is mounted in the ship: resiliently (Fig. 30), solid- or pad-type mounting, or a distributed isolation material.

   Stuctureborne-noise levels are defined in terms of acceleration decibels; adB $= 20 \log_{10} a/a_0$, where adB is the acceleration in decibels, a the measured acceleration in cm/s^2 rms, and a_0 the reference acceleration level of 10^3 cm/s^2.

   Structureborne noise is generated from both magnetic and mechanical forces set up within the motor.

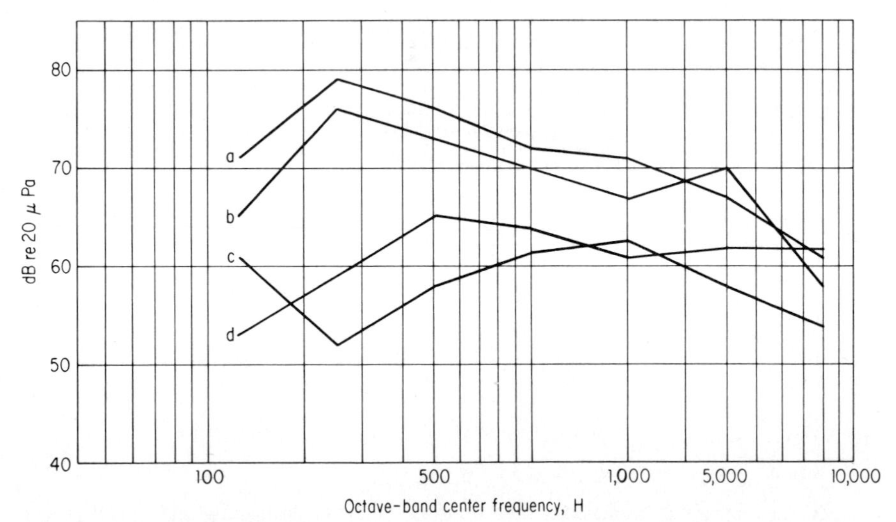

FIG. 29. Typical airborne-noise levels obtained in motors designed for low-noise operation and having totally enclosed fan-cooled enclosures. Curve *a*—150 hp, 3600 r/min; curve *b*—75 hp, 1800 r/min; curve *c*—40 hp, 1800 r/min; curve *d*—7½ hp, 3600 r/min.

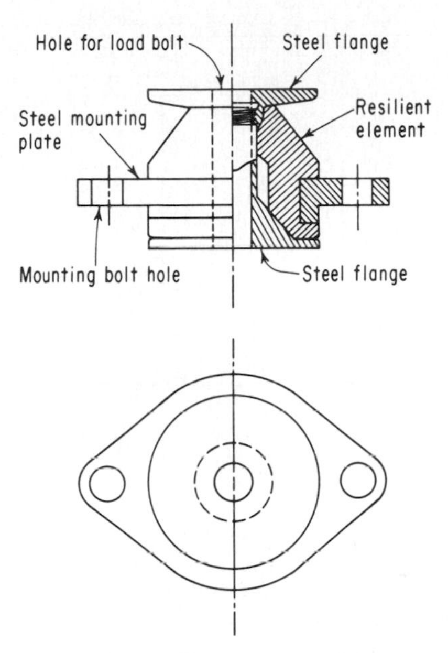

FIG. 30. Typical resilient mount used on navy motors.

The major source of magnetic noise and consequently structureborne noise are the magnetic forces acting across the air gap. To keep these forces at a minimum, consideration should be given to such factors as low flux densities, rotor-bar skew, increased air gaps, slot configurations, stator-coil pitch, and grain orientation of the stator laminations (Fig. 31).

Structureborne-noise reduction may also be designed into the motor by considering the following mechanical features: use of low-noise ball bearings and bearing preload springs, superprecision balance, use of balance rings for final trim balance, minimum eccentricity and a high degree of surface smoothness of the commutator, proper brush selection, correct brush angle, and an effective quality-control system (Fig. 32).

k. Pump motors. Motors for pump applications should be provided with an external shaft slinger (except where friction-type seals are used) on the drive end to prevent entrance of water into the bearing housing. No openings in the pump end shield parallel to the shaft should be permitted in dripproof pump motors.

l. Submarine motors. Submarine motors are Service A motors meeting the additional requirements of submarine service. Illustrative of these requirements are the following.

1. Both structureborne- and airborne-noise limits are imposed on all motors.
2. The feet of motors should be detachable if by that means the motor can be passed through a 25-in-diameter round hatch.
3. Motors driving centrifugal pumps may have a starting torque less than normal to obtain quiet operation.
4. Annular ball bearings for quiet operation are required.

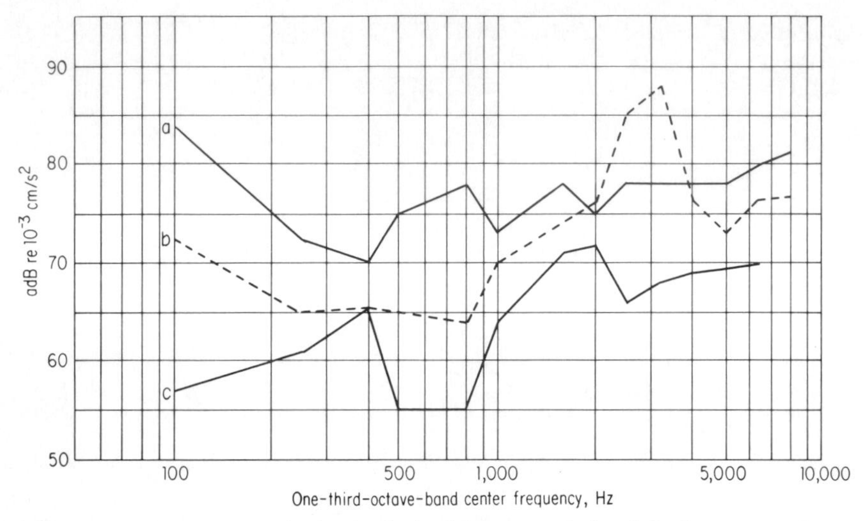

FIG. 31. Structureborne-noise levels obtained in motors designed for low structureborne noise. Curve *a*—250 hp, 3600 r/min, dripproof protected; curve *b*—150 hp, 3600 r/min, totally enclosed fan-cooled; curve *c*—40 hp, 1800 r/min, dripproof protected.

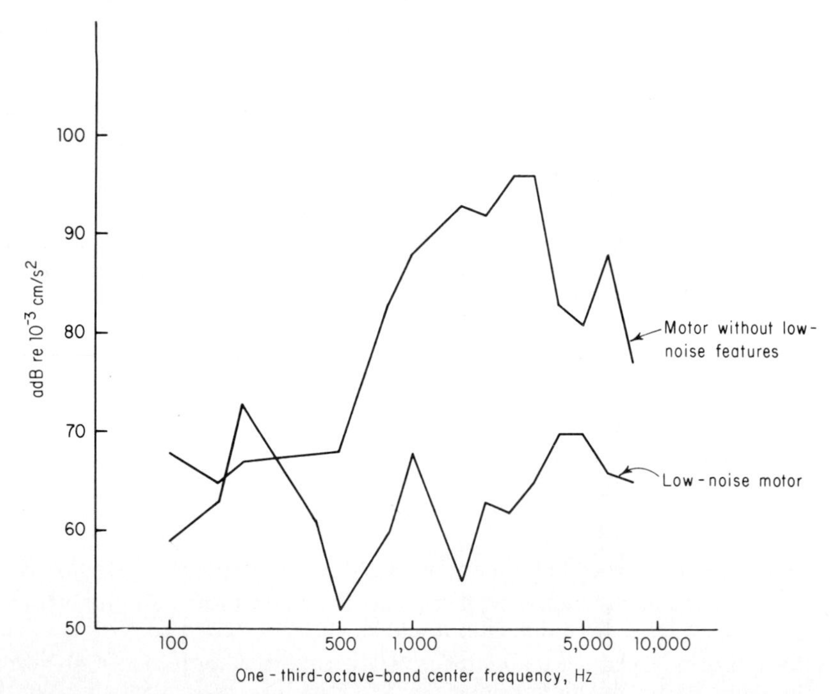

FIG. 32. Noise reduction possible when structureborne-noise-reduction techniques described are applied to a standard motor with dripproof protected enclosure.

55. Service C Motors. These motors may be furnished for those applications not essential to the military effectiveness of the ship. They are marine-type commercial motors meeting the Navy requirements for insulation, lubrication, painting, and the general construction requirements and restrictions spelled out in the motor specifications. Service C motors are found throughout the combatant ship in such nonmilitary duties as machine shop, commissary equipment, and laundry.

REFERENCES

1. "Recommended Practice for Electric Installations on Shipboard," ANSI-IEEE Standard 45-1983.
2. "Motors, 60 Hertz, Alternating-Current, Integral-Horsepower, Shipboard Use," Military Specification MIL-M-17060.
3. "Motors, 60 Hertz, Alternating-Current, Fractional Hp (Shipboard Use)," Military Specification MIL-M-17059.
4. "Motors, Direct-Current, Integral Hp, Naval Shipboard," Military Specification MIL-M-17413.
5. "Motors, Direct-Current, Fractional Hp (Shipboard Use)," Military Specification MIL-M-17556.
6. "Electric Power Equipment, Basic Requirements (Naval Shipboard Use)," Military Specification MIL-E-917.
7. "Airborne and Structureborne Noise Measurements and Acceptance Criteria," Military Standard MIL-STD-740.
8. "Bearings, Ball, Annular (General Purpose)," Federal Specification FF-B-171.
9. "Bearing, Ball, Annular, for Noise Quiet Operation," Military Specification MIL-B-17931.
10. "Requirements for Shock Tests, HI (High-Impact), Shipboard Machinery, Equipment and Systems," Military Specification MIL-S-901.

PART 8

CRANE MOTORS

H. O. Poland*

56. Motor Requirements. The requirements for motors on electric overhead traveling cranes may be compared, to an appreciable degree, with those for traction motors. This comparison is particularly true for bridge and trolley motions, both of which involve movement on rails with attendant critical control of starts, acceleration, and abrupt stops. The third motion involved in crane application, and the real "workhorse" of the group, is the hoist motor, which also must be adaptable to precise control.

Occasionally certain smaller auxiliary motors are used on cranes for such purposes as hook rotation, air-conditioning of control and operator enclosures, and, in more recent developments, air-conditioning of hollow-girder compartments which house control components. However, these are not strictly crane motors since crane motors as such are applied to the three primary crane functions, namely,

1. Load lifting and lowering (hoist motors)
2. Motion along the runway or aisle (bridge motion)
3. Motion transverse to the runway or across the aisle (trolley motion)

In general, one motor is used for each of these functions, except in the case of bucket or grapple cranes wherein two motors may work together on the hoist motion—one of them serving to open the bucket or grapple by appropriate cabling and the other to close it. Both motors share in hoisting or lowering the load.

Certain types of more specialized cranes, such as gantry cranes and tower cranes, may involve other primarily applied motors, examples being individual traction motors to provide rail traction. They are comparable to the overhead bridge or trolley motors. In all these basic functions, however, precise control is a primary requisite, and the motors must therefore be selected for optimum coordination with control.

57. Applications. Obviously, overhead cranes are used in a multitude of applications which affect the severity of duty on both the mechanical and the electrical components of the integrated system. For this reason, and to aid in the selection and standardization of components as much as possible, crane builders have established definite classifications based on the intended usage. The lightest of these is the standby application, such as the powerhouse crane used only for original installation and occasional maintenance of heavy apparatus. The other extreme is the continuous-usage application, as is encountered with cement-mill, incinerator, lumberyard, and other types of bucket or grapple cranes used contin-

*Senior Design Engineer, AC Motor and Generator Engineering, Westinghouse Electric Corporation, East Pittsburgh, Pa.; Senior Member, IEEE.

uously in a production service or otherwise. These classifications of duty assist but do not provide the final solutions in applying motors to crane service.

Other factors of important consideration are ambient conditions such as outside or inside service, clean or dusty conditions, and ambient temperatures. Duty cycles are of major importance in the production-type crane. Therefore it becomes evident that the intelligently applied crane motor must be engineered for its specific type of end usage. A so-called general-purpose motor is never satisfactorily applicable to crane service.

58. Operating Conditions. Such factors as conditions of exposure to weather, ambient cleanliness, and surrounding temperatures are generally readily known. More difficult in analysis and obscure in application is the duty cycle, such as that involved in production-type service. The practice, with production requirements known or established, is usually to compute the repetitive duty cycle in kilowatt-seconds, including energy requirements of acceleration and plug stops (reversal of applied voltage), and to convert the result to rms horsepower, from which the motor rating is selected accordingly with an appropriate margin of safety, involving economic considerations of first cost and overall operating expense. Ambient surroundings are also taken into consideration.

59. Rating Crane Motors. Since, to facilitate the establishment of suitable test and rating criteria, most crane motors are rated on an intermittent basis of time-temperature rise (for example, 2-h 55°C rise), conversion of ratings to satisfy a given duty cycle may become somewhat obscure. This condition is particularly true as the motor-cooling system becomes more involved, as with totally enclosed fan-cooled motors generally used in dusty or weather-exposed ambient conditions. Some builders of motors for crane service have developed methods for analyzing this problem (Ref. 1). The proper application of crane motors involves coordination of the mechanical and service requirements (crane builder) with the motor and control design.

Because of the precise control requirements (variable torque loads, acceleration, stopping, and reversal) crane motors are either dc or ac wound-rotor induction motors. Series dc motors controllable by applied voltage develop increasing torques with increasing loads, thus comparing with traction motors. They are widely used in crane service, particularly in steel mills. The standardized AISE dc mill motors, or 600 series, are used in this application. Aside from steel-mill cranes, ac wound-rotor induction motors are also widely used with resistance or reactance components in the rotor-control circuits to establish required speed-torque characteristics. With the development of more refined ac controls, including static control of stator-voltage reversing, and the establishment of standardized AISE mill motors comparable with the dc line, they are finding acceptance on steel-mill cranes.

60. Safety and Reliability. Obviously, crane-motor requirements may be very severe, entailing the utmost in reliability to assure uninterrupted service as well as safety to personnel and equipment. Construction must therefore be of optimum quality, including ample metal sections with adequate insulation and bracing of coil windings and connections, particularly in rotor windings. Plug stops and overspeed conditions are frequently involved, requiring rotors to be insulated for at least twice rated voltage. These conditions of application create voltage and mechanical stresses not encountered with general-purpose motors or in most other industrial applications.

To minimize failure shutdowns, the protection provided for the motors must be as nearly fail-safe as possible. For example, motor thermal-protective devices should function to warn the crane operator of motor overheating and impending failure and enable the operator to take suitable precautionary measures such as reducing the operating duty cycle. "Nuisance" stopping generally cannot be tolerated because of dangers created in dropping loads or stopping on runways at undesirable positions or in a critical part of the operating cycle. Magnetic brakes on the motor shafts are important features of the crane motor-control system in all three motions to assist in stopping and in holding loads on runway motions. It is essential that these brakes be spring-set for actuation in power-failure situations, and brakes are usually of this type of design.

REFERENCES

1. Harland O. Poland, "Motor Applications for Heavy Duty Ac Crane Service," AIEE paper, Material Handling Conference, Philadelphia, Pa., Apr. 13, 1960.

PART 9

ELECTRONICALLY COMMUTATED MOTORS

Harold B. Harms*

The electronically commutated motor (ECM) is fundamentally a dc motor but with (1) transistors doing the commutation of the armature windings instead of brushes and copper commutator bars as in a conventional dc motor, (2) shaft position sensors substituted for brush position, and (3) roles of stator and rotor reversed. There are two distinguishing similarities, however. One is that commutation is synchronized with shaft position, enabling both motors to be designed for any operating speed, and to be loaded down to zero speed without stalling. The second is that their speed and torque are responsive to voltage and current, respectively. The transistors in the electronically commutated motor offer sparkless, frictionless, and dustfree commutation, an advantage in designing for high speed and for contaminant-sensitive applications. Further, their utility can be extended to the control of speed and torque in response to a commanded load profile. The conventional motor requires add-on electronics to achieve the same degree of control.

The number, arrangement, and voltage rating of the commutating transistors are dependent on the circuit option chosen for the design of the electronically commutated motor, as outlined in Table 7. The circuit option also determines the arrangement and effective use of the motor windings.

61. Shaft-Position Sensing. Shaft position sensing is done by such direct means as hardware sensors mounted in the motor, Hall devices and light interrupters being examples of these, or indirectly by sampling and interpreting the electromotive force at the motor terminals. Either method provides speed as well as position information, although in digital form and with a resolution tied to a commutation interval. Higher, servo-quality resolution for control within a commutation interval will result if, instead, shaft encoders or synchros are used for position sensing.

Permanent magnets are the source of the dc field in the electronically commutated motor. They are mounted on the rotor rather than on the stator as in the conventional dc motor. Still, the geometric options are similar for both motors: (1) radial or axial air gap, (2) rotor inside or outside the stator, and (3) stator laminations shaped with a multiplicity of teeth or with "salient poles." Examples of these options are shown in Figs. 33 and 34.

62. Electronics. The ECM electronics is outlined by function in Fig. 35. Incoming power can be a battery, bulk dc supply, or 110, 220, or 440 V ac, single- or three-phase. The power conditioner converts to direct current, if necessary, and

*Senior Research Engineer, Motor Business Group, General Electric Company, Fort Wayne, Ind.; Registered Professional Engineer (Ind.); Senior Member, IEEE.

TABLE 7. ECM Commutation Circuits

Circuit	Effective use of motor materials, %	Number of transistors required	Transistor voltage rating
Half-bridge			
Single-phase	50	2	2 × dc supply
Two-phase	25	4	2 × dc supply
Three-phase	33	3	2 × dc supply
Full bridge			
Single-phase	100	4	1 × dc supply
Two-phase	50	8	1 × dc supply
Three-phase	67	6	1 × dc supply

also filters and limits current inrush. The transistor bridge, in detail dependent on the circuit option chosen, inverts dc to ac power at the frequency and voltage required by the motor and its load. The frequency is self-generated in real time by the rotor position sensors which feed back their data to the motor control. The primary function of the motor control is to develop a coordinated set of signals, based on the rotor position data, to commutate the transistors in the bridge in such a way as to maintain an optimum relationship of the magnet position with the stator windings in the motor. It can, secondarily, be used to convert a user-generated signal from the machine control to one that controls, by a variety of possible means, the average voltage or current delivered to the motor. If used in this way, the modifying signals are superimposed on the basic commutation signals. A driver stage amplifies these signals and level shifts them when required. The user-generated signals in the machine control can have either analog or digital form, and their origin can cover the range of complexity from a potentiometer to a microprocessor. Some are inherently digital such as on-off or forward-reverse commands, but the speed and torque reference profiles, although they may start out as digital signals, are converted to analog signals to accommodate the analog form of those signals with which they are compared. The torque reference, for example, is compared with actual motor current and the speed reference is compared with average voltage applied to the motor terminals. The machine-control arrangement distinguishes one application from another. It can be physically attached to the motor electronics or to the application electronics.

The majority of electronically commutated motor applications are effectively handled with open-loop control or with delayed speed and/or position feedback. A special case is the machine-tool-servo application, which requires high-gain closed-loop circuits that respond to high-accuracy speed and position measurements. The machine control in this instance is a highly engineered product.

63. Applications. The electronically commutated motor is in the formative stage of its product life cycle, but it is already widely accepted in business-machine and servo-machine-tool applications. The motivations behind the electronically commutated motor are: (1) no brushes, (2) adjustable speed, (3) speed capability

FIG. 33. Representative radial-air-gap configurations of electronically commutated motor. (*a*) Rotor inside stator with conventional tooth and yoke. (*b*) Salient pole. (*c*) Rotor outside tooth and yoke stator. (*d*) Rotor outside salient-pole stator.

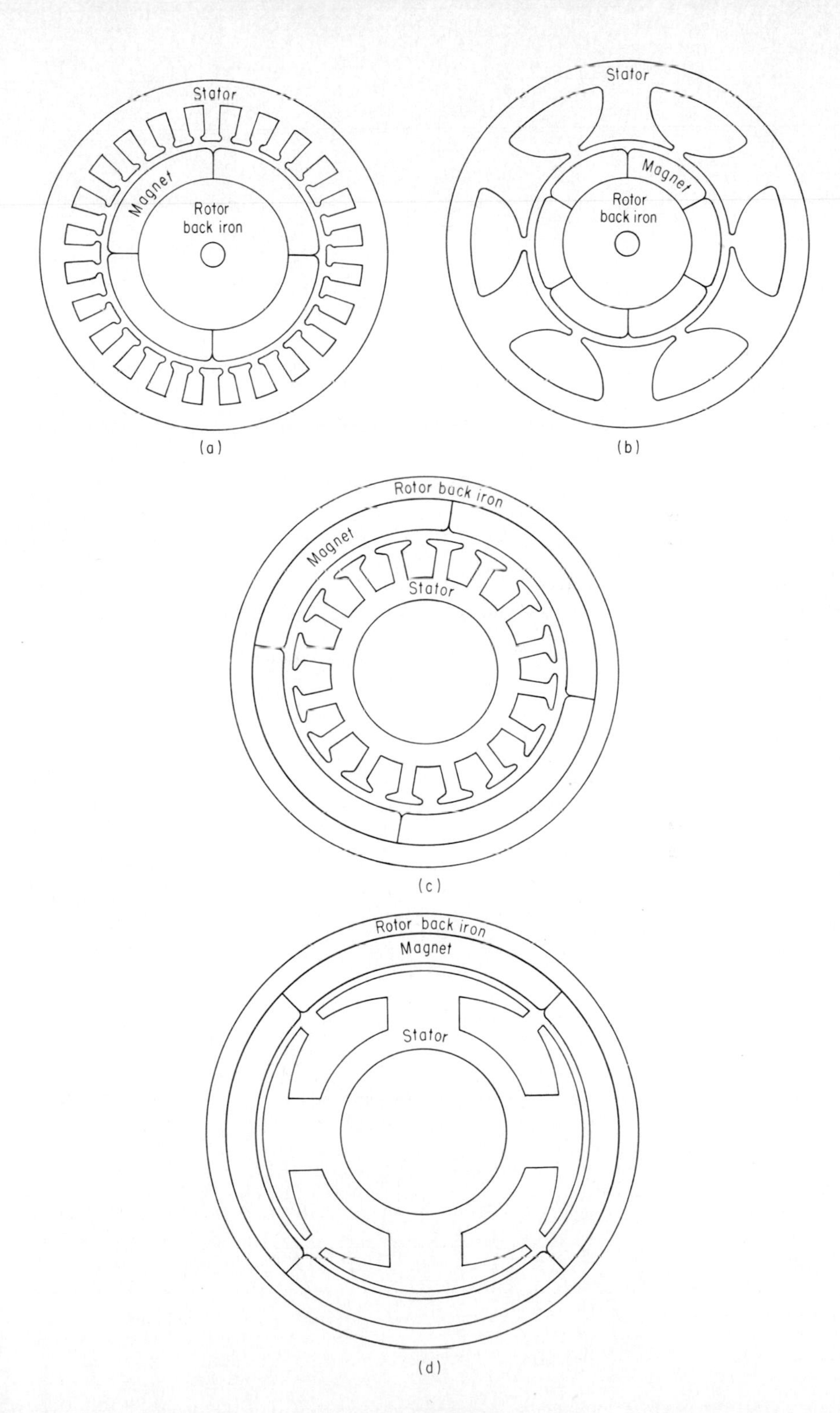
Stator
Magnet
Rotor
back iron
(a)
Stator
Magnet
Rotor
back iron
(b)
Rotor back iron
Magnet
Stator
(c)
Rotor back iron
Magnet
Stator
(d)

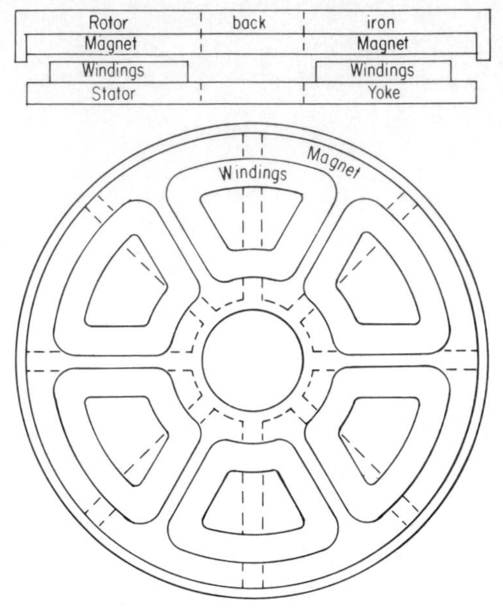

FIG. 34. Electronically commutated motor with axial-air-gap configuration. Stator teeth are not shown, but including them produces a high-output motor.

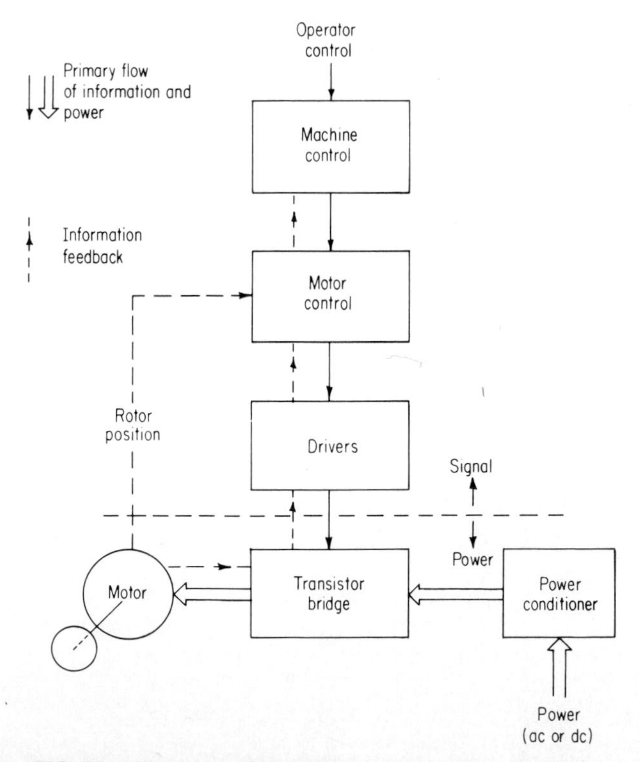

FIG. 35. Electronics of electronically commutated motor, showing basic functional blocks and information flow.

well in excess of 3600 r/min, and (4) a basically high efficiency that is retained over a wide speed range.

The speed-torque profiles in Fig. 36 illustrate the variety of loads the electronically commutated motor can serve and its adjustable-speed properties. Fig. 36*a* is for a ¾-hp, three-speed-range type load where tapped windings are used for the two higher speeds. The ⅛-hp motor in Fig. 36*b* demonstrates a current-torque-type control. Peak currents in that plot were incremented in equal steps. An example of voltage-speed control is shown in Fig. 36*c* for a 5-hp high-speed motor. Voltages in this case were incremented in equal percentage steps. The system efficiencies of these three motors, including losses in both motor and electronics, are plotted in Fig. 37. They illustrate a strong point of the electronically commutated motor: the ability to retain a design efficiency over most of its speed-torque operating region.

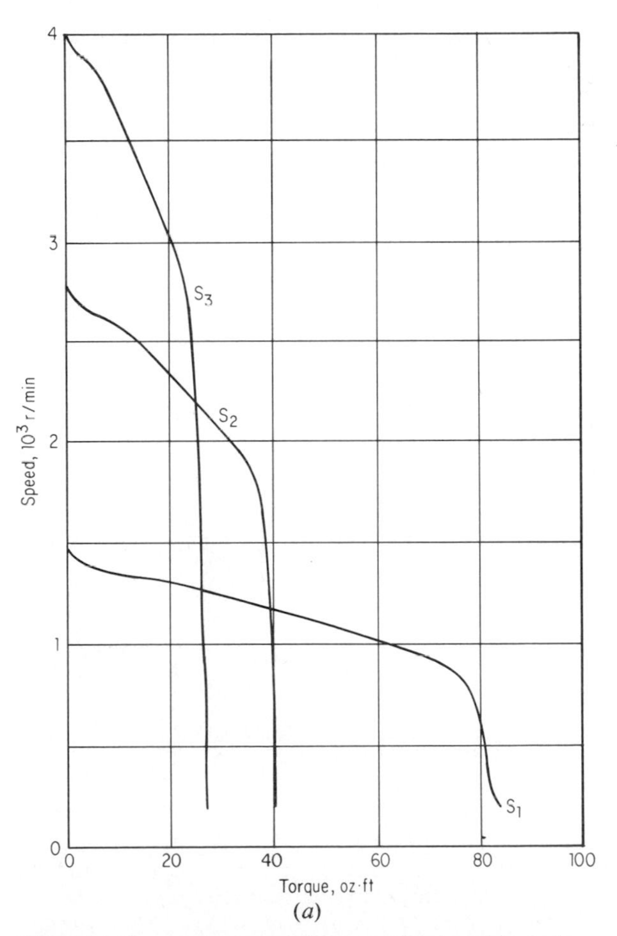

(*a*)

FIG. 36. Dynamometer speed-torque printouts of three-phase full-bridge electronically commutated motors. (*a*) 3-in-stack 40-frame six-pole motor with three optimized speed ranges. (*b*) 1.5-in-stack 30-frame twelve-pole motor under current control. (*c*) 3.5-in-stack 30-frame four-pole motor under voltage control.

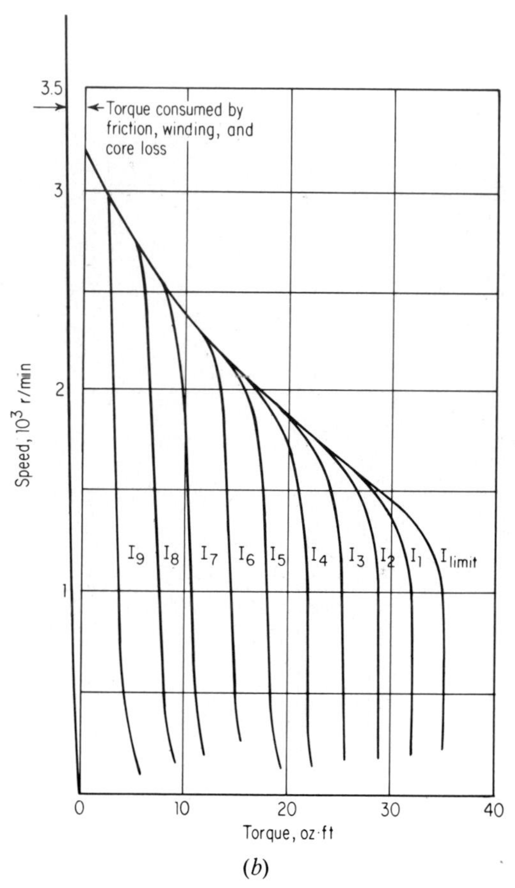

(*b*)

FIG. 36. (*Continued*)

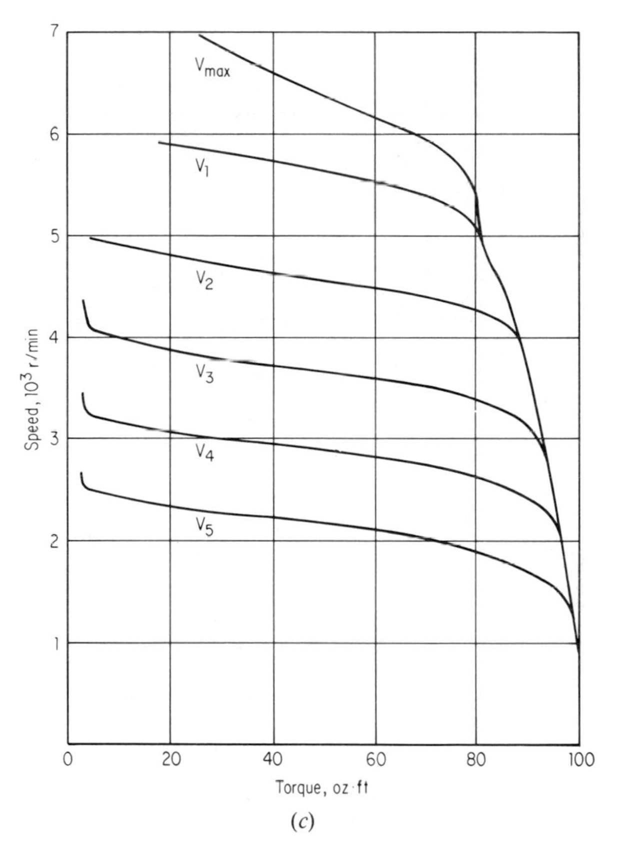

(*c*)

FIG. 36. (*Continued*)

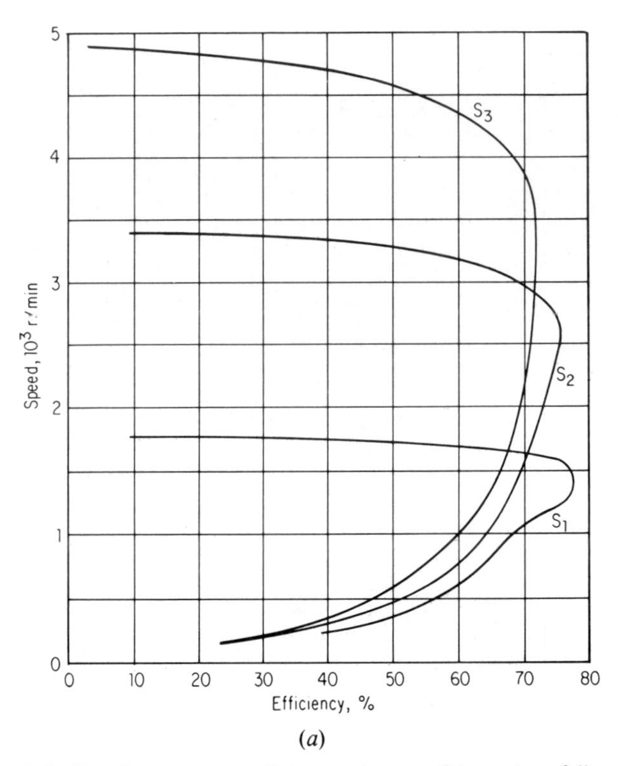

(*a*)

FIG. 37. Dynamometer efficiency printouts of three-phase full-bridge electronically commutated motors shown in Fig. 36.

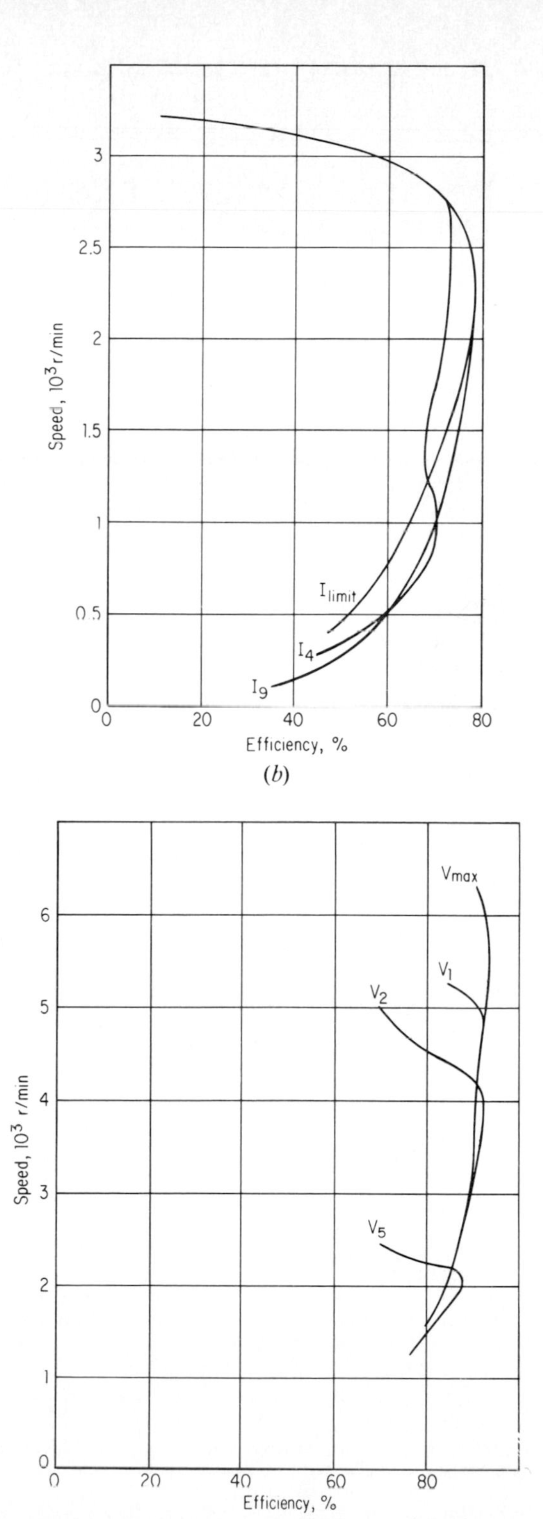

FIG. 37. (*Continued*)

PART 10

PERMANENT-MAGNET-EXCITED SYNCHRONOUS MOTORS

Paul D. Wagner*

The advent of stable, reliable permanent magnets, as well as the need for improved performance, led to the development of permanent-magnet-excited synchronous integral-horsepower industrial motors. The use of large ceramic magnet blocks in the rotors brought slight improvements in performance. Rotor speeds were limited because of the fragile rotor construction (Fig. 38).

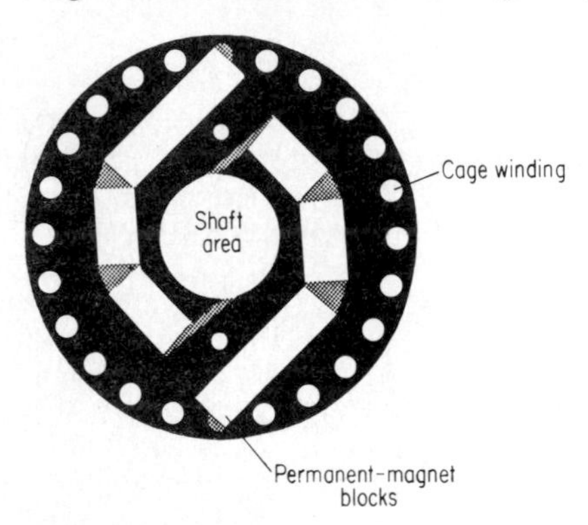

FIG. 38. Ceramic permanent-magnet rotor lamination for 1-hp two-pole three-phase motor. Areas around permanent-magnet blocks and cage are filled with aluminum.

64. Rare-Earth Magnets. Although many times more expensive, these magnets were introduced into industrial motors during the latter part of the 1970s. Prior to this, the rare-earth magnets had only limited use for space and military applications, where light weight and low losses are of paramount importance. Cost in such applications is usually only of secondary consideration.

Table 8 contrasts ceramic and rare-earth permanent-magnet performance. The maximum energy content of a magnet material is found by multiplying the second-quadrant values of the

TABLE 8. Typical Values of Permanent-Magnet-Material Performance

Parameter*	Ceramic material	Rare-earth material
H_c, oersted	3200	8200
B_r, gauss	3400	8000
Maximum energy content, gauss-oersted	2.7×10^6	16×10^6
Temperature coefficient, %/°C	0.2	0.05

*H_c = demagnetizing force corresponding to zero magnetic induction; B_r = residual induction corresponding to zero magnetizing force.

*Manager, Advanced Technology, Industrial Motor Division, Siemens Energy and Automation, Inc., Little Rock, Ark.; Senior Member, IEEE.

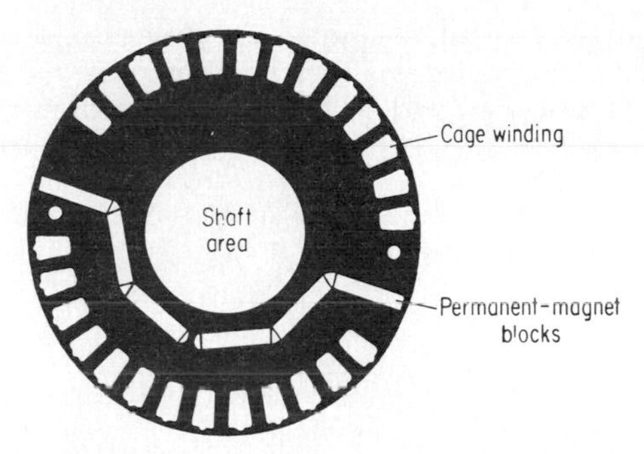

FIG. 39. Rare-earth permanent-magnet rotor lamination for 1½-hp two-pole three-phase motor. Areas around permanent-magnet blocks and cage are filled with aluminum.

magnet material hysteresis curve. The temperature coefficient is important to the motor designer and user, because it tells him how much permanent-magnet flux is lost per degree rise in temperature. Fig. 39 illustrates a two-pole permanent-magnet-motor lamination capable of running in excess of 10,000 r/min.

65. Performance. The permanent-magnet-excited synchronous-motor performance is similar to that of synchronous reluctance motors. They start like an induction motor. Their maximum synchronous torque is designed to be around 150 percent of rated torque. If loaded beyond this point, the motor loses synchronism and will either run as an induction motor or stall. Dynamic braking can be accomplished by either short-circuiting the stator leads or connecting a resistor across the stator leads to obtain a specific braking torque or decelerating time. The rotating permanent magnetic field induces voltages into the unexcited stator windings. This voltage, at rated speed, varies between 60 and 90 percent of rated terminal voltage, depending on the design and magnet material used. This induced voltage decreases linearly with speed and is zero at zero speed. No holding torque is experienced by the rotor at standstill, in contrast to a reluctance motor, which has a small locking torque when its three-phase stator is excited with a dc voltage.

The power factor and efficiencies of the permanent-magnet-excited motor are each 5 to 10 points better than their reluctance-motor counterparts. This factor alone can justify their use despite their higher costs.

Properly designed and used permanent-magnet motors have the same reliability as reluctance or induction motors. In rare instances ceramic permanent magnets have lost their magnetism when exposed to temperatures above 200°C or were inadvertently exposed to very large demagnetizing fields through the erroneous application of voltages in excess of 50 to 100 percent above their rating during normal operation. Rotors of these demagnetized machines can be remagnetized by the manufacturer. Similar problems with the high-coercive-field rare-earth magnets have not been reported.

66. Applications. Permanent-magnet motors are now being applied where precise speed must be maintained to ensure a consistent product. With a constant load, which does not tend to pull the motor out of synchronism after it has been brought up to speed, these motors will precisely maintain the machine speed. A further advantage is that there are no brushes or slip rings to cause sparking, which could be a problem in some atmospheres. Also, brush maintenance is eliminated.

Typically these motors are being applied for synthetic-fiber drawing where constant speeds are extremely important.

SECTION 11
INSULATION

J. L. Kuehlthau* and J. W. Sargent†
Edward J. Adolphson‡

FOREWORD . . . 11-2
1. Insulation Characteristics . . . 11-2
2. Insulation Life . . . 11-3
3. Insulation Classes . . . 11-4
4. Materials . . . 11-4

MAJOR INSULATION . . . 11-5
5. Main Insulation . . . 11-5
6. Strand Insulation . . . 11-5
7. Turn Insulation . . . 11-6
8. Lead Insulation . . . 11-6
9. Crossover Insulation . . . 11-6
10. Ground Insulation . . . 11-6
11. Chemical Aspects . . . 11-6
12. Physics of Insulation Materials . . . 11-7
13. Corona . . . 11-7
14. Stator-Coil Shapes . . . 11-7
15. Voids . . . 11-8
16. Spreaders . . . 11-9
17. Tapes and Sheet Materials . . . 11-9
18. Resins without Solvents . . . 11-9
19. Installing Coils . . . 11-10
20. Coil Connections . . . 11-10
21. Silver-Soldered Joints . . . 11-10
22. Small Stator-Coil Ends . . . 11-10
23. Fluidized and Dip-Type Coatings . . . 11-10
24. Coil Support . . . 11-11

*Consulting Engineer, New Berlin, Wis. (retired); Registered Professional Engineer (Wis.); Member, IEEE.

†Senior Engineer, Motor-Generator Department, Allis-Chalmers Manufacturing Company, Milwaukee, Wis. (retired); Member, ESM and SPE.

‡Senior Insulation Engineer, Motor-Generator Department, Siemens Energy and Automation, Inc., Bradenton, Fla.; Member, IEEE.

NOTE: This section was written by J. L. Kuehlthau and J. W. Sargent for the first edition and reviewed and updated for this edition by Edward J. Adolphson.

25. Field Coils 11-11
26. Mechanical Forces 11-12
27. Reliability 11-12

REFERENCES 11-12

FOREWORD

The electrical-insulation system of an electric motor is one of its most vital components. The satisfactory service life and reliability of a motor depend to a very large degree on the ability of the insulation system to perform its function. While the insulation system is often considered a nonwearing part and thus nonaging, it is probably the part of a motor most easily damaged during handling or maintenance and is the most susceptible to degradation from normal and abnormal operation and from environmental conditions.

The selection of an insulating system for a motor must include consideration of the electrical and mechanical design, the normal and abnormal operating requirements, and the environmental conditions to be encountered. Elements of material selection for a system include their dielectric properties, thermal stability and heat transfer, mechanical characteristics, processability, chemical resistance, and compatibility with related materials in the system.

Data on insulation materials and methods are given to help engineers who specify or apply motors to understand the principles and philosophy of modern motor insulation.

1. Insulation Characteristics. The dielectric-strength characteristics are probably the most important among the dielectric properties of insulation. Dielectric strength is especially important in large industrial machines operating at above 440 V.

As shown in Table 1, materials in their original state have exceedingly high breakdown values. However, experienced motor-insulation engineers cannot use these values. Mica-insulated coils, for instance, seldom are stressed more than 60 V/mil in large motors.

The large variance between the short-time dielectric strength of insulating

TABLE 1. Insulation Strength of Materials in Original State

Material before application onto the substrate	*Avg short-time stress, volts per mil (rms)*
Mica-splitting thin films	3,000–6,000
Mica-splitting plate	400–700
Mica-splitting glass tape	400–600
Mica paper plate	700–850
Mica paper, glass tape	400–600
Varnish-treated fabrics	1,000–1,800
Polyester film	4,000–6,000
Polyethylene film	8,000–10,000
Polyimide film	5,000–9,000
Vinyl film	10,000–12,000
Silicone elastomers	350–650

materials and the voltage stress at which they are commercially used is occasioned by the many variables and imperfections in materials, manufacturing processes, designs, and applications involved. Proper consideration of the voltage-endurance characteristics and the strengths of materials, coordination of voltage-stress limits of the several materials in a system, and experience with the degradable factors listed in Table 2 assist in establishing practical working voltage-stress levels, which assures service reliability and life of an insulation system and of a motor.

TABLE 2. Degradable Factors in Motor Insulations

Operation	*Degradables*
Processing materials into usable forms	Poor-quality elements, damage to elements in preparations, misformulations, misorientations, misgagings, introduction of contaminants, packaging, handling
Processing materials onto coils	Specification and design errors, process distortions, misformulations, introduction of contaminants, process problems, inexperienced and careless applicators; malfunction of equipment, lack of process control
Assembly of insulated coils into motors	Improper transportation and storage, components improperly sized, product damage from winding operations, contaminations, excessive splicing temperatures, misuse of tools, inexperienced personnel, malfunction of equipment, misformulation, destructive testing
Motor application	Misapplication of motor (type, size, duty cycles, enclosure, environments)
Motor operation	Internal and external heat, mechanical forces, electrical disturbances, corona phenomena, abusive operation, chemical atmospheres, abrasive dust, water (humidity, condensations, floods)

Engineers and scientists are presently overcoming many of the degrading factors involved in producing motor-insulation systems. The first steps have been taken in studies involving the theory of dielectric breakdown. Although no one understands this phenomenon completely, the results of research have provided better understanding.

2. Insulation Life. Insulations in large motors are expected to prove serviceable for periods upward of 20 years with minimal maintenance.

To project long life is extremely difficult, primarily because of the inability to anticipate and control the corona activity, the effect of heat, and the other factors found in Table 2. Some engineers claim long life for insulations through accelerated life tests. These tests usually include rigorous thermal cycling coupled with mechanical and electrical abuse. Table 3 gives standard insulation classes based on hot-spot temperatures.

TABLE 3. Motor-Insulation Classes

Class	*Hot-spot temp*
A	105°C (221°F)
B	130°C (266°F)
F	155°C (311°F)
H	180°C (356°F)

3. Insulation Classes. At present most large motors are Class F insulated. A motor may not require a high hot-spot rating; however, insulation systems that can operate at higher temperatures may provide greater reliability for a longer time period because they can more effectively withstand the heat of short-time overloads.

To test a Class F insulation system for assurances of 20-year life seems to demand an accelerated-life-test temperature higher than 155°C. Many materials used in composite insulations are usable dielectrics at 155°C but quickly degrade at, say, 165°C. In fact, a Class F component may provide better insulation for motors in many other qualities than another material capable of resisting higher temperatures. The ideal insulation would have all the requirements listed in Table 4.

TABLE 4. Property Requirements of Motor Insulation

Electrical	Thermal	Mechanical	Environmental
Corona-resistant	Endure heat	Shock-resistant	Resist acid
Arc-resistant	Noncrazing	High compressive strength	Resist alkalies
High insulation resistance	Low weight loss	High internal integrity	Resist water
Long life under continuous electrical stress	Retain strength	Good adherence	Resist solvents
Low power factor	Effects of expansion	Abrasion resistance	Resist salt
High dielectric strength	Conduct heat	High tear strength	Resist fungus
	Temp change shock	High tensile strength	Withstand flame

4. Materials. Advances are being made toward high-temperature-resisting dielectric materials that are physically and electrically strong at room and elevated temperatures. These advances are being made through polymer chemistry. The synthesis of polymers is yielding plastics and elastomers of excellent properties.

a. Thermoplastic materials. Thermoplastic materials, characterized by relatively low temperature softening points, are frequently utilized in smaller motors, such as those used in household appliances. In the future, as a result of continuing technical studies, thermoplastics compounded with fibers are likely to be considered satisfactory for large motors.

b. Thermosetting materials. Thermosetting materials are widespread as major insulation in large motors. Some serve alone and some combined. In the latter, for instance, mica in splittings or platelet size becomes a composite insulation with glass fibers or synthetic fibers, thermosetting saturants, and plastic films.

Materials classified as thermosetting plastics or elastomers are derived from many chemical families. The types most commonly in use are listed in Table 5.

c. Mechanical strength of insulation. The mechanical strength of insulation in many motor components may be considered before its insulation qualities. These uses are usually supports, ties, locking devices, etc. A classic example are the glass-fiber resinous retaining bands used to hold coils in place on rotating equipment such as the armature of a dc motor. The continuous glass filaments surrounded with resin are oriented under tension to yield bands of exceptional strength. This development has replaced steel-wire bands almost entirely. The nonconducting features simplify space designing and add reliability to the product.

TABLE 5. Common Resins and Likely Applications in Motors

Resin	Wire cable	Stators	Field coils	Wound rotors	Arma-tures	Com-muta-tors	Collec-tors	Turn, slot support insula-tion	Fabri-cated pieces	Molded parts	Sleev-ing	Impreg-nants and finishes
Acrylic........		x	x	x	x			x				x
Epoxy.........	x	x	x	x	x	x	x	x	x	x	x	x
Phenolic.......	x	x	x	x	x	x	x	x	x	x	x	x
Polyamide.....	x	x	x	x	x			x	x	x	x	
Polyimide	x	x	x	x	x			x			x	x
Polyester	x	x	x	x	x	x	x	x	x	x	x	x
Silicone........	x	x	x	x				x	x	x	x	x

Other glass-fiber-resin products are either molded or fabricated from sheet into mechanical supports. The most common composite products are polyester glass, phenolic glass, phenolic cotton, melamine glass, and silicone glass. The choice of material is made after consideration of several factors such as:

1. Cost of base material
2. Availability
3. Size and uniformity of product
4. Cost of fabrication
5. Mechanical strength at both room and operating temperatures
6. Arc (or track) resistances
7. Moisture absorption
8. Resistance to environmental contaminants

MAJOR INSULATION

Major insulation systems in motors are found in stator windings and in field coils. The stator-coil insulation is the most critical because of its higher-voltage requirements.

5. Main Insulation. Regardless of the type of coil, the main insulating system is that between the coil and ground. Secondary insulation occurs in strands, turns, leads, and crossovers as the design dictates. Coordination and proper bonding of all component parts of the system, including the magnet wire, are important to produce a serviceable and reliable structure.

6. Strand Insulation. Copper magnet wire is normally used; however, under certain economic and supply conditions, aluminum magnet wire has been used effectively. Insulation is usually preserved on the wire by the supplier and may be any one or a combination of the following basic types:

1. Resinous-coated wire
2. Resinous-film-taped wire

3. Paper-taped wire
4. Fibrous-wrapped wire
5. Mica paper

Strand insulation is frequently of the resinous-coating type because it offers good dielectric strength, occupies a minimum of space (0.003 to 0.005 in), and is low in comparable cost. Other types of insulation are also used depending on design considerations.

7. Turn Insulation. Turn insulation may be any of the five types listed or may be combinations of them if higher dielectric strengths and safety factors are desired. Generally, these insulations are used for smaller lower-voltage equipment of 6600 V and below. For the larger higher-voltage equipment, operating voltages as well as surge voltages may require additional turn insulation such as combinations of those listed or the addition of layers of mica. Often the size or mechanical configuration of the conduction will dictate the use of heavier turn insulation.

The heavy-duty form-wound stator coils often require more insulation between turns than is provided by preinsulated wires. In most coils of this type fibrous-wrapped wire is specified as the substrate.

8. Lead Insulation. Lead insulation is quite similar to the ground-wall insulation. However, less lead insulation is often used because the leads are usually far removed from ground and have ample spacing between them. One problem encountered is joining the lead insulation into the ground insulation without openings. These joints, if not satisfactorily sealed, allow moisture and contamination penetration.

9. Crossover Insulation. Crossover insulation is used to protect wires that cross each other or are transposed in a coil winding. Crossovers are frequently the weakest point in multilayered-type coils such as salient synchronous rotor-field coils. The insulations can be premolded devices, sheet, tape materials, or thin laminates.

10. Ground Insulation. Ground insulation is the main or major part of the insulation system of a motor. It is always subjected to higher voltages and in many cases to the most abuse, in both motor assembly and motor operation. It requires the highest degree of attention in design and manufacture to assure maximum service reliability of the equipment.

To consider ground-wall insulation properly, especially when operating voltages exceed 1000 V, knowledge of dielectric breakdown is necessary. Some of the answers can be found in chemistry and in physics.

11. Chemical Aspects. Chemical aspects are more practical than theoretical. Chemistry, over a period of years, has created new or modified existing insulating materials for improvements as listed below.

1. Molecular structures that are electrical insulations over long periods of time at elevated temperatures and under high electrical stress
2. Casting and potting materials that are compatible with other materials, are noncorrosive, have high bond strength and yield toughness to composites, and provide voidless structures

3. Synthetic films with a high degree of continuity
4. Nonhygroscopic materials
5. Materials resistant to unusual chemicals, solvents, and radiation
6. Materials with characteristics and properties enabling their use even in extremes of room temperature and atmosphere exposure

12. Physics of Insulation Materials The physics of insulation materials provides theory on breakdown or puncture phenomena. It seems quite evident that the complete story of dielectric-puncture-theory conclusions is obscured because of practical imperfections involving the entire insulation structure. These imperfections involve practices and conditions in processing and flaws within composite insulations.

It has been suggested by several authorities that ionic conduction causes internal degradation. Since ionic conductivity varies with the contaminants within the dielectrics, conclusive findings are difficult to obtain. The contaminants are moisture and other conducting impurities. They are found in voids and as inclusions.

13. Corona. Insulation engineers are and should be concerned about corona activity in motor-coil insulations. The erosion by corona discharges can be a serious problem especially when considering long insulation life. Of considerable interest is the study of voids within dielectrics and the shape of these voids. Corona discharge will begin within an odd-shaped void at less voltage than in a spherical void. A small spherical void will have a lower corona voltage start than a large void. Corona is distinguished from a continuous discharge such as an arc in that it consists of relatively low energy intermittent pulse discharges. Corona discharges within an air-filled void produce nitrogen oxides that attack and erode insulation materials.

14. Stator-Coil Shapes. Stator-coil shapes at present resemble a diamond (Fig. 1), which is the most suitable design for compactness. Other coil shapes extend the length of machines and in turn require more materials, longer and larger shafts, bigger bearings, and more floor and cubical space.

The diamond coil has 10 major bends which are relatively acute. Four of the bends occur just at the ends of the coil slots. The ground insulation at these bends must be sound and requires perfect sealing. Because of thermal-expansion forces the insulation in the corner area is stressed in such a way that insulation migration can be a problem. Corners also demand tape materials that drape well and conform. Without these characteristics, insulations "fish-scale" and become bulky during application. Some materials with low tensile and tear strength buckle or rip when applied. Mica-splitting-type tape requires greater attention than a conformable woven-fabric tape. Elastomeric tape, such as silicone rubber, conforms exceptionally well; however, the applicator cannot overtension the rubber.

The wire in the diamond coil has rounded edges to reduce electrical stress and improve insulation applica-

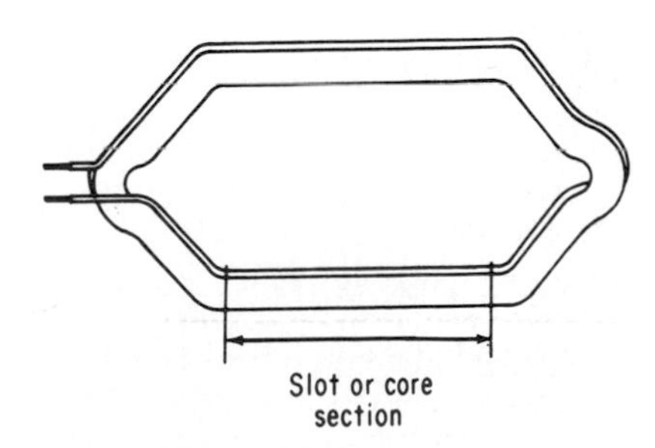

FIG. 1. Diamond-shaped coil.

tion. Residual stiffness in the fibers causes sponge when bent sharply (Fig. 2). Liquid resins and varnish run or pull thin at the relatively sharp corners because of surface tension (Fig. 3). There is also a possibility that the insulation at the corners might be cut as a result of handling the coils.

FIG. 2. Residual stiffness in fibers causes sponge when bent.

FIG. 3. Liquid resins and varnishes run or pull thin at corners.

15. Voids. The voids between the ground wall and the conductor as shown in Fig. 4 can be filled with resins using pressure-vacuum impregnation. Scientific examinations of such systems have shown that in actual production some voids can remain. It is extremely difficult to remove all air, fully replace it with resins, and then retain the resin in the pockets without runout until it solidifies. The resins must be extremely thin, without solvents, and must solidify almost instantaneously to accomplish the objective. Resin that seeps or runs out is replaced with air, leaving voids. With a modern vacuum-pressure impregnation system the size of the remaining voids will be determined by the degree of vacuum and pres-

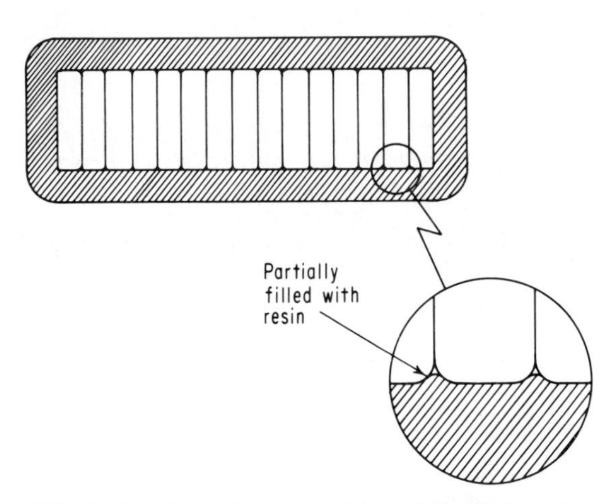

FIG. 4. Pressing coil removes almost all the air from the pockets.

sure used and the resin viscosity. It is possible to virtually eliminate apparent voids.

Ground insulation is usually built up around a stator coil with overlapped layers of tape. Some suppliers use sheet wrappers in the slots (straight sections) with tape at corners and ends. Regardless of the method of application, gas pockets are introduced wherever there is a joint or overlap. Triangular voids are inherent with flap tapes and sheet ends (Fig. 5).

The cross section in Fig. 5 greatly exaggerates the thickness of usual tapes to illustrate the types of voids that can be created. In consideration of high-voltage insulations, which consist of multilayers, it is simple to visualize the host of odd-shaped gas voids that can exist within the structure.

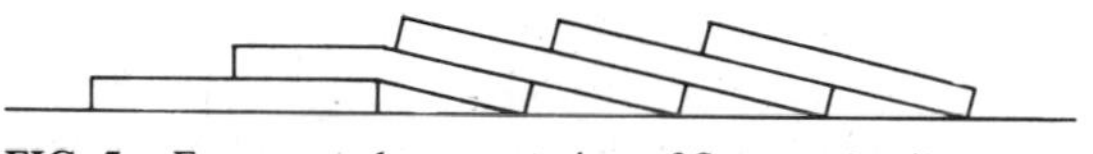

FIG. 5. Exaggerated cross section of flat-taped coil.

Proper selection of material, designs, and manufacturing processes permit the control or elimination of most voids of consequence. However, since it is practically impossible to eliminate all voids, it is desirable to coordinate and limit operating voltage stresses of the insulation system to those that have been shown by experience or test to afford adequate service life and reliability.

16. Spreaders. To force a coil into a diamond shape is difficult. Most production coils are transformed from loops into coils on mechanical devices commonly known as spreaders. It is not uncommon to find damage to the wire insulation if the operator of the spreader is careless or tools are not properly prepared.

17. Tapes and Sheet Materials. Tapes and sheet materials resist bending around a rectangular shape so that "sponge" develops on the flap sides during application. Figure 6 shows a typical taped coil before it is pressed. Usually the pressing is accomplished with the coil sections warmed so that the binders will migrate under the pressure to fill the voids. With the pressure maintained, chilling for thermoplastic or heating for thermosetting binders is introduced. This action solidifies the binders.

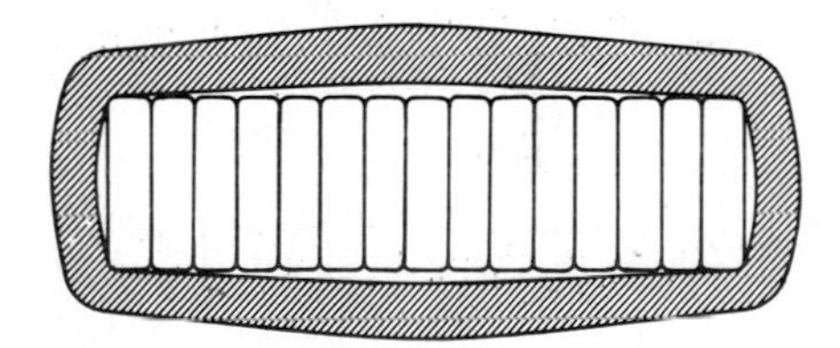

FIG. 6. Unpressed coil cross section showing voids.

Vacuum-pressure impregnated systems will press the coil into the stator-core slot or a form before impregnation, or the individual coil may be impregnated and then pressed in a form. The coils are held in this shape and heated to cure the resin.

18. Resins without Solvents. Resins without solvents (100 percent solids) have greatly improved multitaped insulations. It is not unusual to introduce air into the resin. Vacuum-treating the resin components before applying can reduce the air bubbles in quantity and size. Such action reduces the introduction of air into the sheet or taped insulations with resin binders. Of interest is the fact that the air in the resins is spherical in shape. Spherical air voids have higher corona-starting voltages than laminar voids. With spherical voids pressed flat (such as buttons) the corona activity starts at a lower voltage.

The common resins used for vacuum-pressure impregnation are the epoxies and the polyesters. Both are usually Class F systems. Silicone resin systems are also used to provide for Class H requirements.

19. Installing Coils. Finished coils, especially diamond-shaped stator coils, require some distorting to fit into the stator slots. The distortion is significantly more severe for the throw, or parting, coils. Insulations must withstand winding distortions without cracking or breaking. Half-coils, which are one-half of the diamond, are used where coils are too large or heavy to handle or bend without damage (Fig. 7).

Slot cells are common in small motors having "mush-wound" coils, where they provide ground insulation. They are generally not used with the larger form-wound coils because the ground insulation is applied directly to the coils. Where cells are used with form-wound coils, they primarily provide mechanical protection when inserting the coils.

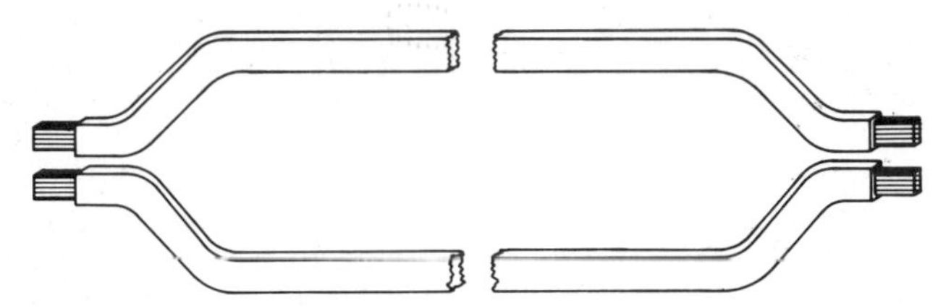

FIG. 7. Coil made in halves permits easier assembly. The ends are connected after assembly in core.

20. Coil Connections. The assembly of lead joints and connectors requires judicious selection of materials, cautious designing, and skillful processing.

At one time tin-lead solders were the usual agent for joinings. They did not require high heat and were easy to assemble and disassemble. Some disadvantages were:

1. Parts required mechanical support in the form of ties, bindings, or clips.
2. Parts to be joined required pretinning to assure a good job.
3. Required fluxes were acids.
4. Flux residue was hard to remove.
5. Solder splatter and remaining fluxes contaminated windings.
6. As motors were designed to work harder, the strength of joints was inadequate.

21. Silver-Soldered Joints. Silver-soldered joints are presently popular. Such joints are termed brazed. Except that high heat (usually about 1200°F) is requried, they overcome many of the disadvantages of soft solders. Proper tooling and setups aid in reducing the hazards of the high heat.

22. Small Stator-Coil Ends. Small stator-coil ends are supported and protected with complete encapsulations. These motors have excellent service records.

23. Fluidized and Dip-Type Coatings. Fluidized and dip-type coatings for small-appliance-size motors are currently in favor for stator-winding ends. These same motors may have similar coatings serving as slot insulation and core-end insulations. The rotating windings are frequently treated in the same manner.

24. Coil Support. Large form-wound stator windings usually require extensive blocking and tying (Fig. 8). The high forces at startup or in fast rotation reversals cause the coil ends to react almost violently. Metal and fabricated rings, spacers, blocks, cords, tapes, and devices are designed to hold the coil ends firm. A large proportion of a motor cost is invested in these intricate lacings.

One manufacturer features stator-coil ends which have cast ring supports (Fig. 9). These rings do not encompass the entire stator ends but just the connections and the U bends. Ventilation is not impaired. This cast ring support serves to insulate and seal out contaminants at the vulnerable lead-coil joints. The mechanical support provided is excellent.

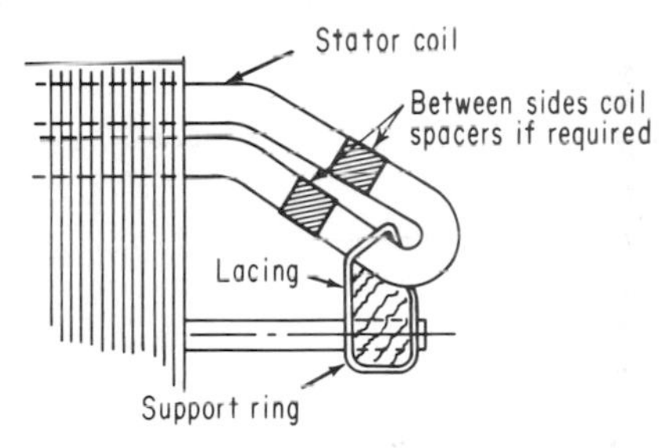

FIG. 8. Coil laced to hold firmly in position.

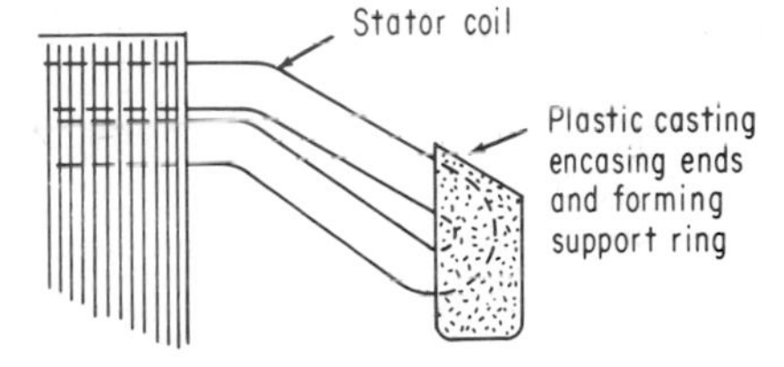

FIG. 9. Coil ends encapsulated for insulation and support.

25. Field Coils. Field coils have been improved considerably in recent years. Although they operate at relatively low voltage, they require insulation systems that are tough enough to resist severely degrading mechanical forces. The strength is derived from glass- or synthetic-fiber-reinforced polymers which are extremely heat-stable. The pole piece and the coil are united into one homogeneous assembly which eliminates contamination between coil and pole. These coils are frequently encased in a thin veneer of continuous oriented-glass fibers saturated with high-strength bonding resins to provide protection from environmental contaminants (Fig. 10).

Field coils for small machines usually have round or square preinsulated wire while machines such as large synchronous motors have strap conductors (Fig. 11). The strap is rectangular in shape and is bent on edge in continuous helical spirals. For cooling purposes the exposed edges are generally bare. Insulating material capable of withstanding high compressive loads (500 lb/in^2) and thermal-expansion forces is "glued" in between turns. The materials usually are mica-splitting

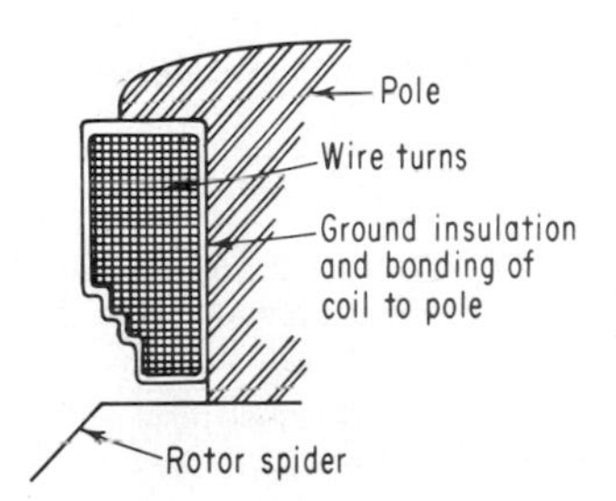

FIG. 10. Wire-wound field-coil insulation.

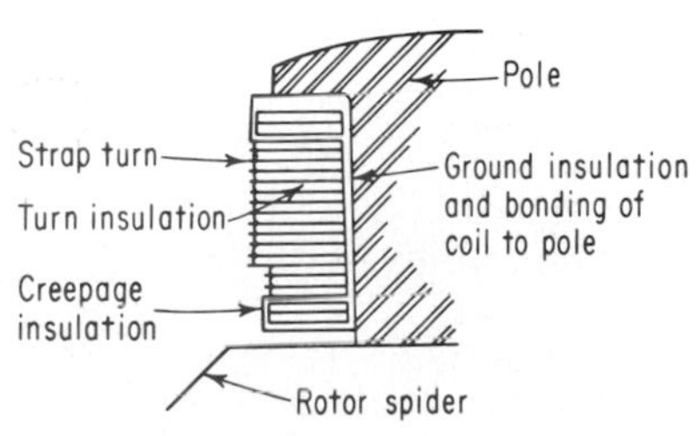

FIG. 11. Strap-wound field-coil insulation.

composites, asbestos-fiber composites, or high-temperature-resisting fiber paper. A high-temperature nylon paper which is capable of at least 200°C continuous is an exceptionally tough material and has largely replaced more expensive mica in this application.

26. Mechanical Forces. Mechanical forces involved with rotating field coils require careful consideration. Leads and windings must be adequately secured to withstand the centrifugal force. Allowances for thermal expansion must be made. The stationary field coils in large dc reversing motors must be secure to withstand the shock of the reversals.

27. Reliability. The reliability of motors can be improved and the life of machines greatly extended by carefully considering the following.

1. Be sure that the motor being purchased is adequate for the exposures at the installation.
2. Inquire into the finer details of the motor, including the insulations.
3. Inquire into the details, such as the mechanical features supporting the windings.
4. Obtain recommended instructions on frequency of inspection and maintenance. See that the instructions are followed.
5. Inquire as to methods of cleaning, what systems to use, where to clean, what solvents to use.
6. Do not repaint windings unless advised because coats of paints and varnishes restrict the flow of heat.
7. Never repaint or revarnish unless all contaminations are thoroughly removed from the surfaces. This operation is usually difficult. It may be better not to recoat.
8. Be sure operators do not abuse the motor in starting too frequently in short periods of time or overloading the motor. Avoid frequent bus transfer and/or chronic low voltage.
9. Install motor control that will adequately protect the windings from overload.
10. Install surge protection at motor terminals if a possibility of severe switching or lightning surges is present.

REFERENCES

1. "General Principles for Temperature Limits in the Rating of Electrical Equipment," IEEE Standard 1-1969.
2. "Recommended Practice for Testing Insulation Resistance of Rotating Machinery," ANSI/IEEE Standard 43-1974 (Reaff 1984).
3. "Standard Techniques for High Voltage Testing," ANSI/IEEE Standard 4-1978.
4. "Guide for Insulation Maintenance of Large AC Rotating Machinery (10,000 kVA and larger)," ANSI/IEEE Standard 56-1977 (Reaff 1982).
5. "Guide for Field Testing Power Apparatus Insulation," IEEE Standard 62-1978.
6. "Recommended Practice for Insulation Testing of Large AC Rotating Machinery with High Direct Voltage," ANSI/IEEE Standard 95-1977 (Reaff 1982).

7. "Recommended Practice for the Preparation of Test Procedures for the Thermal Evaluation of Insulation Systems for Electric Equipment," ANSI/IEEE Standard 99-1980.
8. "Test Procedure for Evaluation of Systems of Insulation Materials for Random-Wound AC Electric Machinery," ANSI/IEEE Standard 117-1974 (Reaff 1984).
9. "Recommended Practice for Thermal Evaluation of Insulation Systems for AC Electric Machinery Employing Form-Wound Pre-Insulated Stator Coils, Machines Rated 6900 V and Below," ANSI/IEEE Standard 275-1981.
10. D. F. Miner, *Insulation of Electrical Apparatus* (McGraw-Hill, New York, 1941).
11. G. L. Moses, *Electrical Insulation, Its Application to Shipboard Electrical Equipment* (McGraw-Hill, New York, 1951).
12. G. A. Van Brunt and A. C. Roe, *Winding Alternating-Current Motor Coils* (McGraw-Hill, New York, 1938).
13. A. Von Hippel, *Dielectric Materials and Applications* (Wiley, New York, 1954).

SECTION 12
MOTOR BEARINGS

H. O. Koons* and J. E. Petermann†

FOREWORD 12-2

BEARING TYPES 12-2
1. Plain Bearings 12-2
2. Antifriction Bearings 12-2

BEARING CHARACTERISTICS 12-3
3. Ball and Roller Bearings 12-3
4. Sliding-Type Bearings 12-9
5. Sleeve-Bearing Load Ratings 12-10
6. Tilting-Pad Thrust-Bearing Capacity 12-12
7. Predicting Bearing Life 12-13
8. Selection of Bearing Type 12-15
9. Protection against Failure 12-16
10. Shaft Circulating Currents 12-16
11. Bearing-Temperature Protective Devices 12-18
12. End Play 12-19
13. Bearings in Fractional-Horsepower Motors 12-20
14. Bearings in Integral-Horsepower Motors 12-21
15. Bearings in Vertical Motors 12-24
16. Bearing Standardization 12-26

REFERENCES 12-26

*Chief Engineer, Waukesha Bearing Corporation, Waukesha, Wis. (retired); Registered Professional Engineer (Ohio); Member, ASME, SNAME, NACE, and ASTM.

†Senior Engineer, Motor-Generator Department, Allis-Chalmers Manufacturing Company, Milwaukee, Wis. (retired); Registered Professional Engineer (Wis.).

FOREWORD

The purpose of a bearing is to support and control the motion of a rotating shaft, while consuming a minimum of power. All bearings can be classified as either plain or antifriction, the latter including ball and roller bearings. Although each type has definite advantages and disadvantages for certain kinds of service, there is considerable overlapping of their suitability for many applications.

In applying, specifying, or maintaining motors it is well to remember that all bearings will wear. Proper application, installation, and care of motor bearings will enable the motor user to get maximum useful life from motor bearings with minimum bearing maintenance. This section provides a rounded background knowledge on bearing application and care.

BEARING TYPES

1. Plain Bearings. The large majority of bearings in use today on motors are of the plain-bearing type. These include all bearings, either sleeve or thrust types, that depend upon a lubricating film to reduce friction between shaft and bearing (see Sec. 14).

When properly designed and lubricated, plain bearings develop tapered, hydrodynamic oil films which have tremendous load-carrying capabilities. In general, bearings of this type are better adapted to supporting heavy loads and shock loads than antifriction bearings.

Under hydrodynamic conditions, the coefficient of friction is usually very low since it is simply a measure of the internal shear within the lubricant film. However, fluid friction increases in proportion to the square of the speed, and in high speeds it may be considerable. When properly designed and applied, plain bearings can operate at speeds of 25,000 r/min or more.

Under certain conditions, the hydrodynamic film may not be developed. When boundary lubrication conditions exist, the coefficient of friction is appreciably higher and life is reduced.

Because of the inherent characteristics of the lining materials and flushing action of the lubricant, plain bearings are not too sensitive to contaminants. Other advantages are that they are relatively inexpensive and operate very quietly. Properly designed hydrodynamic film bearings have an infinite life if properly maintained.

The running clearance, necessary for proper formation of the oil film, must be carefully controlled to hold the shaft centers within close limits, minimize losses, and control vibration. Proper finishes and good design can provide the necessary clearances.

2. Antifriction Bearings. These bearings operate on the principle of rolling contact between elastic circular bodies. The resistance of this rolling action, which replaces fluid boundary friction and plain bearings, is quite low.

At low speeds, ball and roller bearings develop so little resistance through rolling that they are superior to plain bearings operating under boundary conditions. At higher speeds, the resistance to rolling is comparable with the fluid shear in hydrodynamic plain bearings.

Individual antifriction bearings can support, at the same time, both radial and thrust loads in varying degrees—a characteristic not typical of plain bearings.

These bearings require only small amounts of lubricant, and they are relatively insensitive to changes in viscosity. Grease is used as lubricant except in high-speed machines. At exceptionally high speeds they are somewhat difficult to lubricate because the rolling elements have a tendency to jell and throw off the lubricant. When too much lubricant is supplied, the turning action of the rolling element produces excessive fluid friction.

Ball and roller bearings are made of precision-built parts and, as a result, must be treated accordingly. They are sensitive to vibration, poor fits, corrosion, and abrasive dirt. When properly installed, however, they require very little maintenace and have low rates of wear. Bearings can be selected for a definite life with the help of the bearing manufacturer.

The life of ball and roller bearings depends on the fatigue strength of the material and decreases as speed and load increase. At high speeds, a certain amount of noise, the result of the rolling action and windage, is characteristic of this type of bearing.

Ball and roller bearings are manufactured in several grades of precision. They must have a certain amount of internal clearance for proper operation. Thus for some applications they are preloaded to reduce play, provide rigidity of the mounting, and accurately maintain the shaft locations.

BEARING CHARACTERISTICS

When applying electric motors, many power engineers have an understandable tendency to think of them primarily as electric machines. A motor has at least one major moving part; consequently, it is subject also to strictly mechanical problems as applicable to all rotating machinery. Most of these problems involve bearings and applications.

Bearings are designed to carry loads of specific types. Some bearings are designed to carry loads in only one direction, while others can carry loads in two directions.

Thrusts (axial loads) are defined as forces acting parallel with the motor shaft (Fig. 1). Radial loads are defined as forces acting perpendicular to the motor shaft.

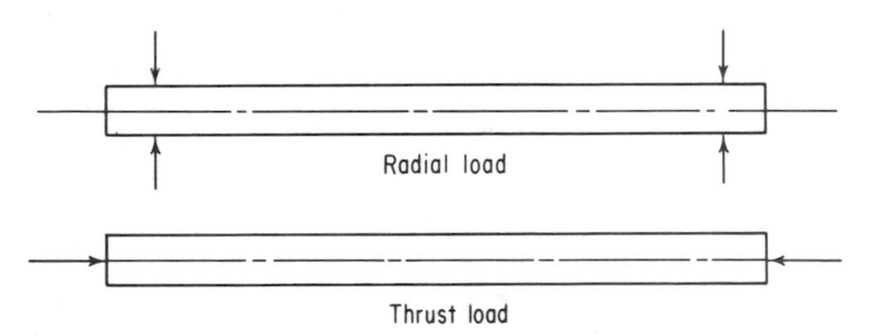

FIG. 1. Both radial and thrust loading must be determined before motor bearings are specified.

3. Ball and Roller Bearings. Normally standard horizontal integral-horsepower motors, in NEMA frame sizes (up through 125 hp at 1800 r/min), are furnished with ball bearings. Except for some of the smallest sizes, sleeve bearings may be obtained as an optional alternate. However, off-the-shelf delivery is not to be expected. The medium-sized machines (150 to 600 hp at 1800 r/min) may readily be obtained with either ball or sleeve bearings. Ball or roller bearings are optional and are selected principally on the basis of the bearing load and speed.

There are seven main types of antifriction bearings: (1) radial deep-groove ball bearings, (2) maximum-type ball bearings, (3) cylindrical roller bearings, (4) double-row spherical roller bearings, (5) angular-contact ball bearings, (6) spherical-roller thrust bearings, and (7) split-inner-race angular-contact-type bearings.

a. Radial deep-groove ball bearings. The most generally used type of antifriction bearing is the radial deep-groove ball bearing. This bearing is designed to carry not only radial loads but also moderate thrust loads in either direction, simultaneous with, or independent of, radial load. The ability of this bearing to carry thrust loads enables the user to mount any motor built on the NEMA 445U frame or smaller in any desired position, that is, horizontal, vertical, with shaft up or down, or at any angle. When the motor is mounted vertically, only small external thrust loading is permissible. The external thrust should be less than the weight of the motor rotor.

Radial deep-groove ball bearings are available in three basic enclosures: (1) open enclosure, (2) sealed enclosure, and (3) double-shielded enclosure.

b. Open-enclosure bearings consist of an inner and outer race, the ball-retainer ring, and balls (Fig. 2). This bearing can be relubricated without diassembling the motor, but the open construction affords slight protection against overgreasing. There is no protection against foreign elements that may be in the grease or in the atmosphere when the motor is dismantled.

Another disadvantage of the open construction is the churning of the grease supply in the reservoir caused by rotating of the balls and retainer ring. The rotation of the ring churns the grease and will increase grease oxidation. Dust particles which may enter along the shaft will mix with the grease and come in contact with the bearing.

c. Sealed-enclosure bearings of the same size and load capacity as an open bearing may have a raceway about 1½ times as wide as the open bearing. The extra width is necessary to install the seals and to provide a grease reservoir. This bearing is lubricated before being installed in the motor, and it is not possible to relubricate, unless the motor is dismantled and the seals are removed (Fig. 3).

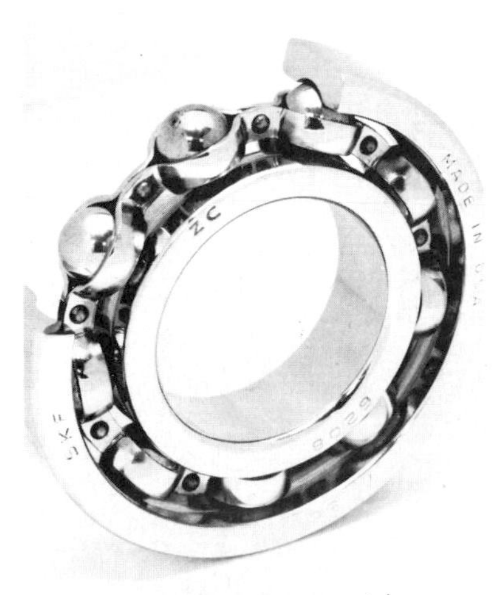

FIG. 2. Single-row open-enclosure deep-groove ball bearing.

FIG. 3. Single-row double-sealed ball bearing.

Sealed bearings with exactly the same raceway dimensions as the deep-groove open type are also available. Bearing manufacturers have found little, if any, difference in operation or life between the wide and the narrow-width sealed bearings. The narrow-width bearings are not relubricatable or repackable, as the seal cannot be replaced. Rubbing seals limit the application speed to 1800 r/min except on the 180 to 210 frames, where 3600 r/min is permissible.

The sealed bearing offers good protection against foreign materials, but other considerations of this type of bearing must be noted.

1. There is no grease on the outside of the bearing, leaving the outside bearing surface unprotected against vapor contamination in the air.
2. No motor is airtight; therefore, any vapor contamination in the air can reach the bearings inside the seals to attack the grease. The amount of grease in the sealed-type bearings is much less than in other types, causing more rapid deterioration through contamination.
3. The length of service of a sealed bearing without relubrication is limited by grease quantity and quality. Tests show that 3600-r/min machines operating under normal conditions for 24 h a day should be relubricated approximately every 3 years because of the imminence of grease failure.
4. The replacement cost of sealed bearings is approximately 40 percent above that for the open type and 35 percent above that for the shielded type for the same size.

d. Double-shielded bearings are simply open-type bearings with a shield rolled into the outer raceway on each side (Fig. 4). The shield has a close running clearance with the inner race. This clearance is sufficient to permit passage of grease but is small enough to protect the balls and raceways from large dirt particles which may enter the grease during assembly or maintenance. This bearing is also prelubricated, but it can be relubricated while in service.

The double shield, while it does not completely eliminate the risk of overgreasing, definitely minimizes this possibility. These shields also serve to keep grease in the ball path when the motor is mounted vertically, by effectively preventing slumping of the grease supply, which results in increased grease life.

The double-shielded bearing gives users the desirable features of both the open and the double-sealed bearings, and also eliminates most of their respective disadvantages. Table 1 summarizes the reasons for selecting the double-shielded bearing.

e. Maximum-type ball bearings. The bearings have the capacity to take 30 percent increased radial load over the radial deep-groove bearings but cannot take thrust loads. The bearing is constructed like the deep-groove ball bearing, except for a notch in the inner and outer race to insert added balls for greater radial load (Fig. 5).

FIG. 4. Single-row double-shielded ball bearing.

TABLE 1. Comparison of Radial Deep-Groove Bearing Features

Features	Open	Sealed	Shielded
Bearing protected from foreign material when removed from its housing enclosure	No	Yes	Yes
Lubrication in service	Yes	No	Yes
Protection against vapor contamination	Yes	Less	Yes
Relative bearing cost	Lower	Higher	Intermediate

f. Cylindrical roller bearings. When the radial load imposed on the bearing is in excess of the capacity of a ball bearing, such as may occur when the load is belt-driven, the cylindrical roller bearing is used. A typical cylindrical roller bearing may be capable of carrying as much as 186 percent more radial load than the same size of ball bearing.

Physically, this bearing is interchangeable with the ball bearing in any given size, but it does have operating limitations. The range of operating speeds for roller bearings is much less than for ball bearings because of noise and friction heat. Recommended maximum speed for the cylindrical roller bearing is 1800 r/min with oil lubrication and 1200 r/min with grease lubrication. The roller bearing can carry no thrust loads and is usually used on the shaft extension end only if the opposite end has a ball bearing to absorb thrust loads. Axial movement of the outer raceway of the roller bearing must be limited to keep it from disengaging from the inner race (Fig. 6).

FIG. 5. Single-row maximum-type ball bearing.

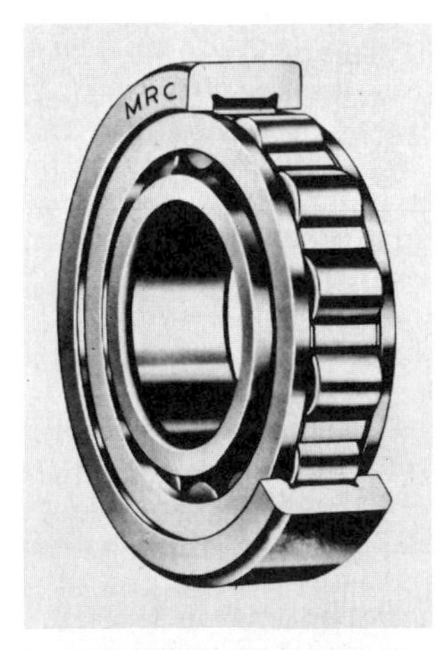

FIG. 6. Cylindrical roller bearing.

g. Double-row spherical roller bearings. These bearings are capable of carrying heavy radial loads. A spherical bearing can carry 400 percent more load than a comparable ball bearing. It can also carry thrust loads in either direction, which differs from the cylindrical roller bearing. This bearing is self-aligning. The width of the race is about 1½ times that of the same size of ball bearing. Because of noise and friction heat developed from the heavy loads, the operating-speed range of the roller bearing is restricted to 720 r/min with grease lubrication (Fig. 7).

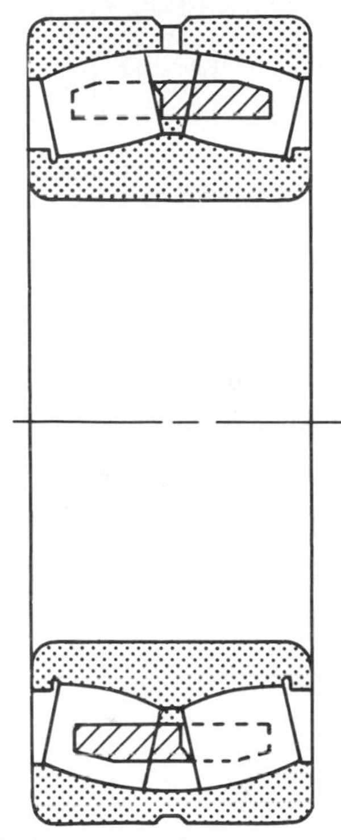

FIG. 7. Cross section of double-row spherical roller bearing.

h. Angular-contact ball bearings. These bearings are designed to carry high thrust loads in one direction only. When the thrust load is too great for a radial deep-groove ball bearing, an angular-contact ball bearing is often used. The load is carried by the wide shoulders, one on the outer race and the other on the inner race, on opposite sides of the bearing. The wide shoulder of the outer raceway designates the direction of thrust.

When thrust loading is in both directions, these bearings are mounted back to back (Fig. 8) rather than face to face because of simplicity of mounting and angular rigidity. For high-down-thrust loads, these bearings may be mounted in tandem, as shown.

i. Spherical-roller thrust bearings. When thrust loads are very high and beyond the capacity of any practical multiple mounting of angular-contact bearings, a spherical-roller thrust bearing is used (Fig. 9). This bearing is self-aligning and is capable of carrying thrust in one direction only. The bearing must be oil-lubricated and normally must be externally forced-cooled. Forced cooling is usually accomplished by running cold water through a coil immersed in the oil reservoir.

Customarily the thrust bearing is located at the upper end of a vertical motor. To hold the rotor shaft assembly centered in the stator, a lower guide bearing is provided. When all the thrust load is downward, thrust is taken entirely by the upper bearing, and the guide bearing carries radial load only. Many pumps, however, are subject to an upward as well as downward thrust due to the pump operation. This condition requires a special consideration in the bearings. One common way of handling this problem is to lock the lower bearing to the shaft and permit it to take the upward thrust. This method is possible because the guide bearing, being a radial bearing, has considerable thrust capacity. The upthrust condition in nearly all cases is considerably less than the downward thrust value and well within the range of the thrust capacity of the radial bearing. Another

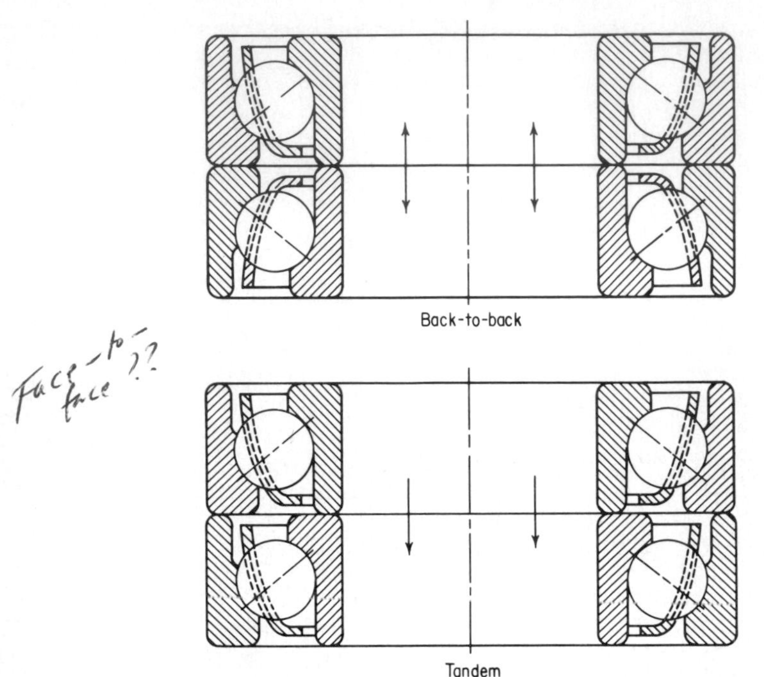

FIG. 8. Mounting arrangements for angular-contact ball bearings.

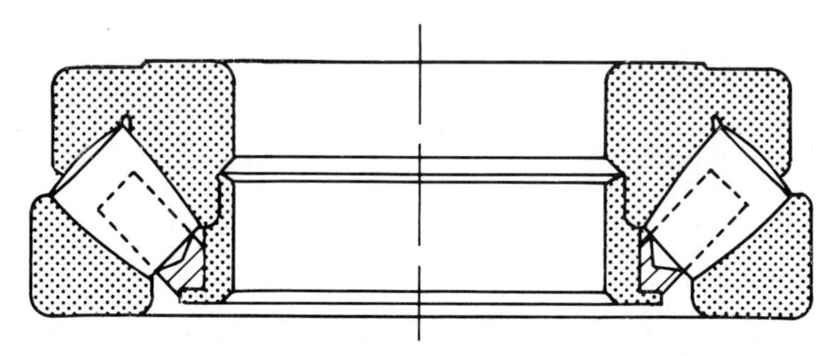

FIG. 9. Cross section of spherical thrust bearing.

factor is that this condition does not occur continuously, and this type of bearing should give sufficient upthrust capacity when required. Antifriction bearing load capacities are compared in Table 2.

j. Split-inner-race angular-contact-type bearings are used to avoid the use of extra parts necessary to lock the guide bearing. Some motor manufacturers prefer this type of bearing since it is designed to handle both the radial and the axial loads and will permit the locking of the bearing to take care of upthrust as well as downthrust forces.

TABLE 2. Summary of Relative Load Capacities, Antifriction Bearings

Bearing type	Thrust capacity	Load radial capacity
Radial deep-groove	Average	Average
Maximum type	None	Medium
Cylindrical roller	None	High
Double row spherical	Medium	High
Angular contact	Medium	None
Spherical roller thrust	High	None

In specifying vertical motors for pump applications, be sure to advise the manufacturer if the pump thrust is expected to be "balanced," or nearly so, for more than a few seconds at a time. This condition occurs on applications such as booster pumps in power-station condensate systems, where pump and motor may run for long periods at light loads. The result is that the rotor may float, being subjected to neither upthrust nor downthrust. If the motor thrust bearing is of the split-inner-race type, such operation may damage the bearing and lead to early failure.

Some deep-well pumps inherently cause a condition whereby the inner and outer races separate at startup. In most cases this condition causes a momentary parting; however, it may be severe enough that when the bearing again reseats itself in its proper manner, brinelling may occur in the races, causing early failure. This condition is most likely to occur when angular-contact or spherical-thrust roller-type bearings are used for this thrust application. Overcoming, or limiting, the amount of upthrust and parting of the thrust bearing can be readily accomplished by the use of springs or other mechanisms to ensure that the outer race would follow the ball or rollers as the inner race rises and tends to part the bearing system.

The above cautions should not be overlooked in specifying vertical motors, as bearing difficulty may result.

4. Sliding-Type Bearings. These bearings frequently are called sleeve bearings. Their application to electric motors is generally understood, and their use has been most successful over a long period of time. On larger motors this kind of bearing consists normally of a babbitt-lined steel or a cast-iron cylinder which is lubricated by oil picked up from a sump or reservoir by oil rings or disks which rotate with the shaft. The babbitt is usually a lead- or tin-base alloy securely bonded to the cylinder and accurately machined to close tolerances to give the optimum operating characteristics. Aluminum-base babbitts have been used under unusual corrosive conditions. However, in modern motors with reasonable maintenance and the proper selection of lubricants, this condition is seldom a problem. The basic early sleeve-bearing designs were developed to be two to three times greater in length than in diameter. The present practice is to maintain a length-to-diameter ratio L/D from 1 to 1.5:1. This ratio reduces the adverse effect on the bearing caused by shaft deflections under both internal and external loads. Figures 10 and 11 show a typical split-sleeve bearing like those used on large motors.

The sleeve bearing is nothing more than a plain cylinder, suitably grooved for proper distribution of the lubricant. The shaft of the motor actually rides on a thin film of oil, having no metal-to-metal contact during running conditions.

FIG. 10. Sleeve bearing in larger motor with cover removed. The fixed-position bearing construction means lower operating temperatures because heat is readily carried away through the broad contact area between bearing and housing. *(Allis-Chalmers.)*

Prior to starting there is metal-to-metal contact because of lack of oil between the motor shaft and the bearing (Fig. 12). As the shaft begins to rotate, it climbs up the bearing wall. As it climbs, the shaft rolls onto a thicker film of oil and begins to slip back. Oil is drawn by this action into the wedge-shaped clearance at the lower right-hand portion of the bearing. More oil is drawn through this converging wedge as the shaft increases speed, thus developing a fluid pressure at the lower right-hand portion of the bearing. This pressure lifts the shaft and pushes it slightly to the left in the final running position.

Oil is carried from the reservoir to the top of the bearing by an oil ring, providing sufficient lubricant to prevent bearing wear. Grooves are cut in the top surface of the bearing to (1) improve oil distribution, (2) increase oil flow to obtain greater cooling, and (3) prevent oil from leaking from the bearing ends by returning it to the reservoir.

Sleeve bearings are normally used only for direct-connected applications. A V-belt drive requires tension between the belts and sheave. To transmit the horse power this tension can become so great that the oil wedge which supports the shaft may not be formed, and the bearing will burn out from heat developed through metal-to-metal contact. A rule of thumb to follow is: *belt drive—antifriction bearings.*

5. Sleeve-Bearing Load Ratings. For sleeve bearings on horizontal-type motors the load-carrying capacity of the bearings is quite simple to calculate. Normally this type of bearing is designed to carry radial loads only. However, some radial sleeve bearings are designed with a small thrust face at one or both ends for the rotor shaft assembly to thrust against momentarily. The amount of thrust capacity actually on this type of bearing is very low. However, if the requirements of the motor are such that the radial bearing must also carry a considerable amount of thrust, then this value must be specified and provisions made at the time the unit is designed. Motors which have thrust requirements are used principally for such drives as paper refiners and Jordon beaters in paper mills. With these loads the driven equipment imposes a thrust load on the shaft, and this thrust must be transmitted by the thrust faces of the bearing. Such thrust loads should be specified in each case because conventional motor bearings will wear out prematurely under these operating conditions.

Radial loads up to 150 lb/in^2 of projected bearing surface may be carried by motor sleeve bearings with little difficulty. Based upon this value, a bearing of 3-

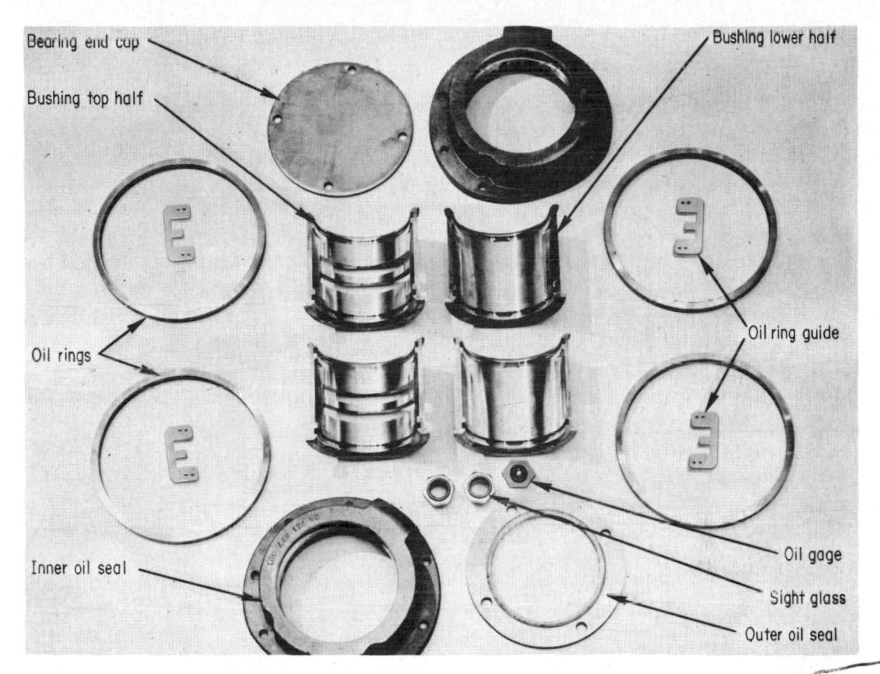

FIG. 11. Parts for bearing in Fig. 10. Bushings are bronze-backed, tin-babbitt-lined typed. Oil rings are made of bearing bronze with jointless construction. Nylon ring guides are nonwearing for long ring life. No grooves are required in the lower half of this bearing, providing more support area. *(Allis-Chalmers.)*

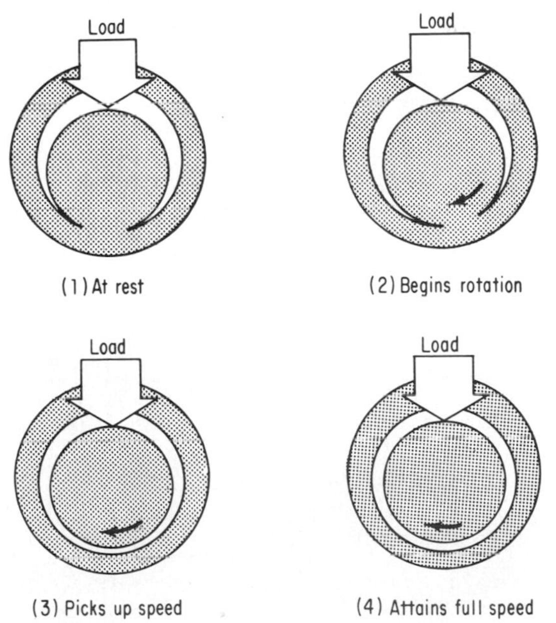

FIG. 12. Changes of oil film in sleeve bearing during starting.

in diameter, 4 in long, should be able to carry a total load of 1800 lb. Several factors that may reduce this capacity must be considered. In the first place not the entire surface of the bearing is available to carry the load. Certain areas or portions of the surface may be cut away to provide oil grooves and reservoirs. This reduction in area means an increase in bearing load.

The location of some of these reservoirs or reliefs is such that the load-carrying capacity is not always the same in both directions of the shaft. This reduction of capacity is a major reason for specifying the direction of rotation and of the belt pull when ordering large sleeve-bearing motors for this type of service. When these factors are taken into consideration, a more conservative figure of from 70 to 100 lb/in^2 of allowable radial load for the motor sleeve bearing is practical. It will be noted that this rating is far below the 1000 lb/in^2 plus allowable for sleeve bearings of other types of machinery and consequently reflects the longer life expected for motor bearings. This extra capacity will allow for slight motor-shaft misalignment and a relatively loose fit as compared with those normally found in such equipment as a conventional automotive engine. Given a reasonable lubrication and protection from circulating electric currents, sleeve bearings will last almost indefinitely at loads within the above limits (see Art. 10).

6. Tilting-Pad Thrust-Bearing Capacity. Figures 13 through 18 show the parts of an equalized thrust bearing. The capacity of a conventional tilting-pad thrust

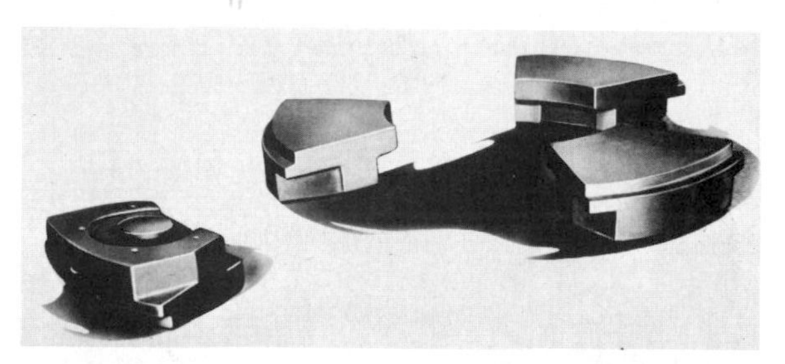

FIG. 13. Pivoted shoes of thrust bearing. Inverted shoe shows hardened-steel shoe support set into its base. *(Kingsbury, Inc.)*

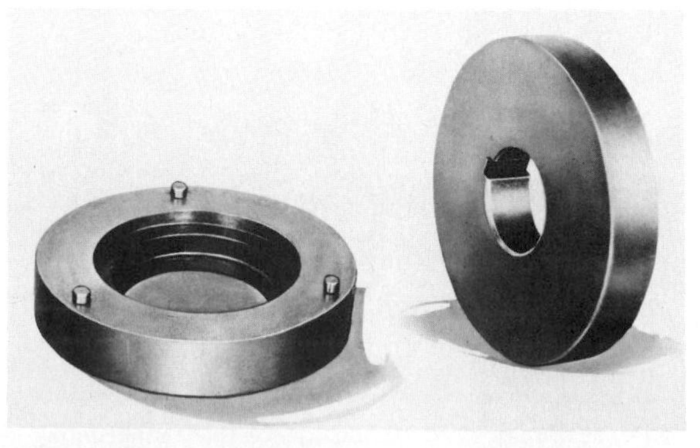

FIG. 14. Bearing runner rides on pivoted shoes in vertical motors. *(Kingsbury, Inc.)*

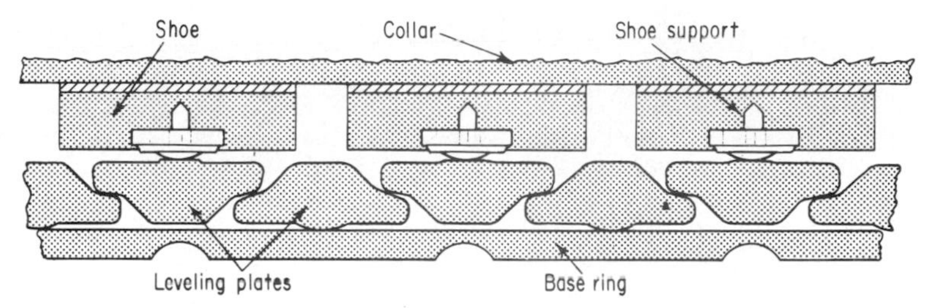

FIG. 15. Principle of equalized support of thrust-bearing shoes. *(Kingsbury, Inc.)*

FIG. 16. Split base ring and leveling plates of small six-shoe bearing. Bearing is held in place by a base ring which also holds shoe against rotation.

bearing, as used in vertical applications, is readily calculated since all the load rests on thrust pads which support the load from the shaft collar (see Art. 15). Good practices will show that the thrust exerted on the thrust pads should be maintained at about 300 lb/in^2. This value is, however, subject to variances due to the speed of the shaft and thrust collar, the design thickness of the oil film, and the heat generated in the bearing system. When these factors are taken into consideration, a rather wide variance in the capacity in any given bearing will be expected. This variance, however, should be within the range of that recommended by the motor and bearing engineers. The physical size of the thrust-bearing housing is also a factor as well as its capacity to radiate the heat generated within the bearing. Most qualified motor engineers are well aware of the limitations of the tilting-pad thrust-bearing designs and have conducted thorough tests in their shops and in the field to substantiate the decision of size for a specific application. As in the case of radial sleeve bearings, thrust bearings, when given a reasonable lubrication and adequate protection for circulating electric currents, will last indefinitely at loads within the above limits.

7. Predicting Bearing Life. The ball- or roller-bearing life is the length of operating time in hours which will pass before the first perceptible sign of failure begins to occur. For bearings of a particular manufacturer, it is possible to predict

FIG. 17. Assembly of six-shoe thrust bearing. *(Kingsbury, Inc.)*

FIG. 18. Vertical runner added to bearing of Fig. 17. The arrows show oil flow. *(Kingsbury, Inc.)*

this life, but only on an average basis as taken from a large number of similar bearings and tested under the same operating conditions.

This life is predicted in two ways. "The minimum life for a bearing that is one of a group of bearings is the time that 90 percent of the group will operate before evidence of failure develops." The average life of a bearing is the average of the life before failure for each bearing in the group, and is roughly five times the minimum life as determined by the standard test methods.

Motors that are not ordered for any specified condition of bearing load are not considered to have a stated bearing life. It is to be assumed, however, that the motor designer has selected the bearing to meet the optimum conditions expected from a conventional motor without consideration of other limiting specifications. Shaft rigidity usually dictates a bearing size which is large enough to ensure the bearing life for many years. However, when specific thrust or radial load conditions are known to be present in the application, it is important to understand what bearing life the motor manufacturer expects. The bearing will normally be chosen to give either 1 or 2 years of minimum life under continuous operation. Larger motors, such as those used in powerhouses for auxiliary service, might be specified to have longer life because of the condition of operation. An average life of 5 to 10 years of continuous operation at specified loading may be selected. If the motor is to run only 12 h a day, the life expectancy would be doubled. Some motor specifications are written calling for a minimum bearing life of approximately 100,000 h, or approximately 12 years or more. When this requirement is applied to vertical motors with high downthrust, particularly at 1800 r/min, the angular-contact ball-bearing combination will usually have to be rejected in favor of the water-cooled spherical roller bearing with an attendant increase in the first cost and maintenance. When this minimum life expectancy is specified, it should be kept in mind that such a minimum bearing life might exceed the life expectancy of the average stator winding. It may therefore be not only costly but unnecessary from a reliability standpoint to insist on extremely long bearing life when other elements of the motor may be expected to fail in a shorter period.

Bearing lubrication is of major importance when considering bearing life. There are many types of lubricants, and a given bearing design will require certain methods of lubrication and maintenance (see Secs. 14 and 17).

8. Selection of Bearing Type. In many motor applications, either the rolling-element type bearing, such as ball or roller bearings, or the slider type, such as sleeve bearings, may fill equally the requirements for the application. The following information is listed so that the user may have a better idea of the merits of various types of bearings from which the one can be selected that will meet a particular application. The facts listed are generally agreed upon by all motor engineers to be valid, and it is intended that the information be used as a general guide in the selection of bearing types.

a. Ball bearings. The merits are as follows:

1. Low friction, especially at starting.
2. High load capacity, particularly thrust loading.
3. Complete standardization. Replacement bearings of many types are available as off-the-shelf items and can be secured all over the world. Those made by one bearing manufacturer are interchangeable for size and rating with bearings of other manufacturers. Remember that the right type as well as the right size is important. A regreasable bearing does not belong in a prelubricated bearing housing, nor does one with a tight internal clearance belong in an application designed for loose clearances.
4. Bearing failure does not usually damage the motor shaft.
5. Grease lubrication is inherently simpler and cleaner than oil and can be used in motors on wall, ceiling, or vertical mountings. Use of prelubricated, sealed bearings helps shorten the tedious round of the maintenance personnel.
6. Loss or deterioration of the lubricant will shorten the bearing life but does not usually result in immediate bearing failure.
7. "Heat-stabilized" ball bearings are available for use at ambient temperatures of above 100°C. Silicone-base greases are used for these applications.
8. Ball bearings may be locked in place to provide the very small shaft end play required for close-coupled drives or where component parts, such as pumps and impellers, are mounted directly on the shaft extension.

b. Sleeve bearings. The merits are as follows:

1. Quiet operation. In some applications this feature is extremely important.
2. The system is relatively insensitive to small amounts of grit and dirt as well as overlubrication.
3. When properly lubricated, indefinitely long life can be expected.
4. When the bearings are badly worn, they can be economically relined and put back into service.
5. Sleeve bearings may be split to permit the inspection and removal of bearings without removing the motor end bells or disturbing the coupling alignment.

c. General factors. When either type of bearing is available in a motor being selected for a given application, the foregoing factors plus the relative cost and delivery date should be considered in making a choice.

The temperatures of standard ball and sleeve bearings should not exceed 85°C. Normally the bearing temperature will level off at least 15 to 20°C below that of the winding. A thermometer on or in the bearing housing is the best check for temperature.

When the ambient temperature for a ball-bearing motor exceeds 40 or 50°C, the motor manufacturer should be consulted so that high-temperature bearings can be supplied. These bearings are especially made to stand operations up to 125°C. Motors so equipped will usually require a long delivery time. Remember that conventional bearing greases will not be satisfactory at this temperature. It should also be remembered that temperatures below minus 25°C require special bearings as well as lubricants.

Sleeve bearings are not rated for operation in ambients above the standard 40°C. Any value in the 30 to 60°C range will probably cause little trouble; however, such applications should be negotiated. Large outdoor motors, subject to below-freezing temperatures, are sometimes equipped with small heaters on the bearing reservoir to prevent the congealing of the oil and to reduce the bearing wear at startup.

On loaded roller thrust bearings, rises of 60 to 70°C or more are not uncommon, especially at higher speeds. Water cooling of the bearing oil is used, where required, to keep the temperature from going higher, and is usually required at 1800 r/min regardless of the thrust. Heating due to bearing speed is excessive at 1800 r/min, even when bearing loading is light. For cooling, water inlet and drain lines must be provided in the motor installations. Water flow usually does not exceed 2 to 3 gal/min. Tilting-pad thrust bearings may also require water for cooling of the lubricant. It is strongly advised in all cases where the ambient temperatures are beyond those stated above that the matter be discussed carefully with the motor manufacturer so that necessary provisions can be included in the design of the equipment to handle these marginal conditions.

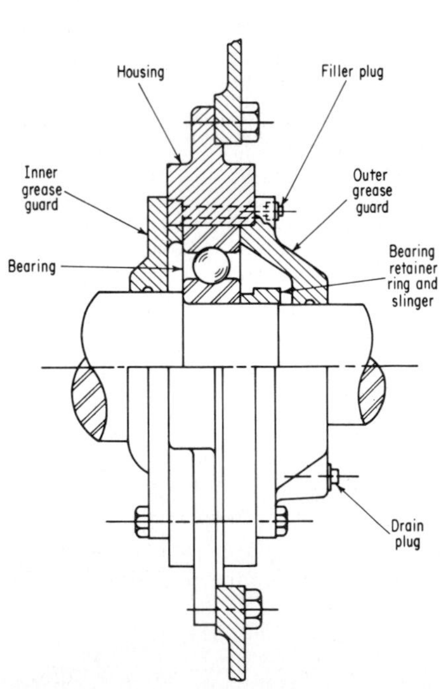

FIG. 19. Typical antifriction-bearing-capsule assembly.

9. Protection against Failure. Assuming that the proper loading and maintenance will be provided, there are additional motor-design features and accessories available to protect against the bearing failures. First to be considered are sealing arrangements to keep contaminants out of the bearing housings (Fig. 19). Totally enclosed motors and outdoor motors meeting the NEMA definition for type 2 weather-protected enclosures are equipped with various types of revolving and stationary seals to exclude water and dirt.

Motors of other enclosure designs may require more complicated sealing in very dirty environments (see Sec. 13).

Where the entrance of dripping and splashing water is a problem, as in the case of dripproof motors mounted with shaft up, neoprene flingers may be used to keep the water out of the bearing.

10. Shaft Circulating Currents. Careful consideration should be given to the use of insulated bearings or bearing housings (Fig. 20). This insulation is generally supplied on only one end of the motor to prevent the passage of shaft circulating currents through the bearings (Fig. 21). The pitting caused by these minute currents is especially destructive to ball or roller bearings. Since the presence of voltage-producing shaft current is a function of the motor electrical design, the decision to furnish insulated bearings is left to the motor manufacturer. The following points are to be remembered in this connection.

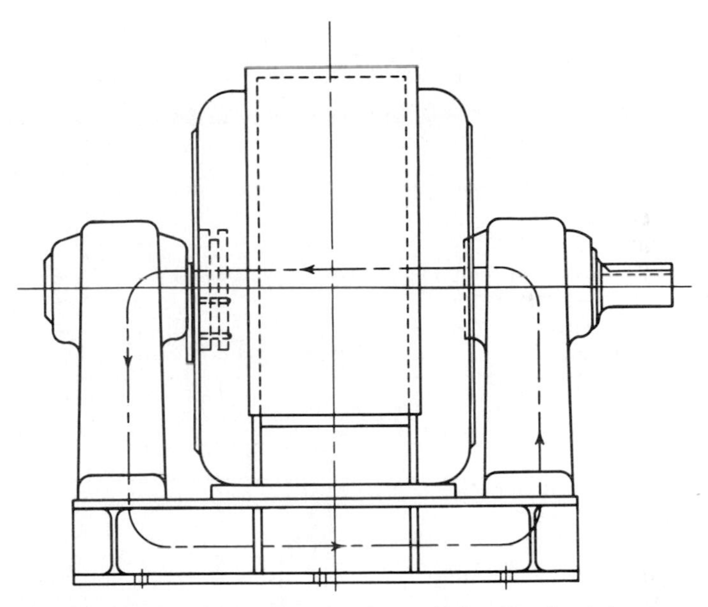

FIG. 20. Arrows show shaft current that could flow if no insulation were provided for bearings. These currents are the result of dissymmetry in magnetic circuit of motor and may be at slip frequency, a multiple of slip frequency, or line frequency.

1. When a motor is furnished with an insulated bearing, it is essential that the insulation be maintained throughout the life of the motor. Excessive contamination by oil or dirt and connections to the bearing housings may short-circuit the insulation and cause a current to flow (Fig. 22). To avoid trouble, the user should clearly understand just how and where the insulation is located.
2. Sometimes, despite the designer's effort, a motor without bearing insulation may exhibit unexpected shaft currents during operation. Should these be detected, it is usually possible to cure the trouble by insulating one bearing. The motor manufacturer should be consulted as to the method to be used. It is further recommended that the motor be shut down until this situation is corrected.

If a motor bearing failure should occur despite all precautions, it is important to know that there is trouble and if possible to get the motor off the line before more damage is done. Bearing failures may ruin the motor shaft, allowing the stator and the rotor to rub and lead to burnout of the windings.

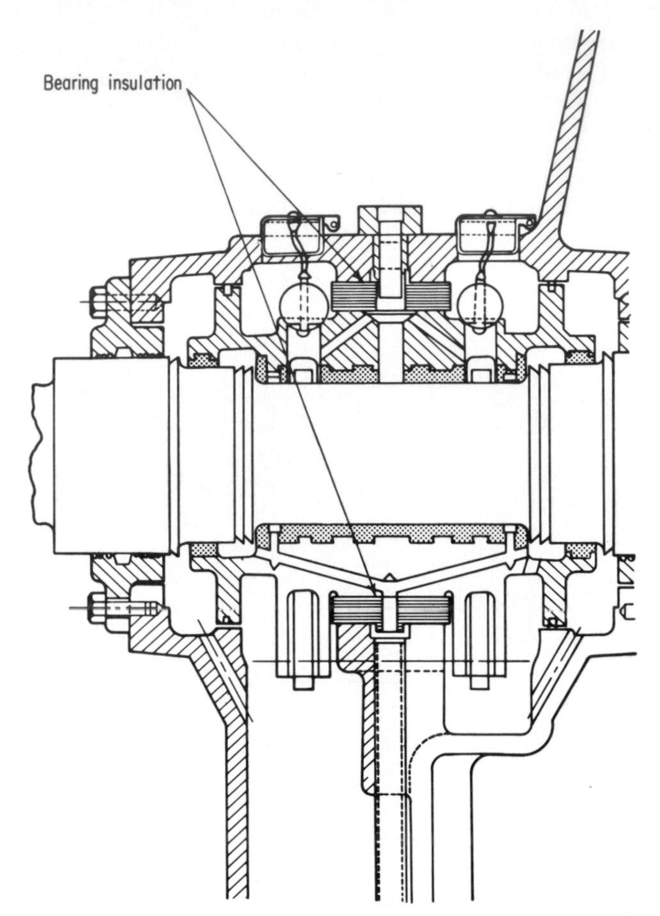

FIG. 21. Insulation of bearing bushing on inside of bearing enclosure provides protection for insulation against dirt accumulation.

11. Bearing-Temperature Protective Devices. Hot bearings can be best detected by the use of temperature detectors mounted into the motor. When bearings overheat, the detectors may sound an alarm, or they may be wired into the motor control circuit to take the motor off the line. This arrangement may be supplied and mounted by the motor manufacturer or mounted later in the field. It is advisable, however, to have this work done on an initial designed basis if possible.

When detectors are to be mounted later, be sure to give enough information on the motor order so that the bearings or bearing housing can be properly drilled. The complete catalog number of the temperature detector will provide this information. When the detectors are to be supplied with the motor, specify the following:

1. Whether a dial thermometer or alarm contacts are to be required.
2. Desired contact arrangement (normally opened or normally closed, etc.).

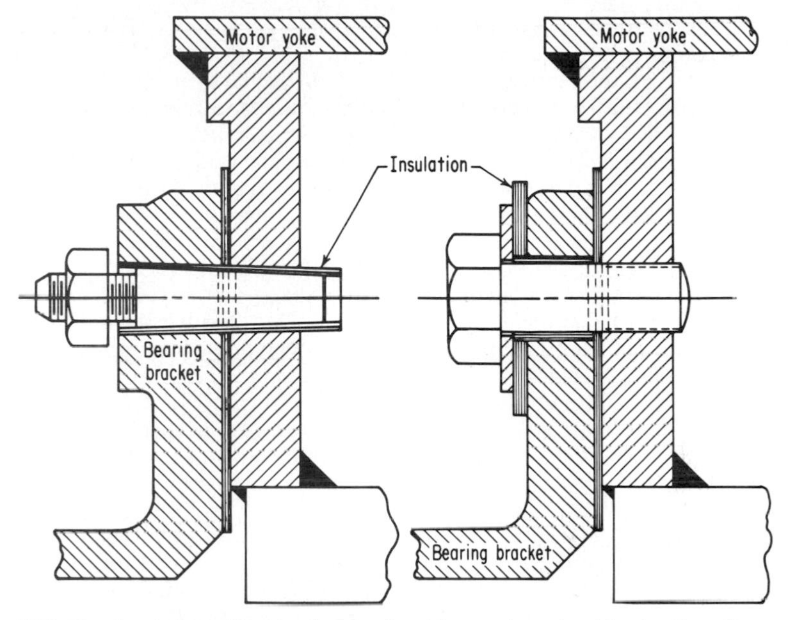

FIG. 22. Insulation of bearing bracket is subject to short-circuiting by dirt, oil, or paint and must be kept clean to be effective.

3. Any coordination of the detectors with similar devices used on driven machine bearings.
4. Some detectors consist of a temperature-probe bulb directly connected by a capillary tube to the relay or thermometer element. If it is desired to mount the control at some distance from the motor, this distance should be so that the proper length of the capillary tube can be ordered. Consideration must be given in the selection of these detectors so that when used they will not impair the function of the bearing insulation.

Detectors are seldom used on motors below 100 hp. The proportionate cost for protection rises rapidly as the motor size decreases. Also, the detectors may be so large that they will not fit effectively on smaller-sized motors.

12. End Play. At least one end of the shaft must be left free to allow for thermoexpansion as the motor warms up during operation. Bearings should not be locked tight against shoulders within the system. The amount required is approximately 0.005 in for each foot of length of shaft between the bearings. This is a minimum, however, and greater amounts may be applied provided the driven units are not affected adversely. Ball bearings for nominal operation will have at least this much clearance in them. Sleeve-bearing motors will have from $\frac{1}{16}$ to $\frac{1}{2}$ in nominal end play, depending upon the rating and the speed. The reason for the large end play is to ensure that despite manufacturing and winding variations, the rotor shaft assembly can be made to "float" during operation, that is, to run without contact between the ends of the bearings and the ends of the journal fits on the shaft. Such contact means thrust is present. The bearing is not designed to withstand this kind of operation.

Since sleeve bearings are not capable of sustained axial thrust, NEMA has standardized on end play for pump motors, as applications most likely to be subjected to such thrust. NEMA Standard MG1-6.11 was written because operating experiences on horizontal sleeve-bearing motors have shown that sufficient thrust to damaged bearings may be transmitted to the motor through a flexible coupling. At operating speeds, the mating parts of the coupling tend to lock together by friction, which does transmit actual thrust to the bearings (see Sec. 16).

Therefore, NEMA recommends ball bearings for all such motors 200 hp and below at 1800 r/min and 100 hp and below at 3000 to 3600 r/min. If sleeve bearings are used, standard end play is ½ in for motors 250 through 450 hp at 3000 to 3600 r/min, and for 500 hp and up at all speeds. Outside of this range, end play is ¼ in. These factors are designed into the motor by the motor builder.

This standard was set up for pump drives, as stated by NEMA, and does not necessarily apply to motors used for other applications. Specifically, large sleeve-bearing motors of speeds below 1800 r/min will normally have ½ in end play. At 1800 r/min the end play may be ¼ in or more, depending upon the horsepower rating.

It is advisable to check with the motor manufacturer beforehand to avoid ordering the wrong type of coupling for the motor. The so-called limited-end-float coupling is recommended for sleeve-bearing motors, in preference to free-floating types, to make sure that the motor rotor cannot move axially to the limit of its end play and thereby impose a thrust load on the motor bearings (see Sec. 16).

When special drives, particularly pumps, require that a total end play be limited to zero or a few thousandths of an inch, locked ball bearings are used. The design modifications necessary to secure this limited end play require that the necessary value of end play be negotiated with the supplier at the time the motor is ordered.

13. Bearings in Fractional-Horsepower Motors. The majority of all the fractional-horsepower motors manufactured are equipped with plain or sleeve-type bearings. They are used in an endless variety of applications. This type of bearing consists of a bushing of antifriction metal which is accurately machined and inserted into the end bell of the motor. The inner surface of the bushing is machined to accept the shaft and to afford a means for lubrication of the bushing. The external shape of the bushing may vary considerably within the limits of the bearing housing to afford a mounting for the bushing. Consequently bushings are frequently designed to fit a spherical seat, an end-flange mounting, or a non-shouldered bored seat. The motor manufacturer has selected the type of bearing bushing to accommodate the housing dimensions as well as for the type of lubricant and environment expected.

The bearing design for this type of motor is well known and is based upon many years of field experience. It should be borne in mind, however, that there are limits to all bearings as applied to motors and other equipment and that, if the motor is abused or misapplied, the bearing system is not immune to failure. The many thousands of fractional-horsepower motors being used daily with assurance of good performance will attest to the proper bearing application and design.

Fractional-horsepower-motor bearings are made from a variety of antifriction metals which lend themselves to the particular requirements of the application. The most common metal used is bronze in its many different alloys. Bearings have also been produced and used successfully made from various sintered materials which afford porosity and a means for lubrication. Some motor manufacturers use bearing bushings impregnated with graphite, which affords protection when lubrication by other means is questionable. In each of the cases, however, it is extremely important that the bearing bushing be accurately machined and inserted into the bearing housing with precision so that the proper running clearances can be ensured when the end bell is assembled into the motor frame.

There are, however, some fractional-horsepower motors which are equipped with ball bearings. When this type of motor is supplied, the user can be assured that careful consideration has gone into the selection of the bearing for the application. Unless the motor is of a special design, the bearings are of the radial type consisting of outer and inner races as well as balls and cages assembled together into an assembly which can be removed and replaced if the system shows evidence of failure. Some earlier designs may have included the inner race as a functional part of the shaft. However, this type of bearing arrangement is seldom found in present-day designs.

In nearly all cases, ball bearings are grease-lubricated with shields and seals as required to prevent the contamination of grease and to retain the grease within the proximity of the rolling elements. The bearings are designed in such a manner that the proper internal clearance is built into the bearing so that when the inner race is pressed upon the shaft, the proper clearances for correct operation of the bearing are assured. It should be borne in mind that a bearing of this type is a precision piece of equipment and is carefully machined and treated to give long trouble-free service. The motor manufacturer has carefully mounted the bearing into the end shield and upon the shaft and has conducted exhaustive tests to guarantee the performance of the motor equipped with these bearings so that the motor will be trouble-free when put into service.

In nearly all cases ball bearings for this size of motor are prelubricated for the life of the equipment during manufacture and no additional provisions are made to permit regreasing. If the motor is to operate under conditions which are not considered to be standard, the motor manufacturer should be notified so that the bearings can be equipped with a proper grade of lubricant and seals. Nearly all motor manufacturers can supply equipment for a variety of environmental conditions, and a motor is selected to meet specific requirements rather than at random.

14. Bearings in Integral-Horsepower Motors. These bearings may be supplied as plain or sleeve-type bearings. Many motors of this size have been furnished with this type of bearing in the past and have given excellent service. On the smaller rated motors one-piece or bushing-type plain bearings have been furnished as standard equipment. On larger-sized equipment a two-piece or split-type bearing has been furnished at the discretion of the motor manufacturer. The decision to use plain bushings or split-type two-piece bearings is left to the motor manufacturer, who has carefully analyzed every bearing application with the size and requirements of the motor involved. Many motor manufacturers will use plain sleeve bushings up to a certain rating, and beyond this size they will use two-piece babbitt-lined bearings.

On the smaller rated integral-horsepower motors, plain or bushing-type bearings are frequently used, which are similar in appearance and design to those used in the fractional-horsepower line, except for their larger physical sizes. Most of these bearings are made from various bearing bronze alloys accurately machined on both the outside and the inside with oil grooves and pockets so that the lubricant can be properly distributed within the bearing. Some plain bushing-type bearings are machined on the outside with a spherical seat to afford self-aligning features. Other types are manufactured with a flange to afford an exact seating of the bearing when mounted into the end bell. In some cases the plain bushings are overcast with babbitt metal to provide the type of riding surface the application requires.

On the larger-size ratings of integral-horsepower motors, bearings of a plain split-type design are frequently furnished (Figs. 10 and 11). This bearing is made in two pieces and split on the horizontal centerline, with provisions for bolting the two halves together, thus forming a complete bearing bushing. This type of bearing is accurately machined on both the outside and the inside, as required, and permits various configurations on the outside to accommodate space limitations and other features desired in the bearing and housing.

Split-sleeve-type bearings are normally made from either cast iron or steel as a shell material, which is lined with a babbitt white metal in the bore. The bore as well as the shell is carefully machined to final dimensions, and the necessary grooves and oil pockets are machined into the bore for proper oil distribution within the bearing. Bearings for this size of motor are normally equipped with oil rings which carry oil to the bearing from the oil reservoir in the housing.

The motor manufacturer has carefully investigated various types of bearing alloys to line the bearing shells. Some of the fundamentals taken into account in the selection of this alloy are worth presenting again for general information regarding motor bearings. The alloy selected should have the following characteristics:

1. Resistance to fatigue
2. Resistance to corrosion
3. Good surface conformance
4. Embeddability
5. Ease in production

Resistance to fatigue is usually the first consideration, as it is a measure of bearing life. The endurance limit, or fatigue, is not easily determined because of such limiting factors as load, distortion, misalignment, clearances, environment, and poor maintenance. All these items have been taken into consideration by the motor-design engineer in the selection of the proper material and bearing design.

There are two basic types of babbitt alloys which are used for this size of equipment. They are best known as tin-base alloys and lead-base alloys.

The tin-base alloy is characterized by its low hardness, and it is necessary to use alloying constituents of copper and antimony to afford this feature. These alloying agents give the tin-base alloy the hardness required as well as wearability and satisfactory surface action and carrying ability.

Since lead alone is too soft to act as a good bearing metal, a lead-base alloy is developed by adding antimony, arsenic, and tin to increase its strength and hardness. While lead and antimonly alone may be used as bearing metals for simple applications, arsenic and tin are added to produce still better mechanical properties. Conventional lead-base babbitts may be affected by weak organic acids, and to prevent this, tin may be added in quantities up to 10 percent, thus making an alloy nearly equal to tin-base babbitt in corrosion resistance.

A great deal has been said about the relative merits of the tin- and lead-base-alloy babbits. It must be realized that the two alloys differ in their casting techniques, and the quality of each alloy may differ for any given casting procedure.

It has been determined that the lead-base alloy may be inferior to tin-base in the presence of some acids, whereas the lead-base alloy is superior to tin-base with regard to load-carrying capacity.

The user of an integral-horsepower size motor which is supplied with babbitt-lined bearings can be assured that careful consideration has been given to the design of the bearings and that the selection of the alloy has been made to give the maximum performance of the equipment.

The horizontal integral-horsepower motors of NEMA size range and furnished with babbitt bearings in nearly all cases are equipped with oil-ring lubrication. This type of bearing is mounted in the end shield of the motor, which may be of one- or two-piece design. In the largest applications, the end shield is made in two sections so that bearing adjustments can be afforded at installation and after wear of the bearings. The bearing housing, which is normally an integral part of the end bell, consists of a mounting accurately machined to accept the bearing as well as a means for securing oil-seal rings at the end of the housings and various auxiliaries. The lower section of the bearing housing contains a reservoir for oil which is lifted to the shaft by means of an oil ring which rides on the shaft. At least one oil ring is used on each of these bearings, and in some cases two oil rings are used on long bearings to supply the proper amount of oil. The housing may include a ventilating duct to afford a means of stabilizing internal air pressures and preventing oil leakage along the shaft. Means may be included on the outside of the housing for indication of the oil level within the housing. Various types of external oilers are available and may be used to maintain the level of the oil within the housing (see Sec. 14).

Other auxiliary equipment can be connected and attached to the bearing housing such as low-level alarms, thermometers, and vibration detectors. Various styles of air baffles and shaft seals may be designed into the housing to prevent the entrance of water and foreign materials into the bearing housing and prevent suction of oil from the oil reservoir. Each of these accessories can be supplied by the motor manufacturer provided it is indicated at the time the order for the motor is placed.

While most of the horizontal integral-horsepower motors furnished in NEMA ratings will be equipped for oil-ring lubrication, there is a possibility that under certain conditions pressure-lubricated bearings may be required. Oil is forced into the bearing for distribution through the oil grooves and pockets to the proper entrance areas to produce the oil film necessary to carry the shaft loads. This type of application would be expected under high-load conditions and should be considered as a special application.

In general, the flow of cooling air being drawn into the motor is sufficient to dissipate the heat picked up by the cooling oil in the bearing housing and to keep the temperature of the bearing within proper limits. Under certain conditions, such as on totally enclosed and externally vented motors without the necessary air movement over a bearing housing, exterior cooling means may be required. The use of cooling coils within the bearing chamber on horizontal machines is not used, as the physical size of the housing is frequently limited by the size of the motor. On some of the larger integral-horsepower motors an external cooling means needs to be afforded to control the temperature of the lubricating oil and the temperature of the bearing. This cooling can be accomplished by circulating the oil from the reservoir through either a water- or an air-cooled heat exchanger. Most large motors which require oil-circulation systems are designed and furnished with a shaft-driven gear pump which assures the movement of the oil as soon as the motor is started. The heat exchangers are normally activated by solenoid valves or blowers which are interlocked with the starting equipment of the motor. If the equipment is to be used in areas of low temperatures, heating means must be provided to prevent the freeze-up of the cooling liquids within the cooling system.

15. Bearings in Vertical Motors. Motors used on vertical applications, such as on deep-well pumps, require special bearing considerations. Both ball and roller as well as sliding-type thrust bearings are used to carry the weight of the motor-rotating element and in some cases the weight of the pump shaft and the impeller. Conventional horizontal-type bearings are not designed for this type of service.

In the smaller rated NEMA motors, some manufacturers furnish a thrust journal bearing as a self-contained bearing with the proper provisions for carrying limited thrust loads as described. The bearing may be either the bushing or the split babbitt-lined type. Accurately machined grooves in the bore and on the thrust faces permit the flow of lubricating oil to the bearing. On larger units, the journal bearing and the thrust bearing may be separate units within the same housing. The thrust bearing, with the provisions for entrance of the oil to the bearing surfaces, may in some cases be mounted in a spherical seat to afford a self-aligning feature.

Thrust bearings are also designed whereby individual segments or pads are used and connected to a self-equalizing system in which any misalignments on the shaft will cause automatic equalization of loads onto each of the respective thrust pads (Fig. 23). This type of bearing, known as the self-equalizing tilting-

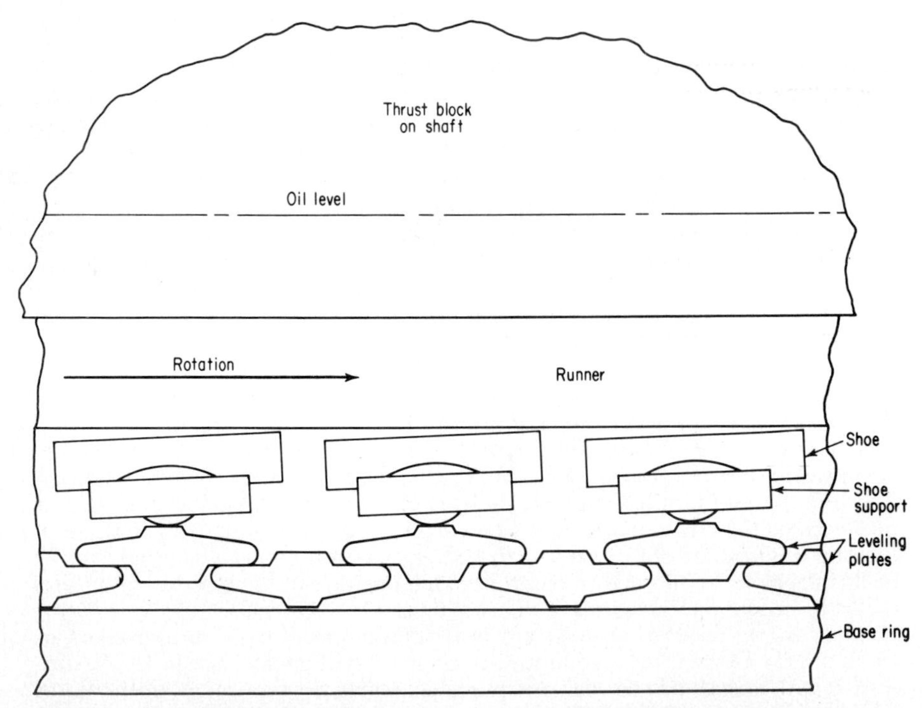

FIG. 23. Thrust-bearing diagram shows how tilting-pad bearings form wedge of oil as motor rotates.

pad-thrust-type bearing, has many advantages over other types of thrust bearings. A similar tilting-pad-thrust bearing is also used without the equalizing feature. However, in this case each of the tilting pads must be adjusted precisely in a vertical manner for each of the pads to carry an equal load. Another design of tilting-pad bearings uses a quantity of springs under each pad to support the load. In this case, all springs must be identical.

Each of the bearings described above must be contained within a bearing housing, which generally includes an oil reservoir. These housings are normally built into the motor and shields. The thrust bearing is either above or below the stator as the motor manufacturer elects. In some cases the frictional losses within the bearing system are so great that an exterior means for cooling of the lubricant and bearing is necessary. This cooling can be accomplished either by circulating the oil through a heat exchanger and returning it to the oil reservoir or by circulating a coolant in the oil reservoir. This latter method is frequently used on large equipment where there is sufficient mechanical room to mount the cooling coils in the oil reservoir. Care must be given to the environment for this type of equipment, as low temperatures may cause freezing of the coolant within the bearing system, causing possible leakage, contamination, and eventually bearing failure.

At least one bearing system must be insulated to prevent the circulation of electric currents across the oil film in the bearing. This can be quite easily done by insulating either one journal bearing or the journal and thrust bearing at the opposite end. Care must be exercised, however, to prevent the short circuiting of this insulation when various accessory equipment is attached to the bearing housings as required.

Some applications of vertical motors require that there be a minimum of clearance between the journal bearings and the shaft itself. This can frequently be accomplished by the use of tilting-pad journal bearings, which run with considerably less internal clearance than the normal journal bearings and will assist in controlling shaft slip and runout. This bearing design includes several babbitt-faced tilting pads equally spaced around the shaft, which are carefully bored to match the diameter of the shaft plus a slight clearance and are retained at their outside diameter by a retaining ring to which the pads are keyed to prevent rotation. The bearing pads are contained within the structure by side plates, and the complete bearing can be installed as a bushing-type bearing or as a two-piece split bearing, as is required. The principal advantage to this design is that normal running clearance can be reduced so that the shaft can be more closely centered with the true center of the bearing. Each individual tilting pad acts within itself, thereby forming its own oil wedge as the shaft revolves.

Thrust-bearing housings and the complementary journal-bearing housings can be equipped with various auxiliary equipment such as oil-level gages, thermometers, flow gages, circulating pumps, alarms, and pressure gages, as may be required for the safe operation of this type of bearing. It is necessary to describe the auxiliary devices when negotiating for this type of motor, as it will be necessary for special matching to be incorporated into the housings to accommodate the mounting of these auxiliaries.

As in the typical horizontal-type motors, various degrees of enclosures are available for protection under various conditions of environment. The bearing housings likewise are adequately protected by the motor manufacturer to include all the features necessary to protect the bearings, as the bearings and bearing structures are critical items of the unit. Motor manufacturers have had sufficient experience and have conducted adequate tests to assure the user of the maximum protection for the bearings.

The user can rest assured that the motor to be delivered will accomplish all that will be needed provided all the requirements of the drive and its operation have been communicated to the motor builder. In this connection, more users are disappointed with the equipment because of inadequate specification than for any other cause. It is advised that with important motors, a carefully considered list of operating conditions be made and that they be covered item by item with the manufacturer so that there is a joint understanding of what is wanted at the time the order for the motor is placed.

16. Bearing Standardization. NEMA and its member organizations have established a set of limiting factors for the design of motors which when applied to the engineering concepts of the motor will give assurance of economical use of materials and good engineering practices. Through this standardization, a system of loading on extended shafts has been established, and when adhered to by the user, it will permit operation of the motor within acceptable bending moments of the shaft between and beyond the bearings. Realizing that a stiff shaft is desirable, these standards establish a minimum shaft diameter as a standard dimension against speed and rating for a given frame size. The exact diameter of the shaft at the bearing seat is at the discretion of the motor manufacturer and the design engineers. The user can be assured that the motors now available are adequately designed and built with the best-quality bearings and bearing materials available to handle the loads which are specified under the limitations of NEMA. It should be recognized, however, that if these limitations are violated, the user has little recourse with the motor manufacturer. The speed at which the shaft turns within the bearing is of course a function of the bearing life. However, if the load is kept within the initially designed limits, the life of the bearing is almost indefinitely long and dependent upon the fatigue limitation of the bearing metals. The user can feel assured that the bearings provided in a motor have had considerable attention in design and testing and that if the equipment is properly used, it will give satisfaction for a long period of time.

The environment in which a motor operates puts a limitation on its life as well as on the life of the bearings. There are several well-known types of enclosures for motors which afford the same protection for hearings. It would be false economy to afford a good insulation system in a motor to prevent deterioration due to unfavorable environmental conditions and then omit consideration of the bearings and housings, which are affected by the same conditions. Labyrinth and rotating seals, fillers, seals, breathers, inhibitors, oxidizers, and other mechanical and chemical devices are used to assist in providing protection against unfavorable environmental conditions and other abuses to which motor bearings are subjected. Each of these items should be considered when purchasing the motors so that a unit can be purchased to meet the exact requirements.

REFERENCES

1. F. T. Barwell, *Bearing Systems: Principles and Practices* (Oxford University Press, New York, 1979).
2. Ian Bradley, *Bearing Design and Fitting* (International Publications Services, New York, 1976).
3. B. J. Hamrock and Duncan Dawson, *Ball Bearing Lubrication* (Wiley, New York, 1981).

SECTION 13
BEARING AND SHAFT SEALS

Robert R. Burke*

FOREWORD 13-2

BEARING SEALS 13-2

1. Clearance Seals 13-2
2. Rubbing Seals 13-3
3. Antifriction Bearing Seals 13-9
4. Shielded Antifriction Bearings 13-10
5. Open Antifriction Bearings 13-11
6. Sleeve-Bearing Seals 13-14
7. Vertical Sleeve and Guide Bearings 13-16

BEARING AND SHAFT SEALS IN SPECIAL ENCLOSURES 13-17

8. Sealing Explosionproof Machines 13-17
9. Sealing Inert-Gas-Filled Machines 13-23

REFERENCES 13-24

*Manager, Mechanical Engineering, Motor-Generator Department, Siemens Energy and Automation, Inc., Bradenton, Fla.; Registered Professional Engineer (Wis. and Fla.).

FOREWORD

Seals used in electric motors can be divided into two broad types: bearing seals, to keep lubricant in and to keep dirt out, and shaft seals for special-purpose machines such as explosionproof and inert-gas-filled types. By far the most common are ordinary bearing seals; these are discussed first, followed by the special bearing and shaft-sealing arrangements for special-enclosure machines.

BEARING SEALS

Sealing bearings of electric motors is a twofold problem:

1. A high rate of leakage of lubricant from a bearing cavity would necessitate frequent maintenance in the form of checking and adding lubricant to bearing cavities for proper performance.
2. Lubricant leakage in the form of mist into a machine promotes an accumulation of dirt and dust in the motor-ventilating passages, as well as attacking some insulation systems.

There are several types of bearing-lubricant seals. Possible divisions are high- and low-speed, clearance and rubbing seals, and sleeve and antifriction bearing seals for horizontal-shaft machines, and the same divisions for vertical-shaft machines. These are also several special types, as for explosionproof and inert-gas-filled machines. All but the explosionproof seals can be either rubbing or clearance seals, depending on the duty to be performed and the surface speed of the rotating part at the point of sealing. Each type of seal will be described, followed by illustrations of the use of each.

1. Clearance Seals. Clearance seals limit leakage by closely controlling the annular opening between the rotating shaft and the stationary housing. There are two types of clearance seals: bushing or ring seals, and labyrinth seals.

Some of the many advantages of clearance seals are ease of design, reliability in continuous operation, fabrication from many materials, and no contribution to loss or heating of components, since there is no rubbing contact.

Leakage of oil vapor is limited when the ring of labyrinth causes the fluid to lose velocity pressure as it is throttled through the radial width of the annular orifice. Ideally, the velocity pressure of the vapor passing through each stage of throttling is entirely reconverted to a pressure potential in the adjacent chamber.

In a practical seal of the straight-through type, the sealing effectiveness is limited because a portion of the kinetic energy associated with each stage is carried over to the succeeding stage. To reduce this carryover effect, grooves or steps are frequently added between stages to deflect the expanded fluid into the space between strips.

The theoretical leakage of any labyrinth or ring seal is a function of the clearance area, which is directly proportional to the radial clearance. Theoretically, leakage is directly proportional to clearance, and clearance should therefore be as small as possible. However, rotating shafts undergo radial motions resulting from an accumulation of bearing tolerances, dynamic deflections and vibration, and thermal distortions.

Radial shaft motions in excess of the clearance between shaft and seal result in either mushrooming the seal or burning either the seal or the shaft, or both, and in high-speed machines they can cause severe vibration.

Figure 1 shows typical recommended seal clearances for various shaft diameters.

It can be readily recognized that if minimum seal clearances are used, nonferrous materials and a sharp-pointed labyrinth design will give minimum difficulties in the form of burning and vibration if a rub should occur, since this type of seal will wear away most readily.

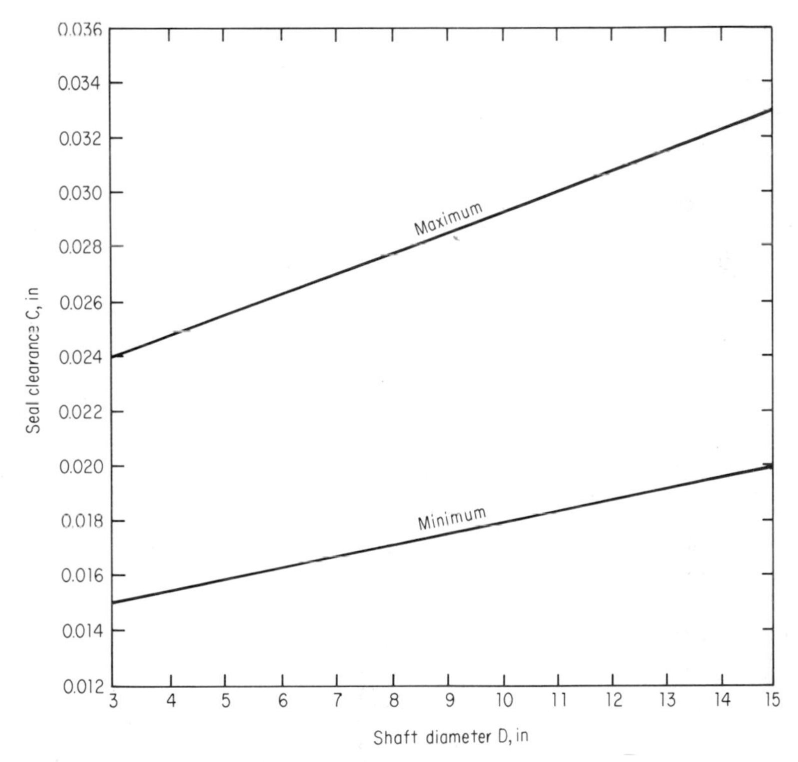

FIG. 1. Recommended diametral seal clearances.

Figures 2 to 5 show types of clearance seals in common use in the motor industry, along with the method of calculating pressure drop across the seal. Because the highest pressure, or head, existing in a large electric motor is probably no more than 10 in of water in the bearing-seal area, this type of seal, although nominally permitting some leakage, will, if designed and applied properly, limit the leakage to small quantities so that maintenance of the oil level in bearing reservoirs is not a frequent or continuing problem.

2. Rubbing Seals. Rubbing seals are applied to motor bearings for both small high-speed and large low-speed motors. The criterion is a rubbing speed of usually

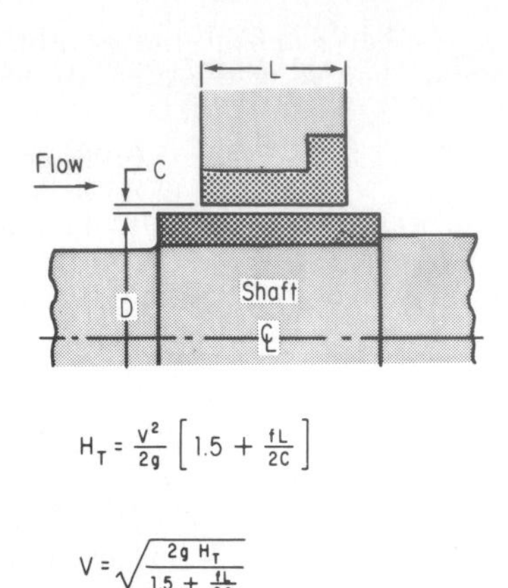

FIG. 2. Sleeve-type clearance seal.

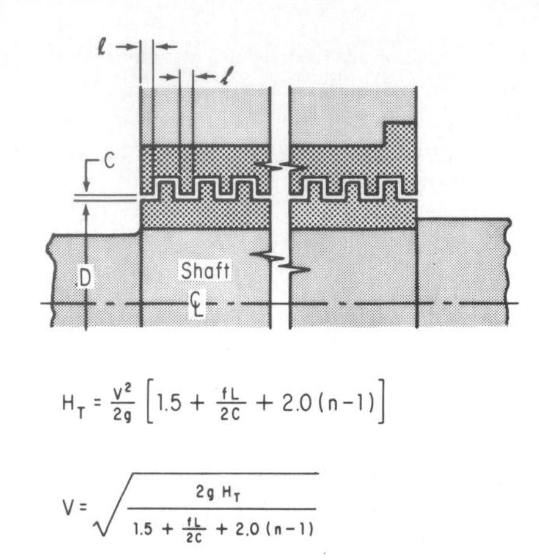

FIG. 3. Interlocking-labyrinth clearance seal.

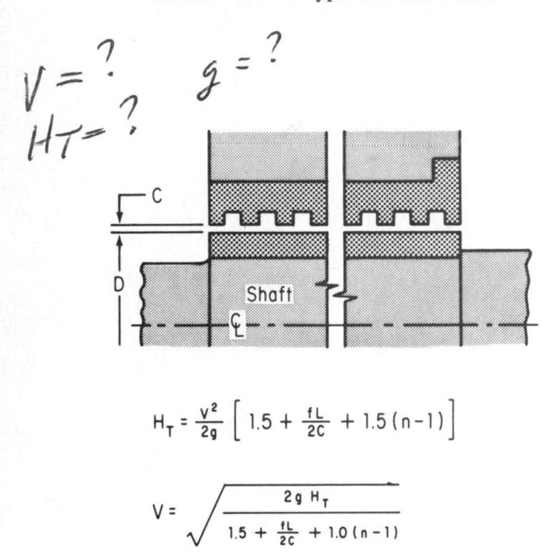

FIG. 4. Noninterlocking-labyrinth clearance seal.

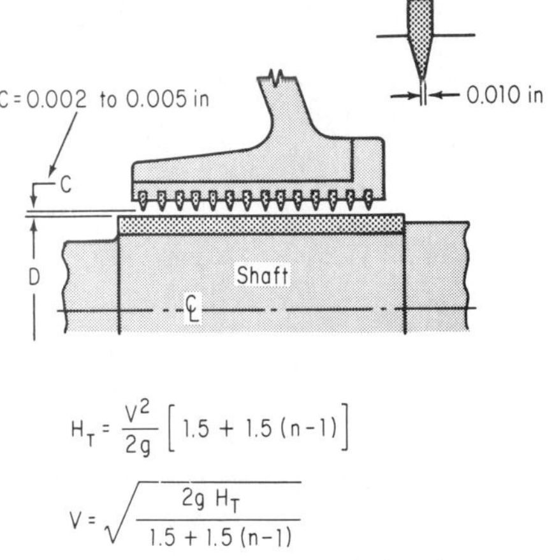

FIG. 5. Tapered-teeth-labyrinth clearance seal.

less than 1000 to 1500 ft/min. Because any friction will produce heat, the speed for the use of rubbing seals must be limited to minimize loss of efficiency and prevent shaft heating. At low rubbing speeds the heat developed is minimal.

Despite the limit of speed at the seal surface, rubbing seals can be used on the majority of small motors built today.

Rubbing seals can be divided into two distinct types: felt seals and lip seals, lip seals being either leather or synthetic seals with or without spring loading.

a. Felt seals. When a felt seal is properly designed of a good-quality wool felt of even density, it is highly efficient. Some advantages include:

1. Ease of installation. Special jigs are rarely needed, and special care to prevent chipping or breakage is generally unnecessary. Distortion is rare, and only rea-

sonable care is necessary to prevent it. Because of the inherent pliability and resiliency of felt, seals are satisfactory even though optimum conditions are not met.

2. Oil absorption and wicking. Although oil-storage capacity is a function of density, felt seals will ordinarily absorb approximately 75 percent of their volume of oil. This absorption assures continuing lubrication, even after long idle periods.
3. Failure is never complete. Resiliency of a felt seal allows it to maintain constant sealing pressure despite wear, end play, minor misalignment, or out of roundness of metal assemblies.
4. Polishing action. Felt seals trap abrasive particles which penetrate into the felt. Thus they protect the steel surface by polishing it rather than scoring it.
5. Filtration. Wool-felt seals, in a dry state, are 99 to 100 percent effective as a filter in removing particles of 0.7-micron size. When saturated in lubricants, even smaller particles are trapped.

Figures 6 to 9 show common types of felt seals used in the motor industry. For effective operation of felt seals, several factors should be considered.

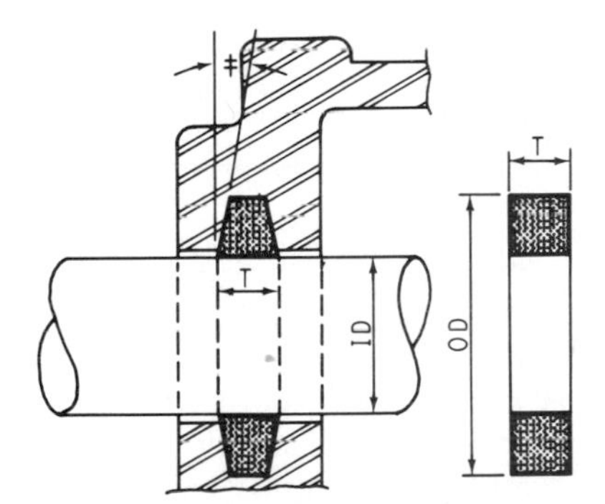

FIG. 6. Machined-groove felt seal.

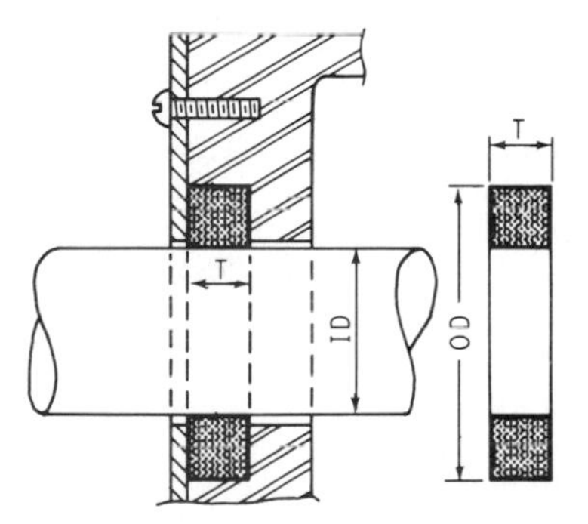

FIG. 7. Recessed-groove felt seal.

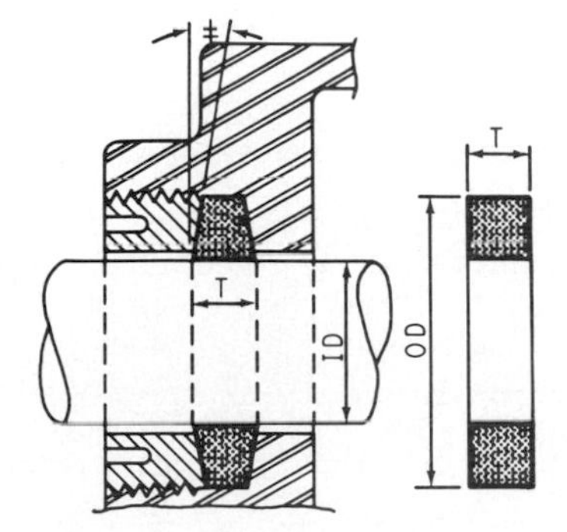

FIG. 8. Stuffing-box-type seal.

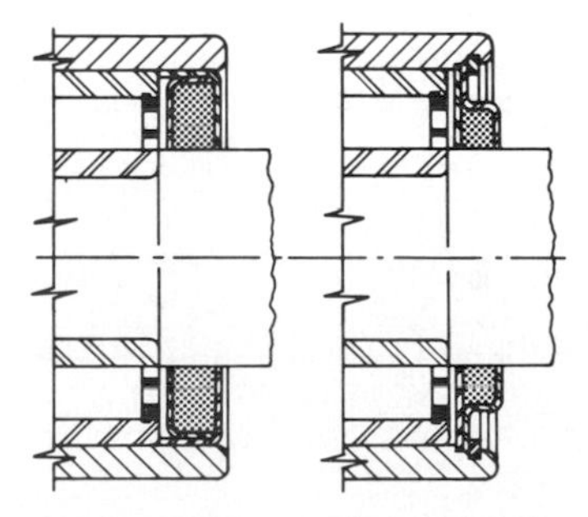

FIG. 9. Metal-stamping-type seals.

1. Avoid excessive stretching when mounting.
2. The felt seal must not be fitted too tightly to the shaft, and the retainer should not exert an excessive compressive force on the felt.
3. In the usual installation, the height of the felt-ring section should be greater than its width. This proportion minimizes seal distortion and permits firm clamping of the felt in its groove.
4. Where possible, the felt should be a solid ring rather than split. However, if a split construction is used, it should be butt-joined at an angle.
5. Felt seals should be replaced when the machine is overhauled. If overhauls are infrequent, an assembly should be used which allows the felt to be replaced easily.

b. Lip seals. Sealing pressure of a lip seal is the result of an interference fit between a relatively flexible sealing element, usually augmented by spring pressure, and a shaft or, in the case of an external seal, a housing bore. In most radial seals, the lip is designed so that increasing pressure in the sealed area causes lip pressure to increase.

Fluid retention of these seals is based on a precise amount of contact pressure so that friction effects, such as heat generation, are minimized.

Conditions of service will determine both effectiveness and life, and should be considered in the seal selection. Most important of these conditions are:

1. Rubbing speed. Tolerances on shaft finish, eccentricity, and end play determine the allowable rubbing speed. The allowable speed may be increased by tightening these tolerances. Shaft eccentricity causes constant flexing of the sealing lip, even at moderate speeds, so that it is unlikely that the sealing contact is continuously maintained without a spring. Under ideal conditions, speeds as high as 2000 ft/min for leather seals and 4000 ft/min for synthetic seals can be attained.
2. Temperature. Heat at the seal lip is generated by rubbing friction, bearing friction, oil agitation, and heat conducted from other parts of the machinery. The upper limit of leather sealing elements is approximately 130°C (200° F), and of molded synthetic elements approximately 157°C (250°F).
3. Shaft surface. Shaft hardness and smoothness are very important factors to consider. Modern seal efficiency and life are obtained with a finely finished shaft surface, usually in the 10-to 20-micron range. The direction of finishing marks is important, and polished or ground finish with concentric marks gives optimum results.
4. Lip diameter. Decreasing the lip diameter of a seal will increase the frictional drag on a shaft, and so increase the temperature.

Thus it is seen that contact pressure is a combined function of lip diameter, interference, spring tension, rubber hardness, and eccentricity. The spring is used to ensure a constant tension or lip load on the shaft, through various temperatures and time conditions. Two types of springs are used, a coil spring, which is very rugged and little susceptible to damage, and finger springs, which with less mass are more responsive, permitting a lighter element with consequent lower friction drag without sacrificing sealing ability. For general service either type is equally suitable.

Two materials are commonly used for motor-shaft seals of this type, leather and synthetic rubber. Leather is absorbent and provides a self-lubricating seal that operates for long periods with little lubrication. It is less sensitive to shaft finish than synthetic rubber, but its speed limit is lower.

Synthetic-rubber seals are used when rubbing speeds or temperatures are high. As compared with leather seals, the synthetic seals withstand shaft runout and eccentricity with less chance of leakage; however, lubrication must always be provided.

Figure 10 may be used to calculate the rubbing speed of a shaft seal, as it relates rubbing speed to shaft diameter and revolutions per minute.

Common lip-seal designs used in the motor industry are illustrated in Figs. 11 to 13, with the most common in Fig. 11.

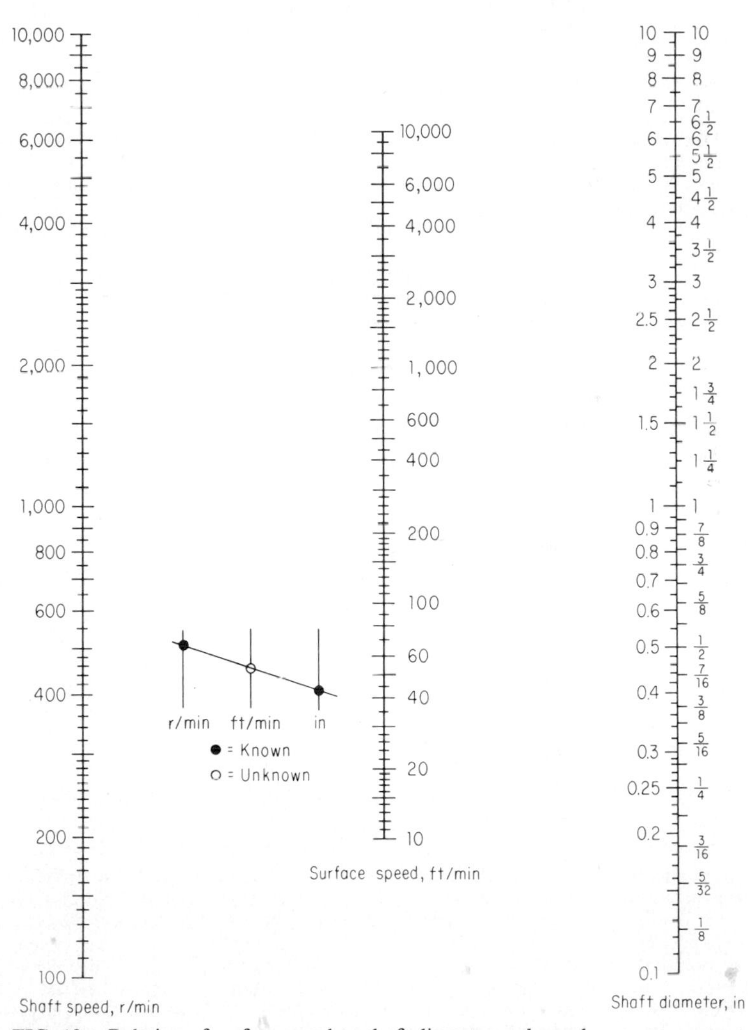

FIG. 10. Relation of surface speed to shaft diameter and speed.

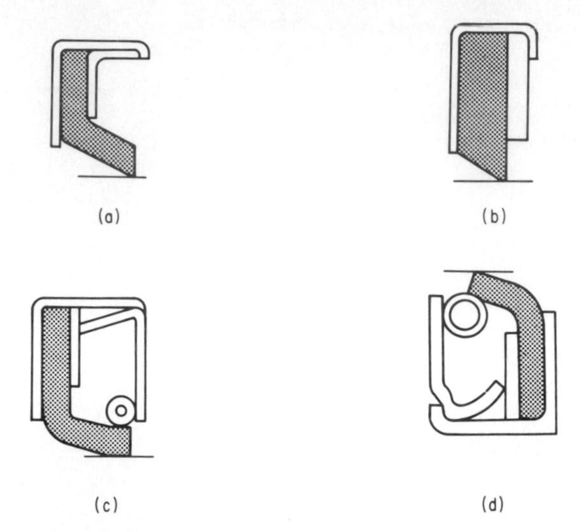

FIG. 11. Cased-type lip seals. (*a*) Flange-type seal with leather or synthetic element. Used primarily to seal against light dust or to retain heavy grease, this type is not suitable for sealing low-viscosity oil. Ideal where space is limited and low friction is required. (*b*) Washer-type leather seal. Adapted for sealing grease or viscous fluid where use of spring-loaded seal is not practicable. Often combined with spring-loaded seal in dual element. (*c*) All-purpose spring-loaded seal with leather or synthetic element. This is the most common type. Pressure range of synthetic type is 0 to 10 lb/in^2. Leather type withstands 0 to 15 lb/in^2. Available with garter or finger spring. (*d*) External seal. Used where speeds are in low range and seal is to be pressed on the shaft. Leather element is suitable for sealing grease or viscous fluids. Synthetic element used with fluids of low viscosity. Available in springless type for stationary shafts where grease is to be sealed.

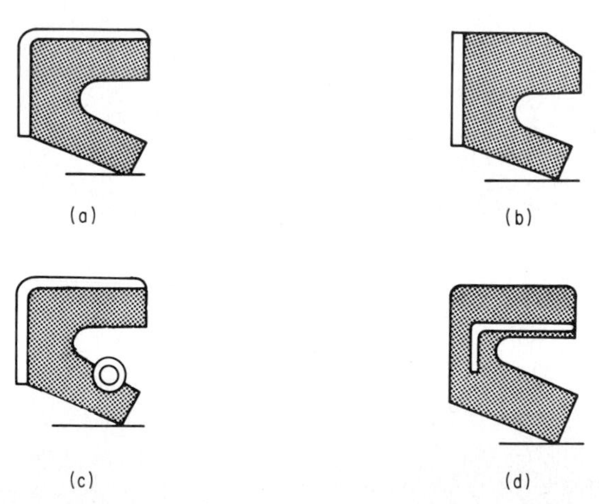

FIG. 12. Bonded-type lip seals. (*a*) Bonded case seal with straight lip, for retaining fluid in rotating-shaft installations. Unit is pressed into straight bore; no groove is required. (*b*) Washer seal with limited-contact lip. Designed for press fitting in straight bore. Limited rubbing-contact area leads to low friction-torque drag on shaft. (*c*) Bonded-case seal with limited-contact lip and spring. Spring accommodates shaft eccentricity, provides initial tension before internal pressure builds up to operating value. Spring can be finger or garter type, or both. (*d*) Embedded-case seal. Synthetic cover on outside of seal compensates for possible unequal expansions of housing and seal. Straight lip suitable for most applications. Also made with springs.

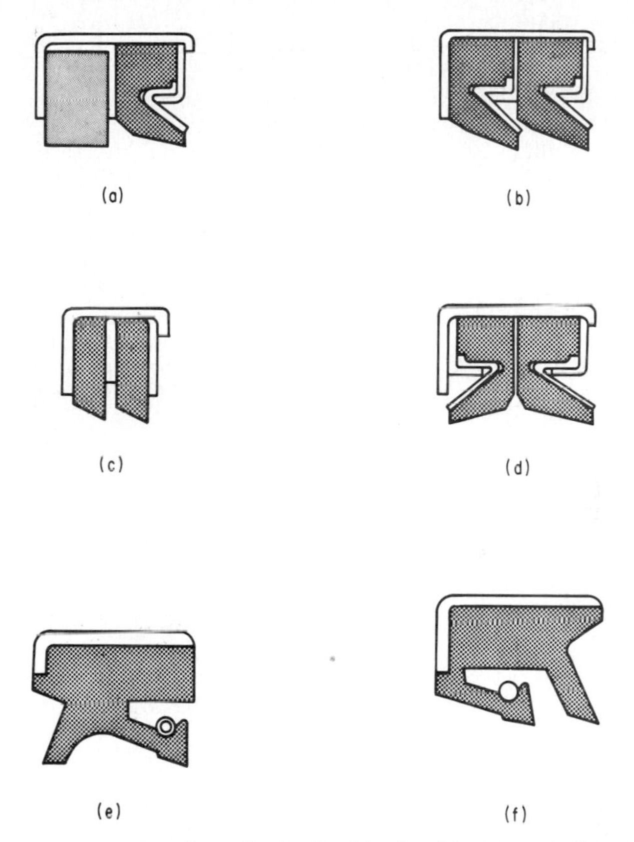

FIG. 13. Double-element-type lip seals. (*a*) Combination felt and synthetic seal. Works well where dusty environmental conditions prevail. Combinations available include either leather or synthetic elements. (*b*) Double-element seal. Used in heavy-duty, medium-speed service where leakage must be held to a minimum. (*c*) Tandem-washer leather seal. Arrangement of elements permits space between washers to be grease-packed when necessary. Leather washers, however, operate satisfactorily with intermittent lubrication. (*d*) Dual-opposed seal. Applicable to service where liquids are present on both sides of seal. Maximum recommended surface speed is 1000 ft/min. (*e*) Spring-loaded seal with opposed auxiliary seal. Loaded seal retains lubricant, auxiliary seal has less contact force and acts as dirt excluder. (*f*) Spring-loaded seal with tandem auxiliary. Since auxiliary member is mounted forward of loaded seal, it permits pressure lubrication without possibility of damaging sealing lip. Acts as dirt excluder when sealed medium contains abrasive particles.

3. Antifriction Bearing Seals. To fit the particular application, either clearance seals or rubbing seals may be fitted to the bearings of an electric motor. Antifriction bearings are classed as open, shielded, or sealed. All three types are used with grease lubrication, but only the open type can be used with oil lubrication.

Sealed bearings usually found in electric motors are single-row bearings, having the width of the raceways corresponding to that of a double-row ball bearing. This width provides a larger reservoir for lubricant and also allows space for an appreciable width of seal. Since these bearings are used where small bearings are required, the bearings and their grease charge are assembled by the bearing manufacturer. The bearing cannot be regreased; so it is foolproof as far as overlubrication or lubrication with an inferior grease is concerned.

Characteristics of typical applications of sealed bearings include one or more of the following: relatively short life requirements, mild operating conditions, necessity for low initial cost, and the difficulty, undesirability, or unlikelihood of relubrication. Since they cannot be readily relubricated, these bearings should be avoided in service where dirt, high temperatures, high speeds, and long-life requirements are encountered. Open bearings in a suitable housing packed with a good grease and with suitable regreasing periods are generally desirable in critical applications.

Because of the simplicity of their bearing-housing designs, sealed bearings have been used increasingly in recent years. A typical housing design for use with a sealed bearing is shown in Fig. 14. It is much less complex than the housing and sealing arrangements for open bearings since it consists essentially of a support

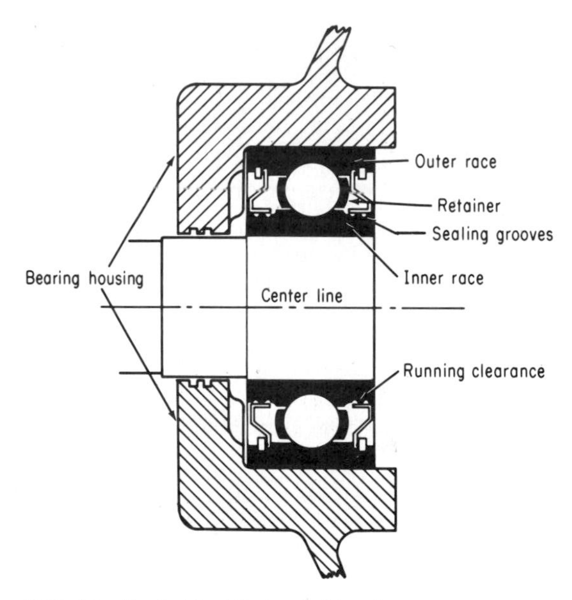

FIG. 14. Typical installed sealed bearing.

and protective shielding over the outside face of the bearing. Its effectiveness against dirt is generally not nearly as good as that possible with a long close-running shaft seal filled with grease. Since it cannot be relubricated, its life is limited to that of the grease contained in the bearing.

4. Shielded Antifriction Bearings. For continuous trouble-free service, shielded bearings are standard on many integral-horsepower-size motors. These bearings allow controlled migration of the grease into the bearings, but protect against overgreasing. The bearing construction allows a proper lubricant balance to be secured by greasing while the unit is fully assembled. If the unit is always shut down before being relubricated, the new grease will be forced into position for migration into the bearing, forcing out old grease when the unit is started.

The design of the inner end cap reservoir is such that several proper relubrications can be accomplished before filling to the extent that old grease will be forced into the motor along the shaft. When this occurs, the motor may be disassembled and the old grease in the inner end cap or seal removed.

The lubrication sequence for double-shielded bearings can be as follows:

1. Stop the unit.
2. Remove the drain plug.
3. Clean the inlet and insert new grease until it comes out of the drain hole.
4. Replace the drain plug.
5. Put the unit back into operation.

The double-shielded bearing, because of its construction, acts as a hydraulic pump when operating. If grease is pumped into the outer grease cavity until it is full, the pressure contact of the grease with the bearing will cause grease to enter the bearing through the small opening between the shield and the inner race. This pressure is created through resistance of the semiliquid grease to flow out the drain hole. If grease is pumped into this cavity while the unit is in operation, the bearing with its inherent pumping characteristic will fill the inner end cap prematurely and then force grease into the motor interior.

There is little danger of overheating a double-shielded bearing by this practice because only the grease inside the bearing is worked. The shields assure that the grease in the bearing housing and end-cap cavities is not worked during bearing operation.

A typical installation of a double-shielded ball bearing in an electric motor is shown in Fig. 15.

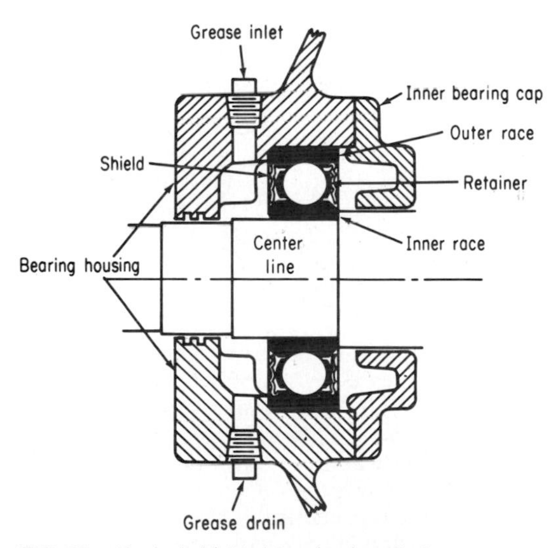

FIG. 15. Typical shielded-bearing installation.

5. Open Antifriction Bearings. The open bearing, with housing enclosures, is designed to provide a large reservoir for grease. This type of sealing arrangement is generally used when the antifriction-bearing size is large. Shielded bearings arc not generally obtainable in large sizes. Also this type of arrangement is suitable for oil lubrication of antifriction horizontal bearings, if this is desired or required by operating conditions, such as operation near the speed limit of the particular bearing.

The open bearing has no protection against foreign matter when removed from its housing enclosure for maintenance purposes. In operation the bearing housing and its grease or oil reservoir provide the only seal against contamination. This type of bearing can be greased in service but has no protection against overgreasing, which could lead to bearing failures.

Figure 16 shows a typical sealing arrangement for open antifriction bearings when used in a horizontal electric motor.

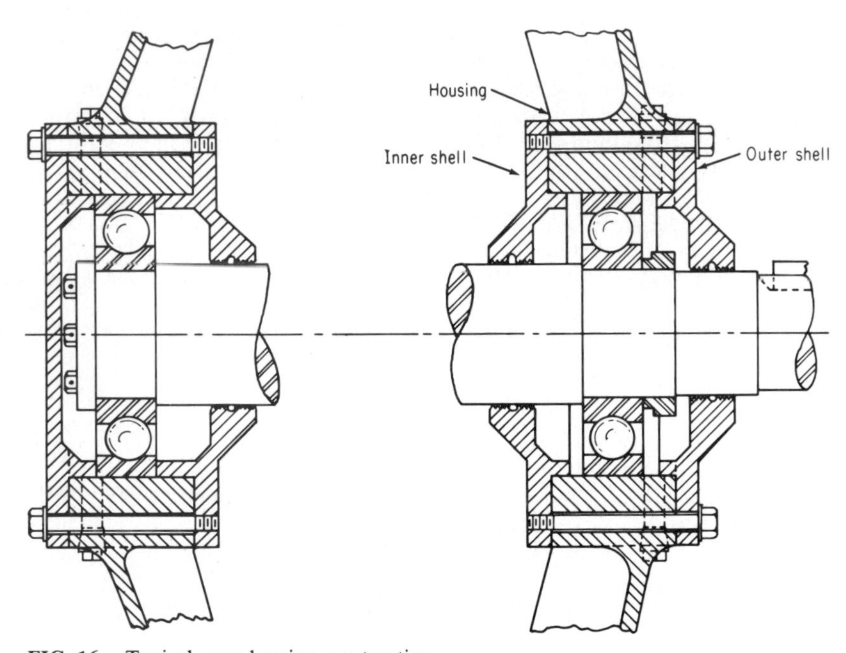

FIG. 16. Typical open-bearing construction.

Vertical-shaft electric motors use the same general sealing arrangements as horizontal machines, having grease-lubricated bearings with adequate capacity for the loads imposed. Often several improvements are made, however, to confine grease to the bearing cavity, particularly in larger vertical machines.

Figure 17 shows a large vertical motor with antifriction bearings, the lower a grease-lubricated ball bearing, while the upper, an oil-lubricated roller bearing, is acting as both a thrust and a guide bearing.

The lower bearing has a rotating seal below the bearing, fitting into a relief in the clearance seal. This arrangement keeps the oil, which may separate from the bearing grease, from leaking out through the clearance seal. Above the lower bearing a rotating slinger is shown, which fits into a relief in the upper clearance seal. This slinger prevents any substance in the airstream—water, dust, etc.—from running down the shaft and into the bearing area.

The oil seal for the upper antifriction thrust bearing is the conventional type

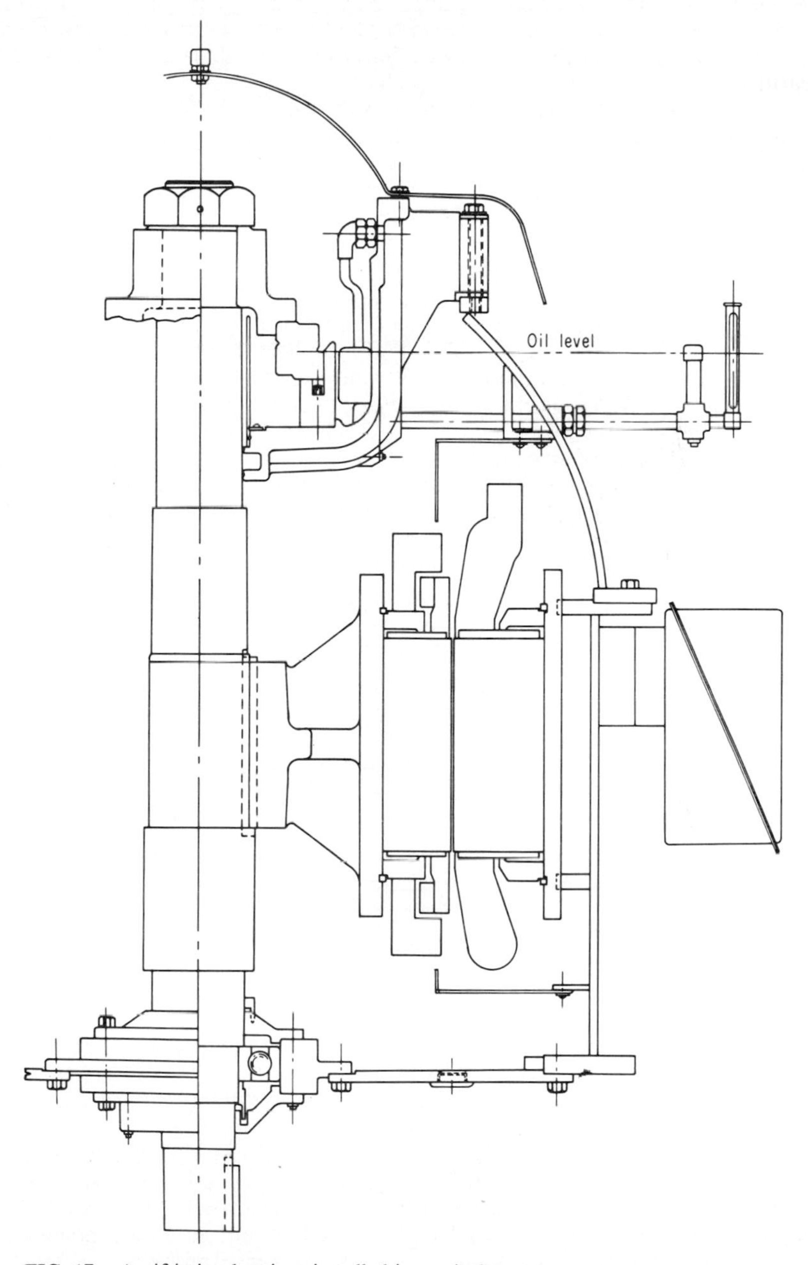

FIG. 17. Antifriction bearings installed in vertical motor.

also commonly used with pad and plate thrust bearings. It consists of an inner tube extending above the oil level, with close-clearance labyrinth seals below. Since the motor-intake air is directly adjacent to the labyrinth seal, a chamber vented to the atmosphere is sometimes placed between two labyrinths, as shown. This arrangement allows the slight vacuum at this region of motor-intake air to draw clean atmospheric air into the machine rather than oil vapor from the bearing reservoir.

Several special types of antifriction bearing seals are used with explosionproof machines. These seals will be described later, together with other seals for special motors.

6. Sleeve-Bearing Seals. The two types of sleeve bearings in general use in free liquid lubricant in the motor industry can be defined, for purposes of sealing, as those with or without free liquid lubricant in the bearing housing.

The type with no free liquid lubricant includes the many sleeve bearings in small intermittent and continuous-duty motors, in which the bearings are either wick-oiled or porous. Sealing bearings of this type is usually accomplished with felt washers, single close-clearance baffles, or other similar devices. When speed and diameter are near the upper limit for this type of bearing, an oil slinger is added to keep oil-vapor pressure at the seal area at an acceptable value.

Figure 18 illustrates the sealing arrangement of a typical wick-oiled sleeve bearing as used on shafts of less than approximately 1-in diameter.

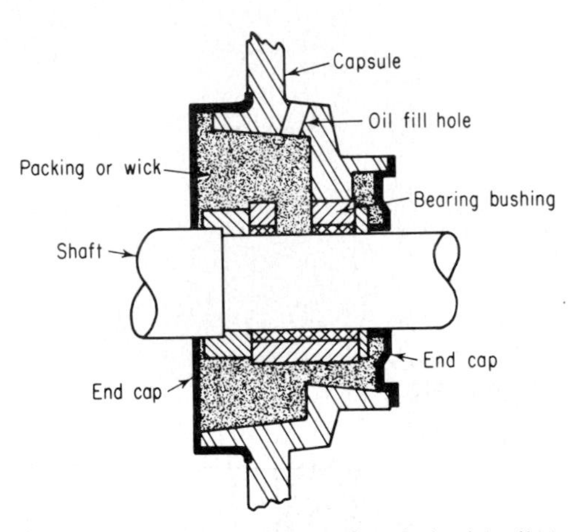

FIG. 18. Seal arrangement of typical wick-oiled sleeve bearing.

Sleeve bearings having free liquid lubricant in their reservoirs include all horizontal bearings larger than approximately 1-in diameter used in modern electric motors. Although in most cases these bearings are lubricated by oil rings, the very large high-speed machines, above 2500 hp at 3600 r/min, require supplementary oil-circulating systems to maintain proper bearing temperatures.

With the free surface of oil in the bearing reservoir, as well as the rapid circulation of oil through a bearing by oil rings or other means, sealing the bearing from leakage of lubricant becomes more difficult. To seal this type of bearing effectively, the common practice is to separate the bearing capsule into two zones, as follows:

1. The center zone is the center portion on the bearing assembly. In this zone are located all the bearing parts that produce oil vapor.
2. The outer zone is an isolation area between the center zone and the surrounding atmosphere, both inside and outside the motor.

The two zones are divided by clearance seals, so that possibly 95 percent of the oil vapor developed remains in the center zone. Thus the outer seals, also clearance seals, have only to restrict the remaining 5 percent of the oil vapor developed.

In addition to the two zones, a third zone is sometimes established between the bearing and the internal parts of an electric motor. This zone is required because the incoming air for motor ventilation usually flows directly past and against the bearing capsule, cooling the bearing but creating a slight vacuum at the inner nose of the capsule. To prevent this vacuum from drawing oil vapor into the motor, a third chamber is vented to the atmosphere. This design removes the vacuum from the second zone, which is retaining 5 percent of the developed oil vapor, and if the vent is sufficiently large, allows nonoily atmosphere air to be drawn into the motor, rather than oil vapor from the second bearing-capsule zone.

Since there is no vacuum-producing air movement near the outside of the bearing capsule, no additional atmospheric chamber is required here.

Figure 19 illustrates the type of bearing sealing used in oil-ring-lubricated bear-

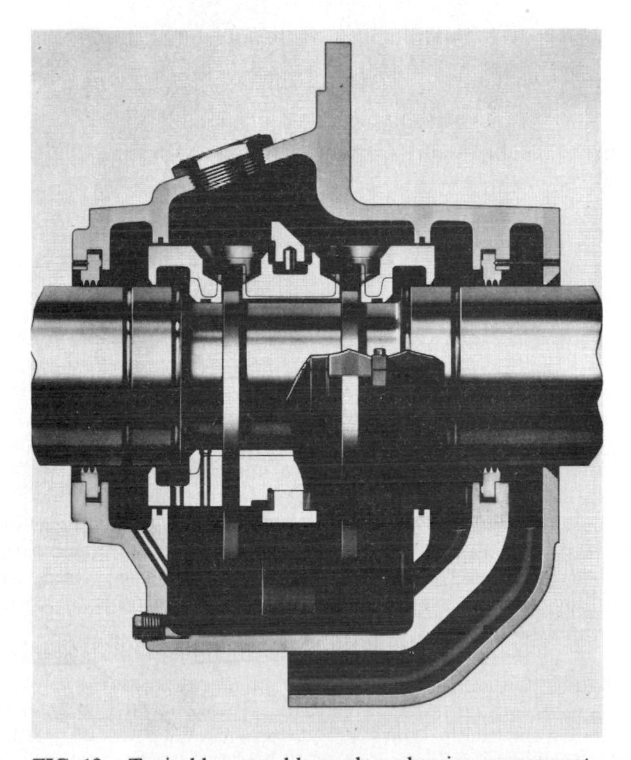

FIG. 19. Typical low-speed large-sleeve bearing arrangement.

ing assemblies for shafts larger than approximately 1-in diameter for speeds to approximately 1200 r/min, when used in bracket-bearing machines.

Figure 20 illustrates the same application when used as a pedestal bearing. The only significant difference between this type and the arrangement in Fig. 19 is that, as a pedestal bearing, both shaft-seal areas are isolated from the motor air-intake stream, so that no atmospheric relief or third chamber is required.

FIG. 20. Typical pedestal bearing arrangement.

Special problems encountered in high-speed bearing operation make necessary a modification from the standard low-speed sealing arrangement. The high-speed machine draws ventilating air into the motor at a higher velocity. This air movement creates a higher suction pressure adjacent to the innermost bearing-capsule seal. If this suction is sufficiently high, the atmospheric vent cannot accommodate the airflow, and oil vapor will be drawn into the motor. For this condition air is bled from the downstream side of the motor fan to pressurize a chamber in the innermost shaft seal. The arrangement assures that an adequate supply of clean air is drawn through the shaft seal into the motor. The atmospheric vent can protect and maintain atmospheric pressure at the secondary seal.

Figure 21 shows this additional oil seal added for high-speed operation. Comparison with Fig. 19 shows the difference between bearing-seal arrangements for high-speed and low-speed operation.

7. Vertical Sleeve and Guide Bearings. Sleeve bearings, as used for guide bearings in standard vertical-shaft motors, may be combined with thrust bearings for

FIG. 21. Typical high-speed large-sleeve bearing arrangement.

sealing arrangements. Figure 22 illustrates the common arrangement used for vertical machines with oil lubrication. The combination thrust- and guide-bearing housing has the inner tube extending above the oil level, with two close-clearance seals in the path that the oil vapor would take to get into the machine. The seals are so arranged that the area between the first and second shaft seals can be vented to the atmosphere, if this is required. In general, however, slow-speed vertical machines do not develop sufficient suction at the shaft-seal area to require an atmospheric vent, because of the construction of the upper housing needed to support the transmitted thrust load. High-speed machines in general do, however, require an atmospheric vent between shaft seals.

The lower guide bearing, being closer to the airflow path, has the normal center zone, outer zone, and atmospheric-relief vent. This design is representative of machines of all speeds. To the exterior, however, the sealing is minimal since there is no area of low air pressure. Thus the inner tube extending above the oil level is by itself an adequate seal in almost all vertical motors.

BEARING AND SHAFT SEALS IN SPECIAL ENCLOSURES

8. Sealing Explosionproof Machines. The sealing requirements of motors to be installed in hazardous locations are governed by Underwriters Laboratories. Although they set guidelines, actual explosive testing of the construction of a similar machine is required for certification.

The types of hazardous locations are grouped as follows:

1. Class I. Locations in which flammable gases or vapors are or may be present in the air in quantities sufficient to produce explosive or ignitable mixtures.

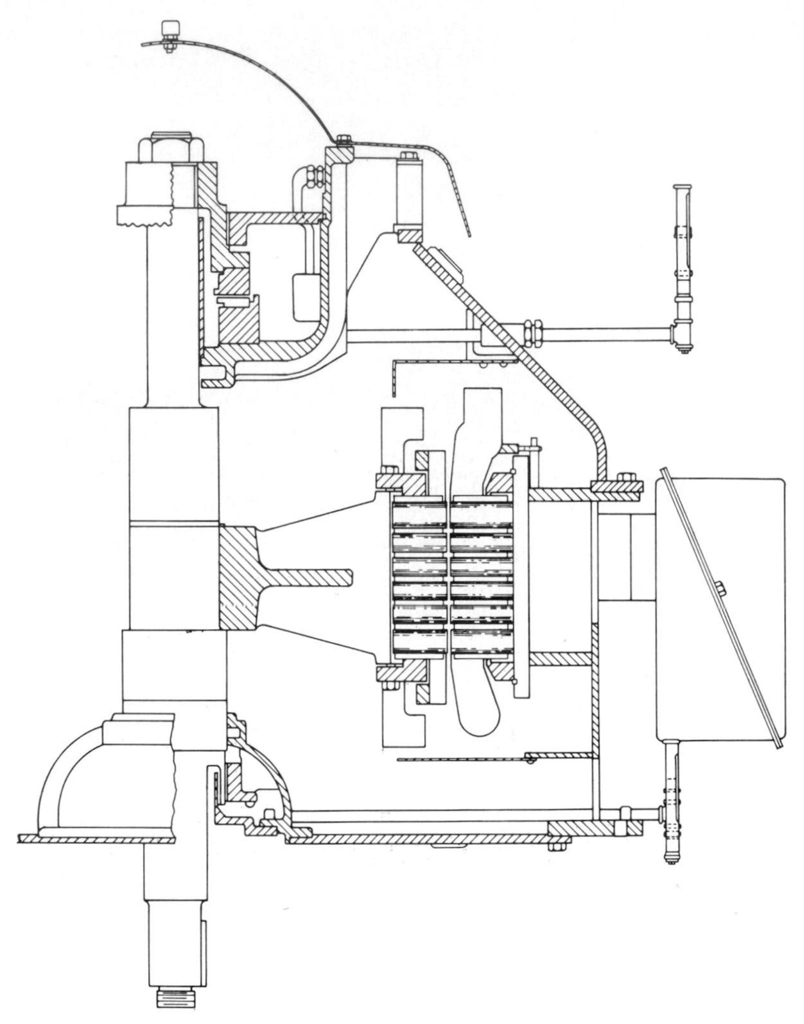

FIG. 22. Oil-lubricated bearings installed in vertical motor.

Group A. Atmospheres containing acetylene.

Group B. Atmospheres containing hydrogen or gases or vapors of equivalent hazards, such as manufactured gas.

Group C. Atmospheres containing ethyl ether vapors, ethylene, or cyclopropane.

Group D. Atmospheres containing gasoline, hexane, naphtha, benzene, butane, propane, alcohols, acetone, benzol, lacquer solvent vapors, and natural gas.

2. Class II. Locations which are hazardous because of the presence of combustible dust.

Group E. Atmospheres containing dusts of aluminum, magnesium, or their commercial alloys.

Group F. Atmospheres containing carbon black, coal, or coke dust.

Group G. Atmospheres containing flour, starch, and grain dust.

3. Class III. Locations which are hazardous because of the presence of easily ignitable fibers or flyings, but in which such fibers or flyings are not likely to be in suspension in the air in quantities sufficient to produce ignitable mixtures.

In addition a breakdown is made as to whether the hazard is present in normal or expected operating conditions (Division I), or whether it will be present only because of failure of equipment or other abnormal operating condition (Division II).

Thus a typical description of an application would be Class I, Group D, Division I.

To indicate the type of sealing required for hazardous locations, the following is taken from the Underwriters Laboratories Standard for Electric Motors and Generators for Use in Hazardous Locations, Class I, Groups C and D, for Integral Horsepower Motors.

> *General.* Requirements of the National Electrical Code with reference to electric motors and their wiring and fittings shall be complied with. The following requirements are supplementary thereto.
>
> The motors shall be of the totally enclosed, or totally enclosed fan-cooled type.
>
> *Casing.* The motor casing shall be of substantial construction and shall be capable of withstanding the internal pressures resulting from explosions without bursting or loosening its joints. Calculated factors of safety of 5 for cast parts, 4 for fabricated steel parts, and 3 for bolts shall be required, based on the maximum internal explosion pressure developed during tests. In case of listing by frame size, the factors of safety shall be on the basis of pressures not less than those indicated by Fig. 23.
>
> A hydrostatic test may be conducted as an alternate to determining the strength of parts by calculations, as referred to above. The motor or parts subjected to hydrostatic tests shall withstand, without rupture or permanent distortion, a pressure of four times in the case of cast parts, and of three times in the case of fabricated steel parts, the maximum internal explosion pressure developed during explosion tests. In the case of listing by frame size, the factors of safety shall be on the basis of pressures not less than those indicated by Fig. 23.
>
> Joints in the casing shall be of the metal-to-metal type having a width of metal-to-metal surface not less than that shown in Fig. 24 and never less than ¾ in. Portable motors having flat joints shall have a minimum width of metal-to-metal surface of not less than 1 in.
>
> *Bolt spacing and location.* If the bolthole is to be counted as part of the flame path, it shall not be more than 1/32 in larger in diameter than the bolt. Bolts may be located approximately in the middle of the joint, but the distance from the bolthole to the inside of the enclosure shall never be less than ½ in. The distance from the joint to the bottom surface of the bolt head shall be equal to at least one-half of the required width of the joint.
>
> The length of shaft openings and clearances for ball-bearing and sleeve-bearing motors and generators shall conform to the following: minimum length of protecting sleeve 1½ in, with maximum diametrical clearance of not more than 0.025 in; and for 2½-in minimum length, maximum diametrical clearance of not more than 0.030 in. Motors for Class I, Group C, will be accepted for tests if each shaft opening is provided with an additional labyrinth flame path of at least ⅛ in.
>
> The length of metal path shall be determined by measuring only the continuous metal-to-metal path, except that oil or grease grooves, if not of sufficient volume to

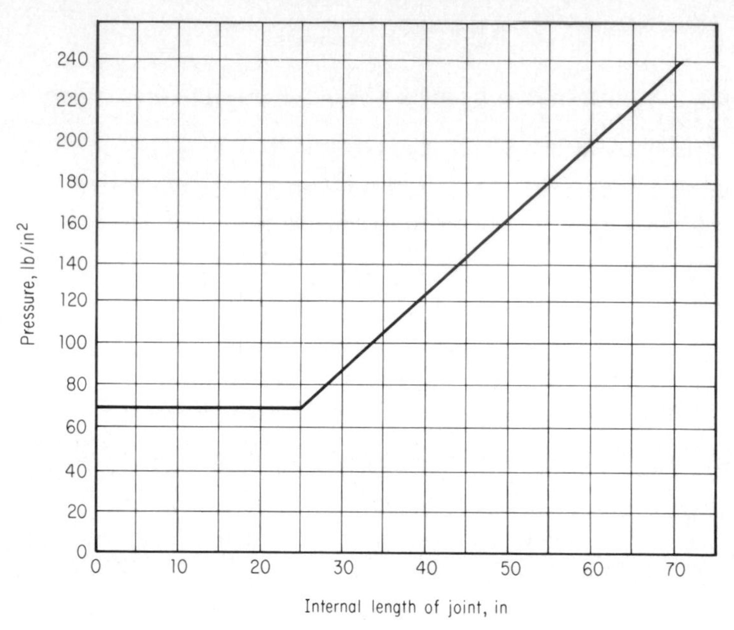

FIG. 23. Design pressure for Class I, Group C and D explosionproof machines.

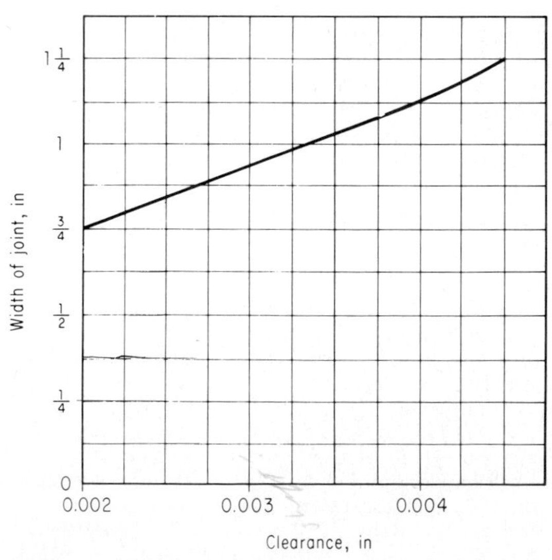

FIG. 24. Required joint dimensions for Class I, Group C and D explosionproof machines.

> nullify in a measure the protective value of the total length of path, are permitted but are not considered as a part of the effective path. Labyrinths, if of substantial form of construction, are considered as equivalent in lengths to straight metal paths. Openings for oil or grease should be located outside the metal-to-metal path in order not to nullify its effectiveness. If particularly deep grooves are used, special tests may be necessary. A removable outer bearing cap is not considered to be part of the metal path.

The foregoing describes the type of sealing arrangement that is adequate for the frame and shaft openings of a motor to be used in a Class I, Division I (gas) type hazardous location, of Group C or D gases.

Construction details of the seal arrangement of a small machine are given in Fig. 25, and a large machine for Class I, Group D, Division I, is illustrated in Fig. 26.

The Class II explosionproof machines—for installation in combustible dust—do not have the strength requirements to resist significant internal explosions but have more stringent restrictions for shaft sealing to keep dust from entering the machine.

The following is quoted from the Underwriters Laboratories Standard for Electric Motors and Generators for Use in Combustible Dust, Class II, Groups F and G.

> *General.* Requirements of the National Electrical Code with reference to electric motors and their wiring and fittings shall be complied with. The following requirements are supplementary thereto.

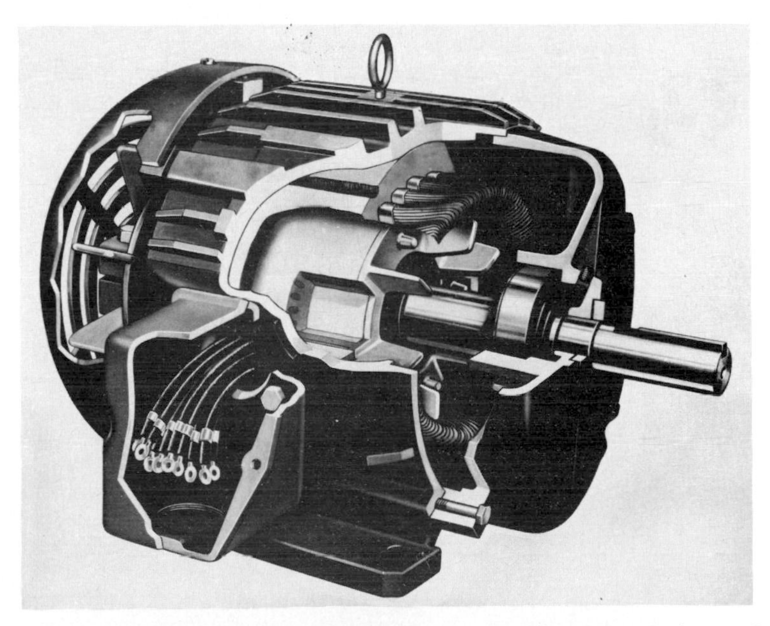

FIG. 25. Typical construction of small Class I, Group C and D explosionproof machines.

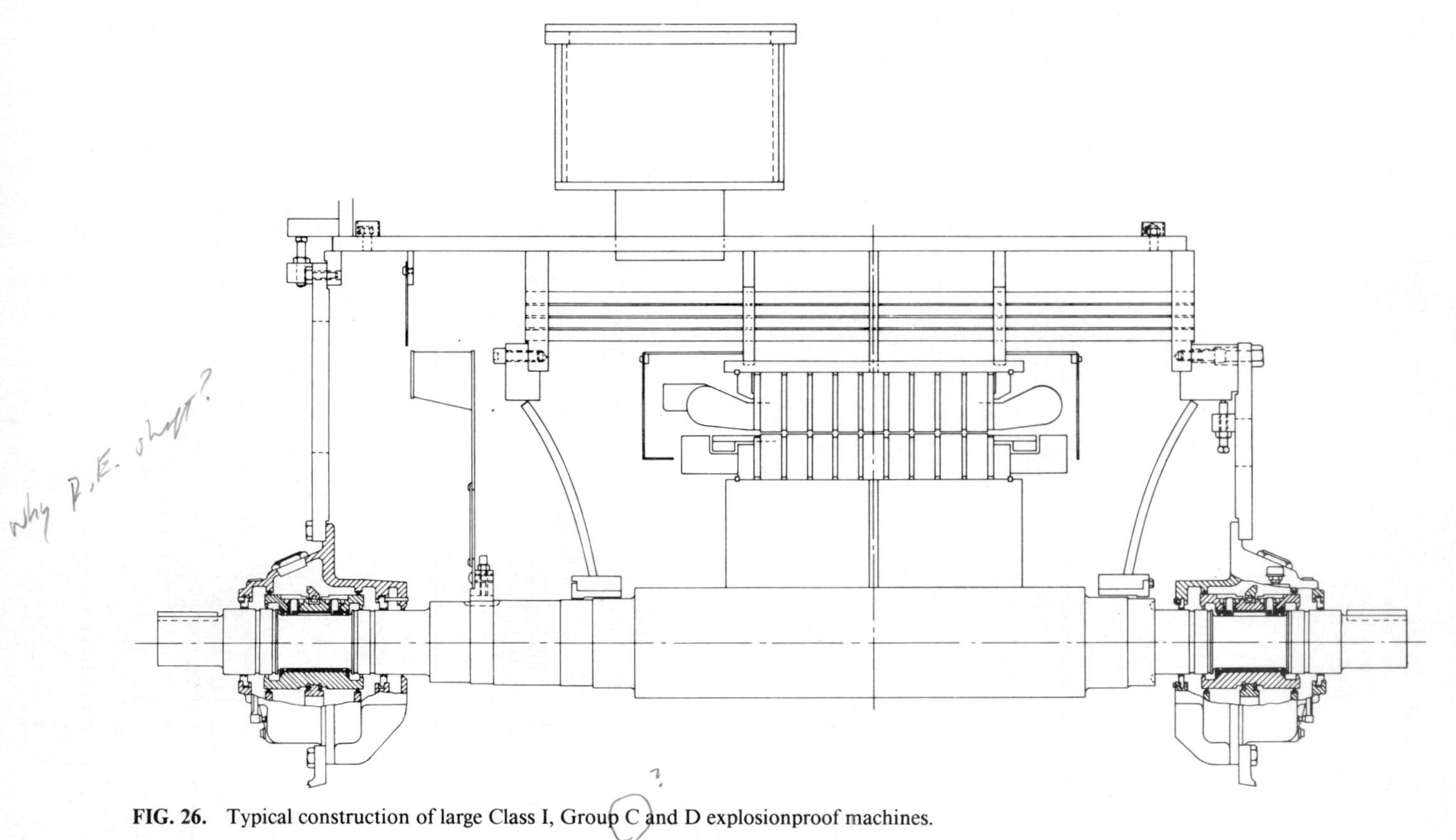

FIG. 26. Typical construction of large Class I, Group C and D explosionproof machines.

Casing. The motor casing shall be of substantial construction and shall be dust-tight as defined in the National Electrical Code, Article 100.

Joints in the casing shall be machined and so designed and constructed as to secure a tight fit at all points. When assembled they shall have a metal-to-metal surface at least 3/16 in in width.

Shaft openings shall have a length of path of not less than ½ in. The radial clearances shall be not more than 0.005 in for ½-in length of path, and not more than 0.008 in for a length of path of 1 in, with intermediate values proportioned.

Thus seals for Class II machines are similar to those of Class I explosionproof machines, but with less radial clearance than Class I. The reason for this is that the Class I seal is to cool and extinguish the internal flame before it reaches the atmosphere, while the Class II seal functions to restrain atmospheric dust from entering the machine.

Division II motors in general do not require special seal arrangements, except to enclose switching mechanisms, sliding contacts, relays, etc., that is, devices that could produce switching arcs in their normal operation. These devices must be enclosed in explosionproof housings.

9. Sealing Inert-Gas-Filled Machines. The totally enclosed inert-gas-filled enclosure has been designed to operate safely in hazardous Class I, Group D, locations, as defined by the National Electrical Code. The pressurized enclosure offers equal protection in any application where airborne contaminants would endanger continued motor operation.

The method by which protection against explosions is obtained is essentially different from that of the explosionproof motor. Explosionproof is defined as "enclosed in a case which is capable of withstanding an explosion of a specified gas or vapor which may occur within it, and of preventing the ignition of the specified gas or vapor surrounding the enclosure by sparks, flashes or explosion of the gas within."

While the inert-gas-filled enclosure is of rugged construction and can operate safely in most hazardous areas, it is not explosionproof by definition of the National Electrical Code.

For inert-gas-filled operation, the motor structure is virtually gastight. Positive internal pressure, usually about 2½ to 3 in water gage, keeps out volatile ambient gases or other contaminants that could cause an explosion or otherwise damage motor parts. Nitrogen gas is generally used for these machines. However, clean, dry, noncontaminated air, such as is used for instrument air in chemical plants and refineries, may also be used.

Since explosion pressure in an explosionproof machine is somewhat proportional to the free internal volume of the machine, it can be seen that, as machines become large, the difficulties of designing and building machines to withstand internal explosion pressures become almost insurmountable. This transition occurs in the range of 2½ to 3 hp per r/min, for instance, 4000 hp at 1785 r/min. The natural solution is to restrain the explosive ambient from entering the machine by maintaining a slight positive internal pressure of an inert gas.

The slight internal pressure denotes a continuous leakage to atmosphere of the inert gas. To reduce gas usage, special seals are built into inert-gas machines. In general, every static opening, such as bearing housings or frame covers, is sealed with O rings or similar gasketing. The only nonstatic opening is the seal around the shaft. This is generally sealed by maintaining an oil film at sufficient pressure to overcome the internal machine pressure. This reduces leakage and replenishment of gas supply to a minimum but does require a flow of oil at all times, non-operating as well as operating time.

Figure 27 shows a typical shaft seal of the clearance-oil-film type to restrict gas flow out of an inert-gas-filled machine.

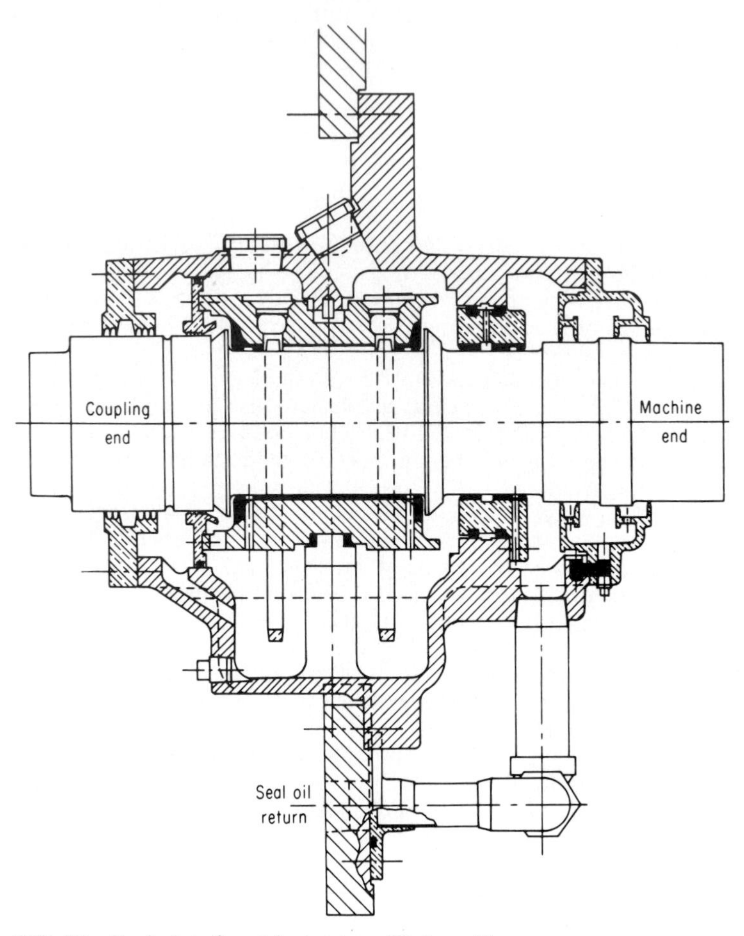

FIG. 27. Typical shaft seal for inert-gas-filled machine.

REFERENCES

1. Louis Dodge, "Labyrinth Shaft Seals," *Prod. Eng.* (Aug. 1963).
2. Edward A. Smith, "Felt Radial Seals," *Machine Design* (Jan. 19, 1961).
3. C. R. McCray, "Radial Positive Contact Seals," *Machine Design* (Jan. 19, 1961).
4. D. F. Wilcock and E. R. Booser, *Bearing Design and Application* (McGraw-Hill, New York, 1957).
5. R. H. Warring, *Seals and Sealing Handbook* (Gulf Publishing, Houston, Tex., 1981).

SECTION 14
LUBRICATION

E. R. Booser*

FOREWORD 14-2

TEMPERATURE 14-2

1. Estimating Bearing Temperatures 14-2

OIL LUBRICATION 14-3

2. Oil Composition 14-3
3. Viscosity 14-4
4. Synthetic Oils 14-6
5. Application Methods 14-6
6. Lubrication Maintenance 14-9

GREASE LUBRICATION 14-11

7. Grease Composition and Properties 14-11
8. Bearing-Housing Arrangements 14-15
9. Regreasing 14-17

GEAR MOTORS 14-19

10. Motor Bearings 14-19

REFERENCES 14-20

*Turbine Technology Laboratory, General Electric Company, Schenectady, N.Y.; Registered Professional Engineer (N.Y.); Member, ASLE, ACS, and ASME.

FOREWORD

Electric-motor bearing systems are usually designed to use commonly available industrial oils and greases. The grade of lubricant and the relubrication schedule are then matched to motor operating conditions, bearing type, and temperature. This section provides a background for both the selection of the proper lubricant and establishing appropriate lubrication maintenance procedures.

Oil selection is covered first. Oil is ordinarily used in small (fractional-horsepower) motors with their wick-oiled sleeve bearings. Oil is again used for heavily loaded sleeve bearings in large motors ranging above about 500 hp. Grease selection then follows. Grease-lubricated ball and roller bearings are used for design and operating simplicity in most industrial motors ranging from about 1 to 500 hp. Gear motors, combining an electric motor and a speed-changing gear set in a single package, involve special lubrication requirements which are covered at the close of this section.

Comments and observations provided here should be useful in considering most normal motor applications. Since lubrication requirements of electric-motor bearings frequently are not critical, subsidiary considerations such as rust protection also become important. Where a common lubrication system is used for the motor and its driven machinery, the requirements of all bearings, gears, and other lubricated parts should be considered. In each case reference should be made to the lubrication instructions supplied by the motor manufacturer. The manufacturer should also be consulted directly for lubrication instructions for unusual operating conditions.

TEMPERATURE

1. Estimating Bearing Temperatures. Insofar as lubrication considerations are concerned, two general types of electric-motor enclosures are in common use: open and totally enclosed. Distinctive features of the two types are discussed in detail in preceding sections of this handbook. Each type has a characteristic effect on the bearing temperature; and this temperature, in turn, strongly influences lubricant selection and relubrication schedules for a motor.

The simplest motor construction, and the one commonly used when the motor is not exposed to outdoor weather or other unusual ambient conditions, is the standard open type with a dripproof enclosure. In these open motors, outside air is usually drawn into the motor enclosure directly over the bearing housings for cooling the motor windings. With this free flow of cooling air, the bearing-temperature rise above ambient is normally about one-quarter to one-half the temperature rise of the electrical windings of the motor. The rated temperatures for various classes of insulation are indicated in Table 1. From this table, an open Class B motor with 1.15-service-factor load capability would be expected to have a bearing-temperature rise of one-quarter to one-half of the 90°C (162°F) maximum rise for the windings, that is, the bearings would be approximately 22 to 45°C (40 to 81°F) above the ambient-air temperature. The higher temperature rises in this range would be experienced with larger 3600-r/min motors. While use of Class B and Class F insulation heavily predominates, the corresponding rises for Class A and Class H windings are 60 to 65°C (108 to 117°F) and 125 to 135°C (225 to 243°F), respectively.

The other general type of motor enclosure uses a totally enclosed construction.

TABLE 1. NEMA Winding-Temperature Ratings for Use in Estimating Bearing Temperatures*

Insulation class	Service factor	Max ambient temp, °C (°F)	Allowable winding temperature rise: Totally enclosed: Dripproof, °C (°F)	Fan-cooled, °C (°F)	Nonventilated, °C (°F)
B	1.0	40 (104)	80 (144)	80 (144)	85 (153)
	1.15	40 (104)	90 (162)	90 (162)	90 (162)
F	1.0	65 (149)	90 (162)	90 (162)	90 (162)
	1.15	40 (104)	115 (207)	115 (207)	115 (207)

*For motors built in frames equivalent to 500 hp at 1800 r/min and smaller.

In this arrangement the motor is completely sealed, with the exception of a very limited clearance space along the shaft extension. With no ventilation, the much more limited cooling for the bearings than in an open motor results in a bearing-temperature rise above ambient of about 75 to 80 percent of the winding-temperature rise.

Supplemental cooling is provided in most larger enclosed motors by blowing air with an external fan along the outside of the motor shell in a totally enclosed fan-cooled (TEFC) construction. A TEFC motor gives a bearing-temperature rise on the order of one-half to two-thirds of the temperature rise of the motor windings. Referring to Table 1, a Class B motor with a nominal temperature rise of 80°C (114°F) for the insulation might experience a temperature rise for the motor bearings of 40 to 53°C (72 to 96°F) above the ambient temperature.

For hazardous locations containing explosive mixtures of gas or dust, a specially designed motor is installed which is either explosionproof or dust-ignition-proof. The bearings in this type of motor will have essentially the same running temperatures as those in the conventional TEFC construction.

Although these temperature approximations are typical for the majority of small- and medium-sized motors and generators built for industrial use, differences will be found in special units and for large mill-type motors where the bearing pedestals are separate from the motor frame.

OIL LUBRICATION

2. Oil Composition. Petroleum oils are normally used for the lubrication of sleeve-bearing motors because of their ready availability, low cost, satisfactory physical properties, good stability, and adequate lubrication ability.

All petroleum oils are complex mixtures of various types of hydrocarbon molecules having about 20 to 70 or more carbon atoms per molecule. With the carbon-atom skeletons primarily linked end to end in chainlike structures, the oil is described as paraffinic. Such an oil will be relatively nonsludging and is the type normally found in high-grade, stable turbine oils commonly used in electric motors. With a large portion of the ring structure in the carbon-atom skeleton, an oil is described as naphthenic. Such oils are normally less stable than the paraffinic type and tend more to soften some types of rubber, but they do have lower

TABLE 2. Suggested ISO Viscosity Grades of Electric-Motor Oils

Bearing type	Speeds under 1500 r/min	Speeds 1500 r/min and above
Sliding type:		
Pressure-fed	68*	32*
Ring-oiled	68	32
Disk	68	32
Wick-oiled fractional-horsepower	32–68	32
Plate-type thrust	68	68
Rolling type:		
Ball	32	32
Cylindrical roller	32	32
Spherical roller	150	150

*Midpoint viscosity, mm^2/s (= cSt) at 40°C (ASTM D 242).

pour points, which brings them into some electric-motor use for refrigeration systems and low-temperature outdoor use in cold climates.

Chemical additives have been used increasingly in lubricating oils since about 1930 to improve a variety of properties. Oxidation inhibitors are now employed at about 0.5 percent concentration in almost all high-quality turbine oils intended for electric-motor service. These oxidation inhibitors slow down the reaction of the oil with air and thus inhibit the formation of varnish, sludge, and organic acids for many months of satisfactory operation. Rust inhibitors are also normally used at low concentration. These commonly are long-chain organic acids which form a protective network by polar attachment to ferrous surfaces. This protective surface layer is formed in such a way as to displace water and prevent rusting.

3. Viscosity. The viscosity of the turbine oil to be used varies with the bearing design, speed, load, and bearing operating temperature. The viscosities listed in Table 2 can serve as a guide. A light turbine oil of 32-centistoke (cSt) viscosity at 40°C (100°F) will be satisfactory for normal application in fractional-horsepower motors and all larger units running at 1500 r/min and higher speeds. For lower speeds, high temperatures, and high loads, a shift is frequently made to an oil of ISO (International Organization for Standardization) viscosity grade 68. To meet the more severe lubrication requirements of some rolling-contact thrust bearings, oils of ISO viscosity grade 150 should be used. Specifications useful in obtaining various viscosity grades of lubricating oils for electric-motor service are indicated in Table 3.

TABLE 3. Typical Electric-Motor Oil Specifications

Grade	Light	Medium	Heavy
Viscosity, cSt at 40°C	28.8–35.2	61.2–74.8	135–165
Viscosity index, min	90	85	85
Pour point, °C, max	−20	−20	−10
Flash point, °C, min	190	200	210
Neutralization number, mg KOH/g, max	0.2	0.2	0.2
Oxidation test life, h, min (ASTM D 943)	1000	1000	1000
Rust prevention test (ASTM D 665)	Pass	Pass	Pass

Actually, lubrication requirements of many electric motors are not severe and one viscosity grade may often be substituted for another. If, for instance, most of the equipment in an industrial plant is being lubricated with a 68-cSt oil, such an oil can also be used in place of a 32-cSt oil in most electric motors running at speeds over 1500 r/min. Slightly higher bearing temperatures will be encountered and power losses will be greater, but the overall motor performance will generally be satisfactory. On the other hand, many motors operating at speeds below 1500 r/min can be lubricated with a light turbine oil of 32-cSt viscosity for low-temperature starting in outdoor service.

Oils other than those of the turbine type are sometimes used. Automotive oils find use because of their ready availability. The viscosities of several SAE grades are compared with the common electric-motor-oil viscosities in Fig. 1. These automotive oils are more expensive, and the detergents and other additives included sometimes cause trouble with foam, emulsions, and other operating difficulties. Uninhibited machine oils used in drip-feed applications for machine tools provide relatively short life in electric motors, may allow corrosion, and will lead to increased maintenance cost and less satisfactory motor operation. Extreme-pressure and antiwear additives as are used in oils for heavily loaded gears function primarily by forming a protective surface film through a mild surface chemical reaction to minimize welding and tearing. The tendency for these surface reactions persists, however, even where extreme-pressure properties are not required. Extreme-pressure additives have reacted with motor insulation to cause winding failures. Such extreme-pressure gear oils provide generally shorter

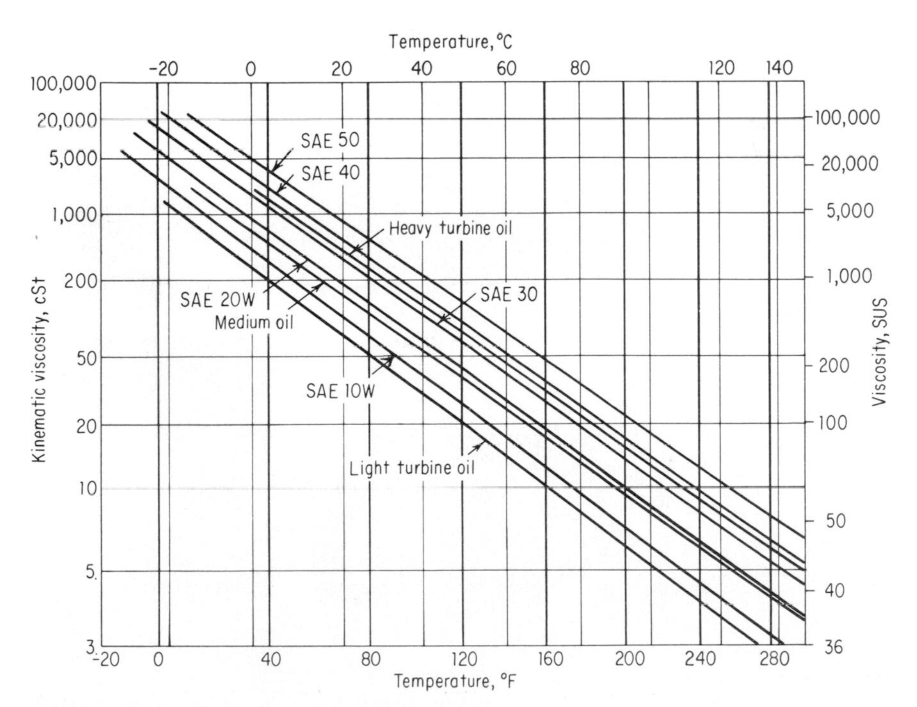

FIG. 1. Oil-viscosity–temperature characteristic.

service life and should be avoided to minimize the possibility of corrosion, except where they are needed in gear motors or for severe sliding conditions in the driven machinery.

4. Synthetic Oils. These oils are now in frequent use in aircraft, for fire resistance in industrial plants, or for other special applications but should not normally be used for electric motors. Almost all synthetics, except synthetic hydrocarbons, will attack paint, rubber, and electrical insulation, and they may not provide satisfactory lubrication. If extreme temperatures or special operating conditions require that a synthetic oil be used, the entire motor should be designed for the application. Even in such cases, first consideration should be given to the use of a synthetic grease in a ball-bearing motor or a circulating petroleum-oil or synthetic-hydrocarbon system with supplemental heating or cooling to maintain a satisfactory oil temperature.

5. Application Methods. Oil-application arrangements vary with both the motor size and the demands of the bearing system.

a. Wick oiling. Fractional-horsepower motors with their minimum oil requirements generally use felt, waste, or yarn packing to feed sleeve bearings up to about 1 in in diameter. A typical wick-oiled bearing is shown in Fig. 2. Oil feeds to the shaft surface through a window at the top of the bearing. Oil discharging from the bearing is then returned by gravity or by centrifugal throw-off to the wick. Tin babbitt, leaded bronze, or porous bronze is the usual bearing material.

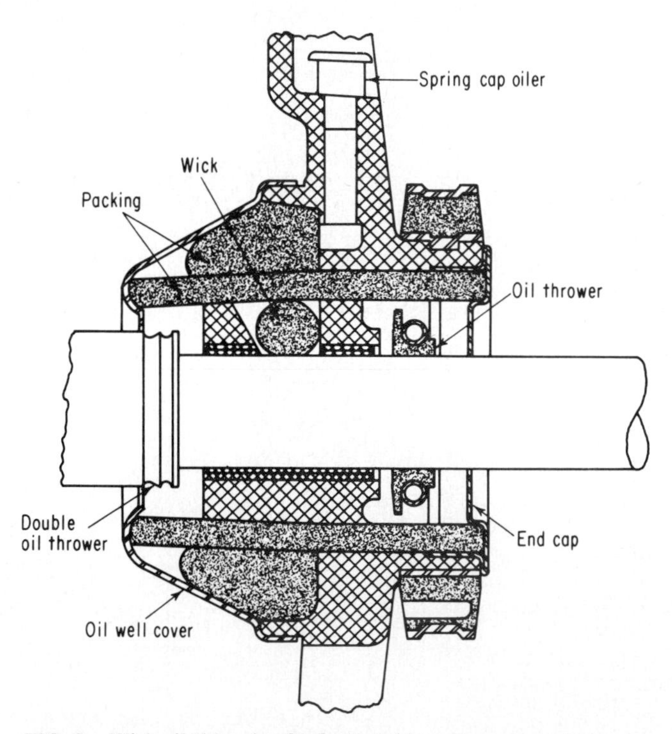

FIG. 2. Wick-oiled bearing for fractional-horsepower motors.

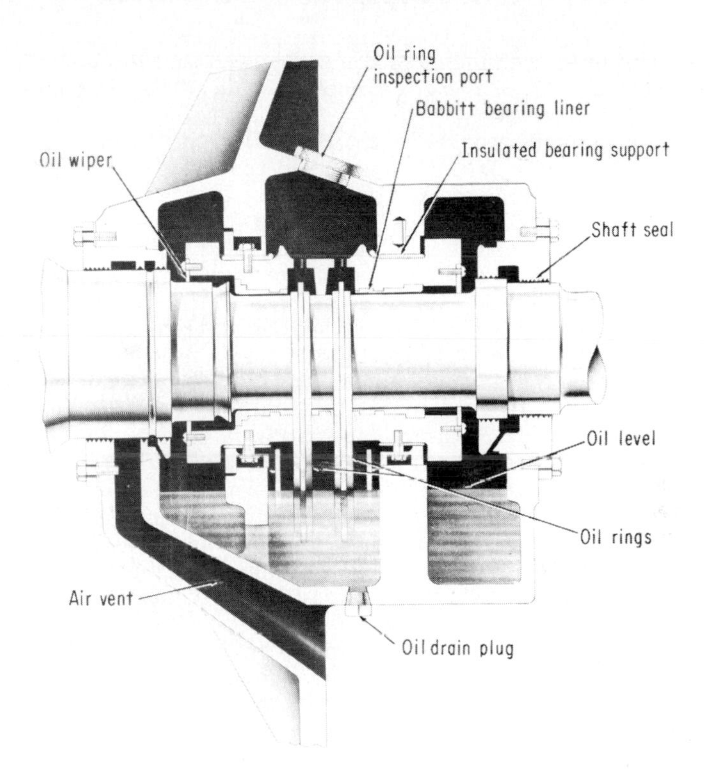

FIG. 3. Ring-oiled sleeve bearing.

Wick-oiled bearings are also used in some mill motors and in railroad traction motors. In these motors, waste packing is used to carry the oil to the journal through a window in a babbitted-bronze or solid-bronze bearing.

b. Ring oiling. Rings are used to lubricate sleeve bearings in motors and generators ranging from about 250 hp in size up to 40,000 kVA. Oil-ring lubrication is ordinarily limited to shaft surface speeds below approximately 3000 ft/min. Either one or two zinc or bronze rings about 1.5 times as large as the journal diameter are usually located in a slot in the upper half of the bearing, as shown in Fig. 3. When the motor runs, the ring will turn at roughly one-tenth the shaft speed. The turning ring or rings then pick up oil at the bottom of their travel and deliver it up to the top of the journal from where it flows to the bearing surfaces through internal grooving. Spiral or racetrack grooving, such as shown in Fig. 4, is useful for belted loads or other applications where load might be applied in any direction. Two axial oil-distributing grooves at the

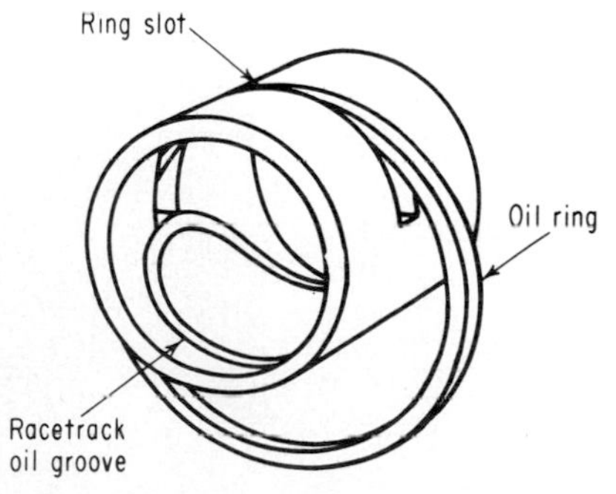

FIG. 4. Racetrack oil groove for belt loads in any direction.

horizontal centerline are commonly used for direct-connected motors. Tin babbitt, lead babbitt, or SAE 660 leaded bronze is frequently used as the bearing material for ring-oiled motors.

A sight glass is normally provided to check the oil level. The oil should always be maintained above the inner surface of the oil ring at its bottom point. Free turning of the rings should be checked on starting a new motor and after maintenance work. Corrosion is occasionally a problem where lead babbitt bearings are used. In such cases, an oxidation-inhibited oil should be used and changed at frequent enough intervals to avoid its acid number exceeding about 1.0 (ASTM D 974).

c. Disk lubrication. A disk either shrunk onto or made integral with the shaft is used to provide lubrication for many large horizontal motors. In operation the rotating disk picks up oil from the reservoir and carries it upward to a scraper at the top of its travel. The removed oil then flows through grooves to the working surface of the bearing. A typical arrangement is shown in Fig. 5. Disk-lubricated bearings are generally either lead or tin babbitt. A 68-cSt turbine oil such as indicated in the specifications in Table 2 is commonly used. The same general maintenance practices should be followed as with motors using ring-oiled bearings.

d. Bath oiling. A surrounding oil bath serves frequently for lubrication of the bearings in large vertical motors and generators. This lubrication scheme is commonly used either with ball and roller bearings or with oil-film sleeve and thrust bearings.

Typical characteristics of a bath-oiling system are illustrated in Fig. 6 for the upper end shield of a vertical motor. Oil slots or metering holes restrict the oil-feed rate in high-speed motors to a quantity that will provide adequate lubrication

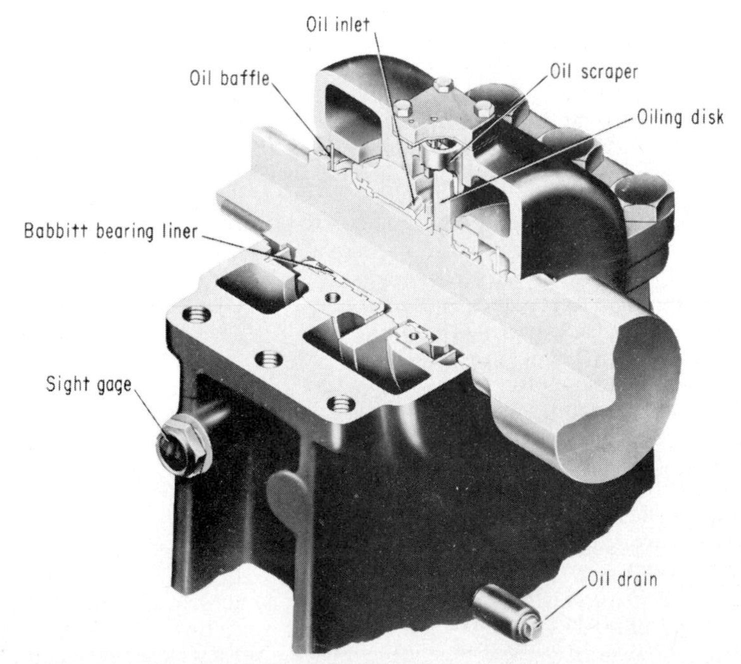

FIG. 5. Disk-lubricated pedestal bearing for large steel-mill drive motor.

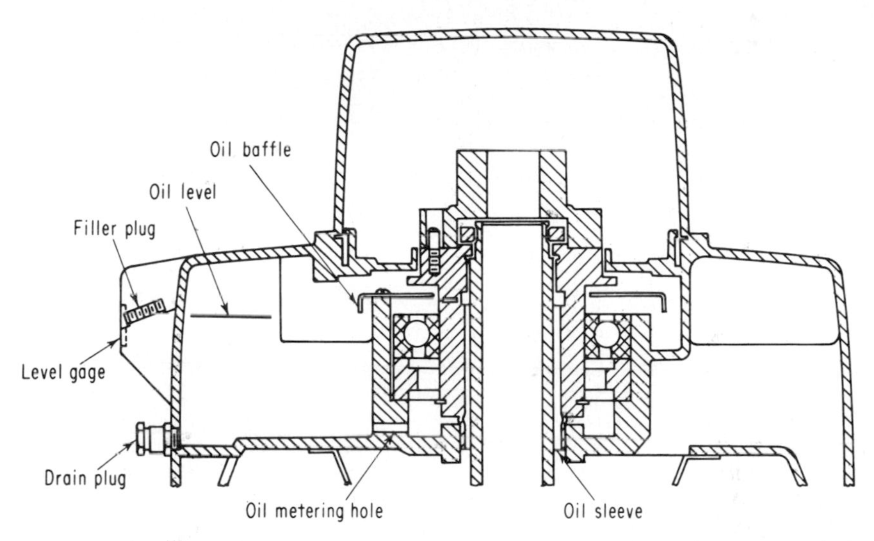

FIG. 6. Bath-oiling arrangement for vertical motor. A second matched ball bearing is added for high-thrust loads.

and yet not involve excessive turbulence and foaming. Baffles or radial fins minimize splashing and aid in the breaking of foam in the oil leaving the bearing. A central oil standpipe prevents oil from running down along the shaft to contaminate the insulation and other motor components.

Although bath oiling is most commonly used with vertical motors, it may also be applied in some horizontal units with ball and roller bearings as shown in Fig. 7. In these, the oil level is set at about the midpoint of a ball or roller at its lowest point of travel. A lower level may result in insufficient lubrication, while the oil churning associated with a higher oil level may cause both oil leakage and excessive friction with a correspondingly high temperature rise.

e. Circulating systems. When lubrication requirements are beyond the capability of an oil-ring or disk arrangement, a circulating-oil system is used to supply electric-motor bearings. Such a system will normally consist of a pump, reservoir, means for cooling, piping, instrumentation, and controls.

A small 2-gal/min oil system for a motor in the size range of approximately 1000 to 5000 hp, for instance, may use a shaft-mounted spiral pump or a separate gear pump driven by a small accessory motor to feed oil at 20 to 40 lb/in^2 manifold pressure. The reservoir capacity would be about 10 gal to provide a 5-min dwell time for the circulating oil. This dwell time is sufficient for avoiding buildup of foam and for the settling out of any contaminants.

In larger motors the pedestal supporting the bearing may serve as the oil reservoir. With some very large motors and generators, the lubrication system may incorporate separate cooling, filtration, and pumping units similar to those provided with steam turbine-generator units in electric generating stations.

6. Lubrication Maintenance. The oil in a motor should be checked once about every 3 or 6 months to ensure that the bearings are receiving an adequate supply of the proper grade of oil free of contamination. This check may be made more or less frequently, depending upon the conditions in which the motor is being

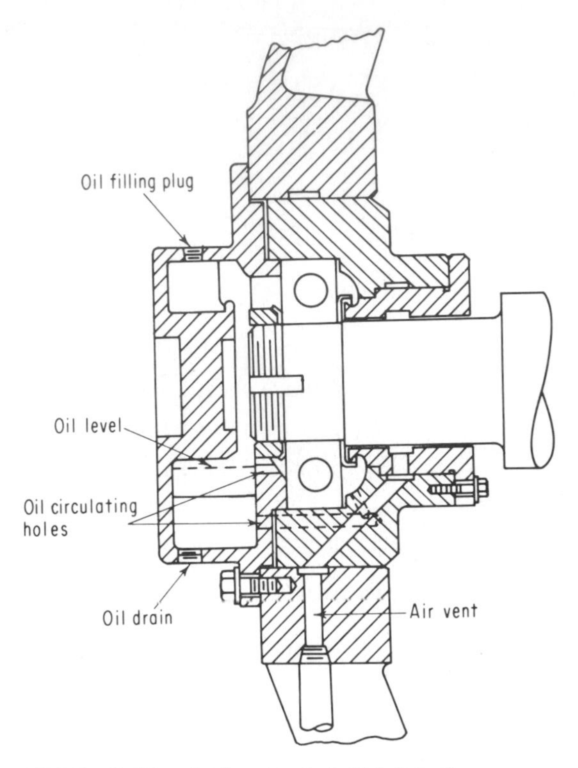

FIG. 7. Ball-bearing housing for bath lubrication.

used. Yearly intervals are usually adequate, for instance, with a combination of infrequent operation, low ambient temperature, and clean surroundings.

The oil should be changed either at regular intervals or whenever a periodic inspection shows oil deterioration. While yearly changes are common, more frequent drains are required for high ambient temperatures and some high-speed motors. On the other hand, operation for a number of years between oil changes is satisfactory with lightly loaded bearings in motors operating at normal indoor ambient conditions. A limiting neutralization number for the oil of 1.0 to 2.0 (ASTM D 974) should serve as a guide. When oxidation of the oil has raised the acidity above this value, bearing corrosion, varnish deposits, and dirt may be expected. Where an acidity check is not practical, a visual observation for oil-color change and contamination will often give a useful indication as to the desirable oil-change time.

If the oil is light in color and contains no dirt, the level should merely be adjusted by addition of the proper grade of oil. Usually the check of oil should be made with the motor at standstill since rotation of an oil ring or other components may cause a change in level. In some cases the motor instructions will indicate whether the level should be set during standstill or while running.

The only cleaning necessary in most oil-lubricated motors is accomplished by removing the drain plug to drain the oil. This will generally flush out any water, dirt, or sediment which has settled to the bottom of a bearing housing or oil reservoir. During motor disassembly for general cleaning, the bearing housing may be washed out with a suitable solvent. After solvent washing, the bearing lining should be dried, and then the shaft and bearing surfaces should be covered with a film of oil before reassembly. Before the drain plug is replaced, it should first be coated with a sealing compound and then securely tightened to prevent oil leakage along the threads.

Fractional-horsepower motors with wick-oiled sleeve bearings should be reoiled once a year, or every 2000 operating hours, whichever occurs first. The amount of oil added should vary from 30 drops of 32-cSt viscosity turbine oil for a 3-in-frame-diameter motor to 100 drops for 9-in-diameter motors. If a turbine oil is not available, SAE 10 automotive oil may be used.

With wick-oiled bearings, the yarn or felt packing should be removed during any other motor-maintenance work. At this time the housing should be washed with a suitable solvent. The bearing housing should then be repacked with clean wool yarn which has been saturated with the proper grade of lubricating oil.

During lubrication maintenance on a motor or generator, a check should also be made on any brushes, commutator, or collector rings in the motor. They should be wiped free of oil with a clean, lint-free cloth and worn brushes should be replaced. The motor fans should also be wiped clean to preserve balance.

GREASE LUBRICATION

Although motors using ball and roller bearings may be designed for either oil or grease lubrication, grease is now almost universally used in motors ranging from about 1 to 800 hp for a number of cost, design, and application reasons. Less lubrication maintenance is normally required. More freedom exists in selecting a motor-mounting position since sealing against leakage along the shaft is much easier with grease. Grease effectively keeps out dirt and moisture when used in combination with well-designed shaft seals. Since there is less tendency for leakage, grease also minimizes contamination of motor windings and any commutator or slip rings and brushes.

The primary function of the grease in a motor is to supply the bearings with oil from the spongelike reservoir of the grease gel structure. Without a protective oil film on the heavily stressed contact zones between the balls and raceways, and at the high-speed rubbing contacts between the balls and the separating cage, excessive friction would result in rapid wear and almost immediate failure. Although this lubricant film is absolutely essential, the amount required is extremely small. Months of satisfactory operation can be obtained in a 5-hp motor, for instance, with as little as a single drop of oil under ideal conditions of cleanliness. A high order of chemical and physical stability is required in a grease so that this small but essential lubricant supply may be available for a period up to many years in electric-motor service.

In addition to providing the lubricant, greases are also required to prevent rust, to protect bearings from dirt, and to minimize maintenance. Since so very little oil is required for lubrication, these secondary functions will often dictate the type of grease to be used. Grease selection should also be integrated with the bearing and housing design for the type of service to be met.

7. Grease Composition and Properties. Greases for electric-motor use consist of either petroleum or synthetic oils thickened with about 8 to 15 percent of any one of several dozen varieties of thickening agents.

TABLE 4. Typical Temperature Limits for Synthetic Greases

Grease type	Max temp for 1000-h life, °C (°F)	Min temp for 1000-g·cm torque in 204 bearing, °C (°F)
Silicone	170 (340)	−35 (−30)
Polyester	150 (300)	−45 (−50)
Synthetic hydrocarbons	145 (290)	−45 (−50)
Silicone diester	140 (280)	<−75 (<−100)
Diester	130 (270)	−55 (−70)
Petroleum	120 (250)	−30 (−20)

a. Oil type. Well over 95 percent of the greases employed in industrial motors use petroleum oil in the SAE 20 to 30 range with 40°C (100°F) viscosities of about 45 to 130 cSt. Oils in the upper end of this range are usually desirable for low evaporation rates and long life while allowing free operation at temperatures down to −29°C (−20°F). Lower-viscosity oils can be used with a sacrifice in life for applications down to about −40°C (−40°F). Somewhat higher viscosities are used in some greases for applications in the 95 to 120°C (200 to 250°F) range. For extreme temperatures, however, it is generally better to consider one of the synthetic greases rather than a petroleum grease with an abnormal oil viscosity. A comparison of limiting temperatures with several different types of synthetic greases and petroleum grease is indicated in Table 4.

b. Thickener type. Typical characteristics of petroleum greases as related to the various common types of thickening agents are indicated in Table 5. There are extreme variations in physical properties and performance characteristics in a group of greases within a given soap type; so care should be given to the selection of specific greases for difficult motor applications.

Lithium-soap greases have given years of excellent performance in electric motors. Good high-temperature stability has also brought polyurea-thickened greases into general use. Both have reasonable resistance to water, good high- and

TABLE 5. Typical Characteristics of Petroleum Greases

Base	Texture	Dropping point, °C (°F)	Max temp for continuous use, °C (°F)	Water resistant	Mechanical stability
Soap:					
Aluminum	Smooth, clear	77 (170)	82 (180)	Yes	Poor
Barium	Buttery	188 (370)	107 (225)	Yes	Good
Calcium	Smooth	82 (180)	77 (170)	Yes	Fair
Complex soaps	Smooth	260 (500)	149 (300)	Yes	Good
Lithium	Smooth	191 (375)	149 (300)	Yes	Good
Sodium	Buttery or fibrous	171 (340)	121 (250)	No	Good
Nonsoap:					
Modified clay	Smooth	None	121 (250)	Yes	Good
Organic	Smooth	260 (500)	163 (325)	Yes	Good

low-temperature characteristics, and good mechanical and oxidation stability. Previously sodium-soap greases were the most widely used for roller-bearing lubrication. Soda-soap greases are generally characterized by good mechanical stability, good rust-prevention characteristics, and wide temperature usability. About 20 percent calcium soap is often incorporated with a soda soap to shorten the fiber length and give a more uniform structure. Although sodium greases have good rust-prevention properties, they are washed away by large quantities of water. In such cases frequent relubrication must be practiced or else a water-resistant grease of the lithium, calcium, barium, or modified-clay type should be used.

The generally poor mechanical stability of inexpensive calcium greases has usually eliminated them from consideration for electric-motor bearings. Older types of calcium greases contained water as a stabilizing agent, and this limited their use to temperatures below 70°C (160°F). Some newer complex calcium-soap greases have been developed, however, which are widely used as multipurpose industrial lubricants with good high-temperature characteristics and other qualities which give satisfactory performance in electric motors.

Modified clay-type thickeners have recently been finding common use for multipurpose applications in industrial plants. These are characterized by good resistance to water and by nonmelting properties. Despite the very high dropping points of these and some other petroleum greases, upper temperature limits for long-time service in electric motors have not been raised by their use. Oxidation, oil bleeding, evaporation, and other properties dictate the life at elevated temperatures, even though no limitation exists insofar as the dropping point itself is concerned.

c. Extreme temperatures. Greases containing synthetic oils provide excellent service below and above the temperature range (−30 to 120°C; −20 to 250°F) at which most petroleum greases are useful. For temperatures down to −73°C (−100°F) synthetic greases of the diester and silicone-diester type can be used. Silicone greases are useful for long-term service at temperatures up to 150°C (302°F) and at the same time can be used down to about −35°C (−30°F). Some of the newer silicone greases thickened with high-melting organic gelling agents other than soaps appear promising for short-time applications up to about 232°C (450°F).

For temperatures above 120°C (250°F) ball and roller bearings of standard 52100 steel should not generally be used. Although practices vary with different bearing manufacturers, final heat treatment or stabilization is often carried out at a relatively low temperature to achieve optimum hardness for long fatigue life. When the bearing operating temperature approaches this stabilizing temperature, permanent increases in bearing dimensions occur. These result from the transformation of small amounts of high-temperature austenite which had failed to transform to the less dense martensite during the prior heat treatment. Bearings of 52100 steel specially heat-treated for use up to 175°C (350°F) are supplied on order by most manufacturers. Tool-steel bearings are available for even higher temperatures.

d. Consistency. In addition to the wide range of grease compositions available, attention should also be paid to physical properties and mechanical characteristics. Grades vary in consistency from those which are fluid at room temperature to others which are so hard that they must be cut with a knife. The depth of penetration in tenths of a millimeter for a cone dropped into a grease sample is the most common measure of consistency (ASTM D 217). This penetration measured after the grease has been worked 60 strokes with a perforated-disk plunger is the basis of the following classification developed by the National Lubricating Grease Institute (NLGI):

NLGI No.	*ASTM worked penetration*
0	355–385
1	310–340
2	265–295
3	220–250
4	175–205
5	130–160
6	85–115

By far the greater proportion of electric-motor greases are NLGI grade 2. Such greases are usually sufficiently stiff so that very little trouble will be encountered with churning, slumping, or leakage. Yet the grease is soft enough to feed readily to the working parts of a bearing. Stiffer greases of NLGI grade 3 are used where good channeling characteristics are required in large high-speed motors or in double-sealed permanently lubricated bearings where the grease must be totally confined and yet stay clear of the churning action in the ball complement.

e. Bleeding and evaporation. In addition to stability under mechanical agitation, the loss of oil from the grease structure by bleeding and evaporation should be low for long bearing life. Generally when a grease has lost about half its initial oil content, it is no longer capable of supplying the lubrication needs of the bearing and failure will result. For most motor use, a bleeding loss of less than 10 percent in 500 h at 100°C (212°F) and an evaporation loss of less than about 3 to 5 weight percent in the same period is desirable for long life. Bleeding rates of greases can be determined by collecting the oil draining from a 10-g sample in a wire-mesh cone into a beaker in a constant-temperature oven. Evaporation loss in such a test is the total weight loss from both the cone and the oil in the beaker.

f. Oxidation. Oxidation of the grease by reaction with air is one of the chief deteriorating factors that limit the life of ball-bearing greases. Oxidation of the oil in a grease results in drying and hardening, while oxidation of the soap usually results in loss of the network gel structure. The combination of these factors often shows up as an initial softening as the soap structure breaks down. This stage is followed by increasing stiffness and darkening, which finally result in a hard, caked mass. Excellent oxidation inhibitors are now available to give extremely long life with high-quality greases. Better greases for electric-motor use will generally not give a pressure drop in a Norma-Hoffmann bomb test of more than 20 lb/in^2 in 750 h.

g. Compatibility. Although compatibility is usually not a serious problem when different electric-motor greases are mixed in a bearing housing, this point should be checked before using greases interchangeably. In general, the mixing of greases of various soap types should be avoided. In such a case, the two different soaps may interact to give either unusual consistency changes or a lower dropping point. The mixing of diester, silicone, or other synthetic greases with each other or with petroleum greases should especially be avoided.

Table 6 lists properties which may be helpful in the initial screening of grease candidates. This is only a guide, and good performance has been obtained with greases that do not provide all of the suggested characteristics. A grease of NLGI grade 1½, for instance, is currently giving satisfactory service in low-temperature ambients. Some large bearings may perform better with a grade 3 grease.

TABLE 6. Recommended Grease Properties for Industrial Motors in 40°C Ambients

Property	Requirement	ASTM method
NLGI consistency grade	2	D 217
Dropping point, °C, min	160	D 566
Oil viscosity at 40°C	110–140 cSt (500–650 SUS*)	D 445
Operating temperature range, °C:		
Low limit (10,000 g·cm)	−30	D 1478
High limit (500-h min life)	150–163	D 3336
Oxidation stability, min h to reach 20-lb/in^2 drop at 99°C	750	D 942
Bleeding in 500 h at 100°C, % max	10	D 1742
Starting torque at room temp, g·cm	150	D 1478
Running torque at room temp, g·cm	150	D 1478

*SUS = Saybolt universal seconds.

Final establishment of suitability can be made only in electric motors performing their function. Lubricant suppliers and motor manufacturers should be checked for their latest recommendations for any unusual operating conditions.

8. Bearing-Housing Arrangements. The simplest housing design for a grease-lubricated bearing can be achieved with a double-sealed ball bearing as shown in Fig. 8*a*. With built-in seals and self-contained grease within the ball bearing itself, the housing functions merely as a protective supporting shell for the bearing. Relubrication is not practical, however; so failures should be expected after a long period of operation, particularly at elevated temperatures. In such a case the motor end shield is removed, and the bearing is pressed from the shaft and replaced with a new bearing.

An open-type ball bearing with a regreasable housing is shown in Fig. 8*b*. Here an inner cap and grease fittings are added to the bearing housing. This arrangement provides a supplemental grease reservoir for the ball or roller bearing, and the unit is regreasable.

Both these types of bearing arrangements have applications for which they are best fitted. Appropriate long-life petroleum greases for each type of enclosure are specified in Table 6. With these, the grease life under ideal conditions is indicated in Fig. 9 for various bearing temperatures.

Various modifications of these two more or less basic designs are in rather common use. A single seal on the inboard side of the bearing housing can replace the inner cap of a regreasable housing in many smaller motors. Such an arrangement is shown in Fig. 8*c*. A double-shielded bearing is sometimes used in place of the open bearing shown in Fig. 8*b*. The close-running shields at the faces of such a ball bearing minimize the leakage of grease to the inboard side of the bearing during regreasing and help exclude dirt from the working parts of the bearing itself. This is done at some sacrifice to the ease of purging old grease from the bearing housing during regreasing.

Various grease entrance and exit locations have been used for regreasable housings. Grease is sometimes introduced at the inboard side of the bearing, as shown in Fig. 8*d*, in an attempt to get a more complete purging action of the old

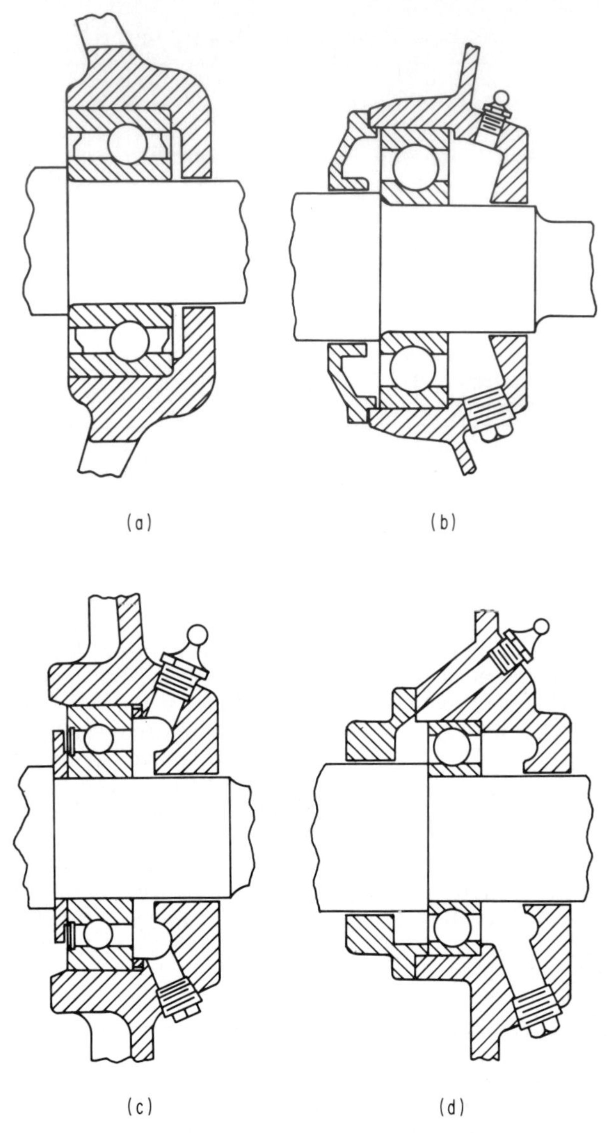

FIG. 8. Housing arrangements for greased ball bearings. (*a*) Self-contained double-sealed bearing. (*b*) Open bearing in regreasable housing. (*c*) Regreasable housing using single-shielded bearing. (*d*) Transverse greasing through bearing.

grease in the inner cap. Rather than provide a separate drain opening, the clearance space between the shaft extension and the housing is sometimes used as the discharge point for purged grease. Various arrangements using a shaft slinger, metering plate, internal baffles, and an enlarged discharge opening have been used in various designs to get a more complete displacement of old grease with a minimum of leakage along shaft seals during regreasing.

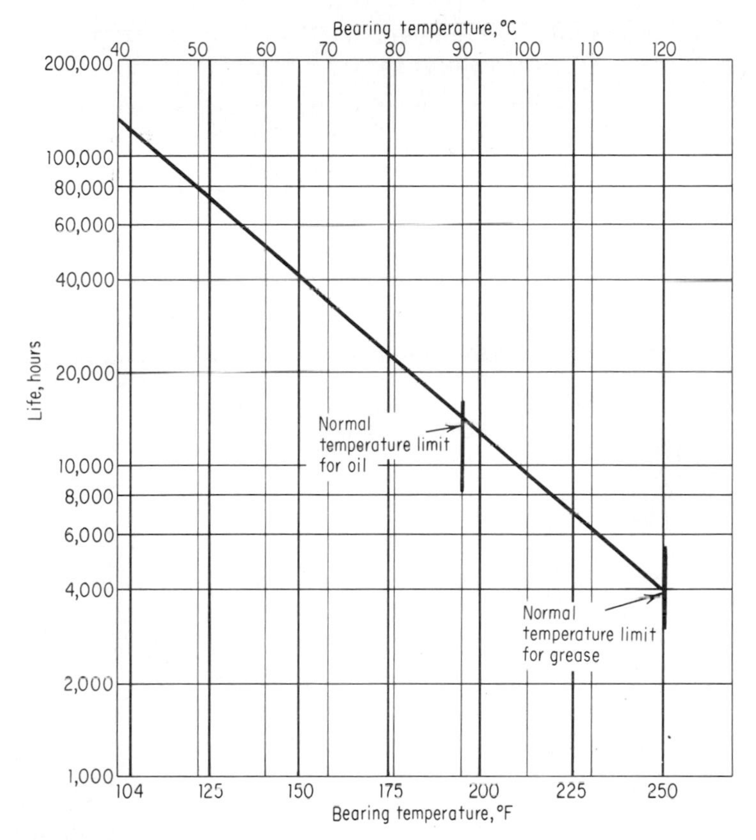

FIG. 9. Variation in ball-bearing grease test life with temperature in 306 size ball bearing at 3600 r/min.

9. Regreasing. Table 7 will serve as a guide in determining regreasing intervals by the type, size, and service of the motor to obtain the most efficient operation and the longest bearing life. Where a variety of motor sizes, speeds, and types of service are involved in a single plant, a uniform relubrication period is sometimes selected. A yearly basis is common, for instance, and such a yearly regreasing might conveniently be carried out on a plant-wide basis during a vacation shutdown.

Motors equipped with grease fittings and relief plugs should be relubricated by the following procedure using a low-pressure grease gun:

1. Wipe clean the pressure-gun fitting and the regions around the motor grease fittings.
2. Remove the relief plug and free the relief hole of any hardened grease.
3. Add grease with the motor at standstill until new grease is expelled through the relief hole. In a great majority of cases it is not necessary to stop the motor during relubrication, but regreasing at standstill will minimize the possibility for grease leakage along the shaft seals. Care should be taken not to overgrease; troublesome contamination of the motor interior could result.

TABLE 7. Guide for Maximum Regreasing Periods

	Motor rating				
	¼–7½ hp 0.2–5.6 kW	10–40 hp 8–32 kW	50–150 hp 40–125 kW	200–300 hp 160–250 kW	>300 hp >250 kW
Type of service	Regreasing interval, horizontal shaft/vertical shaft				
Easy: infrequent operation (1 h per day), valves, door openers, portable floor sanders	NR*/NR	NR/3 years	4 years/1.5 years	2 years/9 months	1 year/1 year
Standard: 1- or 2-shift operation, machine tools, air-conditioning apparatus, conveyors, garage compressors, refrigeration apparatus, laundry machinery, textile machinery, woodworking machines, water pumping	NR/3 years	4 years/1 year	1.5 years/6 months	1 year/3 months	6 months/6 months
Severe: motors, fans, pumps, motor-generator sets, running 24 h per day, 365 days per year; coal and mining machinery; motors subject to severe vibration; steel-mill service	4 years/1.5 years	1.5 years/6 months	9 months/3 months	6 months/1.5 months	3 months/3 months
Very severe: dirty, vibrating applications, where end of shaft is hot (pumps and fans), high ambient temperature	9 months/6 months	4 months/3 months	4 months/2 months	3 months/1 month	2 months/2 months

*NR = regreasing not normally required.

4. Run the motor for about 10 min with the relief plug removed to expel excess grease.
5. Clean and replace the relief plug.

For totally enclosed fan-cooled motors, the above instructions apply for greasing the drive-end bearing. The fan-end housing is frequently equipped with a removable grease relief pipe which extends to the outside of the fan casing. First remove, clean, and replace the pipe. Next, during the addition of new grease from a grease gun, remove the relief pipe several times until grease is observed in the pipe. After grease is once observed to have been pushed out into this pipe, no more should be added. The pipe, after again being cleaned and replaced, will then act as a sump to catch excess grease when expansion takes place during subsequent operation of the motor.

In many vertical motors, the ball-bearing housing itself is relatively inaccessible. In such cases a grease relief pipe is frequently used in a manner similar to that in the totally enclosed, fan-cooled motors. The same regreasing procedures as described above should be used for the totally enclosed fan-cooled motors.

In many small motors no grease fittings are used with bearing housings similar otherwise to that shown in Fig. 8*c*. Such motors should be relubricated by removing the end shields, cleaning the grease cavity, and refilling three-quarters of the circumference of the cavity with the proper grade of grease. In the end shields of some smaller motors, threaded plugs are provided which are replaceable with grease fittings for regreasing without disassembly.

Since regreasing of motor bearings tends to purge the old grease, a more extensive removal of all the used grease is seldom necessary. Whenever a motor is disassembled for general cleaning, however, the bearings and housing should be cleaned by washing with a grease-dissolving solvent. To minimize the chance of damaging the bearings, they should not normally be removed from their shaft for such a washing. After thorough drying, each bearing and its housing cavity should be filled approximately one-half to three-fourths full with new grease before reassembly. Avoid spinning the bearing with an air hose during cleaning, and do not reuse any bearing that has been removed from the shaft by pulling on the outer ring.

GEAR MOTORS

10. Motor Bearings. The electric-motor portion of a gear motor such as that shown in Fig. 10 is usually a conventional 1800 r/min motor in which one end shield has been modified to permit mounting of the gear case directly to the motor frame.

No unusual lubrication requirements are normally encountered in the motor bearings of the gear motor. The gear-end motor bearing either is splash-lubricated by oil from the gear casing or is grease-lubricated. If it is grease-lubricated, both it and the ball bearing at the opposite end of the motor should be greased and maintained in the same fashion as any other electric-motor ball bearing.

The same oil is used in the gear unit for the lubrication of both gears and bearings. Since ball bearings are commonly used with their relatively mild lubrication requirements, oil selection is made to meet the more severe conditions encountered in the gears. The proper grade of oil will be specified either on the nameplate or in instructions provided with the gear motor. This grade of oil will vary with output power, speed, and the type of gear design, which may include spur, helical, herringbone, and worm types. Recommendations on specific oils to be used can be obtained from the gear motor manufacturer, from an industrial-lubricants supplier, or from listings available from the American Gear Manufacturers Association.

Since almost all gear motors are shipped without oil, the proper grade of lubricant should first be added to the gearbox to the indicated level. The unit should then be run for a short period at no load to check the direction of rotation and for free turning. Running at partial load for a day or so is also desirable as a final break-in for the gears.

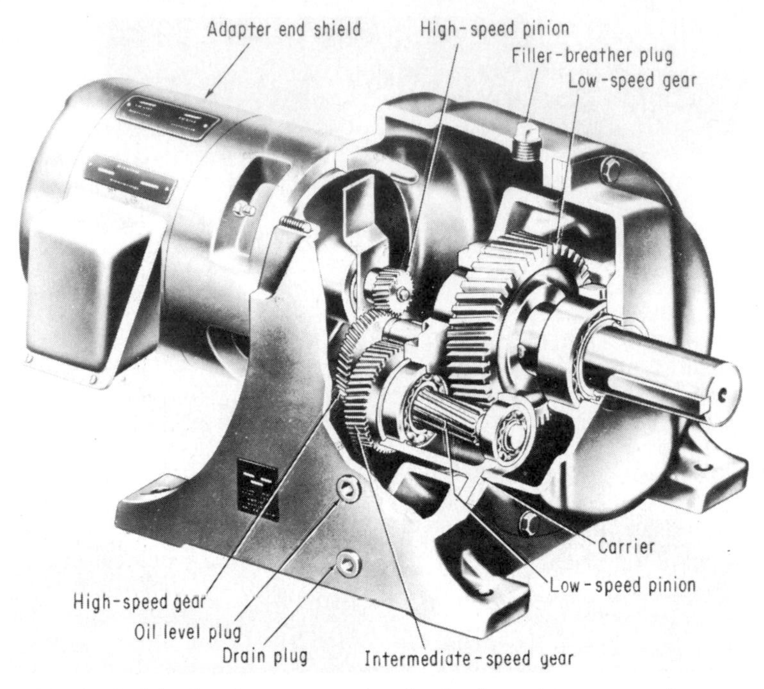

FIG. 10. Lubrication features on triple-reduction integral-type gear motor.

The oil should then be changed after the first 200 h of run-in operation. To do this, the original oil is drained and the gear housing is refilled to the correct level with an SAE 10 oil. The unit should then be run for 3 min for flushing. Finally drain off this light flushing oil and refill the gearbox with the designated grade of gear oil to the correct level.

The oil should then be sampled weekly to check for water and sediment in any moist, dirty, or high-ambient-temperature installations. If water or sediment is present, the oil should be changed. Weekly checks should be continued until the correct interval between oil changes is established. Where problems with moisture, dirt, or high temperatures are not encountered, 3 to 6 months is a reasonable oil-change period. A magnetic drain plug is useful in a gear unit to remove any particles of metal worn from the teeth. If such a drain plug is used, it should be removed periodically to clean away any collected metal particles.

Although the above procedures will be useful for most industrial gear-motor units, a great variety in size and operating conditions leads to unusual designs with different lubrication requirements. A careful check should be made of instructions provided with each unit before it is placed in operation.

REFERENCES

1. T. J. Beebe, *Industrial Electric Motors, Handbook of Lubrication,* vol. 1 (CRC Press, Boca Raton, Fla., 1983), pp. 188–192.

2. D. F. Wilcock and E. R. Booser, *Bearing Design and Application* (McGraw-Hill, New York, 1957).

3. J. J. O'Connor and John Boyd, *Standard Handbook of Lubrication Engineering* (McGraw-Hill. New York, 1968).

SECTION 15
FLYWHEELS

John F. Sellers* and Robert C. Moore†

FOREWORD 15-2

FLYWHEEL APPLICATIONS 15-2

1. Short-Time Sources of Power 15-2
2. Flywheels with Pulsating Motor Loads 15-5

REFERENCES 15-6

*Chief Engineer, DC Machines, Allis-Chalmers Manufacturing Company, Milwaukee, Wis. (retired); Fellow, IEEE.

†Senior Engineer, Motor-Generator Department, Allis-Chalmers Manufacturing Company, Milwaukee, Wis. (retired); Registered Professional Engineer (Wis.); Senior Member, IEEE; Distinguished Lecturer, Milwaukee School of Engineering.

FOREWORD

There are two major groups of motor applications in which flywheels are employed to obtain a desired operating characteristic.

In the first group flywheels are used on motor-generator sets or individual-drive motors when the application requires supplemental power for short periods. These periods are, however, seldom longer than 2 s.

In the second group of applications flywheels are used to obtain additional inertia when a rotating-mass system is operating too close to its critical speed. Such uses of flywheels are especially important when load-torque pulsations are present. Such load pulsations occur with reciprocating compressors, jaw crushers, punch presses, or similar machines.

FLYWHEEL APPLICATIONS

1. Short-Time Sources of Power. There are many applications in which flywheels are used as a momentary source of power.

An inertia constant H has been established for exciter motor-generator sets, on which a momentary loss of ac power may occur for up to 2 s:

$$Wk^2 = \frac{H \times \text{kW} \times 10^6}{0.231 \times n_{fl}^2} \tag{1}$$

where Wk^2 = total inertia for entire motor-generator set, lb·ft^2
kW = rated full-load output of exciter
n_{fl} = speed of set at rated load, r/min
H = arbitrary flywheel constant

The stored energy in kilowatt-seconds given up by a rotating mass between speeds n_{fl} and n_t is

$$0.231 \times 10^{-6} \times Wk^2 \times (n_{fl}^2 - n_t^2) \tag{2}$$

where n_t is the speed after t seconds in r/min. Substituting Eq. (1) into Eq. (2),

$$t_s \times \text{kW} = 0.231 \times 10^{-6} \times (n_{fl}^2 - n_t^2) \times \frac{H \times \text{kW} \times 10^6}{0.231 \times n_{fl}^2}$$

or

$$t_s = H\left(1 - \frac{n_t^2}{n_{fl}^2}\right) \tag{3}$$

Figure 1 shows the relationship between the ratio of minimum speed to rated load speed and the load duration, in seconds, for values of H between 1 and 5. A value of 5 is most commonly used for exciter applications.

On motor-generator sets, such as those used to supply electric power to dc hoist motors and reversing, metal-rolling-mill motors, a flywheel may be used with a wound-rotor induction motor to keep down the maximum peak load drawn from the power line. The energy, in kilowatt-seconds, stored in the flywheel at the synchronous speed of the ac motor is from 20 to 30 times the kilowatt rating of the dc reversing motor. The flywheel gives up energy by speed reduction when resis-

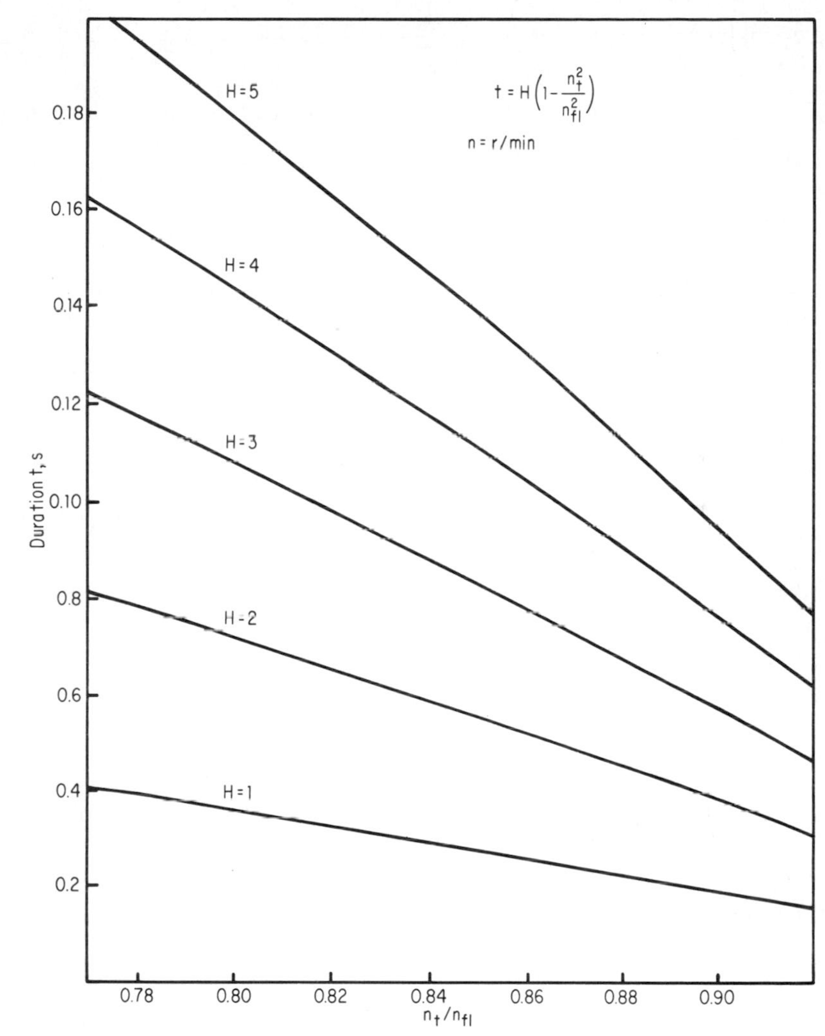

FIG. 1. Ratio of minimum speed to full-load speed as related to load duration for various *H* factors.

tance is increased in the rotor circuit of the induction motor, after the load builds up to approximately 125 percent of rated current. A maximum slip of 15 to 20 percent is considered most practical in these applications.

For duty cycles requiring a short duration of ac or dc electric power followed by a considerably longer period of no load, such as for energizing magnet coils, demagnetizing loops, and special wind tunnels, a flywheel is used with the pulse generator having a driving source of only a fraction of the rating of the generator. The driving source may be any type of prime mover or wound-rotor induction motor. A regulator is used to maintain a nearly constant fuel supply into the prime mover, or constant power input to the induction motor, until a predetermined no-load speed is reached during the speed-recovery portion of the total load cycle.

For example, assume a demagnetizing loop as the intermittent load on a dc generator, having a constant-resistance negligible-inductance characteristic requiring 500 V, 4000 A for 2 s on, followed by 8 s off with the generator field deenergized (Fig. 2). What are the sizes of the flywheel and the drive motor required? A voltage regulator would be used on the output voltage of the dc generator to maintain 500 V down to 80 percent of synchronous speed. The dc generator can be designed with 60°C rise for 200 percent repeated load application, giving a continuous generator rating of 1000 kW, 500 V, 2000 A. This machine could be designed for 1200-r/min synchronous speed, 960-r/min minimum speed.

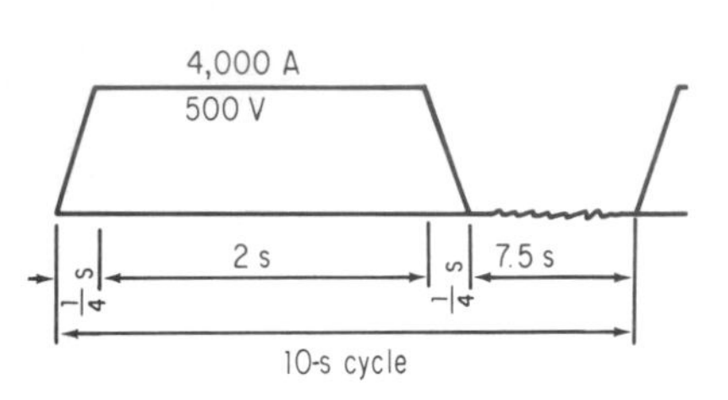

FIG. 2. Load cycle on dc generator driven by motor with flywheel.

With proper field forcing, the generator voltage would build up in ¼ s.

The rms heating value on the generator and loop would be

$$\left[\frac{\frac{1}{4} \times \frac{1}{3} \times 4000^2 + 2 \times 4000^2 + \frac{1}{4} \times \frac{1}{3} \times 4000^2}{10}\right]^{1/2} = 1860 \text{ A}$$

This value is within the 2000-A continuous rating of the generator. The total power, in kilowatt-seconds, required from the induction motor is determined as follows:

Coil load $= 2 \times \frac{1}{4} \times 2000 \times \frac{1}{4} = 250$ kWs during rise and fall

Coil load $= 2 \times 2000 = 4000$ kWs during loading

Generator $i^2r = 2 \times 4 \times 30.6 = 245$ kWs during loading

Generator $i^2r = 2 \times \frac{1}{4} \times \frac{1}{3} \times 4 \times 30.6 = 20$ kWs during rise and fall

Generator brush $ei = 2 \times \frac{1}{4} \times 3 \times \frac{4000}{2} = 3$ kWs during rise and fall

Generator brush $ei = 2 \times 3 \times 4000 = 24$ kWs during loading

Generator core loss $= 2 \times 10 = 20$ kWs during loading

Generator stray-load loss $= 0.01 \times 2 \times 2000 = 40$ kWs during loading

Generator brush friction and winding $= 10 \times 14 = 140$ kWs for total cycle

Flywheel winding $= 10 \times 6 = 60$ kWs for total cycle

Total motor output $= 4802$ kWs in 10 s (1 cycle)

Average motor load $= \frac{4802}{10} = 480.2$ kW $= 643$ hp

Required motor rating $= \frac{643}{0.9} = 750$ hp at 1180 r/min

Flywheel parameters are:

Coil load in 2.5 s = 4250 kWs

Generator loss in 2.5 s = 352 kWs

Generator + flywheel winding and friction = 50 kWs

Motor output $= 750 \times 0.746 \times 2.5 = 1400$ kWs

Net flywheel output = 3252 kWs between 1200 and 960 r/min

From Eq. (2),

$$\text{Total } Wk^2 \text{ in motor-generator set}$$
$$= \frac{3252}{0.231 \times 10^{-6} \times (1200^2 - 960^2)} = 27{,}200 \text{ lb}\cdot\text{ft}^2$$

Generator $Wk^2 = 4650 \text{ lb}\cdot\text{ft}^2$

Motor $Wk^2 = 2350 \text{ lb}\cdot\text{ft}^2$

Wk^2 of generator and motor $= 7000 \text{ lb}\cdot\text{ft}^2$

Required Wk^2 in flywheel $= 27{,}200 - 7000 = 20{,}200 \text{ lb}\cdot\text{ft}^2$

Flywheels for motor-generator sets are usually made of plate steel, using a peripheral speed of 25,000 ft/min at no load. At 1200 r/min this would give a wheel diameter of

$$\frac{25{,}000 \times 12}{1200\pi} = 80 \text{ in}$$

For a solid-disk wheel, the center of gravity is at 70.7 percent of the radius. The flywheel weight is

$$W = \frac{20{,}200}{(0.707 \times 40/12)^2} = \frac{20{,}200}{5.56} = 3630 \text{ lb}$$

The width of the flywheel is

$$\frac{3630}{0.283\pi \times 40^2} = 2.55 \text{ in}$$

A rough plate 7 ft square by 3 in thick would be needed for the flywheel of steel grade ASTM A 113.

2. Flywheels with Pulsating Motor Loads. Obviously flywheels applied to synchronous motors do not drop in speed to release kinetic energy in the same manner as for induction motors.

Constant synchronous-speed operation is characteristic of synchronous-motor operation when the shaft-torque requirement is constant. However, certain types of motor loads, such as reciprocating compressors, may impose a pulsating torque load on the motor. In such cases the motor-rotor angular velocity (Fig. 3) will follow the pulsating torque imposed, although the average motor speed will be synchronous.

Changes in motor-rotor angular velocity are accompanied by variations in motor line current and possibly in supply-line voltage. Periodic supply-line-voltage variations can cause light flicker, speed variations in induction-motor drives, and other problems on the power system supplying the synchronous motor.

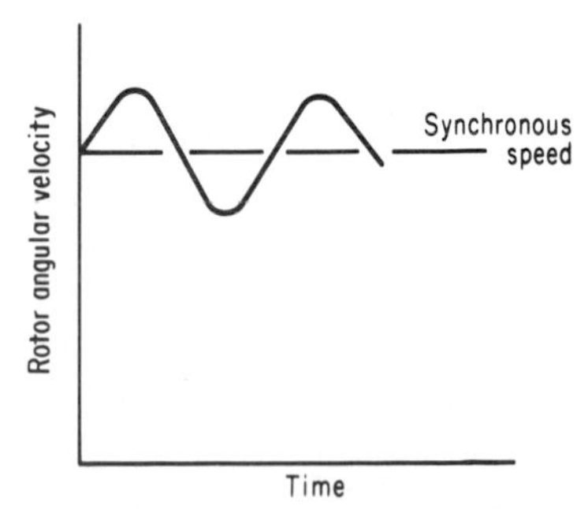

FIG. 3. Load-torque variations cause rotor angular velocity to vary above and below synchronous angular velocity.

A flywheel, correctly applied to a synchronous-motor drive having varying torque requirements, has the effect of reducing angular-velocity variations of the motor rotor, and thus reduces motor current and line-voltage changes. An incorrectly applied flywheel may be worse than none because of resonance between the motor mechanical and electrical characteristics.

To assist in determining proper values of the flywheel effect Wk^2 for a large number of compressor applications, NEMA Standard MG 1-1978 has listed C^2 factors. The numerical C factors relate flywheel Wk^2 requirements and certain other motor-drive characteristics to keep the motor-current variation within limits. For example, C factors for different types and applications of compressors are listed for 60, 40, and 20 percent motor-current pulsation. It is suggested that the standard be consulted for further details.

Flywheels may also be used in induction-motor drives, such as for motor-generator sets or punch presses. In motor-generator-set applications a wound-rotor induction motor with resistance control in the rotor circuit allows the set speed to drop when a large dc load demand occurs. As the set speed drops, flywheel-stored kinetic energy is released to supply the dc load demand partially. Thus the induction motor is relieved of the energy contribution of the flywheel. The motor line current becomes less than would have been the case if the entire peak energy had to be provided by the motor alone. The flywheel energy can be stored by increasing the set speed during light-load periods of the load cycle. Controls are arranged so that the motor will be automatically restored to its original speed.

Punch-press drives may use a high-slip cage-type induction motor. Sudden large amounts of energy may be required by the load. Because of the high-slip nature of the cage-type motor, the drive speed drops, allowing the flywheel to contribute energy to the load. As in the previous case, the induction motor replaces the flywheel kinetic energy during light-load periods of the load cycle.

When flywheels are to be used to obtain additional inertia with pulsating loads, shaft diameters of the machine must be given careful consideration for safe design.

The motor or machine manufacturer should usually be consulted before flywheels are added or the flywheel size is increased on a drive system.

REFERENCES

1. Donald Richardson, *Rotating Electric Machinery and Transformer Technology* (Reston Publishing, Reston, Va., 1978).

SECTION 16
FOUNDATIONS, GROUTING, AND ALIGNMENT

Charles R. Gibbs*

FOREWORD 16-2

INSTALLING MACHINE SUPPORTS 16-2
1. Machine Support 16-2
2. Loading 16-3
3. Slope 16-4
4. Elevation 16-5
5. Horizontal Forces 16-6

SOLEPLATE GROUTING 16-8
6. Important Considerations 16-8
7. Forms, Grout Mix, and Placement 16-9
8. Curing Time 16-13

ALIGNING HORIZONTAL MACHINE SETS 16-13
9. Positioning Machines 16-13
10. Rough Alignment 16-14
11. Checking the Foot Plane 16-15
12. Preparing for Final Alignment 16-15
13. Correct Final Alignment 16-16
14. Making Final Alignment 16-16
15. Rim Check 16-17
16. Face Alignment 16-20
17. Two-Indicator Method 16-20
18. Recording Alignment 16-22
19. Axial Float and Thrust 16-23
20. Limited End-Float Couplings 16-24

REFERENCES 16-24

*Director of Service, Allis-Chalmers Manufacturing Company, Milwaukee, Wis. (retired); Member, ESM.

FOREWORD

Three critical procedures require extreme care when motor-driven machine sets are installed. These procedures are installing machine supports, grouting soleplates, and alignment.

Attention to details in carrying out these three operations ensures:

1. Smoother operating machines and long-term savings on maintenance.
2. That the machine soleplate is solidly anchored and a permanent and integral part of the foundation.
3. Proper alignment of the machines to minimize coupling wear, bearing wear, shaft flexing, and machine vibration.

INSTALLING MACHINE SUPPORTS

1. Machine Support. The machine support on a foundation is frequently considered an integral part of the overall machine design. Provision is made to transmit the static, dynamic, and emergency loading to the foundation through bedplates or soleplates.

Generally, soleplates are large enough to allow the use of permanent-type supports between the soleplates and the foundation supplemented by grout (Fig. 1). Where the size of the soleplate is insufficient to prevent overloading the foundation material, special means of support or grout only may be required (Fig. 2). Experience has shown, however, that permanent supports combined with grout are far superior to grout only, and this support system should be used whenever possible. When using the support and grout method, a sufficient quantity of supports become an integral part of the foundation and are never removed. They are installed between the soleplate and the foundation to position, support, and hold the soleplate adequately to the required degree of flatness with the anchor bolts tight.

Since it is necessary in a typical installation on a foundation to adjust the elevation of the soleplates during the setting process, adjustable height supports should be used to save time (Fig. 3). An exception is marine work, which often demands use of nonadjustable machined-to-size supports or chocks.

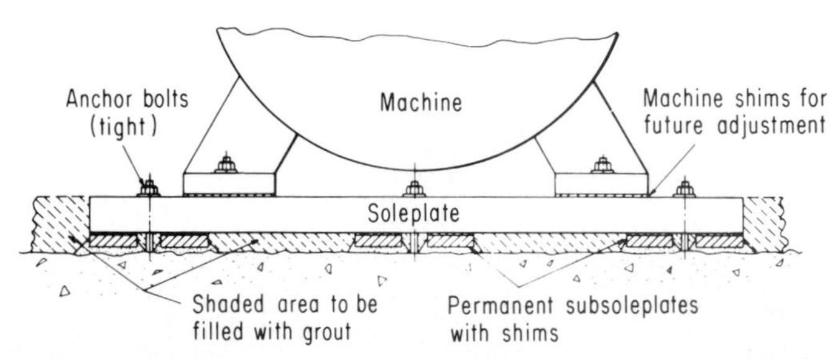

FIG. 1. Permanent supports are frequently used between machine soleplate and foundation.

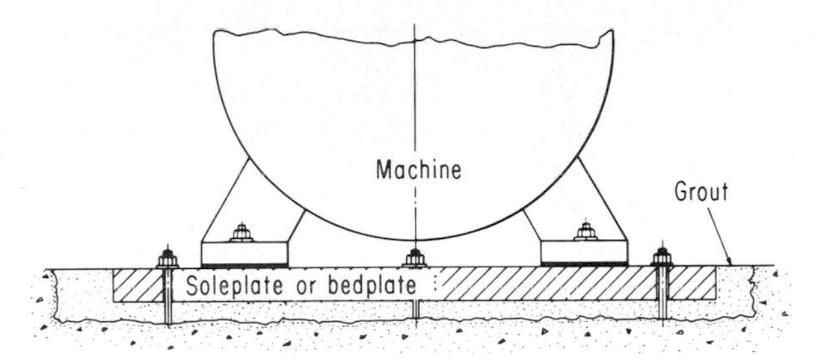

FIG. 2. Grout only is used where more support is required under machine soleplate.

These supports are of a type that will not shift or deform during the setting process or during machine operation and are locked in place by grout or welding to prevent movement. They are arranged to distribute machine and anchor-bolt loading to the foundation. Concentrated loading should not exceed 10 percent of the yield point, or 10 percent of the compressive strength of the weakest material in the support structure. The usual design is 300 lb/in^2 for concrete.

Auxiliary supports are provided under the soleplates at selected points to avoid soleplate deflection when tightening anchor bolts. Very flexible soleplates should be provided with additional support wherever required to ensure contact between them and the machine base or against sagging of the soleplate, which may occur over long periods of time. The proper number, size, and position of supports or subsoleplates can be determined by consulting the machine manufacturer's drawings or from calculation and accepted good practice.

2. Loading. To determine the number of square inches of subsoleplate area required when using concrete foundations, divide the total operating load on the soleplate in pounds by 300 lb/in^2. The answer will usually be less than 50 percent of the total soleplate area, indicating that the subsoleplate method of support is practical. If in excess of 50 percent, grouting may become difficult with the subsoleplate method.

While subsoleplates longer than 20 in can be used, shorter sizes are usually desirable to facilitate shim changes and ensure good contact even with nonparallel conditions between subsoleplate and soleplate. To obtain flatness of within 0.002

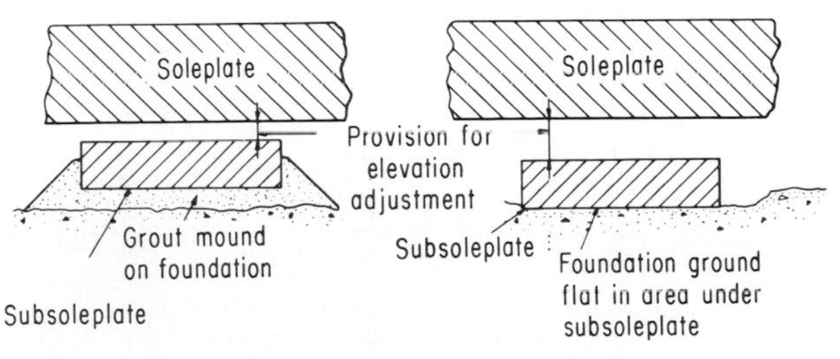

FIG. 3. Adjustment for elevation is provided between machine soleplate and subsoleplate.

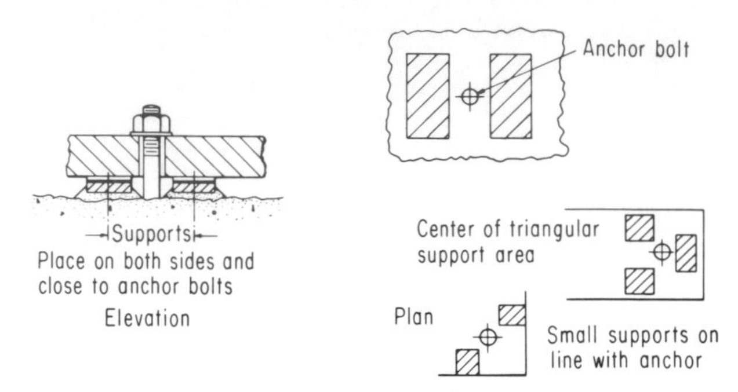

FIG. 4. Supports are placed on both sides of anchor bolts to prevent bending of soleplate.

in, it is usually necessary to grind the subsoleplate top surface. However, grinding both sides is recommended to equalize stresses in the plate.

Poor shim fits, rough surfaces, local loading by soleplate jackscrews, or other factors can cause subsoleplate bending. It is best to use the heaviest material available, but usually not less than ¾ in unless very lightly loaded. The subsoleplates are adjusted to conform to loading, and the anchor-bolt configurations and their number are determined.

Experience, ease of grouting, judgment, and calculations govern the exact placement of subsoleplates. Supports should be provided to prevent adverse bending of soleplates when tightening anchor bolts (Fig. 4). Because machine bases and soleplates have varying degrees of stiffness inherent to their design, all soleplates can be regarded as flexible to some extent. Even though machined flat, they must be checked during erection and the supports positioned as required so that the combined effect of the supports and anchor bolts will return the soleplate to its required degree of flatness.

Supports are positioned and adjusted for elevation so that large stresses in the soleplate are avoided. These stresses could result in upward soleplate movement if the machine load were removed, as when making shim changes.

3. Slope. The setting or positioning of the machine may require placing the soleplates in a predetermined off-level position. Supports are also positioned to this same slope. A kiln may be set at a ⅜-in/ft slope, while long shaft units may require the outboard units to be on a slope of a few thousandths of an inch per foot (Fig. 5). The center soleplates would be set level while the outboard soleplates

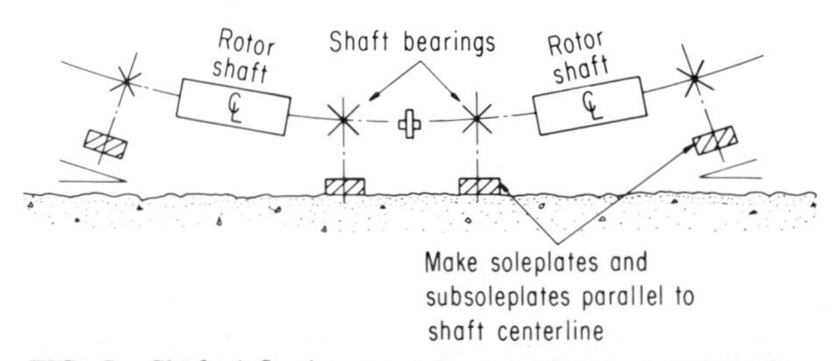

FIG. 5. Shaft deflection must be considered when installing machine with long shaft.

would be sloped approximately parallel to the shaft slope at that point. Long, unsupported soleplate spans should be avoided on at least 18-in centers regardless of loading calculations. Inherent soleplate stiffness is the determining factor governing the maximum spacing of such supports.

The foundation may be ground flat to receive the subsoleplate, or it may be placed on a mound of plain portland-cement dry-pack-type grout and tapped until it is firmly supported and properly positioned. This method makes it easy to adjust the subsoleplates for rough elevation or slope merely by sinking them into the pliable grout mound. A subsoleplate elevation of ⅛ in below the final soleplate elevation with a ±1/16-in tolerance is desirable and may be determined with a carpenter's level. The individual plates must be absolutely level or at a specified slope, and a precision level should be used for individual plate setting.

4. Elevation. For a permanent-type support, the most satisfactory adjusting device between subsoleplates and soleplate is a shim pack the same size as the former. Selected shim thicknesses assembled as a sandwich are inserted between the sole- and subsoleplates. Individual shims should not be less than 0.010 in thick, and thin ones should be sandwiched between those 0.40 to 0.60 in thick. Since a pack of thin shims will be "spongy," it is best to use the heaviest-gage shims to readjust the new pack. Shims should be tack-welded after final setting before grouting (Fig. 6).

Parallel wedges, many forms of which are available, can also be used for elevation adjustment. Since they can shift, they should be locked by tack welding or some other positive method after adjusting. A parallel wedge with a "slow" taper of about ⅛ in/ft provides a fine elevation adjustment as well as a ready means of checking the relative degree of support by merely tapping it with a hammer (Fig. 7).

The use of parallel wedges with a "fast" taper—¾ in/ft or more—is not recommended because of their coarse adjustment and the chance of slippage. They can be used, however, on certain types of equipment not requiring critical alignment, if locked in place by welding.

Some commercially available parallel wedges have a screw adjustment. Others do not need subsoleplates as their base is designed to rest on grout or a smooth foundation top.

Single wedges (Fig. 8) are not recommended as a permanent-type support, but they can be useful as temporary supports while rough-setting machinery.

Jackscrews provide an easy means of adjustment and permanent support when designed for that purpose. Usually the subsoleplates are specified with a recess or

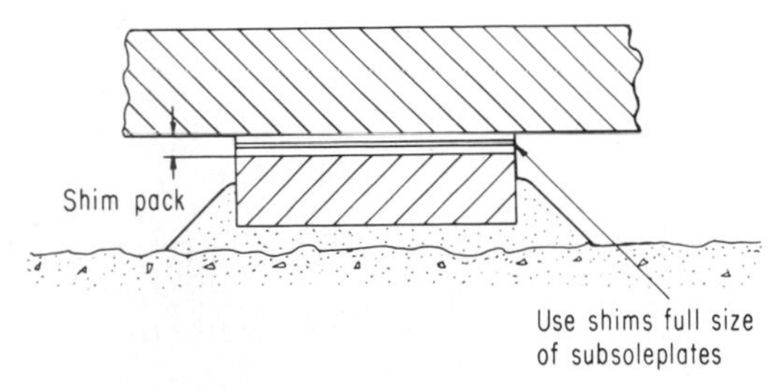

FIG. 6. Shims should be cut to full size of subsoleplate to provide solid support.

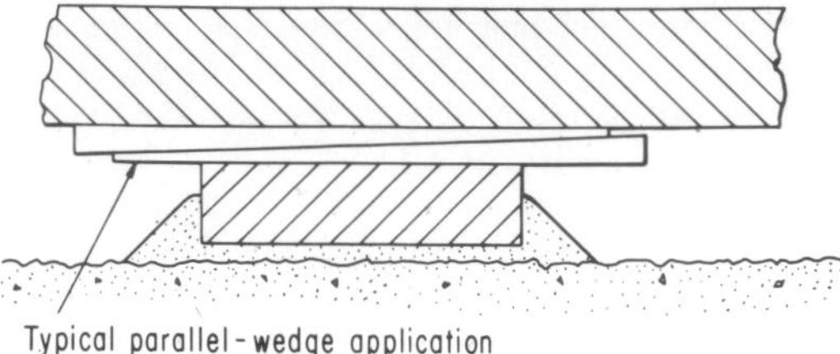

FIG. 7. Slow-taper wedges provide convenient means of adjusting soleplate elevation.

spherical counterbore to receive the jackscrew to help distribute the load. With this relatively local loading on the top of the subsoleplate, the plate must be thick enough to resist bending and transmit the loading to the foundation within 300 lb/in^2 (Fig. 9). A jackscrew used for lifting the soleplate to facilitate elevation changes does not qualify as a permanent support and should be loosened or removed before grouting the soleplate.

Some soleplates are designed to be supported by grout only. Generally these plates are rigid, and they are usually set and grouted before the machine is placed on the soleplate or operating loads are applied. Such a soleplate is usually designed with enough stiffness to permit setting on a foundation and adjustment for elevation with a minimum of hardware. Often jackscrews are provided for elevation adjustment opposing the downward force of the anchor bolts. When jackscrews are not provided, small subsoleplates and shims or wedges can be used (Fig. 10).

When the amount of permanent-type supports exceeds 50 percent of the soleplate area, the use of grout alone without supports may be necessary. While this system is widely used, it is not generally recommended because it is entirely dependent upon the quality of the grout material and its placement. Experience has proved that while the grout-only method can result in cost savings during installation, its use must be carefully weighed.

5. Horizontal Forces. Horizontal loading and operating forces resulting from operating the machine or from expansion of long soleplates with radical temperature differential relative to the foundation should be considered.

While it is assumed that normal machine loading will be in a vertical direction and is transmitted to the foundation by the supports and grout, operating horizontal forces are also transmitted through the soleplate.

Long soleplates should not be exposed to heat which can cause misalignment or heaving of the plate, particularly when its horizontal expansion is restricted. Short soleplates with expansion joints between them may be required if heat is involved.

Horizontal anchors used with soleplates include cast-in soleplates with which the horizontal forces are transmitted directly to the grout (Fig. 11). The different types of grout keys and posts are shown in Figs. 12 and 13.

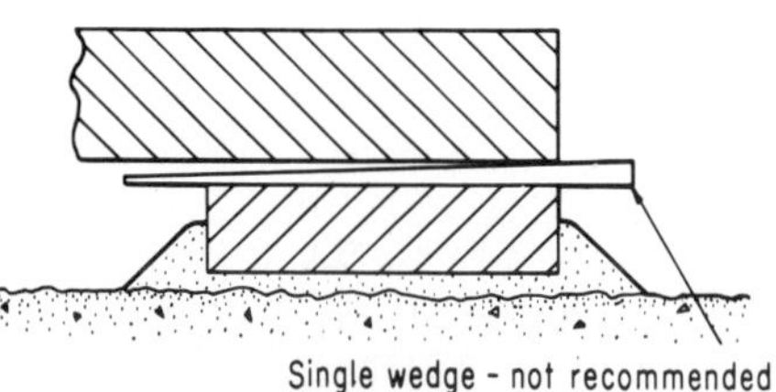

FIG. 8. Single wedges are not recommended for permanent support of a machine.

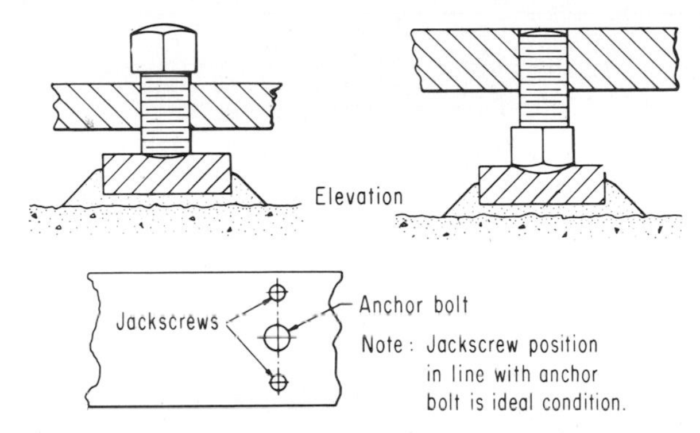

FIG. 9. Jackscrews are used in some installations for adjusting elevation or leveling.

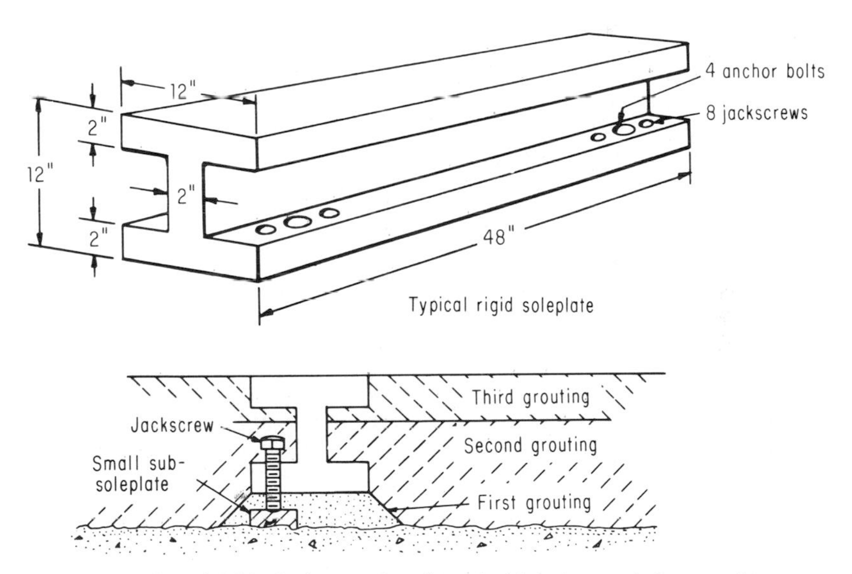

FIG. 10. Elevation of rigid soleplate can be adjusted with jackscrews before grouting.

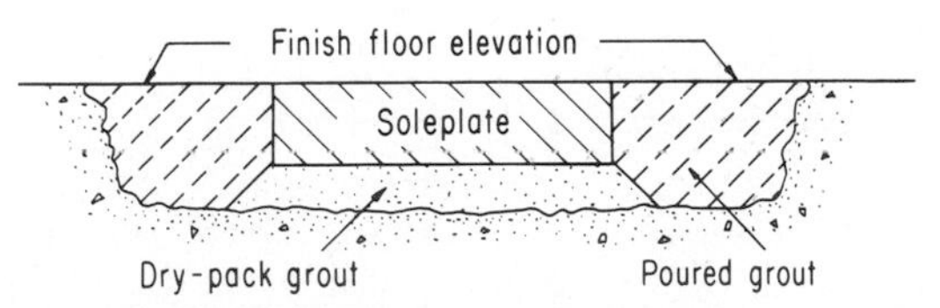

FIG. 11. Poured grout provides horizontal as well as vertical support to machine.

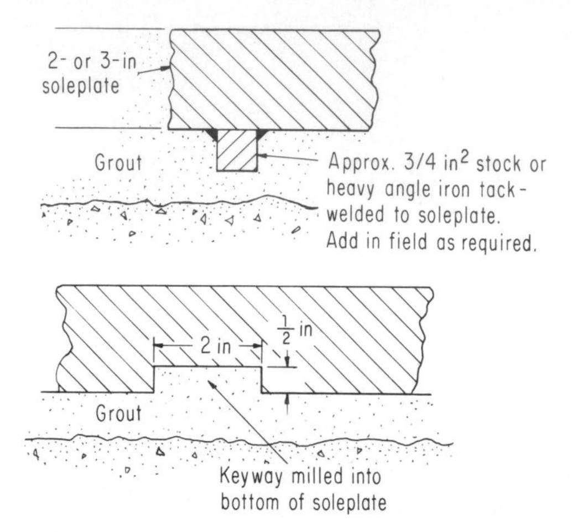

FIG. 12. Grout keys can be installed to transmit horizontal forces to grout.

The care and time applied to support a machine properly provide dividends in longer operating life and greatly reduce maintenance problems.

SOLEPLATE GROUTING

6. Important Considerations. A number of factors must be taken into consideration to assure effective soleplate grouting on concrete foundations. At the outset, it is important that the soleplates be properly positioned and the anchor bolt tightened prior to installation of the grout.

Grout is the filler of plastic consistency which is forced into the void between the machinery base and the main foundation upon which the equipment is to

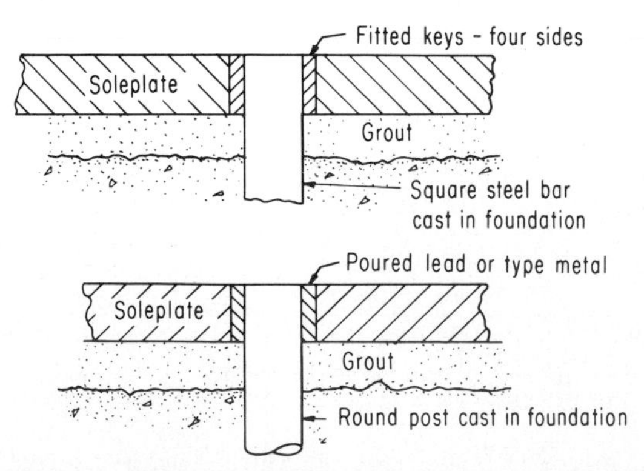

FIG. 13. Posts cast into foundation transmit horizontal forces to foundation.

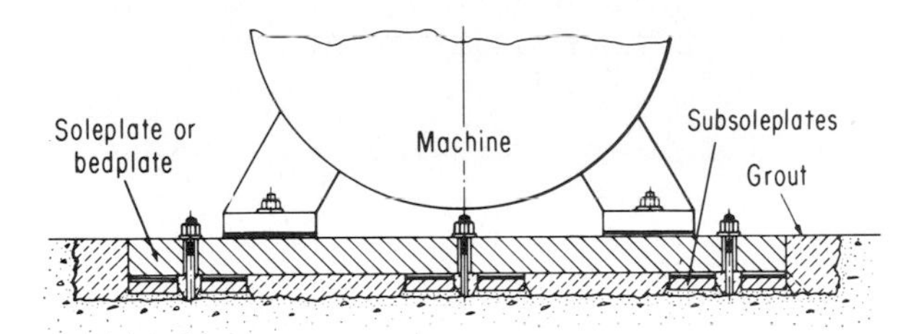

FIG. 14. Grout completely fills space between soleplate and foundation, providing additional support for machine.

operate. It should fill all the air spaces and cavities before it sets or solidifies and becomes an integral part of the principal foundation (Fig. 14). The grout should also be a chemically inert, high-strength, nonshrinking type. Grouting encompasses the materials and processes of application when portland-cement-type mixes are used under machinery soleplates on concrete foundations.

Unless otherwise specified, applications of a plain portland-cement–sand-mix grout 1 to 4 in thick are suitable for noncritical grout areas. In critical grout areas requiring minimum shrinkage, an additive is used with the plain mix. The selection of a suitable additive should be the subject of careful study to make sure that the mixture will suit the job.

When the configuration of the soleplate with its supports and the foundation makes it feasible, an application of dry-pack grout will give minimum shrinkage. Poured grout may also be used in critical areas when use of the dry-pack method is impossible because of physical limitations.

The concrete-foundation surface must be clean and rough to permit a good bond and increase resistance to movement caused by horizontal forces. All dirt, oil, grease, rust, mill scale, concrete dust, imperfect cement, and paint should be removed from the concrete-foundation surfaces and the soleplate. A fiber brush is best suited for cleaning concrete surfaces and a steel-wire brush for metal. The concrete-foundation top always should be chipped down to remove laitance and should be roughened by using a power-driven chipping hammer or handtools. A light film of rust on the soleplate may improve the bond, but a heavy coating of flakes or scale should be removed.

The concrete foundation should be kept wet for at least 24 h, and preferably 48 h, before starting to grout. This precaution will prevent water from being drawn from the grout and hindering the hydration of the cement, the formation of grout plugs which could stop the flow of the grout, and stresses between the foundation and the grout.

The surface of the foundation can be kept moist by using wet burlap or by continuous soaking with a hose. All free water must be removed from the foundation before placing the grout, or the bond and strength will be adversely affected. Compressed air containing no entrained oil can be used to remove the excess water.

Deep depressions in the foundation top and anchor-bolt sleeves should be filled with one part cement and two parts sand and allowed to set and cure for several days before the soleplate's general area is grouted (Fig. 15).

7. Forms, Grout Mix, and Placement. Strong, well-braced, and secure steel or wood forms should be used to prevent bending, slipping, or leakage of grout. On

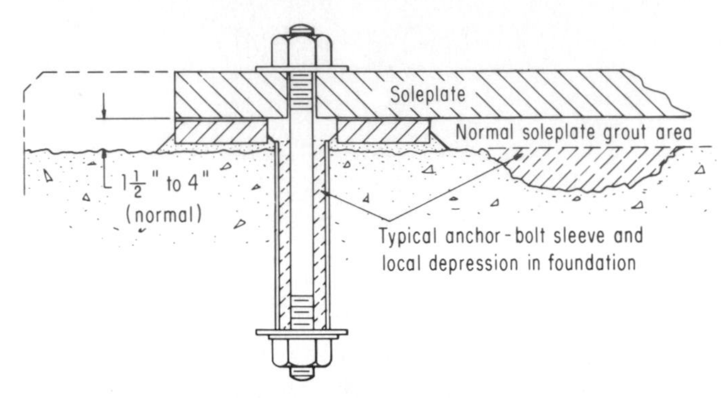

FIG. 15. Grout is placed in anchor-bolt sleeves and depressions and is allowed to set before grouting soleplate.

certain installations it may be necessary to install permanent-type grout forms since they may be inaccessible after the machine is placed on the foundation (Fig. 16).

A controlled amount of aluminum powder mixed with one part of portland cement and one or two parts of well-graded, clean, dry sand is one type of grout used to counteract shrinkage. Other additives and premixed grouts are also commercially available. For dry-pack application, only enough clean water should be added so that when squeezed in the hand, the sample will not crumble or slump when pressure is released. This type of grout must be cast within 45 min after water has been added, or the benefit of the aluminum is lost.

The dry cement-sand-aluminum mix with additional water will provide a poured grout. The water content should be held to a minimum to avoid shrinkage, and it should be poured within 45 min.

Regardless of the grout mixture used, the water-cement ratio should be kept at or below 0.5 to reduce shrinkage and prevent low strength. If the grout must be made more flowable, the water content can be increased as long as the water-cement ratio remains below 0.5 by weight (Fig. 17). No water should ever be

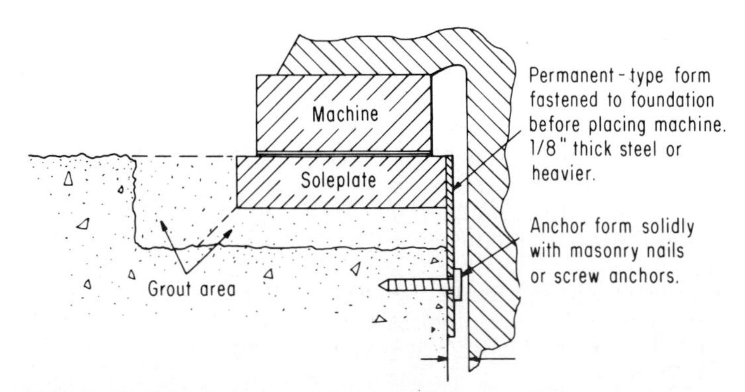

FIG. 16. Permanent forms may be used when machine arrangement does not permit normal wood forms for grouting.

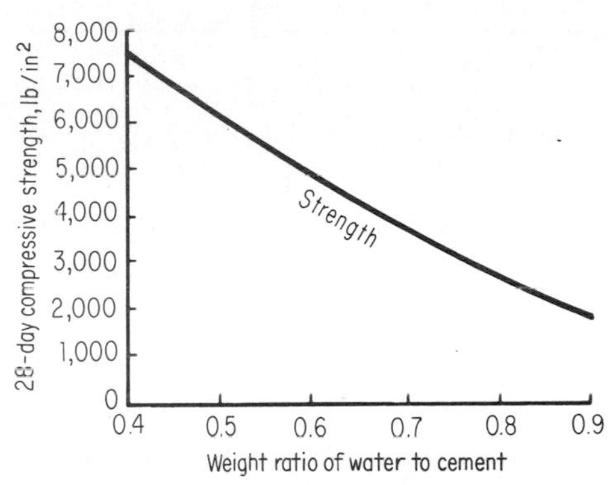

FIG. 17. Concrete grout is strongest when a low water-cement ratio is used when mixing grout for installation.

added to a mix after the initial mixing. If it appears necessary to add water to increase the flowability at a later time, that mix should be discarded.

Two common methods are used to place the grout between foundation and soleplate—dry pack and pressure grouting. Neither should be applied at temperatures below 10°C (50°F) or above 32°C (90°F). At the lower end of the temperature scale, warm water may be used to advantage while iced water may prevent setting too fast at higher temperatures.

The dry-packing method, which is limited to easily accessible areas, is accomplished by ramming the grout material in place with wooden paddles shaped for the particular job. To prevent generating sizable upward hydraulic jacking forces and throwing soleplates out of alignment, the grout should be tamped or, as the name implies, packed rather than pounded in place.

In casting, the excess water of one ramming should be allowed to drain off before more grout is added and another ramming takes place. When packing under a soleplate, one side of the open space should be backed up and the grout placed in from the other side. After the space between the plate and the foundation has been filled, the backup block should be removed and the face of the grout rammed from the opposite side.

Where grout must be pumped or poured into place, sufficiently high forms should be built to hold the grout, flow it into the required area, and provide a static head pressure. Grout must be poured from one side of the soleplate and allowed to flow to the other side, thus avoiding air pockets and voids, or it can be poured from the center with the grout flow outward in all directions. After the grout sets for about 30 min, all air or extra water must be removed by rodding or by pulling plain welded link chain (½ in or larger) through the grout. Do not vibrate the grout or work the chains too vigorously, or settlement of the aggregate may take place in the grout. Holes must be provided in the soleplate to vent any natural air pockets and ensure full contact between soleplate and grout (Fig. 18).

Forms used with dry-pack grout can usually be removed immediately after the grout has been rammed, followed by finishing or pointing. Poured grout should set for 18 to 48 h at a temperature above 21°C (70°F) before the forms are removed. If the temperature is below 21°C (70°F), the forms should remain in place longer (Fig. 19). After the forms have been removed from poured grout, the surface should be pointed.

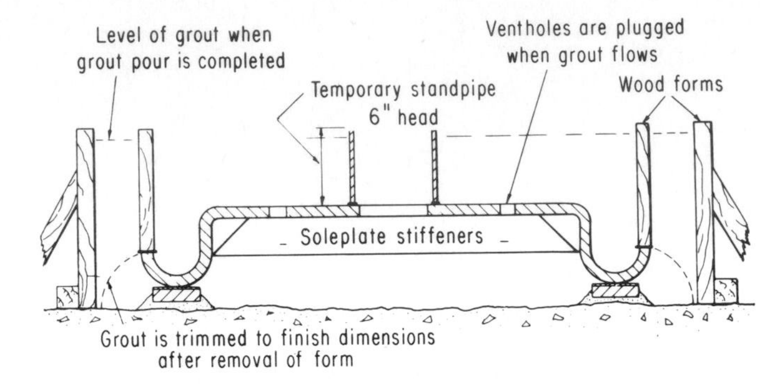

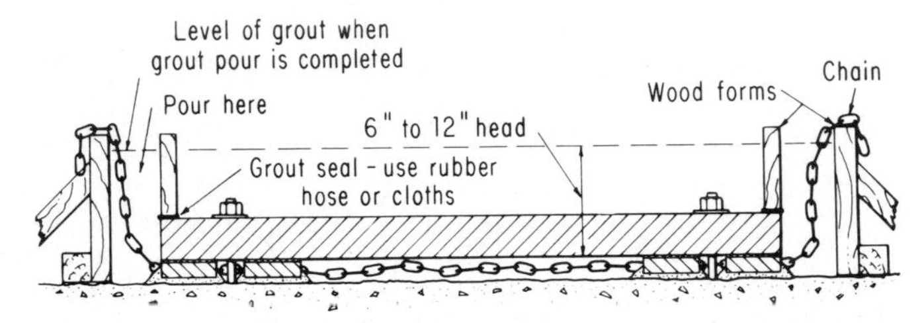

FIG. 18. Grout may be poured or pumped into high forms to fill space under formed or fabricated soleplates (top). Chains or rods are used to compact grout under flat soleplates.

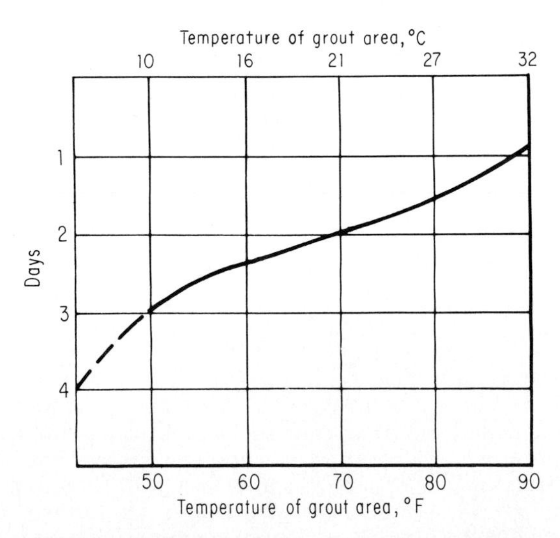

FIG. 19. Temperature in area of grouting determines length of time before forms can be safely removed.

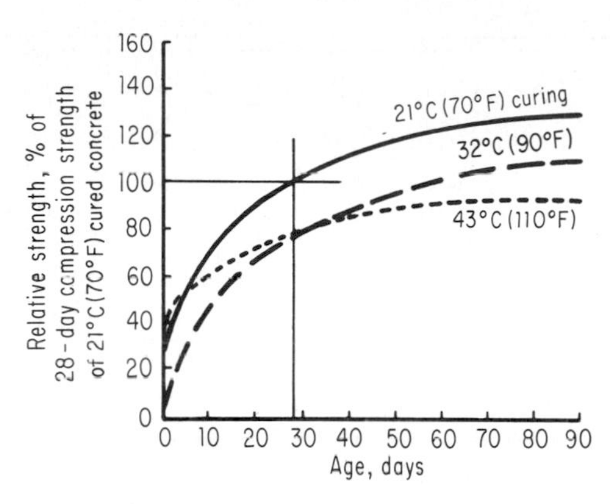

FIG. 20. Reasonable curing temperatures must be used to ensure grout strength.

8. Curing Time. Curing of grout should start as soon as possible after the initial set. Covering it with wet burlap for at least 7 days at temperatures of 10 to 32°C (50 to 90°F) is one of the best methods. An excessively high water requirement with resultant loss of strength prevails at temperatures over 32°C (90°F) (Fig. 20).

Concrete-sealing compounds are often used to provide final curing following a week or more of wet or water curing. Hydration of the grout of hermetically sealed fresh concrete will be arrested if the grout has a water-cement ratio below 0.4 by weight (4½ gal of water per sack). With grouts of higher initial water content, hermetic sealing does not arrest hydration but may slow it down.

Dry-pack or poured grout should set at least 7 days before the equipment is placed in operation, particularly if the machinery is subject to vibration.

ALIGNING HORIZONTAL MACHINE SETS

9. Positioning Machines. Proper alignment starts with one carefully positioned machine. The other machine or machines are then aligned with the first machine following precisely executed steps.

Flexible couplings are not intended to permit permanent misalignment. Any slight misalignment will reduce bearing life and cause other problems, even with good flexible couplings. Flexible couplings permit some temporary slight change in alignment or end play to allow for thermal expansion during startup or during some unusual momentary load conditions.

All machine sets should be carefully checked for alignment when installed, even when mounted on a permanent bedplate and aligned at the factory. Factory alignment can be disturbed during shipment, handling, or installation.

Either the driving or the driven machine is selected and its final position in relation to foundation centerlines, level, and foot plane is established. The type of unit or the local conditions will usually determine which machine is fixed, and this machine is generally the one that is the hardest to reposition. If the fixed machine is mispositioned in relation to the complete set of machines starting from an arbitrary position, it might result in having to revamp connections, slotting boltholes, or reducing bolt diameters. To minimize such problems the fixed unit is set with its tiedown bolts in the center of their respective holes and is shimmed if necessary so that it or units being aligned to it must be raised to obtain alignment.

10. Rough Alignment. Assuming that the coupling halves on the drive and the driven machine are of the same diameter, initial rim alignment is accomplished by laying a straightedge from one coupling to the other. The straightedge is held parallel to one shaft by holding it tight on the outside turn of the respective coupling. Adjust the shaft position of the unit being brought into alignment in both the horizontal and the vertical planes until the rim alignment is reasonably close and a dial indicator can be substituted for the straightedge for final alignment adjustment.

After coupling-gap readings are measured with a rule and angular alignment is adjusted so that the distances F_1 and F_2 (Fig. 21) are reasonably close ($\frac{1}{32}$ to $\frac{1}{64}$ in), dial indicators or micrometers can be used for final alignment purposes. This alignment is made in both planes.

Upon completion of the initial rough alignment, the following should be rechecked before final alignment:

1. The fixed machine should be in a position that will allow the machine being aligned to it to be moved within its normal mechanical limits to obtain correct alignment.
2. The axial distance between the coupling halves should be as specified on the assembly print, taking into account the type of coupling involved (free-floating or thrust-transmitting) and the desired operating position of free-floating shafts. The axial positioning of certain types of machines is critical since expansion of parts, transmission of thrust forces, and adequate clearance of parts must be considered before the machine is located.
3. There must be no abnormal stresses acting on the machines being aligned. Stresses or strains could be caused by connected equipment, piping, etc. If checking indicates excessive stresses acting on the equipment being aligned, the item causing this stress must be corrected.
4. Any piece of machinery can be distorted by improper distribution of weight on the machine feet. This general subject is called foot plane. An example would be a machine supported by four feet having one short leg, or having one support on the base out of plane with the other three. When this machine is bolted down to its supports, the machine will be distorted or out of foot plane.

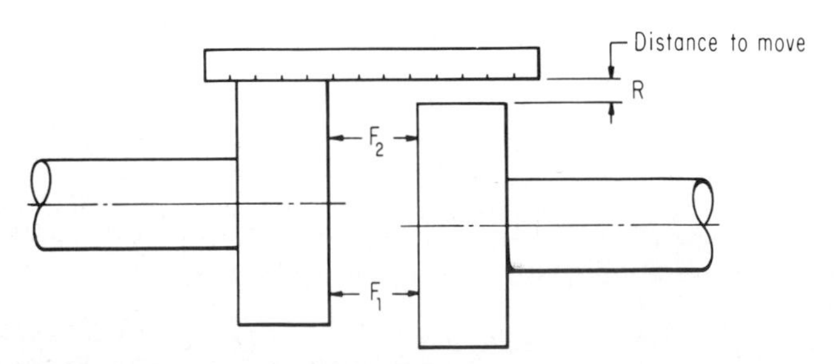

FIG. 21. Initial alignment may be made with a rule to approximate closely the correct position of machines.

This condition, as well as abnormal stress introduced from external sources, could lead to extreme damage to the machine through internal rubbing, vibration, or broken parts. Stresses from all sources must be discovered and rectified before the alignment of the shafts is attempted. In addition, if foot plane or abnormal stresses are overlooked, noise, vibration, and excessive wear can develop later.

11. Checking the Foot Plane. While some special types of machines require a specified loading on each machine foot, most require only normal equalized foot loading. On smaller-sized machines, such as motors, generators, and pumps having a rigid frame or casing, it is important to ensure that each foot conforms to its individual support, that the weight is properly distributed over the total foot area, and that the machine is solidly bolted down. To check for equal support on each foot, one holddown bolt at a time is loosened. A dial indicator or feeler is used to check whether any foot rises off its support as each bolt is loosened. Only two adjacent feet on a four-foot machine need to be checked. If movement is found when the holddown bolt is loosened, the shims under that foot are adjusted until no movement is observed as the bolt is again tightened and loosened.

On machines requiring a measured amount of loading on each foot, crane scales or calibrated hydraulic jacks can be used for weighing the individual foot load. The feet are then elevated or lowered to obtain proper loading.

On many types of machines a temporary support can be used to advantage to equalize foot loading. On a four-footed machine casing, the casing is allowed to rest on the two feet at one end of the machine while at the other end of the machine a temporary support, such as a jack, crane, or wedge, is used on the vertical centerline of the machine. The machine is raised with the temporary support until the feet on its raised end are clear of the supports by a few thousandths of an inch. The relative gap between the two loose feet and their supports is then measured. Shims to make the relative spacing the same are added or removed under each foot. The temporary support is then removed and the loading on each support will be equal.

With the machine now solidly supported by its feet and bolted solidly through its soleplates or bedplates to the foundation, shaft alignment can proceed. The soleplates on new installations will usually be grouted following initial alignment but prior to final alignment.

12. Preparing for Final Alignment. Final alignment is merely the refinement of the shaft positioning to the extent required to ensure a proper operating unit. As the alignment must be accurate to thousandths of an inch, dial indicators and micrometers should be used for the final shaft positioning and alignment checks. Instruments should be carefully checked to be sure they are operating accurately.

The indicator must be mounted solidly. Conventional indicators are mounted so the stem is absolutely radial to the shaft when making rim checks. They should be parallel to the shafts when making gap or face checks.

After the indicator has been mounted, slide a feeler or a piece of shim stock under the button and then remove the feeler. The indicator should return to the same reading each time this is done. The indicator button should rest on a clean, smooth surface.

When reading the indicator, one eye is used, making sure that the eye is in the plane formed by the needle and the needle shaft to avoid introducing error in the reading. When using a mirror to read the indicator, it is particularly important to have your eye in the correct plane and the face of the mirror perpendicular to that plane.

The closest reading possible should be taken. On most indicators this reading is to the nearest one-quarter of a thousandth of an inch.

Before proceeding with alignment it is necessary to differentiate between right and left sides of a machine set for record purposes. The set is arbitrarily viewed from the driving-machine end. For example, a motor-generator set would be viewed from the motor end looking toward the generator.

13. Correct Final Alignment. Consider two rotating shafts connected by a coupling. The shafts are defined as being in correct alignment when the relative position of the two shafts is such that during normal operation no flexing will occur at the coupling.

To accomplish correct final alignment during normal operation the shaft axes at the coupling must be in perfect alignment. If the shaft axis on one shaft is extended in the direction of the mating shaft, the two shaft axes must coincide at the coupling.

In order to show the procedure for accurately checking shaft alignment, the dimensions used in the examples are focused toward a "perfect" zero rim and face alignment. However, perfect, or zero, alignment may not be the ideal goal with the machines cold and at a standstill. Although such an alignment is satisfactory in the majority of cases, factors introduced as the machines are started and when they reach normal operating conditions must be considered. Factors for consideration include foundation and machine dimension changes resulting from thermal changes, radial shaft loading such as experienced on gear or sprocket shafts, change in relative shaft position resulting from change in oil-film thickness, type of couplings involved, or rotational speeds. Compensation for the effect of these factors is made through specifying an offset when the machines are cold and at standstill to achieve a correct final alignment.

Once the procedure for checking alignment is thoroughly understood, it can be readily applied to establish the proper final "cold" alignment position of shafts.

On some machines requiring an offset cold alignment, it is easier to align the shaft to a perfect, or zero, cold alignment and then, after the alignment is complete, change bearing shims or machine shims to the exact amount required to achieve proper final alignment under operating conditions.

14. Making Final Alignment. To make the shaft axes coincide at the coupling point, it is necessary to use an alignment method that will cancel out any conditions that could lead to an inaccurate alignment, such as irregularities of surfaces, end float of the two shafts, or eccentric surfaces. In all cases when taking measurements to check alignment, the measurements must be taken from points that are solid to the shaft.

Two check procedures necessary to prove correct alignment are the rim-alignment check and the face or coupling-gap check.

a. Rim-alignment check. Analyzing the results of the rim check determines how much one shaft is offset in relation to the other shaft. The term "rim check" is used because quite often the machined outside surface of the coupling hub or coupling rim is used as a convenient measurement surface. This check also shows how much one shaft must be moved to make the shaft axes intersect under the indicator. After readings are taken in both the horizontal and the vertical planes, one shaft is moved a measured amount to a new position to make the new shaft position parallel to the original position (Figs. 22 and 23).

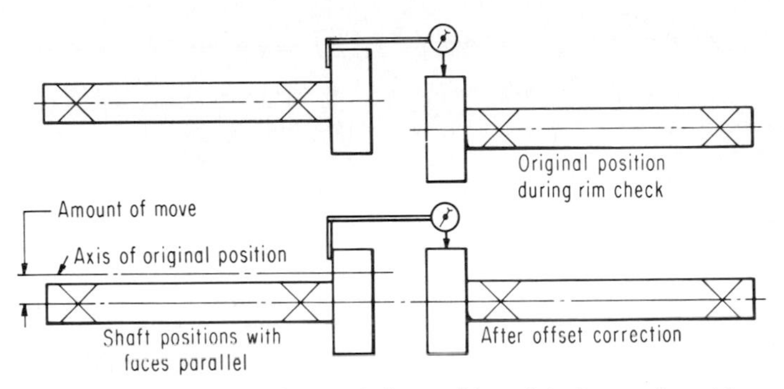

FIG. 22. Rim check indicates relative position of shaft axes of machines with parallel shafts.

b. Face-alignment or coupling-gap check. This check is determined by taking distance measurements between the coupling halves or hubs in both the horizontal and the vertical planes. When the measurements are evaluated, it can be determined how much one shaft must be moved so that its axis becomes parallel to the other (Figs. 24 and 25).

The combination of the rim and face checks makes it possible to evaluate the perfection of alignment attained and provides a basis for predicting further adjustments to accomplish correct alignment. Neither check alone is capable of proving the accuracy of alignment. The status of alignment, therefore, can only be determined when both a rim check and a face check are made and both these checks are made in both planes.

After each alignment check, correction is made by moving the machine that is being aligned in relation to the fixed machine.

15. Rim Check. A dial indicator is mounted on one shaft or coupling half and adjusted so that the indicator button reads radially on the periphery of the coupling half on the adjacent shaft.

An identification mark should be placed on the coupling at the point where

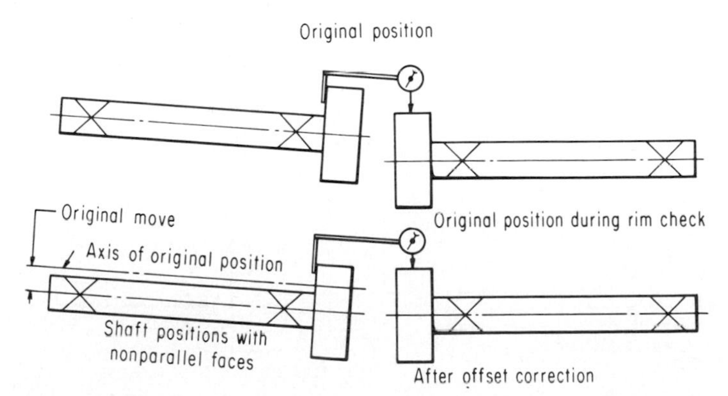

FIG. 23. Rim check also indicates relative position of shaft axes of machines with nonparallel shafts.

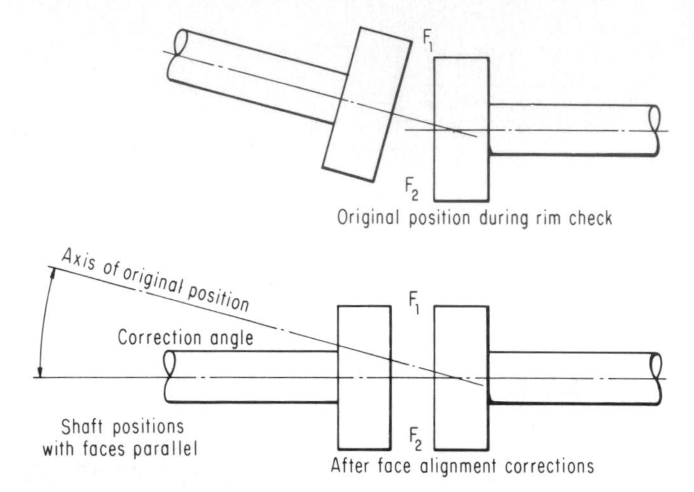

FIG. 24. Face check indicates angular correction needed to make shafts parallel to each other.

the indicator button is resting when the indicator is read. All alignment-indicator readings must be taken with the button on the same mark. To obtain the necessary rim readings at 90° intervals, the two shafts must be rotated the same amount prior to each reading. By using this method all irregularities such as surface imperfections, coupling periphery runout, or even a bent shaft are canceled out and reliable alignment readings can be obtained. This statement is based on the fact that the distance between the point on which the indicator button is resting and the shaft axis will always remain constant, as shown by distance D_1 in Figs. 26 and 27.

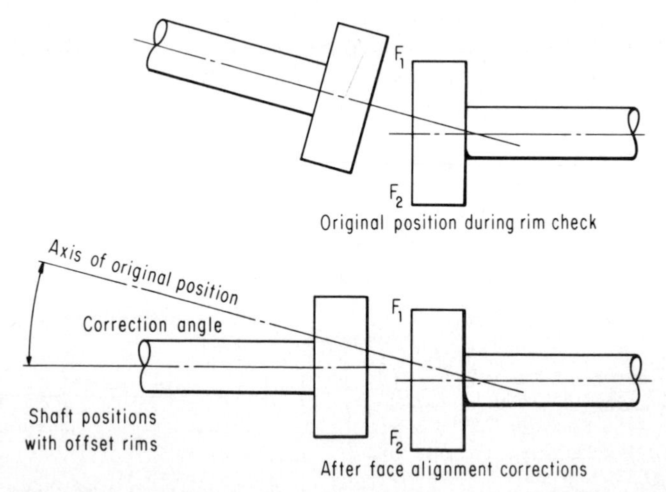

FIG. 25. Angular correction needed can be determined before shaft axes are in perfect alignment.

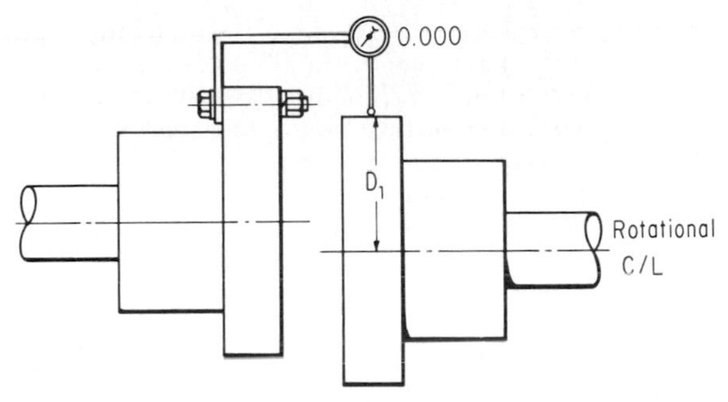

FIG. 26. Setting indicator to zero in top position simplifies making rim-check calculations.

The indicator is mounted as shown, with the button at the top of the coupling and adjusted to zero. Both shafts are turned 180°. Since the reading obtained at the bottom is plus 0.010 in, the centerline of the shaft on which the indicator button is resting is 0.005 in lower than the adjacent centerline (Figs. 26 and 27).

If in the same example we had started with minus 0.003 in on top, then the indicator would have read plus 0.007 in on the bottom. To determine the centerline offset, or amount of eccentricity, we must algebraically subtract one reading from the other, that is, +0.007 in subtracted from −0.003 in = 0.010 in.

This result of 0.010 in must then be divided by 2 to obtain the relationship of one axis to the other; so 0.010 in is divided by 2, and 0.005 in is the offset. When an indicator is used and the subtractions are complete, the plus or minus signs can be disregarded in further calculations to determine the direction of offset. As most indicators move in the plus direction when the button is pushed in toward the indicator, the shaft on which the indicator button is resting will be offset toward the direction of the larger indicator reading. The indicator should, however, be checked for the possibility of reverse reading.

The example illustrates a rim check in the vertical plane using the top and bottom readings. A rim check must also be made in the horizontal plane to obtain left and right readings. Actually the two checks can be combined so that only one full revolution of the two shafts is necessary. The indicator is generally zeroed on the top and the shaft marked. Both shafts are then rotated in the same direction 90° and a reading is taken. Again both shafts are rotated 90° so that the indicator and the mark on the other shaft are on the bottom. Another reading is taken, and the same procedure is followed for the remaining side (270° position) as well as again at the top. The second top indicator reading (360° position) should again be zero. If the indicator does not return to zero, something has happened during the check that makes the check invalid, and the rim check must be repeated.

A complete rim check might show: top 0.000 in, right −0.008 in, bottom −0.015 in, left −0.007 in. Since the

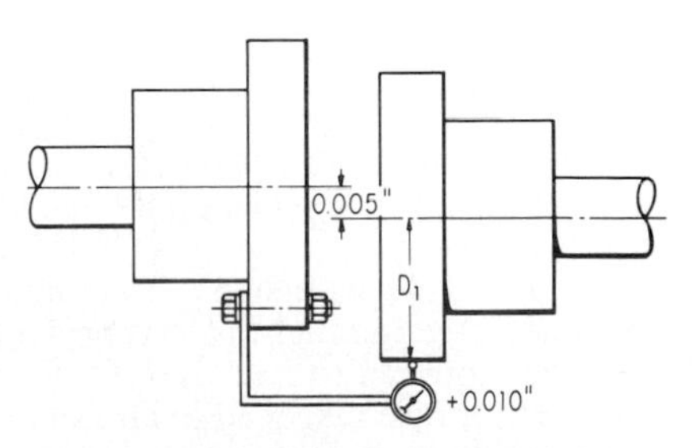

FIG. 27. Both shafts are rotated 180°. New reading is divided by 2 to obtain shaft displacement.

sum of the readings in each respective plane is equal, the readings are valid.

Considering the vertical plane, the top 0.000 and the bottom −0.015 are algebraically subtracted to obtain 0.015. Dividing the difference by 2, the shaft centerline offset is 0.0075 in in the vertical plane. The shaft that is against the indicator button will therefore be 0.0075 in higher than the other because the larger reading is on the top. Considering the horizontal measurements, 0.008 and −0.007 are algebraically subtracted to obtain 0.001 in. Dividing 0.001 by 2, the centerline of the shaft against the indicator button is 0.0005 in to the left of the other centerline since the left reading is larger than the right.

16. Face Alignment. Two methods are available to make this check. The method applied will depend on the tools used. (1) On small coupling gaps, feeler gages or snap gages can be used and on large coupling gaps, inside micrometers. (2) Dial indicators can be used on large or small coupling gaps.

The object of a face-alignment check is to obtain a measurement of the angular relationship between the shaft axes so a correction can be made to make them parallel. Measurements are taken between the coupling hubs. Readings should be taken at a uniform radius from the shaft centerline. The readings should be taken on the largest diameter possible to obtain the most accurate check of angular relationship. From Fig. 28 it can be seen that if distance F_1 were made equal to distance F_2, the alignment of the two axes would not necessarily bring them parallel to each other because the face of the right coupling is not perpendicular to its axis.

To eliminate the effect of this condition, the coupling gap is measured at the top and bottom for a reading in the vertical plane as well as at 90° from the top on both sides for readings in the horizontal plane. Both shafts are then rotated 180° and the gap at each 90° position is again measured. Using this method will also cancel out any end float which may occur when rotating the shafts.

The following are typical data that might be obtained:

	Top, in.	Bottom, in.	Right, in.	Left, in.
Initial or 0° reading	1.000	1.010	1.002	1.002
After rotating both shafts 180°	0.980	1.010	0.990	1.006
	1.980	2.020	1.992	2.008
Average	0.990	1.010	0.996	1.004

The coupling gap in this case is therefore open on the bottom 0.020 in more than at the top, and open on the left 0.008 in more than on the right.

17. Two-Indicator Method. Two indicators rather than one are used when making a face-alignment check to cancel end float of either shaft. The two indicators must be mounted to read 180° from each other on the same surface of the coupling. They must also read in an axial direction parallel to each other and parallel to the shaft centerlines.

Consider first a face check in only one plane using one indicator mounted on the top of the coupling and one on the bottom of the coupling (Fig. 29). The indi-

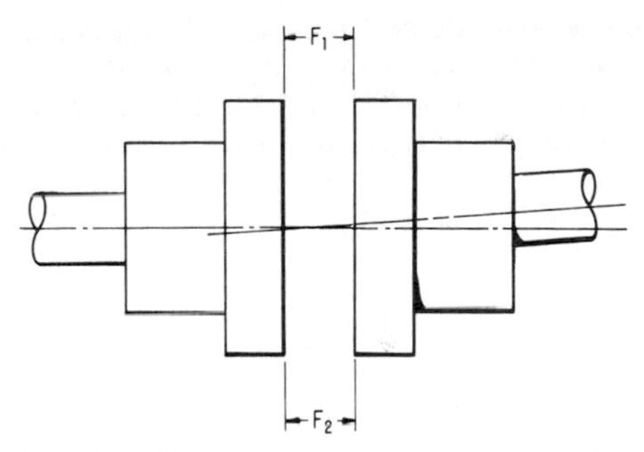

FIG. 28. Coupling faces may not be perpendicular to axes. Readings properly taken will cancel possible error.

cators are set on zero. Both shafts are then rotated 180°. The top reading in this case might be +0.002 and the bottom +0.014. These readings must be algebraically subtracted and the result divided by 2 since two indicators are used. Top 0.002 − bottom 0.014 = −0.012 in. As in the case of the rim check, the minus sign is ignored when making readings. Dividing 0.012 by 2 gives 0.006, the amount of misalignment. The indicator with the larger reading will designate the closest point.

The same procedure can then be used to obtain the face data in the other plane.

To be sure the indicators were not bumped during the check, the shafts should be rotated back to their original positions. If the indicators were set to read the same in the original position, they must read alike in the final position, or the check is not valid. End float may cause a difference between the original two like settings and the final readings; however, the two final readings must be the same. The foregoing method of working one plane at a time is the easiest. Setting the indicators on zero, then rotating the shafts 180°, and then taking the readings does not require any recording of figures during the check and makes the computations very simple.

If, however, it is inconvenient to rotate the shafts or the rotation of the shafts for some reason must be held to a minimum, a face-alignment check can be made during just one full turn of each shaft. The shafts must be stopped at the 90° points to take readings. A third indicator could be used to take a rim check at the same time the face readings are observed. Typical results of the two-indicator one-turn face check might be as follows.

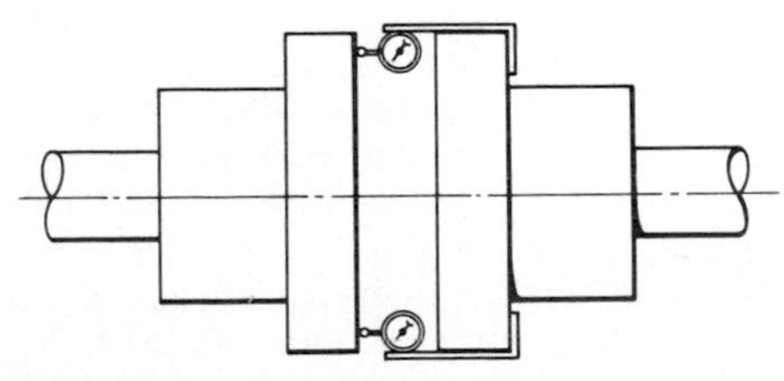

FIG. 29. Two indicators are mounted 180° apart, but mounting for face check depends on make of indicator and attachments required for coupling type and spacing.

The indicators are set at zero at the top and bottom, and recorded as shown by reading 1. Both shafts are rotated 90° and indications recorded, as shown by reading 2. Again the shafts are rotated 90° and the second set of top and bottom readings are recorded, as shown by reading 3.

The shafts are rotated another 90° and recorded as shown by reading 4. The shafts are again rotated 90° back to the original position to check the indicators.

	Top, in.	Bottom, in.	Right, in.	Left, in.
First reading	0.000 (1)	0.000 (1)	−0.002 (2)	−0.008 (2)
After 180° rotation	0.000 (3)	+0.012 (3)	+0.008 (4)	+0.002 (4)

Now to check for the amount of misalignment, the readings at each position must be algebraically added. Then the sums for the respective positions 180° from each other are algebraically subtracted and the answer is divided by 2 to obtain the amount of misalignment.

	Vertical plane		Horizontal plane	
	Top, in.	Bottom, in.	Top, in.	Bottom, in.
First reading	0.000 (1)	0.000 (1)	−0.002 (2)	+0.008
After 180°	0.000 (1)	+0.012 (1)	+0.008	−0.002
Sum for position	0.000	+0.012	+0.006	+0.006
Subtraction	0.012		0.000	
Divide by 2 to obtain amount of misalignment	0.006		0.000	

The above shows the top open 0.006 in more than the bottom since the bottom sum is larger than the top. The right and left gap readings are equal; therefore the shaft axes are parallel in the horizontal plane since the subtraction is zero.

18. Recording Alignment. After the final adjustment of the machine position, a careful record of the shaft positions should be made, recording the results of both the rim-alignment and the coupling-face-alignment data. This information should be kept in the maintenance data file for the machines.

Having an exact record of the alignment that produced satisfactory operation of the machine is essential. In the event machine operation changes, vibration changes, or future maintenance inspections reveal unusual wear, alignment must be rechecked and restored to the alignment figures that gave good operation.

Emphasis is put here on "restoring the machines to the alignment figures that gave good operation." The general case with most classes of equipment calls for an exact alignment with as close to zero rim alignment and face alignment as possible. There are also many units that require a precise amount of offset and angularity between the shaft ends in a cold static condition. These units when at operating temperature and with expected shaft loadings will be in exact alignment when running. An accurate record of such offsets and angularity is essential, as this information is frequently developed to suit local conditions as well as manufacturers' recommendations.

Full bearing, shaft, and coupling life is assured and unnecessary downtime is avoided when care and precision are used in aligning machine sets. Such precision is particularly vital to modern high-speed machinery that is a part of an important process or a continuous plant operation.

19. Axial Float and Thrust. Sleeve-bearing horizontal motors normally are designed and built with a predetermined amount of free rotor float or available axial rotor movement. The amount of rotor float is controlled by shaft shoulders at the ends of the shaft journals and axial bearing surfaces at the ends of the bearing bushings. Depending upon the design of the particular motor, one of the bearings or both may be used to control the float.

The normal axial running position of the rotor in respect to its total available float will be at or near the midpoint. This running position is maintained by electromagnetic forces while in operation and is usually referred to as magnetic center.

The need for axial restraint for the motor rotor is therefore limited to momentary thrust loading during startup periods, or when the motor is deenergized and coasting to a stop. The axial bearing surfaces are more than adequate for these purposes and, in fact, on many drive applications continually demonstrate their ability to accept reliably light continuous loading imposed by the drive system. Nevertheless, to ensure reliability, a motor drive system should be designed to eliminate imposing axial load on the motor bearings.

As the forces acting to center the rotor magnetically are relatively weak, they can easily be overcome by externally imposed thrust forces that could cause motor-bearing damage. These external forces can often occur unexpectedly and are many times unpredictable in magnitude.

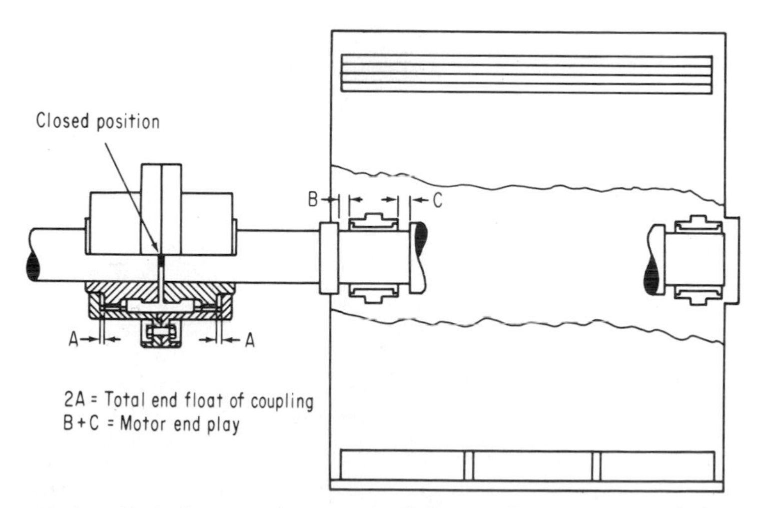

FIG. 30. Shaft alignment of motors with limited end-float couplings requires the same care as with machines without this feature.

One cause of unpredictable thrust loading can be through coupling misalignment which can cause the coupling hubs to pull together or to separate, thus imposing thrust loading.

Coupling locking due to torque can also produce undesirable and unpredictable thrust loads, as can coupling wear.

These causes of excessive thrust loading can be avoided by limiting the end float of the motor shaft so that it does not touch its axial bearing surfaces.

20. Limited End-Float Couplings. Such couplings are available in a wide variety of designs. Figure 30 shows a typical arrangement of clearances for the motor float and coupling. The coupling shown is a typical gear type. In this case a small button on the end of one shaft serves to limit the end float in one direction while the coupling separation is limited by the lips on the coupling cover.

Using a limited end-float coupling and with proper consideration given to the expected thermal-expansion movement of the driven equipment and the motor, the machines can be positioned so that the motor rotor in the running condition is at the midpoint of its end float. The axial position of the motor rotor under these conditions would be determined by the driven-machine thrust bearing. Normally the additional loading on the driven-machine thrust bearing will be negligible and at most it will not exceed the magnetic-centering force of the motor. In most cases the loading will be lower than what the driven machine would be subjected to if a limited end-float coupling were not used.

REFERENCES

1. J. Bowles, *Foundation Analysis and Design* (McGraw-Hill, New York, 1982).
2. Ronald F. Scott, *Foundation Analysis* (Prentice-Hall, Englewood Cliffs, N.J., 1981).

SECTION 17
PREVENTIVE MAINTENANCE

R. C. Blakey*

FOREWORD 17-2

PREVENTIVE-MAINTENANCE PROGRAM 17-2
1. Responsibility and Authority 17-2
2. Planning 17-4
3. Records 17-5
4. Spare and Renewal Parts 17-6
5. Communication 17-10

INSPECTION 17-10
6. Value of Preventive Maintenance 17-10
7. Deteriorating Environmental Conditions 17-11
8. Insulation Systems 17-11
9. Vibrations 17-14
10. Current-Collecting Devices 17-16
11. Stationary Elements 17-18
12. Rotating Elements 17-19
13. General Bearing Care 17-21
14. Care of Ball Bearings 17-24
15. Sleeve-Bearing Care 17-26
16. Testing 17-29
17. Load 17-35
18. Preventive-Maintenance Test Equipment 17-39
19. Safety 17-41
20. Preventive-Maintenance Checks 17-42

REFERENCES 17-43

*Manager, Apparatus Repair, Westinghouse Electric Corporation, Phoenix, Ariz. (retired).

FOREWORD

A study of preventive-maintenance programs in various industries over a period of years has indicated that in most instances those that failed were dropped as a result of frustration and problems involving systems or procedures, and not because of technical problems. Furthermore, those preventive-maintenance supervisors whose programs succeeded and who were further elevated in position or responsibility always understood the importance of these problems and spent the necessary time and effort in selling and arranging an effective system.

Since we are attempting to improve the overall scope of our effectiveness in dealing with preventive maintenance, the first step is to overcome the obstacles which produce failures in this endeavor. For this reason the first part of this section deals with these problems, while the balance of the section deals with the technical aspects of preventive maintenance.

Preventive maintenance encompasses all planning and action necessary to identify and rectify deteriorating influences or conditions before they advance to the stage where the initiative is removed from preventive maintenance and placed in repair maintenance. Logic and experience clearly indicate that preventive maintenance, properly applied, reduces repair costs and increases production as a result of reduced downtime.

PREVENTIVE-MAINTENANCE PROGRAM

1. Responsibility and Authority. Before any effective planning can be done, it is necessary to understand the system of operation and to know the degree of responsibility and authority that management assigns to a given level of supervision as well as the allocation of costs and budgets. It is to be noted that authority is not always commensurate with responsibility, and budgets do not always follow responsibility or authority.

Figure 1 shows a typical management organizational chart, but production supervisor, preventive-maintenance supervisor, and budget costs are not shown

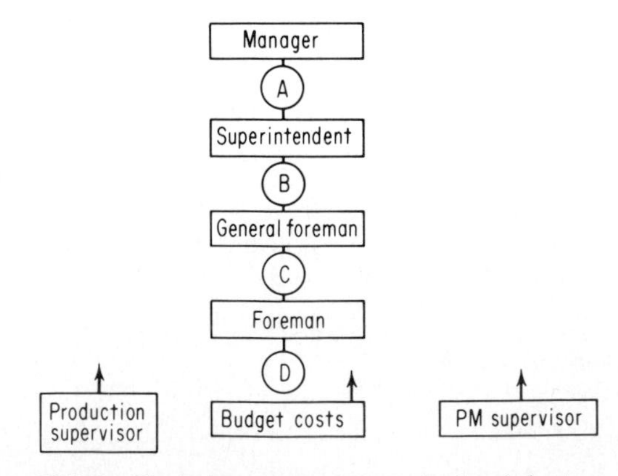

FIG. 1. Typical management organizational chart.

connected. Obviously, they must report to A through D, but not necessarily all to the same level.

To obtain a clear picture of authority, the following job descriptions are given.

a. Manager. Authority to make all decisions for the plant, having different levels of supervision reporting through line organization.

b. Production supervisor. Authority to establish and change machine production schedules.

c. Preventive-maintenance supervisor. Authority to schedule downtime on machines.

d. Budget and costs. Budgets and costs for all maintenance and associated costs may include downtime.

In this scheme of responsibility levels, consider the three following plants:

1. Company X. Utility, having three large turbine generators, each costing $20,000,000.
2. Company Y. Machine-tool builder with 10 departments each having many machine tools, and four departments having individual machines costing $350,000.
3. Company Z. Light manufacturing, many tools and departments, but few machines costing more than $35,000.

It is obvious that only the manager would make the decision in company X to shut down one machine and reduce capacity by one-third for preventive maintenance, or elect to run the machine with the possibility of a failure. The preventive-maintenance supervisor would probably report at A, along with the budget costs. Production (the load dispatcher in this case) might report to B or C.

In the case of companies Y and Z, it is doubtful that the manager would concern himself with the downtime of one machine, and the preventive-maintenance supervisor might report to any position, B through D. The manager might spend more time with production problems, and possibly production might report at A or B. The budget costs might follow production, preventive-maintenance supervisor, or direct to line supervisor.

It is in companies Y and Z where the system is likely to fail since, generally speaking, production is concerned with meeting machine schedules, while preventive maintenance is concerned about keeping machines in operating condition. Under certain conditions, their desires are diametrically opposed.

A typical example of the breakdown of a preventive-maintenance system starts when the preventive-maintenance department reporting at D recommends that the no. 4 motor-generator set be shut down for several hours during the next few days because of increased bearing temperature and vibration. The production department, reporting at A, promptly contacts top management, explaining the importance of meeting a certain schedule, and obtains approval to continue operation. One week later, the bearing fails, wrecking the shaft, pedestals, and bearings. The failure may result in costs 50 times the original necessary costs and a loss of production of several weeks.

Who is responsible for costs and loss of production, and who assigns the penalty? This condition is one of the major reasons for breakdown of the morale in many preventive-maintenance systems. If the production supervisor and the preventive-maintenance supervisor report to levels which are widely separated, the true facts may never be known to all involved. The fallacy of this system is twofold. First, upper management has not been completely sold on an effective preventive-maintenance program, and second, the preventive-maintenance supervisor and production should be on closer or the same levels of supervision.

Aggressive management is vitally interested in eliminating unnecessary and costly failures and in reducing downtime. Frequently management must be sold on such a system and then it must allow production and preventive maintenance to share the responsibility of reducing costs and downtime. There are several ways to reduce costs and downtime.

2. Planning. The system that is usually most effective is to have production and preventive maintenance and their associated budgets report to the same level of supervision. One supervisor can then weigh all probabilities and make a decision. "Before" and "after" facts are then easily obtainable and responsibility is quickly identified. The equal level of reporting sometimes may not be satisfactory in high-producing plants where one preventive-maintenance supervisor may have many production supervisors on the same level and meeting schedules cannot be arranged easily. The obvious failure is the overload of the preventive-maintenance supervisor.

One system which has proved to be quite satisfactory is allocating costs of breakdowns. If the preventive-maintenance department requests downtime on a machine, but the production department refuses to release the machine, then any downtime and associated costs are chargeable to the production department. On the other hand, if the preventive-maintenance department did not foresee the failure approaching, then they are charged with associated costs of repair and production loss. This arrangement, however, could produce unrealistic effects if the program is not applied honestly and conscientiously.

After the preventive-maintenance supervisor is settled in his or her reporting position, two other departments must also be considered. These departments are plant maintenance and plant engineer.

The reporting levels of these jobs can cause as much confusion and system breakdown as the chart outlined in Fig. 1. Again, it is necessary to consider the size of the plant and the mode of operation of the reporting positions. To clarify the responsibility, the following job descriptions are given:

Plant engineer. Responsible for the design of machines and their layout in regard to plant capacity

Plant maintenance. Responsible for repair of machines

Preventive maintenance. Responsible for a preventive-maintenance program

It will be noticed that all three jobs involve machines, and all must be concerned with continuity of operation.

One system which has proved to be effective is shown by Fig. 2. The preventive-maintenance supervisor reports to the plant engineer by direct-line supervision, and equally to production by dotted line. In the smaller plants, all three jobs may be assigned to one supervisor with a title of any of the three. However, in

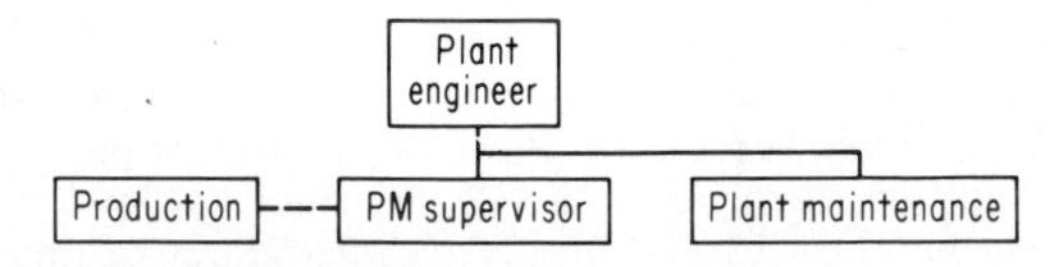

FIG. 2. System of reporting to plant engineer.

larger operations, three separate jobs may be involved and a decision must be reached as to where they report and their area of cross responsibility. For an effective job of reducing machine costs and downtime, all must be closely associated.

There is one great danger of combining preventive-maintenance supervisor with plant maintenance into one job. In times of reduced activity or profits, the tendency is to curtail preventive maintenance. Also, at times when maintenance is extremely busy, there is a tendency to borrow preventive-maintenance employees to do maintenance work, and thus reduce the preventive-maintenance work. Either case eventually produces ineffectual preventive maintenance, and this department deteriorates.

If preventive maintenance is capable of reducing the downtime of machines and increasing production in profit years, then it becomes increasingly more important in lean years. This fact must not be overlooked when trimming costs.

With regard to personnel, a preventive-maintenance employee requires in general considerably more overall knowledge, skill, and training than a maintenance-repair employee. Preventive maintenance involves the machine, control, drive, and protective equipment, and more judgment is necessary as supervision is not so readily available. The personnel is usually drawn from the maintenance staff.

Preventive maintenance is important to any manufacturing operation, but it is only a staff function and must not hinder production in the overall results. Interference to production must be at a minimum, and cooperation by preventive maintenance is a must. By careful planning, much preventive-maintenance work can be done while the machine is in full production. This time is best for observing commutation, vibration, heating, and temperatures. More stress should be placed on safety, however, when examining machines under operating conditions. In general the plant safety engineer should be made aware of all maintenance programs. His or her expertise may prove extremely valuable.

Another important point is that preventive maintenance must know the anticipated schedules of the machines and plan to make inspections when the production requirements are less stringent.

3. Records. Any effective preventive-maintenance system must deal with records and measurements under certain conditions. The fact that a given machine has 20-MΩ insulation resistance today means little unless we know what it has been under past conditions.

The questioning attitude is most helpful in preventive maintenance, provided information is recorded. These necessary records are, in fact, questions answered. A sleeve bearing may have an unusually large clearance, but if records indicate the clearances are not becoming larger, an important question has been answered.

Some of the larger electrical manufacturing companies have sample preventive-maintenance cards available, but these usually do not serve the needs of many preventive-maintenance departments without some revision. A ledger book is sometimes used to record data on a large important machine. Sufficient information must be available on the record card to serve the purpose of the card. If the machine is a spare, it should be indicated on the card, and more information is required to be certain it will fit another application. If the machine is not a spare, then the card may serve only as a ledger to collect information from previous inspections, and possibly costs.

Company policy will dictate whether the card should contain such information as tool number or company number, and whether certain charges are applied against this number. Spare parts or renewal parts available should be indicated, particularly on duplicate machines to reduce stocking of parts.

In large plants, the machine position should be shown to simplify its being located by new personnel and possibly for inventory control. Load capabilities or original specifications should be on the machine card. Was the original saw blade 14 in in diameter and now a 16-in blade is used? Was the original head of water 320 ft, but now the head is 380 ft? Was the original recycling rated at 15 min, but now the recycling is every 8 min? Such information is vital to an effective preventive-maintenance program.

Questions such as "Is the machine so old that the life expectancy of the winding is past; are maintenance costs so high that a new machine would be justified; have the prices of new machines increased to such a value that it would be economical to overhaul completely, or should the machine be scrapped?" should be recorded and recommendations made.

Inspection schedules vary greatly between different machines. By comparing inspection results with previous inspections, the schedules may be revised either upward or downward. Punched cards or computer systems of prescheduled inspections are used in some plants with good success, but good judgment of such prescheduling is a must. The oil of outdoor motors may have to be changed in May and October, which is an ideal computer-card system, while brushes may have to be examined every 3 weeks in high-production times, but perhaps only twice yearly in low-production times. Punched-card systems should be tied in with operating hours. Such scheduling requires not only good judgment but also some idea of the production schedules.

Some punched-card or computer systems become very comprehensive and may produce printouts showing:

1. Company number
2. Location of machine
3. Inspection items due at next inspection (coded items)
4. Whether inspections are to be made at downtime or at load
5. Estimated hours of inspection due
6. Total downtime for a given machine
7. Downtime costs or maintenance cost for a given machine
8. Inspections based upon time only, hours of production, or predetermined time
9. Employee who performed work
10. Hours spent and/or dollar cost (for item 9)

Obviously all these items could be maintained by hand without the aid of a computer.

Caution. Discuss thoroughly with the machine programmer before such a system is started. During periods of low profits, frequently the first item cut is the maintenance of records. It is perhaps better to have fewer records well maintained than massive records with large missing sections due to forced record curtailment.

4. Spare and Renewal Parts. Renewal parts are those parts that have a short life with respect to windings, which determine the life of the machine. Spare parts are those parts that must be replaced as a result of some accident or unusual operation. Renewal parts are usually stocked to keep the machine running while spare parts may be stocked to reduce downtime in case of some premature or accidental failure.

In general, a rather poor job has been done in the past regarding the stocking of spare and renewal parts. This situation has been the result of poor or vague recommendations by some of the manufacturers, lack of planning or knowledge by the purchasing or operating personnel, and a misunderstanding between manufacturer and purchaser.

Experience recalls many situations in which customers thought that they had spare parts available which could quickly get a machine back in service. However, when the spare parts were finally located, they were found to be worthless. A few examples are improperly marked or unmarked boxes; sealed bearings in which the grease had solidified because of age and heat; steel so badly rusted that it could not be used; coils which had been stored in an acid climate so that the insulation had deteriorated; high-voltage coils improperly packed for long storage so that the coils had swelled to such an extent that they could not be used. In one classic example, high-voltage stator coils had been stocked for years. When a failure occurred, it was found that the machine had been redesigned and rewound for different characteristics, but the spare coils were for the original design.

There is no magical formula to determine what spare parts should be carried in stock, but a logical approach to prevent the above examples is to consider the obvious advantages of maintaining spare-parts stock and weigh them against the following disadvantages:

1. Increased investment
2. Increased tax liability
3. Record handling
4. Inventory—audit counting
5. Proper storage facility and space required
6. Obsolescence of parts
7. Possibility of moving and shifting storage due to expansion or moving
8. Possibility of continuing stocking after original machine has been scrapped or retired
9. Deterioration of parts due to aging, improper storage, and poor atmospheric conditions
10. Duplication of spare parts

The factory accountant should be consulted for the first four items.

The reader should not construe that the large number of disadvantages indicate that spare parts should not be stocked. It is only urged that these factors be considered. The following considerations are also important for stocking spare parts.

a. Failure rate. Machines are more susceptible to failure during their early life of several months. This rate results from defects in manufacture, misapplication, improper operation or handling, and damage during construction, shipping, or installation. After this critical period has passed, the failure rate drops sharply until the approach of life expectancy, at which time the failure rate will begin to climb. The life expectancy varies greatly in machines—30 or 40 years is not uncommon on slow-speed motors with low voltage, small shock load, low ambient temperatures, little or no overloads, and ideal climatic conditions devoid of high humidity, acids, and fumes. The opposite of these ideal conditions reduces the life of the machine, in extreme cases to less than several years. Obviously, it is most important to have spare parts during the period of the highest failure rate.

b. Importance of machine. The relative importance of the machine must be carefully considered in the event of failure. Would the machine failure shut down a line or an entire plant? Or could other machines pick up additional load and thus keep production at satisfactory levels? Union contracts containing lost-time or downtime clauses should be carefully considered.

c. Availability of substitute or replacement motor. Consideration should be given to the possibility of a replacement or substitute motor. A 5-hp 1750-r/min three-phase 220/440-V dripproof 60-Hz induction motor with a belted load could easily be substituted. But a 5-hp 370-r/min two-phase 25-Hz 550-V flange-mounted explosionproof motor might be impossible to locate quickly from any supplier.

d. Number of duplicate machines. The original investment and tax liability of spare parts decreases correspondingly when a large number of duplicate machines are involved in one location, and the cost per machine for spare parts or a spare motor is reduced. Even for different company locations, a pool of spare motors might be economical.

e. Availability of repair. Many parts might be made or procured and the motor could possibly be rewound quickly in a metropolitan area; but in certain rural areas, parts stocks or repair facilities may not be readily available.

To determine the actual parts that should be stocked requires some knowledge of the repair facility available, repair procedures, and the design of the motor. Consider, for example, a crane motor that breaks a shaft occasionally. Some motors are designed with a core quill, keyway, and possibly locknuts to center the shaft. A new shaft could be made by a well-equipped shop by the time the motor could be taken down from the crane bridge and dismantled if a drawing is available. Such a shaft could also be stocked.

However, some motors are made with the shaft shrunk in the core laminations and in some instances the shaft cannot be pressed out; it must be drilled out. In this case, the new shaft must be made oversize and a standard stock shaft could not be used.

Most repair shops can rewind stator, rotor, or armature coils of the random- or mush-wound type. However, only the larger shops can wind, pull, or form the form-wound coils used in the larger equipment, and few shops will, or should, make coils for voltages 7200 V or above, where special processing or pressing, vacuum, and pressure are required.

Many shops can wind the field coils of the smaller machine, but the large field coils and particularly rotating-field coils require special techniques and equipment.

Many of the larger shops can rebabbitt sleeve bearings of the split type, where great care must be exercised in processing and final machining.

f. Auxiliary equipment. Frequently, auxiliary equipment is completely overlooked when consideration is given to spare-parts stocking. The failure of a small, special built-in field-rheostat operation motor might shut down a large important drive motor. This auxiliary equipment, costing less, probably would not have the same reliability as the main equipment, because of tighter specifications and a higher degree of quality control on the main equipment.

g. Obsolescence of machines. Some manufacturers will continue to furnish spare parts many years after the original machine was built; others, however, will not furnish spare parts after several years. This situation frequently creates severe problems if the motor has a special mechanical or electrical design where a standard motor would not be suitable.

h. Scarcity of materials. During times of national emergencies, or certain strikes, some materials become extremely difficult to obtain. One of the major items affected is copper wire. During periods of heavy competition, copper wire, drawn to size and insulated, can usually be obtained in less than 1 week. However, during certain critical times, copper wire may require several weeks to draw to size and insulate. Ball and roller bearings may also be affected, particularly for motors built outside the United States, which frequently use a size not readily available.

i. Parts that wear. Renewal parts usually consist of those items which wear and have a shorter life than the actual windings. The windings usually determine the life of the machines, and when new windings are installed, the life of the machine is renewed. Carbon brushes, brush springs, brush fingers, and bearings are parts that wear and must be replaced. The commutator or slip rings also wear, and frequently they have to be turned or ground, but under normal operating conditions there is sufficient material in these parts to last the life of the winding. Under abnormal operating conditions, the commutator, rings, or brushholders may have to be renewed during the life of the machine.

j. Records on spares. Accurate records must be maintained if spare parts or renewal parts are to serve their function. The best practice dictates the use of a master motor card, which lists the parts available and their actual location. The file location of the master motor card should be known to all who will have the responsibility of replacing these parts. The actual boxes should also be plainly marked and cross-indexed with the master motor card. Failure to follow this procedure usually renders the entire stocking program worthless. The master motor card must be kept up to date as present parts are used and new parts are ordered, received, and stored.

As a final word of caution concerning the records, there is a tendency to become lax during low-profit budget-cutting times. If the parts are worth stocking, the effort spent on maintaining records is also worthwhile.

Parts used on duplicate machines, or duplicate parts used on different machines, should be further cross-indexed, to keep stocking to a minimum.

k. Parts storage. When a failure occurs, the parts stocked must be in a usable condition, or all past efforts have been wasted. The boxes must be plainly marked. Paper tags, if used for this purpose, will deteriorate and eventually become illegible. Accumulated dust on top of wood crates will finally obliterate any markings on top of a wood crate.

Steel parts are subject to rusting and must be protected from elements producing this condition. Sludging is usually recommended.

Copper-brass parts are subject to corrosion in acid fumes and can be protected by a coat of grease or sludging.

l. Short-shelf-life items. Certain compounds may have a relatively short shelf life. These items must be dated and replaced as required. Sealed ball bearings also have a limited shelf life as the grease will eventually solidify.

Insulated coils should be stored in a clean, cool, but dry place. The insulation is already preformed; so a dry place is required since humidity must not saturate the coil.

For long-term storage, armature and stator coils, particularly high-voltage stator coils, should be mummified. This operation consists of wrapping the completed stator coil with special tape which will seal out moisture. The coil should then be clamped to hold its shape and keep it from swelling.

5. Communication. It would appear that the simplest problems of preventive maintenance would be intracompany communications, since all personnel within a company should be interested in maintaining machine operation at the lowest possible cost. However, at times this sharing of information will be one of the greatest problems. The old adage "the right hand doesn't know what the left hand is doing" still applies in some instances. Here are some examples.

The operator knows that the deep-well pump had a 200-ft head when the unit was installed, but the water level has dropped to 280 ft, the motor has been running hot, and bearing life has been reduced. However, when the repair purchase order is made out, nothing is said on the purchase order concerning the problem involved and the desired corrective action by the repair shop is not made.

The manufacturing processes were changed from 5 starts per hour to 15 starts per hour, and the motor rewinds have been more frequent. This information should be passed on to the repair shop so corrective action can be taken.

It is imperative that communications between the vendor and the actual operator are not interrupted within the company if the machine is to perform to utmost satisfaction.

All such information should be placed on the master motor card, and care should be exercised in passing on such information, both upward to purchasing and downward to the operator.

INSPECTION

6. Value of Preventive Maintenance. Preventive maintenance follows the inverse curve of diminishing returns. When excessive maintenance is applied to electrical equipment, the expense may become prohibitive; but if maintenance is neglected completely, the result is premature failure and costly repairs. It is therefore necessary to determine the optimum amount of preventive maintenance required to keep electric machinery in proper operating condition. Insurance brokers answer similar questions, such as "How much insurance is needed?" Preventive maintenance is insurance; the maintenance costs are the premiums and the elimination of machine failures is the policy return.

Each individual case must be examined and the following facts determined.

a. Maintenance versus machine costs. What is the relative cost of machine maintenance versus machine cost? The cost of maintenance of a $30 motor must be considered in a different light from the cost of maintenance of a $300,000 turbine if a bearing fails because of improper lubrication. Many plants accumulate maintenance costs and apply them as a ratio to total profits, total billing, or total costs. While this may serve as a guide to comparable industries, the only logical approach and the true effects must be considered as:

$$\frac{\text{Machine maintenance}}{\text{Machine cost}}$$

b. Maintenance versus downtime. To reach this decision on machine maintenance versus downtime, it must be determined what the effect will be if the motor fails. If there are a number of heaters in a building, the loss of a single unit might not be important; if, however, the building has only one heater, then the loss of the unit could result in a costly shutdown during cold weather. Union contracts may have a strong bearing on the downtime costs. Here the decision becomes

$$\frac{\text{Machine maintenance}}{\text{Downtime costs}}$$

c. Maintenance versus hazards. The relative cost of maintenance versus overall hazards should be considered. For example, on a ship's only propulsion motor it would be desirable to have excessive maintenance to prevent serious circumstances which could result from a failure on open water. This question is the most difficult to answer, but at times it must be answered by

$$\frac{\text{Maintenance costs}}{\text{Overall hazards}}$$

7. Deteriorating Environmental Conditions. In a general preventive-maintenance program, two functions must be considered:

1. Identify and correct or minimize the deteriorating condition.
2. Plan repair or replacement before actual failure occurs if the condition cannot be corrected.

The life of a motor is determined, barring mechanical accidents, by the windings, the key point of which is the insulation. Insulation as applied to a given machine can be divided into two major functions: ground insulation, which is the insulation separating live parts from the frame or ground, and turn-to-turn insulation, which separates or insulates turns or conductors of different potentials. An insulation failure of ground insulation produces a grounded winding, and an insulation failure of turn-to-turn insulation produces a short-circuited winding. When insulation fails, the decision must be made to patch the insulation, rewind the motor, or replace the motor. The normally expected life span of the insulation may be for many years (20 years is not unusual), while certain deteriorating conditions may reduce the useful life to a very short period of time.

It therefore becomes imperative that preventive-maintenance operators have a general idea of insulation systems, and particularly the deteriorating elements which produce premature failure.

8. Insulation Systems. Insulation, in general, can be divided into two major kinds: electrical insulation and mechanical insulation (see Sec. 11). The electrical insulation resists the flow of current while mechanical insulation gives the mechanical strength necessary. There are many different types of insulations, designed for one or more of the following conditions: mechanical and electrical stress, compression, tension, abrasion, vibration, extreme temperatures, moisture or high humidity, dirt, grease, oil, and acids or alkalies. To further complicate its selection, during manufacturing operations some must be easily shaped while others must hold their shape. Some insulations have a combination of electrical and mechanical properties, but in general most insulations have one or two strong points with several weak points.

When a machine fails and is to be rewound, preventive maintenance must be certain that the repair plant is informed of any particular deteriorating conditions so that the proper insulation system can be applied.

Ideal motor operation with respect to long insulation life consists of low operating voltages, low temperature, low speed, small shock load, few starts, reduced-voltage starting, and clean, dry cooling air. However, a motor designed for these ideal conditions will satisfy few modern-day applications. So special insulations

TABLE 1. Motor-Winding Temperature Ratings

Insulating class	A	B	F	H
Max hot spot	105	130	155	180
Hot-spot allowance, °C	15	20	25	30
Limiting observable temp, °C	90	110	130	150
Standard ambient, °C	40	40	40	40
Limiting rise over ambient, °C	50	70	90	110
Usual rating, °C rise	40	60	75	90

have been developed to withstand some or all of these deteriorating conditions. Obviously, the more severe applications will call for a premium design, and premium insulation at a premium price.

a. Rated temperature. NEMA has established and standardized temperature ratings and terms (Table 1).

b. Hot spots. Insulation systems have a hot-spot (hottest spot) temperature which must not be exceeded for the optimum life of windings. This maximum temperature is determined by the insulation class (Table 1).

c. Hot-spot allowance. Since this hot spot will be in the center of the windings and cannot be measured (except by embedded temperature-measuring instruments), a hot-spot allowance has been calculated and set up for each temperature rating. This hot-spot allowance is the theoretical difference between the temperature at the hottest spot and the temperature measured with a thermometer at the outside of the coil, or the temperature calculated by resistance test.

Subtracting the hot-spot allowance from the maximum hot spot gives the limiting observable temperature, which is the temperature actually measured by thermometer or resistance method.

d. Ambient temperature. The motor winding will absorb and will be at the same temperature (before running) as the surrounding air (ambient temperature), and NEMA has standardized this normal (and maximum) room temperature at 40°C. All motor ratings are based upon this ambient (or starting temperature) unless otherwise specified for special ratings.

e. Limited rise. Subtracting the standard ambient (40°C) from the limiting observable temperature gives the limiting rise over ambient. This change of temperature is the rise within which a manufacturer may rate a standard machine. If a manufacturer uses all this temperature at full load, there will be no margin for temporary or sustained overloads and the machine will have a service factor of 1.0 (100 percent load only). Therefore, most manufacturers use a usual rating of less than this value; so the machine may carry some overload for short periods.

f. Variations in ambient. Upon this basis all standard motors are designed. However, the greatest problem (uncontrolled variable) is the ambient temperature. The ambient will vary considerably from winter to summer and from north to south, while the motor basic rating is based upon the standard ambient of 40°C. If this ambient goes up to 70°C, the motor rating must be reduced, or the maximum hot spot will be greater than the insulation can endure and premature insulation failure will result. If the ambient is reduced below 40°C, theoretically the windings can carry more load. In actual practice, however, other parts must also be considered such as shaft, bearings, and current-collecting devices, which may limit the overload capacity, even though the ambient may be reduced.

To compensate for a higher than normal ambient, a different insulation class may be used. Assume a 60°C rise motor must be operated in an ambient of 70°C. Since the ambient has increased 30°C, the insulation class should be increased 30°C and Class H insulation should be used.

Note. If the insulation class of a motor is increased, the bearing-grease temperature must also be changed.

The addition of steam pipes, heaters, ovens, or more motors in a small space is a frequent reason why the room ambient might change from the original ambient and produce an insulation-deteriorating influence.

As mentioned previously, the hot-spot temperature is based upon optimum life expectancy. As the hot spot increases, the insulation ages much faster than the increase in temperature.

g. Altitude. Closely associated with ambient temperature is the elevation in altitude. NEMA standard motors are rated at full load up to 3300-ft (100-m) elevation. The manufacturer should be consulted for a possible reduction in rating above this altitude because of the reduced cooling effect of the ambient.

h. Part-time ratings. Special part-time motors such as hoist motors are not designed to run continuously, but only a certain number of minutes out of each hour. Such motors have an intermittent rating, and the nameplate will show how many minutes out of each hour the machine can operate at full load without exceeding the rating and thus producing a deteriorating effect on the life of the winding.

i. Frequent starting. Some operations require recycling or more than the usual number of starts per hour. Check the latest NEMA publications or the manufacturer, as special designs may be necessary for the application.

Excessive starting overheats the windings and reduces insulation life and may cause the cracking of rotor bars or rings in squirrel-cage motors.

j. Moisture. Absorbed moisture reduces the insulation resistance of windings and the insulation may rupture, allowing the line current to flow to ground, thus wrecking the windings; or the reduced insulation may allow a flash between different potentials of the winding, creating a short-circuited winding. Special insulations are available, usually at a premium price, which are completely or partially impervious to moisture absorption.

If a motor has become wet, the insulation should be checked before line voltage is applied, and a dryout may be necessary. Once a winding has flashed to ground or internally, the insulation is ruptured and repairs must be made before further operation can be permitted.

k. Dirt, grease, carbon, dust, etc. These foreign particles all reduce insulation levels on windings and absorb moisture, either of which may produce ground faults or short circuits in windings. In addition, the buildup of these foreign particles reduces ventilation and produces higher operating temperatures. If standard windings cannot be kept clean, special insulations are available which are impervious to the deteriorating action of dirt and moisture, etc.

In addition, if the ventilation is impaired by accumulation of dirt, a higher-temperature insulation may be required.

l. Acids. Acid or alkalies and their fumes attack standard insulations and particularly attack their structural strength. When this strength is reduced to a certain point, no reclamation of the windings is possible, and complete rewinding is necessary. Special insulations are available which are acid-resistant and in some cases acidproof.

Prior to actual winding failure, dipping and baking may forestall failure first by sealing and adhering all weakened insulation and second by sealing off the fumes from the windings.

m. Standard versus special insulation. In many applications normal conditions exist which are devoid of excess temperatures, acids, moisture, and other deteriorating influences, and hence special insulations are not required. However, standard insulations will not stand up under these deteriorating influences, and if such influences exist, special insulation should be used.

n. Vibrations and overspeed. These conditions also exert a deteriorating influence on insulations, but to a certain extent they can be overcome by use of a resilient or stronger insulation. Both conditions produce failure due to mechanical stresses.

When any machine fails and is to be rewound, preventive maintenance should investigate the possibility of special insulations to overcome this effect.

9. Vibrations. Vibrations affect the operation of some loads such as grinding and sawing, and hence cannot be tolerated. However, we are concerned here with vibrations emanating from the motor or the load that affect motor operation.

Excessive motor vibration is detrimental in several respects. First, the vibration tends to produce structural-insulation failure. The insulation is weakened; banding, blocking, wedging, and tying become loose; and coils are allowed to vibrate. The movement causes insulation to flake and wear, and eventual failures will occur. In early stages, dipping and baking will tighten all insulation, but in final stages complete rewinding may be necessary.

Vibrations produce sparking at current-collecting devices; this sparking, in turn, produces burning at the commutator or collector rings and is cumulative in its effect. The problem is thus magnified. Vibrations also tend to "harden and embrittle" copper, and the leads may eventually break. This breakage is particularly apt to happen on rotating windings.

Finally, vibrations cause premature bearing failure. Antifriction bearings "brinell," sleeve bearings will be pounded out of round, babbitt loosens in the shell, and the shell wears the frame housing.

Vibrations may be produced by an electrical unbalance in the motor, a mechanical unbalance in the motor, a mechanical misalignment between motor and load, or by the load itself, as well as by poor foundations that vibrate or amplify a small unbalance into a greater unbalance. End play due to a nonlevel condition or caused when magnetic centers and load centers are not aligned will produce bumping. Worn sleeve bearings may allow the shaft to climb the bearings, and the shaft then falls back into the bearing seat. Damaged ball or roller bearings will often produce vibration, but this can usually be detected by the noise they produce. Bent shafts or journal fits out of round will always produce vibrations. Combinations of these reasons may be extremely difficult to locate and identify.

NEMA standards establish vibration limits on standard machines, but on special or large machines, good judgment plus the manufacturer's recommendations are the only guidelines. NEMA standards (Table 2) may not be satisfactory on certain types of load, and special balancing may be necessary. This extra balancing usually carries a premium price.

Vibrations are dynamic energies and may express themselves in peculiar ways. The dynamic energy tends to expend its energies in an outward motion concentric with the shaft. If the weight of the frame and foundation is great with respect to the out-of-balance energy, the up-and-down movement may be small and the side movement may be larger; if the side bracings are exceptionally strong, the up-and-down movement may be the largest movement. If all areas of direction are well braced, the vibrations may be limited to the rotating member and dissipated in

TABLE 2. Standard NEMA Balance

Frame size	*Total amplitude of vibration, in.*
Up to 220	0.001
Over 220 to 320	0.0015
Over 320 through 500	0.002

the bearings, causing a short bearing life. Such vibrations may not be noticed unless measurements are made directly on the shaft.

If excessive vibrations are present, the motor should be disconnected from the load and only the motor checked. If the motor (only) is found to be unbalanced, then a determination must be made if the unbalance is electrical or mechanical. The motor is brought up to full speed and the vibration is rechecked. If the vibration disappears with the removal of power, it can usually be assumed that the problem is electrical unbalance. If the vibrations continue, the unbalance is caused mechanically. If the motor is still rough with the load disconnected, then the coupling, pulley, or gear is pulled off and the bare motor is checked. Many times a precision-balanced motor is upset by applying a pulley which is not balanced. Frequently the fit between motor shaft and coupling is too loose and vibration is produced.

Vibrations of assembled machines may be checked by mechanical, electronic, or reed-type indicators, usually calibrated in mils (0.001 in) of displacement. The mechanical instruments are usually limited to lower frequencies as they may not be able to follow the higher frequency ranges mechanically. The electronic devices, while more expensive, are able to pick up and follow the higher frequencies. The frequency involved may be a function of the mechanical out-of-balance condition, which occurs once per revolution; or the vibration may be caused by an out-of-round ball in a ball bearing, and the frequency may be many times greater than the speed of rotation. Harmonics may be created or magnified by unstable foundations, shaft deflection at critical speeds, or misaligned couplings, and these harmonics may be greater or less than the frequency of the item actually causing the vibration.

a. Mechanical unbalance is chiefly caused by weight distribution, loose parts, bent shaft, or bad bearings. Some machines pass through a critical speed, usually at about 60 percent of their full speed. Vibrations produced by critical speed may be rather severe but should cause no concern if this is the nature of the motor.

Mechanical vibrations caused by unequal weight distribution are usually status quo, that is, getting no better and no worse. The only remedy is to rebalance the rotating member. Loose parts or parts shifting on the rotating member will cause the out-of-balance condition to change, usually for the worse. Vibrations caused by a bent shaft usually occur after a heavy shock load and after that maintain status quo or possibly get worse as bearings wear. The shafts of large, heavy rotating elements may take a definite bend after long periods of idleness. On prolonged shutdowns, such equipment should be turned over occasionally to prevent this possibility.

b. Electrical unbalance frequently expresses itself as a mechanical unbalance. Magnetically the iron rotor core is pulled to the stator iron across the air gap, but as long as the magnetic effect is equal around the periphery of the rotor, the entire effect is canceled. If, however, the magnetic attraction is unbalanced greatly enough, the shaft will deflect, the rotor leaves its mechanical balance center, and the rotation causes a mechanical vibration. If the iron is demagnetized, the shaft deflects to normal and the mechanical balance is restored to normal.

The electrical unbalance may be due to several causes. If a field coil of a synchronous machine is short-circuited, the coil will produce less than normal flux and shaft deflection may take place. An open rotor or stator winding, particularly where there are a number of parallel paths, will also cause shaft deflection. Open bars in a squirrel-cage motor produce the same effect. If the air gap of a machine varies greatly, as a result of worn bearings or improper centering, the rotor may be deflected to the nearest air gap. The reason is that the magnetic attractive force varies inversely as the square of the distance across the air gap. The total effect of these conditions varies from machine to machine, depending upon the force of deflection and the strength of the shaft.

10. Current-Collecting Devices. Much has been written over the years concerning this subject; most manufacturers are quite generous with descriptive and instruction leaflets, and brush manufacturers also have much free and valuable information available. It is suggested that the reader acquire such data if detailed information is desired. Over years of examining many failures, experience distinguishes between obvious failures and nonobvious failure.

Obvious failures are those failures in which a casual observation of the brush rigging would have foretold a failure, the corrective action being of an obvious nature. Many preventive-maintenance departments, having to do with a preponderance of relatively trouble-free three-phase squirrel-cage motors, simply forget that brush rigging for commutators or slip rings requires more frequent inspection. The result is that a complete failure occurs before the next regular inspection is due.

a. Obvious failures should be ranked with respect to their relative importance.

1. Oil saturation. Because of actual leaking or vapors, the brushes and the commutator mica become saturated with oil. Brush carbon is packed in the undercut mica and begins to conduct from bar to bar. The oil deteriorates the mica, which is further destroyed by the arcing, and eventually the commutator must be rebuilt. This deterioration also is a frequent cause of fire. The commutator film cannot be reestablished in the presence of oil, and further brush wear and commutator wear hasten destruction.
2. Cleanliness. The brush-stud insulators (ground insulators) are subject to line-to-ground voltage, and accumulation of carbon dust and oil reduces the ground insulation to a point where a flashover occurs.
3. Short brushes. Extreme wear causes the brush rivet or brush shunt to contact the commutator. This action not only grooves the commutator but frequently drags the commutator copper across the undercut and results in a short-circuited commutator and armature destruction.
4. Poor condition of commutator. Blackened, burned rough, and nonglazed commutators will continue to get worse and eventually lead to failures.
5. Loose shunts. Vibration may loosen the shunt connection to the brush studs and the load current must then follow the spring or finger. The spring will lose its spring tension, resulting in item 4 above.
6. Equal brush tension. Excessive tension produces heat and excessive wear; insufficient tension allows the brush to jump from the commutator because of commutator roughness, vibrations, or shock loads; unequal tensions cause unequal contact resistance and unequal current distribution. Contact pressures vary because of brush grades and application, but the usual pressure is 3 to 4 lb/in^2 of brush face.

b. Nonobvious failures. These conditions may not immediately be apparent at casual observation, but they will eventually produce failures, and preventive-maintenance operators should be trained to pick out these faults and correct them in the early stages.

1. Brush grade. Only the manufacturer or highly experienced brush personnel should change brush grades on commutators or slip rings. If the grade is to be changed for a specific reason, all brushes should be changed simultaneously. As a rule, different brush grades should not be mixed. All brushes contacting one ring, all positive (+) brushes, or all negative (−) brushes are all operating in parallel with each other. If a different grade of brush is placed in these groups, the different grade will carry more or less than its share of the load. Heating and destruction of the brush film will follow. Because of this parallel circuit, a long brush should not be operated in the same circuit as a short brush.
2. Parallelism of brushes. All brushes must be parallel to the commutator bars, so that all brushes enter a new bar or leave the old bar simultaneously. Some brush boxes are pivoted around the brush stud, and improper adjustment of an individual brush box or holder will allow that brush to lead or lag the other brushes on that brush stud.
3. Equal spacing. Each set of brushes must be the same distance apart around the periphery of the commutator. The manufacturer may specify ±1/64 in tolerance on critical machines. For checking, a continuous strip of paper is wrapped around the commutator under the brushes, and the toe (or heel) of each brush is marked on the paper. The paper is removed and the actual distance measured (see Sec. 18).

   Note. Some special regulating generators may have the brushes at a nonuniform spacing to upset the magnetic circuit deliberately.
4. Box-to-commutator spacing. The bottom of all brushholders should be set the same distance from the commutator. If the brush box is set too far from the commutator, the brushes are apt to chatter, and pivot-type boxes will cause the brushes to extend a greater distance than normal (see items 2 and 3 above).
5. Lightly loaded machines. If a machine is lightly loaded for prolonged periods, the brushes and commutators will not glaze, and excessive heat and friction and wear will take place. Some of the brushes are removed from brush boxes to match the new light-load conditions.
6. Ring polarity. Depending upon the brush grade and the material of the collector rings, some manufacturers recommend that the polarity of rotating dc-field collector rings be reversed occasionally to offset the plating effect of constant polarity.
7. End play. After brushes are reseated, care must be taken to see that end play of the machine does not cause a brush to ride off the collector or commutator.
8. Same-position starting. Geared or direct-drive machines on certain loads may always start with the same commutator bars under a given set of brushes (or slip rings). Since the starting current is greater than the running current, there is a strong tendency for burning or etching the same set of bars. Usually nothing can be done except to provide extra inspection and maintenance.
9. Commutator undercutting and ring grooving. Most industrial commutators are undercut since the mica is harder than the brushes, and high mica will

chip brushes and make them jump and spark. Plating machines (high-current commutators) usually use a grade of brush containing copper and undercutting is not necessary. Some high-speed slip rings are spirally grooved for cooler operation. In either case, the recommendation of the manufacturer should be followed.

10. Concentricity. High-speed commutators and rings should be held to a total concentricity of 0.001 in, medium speed to 0.001 to 0.002 in, and slow speed to about 0.003 to 0.004 in.
11. Good commutation. Good commutation calls for no sparking at any load, no brush vibration (use a round fiber rod for feeling brush vibration), no glazing of the faces of any brushes or glazing of commutator or rings, no signs of excessive heat on any parts, and all parts clean and dry.

11. Stationary Elements. The stationary windings of dc machines and polyphase ac stators are not subjected to a centrifugal force; therefore they are usually less troublesome than rotating windings. However, they are subject to other deteriorating influences (Art. 7), and all inspections should cover these possibilities. Dc series, compensating, and interpole field windings, as well as polyphase stator windings carry full-load current whereas dc shunt and separately excited fields carry a current which is only a fraction of full-load current.

a. Dc fields. In addition to the above-mentioned inspections, all fields carrying full-load current should be examined occasionally for heating caused by poor connections which are produced by the heating cycle or vibrations. In small machines, these connections are usually taped, but in larger machines they are open for inspection. All field coils must be mechanically tight so vibrations or flux surges will not cause the coils to shift on the pole pieces, causing insulation to chaff, wear, and produce an eventual insulation failure. Blocking, typing, roping, or other means of mechanical support must be kept tight.

All leads must be securely braced so they will not break because of vibration, or short-circuit to other leads or to ground.

Care must be exercised not to break an energized shunt-field circuit, as the high inductive voltage is quite dangerous to personnel and will sometimes cause an insulation failure of the shunt fields.

b. Polyphase stators. Severe surge currents may take place when starting, particularly if full-voltage starting or reversing duty is used. The surge currents produce a distorting magnetic effect, the force of which varies as the square of the current. If a machine is started across the line and draws 8 times full-load current, the magnetic distortion is 8^2, or 64, times full-load current distortion. This magnetic effect is an attraction force for all the coils of one group and is a repulsion force between coil groups.

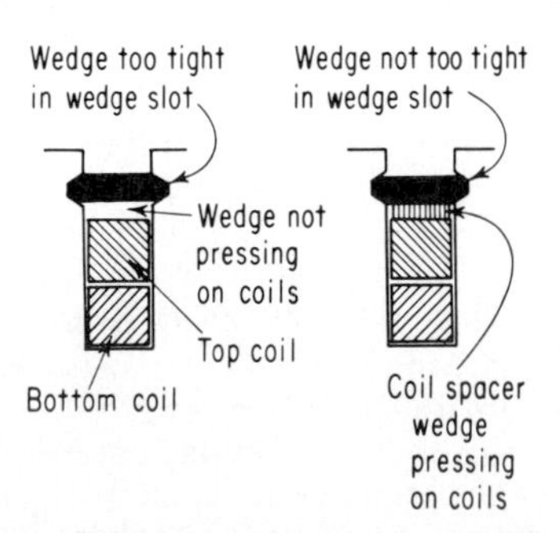

FIG. 3. Wedges of coil windings should have proper tightness in wedge groove and must press down tightly on coils.

Coils should be mechanically tight in the slots to prevent insulation failure caused by vibration. Hairline cracks in the varnish between coils and iron are good indications of loose-fitting vibrating coils.

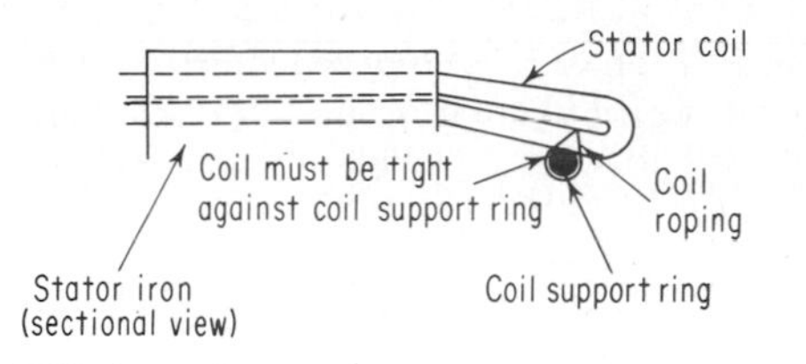

FIG. 4. Coils should be tight against coil-support rings on stator windings. Roping of coils to rings may be necessary.

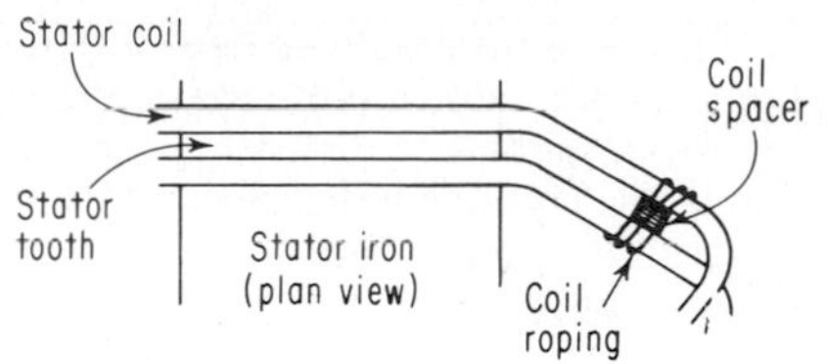

FIG. 5. Roping holds stator spacers tightly in place.

Coil wedges should be tight. Because of improper wedge fit in the wedge groove, the coils might actually be loose even though the coil wedges are tight. Mechanical tests are necessary to determine that the wedges are tight in the wedge groove and that they are pressing down tightly on the coils as shown (Fig. 3).

Coils should be tight against coil-support rings on the stator windings (Fig. 4). Varnish cracking will again indicate coil movement on the support rings, or in extremely bad cases, actual gaps will appear between the coil ends and the support rings. Either of these conditions requires that the coils be reroped firmly to the rings and given proper dipping and baking.

Roping should be in good mechanical condition and should keep the coils firmly against the support rings. Frequently, spacers are used in the diamond portion of the coil to tighten one coil against the other. The spacers should be examined and the roping which holds them in place should be tight (Fig. 5).

Leads should be well braced. The insulation on all leads and lead cleats should be examined to check for condition and for fraying caused by vibration resulting from loose lead plates.

Lead lugs should be free from terminal corrosion caused by loose plates and loose connections.

Bolts, nuts, wedges, welds, key fits, machined fits, and other mechanical connections should be examined for signs of loosening or wear.

12. Rotating Elements. Rotating elements are subject to centrifugal forces as well as other deteriorating actions.

a. Squirrel-cage rotors. Rugged design and carefree operation have been important factors in the widespread use of polyphase motors. However, the rotors are subject to abuse and failure. Usually there is no insulation to cause problems and the only failures that occur consist of open circuits in the bars or end rings.

If a single rotor bar should open, that particular slot bar produces zero torque, and the total torque of the machine is reduced. This reduction may never be noticed (depending upon the total number of bars) except for the vibrations sometimes set up (Art. 9). As more bars open, the noise and vibrations increase and the torque reduces. If all bars open, the rotor flux is only magnetizing flux and the torque is zero.

If the end rings open, an entire rotor pole is idle, and extreme noise and vibration and a greatly reduced torque result.

Bars or end rings may open when the starting torque or starting cycle is too great for the design of the machine. They may also open as a result of distortion or vibration of the rotor (Art. 9).

b. Wound rotors and dc armatures. Besides being inspected as described in Art. 11, these windings are also subject to a centrifugal force. Coils should be tight in the slots, and the wedges tight against the coils.

The banding should be tight with no indication of loosening or shifting because the coils are held in place against the centrifugal force by these bands. The band insulation should be examined to see if it is in good electrical and mechanical condition and has not slipped. Tightness can be observed by hammering lightly on the band and feeling for vibration. The varnish can also be examined for hairline cracks between the bands and coils, indicating coil movement because of loose bands.

The diamond portion of the coils should be tight against the coil-support ring. Again, any mechanical clearance will indicate loose coils, and hairline cracks between the coil-support ring and the coils will indicate coil movement on the support ring. To correct this condition, the winding should be heated and new bands applied.

c. Salient-pole dc fields. Rotating poles, in addition to being subjected to the common inspections listed above, have three additional major inspection points.

1. Loose pole pieces. Pole pieces are anchored in place by means of dovetail slots (Fig. 6) and tapered wedges, or by means of bolts. Mechanically, either system may loosen, allowing the pole to loosen and the coil to move on the pole, thus chafing the insulation. If the pole loosens, the rotor will usually become mechanically unbalanced. Visual mechanical inspections plus feeler gages are used to determine whether pole pieces or coils are loose.
2. Loose coils. Occasionally coils will loosen on pole pieces (Fig. 6) even though the pole pieces may remain tight. This loosening is the result of insulation shrinkage and deteriorating washer insulation. New washers or fillers must be installed to tighten coils.
3. Flared coil turns. The first indications of coil failure are flared turns on the straight portion or side of the field pole. Because of heat and aging, the bonding materials holding the turns together deteriorate, and centrifugal force allows the side turns to flare (Fig. 7). The turns are, or soon will become, short-circuited and rewinding will be necessary.

d. Nonsalient-pole dc fields. This type of rotating field (turbo type) is used on all large two-pole synchronous motors and generators and occasionally on four-pole machines. The core turns are wound in slots and the end turns are covered

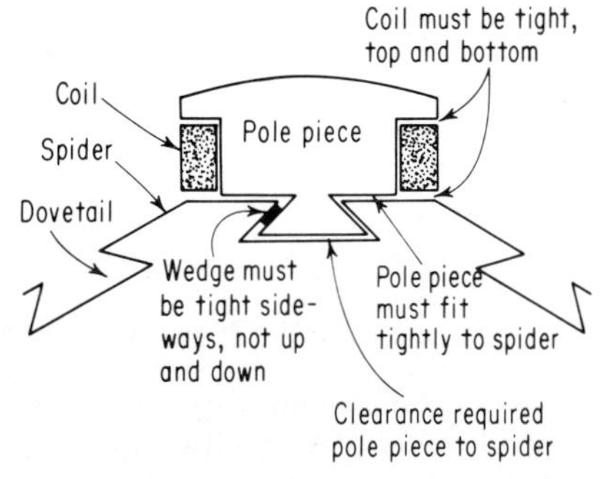

FIG. 6. Coils must be tight on pole pieces. Looseness will cause field-coil vibration on pole piece and eventual coil flashover to ground.

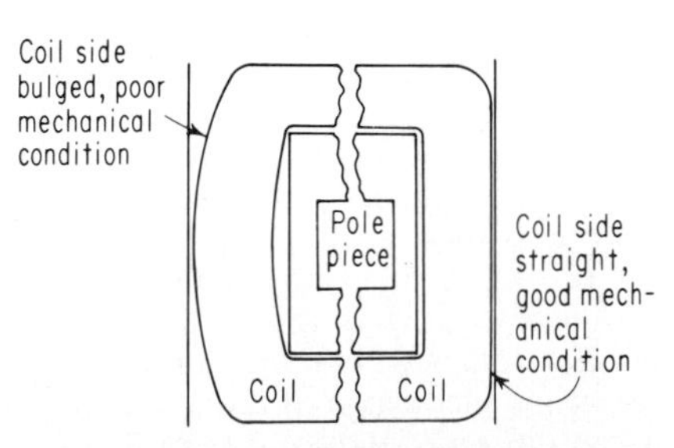

FIG. 7. Heat and age cause binding materials holding coil turns together to deteriorate until wires loosen.

with a steel shrink ring or shell called end bell. No turns or insulation is visible. The core and shaft are usually made from a single forging.

Routine maintenance consists of megger readings and blowing out carbon dust with dry compressed air. Tear-down inspections should be made only by personnel experienced with this type of equipment. Slot wedges are examined for heating and tightness, and the end bells are examined for cracks (usually with X rays), signs of heating, or indications of looseness. End bells are usually removed only if the megger reading falls to a low value because of accumulation of carbon dust not removable by blowing out, or because of short-circuited or grounded windings. They should be removed only by extremely competent shops having extensive experience with this type of equipment.

e. Damper windings. The squirrel-cage damper windings are visible on salient-pole machines and should be inspected the same as other cage windings. The damper windings on nonsalient-pole motors may be covered by the end bell and not accessible for inspection. The damper windings on synchronous motors are used as a squirrel cage for starting. After the machine "pulls into step" (the dc field locks magnetically with the rotating ac field), the damper winding is used only for speed stabilization.

If the field is lost on a motor, the damper winding may sustain the speed (operating as a squirrel-cage motor), but the damper windings are not designed for continuous running and they will usually get hot and melt.

13. General Bearing Care. Ball, roller, and sleeve bearings have many similar and some dissimilar problems (see Sec. 12). Common items of concern are:

a. Temperature. The maximum operating temperature is usually limited by the bearing lubrication. However, for motor application, the insulation class must also be considered. A Class H motor winding operating at 180°C could not use an oil or grease which has a rating of 115°C maximum temperature, and a bearing using a high-temperature grease rated at 175°C could not be operated at maximum temperature when used on a motor with Class A insulation. If the temperature breaks down the lubrication, immediate bearing failure will follow.

Higher operating temperatures will reduce internal bearing clearances because of expansion.

These factors should be taken into consideration if bearing temperatures are to be elevated by changing insulation classes and bearing lubrication.

b. Cooling. The bearings of small- and medium-sized machines are usually self-cooled, but larger or special machines may be force-cooled. This cooling may consist of airflow or water-cooling coils embedded in the bearing housings. Oil-lubricated bearings may use a water-cooled heat exchanger, an oil-feed pump, and oil filters. Bearing temperature relays are frequently used to detect and possibly shut machines down should high bearing temperatures occur.

c. Cleanliness. Care must be exercised to keep oil or grease clean, dry, and free from contaminants. Antifriction bearings are particularly susceptible to failure resulting from foreign particles.

d. Type of oil or grease. Many oils or greases are not compatible with each other and if mixed may form sludge or acids, wrecking the lubricant and pitting the bearings (see Sec. 14). Motors used outdoors may require lubrication changes during extreme cold or hot weather, or perhaps a special lubrication. On special applications, some manufacturers may specify a special brand of lubrication and will not extend warranties unless this brand is used as specified.

e. Loading. Bearing loading factors consist of speed ratings, end thrust, radial thrust, and shock. An increase of any of these factors will shorten life expectancy.

All are a function of design, and changes should not be made without proper consideration. A short-shaft motor is usually made for coupled loads and not for belted loads; merely increasing the shaft length does not necessarily mean the bearing will be suitable for belted loads. For close centers and extra belting a pedestal bearing (outboard) may be required.

f. End thrust. Unless specifically designed, motors may not be able to accept end thrust (particularly sleeve bearings) from thrusting loads, nonlevel foundations, and magnetic thrust resulting from rotor and stator magnetic centerlines being mismatched. Certain vertical motors, such as deep-well pumps, are of course designed for extremely heavy end thrusts.

g. End play. All sleeve-bearing motors must have end play; otherwise, when the shaft expands, an undue amount of pressure is applied between the bearings and the bearing thrust on the shaft. Since the shaft and the bearing are usually designed for very little thrust, any additional expansion would cause the thrust collar to chew up the thrust face of the bearing.

Ball-bearing motors must also have end play for the same reason. However, on certain motors this end play cannot be measured because one bearing may be locked in position to eliminate load end play. There must be room in the opposite end of the bearing housing for shaft expansion caused by heat. This measurement can be checked by loosening the bearing cap bolts on the locked bearing and determining the amount of end play available within the motor.

Each manufacturer has his own standards for end play for ball- and sleeve-bearing motors; therefore, no fixed rule can be given. However, Table 3 gives the usual end play found in typical motors.

TABLE 3. Approximate End Play Required in Sleeve-Bearing and Ball-Bearing Motors

D dimension,* in.	Frame size†	Sleeve-bearing motors				Ball-bearing motors			
		DC motors		AC motors		DC motors		AC motors	
		Min	Max	Min	Max	Min	Max	Min	Max
4½	18	0.070	0.120	0.040	0.065	0.020	0.060	0.020	0.050
5	20	0.080	0.130	0.050	0.075	0.020	0.065	0.020	0.055
5¼	21	0.090	0.140	0.055	0.085	0.020	0.065	0.020	0.055
5½	22	0.090	0.140	0.060	0.085	0.020	0.070	0.020	0.060
6¼	25	0.110	0.150	0.070	0.100	0.025	0.075	0.025	0.060
7	28	0.120	0.170	0.075	0.110	0.025	0.085	0.025	0.065
8	32	0.125	0.180	0.080	0.130	0.300	0.085	0.030	0.065
9	36	0.125	0.180	0.090	0.140	0.030	0.090	0.030	0.075
10	40	0.130	0.190	0.090	0.140	0.035	0.090	0.035	0.080
11	44	0.140	0.200	0.110	0.160	0.040	0.100	0.040	0.085
12½	50	0.160	0.220	0.130	0.180	0.040	0.105	0.040	0.095
14½	58	0.170	0.230	0.140	0.190	0.050	0.110	0.050	0.105
17	68	0.190	0.250	0.160	0.225	0.055	0.115	0.055	0.110

* *D* dimension is the distance in inches between center of shaft and base.

† The full frame designation includes one more digit, such as 203 or 204, which is the same frame diameter.

h. Bearing oiling. Proper oiling techniques should be followed for sleeve-bearing lubrication.

When oil cups have been filled to capacity while the machine is rotating, oil which has collected on the shaft and in bearing surfaces will return to the already full sump when the machine is stopped and overflow—possibly into the motor windings where it can cause winding deterioration.

i. Gear, pinions, and couplings. Maintenance must begin when the machine is first installed through the proper application of pinion, pulley, coupling, and correct lineup. Gears or pinions should never be driven onto a shaft. The shock caused by pounding is transmitted through the shaft to the bearings, which causes brinelling and means shorter life and noisy bearings.

It is always recommended that gears and pinions be heated to expand them, placed on the shaft in the proper position, and then allowed to cool and seize on the shaft.

Gears or pinions should never be placed on the shaft too loosely and then tightened by using an oversize key. This procedure will distort the pinion or coupling, causing it to have poor contact, and this results in a noisy gear with a short life. If the pinion is too loose on the shaft, there is the danger of completely wrecking the key and keyway.

j. Bearing currents. The effect of bearing currents on both ball and sleeve bearings is detrimental and results in premature bearing failure. In sleeve bearings, the shaft becomes rough and will dig and score the bearing until the failure results. In ball bearings, electric arcing on the highly polished surface of the balls or races causes early fatigue and failure.

There are several reasons for bearing currents. One cause of bearing currents is the passage of current from a grounded rotating winding through the bearings to the frame, to be passed off as ground current. Ball bearings should be replaced under this type of failure, and journals and sleeve bearings should be examined for pitting.

When rotors or armatures are ground-tested in the field, the ground-box test leads should be applied to the shaft and the windings, rather than on the frame and rotor windings; then the exciting current will flow directly from the windings to the shaft and not through the bearings.

Voltages may be induced in the shaft by unequal air gaps or stray magnetic fluxes which cause shaft currents to flow in at one bearing and out at another.

Belts sometimes generate static voltages, and if this static voltage is discharged to the motor shaft, it may flow through the motor bearings to ground. Such static charges should be passed to ground without allowing them to pass through bearings.

There are two principal methods of eliminating bearing-current flow: insulated housings and insulation of one pedestal on large dc and ac machines (see Sec. 12, Art. 10). In the first application, the housing of one of the ball bearings is bored considerably oversized and an insulating cup is inserted between the outer race of the bearing and the inside of the housing. The second method separates the pedestal from the base by approximately ⅛-in-thick insulation. The tie bolts and dowel pin must also be insulated so that a short circuit does not exist between the pedestal and the frame. Occasionally, other means of insulating are used.

Some machines are manufactured with one bearing insulated, while some machines have been changed in the field (after manufacture) because of stray or unexplained fluxes which have developed and caused excessive bearing failures.

Bearing currents may be detected in the following ways.

1. Voltage measurements. With the machine running at normal load, and with the aid of long insulated wires, measure for a voltage difference between one end of the shaft and the other. Minimum voltage may be on the order of a few millivolts and maximum voltage probably will not exceed a few volts. If a potential exists between the ends of the shaft, bearing currents will flow unless one bearing is insulated from the frame.
2. Current indication. With the machine running at normal load, and with the aid of a long, heavy insulated cable, hold one end of the cable firmly against the center of one shaft and strike the other end against the opposite shaft end. If bearing currents are available to flow, sparks will be drawn from the shaft.
3. Shaft magnetization. If both ends of the shaft are magnetized, shaft currents have probably been flowing.
4. Inspection. If bearing currents have been flowing, the shaft, balls, races, or sleeve bearing will be pitted.

14. Care of Ball Bearings. There are certain preventive-maintenance considerations that apply especially to ball bearings.

a. Bearing size. Bearing manufacturers have done a wonderful job of standardizing bearing sizes and marking the size on the bearing, but they have not done so well in standardizing and marking other important items, which are necessary if bearing substitution is to be made. Bearing manufacturers are usually very generous with their bearing catalogs, and these catalogs give bearing sizes, recommended sizes of shaft, and housing fits. Most motor manufacturers follow these recommended fits fairly closely except on specially designed motors.

b. Bearing identification and substitution. Because the bearing size, always marked on the bearing, is the only standardization in the bearing industry, it can be extremely dangerous to substitute ball bearings by size only, and many failures result from improperly substituted bearings. Some of the items of importance are discussed in the following.

1. Accuracy. The accuracy of a bearing is designated by its ABEC number. These numbers for electric motors are usually odd numbers from 1 to 9. The higher the number, the greater the accuracy of the bearing. A standard motor usually uses ABEC no. 3, while a precision-manufactured motor may use ABEC no. 7. The greater the bearing precision, the higher the cost of the bearing. The ABEC number may or may not be coded on the bearing marking. It is possible to change vibration characteristics of a motor by substituting a lower-precision bearing. The greater the precision, the more quietly the bearing will run.
2. Internal clearance. Some applications require more internal clearance to permit heat expansion. Substituting a low-clearance bearing for a high-clearance bearing (degree of internal losseness) may wreck the bearing within a very short period of time.

   As mentioned previously, motor manufacturers may or may not follow bearing manufacturers' recommendations as to shaft and housing fits. In general the outer race is a slip fit in the housing (enough friction to hold its own weight). If this fit is too tight, the internal clearance of the bearing will be reduced with possible bearing failure. If the motor manufacturer wants the outer race tight in the housing (possibly due to high shock), a bearing with more initial internal clearance is necessary.

   In general, the light press fit is used between the shaft and the inner race. If the fit is too loose, the inner race will turn on the shaft and the shaft will wear.

If the fit is too tight, the internal clearance of the bearing will be reduced. Motors designed for shock and temperature (diesel-electric traction equipment, for example) usually have tighter shaft and housing fits and consequently use a larger initial internal clearance.

The code marking for internal clearance may or may not be marked on the bearing.

3. Cage material. The cage material spacing the balls or rollers may be pressed steel, brass, or plastic. This material changes the operating temperature and may cause failure if the wrong cage material is used. The type of cage may or may not be code-marked on the bearing.
4. Load rating. Bearings are rated by speed, end thrust, radial load, and shock, which determines bearing life, and any changes in the characteristics may greatly affect the life of the bearing. Most manufacturers have their own coding or numbering system. Bearings may be capable of heavy end thrust and light radial load, heavy radial load and light end thrust, or a combination of either. Double-row bearings may be used for heavy radial thrust, and double bearings may be mounted together, face to face or back to back or in tandem (face to back). Such bearings may have their faces or backs ground specially or may use spacing washers between bearings. Care must be used in replacing such bearings to maintain the original design.
5. Sealed or shielded bearings. Bearings may be sealed or shielded on one or both sides. If both sides are sealed or shielded, they are usually prelubricated for the life of the bearings. These bearings may be designed to be lubricated in other applications, but in general, in electric-motor applications such bearings are not designed to have lubrication added.

   Care must be exercised to obtain grease of the proper temperature to match the operating temperature of the motor (see Sec. 14).

   A shielded bearing has a metal shield which is fastened to the outer race and has a small clearance to the inner race. This shielding will keep the grease in the bearing but will not prevent finely powdered dust from working through the shield to contaminate the grease and wreck the bearing.

   A sealed bearing has a plastic seal which slides tight against the inner race, preventing fine dust from entering the sealed bearing.

   Sealed or shielded bearings do not have an unlimited shelf life as the grease may solidify in hot storage rooms.

   Some applications use a shield on one side (winding side of bearing), which will keep excess grease out of the motor windings, but the bearing can be greased from the other side.

   Bearing manufacturers usually use their own codes for seals or shields.
6. Grease. Open bearings have enough grease to prevent rusting during storage, and at assembly additional grease must be added. As mentioned previously, sealed or shielded bearings are usually nongreasable.

   Special applications may require special grease (brand name), and greasing instructions must be carefully followed or the manufacturer may not back the warranty. Many types of grease are available as far as temperature ratings are concerned.

   The common problem with greasable bearings is overgreasing. An excess amount of grease will "churn" and produce excessive heating, and the excess grease will usually enter the motor and saturate the windings. This leakage usually deteriorates the winding insulation. Ball bearings should be greased while the motor is running, and the sump pipe or plug should be open. After the bearings and grease warm, the excess will flow out the sump. This might require 15 min. The sump can then be closed.

If ball bearings become contaminated with dust, water, or other foreign particles, regreasing will usually flush out the contaminants.

Sealed or shielded bearings may or may not be coded with the type or temperature rating of the grease.

7. General care. If a bearing is pulled from the shaft by pulling on the outer race, brinelling will usually take place and the bearing should be scrapped.

Bearing failure is preceded by increased noise, temperature rise, increased roughness, or any combination; and these items, plus hours of life, form the criteria for rejection of bearings. If a larger bearing is removed for some other reason, an inspection of the races should be made, and in some cases it may be feasible to send it to a laboratory for inspection. Mechanical measurements are worthless (except laboratory), as the bearing will have failed before any measurable wear can be determined by shop measurements.

Some manufacturers recommend that, on special applications, bearings be replaced after a given number of hours.

Usually some of the balls or rollers of oil-lubricated bearings are out of oil when not running, particularly on oil-mist or impeller-oil-fed or oil-pump-fed motors. If such machines are down for an extended time, especially in humid conditions, care must be taken to see that rusting of the bearing does not take place where the balls or rollers squeeze the oil out between them and the race.

On horizontal high-speed motor-generator sets there is no external thrust, and such machines may use a smaller bearing than would be used on an equivalent-size motor, or in reality the weight per ball may be greater. If such machines are shut down and remain in one position for extended periods, undue roughness may be observed when starting. This may be the result of shaft distortion or ball brinelling. It is suggested that these machines be turned to a different position occasionally while in storage. Vertical machines do not have the same condition, as there is no shaft distortion and all balls carry the weight. Slow-speed machines usually have greater shaft and bearing capacity in regard to weight.

15. Sleeve-Bearing Care. Preventive maintenance of sleeve bearings has important considerations that can add considerably to bearing life.

a. Internal clearance. Since there must be clearance between the journal and the sleeve bearing, the journal or the bearing can wear a certain amount before the bearing becomes unusable. Barring accidents, one journal will wear out many sets of bearings before journal wear is noticeable. Excessive bearing wear produces one or more of three effects:

1. Loss of oil pressure and lubrication, particularly on forced-oil-feed bearings.
2. Excessive vibration because the shaft may climb the bearing wall and then fall back in the seat.
3. The air gap is reduced (Art. 9) and the rotor may drag on the stator, or excessive shaft deflections and vibrations may be produced.

While preventive-maintenance operators do not have complete agreement on the clearances given in Table 4, the chart gives a general guideline for maximum clearances for usual applications.

TABLE 4. Maximum Sleeve-Bearing Wear

Shaft size, in.	Approx factory clearance, in.		Approx max wear, in.
	Min	Max	
¾–1	0.0015	0.0025	0.0035–0.004
1 –1¼	0.003	0.004	0.005 –0.006
1¼–2	0.0035	0.005	0.007 –0.008
2 –2½	0.004	0.006	0.008 –0.009
2½–3	0.005	0.007	0.009 –0.0105
3 –4	0.006	0.008	0.010 –0.0115
4 –5	0.007	0.009	0.011 –0.0125
5 –6	0.008	0.010	0.012 –0.014

b. Materials and shapes. Sleeve bearings are usually solid bronze or steel shells babbitt-lined. In smaller-sized machines (fractional-horsepower sizes) the bronze bearing may be spongy and the bronze impregnated with a lubricant so that the bearing becomes self-lubricating. Medium-sized machines may use bronze that is lubricated by oil and waste packing or oil rings. Larger machines and smaller high-speed machines use steel or cast-iron shells with babbitt linings.

One-piece bronze bearings cost less and will stand a shock load better but will not stand up under high-speed operation. Bronze bearings may be made up to 2½- or 3-in-diameter shaft.

There are many grades of babbitt, but in general, a low-speed babbitt is used for speeds up to 1200 or 1800 r/min, and high-speed babbitt is used for 3600 r/min.

Lack of lubrication on bronze bearings will cause the bronze to shrink and seize the shaft, usually wrecking the bearing and the shaft. Lack of lubrication on babbitt bearings will usually cause only the babbitt to melt and "wipe," wrecking the babbitt only. The shell can be rebabbitted. Bearings larger than 3 in are usually split-type bearings. A typical split-type bearing is shown in Fig. 8.

A return-oil groove is machined into the babbitt around the entire inner periphery near both ends so that oil will not leak out the end of the bearing. Holes are drilled in the bottom cap of this oil-return groove so that this excess oil can drain to the oil sump.

A relief is cut into the bearing to about 1/16 to 3/16 in deep. This relief serves as a temporary oil storage or direct-contact reservoir. It is machined, after the bearing has been bored to size, by setting up one cap at a time off center and boring the relief section to a larger diameter. The relief section extends into both sides of the bottom cap by approximately 15°, and the entire top of the bearing may be relieved except for a narrow lip on each side of the top oil groove. The relief section (oil reservoir) must not be run into the oil groove. It will be noticed that the top bearing cap has an actual contact bearing surface of two small sections at each end, which merely steadies the journal by the top cap. If the journal has any appreciable upward radial thrust, the top relief may be omitted to give a larger top-cap bearing surface.

Oil grooves are ground into the babbitt, usually forming an X in the bottom cap. This groove serves as an oil contact reservoir and distributes the oil to the

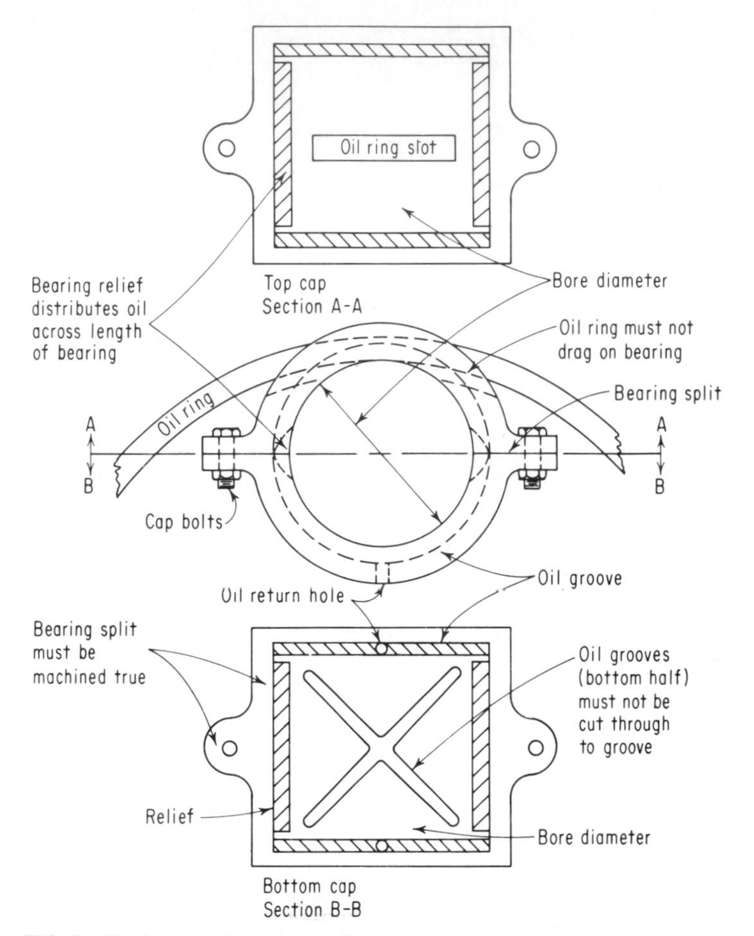

FIG. 8. Typical babbitt bearing split-type steel shell with poured babbitt.

length of the bearing. These oil grooves must not be cut through or into the end oil grooves, as the reservoir oil would flow directly to the sump.

As mentioned above, some bearings are designed for upward radial thrust and the top bearing may not be relieved across the entire top. In this case, the top bearing cap will also have the X oil grooves.

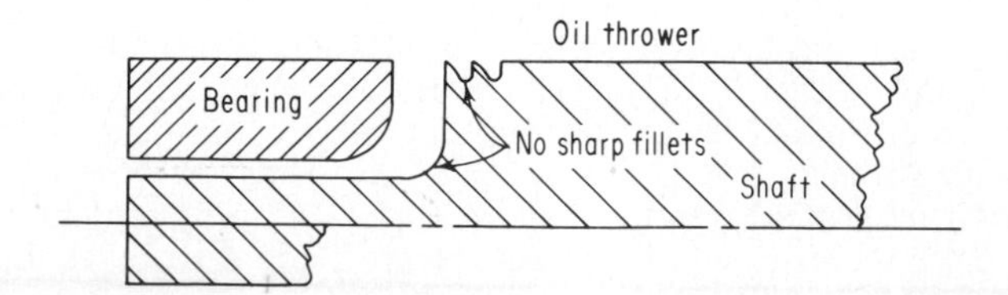

FIG. 9. Single-thrust-face bearing.

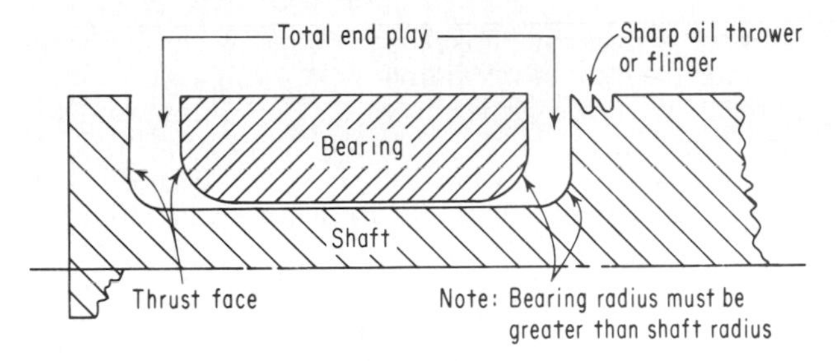

FIG. 10. Double-thrust-face bearing.

The bearing shell must be tight in the housing, usually a few thousandths interference, or a pinch fit.

The thrust face may be single (Fig. 9) or double (Fig. 10). Just behind the thrust face is an oil flinger, or thrower, which throws oil that has escaped from the bearing off the shaft. This oil is thrown into the bearing cage and is returned to the oil sump.

Thrust faces must be smooth and true; otherwise they will chew up the bearing thrust face. The oil throwers must be sharp to ensure throwing off excess oil.

Sharp fillets are never made on journal or shaft fits as the shaft will break at a sharp fillet. Care must be exercised to be sure that the radius on the thrust face of the bearing is larger than the radius of the journal oil-thrower face; otherwise thrust will occur on the radius instead of on the thrust face.

16. Testing. The subject of testing is quite extensive, and it should be understood that this section deals with testing as applied to preventive maintenance or field testing. No effort is made to discuss quality control, factory tests, or field acceptance tests for new installations.

a. Ground-insulation tests. This test is the most important and the easiest to make of the electrical tests, and the one that will prevent the most failures or indicate a possible future failure. It is made to determine the condition of the ground insulation, which is all insulation between live parts and frame or ground.

1. *Megohmmeter test.* Usually the complete machine is tested with all windings connected, and if a low megohm reading is obtained, the various windings are disconnected and each winding is tested separately to determine which winding has the low reading. Such testing is often referred to as megger testing and the readings as megger readings.

The usual and most common test is with a megohmmeter. The test is made in the usual range of 500 or 1000 V dc. The meter may be hand-cranked or motor-driven, and some units employ a rectified and filtered ac circuit. For low-voltage circuits, the hand-cranked 500-V unit is used. The 1000-V units, usually motor-driven, are used for high-voltage machines and in drying processes where a 1-min reading is compared with a 10-min reading.

The test is nondestructive in nature, as the small currents involved (usually 1 mA or less) are not sufficient to burn insulation. Since the open-circuit voltage of the tester is considerably less than the factory ground-test voltage, there is no danger of puncturing the insulation.

The value of the insulation in megohms (million ohms) is reduced as the ground insulation becomes saturated or contaminated with moisture, carbon dust, dirt, oil, and by deteriorated insulation due to age or temperature in excess of that specified for the class of insulation used. NEMA standards, some manufacturers, and many insurance carriers have set a minimum insulation value of 1 MΩ per kilovolt of rating, plus 1 MΩ at normal ambient of 40°C. Practical experience indicates that any medium-sized machine (4-ft diameter or less) will have a megohmmeter reading in excess of 50 MΩ if it is clean and dry and has good live insulation.

A single reading for a given machine means very little if the history of the machine is unknown. A curve should therefore be kept, showing the megohms versus the date of the reading, and the curve will indicate if the ground insulation is deteriorating. Extending the curve will indicate if or when the insulation of the machine will be in a dangerous condition and rejuvenation will be required. Consider two similar machines, *A* having an insulation value of 100 MΩ and *B* having an insulation value of 15 MΩ. Both machines are subjected to moisture, and a megohm recheck is to be made before the machines are started. A recheck shows both machines having a reading of 12 MΩ. This condition indicates that machine *A* has absorbed a considerable amount of moisture and should be dried before starting, whereas machine *B* has absorbed little moisture and may be operated.

Increased temperature reduces the megohm reading, and the temperature should therefore be noted when the megohm test is made. Correction factors are available for different types of insulation, but a simpler method is always to check the machine at shutdown (hot) or before startup (cold). If desirable, two charts can be maintained, one for hot and one for cold conditions.

The frequency of the tests should be determined by operating conditions. Dusty, high-humidity conditions require more frequent testing than clean, dry conditions. Because dc machines generate carbon dust, they should be tested more frequently than ac squirrel-cage machines. A motor in a dry atmosphere may be down for several months without absorbing any appreciable moisture, whereas a mine motor might absorb enough moisture while being down for 1 week to cause a failure when starting. Some operating people feel totally enclosed or explosionproof motors are protected against moisture, but when certain explosionproof motors have been dismantled, several gallons of water have been found trapped in the enclosure as a result of breathing and condensation.

The machine must be electrically dead during megohm testing. All windings under test should be short-circuited, and all other windings should be short-circuited and grounded. This caution prevents any transformer action resulting from a charging current inducing a high voltage in the winding under test to ground and in all other windings in the machine not under test. The test is therefore made between the winding under test and ground and all other windings. One test terminal marked GND or GROUND is connected to a good ground on the machine, and the other terminal marked LINE is connected to the winding under test. The crank (if hand-cranked) is turned at a uniform speed of about 100 r/min until the reading is consistent. About 5 to 15 s will be required to stabilize readings, depending upon the size of the machine. If the speed is erratically changed, different readings may be obtained because of the capacity effect of the insulation. If the megohm reading is extremely low, sometimes the pointer will jump to zero and then climb back again. This spasmodic reading indicates the insulation is so poor that it will not hold the 500 V megohm voltage and the current is jumping to ground over or through the dirt or moisture.

There has been some criticism concerning the use of a megohmmeter in arriving at an estimate of insulation condition. It must be remembered that this test indicates only the condition of ground insulation and not the condition of turn-to-turn insulation. The assumption is frequently made that if the ground insulation is in good condition, the turn-to-turn insulation is also in good condition. This assumption may or may not be correct.

Consider a machine that has mica (Class B) ground insulation and Class A turn-to-turn insulation, and further assume that the machine has been operating at 125°C hot spot for a period of time. Since the temperature was within the safe limits, the ground insulation will be unaffected, but the turn-to-turn insulation will be roasted. In this case, the assumption that the turn-to-turn insulation is equal to the ground insulation is false.

Also consider a machine where a water spray or drip falls on the winding at the end turns but does not get to the iron core. The end turns will be saturated with moisture, but the ground insulation will be dry.

In such cases, more knowledge about the machine and the operating conditions will help to avoid a false decision.

2. *High-potential ground test.* This test consists of applying a higher than normal ac voltage, 25 or 60 Hz, between winding and frame. This stresses the insulation above operating voltage and indicates whether a margin of safety above operating voltage is available.

Consider a 2400-V motor where some damage has occurred to the insulation, breaking the solid insulation and exposing bare copper $\frac{1}{16}$ in from ground. A 500-V megohm meter will not jump the $\frac{1}{16}$-in gap in insulation value. However, when line voltage is applied to the machine, it will jump the $\frac{1}{10}$-in air gap and power current will follow, possibly severely burning the insulation and iron laminations. A high-voltage ground test would have detected this condition.

On the other hand, some operating people feel that a voltage much higher than normal may unduly stress the insulation and possibly cause a failure while testing, which might necessitate rewinding. This controversy will probably never be resolved, but certainly some margin of safety should be provided.

A ground test should never be made on a winding which has a low or minimum megger reading, because if the insulation is wet, the ground-test voltage will rupture the insulation, or if the machine is dirty, the ground-test voltage will creep and flash to ground.

The factory ground-test voltage is usually double the operating voltage plus 1000 V for a period of 60 s, for all machines over 1 hp in rating. Field acceptance tests should be 85 percent of this value, and routine tests in the field should be 65 percent of the original test voltage.

Test voltages of 2400 V and less usually are applied at full value (one step), but test voltages above 2400 V are applied with regulation, increasing from operating voltage to peak test voltage at the rate of 3 kV/s. At the end of the test, the test voltage should be decreased in steps of 3 per kV/s until the operating voltage is reached, when the ground-test voltage can be instantly reduced to zero.

More power is available on this test than on the megohmmeter test, and great care must be taken to protect life when it is made. Both the machine frame and the ground-test equipment must be securely grounded. Static charges may remain on large high-voltage machines if the test voltage is quickly disconnected, and care must be used to drain off this possible static before the windings are touched.

If the ground insulation fails, a short circuit is placed on the test transformer, and the ground-test transformer must be protected against this condition by overcurrent relays. Because of the capacity effect of a winding when tested with alternating current, a large high-voltage machine may draw sufficient charging current to overload the ground-test box and cause the overload relays to trip. Difficult calculations are sometimes required to determine the size of a ground-test box in relation to the size of the machine under test.

A simpler method consists of measuring the charging current at some lower voltage, calculating the necessary charging current at the higher test voltage, and comparing the two with the capacity of the ground-test box. The charging current will vary approximately as the applied voltage. Sometimes the circuits of a winding are individually tested instead of being collectively tested to reduce this capacity effect. All windings not being tested must be short-circuited and grounded.

3. *High-potential dc ground test.* This test is becoming more popular for the following reasons:

a. Since the test is direct current, little charging current is involved, and smaller equipment can be used on large high-voltage machines.

b. The capacity of the equipment is limited to less than 1 mA and the danger to personnel is greatly reduced.

c. The current is limited, and no burning is produced in case of a ground fault.

d. Since the applied voltage and the leakage current are metered, the insulation in megohms can be directly calculated using Ohm's law:

$$\text{Resistance} = \frac{\text{applied voltage}}{\text{current flow}}$$

The equivalent dc test voltage for new windings is double the operating voltage, plus 1000 times 1.6. This voltage is reduced to 65 percent for routine field tests [(2 × rated voltage + 1000) × 1.6 × 0.65].

4. *Repairs.* If previous megohmmeter readings have been recorded and previous operating conditions are known, logic usually dictates what repairs are necessary for a low megohmmeter reading. If a machine has been maintaining a consistently high reading but has been subjected to extreme moisture, a drying operation is necessary. If the reading has been decreasing and the windings have been accumulating carbon, oil, or dirt, a cleaning operation is necessary. If the reading has been decreasing but no moisture or other contaminating influences are present, and if the machine has been heavily overloaded, the life of the insulation could be at its end, and rewinding may be necessary.

If the megohmmeter indicates zero reading on a complete machine, and a dead ground is indicated by means of an ohmmeter or a series test lamp, the grounded part of the machine must be isolated to be located. Raising the brushes on a dc machine will isolate the armature from the other parts of the machine. If the brushholders are grounded, the cross connection must be removed to locate the grounded brushholder. Shunt or series field coils are disconnected and each field winding is tested separately. If a 10-pole shunt field is grounded, the connection between coils 5 and 6 is opened to determine which section is grounded. Further disconnection is necessary to locate the grounded coil.

If the armature is grounded and this condition is not obvious because of flash marks or other signs of damage, further tests must be made by a winder. Locating a grounded armature coil and making repairs require winding skill, which is not a part of preventive maintenance.

b. Short circuits. Many effective means are available in the factory or in modern repair plants for testing for short-circuited windings, but field testing usually is limited by equipment and conditions to voltage-drop tests or resistance testing. The voltage-drop tests use the principle of Ohm's law. Since the various resistances to be checked are normally equal, and the current is the same for a series circuit, the voltage drop will vary directly with the resistance.

1. *Shunt fields.* The field is energized by normal (or less) dc voltage, and the voltage drop is measured across each field. The fields are usually connected in series, and since the resistances of all coils are normally equal, the voltage drop expected across each field is

$$\frac{\text{Applied voltage}}{\text{Number of coils in series}}$$

The voltage should normally not vary by more than 5 percent. The copper wire has approximately 2 percent variation limits, and 3 percent is usually allowed for differences in turns or diameters of the mean turn. It must be remembered that the accuracy of the voltmeter used is based upon full-scale deflection, and therefore for greater accuracy the meter should be operating on a range where it is reading in the upper part of the scale.

Assume a meter with a 2 percent accuracy is used, measuring 96 V on a 100-V scale. The reading could be plus or minus 2 V, or actually 94 to 98 V might be indicated. If the same 100-V-scale meter is used to measure 10 V, the meter could read $10 - 2$ or $10 + 2$ (8 to 12 V) and still be within its rated accuracy, although the percentage error is much greater. The lead insulation can be pierced by a pair of pointed test prods, but care should be taken not to pierce the insulation where the lead rests against the frame, or in close proximity to other live parts. Care must be exercised so as not to open the field circuit while energized, as the high inductive kick voltage could cause the fields to flash to ground and could certainly harm an operator or a voltmeter connected across a field. To reduce the effects of the discharge voltage when opening the field circuit, a field-discharge switch, a variable-voltage supply, or a variable resistance in the field circuit could be used.

Should all the coils be connected in parallel, the voltage drop across each field is equal to the line voltage, regardless of the resistance of each coil, and the coils must be reconnected in series for a voltage-drop test.

If available, a good ohmmeter or wheatstone bridge can be used and the resistance of each coil measured. Since the original factory-specified resistance may vary as much as 5 percent, it is usually impossible to tell whether some fields are short-circuited 5 percent, or whether the combination of meter accuracy, wire resistance, and mean turn diameter, all working together, gives a -5 percent reading. In the usual dc machines this is of little consequence, except on certain exciter generators where a very fast response is necessary. Short-circuited turns will damp quick flux changes. Some machines are designed with some turns actually short-circuited to prevent quick changes in flux, while other special machines must change flux quickly.

2. *Rotating fields (synchronous fields).* Severe vibrations are sometimes set up on synchronous machines if the fields have a few short-circuited turns, and the dc voltage-drop testing or resistance testing will fail to show a percent or two of short-circuited turns. The dc drop test is made as a rough check, and then for greater accuracy 440 V ac is applied to the 125- or 250-V dc field, and an ac voltmeter is used for drop testing. Because of the impedance change with short-circuited turns, a small percentage of short-circuited turns will show up as a much higher percentage on the ac test.

If direct current is applied to the field winding, the induced voltage must be controlled when breaking the circuit (item 1). If ac voltage is applied to the rotor while in place in the stator, care must be taken to be certain that the leads of the machine are insulated against bodily contact, particularly on high-voltage stators.

3. *Series coils, commutating coils, and interpole coils.* The turns of these coils are frequently visible, and inspection can usually detect any short-circuited turns, particularly on the larger machines; then the principal fault is usually a ground. However, if the turns are covered with insulation, a voltage-drop test can be made by circulating up to full-load current through the windings and taking a millivolt drop across similar sections of the windings. For this test, a dc welder is used as well as a variable-scale dc voltmeter or millivoltmeter. Judgment and care must be used to prevent excessive voltage on the meter. The highest scale and lowest current are used, and both are adjusted until a suitable reading is obtained.

4. *Dc armatures.* Except in the factory or a well-equipped repair shop, the test voltages which can be applied for testing short-circuited armature coils will be only a few percent of operating voltage. For this reason the best field test is actual operating conditions.

Short-circuited armatures fail to rotate or the rotation may be jerky, heavy armature current is usually drawn, and heating, either local or general, will always develop. Usually the sparking at brushes will be severe. Direct current, by using a welder, can be circulated through the armature, and brushes are raised from the commutator on all brush arms except two adjacent to the place where the bar-to-bar reading will be taken. The reading should be started near one set of brushes and work toward the other set without turning the armature. Bar-to-bar readings should be compared. An armature without equalizers should give equal bar-to-bar readings. Equalizers will give a reduced reading. If equalizers are used, usually every third or fourth bar will be equalized, and a definite pattern of readings should be established. After all the bars between the brush studs have been tested, the armature should be turned so a new group of bars can be tested. Only the readings between brushes for a set position of the armature should be compared with each other. When the armature is partially rotated, the set of readings will vary from the previous readings. Care must be exercised to keep the armature current constant. Lower readings indicate a possible short-circuited armature coil between the bars under test, a zero reading indicates a short-circuited bar, and a higher reading indicates a poor connection or high-resistance lead connection to the commutator.

Before this test is made, it should be first determined that there are no open circuits in the armature. An open circuit in the armature will give the same voltage reading that exists across the armature brushes. This reading will be as many times greater than the bar-to-bar test as there are bars between brushes. Therefore if the meter is adjusted to give one-half scale deflection on a bar-to-bar test, spanning an open circuit will increase the voltage on the meter to a large enough value to damage the meter.

5. *Stators, polyphase ac motors.* The most effective test that can be made in the field is a phase-balance test, comparing phase currents. The polyphase voltages must be equal for this test to be of value. Resistance testing means very little except in extreme cases where the resistance varies by more than several percent. One turn short-circuited in a 300-turn winding will affect the resistance 0.3 percent, but the resistance of a good winding might vary several percent. One turn short-circuited in a stator winding will cause complete destruction of the winding,

as the induced circulating current will be great, causing the copper to melt. If the motor is in operating condition, full voltage should be applied if possible.

If the motor is dismantled, one-fourth or one-half operating voltage can be applied; 50 percent of operating voltage will force approximately three to five times full-load current through the windings.

6. *Wound-rotor motors.* The brushes can be raised or the rotor leads disconnected from the secondary circuit and insulated. The stator can be phase-balanced at full voltage to test for stator short circuits, and the full ring voltage can be measured for rotor short circuits. The iron noise and the exciting current should be small. Usually there is no tendency for the rotor to turn, and the rotor can be turned in either rotation with the same degree of ease.

c. Open circuits. Open circuits usually occur in operation as a result of some mechanical or electrical fault, and the indications are usually apparent, requiring a shutdown.

Dc motors of the larger size are usually protected against open shunt field by field loss relays, and if the motor is not disconnected by protective relays, the machine will throw itself apart.

1. *Dc armatures.* An open circuit always produces excessive sparking and ultimate destruction of two commutator bars. In the final stages a flashover usually results. No relay protection is possible, and this condition must be located by observation.

2. *Polyphase stators.* Usually an open circuit is produced by one fuse or contact opening, causing single-phase operation. The power output of the machine is reduced to 57 percent. If full load is required of the machine, the remaining fuses or overload devices should open. If the machine has only partial load, it is possible to burn out the winding without having the overloads remove the machine from the line. Large important machines are protected against single-phasing by a three-phase power-loss relay. Polyphase motors will run single-phase but are not capable of starting.

If the open circuit develops in the motor windings, all the above statements are true, but a delta-connected motor will start with an open circuit. The machine will operate open-delta with 86 percent of capacity if connected single-delta. If parallel circuits exist in a star-connected winding, the motor will also start.

3. *Polyphase wound rotors.* The open circuits usually occur in the resistor bank, and the rotor operates with a 57 percent reduction in rating. The stator currents will balance, so no relay can detect this condition unless it is connected in the rotor circuit. Since protection of the rotor circuits is not usually furnished, it is important that all connections in the rotor circuit be examined occasionally for possible open circuits.

4. *Synchronous motors.* Great harm can be produced if the field opens and the motor is allowed to operate on the squirrel-cage winding. The squirrel-cage winding usually melts and the machine usually catches fire. Such failures can be prevented by a field-loss relay or power-factor relay.

17. Load. General terminology of electric-motor load does not necessarily mean the "driven mechanical load" but may refer to the "load current" of the motor. In this section we are not concerned with the mechanical load but only with the electric, or load, current and factors affecting it. Obviously if the mechanical load is too great, the load current will be excessive.

Excessive load current (over nameplate) will increase the heating in the winding involved in proportion to the increase in I^2R (amperes squared times resistance) losses plus additional heating of current-collecting devices. The hot-spot temperature will therefore also increase as the square of the current increases. Since the higher temperature will increase the resistance of the copper R, the total temperature will be greater than the increase of the I^2. The increased temperature will reduce insulation life.

a. Low voltage. The actual effect of lower than nameplate voltage depends upon the type of motor involved.

1. *Induction motors.* The torque will vary inversely as the square of the applied voltage. If 90 percent of nameplate voltage is applied, the torque will be 90 percent squared, or 81 percent of normal. If full mechanical load is applied requiring 100 percent torque, then the motor must draw more current to produce this torque. Excessive heating probably will be produced at lower than nameplate voltage.

2. *Dc series motors.* The speed will vary directly as the applied voltage, so the motor will reduce speed. This reduction in speed may affect the ventilation, and the temperature may increase slightly.

3. *Dc shunt motors.* The shunt-field flux will decrease (depending upon the part of the saturation curve in which the field is operating), and this reduced field strength will reduce torque in proportion to the reduction in field flux. To produce normal torque, the armature current must increase to compensate for the reduced field strength. The shunt fields will run cooler and the armature will run warmer.

b. High voltage. The effect of higher than nameplate voltage on motors varies with the type of motor.

1. Induction motors. Iron saturation may take place and the exciting current may increase faster than the increase in voltage, with the effect of increased heating.
2. Dc series motors. The speed will vary as the applied voltage, and as a result of increased ventilation, the motor may run cooler.
3. Dc shunt motors. The shunt-field temperature will increase as the I^2R losses increase. The armature will increase in speed (depending upon the saturation of the field flux) and will probably be unaffected.

Note. Another factor affecting the load current with increasing or decreasing speed is the type of load. The torque of some loads varies as the speed, and the torque of other loads varies as the square or the cube of the speed.

c. Voltage deviation. Standard motors are guaranteed to operate satisfactorily within −10 percent of nameplate voltage, although not in accordance with any specification guarantees.

d. Frequency variations. Severe disturbances can be produced if frequency and voltage swings occur simultaneously, particularly if the voltage swings in the opposite direction to the frequency. The machine manufacturer should be consulted for maximum variations.

e. Unequal phase voltage or single-phase operation. Unbalanced voltage or single-phase operation of polyphase machines may cause excessive heating and ultimate failure. It requires only a slight unbalance of voltage applied to a polyphase

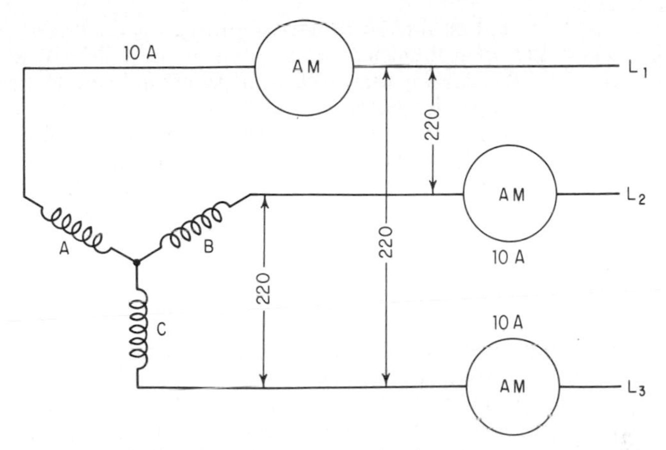

FIG. 11. Simple three-phase motor circuit under normal operating conditions.

machine to cause large unbalanced currents and resultant overheating. In such cases, the power supply should be checked and corrected if even the slightest unbalance (voltage) is found. Single-phase power applied to a three-phase motor will also cause excessive heating from failure to start or from unbalanced currents.

To study the effects of a three-phase motor operating on a single phase, consider Figs. 11 and 12. For simplicity the motor windings are connected one-wye (single-circuit wye).

Now consider the same motor operating single-phase with an open circuit in line 3.

Assuming the line current remains 10 A and ignoring power factor or efficiency, power = 220 W.

Since (maintaining constant motor current) the single-phase circuit (Fig. 12) has only 57.8 percent of the power of Fig. 11 and if the motor and load require

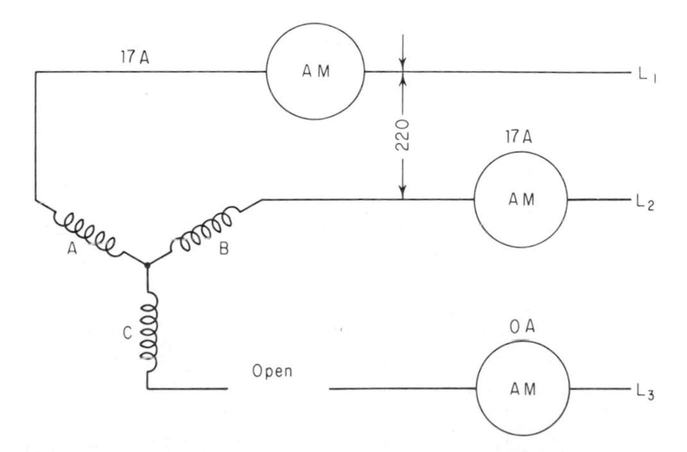

FIG. 12. Simple three-phase motor circuit with one phase open.

100 percent power, the line current will increase in Fig. 12 to a ratio of 1.7, or 17 A (or more, depending upon the design constants). If the full-load rating of the motor has 10 A, phase windings *A* and *B* are overloaded and will heat excessively while phase winding *C* is absolutely idle.

If the motor is at rest, and single-phase voltage is applied to the winding, the motor will not start since there is no rotating magnetic field to produce the starting torque. After the motor is up to speed, the inertia of the rotor will carry the rotor past the dead spots and it will continue to rotate.

At light or no load there is no indication in the operation of the motor that it is operating single-phase. At heavier loads, the magnetic noise and magnetic vibrations will increase, and two of the phase windings will overheat.

The above example indicates how a single motor may burn out, but a more serious condition could occur, affecting many motors, as shown in Fig. 13. Several important and little known phenomena can be learned by a close study of this figure. Assume motor 1 is running on a normal circuit when bus fuse L_3 opens, appling single-phase voltage to motor 1. Also assume that the load has now been removed, and the motor continues to run single-phase at little or light load.

Since bus *C* is dead (from L_3), it might be assumed that the voltage from bus *C* to bus *B* or bus *A* would be zero (which is equal to T_3 to T_2 or T_1). This assumption is incorrect, as the induced flux in the rotor will cause an induced flux and an induced voltage in phase winding *C*, which will almost be equal to normal voltage. Therefore a three-phase voltage will appear across T_1–T_2–T_3 and bus *A*, bus *B*, and bus *C*. The induced voltage in motor phase winding *C* will be in the order of 90 percent of line voltage as

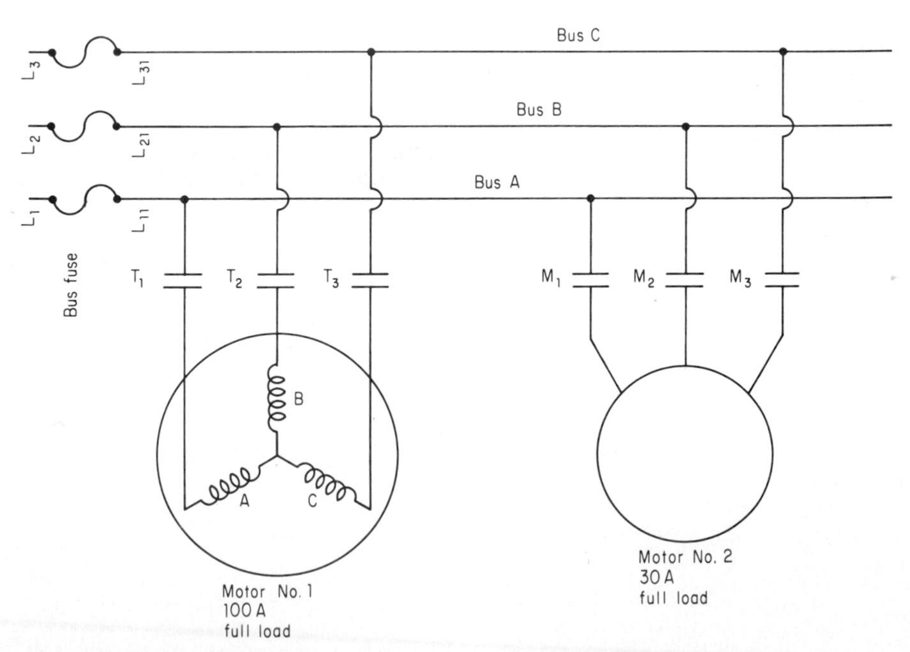

FIG. 13. Two motors of different ratings on one system are susceptible to problems.

$$\text{Voltage } T_1\text{–}T_2 \text{ (bus } A\text{–bus } B) = 100\%$$
$$T_2\text{–}T_3 \text{ (bus } B\text{–bus } C) = 90\%$$
$$T_3\text{–}T_1 \text{ (bus } C\text{–bus } A) = 90\%$$

A test lamp or voltage tester would probably not be able to pick up the difference in voltage (a voltmeter would, of course).

A voltage check at L_1, L_2, or L_3 would show 100 percent voltage, and a voltmeter connected across the open fuse L_3–L_{31} would show the 10 percent "loss" of voltage in phase winding C. It would be dangerous to try to use an ohmmeter to test fuses while in operation, since the 10 percent voltage would damage the ohmmeter. A three-phase power-loss relay would be able to detect the single-phase operation and could open the power circuit.

Now while motor 1 is single-phasing and generating a third leg voltage, assume motor 2 (which is a smaller motor) is connected to the line by contactors M_1, M_2, and M_3.

Since we now have a three-phase supply (10 percent unbalance), motor 2 will start and run. However, it is likely to burn out at part or full load because of the unbalance in phase voltage. A large number of motors could burn out in the same way if they were small in comparison with motor 1.

It is interesting to speculate how long this condition could be occurring before the true facts are known, and how many arguments have been produced when a manufacturer refuses to allow a warranty because a motor was apparently overheated in typical single-phase or unequal-line-voltage operation, but the user's operators insisted that the motors were not single-phasing because they would start! Proper protective control would eliminate those problems (Ref. 1).

18. Preventive-Maintenance Test Equipment. Test equipment for preventive maintenance may, like camera or fishing equipment, vary from the ridiculous (extremely meager) to the sublime (an excess of seldom used equipment). Just as it is difficult to tell another fisherman exactly what equipment he or she needs, it is equally difficult to lay out the test equipment for someone else.

In an attempt to point out the necessary equipment for each category of testing, the first items of each category are considered necessary, while the succeeding items become more exotic.

a. Temperature testing. Since electric equipment is rated in degrees Celsius, all measuring devices should be calibrated in the same scale so that conversion factors will not have to be used. Temperature ratings vary from 40°C (standard ambient) to 180°C (Class H insulation), and measured temperatures will probably vary from 20 to 200°C.

1. Thermometer. Glass thermometers are reasonable in price and sufficiently accurate. They are available with glass eyes for tying. It is not advisable to use a steel-encased thermometer around electric equipment. The bulb is held to the part being tested by putty or a similar plastic. The thermometer is held in place until a constant reading is obtained for a 5-min period.

2. Contact pyrometer. This instrument consists of a thermocouple and a voltmeter. The thermocouple generates a voltage proportional to the temperature and the voltmeter is calibrated in degrees of heat. A reading can be taken in less than a minute. Leads of different lengths can be obtained.

3. Resistance temperature detector (RTD). These devices are usually insulated for high voltages and consist of a resistance element. The resistance varies as the temperature, and is used on a bridge circuit. The balancing arm of the bridge is calibrated in degrees. By suitable switching means, a number of circuits can be measured in sequence.
4. Recording equipment. Occasionally a recorded chart over a period of time is required for varying temperatures. This arrangement consists of a recording mechanism in conjunction with item 1, 2, or 3.

b. Voltage measurements. Since most machines are rather critical with respect to proper operating voltage, every preventive-maintenance department must be able to check voltages.

1. Portable volt-ohmmeter. These meters can be made with 2 percent accuracy, ranging from a few millivolts up to 600 V ac or dc. Some of these portable meters can read up to 6000 V, but they should be limited to electronic circuits and not used on power circuits.
2. Potential transformer (ac only). These unit devices step ac voltages down to a nominal secondary-circuit voltage of 120 V. The nameplate will indicate the ratio between primary voltage and secondary voltage. The iron core or case and one secondary lead should be firmly grounded. High-voltage fuses should be installed in the primary leads.
3. Recording voltmeter. A record is sometimes required of voltage fluctuations over a period of time.

c. Current measurements

1. Clamp-type portable ammeter. These instruments are available in many different current ranges. This type of ammeter should not be on circuits with a voltage higher than the rating of the insulation on the clamps. Accuracies of from 2 to 5 percent are available in these instruments.
2. Current transformer. Used where a greater accuracy is required or where high voltages exist. The primary is connected in series with the circuit being measured, and the secondary is connected to an ac ammeter. The nameplate will indicate the ratio. The secondary circuit is usually 5 A when full load is flowing in the primary. The core or case of the current transformer (CT) and one secondary lead should be firmly grounded.

   Caution. Never use fuses in the secondary of a current transformer. The secondary side may be short-circuited at will, but must never be opened with primary current flowing as extremely high voltages will be induced in the open secondary winding.
3. Recording ammeter. This equipment may be desirable if a chart showing previous readings is required.

d. Megohmmeters. Some type of megohmmeter is a must in any preventive-maintenance department.

1. Portable battery-operated
2. Portable ac rectified and filtered
3. Hand-cranked dc units

4. Motor-operated dc units
5. High-voltage dc units (variable voltage)

Voltages may range from less than 50 to more than 15,000 V, and megohm readings may vary from 25 to 10,000 MΩ. Prices vary according to voltage ratings, megohm scales, and accuracy.

e. Resistance measurements. Scales, accuracies, and voltages vary widely, as well as prices.

1. Portable volt-ohmmeters
2. Kelvin bridge for extremely high resistance
3. Wheatstone bridge for extremely high resistance
4. Megohmmeters for higher megohm values

The Kelvin and Wheatstone bridges are rated as laboratory instruments, and care must be observed in use and handling.

f. Vibration measurements. Such measurements are frequently necessary, as the "hand feel" is extremely inaccurate.

1. Mechanical instruments. These devices are sensitive to the lower frequencies but may not follow the higher frequencies.
2. Electronic instruments. Extremely accurate and will follow the higher frequencies.

g. Speed indicator. This device is extremely useful and is a must.

1. Speed-revolution counter. This counter is a geared device which counts the revolutions. It must be used in conjunction with a watch and will give the revolutions per period of time.
2. Tachometer. This instrument gives an instantaneous reading of revolutions per minute. Several scales are available.
3. Tachometer-generator. The tachometer-generator consists of a generator and a voltmeter. The voltage of the generator is proportional to the speed, and the voltmeter is calibrated in revolutions per minute. These instruments are ideal for remote indicating and may be calibrated for leads of any length.
4. Stroboscope. This unit is a calibrated box using a flashing lamp. It is also used to make rotating or vibrating elements "stand still." It is extremely accurate except that the basic speed must be known. Errors of 100 percent can exist in unskilled hands.

h. Power analyzing. The use of more exotic test equipment such as single-phase and three-phase wattmeters, power-factor meters, and frequency meters may or may not be justified in a preventive-maintenance department.

These and other instruments are fragile and are subject to loss of accuracy by rough or improper use. All instruments should occasionally be checked for calibration and accuracy. Nothing is so useless and frustrating as a meter that reads 20 percent low when it is assumed that the reading is correct.

19. Safety. When machines fail and production lines are down, the pressure may become tremendous for hasty restoration of operation. However, regardless of pressures exerted, unsafe acts are never condoned by responsible management. Any supervisor who performs, allows, or directs his people to assume dangerous positions or acts will be criticized by management. An excellent example of this attitude is the public power companies. Their production flow is seldom interrupted, indicating their desire and determination to maintain operations, yet their safety record is excellent, and "chances" or "hope so" in regard to safety have no place in their thinking. Circuits are held open by locks, the only key being in the pocket of the supervisor. As a further precaution all circuits are grounded before contact is made. If circuits are to be worked hot, safety insulating equipment is regularly checked for adequacy.

Each plant, each industry, and each preventive-maintenance department has its own hazards, and it would be impossible to name or mention all of them. Such a list would even be misleading, as the one item omitted might be the one to cause an accident.

The preventive-maintenance supervisor should, however, observe several recommendations:

1. Convince yourself that your management wants a safe place and will hold you responsible for obtaining it.
2. Convince your people that you want a safe place and that top management insists upon it.
3. At each meeting with preventive-maintenance employees, stress general safety and one or two specific items.
4. Watch your employees and firmly correct them if they perform their work in a hazardous manner. Transfer accident-prone people to a nonhazardous job.
5. Maintain safe equipment.
6. Keep current with new safety information. Much information is available concerning safety, such as films, leaflets, and bulletins (see Ref. 1).
7. Remember, a preventive-maintenance job is well done only when personnel as well as machines are taken care of properly.

20. Preventive-Maintenance Checks. As was pointed out in Art. 1, preventive-maintenance checks and tests must vary in each company and each plant, depending upon the factors involved. A complete check plan must therefore be formulated by the plant personnel. A rough checklist, however, is presented and references to items within this section are made as applicable. Since the section does not deal with certain items such as motor control, no references to these items are included.

The preventive-maintenance checks are made with the motor operating and the motor at rest.

Motor Operating

Item	*Reference, Art. No.*
Review safety procedure	19
Review previous records	3
Vibration	9
Temperature:	
Bearing	13
Windings	7

Motor Operating

Item	*Reference, Art. No.*
Noise:	
Bearing	13
Coupling	13
Other	13
Lubrication	13
Load currents	17
Operating voltage	17
Current-collecting devices	10
Operating conditions (deteriorating influences)	7
Control operation	
Auxiliaries	
Coupling devices	13
Starting	
Accelerating	
Record inspections	3
Communications	5

Motor Not Operating

Review safety procedure	19
Review previous records	3
Cleanliness	7
Megohm reading	18
Current-collecting devices	10
Winding inspections	11, 12
Sleeve-bearing lubrication	15
Sleeve-bearing clearance	15
Tight connections	11
Control inspections	
Inspection of auxiliaries	
Guarding and grounding	
Record inspection data	
Communications	5

When a machine is down and the pressure is on to get it rolling, two problems appear so frequently that a word of caution is in order:

1. Assume nothing or take nothing for granted.
2. Always look for simple solutions first.

REFERENCES

1. Robert W. Smeaton, Ed., *Switchgear and Control Handbook* (McGraw-Hill, New York, 1987).
2. L. R. Higgins and L. C. Morrow, *Maintenance Engineering Handbook* (McGraw-Hill, New York, 1977).
3. Donald V. Richardson, *Handbook of Rotating Electric Machinery* (Reston Publishing, Reston, Va., 1980).

SECTION 18

REFURBISHING OLD MOTORS

Ewald L. Wiedner*

FOREWORD . . . 18-2
1. General Treatment of an Old Motor—Mechanical . . 18-2
2. Cleaning Precautions . . . 18-3
3. Cleaning . . . 18-3
4. Mechanical Inspection . . . 18-3
5. Locked Sleeve Bearing . . . 18-6
6. Scored Shafts . . . 18-6
7. Roller Bearings . . . 18-7
8. Ball Bearings . . . 18-7
9. Lubrication . . . 18-7

DIRECT-CURRENT MOTORS . . . 18-8
10. Dc Motor Problems . . . 18-8
11. Uneven Commutator Surface . . . 18-9
12. Tightening the Commutator . . . 18-9
13. Undercutting the Commutator . . . 18-9

COMMUTATION FACTORS . . . 18-10
14. Carbon-Dust Problems . . . 18-10
15. Discolored Commutators . . . 18-10
16. Open Armature Circuit . . . 18-10
17. Grounded Armature Windings . . . 18-11
18. Function of Commutation in dc Machines . . . 18-11
19. Direction of Rotation . . . 18-11
20. Brush Sparking . . . 18-12

AC INDUCTION MOTORS . . . 18-12
21. Single-Phase, Two-Phase, and Three-Phase . . . 18-12
22. Torque . . . 18-13
23. Starting Torque . . . 18-13
24. Voltage—Rated and Applied . . . 18-14
25. Frequency Changes . . . 18-14
26. Changing Ratings . . . 18-15
27. Changing Voltage . . . 18-16

*Professor Emeritus, Milwaukee School of Engineering, Milwaukee, Wis.

28. Changing Speed 18-16
29. Increasing Torque 18-18
30. Changing Two-Phase to Three-Phase 18-19
31. Scott T Connection 18-19

DC WINDING CALCULATIONS 18-20
32. Formulas for Lap Windings 18-20
33. Formulas for Wave Windings 18-23

REFERENCES 18-26

FOREWORD

Frequently older motors may be repaired or renewed at little cost to provide useful service. When older motors have suitable characteristics for a particular application, a considerable saving in cost may be realized over that of new motors. However, such items as shaft height, mounting-base dimensions, and the available voltage must be considered. If the motor nameplate voltage differs from the available power supply, changes in connections may, in many cases, be made without seriously affecting the motor performance. Some variations in performance may be tolerated; however, only changes which avoid operating difficulties should be considered.

Where motors are to be used to drive production-line machinery, great care should be exercised when using refurbished motors. Failure as a result of a miscalculation may be too costly; however, in many other applications an older machine may give service as satisfactory as a new machine. Old motors usually tolerate a wider voltage range than new ones.

Modern motors are often more efficient than motors built a few years ago. In some applications this gain in efficiency may be important. Also new motors weigh less and are more compact.

1. General Treatment of an Old Motor—Mechanical. The following checks are made to assure good mechanical service from a motor being refurbished.

1. Conduct thorough external cleaning, using an approved solvent.
2. Disassemble machine and clean all internal parts.
3. Carry out careful inspection and micrometer measurement of bearings for sleeve and shaft wear.
4. Check ball-bearing or roller-bearing cage wear.
5. Check ball-bearing races and balls (wear from metal fatigue).
6. Roller bearings usually do not wear but when cages wear, the rollers fall to the bottom when rotation stops and the shaft is not centered on subsequent starts.
7. Oil wells in the bearing housing may be cracked from rough handling of the machine.

8. Sediment accumulation in oil wells prevents oil rings from rolling properly.
9. Oil grooves in bearing sleeve are partly clogged.
10. Oil drain holes are clogged or partially closed.
11. Shaft is bent.

2. Cleaning Precautions. The cleaning methods used depend on the equipment available and the size of the machine. Precautions should be observed since cleaning solvents may be flammable, toxic, or caustic. Proper safety rules must be followed. Careful ventilation and eye protection, as well as skin protection, are important. If flammable solvents are spilled or sprayed on clothing, clothing should be changed to avoid possible skin irritations or fire injury. Static sparks may ignite fumes or liquid. A mixture of combustible fumes and the right percentage of air makes the fumes explosive. Proper grounding should be used as a safety measure.

3. Cleaning. Where cleaning tanks are available for cleaning small parts, there is no problem. A brush or spray is used if no tanks are available. Before a machine is disassembled for cleaning purposes, its method of assembly should be carefully studied and then the reverse procedure followed. For dc machines, brushes, brushholders, and commutators need particular attention during the disassembling operation or process.

4. Mechanical Inspection. After the cleaning procedure has been completed, internal and external, mechanical inspection may begin. Sleeve bearings wear on the belt-pull side as a result of belt pull and shaft pressure. This action may reduce the oil film and increase bearing wear.

Several thousandths of an inch in most machines will not cause any difficulty. However, any wear of the shaft and sleeve changes the air gap and thereby changes the distribution of the magnetic field in the machine, and magnetic and electrical problems may arise. The change in air-gap length in dc machines is perhaps not as critical as it may be in ac motors because the air gap in the design of the former is much greater than it is in the latter. This statement applies particularly to sleeve-bearing machines. Problems arising from unequal air-gap lengths in dc machines cause unequal current distribution in the parallel armature circuits of lap windings (Fig. 1), more so than in wave windings (Fig. 2). The result may be excessive heating and brush sparking.

Polyphase ac motors with unbalanced magnetic fields usually run with too much magnetic noise. Fractional-size motors may not start because of the very strong field on one side compared with the weak field on the long-air-gap side. Once the motor starts, it will run, but bearings will be noisy. In a machine with a small air gap, excessive bearing wear will cause the rotating member to rub on the stator. This rubbing results in excessive friction and heat and could also cause a lamination to be knocked out of place, cutting through the slot insulation and short-circuiting or grounding the winding.

Bearings and shaft should be carefully measured with inside calipers and a micrometer. Carbon or other accumulation of foreign matter in the oil grooves of sleeve bearings should be removed. Oil grooves which are not free of foreign matter prevent proper distribution of the oil on the running surface, and hot oil may result. Oil temperatures of 85°C have been experienced from clogged grooves. At this temperature the lubricating property of the oil may be reduced. Adding oil to

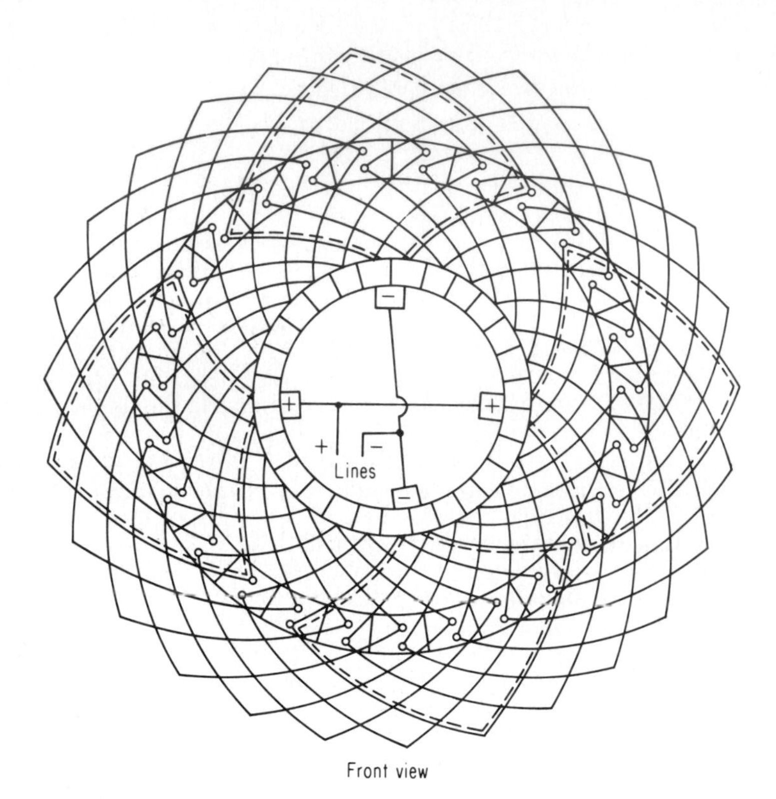

Front view

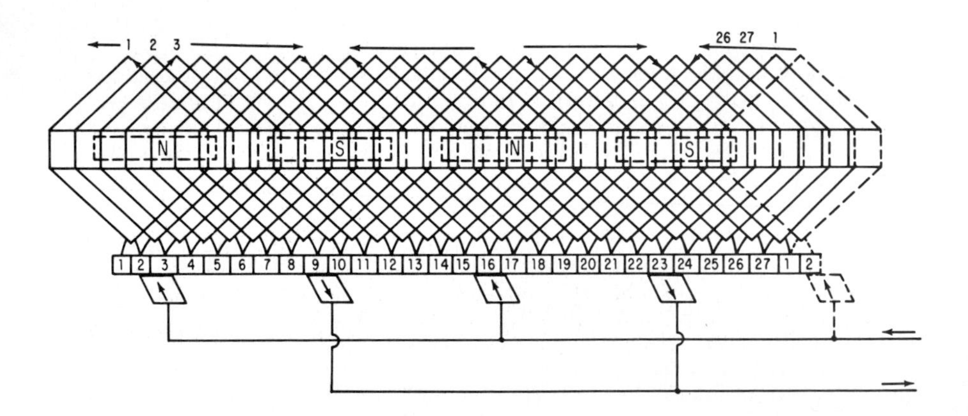

Plane developed view

FIG. 1. Double-layer four-pole 27-slot simplex progressive lap winding with four parallel armature circuits.

$$S = 27 \times 2 = 54 \qquad Y_b = S/P \pm K = 54/4 \pm K = 13\tfrac{1}{2} - \tfrac{1}{2} = 13$$

$$Y_f = 11 \qquad Y_c = 1 \text{ to the right} \qquad N_c = 4 \qquad \text{Coil span} = (13 + 1)/2 = \text{slots 1 to 7}$$

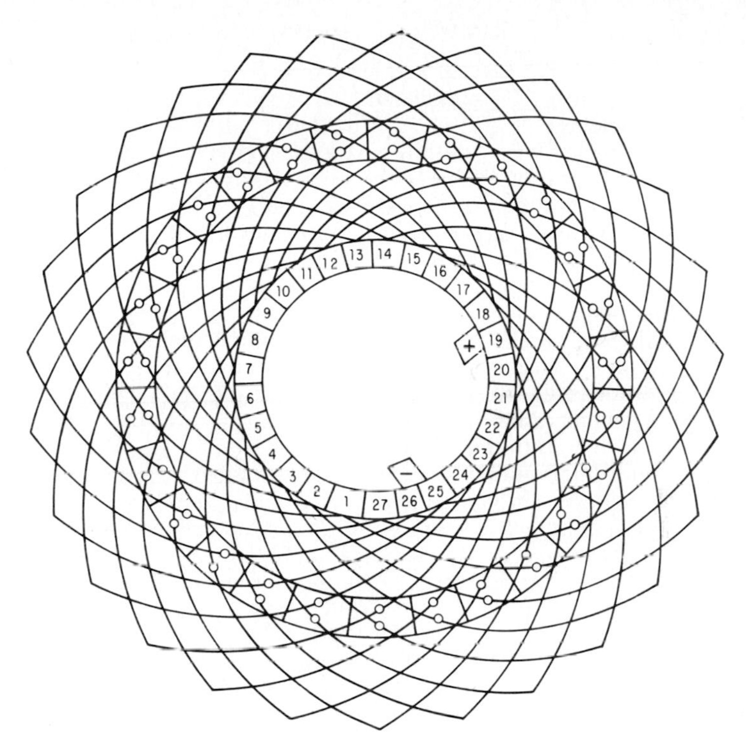

Front view

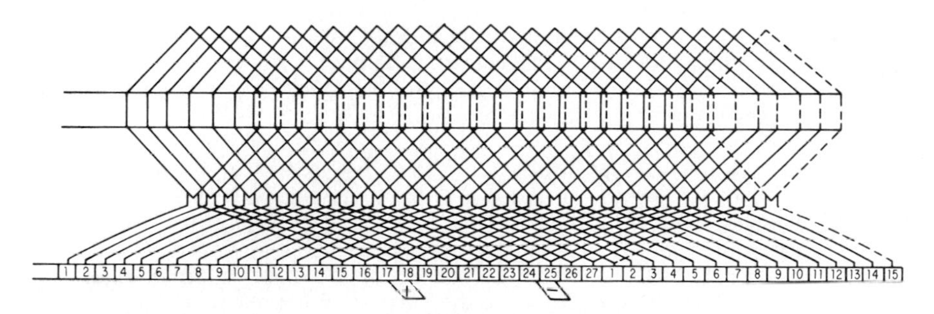

Plane developed view

FIG. 2. Double-layer four-pole 27-slot simplex progressive wave winding with two parallel armature circuits through winding.

$$S = 27 \times 2 = 54 \qquad P = 4 \qquad Y = (54 + 2 \times 1)/4 = 14$$

$$Y_{b1} = 13 \qquad Y_{b2} = 13 \qquad Y_{b3} = 14 \qquad Y_{f1} = 15 \qquad Y_{f2} = 15 \qquad Y_{f3} = 15$$

$$N_c = 2 \qquad \text{Coil span} = (13 + 1)/2 = \text{slots 1 to 7}$$

the bearing will not help. The oil will run out of the overflow or at the shaft and cause further problems.

After the bearing has been carefully cleaned, oil grooves are scraped out clean and scrapings washed away. Oil wells or sumps require inspection for accumulation of sludge or other foreign particles. Sludge and foreign particles prevent the oil rings from rolling and distributing or carrying the oil onto the shaft. The high-speed machines are provided with an oil-ring control lug. This lug is usually a part of the oil-sump inspection cover. Its purpose is to control the motion of the oil ring. Without the lug, the ring will roll so fast that it throws the oil. This action results in a leaky bearing, and the oil level will soon be below the safe limit for proper lubrication. Oil rings with too much freedom of motion produce a distinct ringing sound while being tossed around by the rotating shaft. This ringing sound is the danger signal and indicates that the bearing requires immediate attention because of a low oil level in the sump.

If this low-oil-level condition is not attended to immediately, excessive friction heat begins to melt the soft white metal bearing. Bronze bearing sleeves do not melt so readily but eventually become so hot that bronze adheres to the steel shaft, taking up all the clearance between sleeve and shaft. At this point the motor torque will no longer be sufficient to maintain rotation and the motor stalls from a locked or frozen bearing.

5. Locked Sleeve Bearing. A locked bronze sleeve bearing may or may not be serious. When the bearing has cooled off, the first step is to attempt to crack it loose. Often frozen bearings can be broken loose either electrically by applying power, or mechanically. If a motor-generator set is involved, in some cases it may be necessary to apply controlled power to both machines to get the necessary torque to free the shaft. When the shaft has been freed, it should be rotated slowly with plenty of oil in the troubled bearing. If no trouble is indicated at low speed, the speed may be gradually increased at perhaps 5-min intervals in time and perhaps 10 percent intervals of speed increase to near normal.

If the rotor can be readily turned after perhaps an hour of operation, it may be assumed that not much damage has been done. Depending on the application, machine operation may then be continued under observation, or the machine may be disassembled for careful inspection. If damage is minor, the bearing may give many hours of service without further trouble. Where uninterrupted service is a factor, renewal parts should be provided.

6. Scored Shafts. Bearings which have run dry, particularly where bronze sleeves are involved, may have scored the shaft. The machine may still be used without changing the sleeve when wear is not excessive. With proper care, bearings slightly scored have given satisfactory service for many years. Where the scoring has become bad, the shaft must be machined. A light cut off the shaft and an undersized sleeve bearing may serve the purpose. When the scoring is bad, the shaft may be machined down enough to accommodate a steel sleeve bored to slightly undersize. Heating the sleeve before sliding it onto the shaft and then machining the outside to correct size for the bronze bearing sleeve result in satisfactory performance.

Another method of obtaining a proper fit is sometimes chosen. The shaft may be metal-sprayed and then machined to correct size. When the shaft is machined down to fit the sleeve, the effective shaft diameter is reduced, but the original design safety factor is usually large enough to permit this type of modification.

7. Roller Bearings. Roller bearings, when properly fitted, usually do not develop any wear on the rolling surfaces which carry the weight or belt pressure. The cage lugs that retain the rollers at the proper spacing may wear as a result of lack of lubrication or age. In this case the rollers will no longer be evenly distributed around the circle. This condition changes the rotor centering and brings about its allied difficulties, particularly when the shaft stops. Subsequent starting may result in knocking or excessive vibrations. A new bearing is the only solution.

8. Ball Bearings. In ball bearings the ball-retaining cages wear like roller bearings, usually not enough, however, to permit the balls to get out of position or concentrate at the bottom when the shaft rotation stops. Ball races and balls, however, develop surface wear as a result of metal fatigue. Rolling surfaces disintegrate and become rough. This condition results from repeated pressure deflection.

The elastic life of the metal wears out and the working surfaces become crystallized and brittle, allowing the surface to chip away or disintegrate. The resulting rough rolling surfaces make the bearing noisy. Such a bearing must be replaced.

When ball bearings are installed, proper tolerance is important. Inadequate tolerance when the bearing is pressed into the housing reduces the ball-to-race clearance below permissible values, resulting in increased ball-to-race surface pressure. Reduced bearing life is the result.

9. Lubrication. Proper lubrication requires the right grade of oil or grease in the proper quantities. Local lubrication engineers are always glad to help on any special problem where grades are concerned, and particularly with higher-temperature bearings requiring high-temperature grease (see Sec. 14).

All old grease must be thoroughly cleaned out before new grease is installed. Then the total grease space should be filled to 50 percent to prevent overheating from excessive churning.

A good grade of machine oil will usually supply the needs of oil-lubricated bearings. Some attention must be given to correct oil level and freely rolling rings after the machine has been started.

On older machines the oil-level-inspection fittings are often knocked off during moving and handling. If replaced, the replacements may bring the level too high or too low. If too high, the oil excess will run out and lead to accumulation of dust and dirt, and it could soak up the winding insulation. On dc machines the oil can creep along the shaft on the inside and coat the commutator, brushholders, and brushes, promoting many problems in this area. Rough or careless handling of machines may also result in cracked oil sumps so that oil leaks out in a short time. This condition may be detected if oil leaks out of the bearing, by dripping, when the motor stands idle. Cracks can be found by washing the outside of the bearing housing with a suitable solvent and then wiping the area clean and dry. After machines have been standing idle for several hours, inspection will readily locate cracks if present.

Many fractional-horsepower motors are provided with the stuffing-box type of lubrication. When a motor does not have to operate continuously, the oil well or sump and oil-ring type lubrication are perhaps more elaborate than necessary, and the bearing housing is stuffed with wool yarn or cotton waste. The wool or waste is saturated with oil and the oil-saturated stuffing is held in contact with the shaft through an opening in the bearing sleeve. When an old motor is cleaned, all this yarn or waste should be removed and washed in a solvent and dried before it is replaced in the stuffing box. After the dried stuffing has been replaced, enough oil is applied to saturate the yarn, but not so much as to cause dripping from the vent. In some cases the stuffing box is provided with a light spring to force the oil-soaked waste against the shaft. This pressure or some other means is generally required so that the shaft cannot run dry while an adequate supply of oil is present. Once correctly oiled and assembled, this type of bearing normally needs the addition of a few drops of oil every 6 months if the motor does not operate more than several hours a day.

Overoiling here leads to other problems. In fractional-horsepower ac motors with internal starting switches, creeping oil comes in contact with the breaker points and eventually results in too much arcing or poor contact. Finally total burning away of the points may prevent closing of the auxiliary or starting winding or may cause contact welding. When the starting switch remains open, the motor cannot start, and if improperly protected, the main winding would burn out. In the second case the auxiliary or starting winding would burn out. If the motor is of the capacitor-start type, the starting capacitor can blow out before the starting winding burns. Capacitors for this application are in many cases provided with a safety blow-out plug. If the safety plug is blown out, total destruction of the capacitor is avoided and the capacitor may still be usable.

DIRECT-CURRENT MOTORS

10. Dc Motor Problems. Dc motors and generators have the same mechanical problems as ac machines. Electrically, however, they have additional problems not normally found in ac machines.

Dc motors have commutators which provide problems of their own. The commutator is perhaps the most expensive replacement part of a dc machine. If properly used and taken care of, it will give successful service for many years. Some of the common problems or faults found in old motor commutators are as follows.

1. Uneven surface wear—flat spots on the commutator.
2. Accumulation of brush carbon and dust in undercut mica grooves. This accumulation is hastened by oil creepage.
3. Arcing across mica through the caked brush carbon in the undercut.
4. Short-circuited commutator bars developing after the machine has been operated too long without cleaning out of the undercut mica grooves.
5. High spots and low spots resulting from loose bars shifted in the assembly.
6. Disconnected armature leads at bars, resulting in open armature circuits. This condition is common in fractional-horsepower sizes.
7. Poor soldering jobs where parallel conductors are used on high-current low-voltage machines.
8. Burned brush-stud insulator washers.
9. Bars short-circuited from arcing across the mica.
10. Commutator bars having thrown solder as a result of excessive current circulating in the coil when bars are short-circuited.
11. Brush position incorrect for load applied.

12. Wrong direction of rotation for brush position.
13. Using a motor as a generator or vice versa without considering the relation between the neutral zone and the direction of rotation.
14. Brushes sticking in holder.
15. Brush spring tension too low.
16. Brushes worn too short. Brush spring, in some cases, rests on limit stop instead of brush.
17. Open armature circuit.
18. A grounded armature winding.

11. Uneven Commutator Surface. This condition may result from:

1. Dropping the commutator against the pole face or frame during disassembling of machine or when reassembling.
2. Mica insulating rings not being tight as a result of improper curing.
3. Sufficient heat to bake binder out of mica, leaving assembly no longer a solid mass, and permitting the bars to shift out of position, resulting in high or low bars, or both.

12. Tightening the Commutator. The threaded end ring or cap screws must be tightened after the bars have been tapped into approximate position. The tapping and tightening procedure must be continued until the assembly seems to be solid. The commutator must then be machined to a true circular shape. Commutators should be checked with a dial indicator. Low-speed machines may have an indicator variation of 0.005 in, while the variation may have to be reduced to perhaps 0.001 in for high-speed commutators.

Small commutators which are held together by a rolled tube cannot be rejuvenated and must be replaced. Some smaller commutators wear to an oval shape as the result of a commutation problem. The machine may be put back into service after machining; however, the commutator must be undercut for low-voltage high-current machines.

13. Undercutting the Commutator. On higher-voltage machines the undercutting is not quite as important as on lower-voltage machines. When commutation is not good, the copper wears down faster than the mica, resulting in objectionably high mica. While undercutting is essential, there is always the possibility that brush carbon will accumulate in the undercut, which will eventually cause sparking or arcing across the undercut. Mica burns away and bars are short-circuited. Short-circuited bars may be observed by discoloration of the copper or by solder being thrown off the bars at the neck of the bar where winding or coil ends are soldered to the bars.

On commutators where mica between bars or segments is not so thick undercutting is not as important because the mica wear is equal to the copper wear. The softness or hardness of the brush used determines the rate of wear in this case. Higher-resistance brushes have a self-cleaning effect on the commutator, and softer brushes have reduced voltage drop, more short-circuit commutation current, and less cleaning effect.

COMMUTATION FACTORS

So many factors are involved in satisfactory commutation that each case must be analyzed on its own merits.

14. Carbon-Dust Problems. Carbon-dust accumulation in undercut mica is more likely to occur in machines which have been idle for some time. Dust, creeping oil, and brush carbon dust fill in the undercut spaces, causing ring fire when the machine is put back into service. Ring fire represents arcing across the mica through the carbon and dust accumulation in the cut between the bars. The remedy is to clean first the commutator with a suitable solvent and then scrape the cuts clean with a properly ground piece of hacksaw blade. If the machine is operated without properly removing the accumulation of carbon and if the accumulation is not serious, the carbon will burn away. This burning, however, introduces another bad feature: the mica burns and gets a charred surface across which more arcing takes place. If not properly cleaned out, the bars involved become overheated and show a definite discoloration from excessive heat caused by current circulating in the coil or coils involved. Coils show overheated varnish and bars throw softened solder. On lap-wound machines only one coil is involved for one short circuit, but on wave windings there will be two, three, four, or more, depending on the number of poles in the machine (Figs. 1 and 2). In case of one short-circuited coil, the coil may be cut to prevent circulating currents and a wire of suitable size may be soldered across the bars involved. If it is a simple wave winding, all series coils must be cut and the two bars involved must be bridged.

When coils are cut and bars bridged the speed of the motor is increased. When an emergency repair of this type is made, circumstances will dictate whether this method is permissible. The change in operation will be approximately according to the percentage of conductors involved.

15. Discolored Commutators. Commutators become discolored from overheating. Overheating may result from too much load for the brush position, or the direction of rotation may have been changed without changing the brush position to correspond to the new neutral zone. When only two adjacent bars are discolored, arcing across the mica below the surface is indicated. Using a motor as a generator or a generator as a motor may also cause overheating of the commutator and winding, although line current is still within reasonable limits.

In older machines, oil creepage and dust accumulation gum up brushes and brushholders. This condition in conjunction with out-of-round commutators causes sparking because of poor or improper brush contact. A machine operating in this condition too long causes flat or burned spots to develop on the commutator. When brushes have worn to too short a length without proper spring adjustment, this or a similar condition may develop. If brushes have been worn so short that the spring can no longer exert the proper pressure, poor contact between brush and commutator causes too much arcing, overheating the commutator. The commutator bars will then throw the softened solder and cause open armature circuits.

16. Open Armature Circuit. A motor with one open armature circuit will start and run, but when a given load is applied, the current per armature circuit will be higher according to the portion of the winding involved. The open circuit will be closed by the brush when the brush bridges the break in the circuit. When load is applied, the speed will drop off faster and the remaining windings will heat at a higher rate. The commutator bar at the open circuit will begin to burn black because the brush bridges the open circuit.

When the commutator bar or segment leaves the brush, the breaking of the circuit produces an arc, and this continuing arcing burns the involved bar black. Further operation will burn the bar or segment away on the trailing edge. The total effect which one open circuit has on the operation of the machine depends on whether it has a wave or a lap winding since there are only two parallel circuits through the simple wave winding for any number of poles, while the simple lap winding has as many parallel circuits as poles.

17. Grounded Armature Windings. When the frame of the machine is grounded as required by code for permanently installed machines and the armature winding contacts the iron core in one place, a commutator bar will burn similar to an open circuit.

Portable machines should also be grounded as a safety measure. When the frame is not grounded, one winding ground will not show up on the commutator as it does on a grounded-frame motor. This condition can be dangerous to the personnel operating the machine.

When insulation resistance is checked as a safety measure, 2000 Ω/V may be considered satisfactory. Measurement may be made with a Vibrotest unit or a megohmmeter, or by the series-voltmeter method.

18. Function of Commutation in dc Machines. Commutation is a very important factor involved in the operation of dc motors and generators. It consists of reversing the current in an armature coil when the coil (ends) moves from one side of the brush to the other side of the same brush, also to complete the connection between the armature winding and the external circuit.

Reversing the current in the coil while it moves from one side of the brush to the other side must take place in a small fraction of a second. Therefore this reversal involves coil inductance as the current is reduced to zero and then again builds up to normal since the coil has moved to a circuit of opposite polarity. Any short-circuit current flowing in the coil when it moves away from the brush causes a spark. Also if a voltage is induced in the coil when it is short-circuited by the brush, the low-resistance coil has a considerable current flowing. The intensity of the sparking depends on the current in the circuit which is broken. The voltage induced in the coil being commutated varies with the load applied since the magnitude of the load affects the flux distortion in noninterpole machines. Proper brush shifting within reasonable limits may induce a reversed voltage to overcome some of the objectionable sparking.

19. Direction of Rotation. If the brush position for the direction of rotation and the interpole magnetomotive force is correct, in machines provided with interpoles, the normal flux distortion is counteracted and a few additional ampere-turns provide flux for reverse voltage to counteract the effect of self-inductance in the coil being commutated.

If the direction of rotation is not correct for the interpole polarity, or the direction of brush shift is wrong, the machine will work fairly well at light loads but will spark excessively at rated line current and may become unstable under overload conditions. The instability here means that without any change in load the motor will accelerate and decelerate alternately. A change in the direction of rotation usually restores stability even when extreme overloads are temporarily applied.

20. Brush Sparking. Some reasons for sparking at the brushes and commutator have been given in Art. 10. However, the most important reasons for sparking are covered here. At times there will be sparking at one or more brushes while there is little or no spark at the others. This condition results from brushes that are not equally spaced or that have improper brush-contact pressure. If brushes are not equally spaced, shifting brushes may improve the bad ones but causes undesirable sparking at the previously good ones.

Unequal spacing of the brush is often caused by rotating the brushholder on the brush stud. This position cannot be determined by looking at the brushes, but a piece or strip of thin insulating paper the width of the brushes should be laid around the commutator with the brushes resting on the paper. The paper may be held in place with a piece of masking tape. The heel and toe of each brush are then marked with a sharp pencil on this ring of paper. Once the brush location has been marked on the paper, the paper ring may be cut with a sharp knife and carefully removed from the commutator. The brush spacing is then checked with a pair of dividers or calipers after the paper has been laid on a flat surface. Any difference in spacing may be corrected by rotating the brush stud or the brushholder. The amount of correction required may be marked on the original paper ring, which was marked off and replaced. Rotating the brushholders or the brush stud to obtain equal spacing results in poor brush seating for the affected brush or brushes. The brush must be reseated with a suitable seating stone or with fine-grained sandpaper taped onto the commutator with the grit side up to contact the brush or brushes requiring this treatment. After the sandpaper has been properly installed, the armature may be moved back and forth by hand on smaller machines, until the whole end of the brush or its face has the contour of the commutator. On large machines, which cannot be moved back and forth, it may be necessary to drive or to run the machines and use the commutator seating-stone method to obtain the proper brush contact.

When a brush or brushes have been seated by either the stone or the sandpaper method, the brush face is rough. This condition will affect the performance of the machine for about a 24-h period. During the "running in" of the brush, its face becomes somewhat smooth, and the change will affect the output of high-current low-voltage machines more than that of low-current high-voltage machines. Motors are not affected quite as much as generators.

AC INDUCTION MOTORS

In the following some information is provided on what to expect when an old motor is available and the use for which it is intended is perhaps different from what the designer had in mind. Performance will be different but it may still be within reasonable limits.

21. Single-Phase, Two-Phase, and Three-Phase. Operation is based on applying a voltage to the primary and an inducting voltage to the secondary, similar to a transformer. The reaction between primary and secondary magnetic fields produces torque. The value of torque depends on the magnetic-field strength and the angular and time displacements between primary and secondary flux.

The primary magnetic field rotates or shifts at a speed of $120f/P$, or

$$\text{Speed} = \frac{60 \times \text{frequency}}{\text{pole pairs}}$$

and in so doing induces a voltage in the secondary conductors. Circulating current in the rotor provides the secondary field, which reacts with the stator field. The magnitude of the voltage induced in the rotor depends on the relative motion as the stator field sweeps past the rotor conductors.

The rotor speed of the induction motor is somewhat below that of the stator field, as indicated above.

The torque developed and the countertorque (load applied and frictional losses) relationship determines the actual operating speed. The nameplate speed in most cases is the standard expected speed when rated voltage at indicated frequency and rated load are applied. An exception is an accepted standardized speed as indicated on the nameplate. This value is usually below the actual operating speed.

22. Torque. Since the primary function of a motor is to develop torque, the factors that influence it must be considered. When a designer makes the calculations for a proposed motor, he or she has direct control. Once the motor is built, conditions outside the original plan may change the actual performance.

A change in applied voltage also changes the current for a given applied load. Under these conditions, current and voltage changes are opposite to each other. Minor changes in applied voltage result in minor speed changes for a fixed load. A convenient rule to use in connection with voltage changes is that the applied voltage may be changed 10 to 15 percent above and below the nameplate value without producing operating difficulties.

Under running conditions, a change in the applied load changes the induced rotor voltage, rotor frequency, and rotor current. The change in rotor frequency is not enough, over the normal load range, to become an important factor in the phase relationship between primary and secondary flux values. Under excessive overload conditions, this change becomes important in the motor performance.

23. Starting Torque. When the starting torque is to be increased, the voltage is raised above normal or the motor can be rewound to bring about this new performance. The number of turns in the winding must be reduced so as to increase the starting current. With fewer turns in the coil of the winding, the iron will be worked at a higher point on the magnetization curve. This change is permissible within reasonable limits. Higher flux density results also in increased core losses and increased iron temperature.

When coils of the stator winding are to be rewound, a comparison should be made between slot room and wire size. If slot capacity permits the use of the next larger wire size, it should be used because of the increased current. Reducing the number of turns in the winding increases the current from the no-load value all along the load range. It also lowers the operating power factor.

In fractional-horsepower motors it is often found that the no-load current is too high, resulting in high operating temperatures. If the starting torque is not an important factor, the motor may be rewound using perhaps a 5 to 7 percent increase in turns per coil with the same size or one gage number smaller wire. From this increase in turns the no-load input, flux density, and operating temperature are reduced. With the reduction in no-load current input, the motor output may be substantially increased before the rated line current is reached. The gain in output applies more to fractional-horsepower motors than to integral sizes, as a given amount of iron is good for a limited amount of magnetism if the iron is held within a reasonable operating flux density.

The starting torque of slip-ring motors may be increased by using external resistors or by rewinding the rotor with the next smaller size of wire with an increased number of turns. The increased resistance of the rotor improves the phase angle between rotor and stator flux. When too much resistance is used, the effect of improved phase angle is lost by excessive reduction in current and decreased torque.

24. Voltage—Rated and Applied. Older motors are usually rated at lower voltage than is supplied in many metropolitan areas. Voltages in these areas also vary with the system load and the time of day. Old standard voltage ratings are 110 and 220 V, and the newer ratings are 115 and 230 V as well as 440 and 480 V. If the applied voltage does not exceed 15 percent of the rated values, there should be no operating difficulties. This 15 percent does not, however, apply to motors built at the time of this writing. In a metropolitan area voltages may change from 236 to 246 V during the regular working day. At night and on Sundays the voltage increases to 256 V. On the lower-voltage circuit this value would be 128 V, or 117 percent of 110 V. Using an older 110-V motor on 128 V or a 220-V motor on 256 V is not objectional but it will produce the following results:

1. A small increase in no-load current (depending on the degree of saturation of the iron core)
2. Better speed regulation
3. The same load carried at slightly lower current
4. Increased starting torque
5. A small reduction in the insulation safety factor

Present-day motors will overheat when operated at the above increase in voltage over rated value. Overheating occurs because of the difference in iron-core saturation characteristics and the higher flux densities used.

25. Frequency Changes. A change in frequency results in a speed change. Related to the stator and rotor windings, a frequency change represents a change in inductive reactance. To maintain the original current, the applied voltage should be changed according to the frequency. A 110-V 25-Hz motor may be operated on a 240-V 60-Hz power supply. In doing so, several changes in performance take place. Synchronous speed changes by a factor of 2.4. The higher frequency increases the rate of producing heat in the iron core. The higher core loss would normally increase the rate of producing heat in the iron and seemingly raise the operating temperature. However, the increased speed increases the cooling air blown through the machine, and the cooling effect would be increased at a greater rate than the increase in heating. Normal safety factors provided in the smaller machines will tolerate the higher speeds. In the larger sizes, increased vibration and centrifugal forces must be considered. For fractional-horsepower split-phase motors, the centrifugal starting-switch operation must also be considered.

Operating a 60-Hz motor on 25 Hz involves fewer problems. The applied voltage must be reduced by an approximate factor of 2.4 to limit the current in the winding to a safe value according to the reduced reactance. Since the speed is a function of the frequency, the horsepower output will change directly with the speed at the proper applied voltage. A 1-hp 25-Hz motor will become a 2.4-hp motor on 60 Hz and a 1-hp 60-Hz motor becomes a 0.4-hp motor on 25 Hz.

When operated at reduced speed, single-phase machines with starting switches may require replacement or modification of the starting switches. When a 60-Hz motor is operating at a 25-Hz speed, the starting switch may not open. The starting winding will burn out if the motor is run under these conditions. A reduced rate of iron heating will partly be compensated for by reduced air circulation.

Another frequency not so common is 50 Hz. A 50-Hz motor may usually be connected to a 60-Hz power supply without getting into operating difficulties. If the rated voltage is applied, the starting torque will be lower but the horsepower output is increased. Older motors usually have a voltage rating below the present power-system values. This lower rating will aid in improving the performance of 50-Hz motors used on 60-Hz systems.

More modern machines may be rated at 50 and 60 Hz. These machines may be designed for 50 Hz and a service factor of 1 at rated voltage. At 60 Hz the service factor may be more than 1, although not indicated. Another plan used in the design is to calculate the winding for 55 Hz and then using the machine on either 5 Hz above or 5 Hz below design value and still obtain satisfactory performance.

Shortly after the end of World War II some communities were flooded with small 220-V 180-Hz motors. Since no 180-Hz power supply was available, the use of these motors was severely limited. Frequency converters were too costly to use in this case. A simple phase-splitting device could be rigged up, however, so that the motors could be used on 120 V, 60 Hz, single-phase. The load applied and the operating time, however, had to be regulated according to the heating of the motor.

26. Changing Ratings. Sometimes it may be desirable to change the ratings of a machine. Possible changes of frequencies with resulting speed changes plus corresponding voltage and horsepower-output changes were discussed in Art. 25. Simple voltage changes will be considered here. These involve series-to-parallel or parallel-to-series connections and delta-wye or wye-delta connections. The parallel-to-series or series changes involve only voltage and current ratings, but not speed or horsepower output. The delta-wye or wye-delta changes introduce changes of 100 to 58 percent or 58 to 100 percent of rated voltage. This type of conversion is popular on six-lead dual-voltage three-phase machines. A 440-V series-wye-connected in series-delta would change the rated voltage to 254 V, but this winding would work on a regular 240-, 230-, or 220-V circuit. The horsepower output at the original phase current would be reduced, or it could be considered at the same rated horsepower output with a slight increase in phase current. The line-current rating for the delta-connected winding is 173 percent of the original. Change from a 240-V delta to a 240-V wye connection would change the line current to 1/1.730 of the original value. The 416-V winding will provide good performance on a 440-V circuit. Other convenient voltages possible are perhaps the 208-V two-parallel, connected series, used on 440 V, or vice versa. A 660-V series-wye-connected winding could be connected two-parallel delta for a rated voltage of 191 V and used on a 208-V system. The 208-V supply would be slightly above the normal permissible overvoltage, and the operating temperature may also be high but not dangerously high. Assuming the original current rating to have been 10 A, the new current rating would now be 34.6 A. Since the heating can, to a certain extent, be controlled by applied load, this current could be controlled by limiting the load. The flux density would be increased as a result of the increased applied voltage, but not enough to result in excessive core loss.

27. Changing Voltage. It is frequently desirable to be able to change the voltage connection of a motor winding, so that an otherwise useless machine can be used without rewinding. Rewinding will in most cases produce more nearly the rated performance, but this operation requires both material and labor cost. Not all winding connections lend themselves to reconnection for other voltages. Establishing the present connections is required before it is possible to determine which reconnections are permissible for existing voltages. To change from higher to lower voltage usually involves no insulation problem, but when changes are made from lower to higher voltage, insulation resistance to ground or between phases should be considered. However, insulation resistance alone may not be enough because the dielectric strength is also important when the voltage is more than doubled. Assuming 1000 Ω/V, for a good approximation, might be considered adequate. The presence of moisture in the insulation reduces the resistance, and heating to evaporate excess moisture would be an improvement. A comparison of values of insulation resistance, rated voltages, and operating temperature may show considerable difference between the design values from the engineering department and the values from the maintenance department, when shop production is the important consideration.

Simple changes are from series to two-parallel to reduce from 440 to 220 V or from 240 to 120 V. Other permissible changes are readily made possible by the wye-delta combinations. Some common permissible connection changes are:

1. 220-V wye to delta to provide a 127-V rating
2. 440-V wye to delta to provide a 254-V rating for a 240-V system
3. 240-V delta connected wye to provide a 416-V rating which, when compared with 440 V, is slightly below 95 percent

28. Changing Speed. The primary speed-determining factors for induction motors are pole pairs and the frequency of the applied voltage. A change in speed also changes the rated voltage of the winding. If the change is to be accomplished by reconnecting and regrouping the coils, the new voltage rating must be calculated according to the new distribution factor and the coils or turns in series. A difference of not over 15 percent of the original voltage may be tolerated. Raising the speed increases the vibration and stresses resulting from centrifugal forces and reduces the equivalent amount of iron per pole flux. A second change in speed may be accomplished by reconnecting from whole-coiled to half-coiled winding, which reduces the voltage and speed to half. For a single-phase 220-V series-connected winding this change works out just right, but for three-phase windings the spread or coil span should be 90 or 120°. When the coil span is 120°, there is some cancellation of flux and the corresponding voltages, but the effectiveness of the coil is somewhat higher. Changing from six poles to four poles will change the original iron available for six parallel paths of pole flux to 4 parallel paths. This change raises the flux density in the core considerably.

Reducing the speed by increasing the number of poles seldom brings about a flux problem if the proper voltage is applied to the winding. Changes involving six and eight poles, eight and ten, or ten and twelve poles usually do not present serious problems.

When planning a speed change by regrouping the coils for a different number of poles, a few examples might be helpful. Assume that a 72-slot six-pole three-phase stator with the coil installed in slots 1 and 11 is to be reconnected or regrouped for eight poles. 72 slots ÷ (6 × 3) = four coils per pole per phase; coil spans 150°. 72 slots ÷ (8 × 3) = three coils per pole per phase; coil spans 200°.

$$\text{Chord factor for six-pole winding} = \sin\frac{150^\circ}{2} = \sin 75^\circ = 0.965$$

$$\text{Chord factor for eight-pole winding} = \sin\frac{200^\circ}{2} = \sin 100^\circ$$

or

$$\sin\frac{160^\circ}{2} = \sin 80^\circ = 0.985$$

Distribution factor for four coils per group at 15° = 0.956

Distribution factor for three coils per group at 20° = 0.96

Voltage-inducing effectiveness per turn for six-pole grouping = 0.965 × 0.956 = 0.923

Voltage-inducing effectiveness per turn for eight-pole grouping = 0.985 × 0.96 = 0.946

Comparing the latter two values,

$$0.946 \div 0.923 = 1.025$$

or an improvement of 2½ percent

The loss in induced voltage resulting from speed reduction from six poles to eight poles is approximately 25 percent. The combined change represents approximately 77.5 percent effectiveness, or for a 440-V winding, the rated voltage will be 440 × 0.775 = 340 V. A wye-connected winding changed to delta would correspond to a rating of 197 V. This value would be suitable on a 208-V system, or even 220 V. The 220-V value is 11.5 percent above rating and may be permissible.

Assuming another possibility, a two-parallel 440-V delta connection, changed to a series-delta multiplied by 0.775, makes the rating equal to 682 V, or suitable for a 660-V service. A 660-V series-wye winding reconnected for the eight-pole corrected voltage would have a rating of 513 V and should work on a 550-V power supply. Or a 550-V series-wye winding reconnected for eight-pole series-wye would change its rated voltage to 427 V, making it suitable for a 440-V supply. Other combinations may be worked out, once the voltage-inducing effectiveness has been determined and the winding connection has been traced out. Some changes are possible and others have no suitable voltage available.

Another problem concerns the possibilities of a ten-pole 440-V three-phase series-wye-connected 120-slot stator to be reconnected for eight poles. Coil sides are laid in slots 1 and 11. Ten-pole, 440-V, three-phase, 120-slot = 120/10 × 3 = four coils per pole per phase, and degrees per slot = 180 × 10/120 = 15°. Span = 150°. For the eight-pole winding, 120 slots = 8 × 3 phases = five coils per pole per phase. Eight-pole, three-phase, 120-slot = 180 × 8 poles/120 slots = 12° per slot. Coil span = 120° for the contemplated new connection. Chord factor for the ten-pole winding = sin 150°/2, or 75°, = 0.965. Chord factor for the eight-pole winding = sin 120°/2, or 60°, = 0.866. Distribution factor or vector sum of five coils in a series-group-induced voltage at 12° = 0.956. Speed relation in changing from ten to eight poles is approximately 1.26. Ten-pole chord factor times distribution factor = 0.965 × 0.956 = 0.828. Voltage-inducing effectiveness of eight poles compared with ten poles = 0.828/0.922 = 0.9. Combining this factor with the increased speed factor of 1.26 = 0.9 × 1.26 = 1.134 for each 1 V induced in the ten-pole original.

Using the original assumed wye connection, the winding would be good for 595 V as wye-connected. This value is 90 percent of a possible 660 V across which the eight-pole winding could be used. The starting torque would be about 81 percent of the 660-V torque, and the increased speed should take care of additional heat resulting from the increased current. The horsepower output would be increased in the approximate ratio of increased speed.

As another example, assume the original ten-pole winding series-delta-connected. As an eight-pole winding its rated voltage would be 500 V. If reconnected for two-parallel delta, its new rating would be 250 V. With this connection the motor would be suitable for any commercial voltage from 220 to 240 V without operating problems.

Where the calculated voltage for the reconnected winding seems somewhat high for the power-supply voltage, coils could be left out of the series connection. This omission must be made with careful planning so that the coil grouping is balanced diametrically opposite in the stator circle. Otherwise the motor may make too much magnetic noise.

29. Increasing Torque. The torque developed by a particular motor may be stated as starting, maximum, and rated load. The torque depends on the mechanical constant, the stator flux, the rotor flux, and the phase angle between the two flus values. The stator-flux magnitude may, in many cases, be increased up to 20 percent in older motors, depending on the air gap and the magnetization characteristics of the magnetic circuit. The degree of magnetization or the stator flux at a fixed frequency is controlled by the applied voltage. If the rated voltage of a motor is 220 V, using a more modern system value of 240 V, the increase is approximately 9 percent. Since the starting torque varies according to the square of the applied voltage, the starting torque would be increased by about 19 percent. The load torque would be determined by the value of load applied, or the rated load would be carried at a slight increase in speed and a reduction in rated current.

From another example for possible reconnection, consider a 220-V series-delta winding rated at 17.3-A line current. Reconnecting this winding into a two-parallel wye connection would make the rated voltage 100 × 1.73 = 191 V and the rated line current 20 A. This reconnected winding would be suitable for use on the new system voltage of 208 V (109 percent) or 220 V (115 percent). A newer standard voltage is 230 V (120 percent). The 120 percent of rated voltage could introduce a temperature problem and should not be used without test. The starting torque at 208 V is about 118.5 percent and at 220 V it is 132 percent. At 230 V the starting torque becomes about 145 percent of the original. This voltage should, however, not be used unless a test has determined that the iron is not operated too high on its magnetization curve.

If reconnecting the old winding is not suitable, a new winding may have to be installed. A reduction of the number of turns per coil by about 8 percent would be within the reasonable limits of permissible change in magnetizing ampere-turns for older motors. If the slot room permits, the next larger size wire should be used because of the increased magnetic flux density. It is important that the same connection be used as before rewinding. Any of the foregoing changes will increase the operating temperature at rated load. However, the increase is not enough to rule out the changes.

Another change possible is to improve the rotor-flux phase angle. This change can be made by increasing the rotor resistance. In die-cast aluminum rotors the end rings may be machined to reduce the cross-sectional area. If too much material is machined away, however, the reduced rotor current may result in a greater loss than the gain from improved flux relation.

On larger rotors where the end rings are fastened to the bars by bolts or cap screws, higher-resistance-material rings may be substituted, or rings of smaller cross-sectional area may be used. No exact information can be given on this subject because each particular motor is a separate problem. Test information may have to be obtained before a plan of procedure is selected.

30. Changing Two-Phase to Three-Phase. At times it may be necessary to change a two-phase winding to operate on three phases. The first possibility would be to regroup the coils, as if, for example, a 48-slot four-pole two-phase winding were to be changed to four-pole three-phase. In the two-phase winding there would be 24 coils per phase and six coils per pole phase group. Assuming all coils connected in series, some of the series connections would have to be opened and the coils reconnected with four coils per group because 48 coils grouped for three-phase four-pole would be 48 coils divided by 3, or 16 coils per phase; and 16 coils grouped for four poles results in four coils per pole phase, instead of the original six. This method works out well for a 48-slot four-pole machine, but if it has 48 slots and six poles, the problem is different. The 48-slot, six-pole two-phase works out be be four coils per pole phase group, but for three-phase it will be 16 coils per phase for six poles; and 16 coils divided by six poles results in $2\frac{2}{3}$ coils per group. This arrangement requires odd grouping and the coils would have to be grouped as 3-2-3-3-2-3 in the respective poles. Each pole must have eight coils, so that the second phase grouping might be 2-3-3-2-3-3 and the third phase 3-3-2-3-3-2. The most important factor in odd-coil-grouped windings is that coil groupings must be balanced diametrically across the stator to avoid excessive magnetic noise. Other pole and slot combinations may be worked out in a similar manner. Those that work out as odd groupings must have the same number of coils per phase, and the three phases must have the same number of coils in each pole. In the above example there are eight coils in the three phases for each pole.

When making winding changes after the desired coil connections have been planned, the next problem is voltage. Assuming a 220-V series-connected two-phase winding, and since the phase consists of 24 coils, in the above motor, the voltage per coil in the original connection was 220/24 coils = 9.17. The reconnected three-phase winding would have 16 coils per phase when reconnected. This arrangement would work out to a new phase voltage of $16 \times 9.17 = 147$ V. When wye-connected, the rated voltage would work out to 254. Experience has shown that older motors may be operated at as much as 15 percent above or below rated value without experiencing operating difficulties. No definite statement can be made as to what the limitations are going to be because of the absence of definite details.

31. Scott T Connection. A second change possible is to use the Scott T connection. One of the two phase ends of phase *A* is connected to the center of phase *B*, and the remaining three leads are connected to the three-phase line. For the *B* phase there will be no voltage problem, but for the *A* phase the voltage will be too low, or the *A* phase has too many turns for a balanced condition of voltage and current. To balance the voltage more suitably, selected coils may be removed from the circuit. To balance the voltages, only about 87 percent of the *A*-phase turns are required. Using the original two-phase winding with 24 coils per phase, 87 percent of 24 coils equals about 21. To obtain as near as possible a balanced magnetic field around the stator, one coil in every other pole of the *A* phase should be left out of the circuit (omit the corresponding coil in each group).

The current in a three-phase motor for the same horsepower output, same efficiency, and same power factor is slightly higher than the two-phase motor. An example might be to assume two-phase line current of 10 A; by comparison the three-phase line current would be 11.5 A. Since the relationship between these two currents is smaller than the difference in area between two gage numbers of copper wire, it is possible that the same wire size would satisfy the original design of a 10-hp two-phase motor and a 10-hp three-phase wye-connected winding. In any event, a winding normally intended for 10 A will operate at a higher temperature at 11.5 A but should not be excessive when the two-phase winding is connected for a balanced three-phase at the same load.

In the above two- to three-phase conversion there is no change in coil-span degrees between coil sides or chord factor; but in changing from six coils to four coils per pole phase group the distribution factor, which is the space-vector summation of the coil voltage per group, is improved for the resultant three-phase winding. In the suggested change in coil groupings the differences would be about 6 percent. Since most modern power-supply voltages may be closer to 240 V than to 220 V, the above indicated differences in actual operation are perhaps somewhat less than the figures shown.

DC WINDING CALCULATIONS

32. Formulas for Lap Windings

a. Symbols and basic formulas

S = coil sides

P = poles

Y_b = back pitch (coil side count on back end of armature)

Y_f = front pitch (coil side count on front end or commutator end)

Y_c = commutator pitch

N_c = number of armature circuits when armature is in normal operation

m = number of windings (1 for simplex, 2 for duplex, etc.)

Coil span = spread of coils installed in slots

S = slots × 2 for a one-wire double-layer winding

$$Y_b = \frac{S}{P} \pm K$$

where K must be an odd whole number. K is a convenient number used to determine the required or desired value for back pitch.

$$Y_f = Y_b \pm 2m$$

Here the minus results in a progressive winding, the plus in a retrogressive winding.

$Y_c = m$ to the right of the start for a progressive winding, to the left for a retrogressive winding

$$N_c = Pm$$

$$\text{Coil span} = \frac{Y_b + 1}{2} \quad \text{for one-wire windings}$$

$$= \frac{Y_b + 3}{4} \quad \text{for two-wire windings}$$

$$= \frac{Y_b + 5}{6} \quad \text{for three-wire windings}$$

Figure 1 shows a four-pole simplex progressive lap winding together with its calculations.

b. Examples of applications

1. 24-slot four-pole simplex progressive lap winding:

$$S = 24 \times 2 = 48$$

$$P = 4$$

$$Y_b = \frac{S}{P} \pm K = \frac{48}{4} - K = 12 - 1 = 11$$

$$Y_f = Y_b - 2m \text{ (for progression)} = 11 - 2 \times 1 = 9$$

$$Y_c = 1 \text{ to the right of the start of the coil}$$

$$N_c = Pm = 4 \times 1 = 4$$

$$\text{Coil span} = \frac{Y_b + 1}{2} = \frac{11 + 1}{2} = \text{slots 1 to 6}$$

Commutator connections are determined by the brush position with respect to the poles. The two common brush positions used are in line with poles or between poles. Minor adjustments are required for motor or generator operation, and the direction of rotation of the machine.

2. 35-slot six-pole double-layer simplex retrogressive lap winding:

$$S = 35 \times 2 = 70$$

$$P = 6$$

$$Y_b = \frac{S}{P} \pm K = \frac{70}{6} - K = 11\frac{2}{3} - \frac{2}{3} = 11$$

$$Y_f = Y_b + 2m \text{ (for retrogression)} = 11 + 2 \times 1 = 13$$

$$Y_c = 1 \text{ to the left of the start of the coil}$$

$$N_c = Pm = 6 \times 1 = 6$$

$$\text{Coil span} = \frac{Y_b + 1}{2} = \frac{11 + 1}{2} = \text{slots 1 to 6}$$

3. One-wire duplex winding (four-pole 32-slot progressive):

$$S = 32 \times 2 = 64$$

$$P = 4$$

$$m = 2 \text{ because the winding is double}$$

$$Y_b = \frac{S}{P} \pm K = \frac{64}{4} - K = 16 - 1 = 15$$

$$Y_f = Y_b - 2m = 15 - 2 \times 2 = 11$$

$$Y_c = 2$$

$$N_c = Pm = 4 \times 2 = 8$$

$$\text{Coil span} = \frac{Y_b + 1}{2} = \frac{15 + 1}{2} = \text{slots 1 to 8}$$

A duplex winding has half as many conductors in series as a simplex and twice as many parallel circuits. When all other conditions are the same, the duplex winding is good for half the voltage and twice the current of a simplex winding.

To change from a simplex to a duplex winding, the finish end of the armature coil is moved over one commutator bar (to the right for a progressive or to the left for a retrogressive).

c. Two-wire windings. To improve commutation, the armature coils may be wound with two, three, or more wires, depending on the voltage. A two-wire coil has the same number of wires as the one-wire coil, but it has four wire ends instead of two because the coil is divided into two parts. The two parts are connected in series at the commutator bar. The commutator in this case has twice as many bars as slots.

21-slot four-pole two-wire simplex progressive lap winding, 42 commutator bars:

$$S = 21 \times 4 = 84$$

$$P = 4$$

$$m = 1 \text{ since it is still simplex}$$

$$Y_b = \frac{S}{P} \pm K = \frac{84}{4} \pm K = 21 \pm 0 = 21$$

To sample the first calculation it is substituted in the span formula. The answer must be a whole number, or the value of K must be readjusted.

$$\text{Coil span} = \frac{Y_b + 3}{4} \quad \text{for two-wire winding}$$

$$= \frac{21 + 3}{4} = \text{slots 1 to 6, which is suitable}$$

$$Y_f = Y_b \pm 2m = 21 - 2 \times 1 = 19$$

$$Y_c = 1 \text{ to the right}$$

$$N_c = Pm = 4 \times 1 = 4$$

33. Formulas for Wave Windings. Wave windings are often referred to as high-voltage windings. The higher voltages result from a reduced number of parallel circuits compared with the lap windings.

Referring to the lap-winding calculations, $N_c = Pm$ indicates that the greater the number of poles, the more parallel circuits there are through the winding.

For the wave winding, $N_c = 2m$, from which it is observed that the parallel circuits are independent of the poles, or that any simple wave winding has only two parallel circuits. The smaller number of circuits means that there are more conductors in series, which increases the voltage of the armature winding.

Armatures for wave windings do not work out as simply for various slot and pole combinations as for lap windings because a different method of calculation must be used. Also the coil-side count for planning the winding circuit is made differently. While the front-end coil-side count on a lap winding is made backward with respect to the back pitch, the count on the wave winding is always made forward for both front- and back-pitch values, as well as for the commutator-bar count.

a. Symbols and basic formulas

S = coil sides

P = poles

$$Y = \text{average pitch} = \frac{\text{all values of } Y_b + \text{all values of } Y_f}{\text{their number}}$$

All values of Y_b and all values of Y_f used here are those required to count the coil sides to get around the armature once. Figure 2 shows a 27-slot four-pole simplex progressive wave winding and its calculations.

b. Examples of applications

1. 21-slot four-pole simplex progressive double-layer wave winding:

$$S = 21 \times 2 = 42$$

$$P = 4$$

$$Y = \frac{S \pm 2m}{P} = \frac{42 + 2 \times 1}{4} = 11$$

If $Y = (S + 2m)/P$, winding is progressive; if $Y = (S - 2m)/P$, winding is retrogressive.

In this case since Y is an odd whole number, all values of Y_b, Y_f, and Y_c may be selected as 11. To complete the count of the coil sides to get around the armature once, Y_b, Y_f, and Y_c are counted twice. Therefore the pitch values may be written as

$$Y_{b1} = 11 \qquad Y_{b2} = 11$$

$$Y_{f1} = 11 \qquad Y_{f2} = 11$$

$$Y_{c1} = 11 \qquad Y_{c2} = 11$$

$$\text{Coil span} = \frac{Y_b + 1}{2} = \frac{11 + 1}{2} = \text{slots 1 to 6}$$

If the winding is retrogressive,

$$Y = \frac{S - 2m}{P} = \frac{42 - 2 \times 1}{4} = 10$$

Since this is average pitch and Y_b must always be an odd whole number, the values become

$$Y_{b_1} = 11 \qquad Y_{b_2} = 11$$

$$Y_{f_1} = 9 \qquad Y_{f_2} = 9$$

$$Y_{c_1} = 10 \qquad Y_{c_2} = 10$$

$$\text{Coil span} = \frac{Y_b + 1}{2} = \frac{11 + 1}{2} = \text{slots 1 to 6}$$

2. While a 21-slot four-pole winding works out for either progressive or retrogressive, not all other slot and pole combinations work out so conveniently. As an example a 31-slot 6-pole armature may be considered:

$$S = 31 \times 2 = 62$$

$$Y = \frac{S \pm 2m}{P} = \frac{62 - 2}{6} = 10 = \text{retrogressive}$$

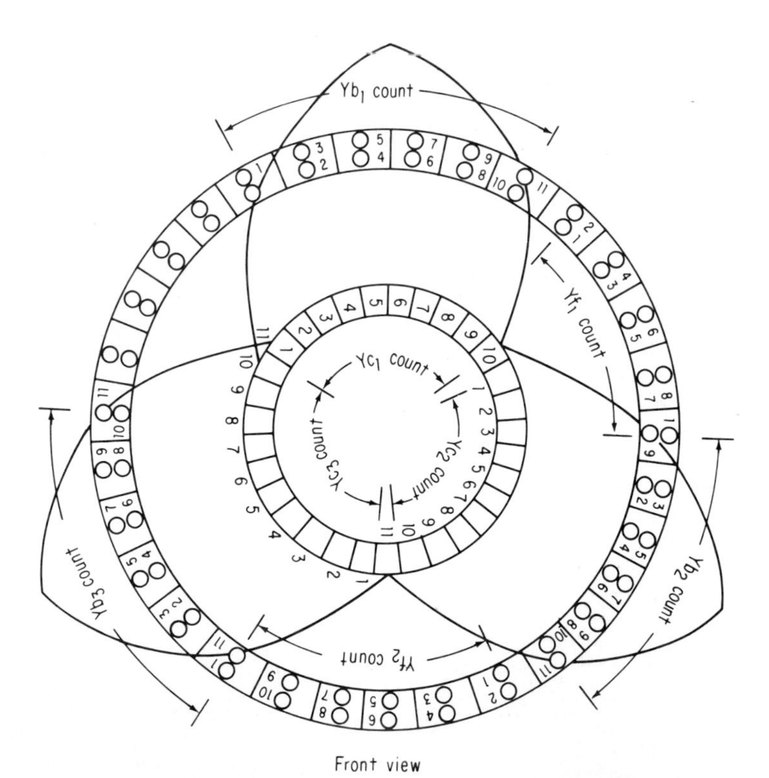

FIG. 3. Simplex progressive one-wire 31-slot six-pole wave winding. Y_{b1}, Y_{b2}, Y_{b3} and Y_{f1}, Y_{f2}, and Y_{f3} are counted in coil sides. Y_{c1}, Y_{c2}, and Y_{c3} are counted in commutator bars.

However, if

$$Y = \frac{S + 2m}{P} = \frac{62 + 2}{6} = 10\tfrac{2}{3}$$

dead coils are indicated.

When Y (the average pitch) for the progressive winding results in a mixed number, the fraction indicates that not all 31 coils fit in the wave winding circuit. All coils, however, are installed as a convenient method to balance the armature assembly mechanically, but coil ends are properly taped to avoid short circuits. The fraction $\tfrac{2}{3}$ shown in the calculations represents two dead coils and two vacant commutator bars. The unused bars are electrically connected to adjacent bars to maintain proper continuity of the circuit.

An alternative would be to use a special commutator, one with 29 bars instead of 31. The commutator-bar count for coil connection would then be 10 for all three values.

The completed calculation would become

$$Y_{b1} = 11 \qquad Y_{b2} = 11 \qquad Y_{b3} = 11$$

$$Y_{f1} = 9 \qquad Y_{f2} = 11 \qquad Y_{f3} = 11$$

$$Y_{c1} = 10 \qquad Y_{c2} = 11 \qquad Y_{c3} = 11$$

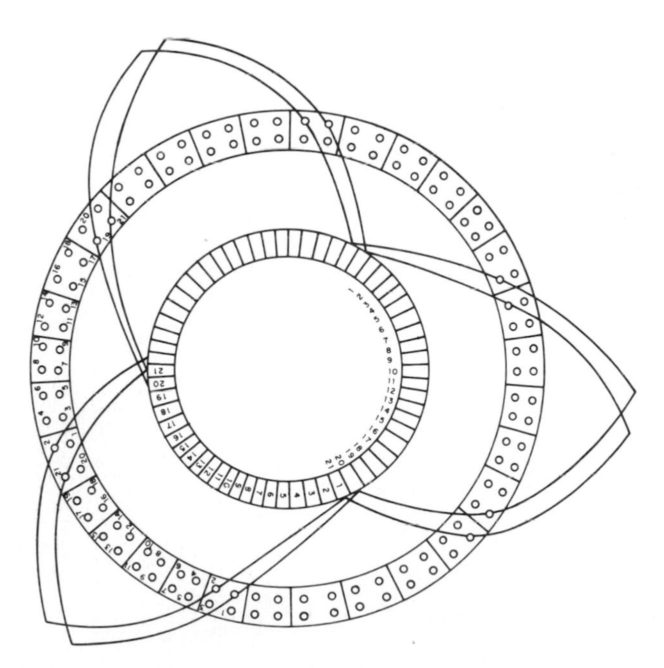

Front view

FIG. 4. Simplex progressive two-wire 31-slot six-pole wave winding. Numbers indicate counting procedure.

$$\text{Parallel circuits} = 2$$

$$Y = \frac{11 + 11 + 11 + 9 + 11 + 11}{6} = 10\tfrac{2}{3}$$

For the commutator count,

$$Y_{c\,(\text{av})} = \frac{10 + 11 + 11}{3} = 10\tfrac{2}{3}$$

$$\text{Coil span} = \frac{Y_b + 1}{2} = \frac{11 + 1}{2} = \text{slots 1 to 6}$$

The coil-side count must be done carefully in the order given when drawing the circuit to see whether the winding will work out. The last two coils left over are the ones that do not fit into the circuit, and they become the dead or inactive ones. Figure 3 shows a 31-slot six-pole one-wire progressive wave winding. Figure 4 shows a 31-slot six-pole two-wire progressive wave winding. The latter arrangement requires no dead coils.

$$S = 31 \times 4 = 124$$

$$Y = \frac{S + 2m}{P} = \frac{124 + 2}{6} = \frac{126}{6} = 21$$

When Y is an odd whole number, all values of Y_b, Y_p, and Y_c may be selected as 21.

REFERENCES

1. L. J. Rejda (handwritten correction; "Redja" struck out) and K. Neville, *Industrial Motor Users' Handbook of Insulation for Rewinds* (Elsevier, New York, 1977).
2. Robert Rosenberg, *Electric Motor Repair* (Holt, New York, 1980).

SECTION 19

EMERGENCY CARE OF DAMAGED MOTORS

Commander Leonard T. Daley, USNR*
Paul J. Dobbins†

FOREWORD 19-2

MOTOR PROBLEMS AND CORRECTION 19-2

1. Catastrophic Failures 19-2
2. Cleaning 19-4
3. Dry-Out Methods 19-7
4. Polarization Index 19-8
5. Troubleshooting 19-8
6. Instruments for Motor Inspection 19-8

*Philadelphia Naval Ship Yard, Naval Ships System Command, Philadelphia, Pa.; Registered Professional Engineer (Wis. and N.Y.); on leave from General Electric Company.

†Electrical Design and Development Engineer, Large AC Motor and Generator Department, General Electric Company, Schenectady, N.Y. (retired); Member, IEEE.

NOTE: This section was written by Leonard T. Daley for the first edition and reviewed and updated for this edition by Paul J. Dobbins.

FOREWORD

Electric motors are unique in that they are capable of being repaired economically to an extent not usually feasible with most prime movers. Further, when such repairs have been properly made, there is little or no operational difference between the repaired motor and a modern replacement unit. Characteristics of speed, overload, vibration, heating, noise, and even efficiency remain essentially the same.

The extent of repairs that can be made to a motor to restore it to operating condition is often limited only by the experience of the person making the repairs. Loose core iron may be tightened, coils with damaged or severely aged insulation patched, and the motor will run with no change in characteristics.

The art of rewinding motors is not dealt with here because of the specialized skills involved and the limited amount of rewinding now being done in modern industrial plants. Many texts are available on this subject, however.

Those repairs are covered which may be made, often without removing the motor from the plant, to get a motor back into service in minimum time, following such catastrophes as flooding and fire. Emphasis has been placed on minimizing the time a plant is out of production, through inspection, troubleshooting, and expeditious repair.

MOTOR PROBLEMS AND CORRECTION

1. Catastrophic Failures. These types of failure affect large numbers of motors, often simultaneously. The entire plant may be inoperative, and the first consideration is to restore production as quickly as possible. Included are floods, fires, lightning, and system low voltage and/or frequency due to power failure.

a. Flooding. Electric motors that have been subjected to flooding can usually be salvaged without rewinding and, if handled promptly, without replacing the more expensive ball and roller bearings. Parts replacement, including such normally available items as bearings, may be difficult after a flood because of the demand in an area. Simple precautions rapidly executed can make a difference of days or weeks in getting back into service.

While perhaps obvious, a hasty evacuation of all electric motors in a threatened area may be the best answer. Cut off cables if necessary (rewiring is easier than repairing the motor itself), but get the motors to a higher area. Shelter is not necessary. A simple dry-out is usually sufficient to return the motor to service. It is usually much faster to remove surface moisture by oven dry-out or heat lamps than to provide the complete cleaning and baking necessary after flooding.

In recent years as the major rivers have been subjected to flood control, those areas that historically flooded have seen less damage, while the major floods have been associated with hurricanes along the East Coast and the Gulf Coast. The result has been an increase in saltwater damage to electric equipment.

Even if no salt is present, however, a flood will often bring chemical contaminants in contact with the insulation, and therefore the motor should be thoroughly cleaned. Experience has shown that while it is possible to dry out and reuse equipment without thoroughly cleaning it, salt and certain contaminants may absorb moisture on subsequent shutdowns. The resultant chemical reaction will lead to early failure.

Even after considerable periods of immersion, however, machines that have been properly cleaned and varnish-impregnated will provide years of service.

b. Fire damage. Damage to motors in a fire ranges from the extreme of a meltdown to at the very least a thorough soaking. Nothing need be said about the molten mass of the first case, but for damage short of this, many motors can be repaired economically.

With respect to fire damage, we should recognize that one of the most common means of removing insulation during a normal rewind is to burn out the insulation with an open gas flame. Little appreciable effect of heat has been observed on the interlamination insulation of the stator or armature-core iron, provided that the temperature of the iron did not exceed about 400°C (750°F). This temperature may not be any hotter than that experienced by a motor in a fire.

A motor armature, or frame, is a large heat sink that usually takes several hours to come up to a uniformly high temperature. An example of the thermal-absorption capability of electric machinery is a large dc machine that was involved in an accident in which several thousand gallons of oil caught fire directly under the machine. The brush rigging was melted, as was the interconnecting cable between some field coils. After the armature was removed and cleaned, however, it was discovered that the coil insulation was in satisfactory condition. It was not necessary to rewind it. The insulation material was rated Class H and therefore suitable for high temperature. Many materials are common to Class B, F, and H insulation systems, but the system is only as effective as the temperature limitation of the lowest rated material permits.

While the ambient in this instance was enough to melt brass, the large mass of iron coupled with the thermal capability of the insulation system apparently soaked up the heat fast enough to preclude conductor or coil-insulation damage. In the more common situation, however, fire damage means a complete rewind. If there is frame distortion, the motor fits must be machined to restore dimensions.

c. Lightning damage. Damage caused by lightning is one of the most elusive causes of motor failure, which often shows up sometime after the damage occurs. On occasion it has appeared within a few hours or even days after the storm, and usually after a subsequent shutdown of the motor. Such failures may be explained by considering the logical sequence of events.

Assume the motor insulation suffers a small puncture from the overvoltage spike during the storm and while rotating. Being warm and dry, the insulation value remains satisfactory. After a subsequent shutdown, however, the motor may fail on start up or even after being in operation for a short time because of moisture absorption, which leads to a tracking of current to ground. It is difficult to prove lightning damage, because the evidence of insulation puncture is destroyed in the subsequent failure to ground.

Lightning may cause an extremely high voltage wavefront to move across the motor terminals and into the windings, particularly if the power system has little impedance between the substation and the motor. This natural phenomenon is an uncontrolled version of the commonly used surge-testing method used by motor manufacturers to locate insulation defects. If there is a weakness in the insulation, it will probably cause a turn-to-turn failure, followed usually by failure to ground (Fig. 1). Such failures have happened to several motors on a system simultaneously, even though one or more have operated for several days after others have failed.

Such damage will usually require a stator or armature rewind and may even

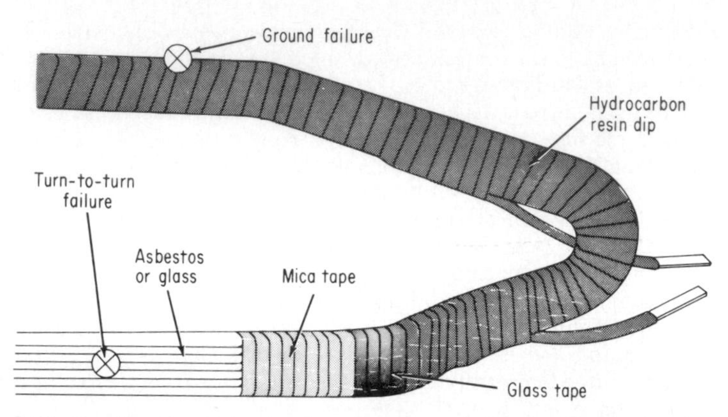

FIG. 1. Class F or B insulated motor stator coil showing likely points for turn failure and ground failure.

require replacement or restacking of stator-core laminations if there is core damage. If, however, the damage is minor and of such a nature that one can locate and isolate the failure, it may be possible to patch the winding.

Installation of protective capacitors and surge arresters at the terminals of large machines is a wise investment for the prevention of lightning damage.

2. Cleaning. The first step to be taken in the event of flooded equipment is to hose down the motor thoroughly to remove all accessible dirt and foreign material. Disassemble completely the stator, rotor, all inspection covers, bearings, and brush rigging, and open the connection box.

Wash down thoroughly with hot water and commercially available detergent cleaner. If possible obtain a steam-cleaning unit. Use reasonable care, however, not to damage insulation further by excessive pressure.

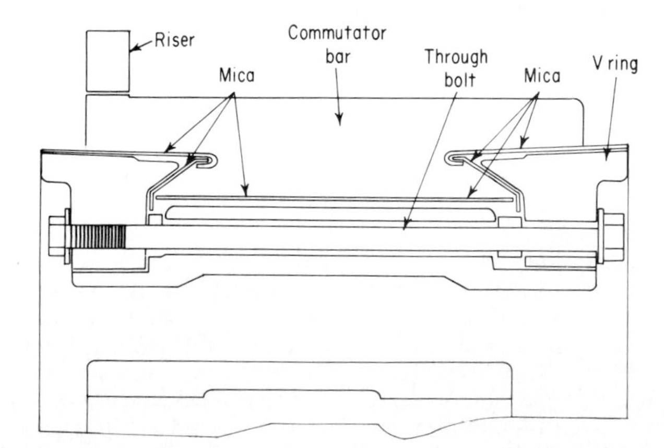

FIG. 2. Commutator cross section showing V rings and mica insulation. Silt from flooding may settle between riser and armature windings.

The hot water and detergent will remove dirt and grease. The water will also reduce the concentration of salt and other contaminants to a level at which they will not cause further damage. Water should be used copiously. If possible, submerge the motor in a large open tank filled with water. An open steam line placed in the tank will increase agitation and warm the water.

a. Synchronous motors. On large synchronous machines the cleaning can be improved and the drying hastened by placing the axis of the stator in a vertical position. This allows the dirt and contaminants to be flushed off rather than washing them into the lower coil slots.

b. Induction motors. To do a thorough cleaning job on an induction motor having long, deep slots, the slot wedges and top sticks should if possible be removed to assure adequate flushing. Some caution however is in order. (1) Rein-

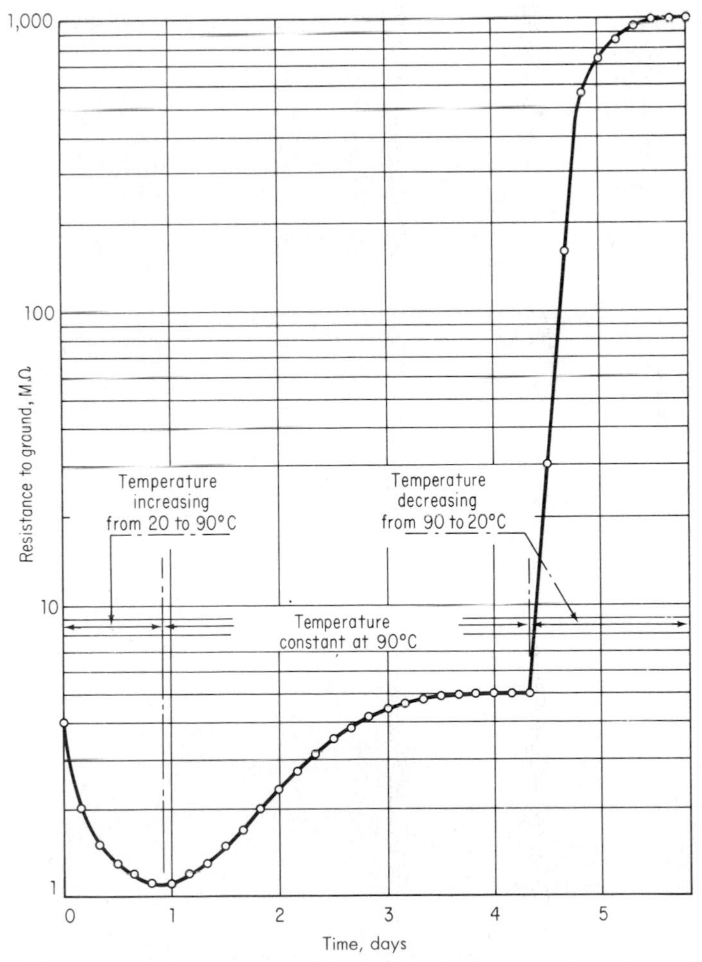

FIG. 3. Typical drying curve for ac motor armature. 1-min readings of insulation resistance were taken every 4 h. *(James H. Biddle Company.)*

stallation of wedges is a job for experienced motor winders, as there is a possibility of damaging aged insulation. (2) New wedges should be used, and these must be either machined in a special wedge cutter or obtained from the motor manufacturer, which may take excessive time.

c. Dc motors. On dc motors similar procedures should be followed. Because of the open construction, the stationary field assembly is easier to clean and flush.

(*a*)

(*b*)

FIG. 4. Typical vacuum-drying tank designed specifically for rapid dry-out of equipment.

The armature is much more difficult to clean, however, as dirt and contaminants may be trapped behind the commutator riser (Fig. 2).

d. Cleaning lubrication systems. When cleaning lubrication systems, be certain that sumps and piping are flushed thoroughly with water to remove all sand after a solvent has been used to remove the grease, which is not water soluble. Silt will penetrate even the most minute aperture, in seals, joints, and other openings. You cannot be overly meticulous in getting rid of silt. Failure to do so may well result in premature bearing failures and no end to mechanical problems.

3. Dry-Out Methods. Methods used for drying out motors that have been saturated with water are limited only by the equipment and the facilities available. The necessity of removing moisture is the same regardless of the method or equipment used. Moisture within the windings of the motor must be vaporized and the vapor removed. The higher the temperature the faster will be the moisture removal provided there is sufficient airflow to get the moisture-laden air out of the ambient. The apparatus should be brought up to temperature slowly because excessive moisture may be present in the windings. If heated too rapidly, this moisture may vaporize quickly enough to rupture the insulation.

a. Bake-out oven. After motors have been thoroughly washed and cleaned with water, the easiest and by far the most common method of drying is to place them in a suitably sized forced-circulation oven. The temperature is then raised to about 125°C and held there for 2 to 4 h, after which it is reduced to 115°C maximum.

An excellent way to check the ground resistance of the windings during dry-out is to run leads from the motor connections of all the units to a point outside the oven where they may be checked periodically by a megohmmeter.

Because the insulation resistance to ground can be read in this manner, it will not be necessary to open the oven doors as often, with a resultant reduction in drying time. The characteristic drying curve for insulation found in electric motors is shown in Fig. 3 (see Art. 5).

b. Vacuum dry-out. The alternative to high temperature is lowering the atmospheric pressure in a vacuum tank, which in turn lowers the vaporization pressure of the moisture (Fig. 4). If available, a vacuum tank with heat can cut dry-out

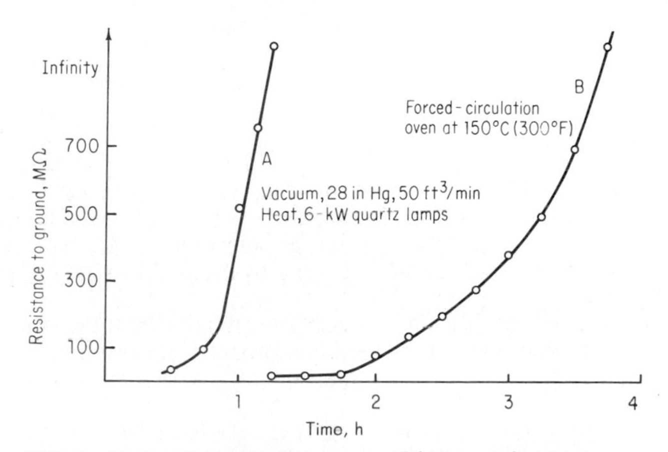

FIG. 5. Drying time. *A* under vacuum, *B* in conventional oven.

time by 50 to 70 percent or to 1 day instead of 4 days. Prolonged high temperatures are not required, although heat will accelerate the dry-out (Fig. 5).

c. Dry-out of dc motors. Drying out the armature of a dc machine may take several days. A plot of insulation resistance versus time will follow a curve similar to that shown in Fig. 3. Note that the highest readings are after the armature has cooled to room temperature. The dry-out time was over 4 days at an oven temperature of 90°C.

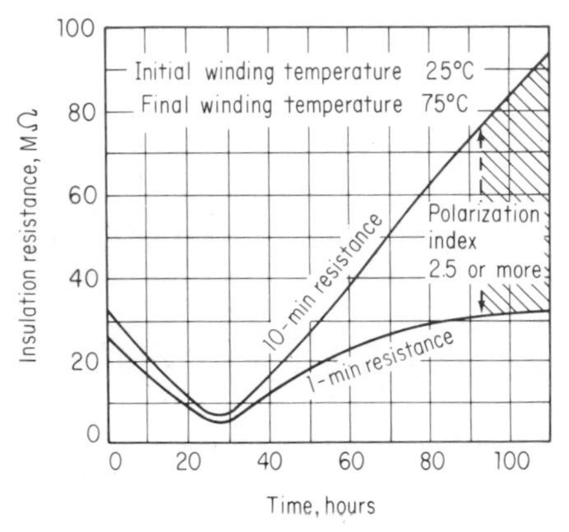

FIG. 6. Change in 1- and 10-min insulation resistance during drying process of Class B insulated ac armature winding.

d. Dry-out of bearings. As soon as possible after motor disassembly, wash all mechanical parts in a hot-water detergent solution, or with one of the commercially available nontoxic cleaning solvents suitable for electric equipment. Dry all bearings and journals as soon as possible after motor disassembly, and coat with moisture-free oil to prevent rust (heat the oil for a short time to remove moisture). Protect the bearings from any dirt, and store them in a clean, dry place until needed for assembly.

4. Polarization Index. If a constant-potential (motor-driven) megohmmeter is available, a ratio of the 1-min ground-insulation-resistance reading to that obtained after 10 min is a very useful guide to insulation dryness. This test should be run only when the winding is above the dew point.

Values of 1.5 and over are obtained from a clean, dry winding (Fig. 6). The lower the ratio the greater the leakage path of the test voltage to ground. This method in effect measures the capability of the insulation to hold a capacitive charge.

There are no absolute "go, no-go" values to look for, but there are guides. New Class B or F systems will give readings of 2.5 and higher with form-wound coils (Fig. 1). Class A systems will usually be closer to 1.5 and up. No vaild data are available covering random, or mush, wound motors, but the writer has found values of 1.25 to 2.5 with motors in good condition. The best guide to a good dry winding is a constant ratio over two or three successive readings, taken periodically, with the megohmmeter removed long enough between readings to let the charged insulation dissipate.

5. Troubleshooting. Effective troubleshooting is an analytical approach to motor repair and maintenance. Experience is a valuable asset but no substitute for a logical approach coupled with a basic knowledge of electrical theory and machine design. Table 1 will assist in getting to the problem area as quickly as possible.

A large number of motor problems actually are failures in the control circuit or improper mechanical alignment of either motor components or the motor with connected apparatus.

6. Instruments for Motor Inspection. Following is a minimal suggested list of instruments for plant electric-motor maintenance and inspection, which will

TABLE 1. Troubleshooting Chart

Trouble	Cause	Remedy
	Induction Motors	
Motor will not start	Overload control trip	Wait for overload to cool. Try starting again. If motor still does not start, check all the causes as outlined below
	Power not connected	Connect power to control, and control to motor. Check clip contacts
	Faulty (open) fuses	Test fuses
	Low voltage	Check motor-nameplate values with power supply. Also check voltage at motor terminals with motor underload to be sure wire size is adequate
	Wrong control connections	Check connections with control wiring diagram
	Loose terminal-lead connection	Tighten connections
	Driven machine locked	Disconnect motor from load. If motor starts satisfactorily, check driven machine
	Open circuit in stator or rotor winding	Check for open circuits
	Short circuit in stator winding	Check for shorted coil
	Winding grounded	Test for grounded winding
	Bearings stiff	Free bearings or replace
	Grease too stiff	Use special lubricant for special conditions
	Faulty control	
	Overload	Reduce load
Motor noisy	Motor running single-phase	Stop motor, then try to start. It will not start on single phase. Check for "open" in one of the lines or circuits
	Electrical load unbalanced	Check current balance
	Shaft bumping (sleeve-bearing motors)	Check alignment, and condition of belt. On pedestal-mounted bearing, check cord play and axial centering of rotor
	Vibration	Driven machine may be unbalanced
	Air gap not uniform	Center the rotor and if necessary replace bearings
	Noisy ball bearings	Check lubrication. Replace bearings if noise is persistent and excessive
	Loose punchings, or loose rotor on shaft	Tighten all holding bolts
	Rotor rubbing on stator	Center the rotor and replace bearings if necessary
	Objects caught between fan and end shields	Disassemble motor and clean it. Any rubbish around motor should be removed
	Motor loose on foundation	Tighten holding-down bolts. Motor may possibly have to be realigned
	Coupling loose	Insert feelers at four places in coupling joint before pulling up bolts to check alignment. Tighten coupling bolts securely
At higher than normal temperature or smoking	Overload	Measure motor loading with ammeter. Reduce load

TABLE 1. Troubleshooting Chart (*Continued*)

Trouble	Cause	Remedy
	Electrical load unbalance (fuse blown, faulty control, etc.)	Check for voltage unbalance or single phasing. Check for "open" in one of the lines or circuits
	Restricted ventilation	Clean air passages and windings
	Incorrect voltage and frequency	Check motor-nameplate values with power supply. Also check voltage at motor terminals with motor under full load
	Motor stalled by driven machine or by tight bearings	Remove power from motor. Check machine for cause of stalling
	Stator winding shorted	Cut out coil or rewind
	Stator winding grounded	Test and locate ground
	Rotor winding with loose connections	Tighten, if possible, or replace with another rotor
	Belt too tight	Remove excessive pressure on bearings
	Motor used for rapid-reversing service	Replace with motor designed for this service
Bearings hot	End shields loose or not replaced properly	Make sure end shields fit squarely and are properly tightened
	Excessive belt tension or excessive gear side thrust	Reduce belt tension, or gear pressure, and realign shafts. See that thrust is not being transferred to motor bearing
	Bent shaft	Straighten shaft
Sleeve bearings	Insufficient oil	Add oil—if oil supply is very low, drain, flush, and refill
	Foreign material in oil or poor grade of oil	Drain oil, flush, and relubricate using industrial lubricant recommended by a reliable oil company
	Oil rings rotating slowly or not rotating at all	Oil too heavy; drain and replace Oil ring has worn spot; replace with new ring
	Motor tilted too far	Level motor or reduce tilt and realign, if necessary
	Rings bent or otherwise damaged in reassembling	Replace rings
	Ring out of slot (oil-ring retaining clip out of place)	Adjust or replace retaining clip
	Motor tilted, causing end thrust	Relevel motor, reduce thrust, or use motor designed for thrust
	Defective bearings or rough shaft	Replace bearings. Resurface shaft
Hot bearings	Too much grease	Remove relief plug, and let motor run. If excess grease does not come out, flush and relubricate
	Wrong grade of grease	Add proper grease
	Insufficient grease	Remove relief plug and regrease bearing
	Foreign material in grease	Flush bearings, relubricate; make sure that grease supply is clean
	Bearings misaligned	Align motor and check bearing-housing assembly. See that races are exactly 90° with shaft
	Bearings damaged (corrosion, etc.)	Replace bearings
Wound-rotor motor problems:	Wires to control too small	Use larger cable to control

TABLE 1. Troubleshooting Chart (*Continued*)

Trouble	Cause	Remedy
Rotor runs at low speed with external resistance cut out	Control too far from motor	Bring control nearer motor
	Open circuit in rotor circuit (including cable to control)	Test by ringing out circuit and repair
	Brushes sparking	Adjust commutation
	Dirt between brush and ring	Clean rings and insulation assembly
	Brushes stuck in holders	Use right size brush
	Incorrect brush tension	Check brush tension and correct
	Rough collector rings	File, sand, and polish
	Eccentric rings	Turn in lathe or use portable tool to true up rings, without disassembling motor
	Excessive vibration	Balance motor
	Current density of brushes too high (overload)	Reduce load (if brushes have been replaced, make sure they are of the same grade as originally furnished)
		Check shaft for straightness
Synchronous Motors		
Motor will not start	Faulty connection	Inspect for open or poor connection
	Open circuit one phase	Test, locate, and repair
	Short circuit one phase	Open and repair
	Voltage falls too low	Reduce the impedance of the external circuit
	Friction high	Make sure bearings are properly lubricated
		Check bearing tightness. Check belt tension. Check load friction. Check alignment
	Field excited	Be sure field-applying contactor is open and field-discharge contactor is closed through discharge resistance
	Load too great	Remove part of load
	Automatic field relay not working	Check power supply to solenoid. Check contactor tips. Check connections
	Wrong direction of rotation	Reverse any two main leads
Fails to pull into step	No field excitation	Check circuit connections. Be sure field-applying contactor is operating. Check for open circuit in field or exciter. Check exciter output. Check rheostat. Set rheostat to give rated field current when field is applied. Check contacts of switches
	Load excessive	Reduce load
		Check operation of unloading device (if any) on driven machine
	Inertia of load excessive	May be a misapplication—consult manufacturer
Motor pulls out of step or trips breaker	Exciter voltage low	Increase excitation. Examine exciter as shown in dc motors. Check field ammeter and its shunt, to be sure reading is not higher than actual current
	Open circuit in field, and exciter circuit	Test with magneto and repair break
	Short circuit in field	Check with low voltage and polarity indicator and repair field

TABLE 1. Troubleshooting Chart (*Continued*)

Trouble	Cause	Remedy
	Reversed field coil	Check with low voltage and polarity indicator and reverse incorrect leads
	Load fluctuates widely	See Motor "hunts" below
	Excessive torque peak	Check driven machine for bad adjustment, or consult motor manufacturer
	Power fails	Reestablish power circuit
	Line voltage too low	Increase if possible. Raise excitation
Motor "hunts"	Fluctuating load	Correct excessive torque peak at driven machine or consult motor manufacturer
		If driven machine is a compressor, check valve operations
		Increase or decrease flywheel size
		Try decreasing or increasing motor field current
Stator overheats in spots	Rotor not centered	Realign and shim stator or bearings
	Open phase	Check connections and correct
	Unbalanced currents	Loose connections. Improper internal connections
One or more coils overheat	Short circuit	Cut out coil as expedient (in motors up to 5 hp); replace coil when the opportunity arises (rewind)
Field overheats	Short circuit in a field coil	Replace or repair
	Excessive field current	Reduce excitation until stator current is at nameplate value
All parts overheat	Overload	Reduce load or increase motor size
		Check friction and belt tension, or alignment
	Over- or underexcitation	Adjust excitation to nameplate rating
	No field excitation	Check circuit and exciter
	Reverse field coil	Check polarity and, if wrong, change leads
	Improper voltage	See that nameplate voltage is applied
	Improper ventilation	Remove any obstruction and clean out dirt
	Excessive room temperature	Supply cooler air
	DC Motors	
Motor will not start	Open circuit in control	Check control for open starting resistor, open switch, or burned fuse
	Low terminal voltage	Check voltage with nameplate rating
	Bearing frozen	Recondition shaft and replace bearing
	Overload	Reduce load or use larger motor
	Excessive friction	Check lubrication in bearings to make sure that the oil has been replaced after installing motor. Disconnect motor from driven machine, and turn motor by hand to see if trouble is in motor. Strip and reassemble motor; then check part by part for proper location and fit
Motor stops after running short time	Motor is not getting power	Check voltage at the motor terminals; also clips, and overload relay
Motor attempts to start but overload relays operate	Motor is started with weak or no field	If adjustable-speed motor, check rheostat for setting. If correct, check condition of rheostat
		Check field coils for open winding
		Check wiring for loose or broken connection
	Motor torque insufficient to drive load	Check line voltage with nameplate rating, larger motor, or one with suitable characteristic match load

TABLE 1. Troubleshooting Chart (*Continued*)

Trouble	Cause	Remedy
Motor runs too slowly under load	Line voltage too low	Check and remove any excess resistance in line, connections, or control
	Brushes ahead of neutral	Set brushes on neutral
	Overload	Check to see that load does not exceed allowable load on motor
Motor runs too fast under load	Weak field	Check for resistance in shunt-field circuits. Check for grounds
	Line voltage	Correct high-voltage condition
	Brushes back of neutral	Set brushes on neutral
Brushes:		
Sparking at brushes	Commutator in bad condition	Clean and reset brushes
	Eccentric or rough commutator	Grind and true commutator, also undercut mica
	Excessive vibration	Balance armature
		Check brushes to make sure they ride freely in holders
	Broken or sluggish acting brushholder spring	Replace spring, and adjust pressure to manufacturer's recommendations
	Brushes too short	Replace brushes
	Machine overloaded	Reduce load, or install larger motor
	Short circuit in armature	Check commutator, and remove any metal particles between segments. Check for short between adjacent commutator risers
		Test for internal shorts in armature and repair
Brush chatter or hissing noise	Excessive clearance of brushholders	Adjust holders
	Incorrect angle	Adjust to correct angle
	Incorrect brushes for the service	Get manufacturer's recommendations
	High mica	Undercut mica
	Incorrect brush-spring pressure	Adjust to correct value
Motor will not come up to speed	Excessive load	Decrease the load
		Check operation of unloading device (if any) on driven machine
	Low voltage	Increase voltage
	Field excited	Be sure field-applying contactor is open, and field-discharge contactor is closed through discharge resistance
Poor commutation	Insufficient brush-spring pressure	Adjust to correct pressure, making sure brushes ride free in holders
One brush takes more load than it should	Unbalanced circuits in armature	Eliminate high resistance in defective joints by inserting armature or equalizer circuit or commutator risers. Check for poor contacts between bus and bus rings
Excessive sparking	Poor brush fit on commutator	Sand in brushes, and polish commutator surface
	Brushes binding in the brushholder	Remove and clean holders and brushes with solvent (nontoxic). Remove any irregularities on surfaces of brushholders or rough spots on the brushes
	Insufficient or excessive pressure on brushes	Check and set brush arm for correct pressure (varies with motor design)
	Brushes off neutral	Set brushes on neutral
Sparking at light loads	Paint spray, chemical, oil or grease, or other	Use motor designed for application. Clean commutator, and provide protection

TABLE 1. Troubleshooting Chart (*Continued*)

Trouble	Cause	Remedy
	foreign material on commutator	against foreign matter. Install an enclosed motor designed for the application
Field coils overheat	Short circuit between turns or layers	Replace defective coil
Commutator overheats	Brushes off neutral, or overload	Adjust brushes Reduce load or increase motor size
	Excess spring pressure on brushes	Decrease brush-spring pressure but not to the point where sparking is introduced
Grooving of commutator	Brushes not properly staggered	Stagger brushes
Brushes wear rapidly	Rough commutator	Resurface commutator and undercut mica
	Excessive sparking	Make sure brushes are in line with commutating fields
Armature overheats	Motor overloaded	Reduce load to correspond to allowable load
	Motor installed in location where ventilating is restricted	Arrange for free circulation of air around motor
	Armature winding shorted	Check commutator, and remove any metallic particles between segments. Test for internal shorts in armature and repair

enable a plant engineer or electrician to perform a thorough electrical test and check alignment and vibration:

Portable indicating ammeter

Portable indicating voltmeter

Megohmmeter (500-V hand-crank type)

Hand revolution counter

Spring balance (0 to 20 lb scale)

Thermometers (centigrade)

Portable analyzer or multipurpose testing instrument (volts ac; dc; ohms)

Hook-on ammeter (voltmeter)

Portable phase-sequence indicator

Vibration indicator (hand type)

Dial indicator and magnetic mount

SECTION 20
MOTOR NOISE

Robert G. Bartheld*

FOREWORD 20-2

TERMINOLOGY 20-2

1. Decibels 20-2
2. Frequency 20-2
3. Octave Band 20-3
4. One-Third-Octave Band 20-3
5. Band Center 20-3
6. Narrow Band 20-3
7. Free Field 20-3
8. Reverberant Field 20-3
9. Semireverberant Field 20-5

MEASUREMENT METHODS AND INTERPRETATION 20-5

10. Sound Levels 20-5
11. Background Noise 20-5
12. Sound-Pressure Levels 20-7
13. Averaging Sound-Pressure-Level Readings 20-8
14. Sound-Power-Level Calculations 20-8
15. Additional Testing Information 20-9

NOISE SOURCES 20-9

16. Magnetic Noise 20-10
17. Rotor and Stator Slotting 20-11
18. Skew of Rotor Bars 20-12
19. Windage Noise 20-12
20. Fan-Blade Noise 20-13
21. Siren Effect 20-13
22. Bearing Noise 20-14
23. Preloading 20-16
24. Friction Noise 20-17

*Manager, Technology Transfer, Siemens Energy and Automation, Inc., Atlanta, Ga.; Member ASME, IEEE, ASA; Representative to NEMA Motor Generator Section; Chairman of IEEE WG 85, Test Procedure for Airborne Noise Measurements on Rotating Electrical Machines; U.S. Expert to two international standards working groups covering noise of rotating electric machines.

25. Vibrating Surfaces 20-17
26. Sound Absorption 20-17
27. Brush Noise 20-18

REFERENCES 20-18

FOREWORD

In applying or installing motors careful thought should be given to motor noise. Airborne noise is defined as unwanted sound in air. Motors are therefore potential sources of noise problems. Such problems may arise when a new motor is installed with a machine in a quiet area or when an older motor is moved to a new location. In other situations noise from a machine, such as a pump or compressor, may be blamed on its drive motor, or the motor noise is blamed on the machine it drives.

Since it is impossible to eliminate noise completely, this section will deal with the basic terminology of noise, the methods of measurement, its sources, and means of decreasing the final noise level.

TERMINOLOGY

1. Decibels. Measuring sound levels in their normal term of reference yields very large values. The decibel is used to reduce these magnitudes to a usable level. The decibel is a nondimensional ratio of two power values expressed as dB = $10 \log_{10}$ (power I/power II). The term, as generally used in sound work, has several specific relationships, depending upon the reference value. Normally two kinds of levels are used.

a. Sound-pressure level is the rms sound pressure in decibels relative to the standard sound-pressure reference expressed by L_p (dB) = $20 \log_{10} (P/20\ \mu\text{Pa})$, where P is the measured sound pressure in pascals, and 20 μPa is the standard sound pressure reference. When no other reference is stated, the standard reference is used.

b. Sound-power level is the sound power emitted by a sound source relative to a sound-power reference expressed by L_w (dB) = $10 \log_{10} (W/10^{-12}\ \text{watt})$, where W is the source sound power in watts, and 10^{-12} watt is the reference sound power for rotating electrical equipment. Sound-power levels cannot be measured directly.

2. Frequency. The quality of sound is determined by its frequency distribution. Rotating equipment generates sounds that have a multitude of frequencies. A plot of the sound-pressure level or the sound-power level against frequency is called frequency spectrum. Sound expressed by a single frequency is termed pure tone.

Frequency is defined as the number of times a pattern repeats itself during a given time interval. In sound this is expressed in hertz.

A human being has a normal audible frequency range of 20 to 10,000 Hz. For analysis purposes it is convenient to divide the frequency range into bands, defined by octaves.

3. Octave Band. The term "octave" means that the frequency of the upper limit of the band is twice that of the lower limit, that is, $f_h/f_l = 2$. The octaves normally associated with sound measurements are shown in Table 1.

4. One-Third-Octave Band. A one-third octave is a band of frequencies where the ratio of the upper to lower frequency limits is equal to the cube root of 2, that is, $f_h/f_l = \sqrt[3]{2}$. The preferred one-third-octave bands are those bands whose centers have been adjusted so that the ratio of successive centers is 1.259. The one-third octaves consistent with the preferred frequencies are shown in Table 2.

5. Band Center. Octave and one-third-octave bands are referred to by their center frequency, the geometric mean between the extreme frequencies of the band, that is,

$$f_c = (f_h f_l)^{1/2}$$

Frequency limits for one-third octaves are shown in Table 2 and those for full octaves in Table 1.

6. Narrow Band. A narrow band is a band whose width is less than one-third octave but not less than 1 percent of the center frequency.

7. Free Field. Free-field conditions exist if, at any point of measurement surrounding the sound source, appreciably no sound energy is present because of reflections. Free field is characterized by a 6-dB reduction in sound-pressure level for each doubling of distance from the source. Free-field, or nearly free-field, conditions can be obtained indoors by satisfying the following requirements:

1. A laboratory room which provides a free field over a reflecting plane
2. A room in which the contributions of the reverberant field to the sound pressures on the measurement surface are small compared with those of the direct field of the source

Conditions described under requirement 2 are usually met in very large rooms as well as in smaller rooms with sufficient sound-absorptive materials on their walls and ceilings.

Free-field rooms are also called anechoic rooms.

8. Reverberant Field. A reverberant field is characterized by the great reflection and diffusion of the generated sound field. Reverberant rooms have volumes large

TABLE 1. Preferred Octave Bands

	Frequency, Hz							
Midfrequency	63	125	250	500	1,000	2,000	4,000	8,000
Approx frequency limits:								
Lower	45	90	180	355	710	1,400	2,800	5,600
Upper	90	180	355	710	1,400	2,800	5,600	11,200

TABLE 2 Preferred One-Third-Octave Bands

	Frequency, H_2										
Midfrequency	16	20	25	31.5	40	50	63	80	100	125	
Approx frequency limits:											
Lower	14	18	22.4	28	35.5	45	56	71	90	112	
Upper	18	22.4	28	35.5	45	56	71	90	112	140	
Midfrequency	160	200	250	315	400	500	630	800	1,000	1,250	
Approx frequency limits:											
Lower	140	180	224	280	355	450	560	710	900	1,120	
Upper	180	224	280	355	450	560	710	900	1,120	1,400	
Midfrequency	1,600	2,000	2,500	3,150	4,000	5,000	6,300	8,000	10,000	12,500	16,000
Approx frequency limits:											
Lower	1,400	1,800	2,240	2,800	3,550	4,500	5,600	7,100	9,000	11,200	14,000
Upper	1,800	2,240	2,800	3,550	4,500	5,600	7,100	9,000	11,200	14,000	18,000

enough to contain at least 20 modes of vibration in the lowest frequency band used. For octave bands, the minimum volume is given approximately by the equation

$$V = \frac{4}{3} \lambda^3$$

where λ is the wavelength of the center frequency of the lowest frequency band of interest.

The average sound-absorption coefficient for all surfaces of the room should not exceed 0.06.

Reverberant-field conditions are assumed for a room when sound-pressure measurements are observed to be independent of the distance from the sound source.

9. Semireverberant Field. Rooms having a room constant neither high enough for free-field conditions nor low enough to satisfy reverberant-field conditions have a semireverberant field and are called semireverberant rooms. Most acoustical environments possess semireverberant fields, unless they were designed for the purpose of making acoustical measurements.

MEASUREMENT METHODS AND INTERPRETATION

Individual or community reaction to noise is based upon psychological and physiolgical responses to air-pressure variations at an individual's car. The pressure level experienced is a function of distance, field conditions, and sound-power level of the noise source, as shown in Fig. 1. Because of this relationship, sound-pressure levels do not completely describe the actual noise source nor do sound-power levels adequately reflect the degree of annoyance.

10. Sound Levels. A sound-level meter reads directly in sound-pressure levels. Readings of this type must be taken at a defined distance from the unit being measured. This distance is generally 1 m (3 ft).

Broadband levels may be measured directly on one of three weighted networks, specified as *A*, *B*, and *C*. A comparison of different network readings taken from the same source can be used to indicate whether the source contains predominantly high- or low-frequency noise. The three curves representing the weighting characteristics are shown in Fig. 2.

Network *C* has an almost flat response and may be used for making frequency-spectrum analyses. The weighting may be used to approximate the response to the human ear. When this is done, the networks are generally associated with a decibel range. Network *A* is used for levels below 55 dB, network *B* between 55 and 85 dB, and network *C* for levels above 85 dB.

Sound levels taken of a noise source generally reflect an average of five microphone positions. The positions commonly used are one directly above the motor and four in the horizontal plane of the shaft at 90° intervals. One position is usually in line with the shaft. Care must be taken not to put a microphone in the airstream.

11. Background Noise. Background noise is sometimes referred to as ambient noise. Measurements made under field conditions may be subject to rather high background noise levels.

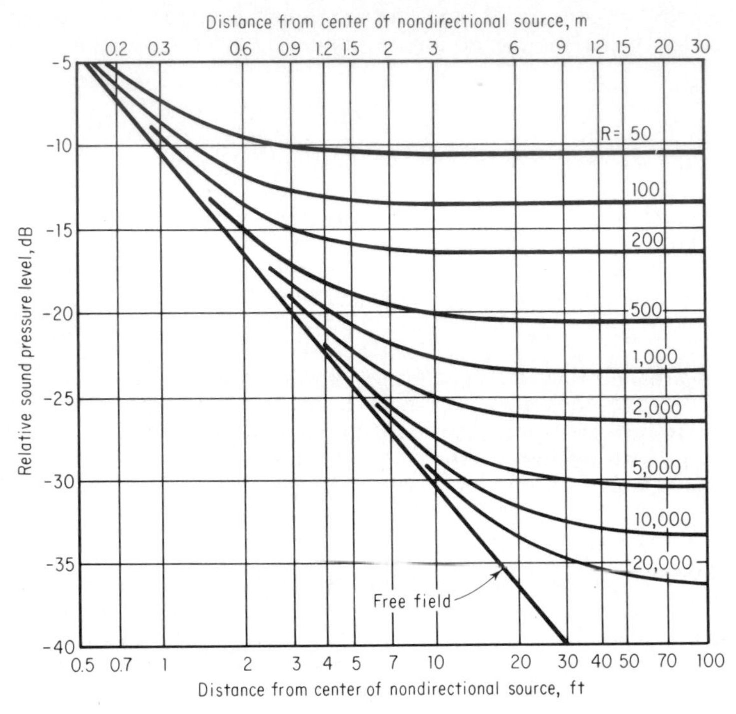

FIG. 1. Relationship of sound-pressure level to sound-power level of noise as a function of distance and enclosure-room constant *R*.

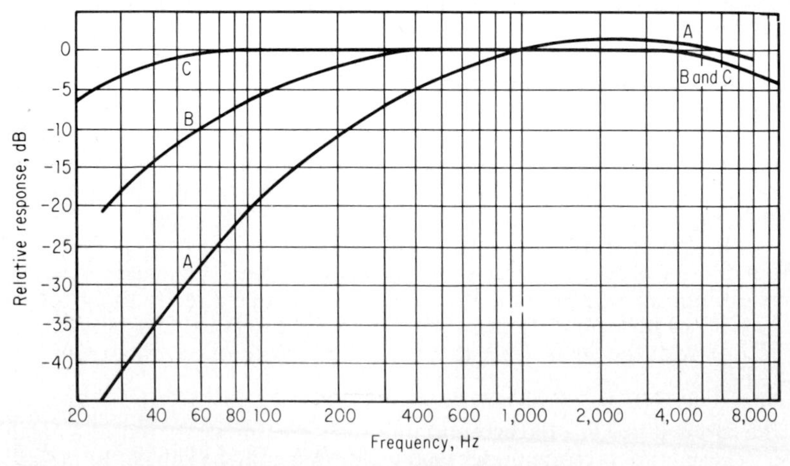

FIG. 2. Response of sound-level meter for different networks.

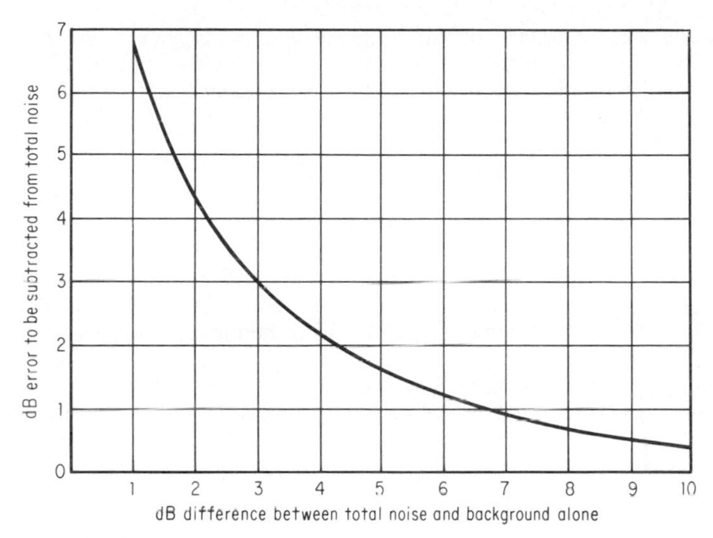

FIG. 3. Corrections for ambient noise level.

The readings given by the measuring instruments may be corrected to obtain sound-pressure levels that more accurately describe the motor noise level. Figure 3 is used to determine the corrective values based on the difference in meter readings between conditions of motor running and motor off for each microphone location.

When the difference exceeds 10 dB, no correction has to be made. If the difference is less than 3 dB, it indicates that the motor noise level is less than the background noise and the readings are of little value, except to establish the upper limit of motor noise.

12. Sound-Pressure Levels. All noise measurements originate with the taking of sound-pressure-level readings. Interpretation and reporting of these readings do not, however, lend themselves to any single method to be employed when making the measurements.

When it is required that an area be classified as to the degree of distress, readings should be taken at positions normally occupied by persons affected by the noise. Readings should be taken for the bandwidths of interest as required by the mode of interpretation. The reading, or set of readings, representing the maximum noise condition should be used to calculate the noise level in the terms desired.

Small motors and motor-driven equipment that produce low noise levels in the annoyance range may be tested and noise-rated prior to installation. The units should be mounted in a manner similar to how they will be mounted in service. Test environment conditions should be as near reverberant as possible. This will permit installation in any environment with the assurance that the resultant noise level will not exceed the test value. Readings should be taken for the bandwidths of interest as required by the mode of interpretation. Corrections should be made for background noise.

Larger motors should be tested in a manner that permits conversion into sound-power levels. The main purpose of these tests is to provide information that can be used to predict ultimate noise levels in the final environment or in making a comparative evaluation of the motor on test.

Motors tested under free-field conditions should have measurements taken on a hypothetical surface at points that represent equivalent surface areas. Several possible arrays are described in Ref. 1. The microphone locations are selected so that they are not closer than 1 m (3 ft) from reflecting surfaces of the motor to be tested. At least three locations are necessary for reverberant fields, with six evenly spaced points being sufficient for semireverberant-field conditions. Measurements should be taken of the test motor, the calibrated sound source, and the background noise at all locations for each octave band of interest.

Whenever sound-pressure levels are reported, they should include the field condition and the measurement distance from the motor.

13. Averaging Sound-Pressure-Level Readings. When it is necessary to average n sound-pressure-level readings, it should be done on a power basis in accordance with the following equation:

$$\overline{L}_p = L_{p\,(\mathrm{av})} = 10 \log_{10} \times \frac{1}{n} \left(\mathrm{antilog}_{10} \frac{L_{p1}}{10} + \cdots + \mathrm{antilog}_{10} \frac{L_{pn}}{10} \right)$$

14. Sound-Power-Level Calculations. The calculation of sound-power levels is dependent upon the testing environment. This environment is subdivided into three field conditions: Free, reverberant, and semireverberant field.

a. Free field. In a free field over a reflecting plane,

$$L_w = \overline{L}_p + 10 \log_{10} (S/S_0)$$

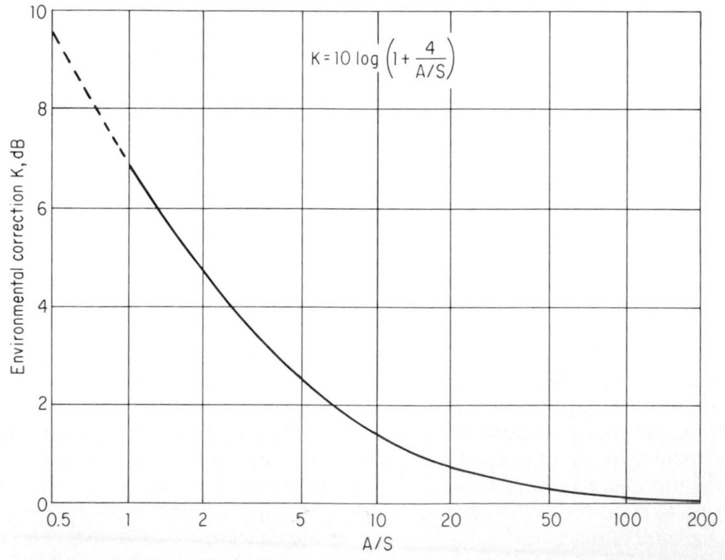

FIG. 4. Environmental correction factor K versus A/S, where A is area of sound absorption of test room and S is area of measurement surface.

where L_w = sound-power level, referred to 10^{-12} watt
$\overline{L}_p$ = average sound-pressure level, referred to 20 μPa
S = area of measurement surface (Ref. 1)
S_0 = 1 m^2

b. Reverberant field. In a reverberant field, using a calibrated sound source, the sound-power level of the motor for each frequency band is obtained as follows:

$$L_w = \overline{L}_p + (L_{wr} - \overline{L}_{pr})$$

where L_w = sound-power level, referred to 10^{-12} watt
$\overline{L}_p$ = average sound-pressure level, referred to 20 μPa
L_{wr} = sound-power level of reference, referred to 10^{-12} watt
$\overline{L}_{pr}$ = sound-pressure level of reference, referred to 20 μPa

c. Semireverberant field. In a semireverberant field, using a calibrated sound source, the sound-power level of the motor for each frequency band is obtained in the same manner as for reverberant-field conditions.

When a calibrated sound source is not available, the sound-pressure levels measured in a semireverberant field must be corrected to free-field conditions. A correction factor K from the curve of Fig. 4 is subtracted from the average sound-pressure level (Ref. 2). The average corrected sound-pressure level is then used in the free-field computation to obtain the sound-power level.

15. Additional Testing Information. Further and more complete testing information may be obtained from Refs. 3 to 5.

NOISE SOURCES

Airborne induction-motor noise can be categorized as magnetic noise, windage noise, and mechanical noise, according to its source.

Magnetic noise that is objectionable usually has loud single-frequency components in the high-frequency range.

Windage noise is largely broadband in character, but single-frequency components associated with resonant volumes within the frame of the motor, the flow of the ventilating air through or over the motor, the number of fan blades, or a siren effect caused by the rotor bars periodically interrupting the flow of air may also appear.

Many mechanical noises are associated with the bearing assembly and can be appreciable if, for example, the bearing parts are deformed in some manner, or if excessive clearances permit axial travel of the shaft.

Unbalance or magnetic forces can excite natural frequencies of the motor structure which radiate noise to the surrounding air.

Some of the noise sources in electric motors and possible noise controls are shown in Table 3. The noise sources are normally identified by a frequency analysis.

No general rule can be given for induction motors as to the relative importance of these three types of noise. Individual design characteristics set the pattern which makes one or the other a major noise source.

During the initial design stage of a motor, the designer should keep the possi-

TABLE 3. Electric-Motor Noise Sources and Controls

Noise source	Cause	Component causing noise	Noise-control remedy
Mechanical	Impact	Sleeve bearings	Correct end play of shaft, decrease clearance between shaft and sleeve, modify oil grooves
		Ball bearing	Reduce radial play, by preloading, reduce shaft or housing tolerances; natural frequency of end shield should not equal characteristic ball-bearing frequency
		Brushholders and brushes	Tighten brushholder, change natural frequency, bias brush, and polish commutator or rings
		Loose laminations	Improve clamping of laminations
	Friction	Sleeve bearings	Increase clearance, increase oil viscosity, scrape sleeve surface
		Ball bearings	Increase preload pressure, change type of grease
	Unbalance	Rotor	Mechanical balance required
	Instability	Bearings	Change oil grooves in bearings, change oil viscosity
Windage	Modulation	Siren effect	Displace rotor ducts axially or acoustically treat interior surfaces
		Fan blades	Change number of fan blades, remove stationary obstacles from airstream, or use aerodynamic directional fan
	Turbulence	Fan	Redesign fan and airstream path, add acoustic filters or acoustically treat the ventilating passages
Magnetic	Radial	Air gap	Correct eccentricity of rotor or stator by machining and adjustment
	Variable force field radial in direction	Slotting of rotor and stator punchings	Skew stator or rotor slots, use closed rotor and semiclosed stator slots
		Dissymmetry harmonics due to higher harmonics in stator current	Reduce dissymmetry by improved magnetic-circuit design; avoid resonance of rotor frame, particularly at lower modes, which are more effective sound radiators

bilities for excessive sound generation to the minimum. Once a motor is designed and manufactured, it is much more costly to effect a suitable noise reduction.

16. Magnetic Noise. Magnetic noise of induction motors has two predominant sources: (1) the radial force waves created by the air-gap flux density and (2) the magnetostrictive expansion of the core steel. Whereas the force waves are imposed on the air-iron surface of the air gap, the magnetostrictive expansion is an internally generated variation in dimension. Both these sources of magnetic noise produce noise that is twice the line frequency.

The first part of the magnetic noise is caused by periodic forces which exist primarily in the air gap between stator and rotor. These forces, which are proportional to the square of the flux density, produce radial components that appear at harmonics of the line frequency and at frequencies related to the slip and line frequency. *Flux densities should be kept low to minimize these double-line-fre-*

quency noises. In some commercial motors it is the practice to employ unbalanced fractional-slot windings to obtain maximum utilization of standard laminations. Such unbalanced windings are not admissible in motors which are to be quiet, since unbalanced line currents and hence double-line-frequency vibrations will result.

The forces in the air gap can be reduced effectively by altering either the electromagnetic or the mechanical characteristics of the rotor-stator assembly. It is not always practical to make these changes, since they also affect the torque characteristics, core losses, and other electrical characteristics.

In well-designed induction motors the noise source, in addition to the induction-motor hum, is from the basic slot harmonic force field, which occurs at the following frequencies:

$$f_r = Rn$$

$$f_r = Rn + 2f$$

$$f_r = Rn = 2f$$

where R = number of rotor slots
n = rotor speed, r/s
f = line frequency, Hz

Appropriate skewing of either the rotor or the stator slots will decrease the magnitude of these slot harmonics.

17. Rotor and Stator Slotting. The most important single parameter in the electrical design of a quiet motor is the selection of the rotor and stator slot combination. In general the slot combination should be chosen so that the number of exciting-force pole pairs is as high as possible since the motor frame will usually be stiffer in these modes. Factors other than noise, such as elimination of hunting, crawling, and cogging, eliminate many possible combinations. Table 4 lists the best slot combinations resulting from the computation of rotor and stator harmonics whose interaction produces the greatest number of force pole pairs.

Squirrel-cage induction motors have either closed or semiclosed rotor slots and semiclosed or open stator slots. The partial or complete steel bridge over the slots reduces the abrupt permeance variation as sensed by the opposing member. A decrease in permeance variation will result in lower magnitudes of slot harmonics.

A closed rotor slot, however, causes less permeance variation than a semiclosed slot unless a relatively small bridge thickness becomes saturated. As the steel bridge saturates, the permeance variation approaches that of a completely open slot. Under such situations the closed slot may not appreciably reduce magnetic noise. Since the use of closed slots never causes an increase in noise, it is recommended that rotors with closed slot configuration be used.

For similar reasons, stators should utilize semiclosed rather than open slot configuration. The width of the slot opening should be kept at a minimum. The ability to wind the motor will determine the minimum opening that may be used.

An increase in the air-gap radial length will also decrease the magnitude of the permeance variations.

Slip noise is a low-frequency beating of higher-frequency components. Because of its intermittence, this noise may be objectionable even though its level is relatively low. Being a function of slip, it is more noticeable under load, with the frequency varying directly with the slip. The noise is often associated with a defect in the uniformity of the squirrel-cage rotor, in which case a new rotor is required.

18. Skew of Rotor Bars. The practice of skewing either the rotor or the stator slots with respect to the axis of rotation to reduce noise and provide for smooth acceleration has become so prevalent that the vast majority of induction motors are skewed. Authorities agree that skewing reduces magnetic noise, but there is no clear agreement on the optimum amount of skew. The noise produced by varying amounts of skew cannot be calculated accurately. One test made is shown in Table 5.

It is suggested that the rotor be skewed at least one rotor or stator slot, whichever has fewer slots. Skewing less than one slot is not effective in appreciably reducing the unit's noise level. Larger skews may be used, but these generally sacrifice motor performance.

19. Windage Noise. Windage noise is most common in high-speed rotating machines. It is caused by the presence of obstructions in the vicinity of the rotating part which moves air and creates turbulence. The acoustic power generated varies approximately as the fifth power of the airflow rate, and the fundamental frequency is a function of the rotational speed. Windage noise differs from the majority of motor noise in that it is created in the airstream rather than in the motor components. In most cases it is broadband noise with essentially no significant pure-tone components. In large motors which use radial ventilating ducts, and in some motors with large cooling fans, discrete frequencies manifest themselves as a series of peaks superimposed on the broadband noise.

Air flowing around or against surfaces produces turbulence, a potential source of sound. The turbulence energy content determines the magnitude of the radiated sound. For this reason wire cloth is preferable to expanded metal for screening purposes.

While the housing, fan, and other components are designed to minimize noise due to turbulence, the constrictions, bends, and obstructions are usually not aerodynamically designed and may still produce severe turbulence. Minimizing this noise calls for the elimination of sharp edges and burrs on all parts in contact with the airstream. It requires aerodynamically designed fans and an airstream path with the following features:

TABLE 4. Preferable Rotor-Stator Slot Combinations for Three-Phase Motors

Stator slots	No. of rotor slots			
	2-pole	4-pole	6-pole	8-pole
24	32 34	32 34	32 36	
36	26 28 44	26 28 44	46 48	48 52
48	38 40 56	38 40 56	58 60 64 68	58 64
54	46 62	46 62	42 66	38 40 66 70
60	52 68 78	44 46 52 76	44 48 72	44 76
72	....	....	60	56 58

TABLE 5. Skew-Test Sound Levels

Band	0 slot	1 slot	1.3 slot	1.75 slot
125	35	29	30	28
250	39	34	33	38
315	49	42	40	43
800	52	42	42	48
1,250	49	41	42	45
4,000	44.5	31.5	33	36
5,000	40	34.5	36.5	34.5
Overall	59	53	52	55

The skew is referenced to the component having the fewest slots.

1. Use a short, streamlined flow path in which the motion of air is orderly and predictable.
2. Keep the flow-path cross section as large as possible throughout as much of its length as possible.
3. Minimize abrupt discontinuities and changes in flow direction.
4. Eliminate unnecessary obstructions in the flow path.
5. Keep boundary surfaces smooth.
6. Provide gradual changes in flow-path cross section.

Reducing fan-tip speed (the product of rotational speed and fan diameter) produces low maximum air velocity with respect to the rotating fan blades and thus minimizes the energy content of turbulence in the wake of the blades.

20. Fan-Blade Noise. Fan-blade noise is the result of an air impulse every time a blade passes a protruding stationary member, or a standing-wave resonance point. The blade frequency is

$$f_b = Nn \quad \text{Hz}$$

where N = number of blades in fan
n = rotational speed, r/s

Totally enclosed motors with external fans produce discrete fan-blade frequencies superimposed on a broadband noise level. The fan-blade-frequency noise level may be reduced by increasing the clearance between fans and stationary parts, or by using a nonsymmetrical group of blades in relation to the stationary protrusions. A directional fan that decreases the effective air impulse leaving the blades will also decrease noise levels.

Large open motors may exhibit standing-wave resonances that can be reduced at the source by changing the number of fan blades, or in the motor frame by changing the critical dimension.

21. Siren Effect Open motors with radial air ducts through the rotor and stator may produce pure-tone components of airborne noise that are highly irritating. The frequency is usually above 1000 Hz and is called siren effect. The noise itself is produced by the sudden interruption of the flow of air from the radial ducts in the rotor. The frequencies associated with this type of noise are

$$f_s = Rn \quad \text{Hz}$$

$$f_{s2} = 2Rn \quad \text{Hz}$$

where R = number of slots in rotor
n = rotational speed of motor, r/s

The offsetting of rotor ducts with respect to stator ducts can be beneficial in reducing noise levels. The elimination of stator or rotor ducts by design considerations is the best approach for low-noise-level motors.

22. Bearing Noise. Bearings used in induction motors are a source of noise and vibration because of the sliding or rolling contact of bearing components. Sleeve or ball bearings are the types usually used, with the former having a sliding contact between components, while the latter has a combination of rolling and sliding contact.

a. Sleeve bearings. Induction motors may be equipped with sleeve bearings when low noise levels are required and when the bearing noise is predominant.

Sleeve bearings do not cause objectionable motor noise unless they are unstable. If there is large clearance, the oil film in the bearing may lift the shaft, unloading the oil film, thereby reducing the pressure. Consequently the shaft drops, hitting the bearing material, causing noise. This instability may be rectified by decreasing clearances in the bearing, by changing the location of the oil grooves, or sometimes by changing the viscosity of the oil.

Bearing oil film is based on the eccentric pump action inherent in a sleeve bearing. Just as a common pump must turn to produce a pressure, the shaft must turn to pressurize the oil. The pressurized oil is capable of transmitting sufficient force to lift the shaft from the mating sleeve. The lubricant film may be considered as a series of laminar layers. The innermost layer of the oil will whirl around with the shaft; the outermost layer will be stationary on the sleeve surface. The viscosity of the oil will determine the ease of relative motion of layers between these two. Note that without relative motion, there is no pressure and thus no support from the lubricant. At standstill, the oil is pressed from between the two bearing members, and during starting there will be metal-to-metal contact and therefore wearing noise. Oil with insufficient viscosity will permit intermittent metal-to-metal contact during shaft rotation. In addition to wear, the contact generates an impact noise.

For very quiet operation, bearings made from plastics, graphite, or similar materials may be used. These bearings are limited to small motor applications where the surface velocity is not too great.

Axial bumping may be another impact-noise source. The bumping occurs when the shaft shoulder oscillates axially, with one boundary being the bearing face. Motors whose sleeve bearings are designed to take axial-thrust loads should have the shaft shoulder in continuous contact with the bearing. When thrust capabilities are not present, the rotor shaft should "float" between bearing faces without contact. Close control of design clearance and coupling type generally overcomes this problem (see Sec. 16).

b. Antifriction bearings. Excessive noise in antifriction ball bearings may be caused by nonuniform balls, poor surface finish, ball-retainer rattle, and eccentricity, which result in impact noise or resonance excitation of the bearing housings, air baffles, and other parts that are efficient noise radiators.

Precision-grade bearings are usually quieter than the more common industrial grade because of closer tolerance and better surface-finish control during manufacture. Bearings made as "electric-motor quality" possess the necessary features for most induction-motor applications.

The most common sources of noise in antifriction ball bearings of induction motors are listed below, along with their possible causes.

1. Noise due to bearing configuration:

   Mounted-bearing radial clearance too tight or too loose

   Retainer too tight and rubbing or too loose and rattling

   Poor race finish

   Excessive eccentricity and runout

   Shield or retainer resonance

   Shield interference with retainer or race

   Improper retainer material for application

2. Noise due to lubricant:

   Grease tacky (remedy: use softer grease)

   Ball skidding in race (remedy: add lubricant)

   Excessive oil present (remedy: clean bearings)

   Residue present on races because of long storage

3. Noise due to machining:

   Housing or shaft fit too tight or too loose

   Out-of-round housing

   Poor shaft or housing finish (shaft finish should be 16 to 32 μin)

   Bent shaft

   Misalignment (shoulder against which bearing rests is not perpendicular to axis)

   Interference from adjacent parts

4. Noise due to abuse or neglect:

   Brinelled races and damaged balls

   Dirt and dust inclusions

   Corrosion

   Damage to shield

   Ball skidding and fatigue damage

Increasing the bearing size and the rotating speed of the induction motor has an increasing effect on the bearing noise level generated.

The noise due to some of these sources is distinct enough to be identified easily. For example, brinelling can be recognized by a low-pitched noise; the presence of dirt in the bearings causes a shrill noise; and skidding at low temperatures with insufficient lubrication results in destruction of the surface, causing a high-frequency noise. An intermittent crackle is often due to the grease. A noise in the frequency range of 100 to 300 Hz may be caused by the passage of the balls or rollers and is characteristic of antifriction bearings. This noise is generally of low level and not disturbing unless it excites motor parts to vibrate at their natural frequencies.

The discrete frequencies of ball-bearing noise are produced by irregularities of the ball-bearing components and may be determined if the bearing geometry and shaft speed are known.

The following symbols are defined:

d = diameter of rolling element, inches
E = pitch diameter, inches
N = number of rolling elements
n = speed of rotating member, r/s
B = contact angle, degrees

The fundamental rotational frequency n is apparent with the slightest unbalance or race eccentricity. Any irregularity of a rolling element or the cage causes noise with a train frequency of

$$f_t = \frac{n}{2}\left(1 - \frac{d}{E}\cos B\right) \quad \text{Hz}$$

The speed of the train of rolling elements f_t is roughly one-half the rotational speed of the inner race n.

The spin frequency of a rolling element is

$$f_s = \frac{E}{2d}n\left[1 - \left(\frac{d}{E}\right)^2 \cos^2 B\right] \quad \text{Hz}$$

A rough spot or indentation of an element causes a frequency component

$$f_e = 2f_s \quad \text{Hz}$$

because the spot contacts the inner and outer races alternately.

Another frequency occurs if there is an irregularity (high spot or indentation) on the inner raceway:

$$f_i = N(n - f_t) \quad \text{Hz}$$

The frequency due to an irregularity on the stationary raceway is

$$f_o = Nf_t \quad \text{Hz}$$

In the case of many spots, the harmonics of f_i or f_o will be more pronounced.

23. Preloading. One of the best-known methods in the suppression and damping of bearing noise is the use of a thrust washer to preload the bearings axially. Wavy-spring washers are one of the more common types of thrust washer. This washer acts as a spring to exert a force, usually on the outer race of the ball bearing. The reactive force is supplied by the shaft pressing against the inner race of the bearing. This force couple across the bearings takes up the internal clearances and causes each ball to follow the same path on each bearing raceway.

Preloading bearings decreases noise caused by the balls rattling within the raceway and cage, decreases the generation of high-frequency vibration, and improves

improves balance by the removal of bearing looseness. But preloading also causes the balls to follow closely the surfaces of the raceways. Good surface finish of the raceways and of the balls will permit the bearing to run smoothly and thus more quietly.

Too much bearing preload produces low-frequency noise and possibly bearing overheating.

24. Friction Noise. Friction noise arises from a lack of sufficient lubrication between two sliding surfaces and is caused by a high-impact type of vibration resulting from rapid intermittent contacting of the surfaces. Sleeve bearings exhibit this phenomenon when there is insufficient oil film, whereas ball bearings commonly experience sliding contact due to a lack of preload. The noise at the point of contact occurs at a high frequency, like that of hissing air, and when the impact vibration is transmitted to a resonant point in the adjacent structure, the audible sound is best described as a screech. The frequency at which the noise occurs cannot be described by an equation, but it is related to the surface finish of the areas in contact. Both surface finish and normal loading values affect the noise level. An additional effect of the high vibration rate is the breakdown of the lubricant at the surfaces.

25. Vibrating Surfaces. Any structural part of a motor may act as a source of airborne noise if it is excited with sufficient energy. For example, rotational unbalance itself may not emit airborne noise of any magnitude, but it may act as an energy source for vibrations which are then transmitted through the support structure to be converted to airborne sound waves at some resonant point. Thus the vibrating member or housing sets the air into motion, making it appear as if it were the noise source.

Air baffles, drip covers, and similar parts may perform their particular function but tend to resonate. Under such conditions two remedies can be effected:

1. Deadening or damping material can be applied to change vibratory motion into heat energy by making use of the internal friction of the material.
2. The stiffness characteristics of the structure can be changed so that resonance will occur at another frequency.

Usually, rigid parts produce less vibration and noise than parts that are more flexible. Consideration should also be given to the use of cast and molded plastic parts to decrease vibration-originated noise.

26. Sound Absorption. The emission of airborne noise generated within the motor can be reduced by using sound-absorbing materials.

Sound-absorbing materials are porous structures which absorb the energy from the sound waves passing into their pores and convert it to heat energy. The absorption capability of this type of material increases with its density, tightness or pore structure, and thickness. For best results, the barrier should completely enclose the source, or at least intercept the direct path between the source and the receiver.

Introduction of 90 or 180° bends into the ducts is further advantageous, in particular for the control of low frequencies (up to 400 Hz), where the absorption coefficients of commercially available sound-absorptive materials are low.

27. Brush Noise. Brush noise is caused by the sliding contact of brushes against a slip ring or commutator. The noise produced by the brushes running over a segmented commutator is considerably greater than that of brushes sliding on a continuous slip ring. Brushes produce a high-frequency squeal from sliding, an electric-discharge or sparking noise, and, when used with commutators, a commutator-bar frequency

$$f_b = Bn \quad \text{Hz}$$

where B = number of bars
n = speed, r/s

Brush noise is normally not a serious factor in airborne noise, except at high speeds. Careful selection of brush pressure, angle, and material as well as surface finish of the opposing surfaces will tend to minimize this source of noise.

REFERENCES

1. "Engineering Methods for the Determination of Sound Power Levels of Noise Sources for Essentially Free-Field Conditions over a Reflecting Plane," ANSI Standard S1.34–1980.
2. "Survey Methods for the Determination of Sound Power Levels of Noise Sources," ANSI Standard S1.36-1979.
3. "Test Procedure for Airborne Noise Measurements on Rotating Electric Machinery," IEEE Standard 85-1973 (Reaff 1980).
4. "Airborne and Structureborne Noise Measurements and Acceptance Criteria," Military Standard MIL-STD-740.
5. "Bearings, Ball, Annular, for Noise Quiet Operation," Military Specification MIL-B-17931.

INDEX

Ac mine motor characteristics, **10**-36
Ac motors, **3**-37
  construction of, **3**-38
Acceleration, **1**-3, **3**-15, **8**-21
Active iron, **3**-5
Adjustable speed, **3**-8
Adjustable-speed motor, **7**-13
Adjustable varying speed, **3**-8
Adjustable varying-speed motor, **4**-7, **7**-13
Alignment:
  coupling face, **16**-21
  dial indicator, **16**-17
  final, **16**-15
  machine sets, **16**-13
  rim check, **16**-17
  rough, **16**-14
  two-indicator method, **16**-20
Altitude:
  above 3300 ft (1000 m), **6**-17
  high, **1**-15
Amortisseur-winding protection, **3**-32
Antifriction bearing seals, **13**-9
Antifriction bearings, **12**-2
Application analysis, **6**-57
  classification, **7**-3
  data, **4**-12, **6**-24, **7**-17
Armature, **3**-33
  grounded, **18**-11
Armature circuit, open, **18**-10
Armature current:
  initial rate of rise, **2**-39
  peak calculations, **2**-40
Axial float, **16**-23
Axial thrust on bearings, **6**-47

Ball and roller bearings, **12**-3
Ball bearings, **18**-7
  care of, **17**-24
Basic motor power formulas, **3**-14
Battery-powered motors:
  characteristics, **10**-20
  designs, **10**-22
  enclosures, **10**-16
Battery-powered motors (*Cont.*):
  power source, **10**-15
  ratings, **10**-19
Bearing care, **17**-21
Bearing end play, **12**-19
Bearing-housing arrangements, **14**-15
Bearing load ratings, **12**-10
Bearing noise, **20**-4
Bearing protection, **2**-58
Bearing seals, **13**-2
  for special enclosures, **13**-17
Bearing selection, **12**-15
Bearing standardization, **12**-26
Bearing-temperature detectors, **3**-30
Bearing-temperature protective devices, **12**-18
Bearing temperatures, **14**-2
Bearing types, **12**-2
Bearings, **3**-5
  ball and roller, **12**-3
  in fractional-horsepower motors, **12**-20
  in integral-horsepower motors, **12**-21
  plain, **12**-2
  sliding-type, **12**-9
  thrust, **12**-7
    tilting-pad, **12**-12
  in vertical motors, **12**-24
Brake selection, **6**-47
Braking, **3**-16, **3**-37
  capacitive, **10**-7
  dc, **10**-7
  dynamic, **3**-20
  external, **3**-16
  internal, **3**-17
  mechanical, **10**-7
  plugging, **10**-7
  regenerative, **3**-18, **10**-7
Braking torque, **3**-7
Breakdown torque, **3**-6
Brush noise, **20**-18
Brush-shifting motors, **3**-52
Brush sparking, **18**-12
Brushless dc-excited synchronous motors, **10**-10
Brushless synchronous-motor control, **10**-14

Brushless synchronous-motor systems, **10**-14

Capacitor industrial-instrument motors, permanent-split, **7**-20
Capacitor instrument and gear motors, **7**-20
Capacitor limitations for unit switching, **2**-44
Capacitor motors, permanent-split, **7**-10
Capacitor ratings for low-voltage induction motors, **2**-47
Capacitor-start motors, **7**-9
  dual-voltage, **7**-10
  induction-run, **3**-41, **7**-11
    single-phase, **3**-39
Capacitor values, suggested maximum, **2**-46
Capacitors, **2**-44, **4**-28
  for induction motor, **2**-44
  switched, **2**-47
Carbon-dust problems, **18**-10
Catastrophic failure, **19**-2
Cellar drainers and sump pump motors, **7**-26
Ceramic permanent magnet rotor, **10**-74
Chain sprockets, **6**-20
Circuit breakers, **2**-48
  for motor starting, **2**-12
Circulating currents, shaft, **12**-16
Classification by variability of speed, **7**-13
Cleaning, **19**-4
Code-letter designations, **4**-12
Codes, **3**-25
  commercial, **8**-25
Coil support, **11**-11
Commercial marine motors, **10**-53
  regulatory bodies, **10**-51
Commutating field windings, **3**-33, **3**-34, **3**-36
Commutation, **18**-11
Commutator, **3**-33
  discolored, **18**-10
  tightening, **18**-9
  undercutting, **18**-9
Compensated series single-phase motor, **3**-39
Compound-wound motors, **3**-35
Compounding:
  differential, **3**-36
  indirect, **3**-36
Constant speed, **3**-8
Constant-speed motor, **4**-7, **7**-13
Control systems, **3**-37
Controller, electric-motor, **3**-16
Controllers, **3**-37, **3**-55, **3**-66
Coolant-pump motors, **7**-30
Corona, **11**-7
Couplings, limited end-float, **16**-24
Crane motors, **10**-62
Critical speed, **3**-9
Current-collecting devices, **17**-16
Current-limiting power fuses, **3**-32
Current-responsive protectors, **3**-28
Cylindrical-rotor synchronous motor, **5**-2

Dc-excited synchronous motors, **3**-57
Dc motor, **2**-23, **3**-32
  acceleration and deceleration, **8**-21
  applications of, **8**-2
  classification of, **8**-4
  construction, **3**-33
  groups of, **2**-40
  horsepower and torque capabilities, **8**-18
  insulation systems, **8**-10
  output-factor application, **8**-25
  overhauling loads, **8**-2
  overload capability, **8**-20
  protective covers and method of cooling, **8**-11
  speed range, **8**-16
  temperature rise and ventilation, **8**-18
  time and service conditions, **8**-12
  transient-speed or impact-speed drop, **8**-23
Dc motor behavior during faults, **2**-38
Dc power source, **8**-2
Dc power systems, **2**-11
Deceleration, **3**-15, **8**-21
Definite-purpose fractional-horsepower ac motors, **7**-20
Definite-purpose motor, **7**-3
Definitions, miscellaneous useful, **7**-14
Definitive motor speed terms, **3**-9
Definitive motor torques, **3**-7
Differential compounding, **3**-36
Differential protection, **2**-53
Differential relays, **3**-31
Dimensions and frame assignments, **7**-17
Dry-out methods, **19**-7
Duty classification, **6**-43
Duty cycle, **1**-5, **6**-46

Dynamic braking, **3**-20, **5**-18, **9**-3
by capacitor-rectifier-resistor combination, **3**-22
by capacitors, **3**-22
by dc excitation, **3**-20
with resistors, **3**-21

ECM (electronically commutated motor) electronics, **10**-65
Efficiency, **4**-9, **6**-50, **7**-14
of dc motors, **3**-36
motor, **3**-10
of polyphase induction motors, **3**-44
Efficiency economics, **6**-53
Electric power, **3**-10
Electronically commutated motor (ECM), **10**-65
Emergency switching, **4**-26
Enclosure, classification by method of, **7**-13
Enclosures, **3**-25, **4**-8
End-thrust damage, **3**-31
Energy considerations, **6**-49
Engine-cranking motor, **10**-26
construction, **10**-30
performance characteristics, **10**-27
Engine-cranking requirements, **10**-26
Environment, **1**-15
Essential-service motors, **2**-58
Excitation, **3**-12

Failure, catastrophic, **19**-2
Fan-blade noise, **20**-13
Fault-current contribution, small motors, **2**-38
Fault protection:
circuit breakers, **2**-48
motor starters, **2**-48
stator winding, **2**-53
Faults, system, **2**-36
Field-coil insulation, **11**-11
Field-discharge resistors, **3**-32
Field-failure relays, **3**-32
Field-protective relays, **3**-32
Field winding, **3**-35
Flat-belt pulleys, **6**-19
Flywheel machines, **1**-20
Flywheels, **15**-1
with pulsating loads, **15**-5
Formulas, handy, **6**-48
Foundations, **16**-1
Fractional-horsepower motor, **7**-3
Frame, **3**-33
dc motor, **3**-33
Frame assignments, **6**-22
Free field, **20**-3
Frequencies, **7**-16
Frequency, **7**-19
higher, **2**-10
of power supply, **4**-3
relays, **3**-31
Frequency-change effects, **4**-33
Frequency changes, **18**-14
Frequency stability, **2**-10
Frequency variation system, **2**-10, **4**-13, **6**-18, **7**-19
Friction noise, **20**-17
Full-load speed, **3**-8
Full-load torque, **3**-7
Full-voltage starting, **4**-15
Fuses:
current-limiting, **3**-32
for motor-starter fault protection, **2**-48

Gasoline-dispensing pumps, motors for, **7**-26
Gear motors, **7**-21, **14**-19
General purpose motors, **6**-10, **7**-3
Grease composition and properties, **14**-11
Grease lubrication, **14**-11
Ground current, **2**-53
Ground insulation, **11**-6
Grout, curing time of, **16**-13
Grout keys, **16**-8
Grout mix, **16**-9
Grout placement, **16**-9
Grouting, **16**-1

Handy formulas, **6**-48
Hazardous atmospheres, **6**-22
Hermetic refrigeration compressors, motors for, **7**-23
Home-laundry equipment, motors for, **7**-28
Horsepower ratings, **4**-3
Humidity, **1**-15
protection from, **3**-30
Hysteresis motors, **3**-57, **7**-7

Impulse motor, **9**-29
performance, **9**-30

Inactive iron, **3**-5
Incomplete-sequence protection, **3**-32
Indirect compounding, **3**-36
Induction motors, **4**-2, **4**-9, **4**-12
  combined characteristics of several, **4**-24
  design classification, **6**-3
  fault behavior, **2**-37
  fault-current contributions, **4**-30
  multispeed, **6**-7
  polyphase, **7**-3
  power factor, **2**-41
  single-phase, **3**-41, **7**-6
  speed variation of, **4**-21
  starting methods for, **4**-15
  wound-rotor, **2**-22, **6**-8
Instrument motors, industrial, **7**-21
Instruments for motor inspection, **19**-8
Insulation, **4**-5, **11**-1
  breakdown or puncture, **11**-7
  field coils, **11**-11
  fluidized and dip-type coatings, **11**-10
  ground, **11**-6
  materials, **11**-4
  resins without solvents, **11**-9
  tapes and sheet materials, **11**-9
Insulation characteristics, **11**-2
Insulation life, **11**-3
Insulation systems, **17**-11

Jet pump motors, **7**-29

Lettering dimensions and outline tables, **6**-22
Lightning protection, **2**-**60**
Lint-free motors, **10**-43
Liquid-level relays, **3**-31
Load characteristics, **1**-4
  constant horsepower, **1**-4
  duty cycle, **6**-49
  inertia, **4**-13
  machine characteristics, **1**-4
  motor load, **1**-2
  overhauling loads, **1**-20
  speed range, **1**-15
  torque: constant, **1**-4
    variable, **1**-4
  Wk^2, **4**-13
Locked-rotor current, **4**-11
  and torque values, **4**-12
Loss of excitation, **2**-57
Loss of oil pressure, **3**-30
Loss of synchronism, **2**-57
Losses, motor, **3**-10, **6**-52
Lubrication, **14**-1, **18**-7
  disk, **14**-8
  gear motor, **14**-19
  oil, **14**-3
Lubrication maintenance, **14**-9

Machine:
  checking foot plane, **16**-5
  repetitive, **1**-19
Machine supports, **16**-2
Magnetic noise, **20**-10
Magnetic vibrating motor, **9**-23
  characteristic curves, **9**-25
Main-field current, **3**-33
Main-field winding, **3**-33
Maintenance, preventive, **17**-1, **17**-43
  checks, **17**-43
  inspection, **17**-10
  planning, **17**-4
  program, **17**-2
  records, **17**-5
  test equipment, **17**-39
  testing, **17**-29
Marine motors, **10**-51
Mechanical arrangements, **6**-19
  parts protection, **3**-30
  power, **3**-10
Magohmmeters, **17**-40
Meters, water-flow, **3**-31
Mine:
  ac motor characteristics, **10**-36
  dc motor characteristics, **10**-35
  duty requirements, **10**-36
Mine motor:
  auxiliary, **10**-34
  enclosures, **10**-37
  explosionproof requirements, **10**-37
  face, **10**-34
  lubricants, **10**-39
  power supply, **10**-34
  service requirements, **10**-37
Miniature motors, **9**-39
Motor loss components, **6**-52
Mounting assembly symbols, **6**-32
Multiple drives, **1**-16
Multispeed, **3**-8
Multispeed induction motors, **3**-52
Multispeed motor, **4**-7, **7**-14

Nameplate marking, 4-9
Natural frequency oscillations, 3-60
Navy motors, 10-53
  service C motors, 10-61
NEMA rerate, 6-5
NEMA standard code letters, 6-22
Noise, 20-1
  airborne, 10-57
  bearing, 20-14
  brush, 20-18
  fan-blade, 20-13
  frequency, 20-2
  friction, 20-17
  magnetic, 20-10
  quality of sound, 20-2
  stuctureborne, 10-58
  windage, 20-12
Noise measurement, 6-19
Noise sources, 20-9
Nonexcited synchronous motors, 3-57

Octave band, 20-3
Oil:
  synthetic, 14-6
  viscosity of, 14-4
Oil-application methods, 14-6
Oil burners, motors for, 7-26
Oil composition, 14-3
Oiling:
  bath, 14-8
  ring, 14-7
Old motors, refurbishing, 18-1
Open-phase relays, 3-31
Outline tables and lettering dimensions, 6-22
Overexcited synchronous motors, 5-17
Overheating, causes of, 3-27
Overheating protection:
  circuit breakers, 2-52
  motor starters, 2-50
  rotor, 2-56
Overload relays, 3-31, 4-23
Overspeed, 2-58, 3-8, 3-31, 6-18
Overspeed protection, 2-58

Payback analysis, 6-54
Permanent-magnet motor, 3-57, 7-7, 9-27
  performance curves for, 9-28
Permanent-split-capacitor single-phase motor, 3-40
Phase changes, 18-19
Phase overcurrent, 2-53
Phase-sequence or phase-reversal relays, 3-31
Plugging, 3-17
Pole-changing polyphase motor, 7-13
Pole-changing single-phase motor, 7-14
Polyphase induction motors, 3-44
Positioning drives, 1-20
Power, 3-10
  electric, 3-10
  mechanical, 3-10
  synchronizing, 3-10
Power factor, 2-41, 3-12, 4-10, 6-55, 7-14
  of induction motor, 3-45
Power factor controllers, 6-60
Power supply, 2-1
Power-supply system, 3-24
Power systems, dc, 2-11
Present-worth life-cycle analysis, 6-54
Primary reactor starting, 2-21
Printed-circuit motor, 9-37
Protection, 3-37, 3-66
  amortisseur-winding, 3-32
  bearing, 2-58
  differential, 2-53
  electrical, 4-22
  fault, 2-48
  ground-fault, 2-56
  from humidity, 3-30
  incomplete-sequence, 3-32
  inherent, 6-48
  lightning, 2-60
  loss of excitation, 2-57
  loss of synchronism, 2-57
  of mechanical parts, 3-30
  motor, 2-46
  motor parts, 3-23
  open phase, 2-57
  from overheating, 3-28
    circuit breakers, 2-52
    motor starters, 2-50
  overload, 4-23
  overspeed, 2-58
  phase sequence, 2-57
  pull-out, 3-32
  rotor, 2-55
  stator-winding overheating, 2-55
  surge, 2-59
  thermal, 4-22, 6-47
  undervoltage, 2-52, 2-57
  winding-insulation, 3-27

Protective features of motor-starting equipment, **2**-46
Protectors:
current-responsive, **3**-28
temperature-responsive, **3**-29
thermal, **3**-30
Pull-in torque, **3**-7
Pull-out protection, **3**-32
Pull-out torque, **3**-7
Pull-up torque, **1**-3, **3**-6
Pulsations, **3**-59

Rare-earth permanent-magnet rotor, **10**-75
Rating, changing, **18**-15
Ratings and performance characteristics, **7**-16
Reactance values, motor, **2**-37
Reactor starting, **4**-15
Rectifiers, **8**-2
Reduced speed, **3**-8
Refurbishing old motors, **18**-1
Regenerative braking, **3**-18
Relays:
differential, **3**-31
field-failure, **3**-32
field-protective, **3**-32
frequency, **3**-31
liquid-level, **3**-31
open-phase, **3**-31
overload, **3**-31
phase-sequence or phase-reversal, **3**-31
Reluctance motors, **3**-57, **7**-7
Reluctance torque, **3**-7
Renewal parts, **17**-6
Repulsion induction motor, **3**-42, **7**-12
Repulsion motor, **7**-11
Repulsion single-phase motor, **3**-42
Repulsion-start induction motor, **3**-42
Repulsion-start induction-run motors, **7**-11
Reverberant field, **20**-3
Reversed speed, **3**-8
Roller bearings, **18**-7
Rotating elements, **17**-19
Rotation, direction of, **3**-14, **4**-13, **7**-19, **18**-11
Rotor, **3**-38
Rotor bars, skew of, **20**-12
Rotor-overheating protection, **2**-56
Rotor protection, **2**-55
Rules of thumb, **6**-48

Safety, **17**-41
Salient-pole synchronous motor, **5**-2
Scott T connection, **18**-19
Seal types, **13**-3
Sealing explosionproof machines, **13**-17
Sealing inert-gas-filled machines, **13**-23
Seals:
antifriction bearing, **13**-9
bearing and shaft, **13**-2
clearance, **13**-2
rubbing, **13**-3
sleeve-bearing, **13**-14
for special enclosures, **13**-17
Secondary data, **6**-40
Selection factors, motor, **6**-42, **6**-49
Self-excitation overvoltage, **2**-45
Series field winding, **3**-34
Series-type single-phase motors, **3**-39
Series-wound motors, **3**-34
Service C motors, **10**-61
Service conditions, **3**-24, **4**-14, **6**-10
unusual, **6**-16
usual, **6**-14
Service factor, **4**-11, **6**-47, **7**-14, **7**-19
general-purpose ac motors, **6**-14
synchronous motor, **5**-8
Servo motors, **7**-21
Shaded-pole motor, **7**-12, **9**-19
performance characteristics, **9**-19
single-phase motor, **3**-43
speed control of, **9**-21
Shaft currents, **3**-30, **12**-16
Shaft-mounted fans and blowers, motors for, **7**-23
Shafts, scored, **18**-6
Shim pack, **16**-5
Shipboard motors, **10**-52
Short-circuit contributions from small motors, **2**-38
Short-time-rated electric machines, **6**-18
Shunt-field current, **3**-33
Shunt-field winding, **3**-33
Shunt-wound motors, **3**-33
Silicon-controlled rectifier, **9**-9
Single-phase repulsion motor, **3**-42
Sleeve bearing, locked, **18**-6
Sleeve-bearing care, **17**-26
Sleeve-bearing load ratings, **12**-10
Slip speed, **3**-8
Soleplate, **16**-2
Soleplate grouting, **16**-8
Solid-rotor induction motor, **4**-3

Solid-state-type controls, **9**-9
Sound absorption, **20**-17
Sound levels, **20**-5
Sound-pressure levels, **20**-7
Space heaters, **4**-23
Spare parts, **17**-6
Special-purpose motor, **7**-3
Specialty-motor types, **9**-3
Specifications, commercial, **3**-25
Speed, **3**-8
  adjustable, **3**-8
  adjustable varying, **3**-8
  changing, **18**-16
  classification by, **4**-7
  constant, **3**-8
  critical, **3**-9
  definitions, **3**-8
  definitive terms, **3**-9
  full-load, **3**-8
  multispeed, **3**-8
  overspeed, **3**-8
  reduced, **3**-8
  reduction, **3**-37
  reversed, **3**-8
  slip, **3**-8
  synchronous, **3**-8
  varying, **3**-8, **3**-37
Split-phase motors, **3**-44, **7**-7
Squirrel-cage induction motor, **3**-50, **4**-3
Stabilizing field windings, **3**-34
Standard motor definitions, **7**-15
Starter classes, **2**-12
  overload protections, **2**-50
  ratings, **2**-12
Starters, **3**-37, **3**-54, **3**-66
  combination, **2**-12
  magnetic, **2**-11
  manual, **2**-11
Starting:
  autotransformer, **2**-18, **4**-16
  with circuit breakers, **2**-12
  full-voltage, **2**-18, **4**-15
  of large polyphase wound-round motors, **4**-20
  part winding, **2**-22, **4**-19
  power factor, **2**-32
  primary reactor, **2**-32
  primary resistor, **2**-18
  reactor, **4**-15
  synchronous motors, **5**-14
  wye-delta, **2**-22, **4**-19
Starting torque, **18**-13
Static exciters, **5**-22
Stationary elements, **17**-18
Stationary windings, **17**-18
Stator-coil shapes, **11**-7
Stator-fault protection on essential-service motors, **2**-58
Stator frame, **3**-38
Stator-overheating protection on essential-service motors, **2** 58
Stator-winding fault protection, **2**-53
Stator-winding overheating protection, **2**-55
Stepping motor, **9**-31
  performance of, **9**-33
Submarine motors, **10**-59
Submersible motors for deep-well pumps, **7**-31
Successive starts, **4**-13
  synchronous motor, **5**-11
Supports, machine, **16**-2
Surge arresters, **3**-32
  and capacitors, **2**-60, **3**-32, **4**-24
Surge protection, **2**-59
Synchronization, **3**-58
Synchronizing power, **3**-10, **3**-58
Synchronous-inductor motors, **7**-7
Synchronous motor, **2**-23, **3**-55, **5**-2, **7**-7
  application, **5**-9
  construction, **3**-55
  dc-excited, **3**-57
  dynamic braking, **5**-18
  electronically commutated, **10**-65
  excitation, **5**-21
  ground-fault protection, **5**-23
  incomplete sequence protection, **5**-23
  for kvar compensation, **5**-17
  locked-rotor current, **5**-10
  locked-rotor torque, **5**-9, **5**-12
  loss-of-field protection, **5**-23
  nameplate data, **5**-7
  noise levels, **5**-12
  nonexcited, **3**-57
  number of starts, **5**-11
  overexcited, **5**-17
  permanent-magnet, **10**-74
  for power-factor correction, **2**-43
  protection, **5**-22
  pull-in torque, **5**-10, **5**-13
  pull-out protection, **5**-23
  pull-out torque, **5**-11
  ratings, **5**-5
  reactor starting, **5**-14
  rotor, **3**-56

Synchronous motor (*Cont.*):
service conditions, **5**-12
service factor, **5**-8
small specialty, **9**-34
speed, **3**-8, **4**-5
starting, **5**-12
starting-winding protection, **5**-23
stator, **3**-56
torques, **5**-7
Synchronous-motor behavior during faults, **2**-36
Synchronous-motor characteristics, **5**-7
Synchronous-motor control, **3**-66
Synchronous-motor power factor, **2**-42
Synchronous reluctance motor, **10**-4
performance of, **10**-5
Synchronous torque, **3**-7
Synthetic oil, **14**-6
System capacity, released, **2**-46
System faults, **2**-36
System-frequency variations, **2**-10

Temperature:
cold, **1**-15
high, **1**-15
of synchronous motors, **5**-6
Temperature-responsive protectors, **3**-29
Temperature rise, **4**-5
Test equipment for preventive maintenance, **17**-39
Testing, **17**-29
Tests, motor, **4**-11
Textile motors, **10**-42
Thermal protectors, **3**-30
sensing detectors, **4**-22
Thrust bearings, **12**-7
Tilting-pad thrust-bearing capacity, **12**-12
Time and temperature ratings, **7**-17
Time constants of induction motors, **3**-52
Torque, **3**-6, **4**-10, **7**-14, **8**-13
accelerating, **1**-3, **3**-6
braking, **3**-7
breakaway, **1**-2
breakdown, **3**-6
definitions, **3**-6, **7**-15
definitive, **3**-7
full-load, **3**-7
increasing, **18**-18
locked-rotor, **3**-6
peak, **1**-3
pull-in, **1**-3, **3**-7, **10**-6

Torque (*Cont.*):
pull-out, **3**-7, **10**-6
pull-up, **1**-3, **3**-6
rated, **10**-6
reluctance, **3**-7
starting, **18**-13
synchronous, **3**-7
synchronous motor starting and accelerating, **3**-59
transient, **2**-45
Torque angle, **10**-8
Torque pulsations, **1**-18
Torque requirements, **1**-17
Troubleshooting, **19**-8

Undervoltage protection, **2**-52, **2**-57
Undervoltage relays, **3**-31
Ungrounded systems, **5**-24
U.S. Navy specifications, **10**-51
Universal motor, **3**-39, **7**-7, **9**-3
performance curves of, **9**-8
Universal-motor parts, **7**-21
Usual service conditions, **6**-50, **7**-17

V-belt sheaves, **6**-20
V curves for synchronous motors, **5**-9
Varying speed, **3**-8
Varying-speed motor, **4**-7, **7**-13
Ventilation, **8**-20
Vibration, **17**-14
Voltage, **7**-19
changing rating, **18**-16
minimum generator, **2**-26
restored, **2**-30
Voltage-drop determination, **2**-24
in cables and overhead lines, **2**-33
calculations, **2**-24
combined system, **2**-36
distribution-system, **2**-32
due to motor starting, **2**-25
of power systems, **2**-35
of reactors, **2**-35
transformers, **2**-32
Voltage flicker, **2**-5
Voltage level, **2**-4
Voltage nomenclature, **2**-10
Voltage ratings, induction motor, **3**-46
Voltage regulation, **2**-4
Voltage spread for motors, **2**-9

Voltage variation, **2**-4, **3**-36, **4**-13, **6**-18, **7**-19
  effect on motors, **2**-6
Voltages, **7**-16
  standard, **4**-3

Water-flow meters, **3**-31
Windage noise, **20**-12
Winding-insulation protection, **3**-27
Windings, **3**-5
Wound-rotor induction motors, **2**-22, **3**-50, **4**-3

ABOUT THE EDITOR

For over 25 years, Robert W. Smeaton was editor of the *Allis-Chalmers Engineering Review*. This magazine provided practical application information on a wide range of electrical, mechanical, and hydraulic products and served over 30,000 engineers in all industry.

Prior to being an editor, he held positions as a test and development engineer with the Square D Company and as a switchgear and control design and application engineer with Allis-Chalmers.

While in control engineering, Mr. Smeaton worked with many experienced motor design and application engineers as well as a range of mechanical engineers to solve difficult control problems. Such problems involved special types of controls for both large and small industrial motors and for marine and Navy motors. This experience gave him an unusual knowledge of the complete range of motor types and sizes and their application.

Robert W. Smeaton is also the editor of the SWITCHGEAR AND CONTROL HANDBOOK also published by McGraw-Hill. The work of over 30 nationally and internationally known contributors, this authoritative handbook covers today's best approaches to selecting, applying, and maintaining switchgear and control equipment.